PROGRESS IN PLANT CELLULAR AND MOLECULAR BIOLOGY

Current Plant Science and Biotechnology in Agriculture

VOLUME 9

Aims and Scope
The book series is intended for readers ranging from advanced students to senior research scientists and corporate directors interested in acquiring in-depth, state-of-the-art knowledge about research findings and techniques related to plant science and biotechnology. While the subject matter will relate more particularly to agricultural applications, timely topics in basic science and biotechnology will also be explored. Some volumes will report progress in rapidly advancing disciplines through proceedings of symposia and workshops while others will detail fundamental information of an enduring nature that will be referenced repeatedly.

The titles published in this series are listed at the end of this volume.

Progress in Plant Cellular and Molecular Biology

Proceedings of the VIIth International Congress on Plant Tissue and Cell Culture, Amsterdam, The Netherlands, 24-29 June 1990

edited by

H. J. J. NIJKAMP
Free University, Amsterdam, The Netherlands

L. H. W. VAN DER PLAS
Agricultural University, Wageningen, The Netherlands

and

J. VAN AARTRIJK
Bulb Research Centre, Lisse, The Netherlands

KLUWER ACADEMIC PUBLISHERS
DORDRECHT / BOSTON / LONDON

ISBN 0-7923-0873-5 (HB)

Published by Kluwer Academic Publishers,
P.O. Box 17, 3300 AA Dordrecht, The Netherlands.

Kluwer Academic Publishers incorporates
the publishing programmes of
D. Reidel, Martinus Nijhoff, Dr W. Junk and MTP Press.

Sold and distributed in the U.S.A. and Canada
by Kluwer Academic Publishers,
101 Philip Drive, Norwell, MA 02061, U.S.A.

In all other countries, sold and distributed
by Kluwer Academic Publishers Group,
P.O. Box 322, 3300 AH Dordrecht, The Netherlands.

Printed on acid-free paper

Printed in the Netherlands

Preface

Plants are an important source of food and of valuable products for industry, agriculture and medicine. They are unique in many aspects of metabolic processes, development and reproduction. Most of these aspects can now be studied by the modern methods and technologies of molecular and cellular biology. Such studies are also encouraged as to improve plant yield and quality.

During the past decade research in plant sciences has demonstrated the feasibility of plant cell and tissue culture techniques as major tools in biology and agriculture. These techniques are also essential in strategies for engineering of biological systems.

The proceedings of the VII International Congress on Plant Tissue and Cell Culture in Amsterdam show that in recent years an impressive progress has been achieved.

The papers of the congress, with more than 2000 participants, include the full text of plenary lectures, keynote lectures and presentations of speakers who have been selected out of more than 1400 abstracts. This combination, which provides readers with reviews as well as recent findings and future developments, captures an important part of the scientific exchange during the congress.

The papers in these proceedings are a reflection of the role of plant cell and tissue culture in disciplines varying from plant breeding to molecular biology. Basic as well as applied studies in a variety of plant disciplines are presented in 4 sections: (1) Genetic manipulation and propagation, (2) Morphogenesis and metabolism, (3) Secondary metabolites and (4) Biotechnology and developing countries.

It was the intention of the organizing committee to bring together outstanding workers in the field of plant sciences from throughout the world and to provide a forum for scientific exchange.
The high quality of the contributions to these proceedings attest to the highly successful outcome of the VII International Congress of The International Association for Plant Tissue Culture.

H.J.J. Nijkamp
L.H.W. van der Plas
J. van Aartrijk

TABLE OF CONTENTS

Section A / Genetic Manipulation and Propagation

Section C / Secondary Metabolites

Section D / Biotechnology and Developing Countries

Section A
Genetic Manipulation and Propagation

MICROPROPAGATION IN CROP SPECIES

Amsterdam, The Netherlands, 24-29 June 1990

W.D. Beversdorf, Crop Science Department,
University of Guelph, Guelph, Canada N1G 2W1.

INTRODUCTION

Plant organ, tissue and cell culture procedures have developed rapidly in the half century since the pioneering efforts of Laibach, Gautheret, Nobecourt and White (1). Micropropagation may generally be defined as *in vitro* regeneration of plants from organs, tissues, cells, or protoplasts. Availability of micropropagation techniques vary widely among species. Within species, micropropagation protocols vary with objectives.

Applications of micropropagation technologies have been reviewed frequently during the past decade (1,2,3). Numerous examples of successful applications have been reported for a) meristem culture, b) organogenic micropropagation from undifferentiated tissues, cells or protoplasts, c) zygotic embryo culture, d) somatic embryogenesis and e) gametic embryogenesis.

Hu and Wang (4) recently reviewed meristem, shoot tip and bud cultures. Meristem cultures have been employed to eliminate virus infection in some asexually propagated species. Meristem and bud cultures have been used commercially for multiplication of some high value genotypes.

Zygotic embryo culture has been successfully employed to circumvent post-fertilization, cross-incompatibility during interspecific transfer of genes among related plant species for many years (5). More recently, embryo rescue techniques have also proven useful in recovery of haploid plants that emerge from chromosome elimination in zygotic embryos following interspecific crosses involving *Hordeum bulbosum* (6).

Regeneration of plants from callus, individual cells and protoplasts have involved organogenic or embryogenic differentiation. Cell and tissue culture regeneration technologies have been useful in providing both spontaneous (7) and mutagen-induced (8) genetic variation (somaclonal/gametaclonal/protoclonal variance). Both embryogenic and organogenic differentiation have been involved among examples of plant regeneration following protoplast fusion, genetic transformation, anther culture and microspore culture. Embryogenesis has played a significant role in recovery of haploid plants from cultured anthers or microspores.

The utility of any technology will ultimately be reflected in the relative value of information or products derived there from. The remainder of this discussion will explore utility of several micropropagation technologies through recent and potential applications as related to crop species.

ISOLATION AND MAINTENANCE OF DISEASE-FREE STOCKS

Meristem tip culture have been very useful in the elimination of virus from valuable genetic stocks normally multiplied through vegetative cuttings or tubers.

The successes of Morel and Martin (see 9) in eliminating virus from infected dahlia plants in 1952 and from potato plants in 1955 stimulated stem tip culture in many horticultural species. Frequencies of virus-free 'mericlones' from infected plants have generally been low and variable, in part, dependent on pre-cloning plant growth environments (10). Once isolated, virus-free clones can be maintained *in vitro* to provide disease-free stocks for multiplication through conventional or micropropagation (4).

PRESERVATION

Techniques for establishing cryopreserved micropropagation stocks (meristem, callus, adventitious shoots) in liquid nitrogen have been successfully demonstrated for a number of species (4). Recently Charne (11) demonstrated maintenance of embryogenic capacity of cryopreserved microspores. Decline in viability of pea meristems over a 6-month storage period in liquid nitrogen (10) suggest that long-term maintenance of valuable germplasm via meristem cryopreservation of *in vitro* propagules maybe pre-mature.

PROPAGULE MULTIPLICATION

Meristem and bud cultures have achieved propagule multiplication rates competitive with conventional asexual propagation based on vegetative cuttings in some species (4). Levin and Vasil (12) recently reviewed progress and potential for reducing costs associated with the multiplication phase of *in vitro* propagule production. They suggested labour represents 60-85% of production costs for micropropagation utilizing *in vitro* meristem proliferation, due to manual shoot separation and transfer. Improved bioreactor technology and automated or robotic handling systems may reduce propagule costs significantly (12).

The value per unit propagule from *in vitro* meristem or bud multiplication (Table 1) has been competitive with multiplication based on vegetative cuttings for unique genotypes of some high value horticultural species. Stem and bud-based *in vitro* multiplication of most crop species have not been commercially competitive with seed-propagated genotypes of field crops (Table 1).

Somatic embryogenesis is inherently more efficient as an *in vitro* multiplication technology than organogenic shoot proliferation (eg. meristem culture). Somatic embryos may serve as independent propagules from cell cultures without laborious hand separation or subculture to initiate root differentiation (12). Improvements in preservation of somatic embryos through gel incapsulation (13) or desiccation (14) allow micropropagated somatic embryos to more closely reflect "artificial seeds". By comparison however, problems persist regarding: a) applicable protocols for induction of somatic embryogenesis is many important crop species and/or crop genotypes, b) poor synchronization of somatic embryo development, c)incomplete somatic embryo development resulting in poor propagule vigour, d) inadequate uniformity of plant regeneration from somatic embryos, and e) poor viability of somatic embryos following storage, handling and/or transportation.

Table 1. Relative Commercial Propagule Prices (1990).

Species	Unit Retail Price [1]
Certified seed: (15)	
Alfalfa (*Medicago sativa*)	0.00001
Barley (*Hordeum vulgare*)	0.00002
Corn (*Zea mays*)	0.001
Soybeans (*Glycine max*)	0.00007
Canola (*Brassica napus*)	0.00002
S.W.W. wheat (*Triticum aestivum*)	0.00001
Rooted propagule: (16)	
Red Maple (*Acer rubrum*)	0.80
River Birch (*Betula nigra*)	0.75
Mountain-Laurel (*Kalmia latifolia*)	1.00
Star Magnolia (*Magnolia stellata*)	0.81
Heavenly Bamboo (*Nandina domestica*)	1.50
Rose (*Rosa sp.*)	0.60
Non-rooted propagule: (17)	
Mericloned orchids	1.00
Ferns	0.45

1 Unit propagule prices have been adjusted to 1990 USA dollar values using an annual rate of inflation of 7%

EMBRYO RESCUE

In vitro zygotic embryo culture continues as a means of rescuing hybrid embryos during interspecific gene transfer. Recently, Parker and Michaels (18) employed this technology to recover hybrids of *Phaseolus vulgaris L.* and *P. acutifolius* A. Gray. *P. acutifolius* has improved tolerance to common bean bacterial blight (*Xanthomonas campestris* p.v. *phaseoli*) than *P. vulgaris* (common field bean) but hybrid and early backcross embryos from crosses between these species frequently abort. Embryo rescue assisted the transfer of improved bacterial blight tolerance to economically important common field beans. In another recent example, Ayotte, Harney and Souza-Machado (19) utilized embryo rescue in the transfer of cytoplasmically-inherited herbicide tolerance from *Brassica napus* to *B. oleracea*. As with the many earlier examples, interspecific hybrids between the recipient and donor species were normally aborted as F_1 embryos. This technology is currently being employed in our laboratories to assist in the interspecific transfer of previously defined self-incompatibility alleles from *B. oleracea* to amphiploid *B. napus*.

Embryo culture technology also continues to contribute to the recovery of haploids, particularly in cereal species where hybridizations with *Hordeum bulbosum* frequently results in haploid zygotic embryos through elimination of *H. bulbosum* chromosomes during early embryo development (6).

GENERATING AND ISOLATING GENETIC VARIATION *IN VITRO*

In vitro cell and tissue cultures of plant give rise to genetic variation spontaneously (7). While spontaneous variation is not desired during propagule multiplication, it has been useful in providing some genetic variants among crop species. The observed frequencies of such 'somaclonal variants' varies widely and is probably related to both time in culture and the observers ability to detect genetic variants. Several 'point mutations' have been observed, the majority of analyzed spontaneous genetic variants from somatic cultures appear to have resulted from induction of aneuploidy (7), loss or alteration of mitochondia-associated plasmids (20,21,22), chromosome interchanges or intrachromosomal segment duplications, deletions or structural re-arrangements (7).

Spontaneous or mutagen induced genetic variation in somatic cell cultures coupled with *in vitro* selection techniques have been effective in isolating desired novel genetic variants (23) with cellular level expression while in a heterozygous condition (23). Haploid cell cultures were reported to have higher frequencies of detectable variants that diploid cultures.

Among the numerous reports of *in vitro* variants, several may provide genetic variation useful for crop improvement. Modified tolerance to specific herbicides (23,8), altered composition (24), and modified pest/stress tolerance (g), if not associated with deleterious pleiotrophic effects, will likely emerge in commercial crop varieties in future.

IN VITRO HYBRIDIZATION

Hybridizations, involving both sexual fertilization and a variety of protoplast-protoplast and cytoplast-protoplast fusions have been successfully demonstrated *in vitro* in a few crop species. 'Test tube fertilization' has been employed to help overcome pre- and post-fertilization cross incompatibilities (see 25), while parasexual fusions have been employed to produce interspecific and intergenomic hybrids (26,27,28), and recombine cytoplasmically inherited traits (29,30,31,32).

Utilization of *in vitro* hybridization and fusion has been of limited value in crop improvement, to date. Most applications have been related to synthesis or resynthesis of amphiploids (26,28) or transfer of cytoplasmic organelle encoded traits (eg. triazine herbicide tolerance, cold tolerance) into otherwise potentially useful crop germplasm (30,31,33), almost exclusively in Brassicaceae and Solanaceae.

PRODUCTION OF GAMETE-DERIVED HAPLOID *IN VITRO*

In vitro gametic embryogenesis has become an efficient means of producing haploids in a growing number of species since the successes of Guha and Maheshwari with cultured anthers of *Datura* (34). Successes in anther culture, and more recently in isolated microspore culture have stimulated interest in this technology for plant breeding applications in both monocot and dicot species of crops.

The utility of haploids in the rapid production of homozygous breeding lines from highly heterozygous parents is well recognized including reduced time requirements per cycle of plant breeding (hybridization/selection/evaluation) due to immediate phenotypic expression of recessive traits in haploid and haploid-derived homozygous lines (35,36). Potential theoretical problems associated with haploid breeding approaches including reduced potential for recombination per breeding cycle have also been reported (37).

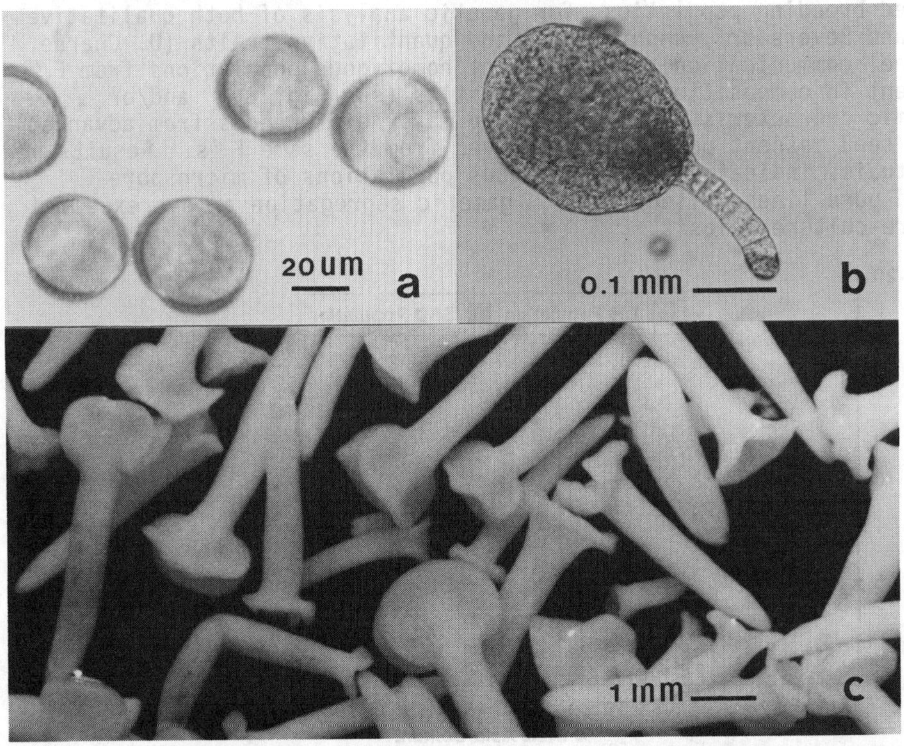

FIGURE 1. Developmental sequence of microspore-derived embryos of *Brassica napus*. a) Microspores in culture at isolation. b) developing embryo with prominent suspensor at day 7. c) mature cotyledonary embryos day 21.

Anther culture in *Nicotiana spp.* and cereal species, and isolated microspore culture in *Brassica spp.* have provided sufficient populations to evaluate genetic and breeding properties of gamete-derived populations (38). In *Nicotiana*, anther culture-derived doubled haploids had poorer performance than homozygous lines or maternal haploids (39,40), suggesting that culture of anthers may provide deleterious effects not expected from gametic segregation. DNA amplification in anther-derived haploids have also been related to reduced agronomic performance of anther culture-derived lines in *Nicotiana* (40). Chlorophyll deficiencies are common among anther-derived haploids in cereals. Day and Ellis (41) detected deletions in plastid DNA of anther-derived wheat albinos while independent analysis of green anther-derived doubled-haploids detected no cytoplasmic organelle DNA changes (42).

EMBRYOGENESIS IN ISOLATED MICROSPORE CULTURES

Efforts in the culture of isolated microspores (Fig. 1.) of *B. napus* (43,44) have improved understanding of microspore embryogenesis (45). An efficient microspore isolation and culture system has provided breeding populations for genetic analysis of both qualitative (Chen and Beversdorf, unpublished) and quantitative traits (D. Charne, personnel communication). Analysis of homozygous populations from F_1's divergent in compositional characteristics (see Fig. 2.) and/or agronomic characteristics indicated no major differences from advanced single seed descent populations derived from the same F_1's. Results of both studies indicated that homozygous populations of microspore-derived pure lines reflect the F_1 gametic segregation arrays expected from pre-culture meiosis.

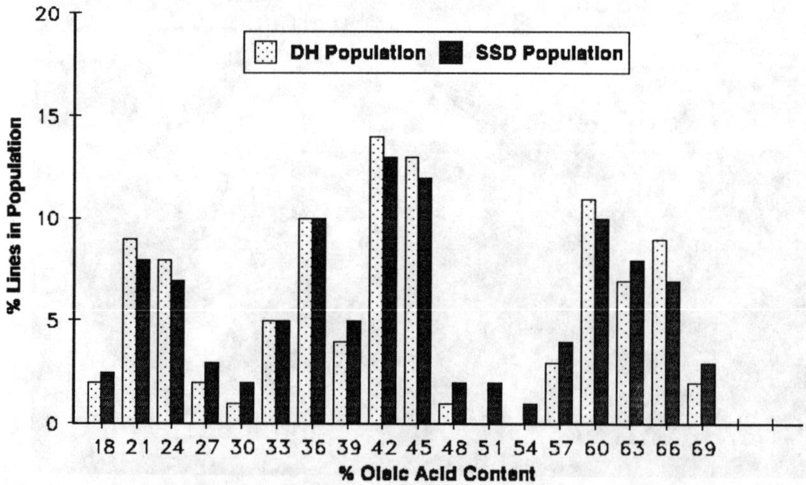

FIGURE 2. Comparative frequency distributions for oleic acid contents within microspore versus single seed descent populations in *B. napus*. (Chen and Beversdorf, unpublished)

In vitro selection of embryogenic microspores have been proposed (46) as a procedure for generating and isolating useful mutants. Swanson *et al.* (8) coupled large scale haploid embryo culture techniques (47) with *in vitro* selection to isolate a series of herbicide tolerant variants in *B. napus*. Kott and Polsoni (unpublished) have mutagenized large scale cultures of similar microspores coupled with subsequent *in vitro* selection for several herbicides to isolate an array of mutants in currently under field evaluation.

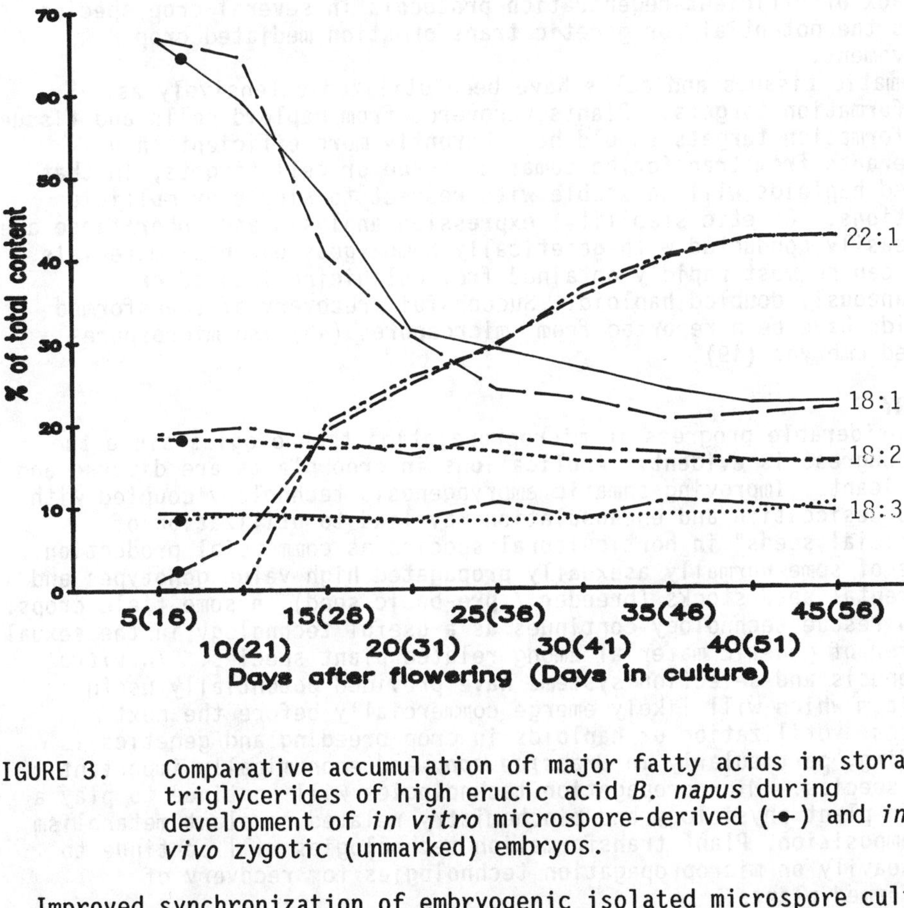

FIGURE 3. Comparative accumulation of major fatty acids in storage triglycerides of high erucic acid *B. napus* during development of *in vitro* microspore-derived (-●-) and *in vivo* zygotic (unmarked) embryos.

Improved synchronization of embryogenic isolated microspore culture (Kott unpublished) stimulated Chen to evaluate storage lipid fatty acid accumulation comparatively with *in vivo* and in vitro gametic embryos (see Fig. 3.). From plants with high or low erucic acid, both types of embryos accumulated fatty acids very similarly. Both types of embryos had similar lag periods before accumulating erucic and eicosenoic acids and both had remarkable synchrony of development. *In vivo* embryos accumulated slightly more oil than *in vitro* embryos (Chen and

Beversdorf, unpublished). These results suggest that lipid biosynthesis in seeds could probably be studied directly in gametic embryos.

MICROPROPAGATION AND TRANSFORMATION

Micropropagation technologies have played an obvious role in the selection of transformed cells or tissues and recovery of plants following genetic transformations involving both co-cultivation with *Agrobacterium spp.* or various direct DNA uptake methods (electroporation, micro-injection or ballistic particles). Currently, the lack of efficient regeneration protocols in several crop species limits the potential for genetic transformation mediated crop improvement.

Somatic tissues and cells have been utilized extensively as transformation targets. Plants recovered from haploid cells and tissue transformation targets should be inherently more efficient than regenerants from transformed somatic tissue or cell targets, in that doubled haploids will be stable with respect to single or multiple insertions. Genetic stability, expression analysis and inheritance are more easily conducted with genetically homozygous parental materials, which can be most rapidly obtained from colchicine induced or spontaneously doubled haploid. Successful recovery of transformed haploids have been reported from microspores (48) and microspore-derived embryos (49).

SUMMARY

Considerable progress in micropropagation technologies since the last Congress is evident. Applications in crop plants are diverse and significant. Improving somatic embryogenesis technology coupled with embryo desiccation and encapsulation may lead to utilization of "artificial seeds" in horticultural species as commercial production stocks of some normally asexually propagated high value genotypes and as parental seed stocks (breeder / pre-basic seed) in some field crops. Embryo rescue technology continues as a useful technology in the sexual transfer of genetic material among related plant species. *In vitro* mutagenesis and selection systems have provided potentially useful germplasm which will likely emerge commercially before the next Congress. Utilization of haploids in crop breeding and genetics is expanding, particularly in *Brassica spp.* and economically important grain species. Micropropagation technologies will continue to play a role in plant physiology and biochemistry related to plant metabolism and composition. Plant transformation technologies will continue to rely heavily on micropropagation technologies for recovery of transformed plants.

1 Pierik, R.L.M. (1987) *In Vitro* Culture of Higher Plants pp. 344, Martinus Nijhoff, Dordrecht.
2 Tomes, D.T., Ellis, B.E., Harney, P.M., Kasha, K.J. and Peterson, R.L. (eds) (1982) pp. 231, University of Guelph, Guelph.
3 Evans, D.A., Sharp, W.R. and Bravo, J.E. (1984) in Handbook of Plant Cell Culture Crop Species (Sharp, W.R., Evans, D.A., Ammirato, P.V. and Yamada, Y., eds.), Vol. 2, pp. 47-68, Macmillan, New York.
4 Hu, C.Y. and Wang, P.J. (1983) in Handbook of Plant Cell Culture Crop Species (Sharp, W.R., Evans, D.A., Ammirato, P.V. and Yamada, Y., eds.) Vol. 1, pp. 177-227, Macmillan, New York.
5 Rappaport, J. (1954) Bot. Rev. 20, 210-225.
6 Kasha, K.J. and Kao, K.N. (1970) Nature 225, 874-876.
7 Scowcraft, W.R., Brettell, R.I.S., Ryan, S.A., Davies, P.A. and Pallotta, M.A. (1987) in Plant Tissue and Cell Culture, Plant Biology Vol. 3 (Green, C.E., Somers, D.A., Hackett, W.P. and Biesboer, D.D. eds.) pp. 275-286, ARL, New York.
8 Swanson, E.B., Coumans, M.P., Brown, G.L., Patel, J.D. and Beversdorf, W.D. (1988) Plant Cell Rep. 7, 83-87.
9 Morel, G.M. and Martin, C. (1952) C.R. Acad. Sci. (Paris) 235, 1324-1325.
10 Kartha, K.K., Gamborg, O.L. and Leung, N.L. (1979) Plant Sci. Lett. 15, 7-15.
11 Charne, D.G., Pukacki, P., Kott, L.S. and Beversdorf, W.D. (1987) Plant Cell Rep. 7, 407-409.
12 Levin, R. and Vasil, I.K. (1989) IAPTC Newslett. 59, 2-12.
13 Redenbaugh, K., Viss, P., Slade, D. and Fujii, J.D. (1987) in Plant Tissue and Cell Culture, Plant Biology Vol. 3, pp. 473-493, ARL, New York.
14 Senaratna, T., McKersie, B.D. and Bowley, S.R. (1989) Plant Sci. 65, 253-259.
15 Crop Budget Committee 1990. 1990 Crop Budgets. Publ. 60, Ontario Ministry of Agric. and Food.
16 Dirr, M.A. and Heuser, C.W. (1987) in the Reference Manual of Woody Plant Propagation. pp. 65-78, Varsity Press, Athens, USA.
17 George, E.F. and Sherrington, P.D. (1984) in Plant Propagation by Tissue Culture. pp 382-386. Exegetics Ltd, Eversly, England.
18 Parker, J.P. and Michaels, T.E. (1986) Plant Breed. 97, 315-323.
19 Ayotte, R., Harney, P.M. and Souza-Machado, V. (1987) Euphytica 36, 615-624.
20 Earle, E.D., Gracen, V.E., Best, V.M., Batts, L.A. and Smith, M.E. (1987) TAG 74, 601-609.
21 Kemble, R.J. and Mans, R.J. (1983) J. Mol. Appl. Genet. 2, 161-171.
22 Wilson, A.J., Chourey, P.S. and Sharpe, D.Z. (1985) in Tissue culture in Forestry and Agriculture (Henke, R.R., Hughes, K.W. and Constantin, M.J. and Hollaender, A. (eds.) pp. 368-369, Plenum, New York.
23 Chaleff, R.S. and Ray, T.B. (1984) Science, 223, 1148-1151.
24 Hibberd, K.A. and Green, C.E. (1982) Proc. Natl. Acad. Sci. USA 79, 559-563.

25 Rangan, T.S. (1982) in Experimental Embryology of Vascular Plants (Johri, B.M. ed) pp. 105-129, Springer-Verlag, Berlin.
26 Carlson, P.S., Smith, H.H. and Dearing, R.D. (1972) Proc. Nat. Acad. Sci. (Wash.) 74-5109-5112.
27 Power, J.B., Frearson, E.H., Hayward, C., George, D., Evans, P.K., Berry, S.F. and Cocking, E.C. (1976) Nature 263, 500-502.
28 Sjodin, C. and Glimelius, K. (1989) TAG 77, 651-656.
29 Belliard, G., Vedel, F. and Pelletier, G. (1979) Nature 281, 401-403.
30 Pelletier, G., Primard, C., Vedel, F., Chetrit, P., Remy, R., Rouselle, P. and Renard, M. (1983) Mol. Gen. Genet. 191, 244-250.
31 Chuong, P.V., Beversdorf, W.D., Powel, A.D. and Pauls, K.P. (1988) Plant Cell, Tissue and Organ Culture. 12, 181-184.
32 Medgyesy, P., Fejes, E. and Maliga, P. (1985) Proc. Natl. Acad. Sci. USA. 82, 6960-6964.
33 Binding, H., Jain, S.M., Finger, J., Mordhorst, G., Nehls, R. and Gressel, J. (1982) TAG 63, 273-277.
34 Guha, S. and Maheshwari, S.C. (1964) Nature 204, 497.
35 Chase, S.S. (1974) in Haploids in Higher Plants, Advances and Potentials (Kasha, K.J. ed) pp. 211-230. Proc. of the First Internat. Symp. University of Guelph, Canada.
36 Kasha, K.J. and Reinbergs, E. (1975) in Barley Genetics III. pp. 307-315. Proceed. of 3rd Internat. Barley Genetics Symposium, Garching.
37 Walsh, E.J. (1974) in Haploids in Higher Plants, Advances and Potential (Kasha, K.J. ed.) pp. 195-209. Proc. of the First Internat. Symp., University of Guelph, Canada.
38 Keller, W.A., Arnison, P.G. and Cardy, B.J. (1987) in Plant Tissue and Cell Culture, Plant Biology Vol. 3 (Green, C.E., Somers, S.A., Hackett, W.P. and Biesboer, D.D. eds.) pp. 223-241. ARL, New York.
39 Collins, G.B., Legg, P.D. and Litton, C.C. (1974) Tob. Sci. 18, 40-42.
40 Reed, S.M. and Wernsman, E.A. (1989) Crop Sci. 29, 1072-1076.
41 Day, A. and Ellis, T.H.N. (1984) Cell 39, 359-368.
42 Rode, A., Hartmann, C., Dron, M., Picard, E. and Outier, F. (1985) TAG 71, 320-324.
43 Lichter, R. (1982) Z. Pflanzenphysiol. 105, 427-434.
44 Chuong, P.V. and Beversdorf, W.D. (1985) Plant Sci. 39, 219-226.
45 Kott, L.S., Erickson, L.R. and Beversdorf, W.D. (1990) in Rapeseed/Canola: Production, Chemistry, Nutrition and Processing Technology (Shahidi, F. ed.) Van Nostrand Reinhold, New York.
46 Beversdorf, W.D. and Kott, L.S. (1987) Iowa State J. of Res. 61, 435-443.
47 Polsoni, L., Kott, L.S. and Beversdorf, W.D. (1988) Can. J. Bot. 66, 1681-1685.
48 Neuhaus, G., Spangenberg, G., Mittelsten Sheid, O. and Schweiger, H.-G. (1987) TAG 75, 30-36.
49 Swanson, E.G. and Erickson, L.R. (1989) TAG 78, 831-835.

GENETIC PRESERVATION *IN VITRO*

BRIAN W.W. GROUT, CELL SYSTEMS LTD., ORWELL HOUSE, COWLEY ROAD, CAMBRIDGE CB4 4WY, UK

1. INTRODUCTION

The purpose of any technique designed for the conservation of plant genetic resources is to preserve the greatest possible diversity for future use. This diversity is found at the level of species, variety, cultivar and ecotype,together with the new manipulated genetic combinations that are being derived from laboratory studies in many different fields of plant science. There are plant populations from natural habitats to be considered as well as those from cultivation at all levels of sophistication, and recent estimates suggest that 9000 wild plant species are in some way threatened and should be considered for active conservation (1), together with an unknown number of agricultural and horticultural subjects that are also vulnerable to erosion and loss. Varieties and cultivars of the past, together with the selections of simple agricultural systems represent a valuable history of observation, selection and breeding that requires investigation and characterisation, and should not be discarded lightly, even though it does not seem immediately useful.

Effective genetic conservation will help to meet the needs of the future in terms of the management of both natural and cultivated plant populations, and one such need may be to respond to global climatic change. Particularly with regard to agriculture it may be that new strategies for crop production will have to be devised and implemented swiftly to sustain habitats and crop production, and the availability of the maximum genetic diversity for critical plant species will be crucial. Similarly, it may be that recovery of forest populations on a global basis will depend on extensive replanting schemes, and the genetic material necessary should be collected and conserved as a matter of urgency , before replanting has to be based on material with worryingly restricted genetic diversity.

Conservation of plant material *in vitro* is appropriate where conventional seed storage cannot, or is not, to be used and where *in situ* conservation may prove inadequate. It is of particular value for vegetatively propagated species (2), species with recalcitrant seed (3) and material modified by one of many techniques in molecular biology (4). The material to be stored may be protoplasts, isolated cells grown in suspension or on semi-solid medium, meristem cultures at various stages of development, somatic embryoids or organised plantlets. It can be assumed that genetic stability will increase with increasing organisation of the culture, with regenerated plantlets being the most stable material of all

(5). If the conservation conditions to be employed are such that growth and development will still occur, albeit at reduced levels, then organised plantlets must be the material of choice, if genetic change during a prolonged *in vitro* history is to be avoided. Where growth is completely inhibited, as in cryogenic storage, shoot meristems may be the more suitable explant (6,7) as they will grow out directly into plantlets, avoiding the non-differentiated stages most susceptible to somaclonal variations (8). When non-organised cultures (callus, suspension cells) or simple materials such as pollen must be preserved it is only by resorting to cryopreservation that the maximum of genetic stability will be assured.

2. PRESERVATION WITHOUT FREEZING

At an elementary level conservation *in vitro* is achieved by simple subculture, maintaining the nutrient supply and providing continued protection from an unfavourable environment and pathogens. This technique for conservation is demanding of labour and physical resources and carries a sizeable risk of loss of material due to human error or breakdown of *in vitro* security eg. a pathogen invasion following subculture or mechanical/electrical failures in an environment room. The risks of genetic change during serial subculture (somaclonal variation) are reduced as the level of organisation of the cultures increases ie from callus to plantlets, and is very real where a prolonged, undifferentiated phase of growth persists before regeneration (9,8,10,11,5). Somaclonal variation has been reported over a range of genera and ploidy levels (8) and often at high frequency eg in the legume *Stylosanthes* some 40% of the regenerants from leaf callus were shown to be variant (12).

Clearly, it is unwise to maintain non-differentiated cultures in this way, and differentiated plantlets propagated by nodal cuttings are the conservation materials of choice. The labour and expense of repeated subculture can be reduced by lowering the growth temperature, to slow metabolism and development (13,14). The security of the cultures is also increased as a simple consequence of the number of interventions being reduced. Similarly, the use of growth retardants and hypertonic osmotica in the growth medium have effectively prolonged the period between subcultures (15,7). Modification of the gaseous environment to retard culture growth has also been considered, with particular regard to compounds such as ethylene and carbon dioxide (16,17,18). The physical dimensions of the gas space within a culture vessel will also affect culture growth, the critical parameter being the diffusion distance from the growing culture to the point of gas exchange between the vessel and the external environment. In a study using regenerating carrot plantlets a range of culture vessels were constructed with identical above-culture gas volumes but differing diffusion distances (19). Biomass measurements showed that the increase during culture was halved by increasing the diffusion distance from 50 to 150mm, which effectively doubled the period between subcultures (Fig.1). This technique may form the basis of a very simple extended storage system for *in vitro* material.

The use of gas-permeable culture vessels may also influence non-frozen conservation by providing an *in vitro* environment in which growth can be

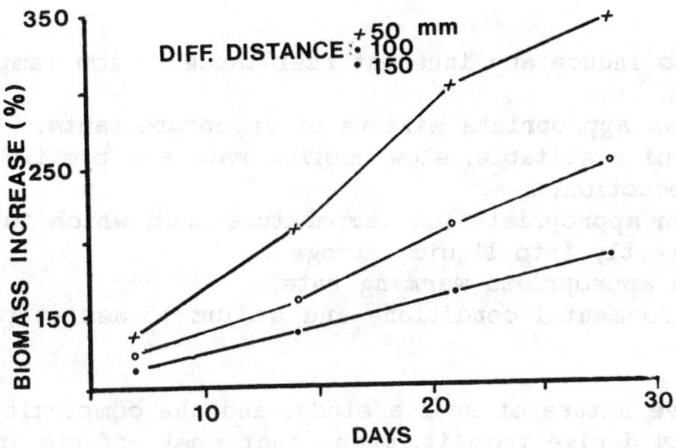

FIGURE 1. The growth of regenerating callus cultures of Daucus carota with differing diffusion distances and a 350ml gaseous volume in vitro. Redrawn from (19).

maintained autotrophically (20). If differentiated plantlets were being cultured the use of these vessels would allow the medium to be reduced to a basal level, which could be replaced or supplemented without removing the plantlet, effectively extending the transfer period. This technique could be very successful when used in combination with chemical growth inhibitors, particularly where relatively short-term storage is envisaged.

Progress in the induction and manipulation of somatic embryogenesis (21) may contribute another technique for in vitro conservation, with the potential of a high degree of genetic stability. Techniques have been developed that result in the dessication of the somatic embryo in a way that appears to reflect normal seed physiology, using pretreatments with compounds such as ABA, IBA and sucrose, and have been successfully demonstrated for species such as grape (22), wheat (23) and soybean (24). If this dessication does accurately parallel the dried condition of the zygotic embryo, with the same level of viability , then the potential exists to store such material at conventional temperatures (-20°C) with the level of genetic stability required for effective in vitro conservation (25). The system becomes even more useful if it could be coupled to a coating technique to produce an artificial seed (26).

3. CRYOGENIC STORAGE

The greatest genetic stability for conserved in vitro plant material will be achieved by storage in liquid nitrogen, with the stability being a consequence of the ultra-low temperature [-196°C] (27,28,29,4,30,7). The design of a successful cryopreservation protocol for cultured plant cells and tissues is still largely an empirical process, although there are some approaches and procedures that seem more broadly applicable than others(31). The factors that most commonly have to be taken into account are :

i) the stage of development and cellular activity of the starting material,

ii) pretreatment to induce any inherent resistance to low temperature or dessication,

iii) incubation in an appropriate mixture of cryoprotectants,

iv) determination of a suitable, slow cooling rate and profile for temperature reduction,

v) selection of an appropriate low temperature from which to quench the sample directly into liquid nitrogen,

vi) adoption of an appropriate warming rate,

vii) design of environmental conditions and medium to assist in recovery.

The broad, investigative nature of such a study, and the complexity of the cooling regime that may derive from it, means that most efforts in cryopreservation will require programmable freezing equipment, which restricts activity to a limited number of specialist laboratories. In order to make cryopreservation more accessible it must be the goal of these laboratories to find unifying themes in their various studies and to attempt to simplify the cryopreservation process. Once reliable, routine protocols are available there are relatively simple pieces of equipment that can be constructed (31), or are commercially available, to achieve a single cooling profile. Techniques of cryopreservation that work well in such equipment would speed up transfer of the technology to a great many more laboratories, particularly in areas where financial resources are limiting.

Encouragingly, cryopreservation by vitrification does seem to be feasible for plant cells, and would reduce the need for complex equipment if it were found to be of broad application (see 4. below). Further, the successes of cryopreservation may be increased if more care is taken of the cells or organs immediately upon their recovery from liquid nitrogen, as it may be naive to expect them to undergo the stresses of the cryopreservative procedure without being in some way damaged. Some laboratories are beginning to consider aspe cts of repair and recovery, and this approach is to be encouraged (see 5. below)

4. VITRIFICATION

As discussed above the majority of successful protocols for cryopreservation of *in vitro* higher plant cells and organs have been based around slow cooling rates (29,7), with fewer successes being attributed to more rapid rates (32,4,33). At slow cooling rates unfrozen cells are exposed to extracellular ice and are liable to lose water as a result of cryodehydration and aqueous solutions within the cell become progressively more concentrated as this water loss continues , for most of the water leaving the cell contributes to ice formation and not an osmotic equilibrium. Given this concentration of cytoplasmic content it is probable that quenching into liquid nitrogen that typically follows slow cooling to temperatures between -30 and -60°C results in the vitrification of the residual solutions.

Vitrification is the transition of water from the liquid phase directly into an amorphous phase, or glass, at low temperature whilst avoiding the formation of crystalline ice. Vitrification of pure water in any significant volume is rarely achieved (33,34,35) and requires an ultra-rapid cooling rate whereas solutions such as those found within, and surrounding, cells during cryopreservation will vitrify more readily as their concentration increases, and at lower cooling rates. In the simplified phase diagram below (Fig.2) vitrification will occur when the curve representing the homogeneous nucleation temperature intersects the curve representing the glass transition temperature (36). For common cryoprotectant compounds such as dimethyl sulphoxide and glycerol it has been calculated that this will occur at concentrations approximately between 50 and 60% w/w. If the content of plant cells could be manipulated at temperatures significantly above that of liquid nitrogen (-196°C) such that vitrification were possible following apporpriate cooling, a siginificant benefit for conservation _in vitro_ might be derived. This benefit would arise because, following the pretreatments needed to ensure vitrification in the cells there is no need for slow, regulated cooling and the subject material may be immersed directly into liquid nitrogen. The devitrification curve is also shown in Fig.2, indicating the way in which a previously vitrified solution freezes, forming crystalline ice upon warming. To produce viable material after cryopreservation by vitrification it is necessary to warm rapidly such that the amount of crystallisation that occurs before melting is insignificant.

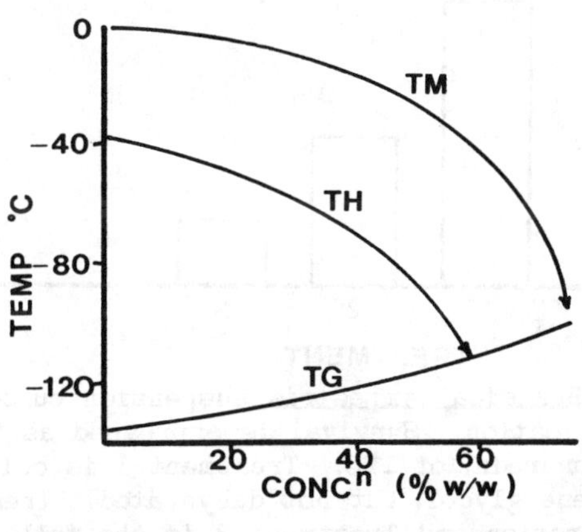

FIGURE 2. A simplified phase diagram for a hypothetical compound with cryoprotectant function. TM-melting point curve, TH-homogeneous nucleation curve; TG-glass transition temperature. redrawn from (36).

Recently, vitrification has been successfully applied to the cryopreservation of cell suspensions of _Asparagus_ and _Brassica_, demonstrating the potential of the technique as one that does not require complex controlled-rate freezing apparatus (37,38). In the latter case _Brassica campestris_ cells were incubated in a 1.5M glycerol solution and then treated with a cryoprotectant mixture estimated at c.19 osmolal. The cells were placed in a 0.5ml plastic straw for quenching into liquid nitrogen, and then recovered by warming at a rate in excess of 60°C per minute. Survival was estimated by a tetrazolium reduction assay (Fig.3) and by subsequent regrowth. The use of plastic straws, as used in semen freezing, ensures a rapid temperature response to the environment. A further benefit of this type of container is that, as a result of their long and widespread use in semen freezing, there is a great deal of equipment available for their manipulation, labelling, storage and transport.

5. POST-THAW INJURY AND SUB-OPTIMAL PROCEDURES

It is probable that many cryopreserved cells are returned to physiological temperatures with some degree of injury, and have to undertake active repair to recover their original structure and function. The ability to complete this repair will depend upon the nature and extent of the damage, the level of cellular reserves available and the degree of metabolic integrity remaining to undertake the necessary synthesis, and may be facilitated by appropriate modification of the physical and chemical post-thaw environments (39,40,41,42,43 and see Table 1.

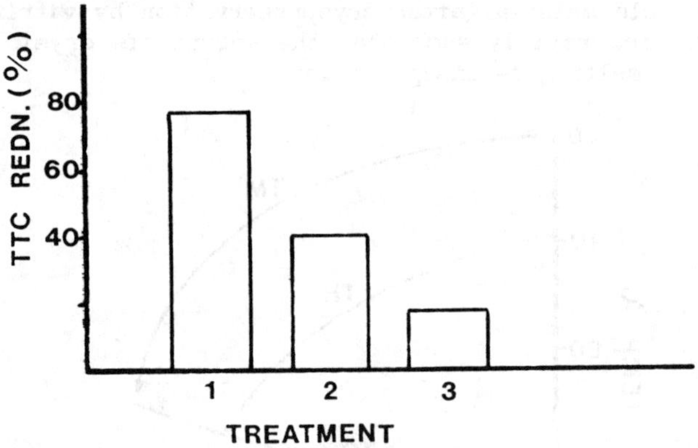

Figure 3. The survival of Brassica campestris suspension cultures following vitrification. Survival is expressed as % of the control value for redn. of TTC. Treatment 1 is cells loaded with 1.5M ethylene glycol, but not dehydrated. Treatment 2 is as 1 plus dehydration and Treatment 3 is the full procedure including quenching in LN. Redrawn from (38).

It is also possible that a range of different lesions might be generated by different freezing protocols and therefore the required recovery media and patterns of repair might differ also. Any occurence of permanent damage or changes following repair in surviving cells would be unwelcome as it might impair the required functions of the material, and would be of particular significance if it had a genetic basis. Whilst it is unlikely that such a situation would affect the overall activity of multicellular organs such as meristems it would be of greater importance where suspension cells, pollen or genetically manipulated material was concerned, and where populations are required to be raised from single cells. The occurence of such genetic injury resulting from freezing procedures with cultured plant tissues is, at present, speculation and no evidence exists to support it. However, there is a body of data from microbial studies which suggests that sub-optimal freezing protocols, whilst producing acceptable levels of viable recovery, may cause genetic alterations in a proportion of the surviving cells. Chromosomal damage has been observed in E. coli in situations where it seems that DNA repair mechanisms are an integral part of cell recovery following cryopreservation (44). Loss of plasmid expression in bacterial systems has also been associated with freezing injury under sub-optimal conditions (45). Similarly, in studies with yeast the extent of plasmid expression and proportion of rspiratory mutants in populations recovered from liquid nitrogen has been shown to be a function of cooling rate, and not to have the expected, simple relationship with viable recovery (Fig.4;46).

Table 1. The percentage of explants producing shoots directly from cryopreserved shoot meristems of tomato, on a range of growth media. Callus production at the cut surface is also indicated.

Medium (MS +)	Frozen Shoots	Frozen Callus	Non-frozen control Shoots	Non-frozen control Callus
+ NAA	0	40	76	82
+ GA	45	22	86	90
+ GA/Cas.Hyd.	40	11	78	94

This type of information must raise a note of concern where cryopreservation is used for genetic conservation and where techniques for genetic manipulation that involve microbial components have been used. It may be that injury is sustained before the material reaches liquid nitrogen, under sub-optimal conditions, and that this injury is preserved in the frozen state to be expressed upon thawing. It is essential, therefore, that cells or organs be characterised as precisely as possible with respect to the function required of them following storage before they are subjected to cryopreservation, and that they be characterised again in the same way after recovery.

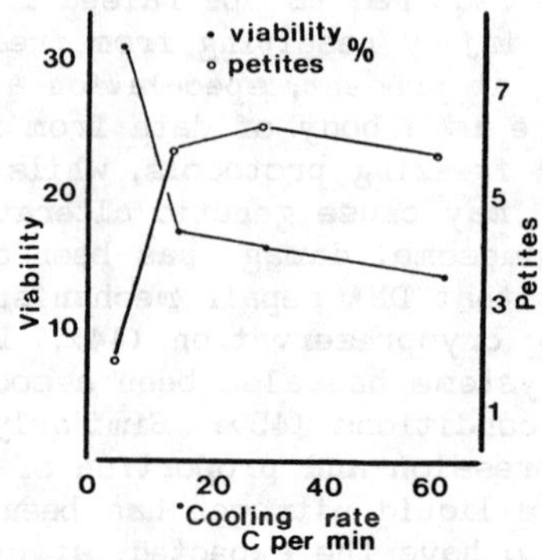

Figure 4. Viability and respiratory deficient cells in a yeast population following different rates of cooling. Redrawn from (46).

REFERENCES

1 Mlot, C., 1989. Bioscience, 39(6): 364-368.
2 Withers, L.A., 1986. In: L.A. Withers and P.G. Alderson (Editors), Plant Tissue Culture and its Agricultural Applications. Butterworths, London, pp. 261-276
3 Grout, B.W.W., 1986. In: L.A. Withers and P.G. Alderson (Editors), Plant Tissue Culture and its Agricultural Applications. Butterworths, London, pp. 303-309.
4 Kartha, K.K. 1987. In: C.E. Green, D.A. Somers, W.P. Hackett and D.D. Biesboer (Editors), Plant Tissue and Cell Culture. Alan R. Liss, New York, pp. 447-458.
5 Karp, A. 1989. IAPTC Newsletter, 58: 2-11.
6 Ford-Lloyd, B. and Jackson, M., 1986. Plant Genetic Resources: an introduction to their conservation and use. Edward Arnold, London, v + 152 pp.
7 Withers, L.A., 1987. In: B.W.W. Grout and G.J. Morris (Editors), The Effects of Low Temperatures on Biological Systems. Edward Arnold, London, pp.389-409.
8 Larkin, P.J., 1987. Iowa State Journal of Research, 61: 393-434
9 Larkin, P.J. and Scowcroft, W.R., 1981.Theor. Appl. Genet., 60: 197-214.
10 Sala, F. and Biasini, M.G., 1987. G.Bot.Ital, 120: 43-54.
11 Scowcroft, W.R., Brettell, R.I.S.. Ryan, S.A. Davies, P.A. and Pallotta, M.A., 1987. In: C.E. Green, D.A. Somers, W.P. Hacket and D.D. Biesboer (Editors), Plant Tissue and Cell Culture. Alan R. Liss, New York, pp. 275-288.
12 Godwin, I.A., Gordon, G.H. and Cameron, D.F., 1987. Plant Breeding, 98: 220-227.
13 Staritsky, G. Dekkers, A.J. Louwaars, N.P. and Zandvoort, E.A., 1986. In: L.A. Withers and P.G. Alderson (Editors), Plant Tissue Culture and its Agricultural Applications. Butterworths, London, pp. 277-284.

14 Wanas, W.H., Callow, J.A. and Withers, L.A. (1986) In Plant Tissue Culture and its Agricultural Applications (Withers, L.A. and Alderson, P.G., eds.), pp285-290, Butterworths, London.
15 Pritchard, H.W., Grout, B.W.W. and Short, K.C., 1986. Ann. Bot., 57: 41-48.
16 Deproff, M.P., Maene, L. and DeBergh, P., 1985. Physiol. Plant., 65: 375-379.
17 Gould, J.H. amd Murashige, T., 1985. Plant Tissue, Cell and Organ Culture, 4: 29-42.
18 Belcher, A.R., Abbott, A.J., Hall, K.C. and Jackson, M.B., 1987. Gaseous constituents of cultures in vitro. Report of the Long Ashton Research Station 1986. AFRC, UK.
19 Bateson, J.M., Grout, B.W.W. and Lane, S., 1987. In: G. Ducote, M. Jacob and A. Simeon (Editors), Plant micropropagation in the Horticultural Industries, Belgian PTC Group, Arlon, pp. 275-277.
20 Tanaka, M.K.,Jinno, M.Goi and Higashiura, T., 1988. Acta. Hort., 230: 73-80
21 Ammirato, P.V. (1989) IAPTC Newsletter 57, 2-16.
22 Gray, D., 1987. Hortsci., 22: 810-814.
23 Carman, J.G., 1988. Planta, 175: 417-424.
24 Obendorf, R.L. and Slawinska, J., 1988. In Vitro Cell Dev. Biol., 24: 71A.
25 Roberts, E.H., King, M.W. and Ellis, R.H., 1984. In: J.H.W. Holden and J.T. Williams (Editors), Crop Genetic Resources: Conservation and Evaluation. George Allen and Unwin, London, pp. 38-52.
26 Redenbaugh, K., Fugii, J.A. and Slade, D., 1988. In: A. Mizrah (Editor), Biotechnology in Agriculture. Alan R. Liss, New York, pp. 225-248.
27 Ashwood-Smith, M. and Grant, E., 1977. In: K. Elliot and J. Whelan (Editors), The Freezing of Mammalian Embryos. Elsevier, Amsterdam, pp. 251-268.
28 Finkle, B.J., Zavala, M.E. and Ulrich, J.M., 1985. In: K.K. Kartha (Editor), Cryopreservation of Plant Cells and Organs. CRC Press, Boca Raton, pp. 75-113.
29 Kartha, K.K., 1985. Cryopreservation of Plant Cells and Organs. CRC Press, Boca Raton, 276 pp.
30 Grout, B.W.W. and Morris, G.J., 1987. The Effects of Low Temperatures on Biological Systems. Edward Arnold, London, 500 pp.
31 Withers .L.A. and King, P.J., 1980. Cryoletters, 1: 213-220.
32 Grout, B.W.W. and Henshaw, G.G., 1978. Ann. Bot., 42: 1227-1229.
33 Angell, C.A. (1982) In Water: A Comprehensive Treatise (Franks, F., ed.),pp.1-81, Plenum Press, New York.
34 Reid, D.S., Rall, W.F. and Fahy, G.M., 1985. Cryobiology, 22: 60
35 Taylor, M.J., 1987. In: B.W.W. Grout and G.J. Morris (Editors). The Effects of Low temperatures on Biological Systems. Edward Arnold, London, pp. 3-71.
36 Fahy, G.M., MacFarlane, D.R., Angell, C.A. and Meryman, H.T. (1984) Cryobiology 21, 407-426.
37 Uragami, A., Sakai,A., Nagai, M. and Takahashi, T., 1989. Plant Cell Rep., 8(7): 418-421.
38 Langis, R., Schnabel, B., Earle, E.D. and Steponkus, P., 1989. Cryoletters, 10(6): 421-428.
39 Grout, B.W.W., Westcott, R.J. and Henshaw, G.G., 1978. Cryobiology, 15:

478-483
40 Benson, E.E. and Withers, L.A., 1987. Cryoletters, 8(1): 35-46.
41 Benson, E.E. and Noronha-Dutra, A.A., 1988. Cryoletters, 9(2): 120-131.
42 Benson, E.E., Harding, K. and Smith, H., 1989. Cryoletters, 10(5) : 323-344.
43 Withers, L.A., Benson, E.E. and Martin, M., 1988. Cryoletters, 9(2): 114-119.
44 Williams, D.L. and Calcott, P.H., 1982. J.Gen. Microbiol. 123: 215-218.
45 Calcott, P.H. and Macleod, R.A., 1975. Can. J. Microbiol., 21: 1724-1732.
46 Pearson, B.M., Jackman, P.J.H., Painting, K.H. and Morris, G.J., 1990. Cryoletters, In Press.

CURRENT RESEARCH IN BIOTECHNOLOGY WITH APPLICATION TO PLANT BREEDING

Dwight T. Tomes
Pioneer Hi-Bred Int'l Inc., P.O. Box 38, Johnston, IA, USA

1. INTRODUCTION

A common rationale for research in new technologies is the flow of new products and knowledge which will result in significant economic development for the governmental sponsors or significant increases in profit for the company sponsor. Short term expectations are rarely met although long term consequences often have dramatic impact on product research and development. Viewed in the short term, plant cell and tissue culture has had a rather minimal effect on plant breeding; but long term impact will be substantial and will increase as technologies mature. The acceptance of plant cell and tissue culture as a unifying discipline between plant breeding and molecular biology will insure the continued health of our discipline, and have a positive impact on agriculture and humankind. The applied science of plant breeding, perhaps the most important economic use for plant cell and tissue culture, critically measures the impact of any technology by the speed with which new and improved cultivars and/or hybrids move into the marketplace or into widespread use in agriculture. New cultivars and/or hybrids most often reflect tissue culture technology that is relatively easy and widespread in its use. This brief review will focus on a sample of the applied successes as measured by germplasm and/or cultivar releases in the broad areas of haploidy, variation from cell and tissue culture, and gene transfer. Length precludes a complete listing of cultivars/germplasm, and only a portion of the significant accomplishments in each of these technologies will be mentioned. Many important technology improvements have not yet had sufficient time to impact economic development or research progress.

2. HAPLOIDY

2.1. Anther culture

The success of haploid procedures could be anticipated because of the development of the technology in economic species and because plant breeders could more readily anticipate advantages of the its use (1).

The largest number of cultivars has come from Hordeum vulgare using the "bulbosum" maternal haploid method. Seven barley cultivars have been released in Canada alone (Table 1). Cultivars either are in test or have been released in all the major cereals including barley, wheat, rice and corn. These

TABLE 1. A SAMPLE OF CULTIVARS/GERMPLASM RELEASED OR IN FIELD TRIALS FROM ANTHER/MICROSPORE DERIVED TISSUE CULTURE

SPECIES/CULTIVAR NAME	SOURCE	COUNTRY/YEAR	NOTED CHARACTERISTICS (REFERENCE)
Brassica napus			
N/A	M*	Canada/1990	agronomics (2)
B. napus			
Granite	M	Denmark/1990	(3)
Hordeum vulgare			
Mingo	B	Canada/1979	agronomics (4)
Rodeo	B	Canada/1983	"
Craig	B	Canada/1988	"
Etienne	B	Canada/1989	"
Winthrop	B	Canada/1987	"
TBR 579-5	B	Canada/USA 1990	"
TBC 555-1	B	USA/1990	
Gwylan	B	New Zealand/1989	" (5)
Nicotiana tabacum			
CTDH38	B	Chile/1988	Verticillium wilt (6)
CTDH40	B	Chile/1988	
HI,LI	A	USA/1988	nicotine levels (7)
Orzya sativa			
Texmont	A	USA/1990	early maturity yield (8)
Triticum aestivum			
Florin	A	France/1985	agronomics (9)
Zea mays			
AC4115	A	USA/1987	agronomics (10)

*A=ANTHER CULTURE, M=MICROSPORE, B=MATERNAL HAPLOID

data include specific information from China, Japan or Korea because specific cultivar descriptions were not available.

2.2. Technology advances

The first reports of tobacco anther culture were soon improved by a relatively simple procedure such as staging of microspores prior to culture, and by the use of temperature pretreatments (11). At first, anther culture progress was quite modest in economic species such as barley, rice, wheat, corn, and rape. Other techniques such as the "bulbosum" maternal haploid system were somewhat easier to use because virtually any cultivar could be used in a haploid breeding strategy (12).

Hordeum anther culture was significantly improved by the use of pretreatments, responsive genotypes, culture temperature (13) and anther orientation (14). Anther culture of other species such as maize required the use of different pretreatment conditions, culture temperatures, specific media modifications, and the use of responsive genotypes (15).

The first positive reports with corn anther culture were from Chinese genotypes. Subsequent screening of elite germplasm has revealed that responsive genotypes are present (10). It is especially significant that the repeated use of anther culture derived genotypes dramatically improves the response level (10, 15).

Anther culture of Brassica was dramatically improved by careful attention to growth conditions for the donor plants, and the use of stress to induce microspore development into an embryogenic pathway (16). These stresses can either be in the form of incubation at 37°C for several days, low doses of high energy radiation, or sub-lethal concentrations of ethanol (17).

2.3. Microspore culture

Improved efficiencies in anther culture brought about the possibility of direct culture of microspores in several species such as Brassica, Hordeum and Zea. The knowledge and application of environmental modifications such as media constituents and environmental conditions in rape seed (18) and the manipulation of responsive genotypes has dramatically improved culture systems (19). The observation that individual plants from responsive genotypes that have been anther cultured and selected for responsiveness respond more efficiently is noteworthy (20, 21). For example, the culture of single inflorescences can produce large numbers of haploids in Brassica and Zea as well as providing the large numbers of cells from single microspores for further in vitro manipulation, selection, and as potential targets for genetic engineering (22, 23).

3. GENETIC VARIATION FROM CELL AND TISSUE CULTURES

3.1. Somaclonal variation

As plant regeneration became routine for more species, a curious and somewhat surprising observation was made. Not all plants recovered resembled the original plant donor. The more optimistic view was that a large number of useful new genetic traits would be available for plant breeders with eventual use in new and improved versions of older cultivars (24, 25). A pessimist viewed the same data with alarm since the variation could not be controlled, and general variation was observed similar to that observed in either the same germplasm, or other germplasm within the species (26, 27, 28). The widespread observation of variation from tissue culture, generally more severe with longer periods of culture, is now commonly accepted. Its value in applied agriculture is still controversial, and will likely remain so for the next several years.

Several interesting and potentially useful traits have been recovered from cell cultures including the widely publicized high solids tomato (25, Table 2). A white

TABLE 2. A SAMPLE OF CULTIVARS/GERMPLASM RELEASED OR IN FIELD TRIALS FROM GENETIC VARIATION FROM TISSUE CULTURE

SPECIES/CULTIVAR	SOURCE	COUNTRY/YEAR	NOTED CHARACTERISTICS (REFERENCE)
Brassica napus	S*	Canada/1988	chlorsulfuron, imidazilinone res. (29,30)
Capsicum annum			
Bell Sweet	G	USA/1987	low seeded (31)
Lycopersium esculatum			
"DNAP-9"	S	USA/1984	high solids in fruit (32)
"DNAP-17"	S	USA/1987	*Fusarium* Race 2 (33)
Medicago sativa			
C2-M4	S	USA/1988	White flower, transposable element (34)
Nicotiania tabacum			
NCT G-51, NCTG-52	G	USA/1990	potato virus Y (35, 36)
Sorghum bicolor	S	USA/1989-89	fall armyworm resistance (37)
Zea mays	S	USA/1990	imidazilinone res. (38, 39)

*S=SOMACLONAL, G=GAMETOCLONAL, S=SELECTION

flowered trait that behaves as a transposable element in tissue culture in alfalfa has been described by Bingham (33). Resistance to Potato Virus Y has been recovered from anther culture of tobacco (Witherspoon, 34 personal communication). Other traits have been reported, but have not been released either as germplasm or in new varieties (Vasil, 28).

3.2. Directed selection

Selection at the cellular level has not resulted in as many useful genes for applied agriculture as initially suggested (28). Herbicide resistance, in which specific biochemical pathways condition susceptibility or resistance, have given rise to mutants with clear levels of resistance or susceptibility in vitro and in field plantings (38, 39). Chlorsulfuron resistance was obtained from cell cultures of tobacco in which dramatic differences between unselected and selected mutants with chlorsulfuron was observed (40). Mutants with 10-100 fold increases in tolerance to the imidazilinone herbicides in maize have been obtained (38), and are in the advanced testing stage by Pioneer Hi-Bred International, Inc. (39, see also 29, 30). Despite the early recovery of mutants for herbicide resistance in tobacco, the additional cost of labeling herbicides for use in that crop have precluded commercial use (E.A. Wernsman, personal communication).

Unfortunately, other selection strategies have not produced mutants which possess clear expression *in vitro* and *in planta* (28). Selection against phosphinothricin resistance in alfalfa cell cultures resulted in modifications in tolerance, but expression in regenerated plants was only modestly improved over non-selected germplasm (41). Similar results have been observed for 2,4-D (42) and several disease resistant factors except for virus resistance in tobacco (36, 43).

The advantages of single cell selection and mutagenesis from microspore cultures has resulted in chlorsulfuron and imidazolinone tolerant plants in *Brassica* (29, 44). Such a microspore selection system partially off sets the disadvantage of interference of neighboring cells in mutant expression. The most significant limitation for directed selection is the availability of traits which are understood well enough at the biochemical level to warrant extensive research in mutant isolation (28).

4. GENE TRANSFER

4.1. Somatic hybridization

The advent of protoplast technology brought about a revolution in plant biology. Non-sexual gene transfer became a reality for the first time. The first parasexual hybrids among *Nicotiana* species by Carlson and Dearing (45) were the first of many somatic hybrids of both distant and relatively closely related species. Although the wide somatic hybrids gained widespread attention, hybrids between more closely related species have been of more immediate use to applied plant biology (28). A somatic hybrid between *N. tabacum* and *N. rustica* gave rise to a wide range of variation which included genes for nicotine content, blue mold and black root rot disease resistance. Interestingly, the sources of nicotine genes are unique to these lines compared to other tobacco lines (46). Somatic hybrids between

TABLE 3. A SAMPLE OF CULTIVARS/GERMPLASM RELEASED OR IN FIELD TRIALS AS A RESULT OF GENE TRANSFER USING EITHER SOMATIC HYBRIDIZATION OR GENETIC ENGINEERING.

SPECIES/CULTURE	SOURCE	COUNTRY/YEAR	NOTABLE CHARACTERISTICS
Brassica napus	S*	Canada/1987	CMS, atrazine res (47)
	G	Canada/1989	Glyphosate res (48)
Glycine max	G	US/1989,1990	Glyphosate tolerance (48)
Linnium	G	Canada/1989	Glyphosate (48)
Lycopersium esculatum			
	G	USA/1988,89	Virus (49)
	G	USA/1988,89	Insect-Bacillus thuringensis (50)
Medicago sativa	G	USA/1988,89	Glyphosate (48)
	G	USA/1990	Alfalfa Mosaic Virus (50, 51)
Nicotiana tabacum	G	USA/1988	AMV (53)
Solanum tuberosum	G	Europe/1989	Phosphinothricin (54)
	G	USA/1988	Virus (48)
		Europe/1988	Phosphinothricin (54) Reslst.

*S=SOMATIC HYBRIDIZATION, G=GENE TRANSFER

Brassica species combined cytoplasmic and nuclear organelles in all possible combinations including CMS genes derived from mitochondria, and atrazine tolerance derived from chloroplasts while retaining the desirable nuclear composition of B. napus (47).

The major limitations for more widespread use of somatic hybridization are the restrictions imposed by the difficulties of the protoplast technique in most of the major crop species and inability to selectively use portions of the genome from one species in a recipient species (28).

4.2. Genetic engineering

Genetic engineering has reduced or eliminated many of the barriers that once existed in transferring genes from virtually any source into plants. This form of gene transfer has also changed the way those in plant cell and tissue culture view their discipline. Many cell biologists now consider their collaborations with molecular biologists and biochemists as an inseparable part of their research. Gene availability and more importantly, function in the desired recipient plant at the correct time and place are the product of such collaborations. A great deal of suspicion and mistrust has gradually given way to spectacular scientific accomplishments and equally spectacular results in both greenhouse and field trials (Table 3).

Cell and tissue culture, especially developmental biology, has never been more relevant than now. Each of the major methodologies for genetic transformation (see below) are dependant on accurate information about the location and frequency of responsive cells to accept DNA and which can also continue their development into plants. Three major DNA delivery mechanisms have been used with plant cells: 1. direct through protoplasts, 2. Agrobacterium, and 3. direct into intact cells via DNA-coated microprojectiles.

4.2.1 Direct through protoplasts. The first demonstration of transfer of biologically active DNA into a plant was accomplished in *Petunia* (55, 56). Dramatic improvements in the efficiency of DNA uptake with chemical treatments and by electroporation (see 57 for review) have been shown. Electroporation has also been used to predict gene function through transient assays in several species (58). The major limitation, as in somatic hybridization, has been the difficulty with which plants are recovered from protoplasts in many species, and the unwanted variation which comes from extended culture periods (26, 28). Despite the recovery of plants from maize by several labs (59, 60, and Maddock unpublished data) the source suspension cultures often give rise to abnormal plants which set seed only with difficulty. Continued improvement in plant regeneration from protoplasts of a wide range of species should improve the prospects for this method of gene transfer (see others papers this section).

4.2.2 *Agrobacterium*: The development of the leaf disk procedure with *Agrobacterium* in *Petunia* by Horsch et al (61) has allowed routine production of transgenic plants in laboratories throughout the world. Virtually all of the genes now in test have been obtained using variations of the leaf disk procedure (Table 3). The most notable modifications in accomplishing gene transfer has occurred by finding the most suitable source cells which interact with *Agrobacterium* and which also regenerate with high efficiency (62). In addition, the use of wide host range *Agrobacterium*, or identification of *Agrobacterium* strains which are specific to the species and plant genotypes in use has been very important (63). The major limitation of *Agrobacterium* is a host range which excludes most monocots.

4.2.3 Microprojectile bombardment: Several plant species have recently been transformed by the use of high velocity microprojectiles to deliver DNA directly into intact plant cells. (see Klein, et al. this volume). Transient activity was demonstrated in maize in 1987 (64), with stable transformation obtained in non-regenerable tobacco and maize cultures soon after (Ross and Tomes, unpublished). Stable plant transformation with seed transmission and DNA integration patterns consistent with other methods of direct DNA uptake has been reported in tobacco (65 and Klein et al. in this volume). Current, but unpublished results, demonstrate transformation of maize by microprojectile bombardment using embryogenic suspension cultures and either the selectable marker gene for chlorsulfuron (ALS), (Fromm M.E., personal communication, Maddock, S.E. unpublished data or phosphinothricin (BAR) Gordon Kamm W., personal communication). Transformed seed progeny have been obtained for both the ALS (Fromm, M.E. personal communication) and BAR transformants respectively (Fromm, M.E. and Gordon Kamm, W., personal communication).

The LC gene for anthocyanin has been used as a visible marker gene in a W22 R-G marker stock to examine gene function in aleurone cells (66) and also in stable transformed plants (Ross, M. et al. unpublished data.)

The use of the microprojectile bombardment method requires location of potentially responsive cells which are accessible to DNA-coated particles. In addition, selectable marker genes that are expressed at high efficiency in the targeted tissue of the host species is prerequisite (Tomes, D.T., unpublished observations) unless procedures such as that described for soybean (67) can be used.

5. CONCLUSIONS

The art and science of tissue culture has matured, both in the sheer volume of knowledge and with the cellular and molecular tools now available. The knowledge base of plant cells and their developmental biology, though rudimentary, offers a starting point for both basic and applied biology pursuits. A quote attributed to Henry Kissinger (Borman, M, personal communication) at the start of a press conference best applies, "Which of you have questions for the answers I have prepared?" For tissue culturists this could be paraphrased as "Which problems in biology and plant breeding are ready to be solved with the solutions we have now."

6. ACKNOWLEDGEMENTS

The author sincerely appreciates the unpublished results and information from the following individuals K.J. Kasha, G. RaKow, L. Shugar, E.A. Wernsman, G.B. Collins, C.N. Bollich, C. Magill, A.D. Genovisi, E. Reinbergs, W.A. Keller, E.B. Swanson, E.T. Bingham, I. Grant, D.J. Isenhour, R.E. Fincher, X. Delannay, R. Townsend and critical reading of the manuscript by Nancy Tomes.

REFERENCES:

1. Kasha, K.J. 1974. Hapoids in higher plants. (Ed. Kasha, K.J.) University of Guelph, Guelph, Canada
2. RaKow, G. 1990. personal communication. Ag. Canada Research Branch, Saskatoon, Canada.
3. Keller, W.A. 1990. personal communication. From Manbo Seeds, Holesy, Denmark.
4. Shugar, L., 1990. W. G. Thompson & Sons Ltd., Box 130, Blenheim, Ontario, Canada.
5. Kasha, K.J., 1990. personal communication.
6. Wernsman, E.A. 1990. personal communication. From Chili Tabacos.
7. Nielson, M.T., Legg, P.D. and Collins, G.B. 1988. Crop. Sci 29, 242.
8. Bollich, C.N. & Magill, C. 1990. personal communication.
9. Buyser, J., et al. 1987, Plant Breeding 98, 53-56.
10. Genovisi, A.D. 1990. In. Biotechnology in Agriculture and Forestry: Haploids in Crop Improvement. I. (Y.P.S. BAJAJ, ed) Vol 12. pp. 176-203. Springer Verlag.
11. Dunwell, J.M. 1976. Env. and Exp. Botany 16, 109-118.
12. Kasha, K.J. and Reinbergs, E. 1979. Proceedings 2nd International Haploidy Symposium pp. 215-230.
13. Dunwell, J.M. 1983. Annals of Botany 48, 535-542.
14. Shannon, P.R., et al. Plant Cell Tiss Org Cult 4, 271-280.
15. Petolina, J.F., et al. 1988. Theor Appl Genet 76, 157-159.
16. Keller, W.A., and K. C. Armstrong. 1978. Zeitschrift Fur Pflanzenzuchtung 80, 100-108.
17. Pechan, P.M. and Keller, W. A. 1989. In Vitro Cell & Dev Biol 25, 1073-1074.
18. Dunwell, J.M., et al. 1985 J. Exp. Bot. 36, 1-11.
19. Coumans, M.P., et al. 1989. Plant Cell Reports 7, 618-621.
20. Pescitelli S.M., et al. 1989. Plant Cell Rep. 7, 673-676.
21. Petolina, J.F., et al. 1988. Theor. Appl. Genet. 76, 157-159.
22. Swanson, E.B., et al. 1987. Plant Cell Rep. 6, 94-97.
23. Polsoni, L., et al. 1988. Can. J. Bot. 66, 1681-1685.
24. Larkin, P.J. and Scowcroft, W.R. 1981. Theor. Appl. Genet. 60, 197-214.
25. Evans, D.A., and Sharp, W.R. 1983. Science 221, 949-951.
26. Bingham, E.T. and McCoy, T.J. 1986. Plant Breeding Rev 4, 123-152.
27. Maddock, S.E. and Semple, J.T. 1986. J. Exp. Botany 37, 1065-1078.
28. Vasil, I.K. 1990. Biotechnology 8, 296-301.
29. Swanson, E.B., et al. 1989 Theor Appl Genet 78, 525-530.
30. Grant, I. 1990. personal communication. Allelix Crop Technologies, Mississauga, Ontario, Canada.
31. Plant Variety Protection Certificate #8700124. 1988. USDA Plant Variety Protection Office, Beltsville, MD.
32. Plant Variety Protection Certificate #8400146. 1984. USDA Plant Variety Protection Office, Beltsville, MD.
33. Plant Variety Protection Certificate #8700123. 1987. USDA, Plant Variety Protection Office, Beltsville, MD.

34. Bingham, E.T., et al. 1988. Plant Transposable Elements (Nelson, O. ed.) pp. 325-337. Plenum Publishing.
35. Witherspoon, W.D., et al. 1990. unpublished results.
36. Wernsman, E.A. 1990. personal Communication. N.C. State University, Raleigh, N.C.
37. Isenhour, D.J., et al. 1990. J. Econ. Entomol. in press.
38. Anderson, P.C., and Hibberd, K.A. 1988. U.S. Patent #4,761,373.
39. Fincher, R.E. 1990. personal communication, Pioneer Hi-Bred Intl. Inc., Johnston, IA.
40. Chaleff, R.S. and T.B. Ray. 1984. Science 223, 1148-1151.
41. Donn, G., et al. 1984. J. Mol. Appl. Gen. 2, 621-635.
42. Swanson, E.B. and D.T. Tomes. 1983. Plant Sci. 29, 19-24.
43. Toyoda, H., et al. 1989. Plant Cell Reports 8, 433-436.
44. Swanson, E.B., et al. 1988. Plant Cell Reports 7, 83-87.
45. Carlson, P.S., et al. 1972. Proc Nat Acad Sci (Wash) 69, 2292-2294.
46. Pandeya, R.S., et al. Pflanzenzuctung 96, 346-359.
47. Barsby, T.L., et al. 1987. Theor Appl Genet 73, 809-814.
48. Delannay, X. 1990. personal communication, Monsanto Company, St. Louis, MO. USA
49. Nelson et al. 1988. Biotechnology 6, 403-409.
50. Delannay, X. 1989. Biotechnology 7, 1265-1269.
51. Loesch-Fries, S.L., et al. EMBO 6, 1845-1851.
52. Townsend, R. 1990. personal communication. Pioneer Hi-Bred Int'l., Inc., Johnston, IA. USA.
53. Tumer, N.E., et al. 1987. EMBO 6, 1181-1188.
54. De Greef, W., et al. 1989. Biotechnology 7, 61-64.
55. Draper, J., et al. 1982. Plant Cell Physiol 23, 451-458.
56. Krens, F.A., et al. 1982. Nature (London) 296, 72-74.
57. Paszkowski, J., et al. 1989. in Cell Culture and Somatic Cell Genetics of Plants (Schell, J. and Vasil, I.K., eds.) Vol 6, pp 52-65. Academic Press, Inc. San Diego.
58. Callis, J., et al. 1987. Genes Dev 1, 1183-1200.
59. Prioli, L.M. and Sondhal, M.R. 1989. Biotechnology 7, 589-595.
60. Shillito, R.D., et al. 1989. Biotechnology 7, 581-587.
61. Horsch, R.B., et al. 1985. Science 227, 1229-1231.
62. Fillatti, J.J., et al. 1987. Biotechnology 5, 726-730.
63. Hinchee, M.A.W., et al. 1988. Biotechnology 6, 915-922.
64. Klein, T.M., et al. 1988. Proc Natl Acad Sci 85, 4305-4309.
65. Tomes, D.T., et al. 1990. Plant Mol Biol 14, 261-269.
66. Ludwig, S.R., et al. 1988. Biotechnology 247, 331-334.
67. McCabe, D.E., et al. 1988. Biotechnology 87, 923-926.

PLANT REGENERATION FROM EMBRYOGENIC CALLI, CELL SUSPENSION CULTURES AND PROTOPLASTS OF *Triticum aestivum* L. (WHEAT)

INDRA K. VASIL, VIMLA VASIL AND FLONA REDWAY

Laboratory of Plant Cell and Molecular Biology, Department of Vegetable Crops, University of Florida, Gainesville, FL 32611, U.S.A.

1. INTRODUCTION

Efficient and reproducible methods were developed during the 1980's for plant regeneration from tissue cultures of all of the important species of the Gramineae (1,2). Regeneration of plants has been shown to be predominantly by the process of somatic embryogenesis (3). Embryogenic cultures can be initiated mostly from explants such as immature embryos, young inflorescences and bases of young leaves which contain meristematic and undifferentiated cells which are not yet committed to specific developmental pathways. Embryogenic cells are relatively small, isodiametric, richly cytoplasmic, strongly basophilic, and contain prominent nuclei, amyloplasts, lipid droplets and only very small vacuoles. Like meristematic cells, they maintain their genetic fidelity during culture and have a selective advantage over cytologically aberrant cells in forming somatic embryos and plants (4-7). Furthermore, it has been shown that somatic embryos develop - directly or indirectly - from single cells (8,9). Consequently, plants derived from embryogenic cultures are neither chimeras nor variants, but true clones. The single cell origin, non-chimeral nature and genetic fidelity of plants derived from somatic embryos are very attractive features for any genetic transformation system.

Genetic transformation and the production of transgenic plants is one of the most useful and powerful tools of plant biotechnology. In most instances it can be achieved by the use of *Agrobacterium tumefaciens*, a natural vector for the transfer of genes into plant cells (10). Unfortunately, this well defined, efficient and relatively simple method has not been useful thus far for the transformation of gramineous species which are outside the natural host range of *Agrobacterium*. Accordingly, much of the research on transformation of grass and cereal species has depended on techniques which allow the direct delivery of DNA into cells (11-13). This necessitates the use of regenerable single cells or protoplasts for obtaining transgenic plants. In spite of extensive efforts, there is as yet no convincing evidence of plant regeneration or even sustained cell divisions in mesophyll protoplasts - which have been used extensively for the regeneration of plants in dicotyledonous species - of the Gramineae. Consequently, embryogenic cultures, in which plants arise from single cells through the formation of somatic embryos, and which are also the only current source of totipotent protoplasts in the Gramineae (1-3,14-16), are of critical importance in the genetic transformation of this important group of food plants.

Wheat is one of the most important crop species in the world. Earlier, we have described plant regeneration from embryogenic cultures derived from immature embryos and inflorescences of wheat (17-19). Culture of wheat protoplasts isolated from suspension cultures has been described previously. Only in one instance, where a suspension culture was established from a single anther-derived callus, rare plantlets were obtained from protoplasts dividing at very low frequencies (20). More recently, we have identified the types of calli (21) that can be used to establish embryogenic suspension cultures (22), and have recovered green plants from protoplasts isolated from such suspension cultures (23). In this paper we provide a summary of these results and also report on our genetic transformation experiments.

2. MATERIALS AND METHODS

Only a brief description of the materials and methods used are described below. Detailed protocols have been published elsewhere (21-23).

2.1. Plant Regeneration from Embryogenic Calli

Eight commercial cultivars of wheat were used for the experiments: Anza, Chris, Coker 983, FLA301, FLA302, Fremont, Hunter and Pavon. Immature embryos (0.5-1.5 mm long, 10-12 days post-anthesis), and young inflorescences (5-20 mm long) were cultured under aseptic conditions on MS (24), MSAA (25) and C_8 (26) media for the induction of callus and plant regeneration. Anthers were obtained from spikes which were pre-treated at 4°C for 4-7 days, and cultured on solid or liquid N_6 (27), P_2 (28) and WM2 (29) media. The media were supplemented with a variety of substances, including 2,4-dichlorophenoxyacetic acid (2,4-D), dicamba, casein hydrolysate, glutamine, silver nitrate and cefotaxime. Immature embryo and inflorescence cultures were incubated at 28°C, and anther cultures at 30°C, in the dark.

Immature embryo derived embryogenic calli were transferred to MS medium with 1 mg/l indole-3-acetic acid (IAA) and 1 mg/l zeatin for plant regeneration (28°C, 16 h photoperiod). Calli from inflorescences were transferred to MS medium with 0.2 mg/l 2,4-D and 1 mg/l 6-γγ-dimethylallylaminopurine (2iP) at 28°C (16 h photoperiod), and those from anthers to MS medium with 0.5 mg/l naphthaleneacetic acid (NAA) and 0.5 mg/l kinetin (30°C in the dark).

Regenerated plantlets were hardened by transfer to MS medium without any plant growth regulators in Magenta jars, and then to Terra-lite Metro-mix in Conetainers, and finally to soil in pots in the greenhouse. Detailed statistical analysis of data was carried out (21).

2.2. Plant Regeneration from Embryogenic Suspension Cultures

Approximately 1 gm of embryogenic callus was transferred to a 125 ml Erlenmeyer flask containing 15 ml of liquid MS medium with 2 mg/l 2,4-D. Flasks were placed on a gyrotory shaker at 110 rpm in the dark at 27°C. For 2-3 months, about every 10 days, 10 ml of the supernatant medium was replaced with an equal volume of fresh medium. The established suspension was maintained by adding 15 ml of fresh medium to 15 ml of the culture once a week.

One ml of the cell suspension culture of FLA302 was plated onto MS medium (with various hormones) for plant regeneration. Somatic embryos and shoots formed were transferred to MS medium containing various concentrations and combinations of IAA, zeatin, benzyladenine, kinetin, gibberellic acid (GA_3) and NAA at 28°C with 16 h photoperiod.

2.3. Plant Regeneration from Protoplasts

Protoplasts were isolated from 5 ml settled cells obtained from three, five and seven days old suspension cultures, in a solution containing Cellulase Onozuka RS, Pectinase Serva and Pectolyase Y23. The cleaned and washed protoplasts were cultured at a density of 1-1.5 x 10^6/ml in liquid or Seaplaque agarose MS or KM (30) media, at 27°C in dark. For regeneration of somatic embryos and plants, the protocolonies were transferred to various media supplemented with plant growth regulators. Regenerated plants were first grown in Magenta jars and then transferred to soil.

3. RESULTS

3.1. Plant Regeneration from Embryogenic Calli

Embryogenic calli were obtained from cultured immature embryos on MS, MSAA and C_8 media (each with 3% sucrose and 2 mg/l 2,4-D), but the MS medium was clearly superior. Many modifications of the MS medium were used, but the best results were obtained when it contained 3% sucrose, 2 mg/l 2,4-D, 100 mg/l casein hydrolysate and 500 mg/l glutamine.

Immature embryos proved to be the best, and anthers the worst, explants for the formation of embryogenic calli. The response of inflorescences was intermediate. The developmental stage of the embryos was the most important factor for the formation of embryogenic callus.

Within a month after the culture of immature embryos, the following three types of calli were formed: (i) a soft, watery and non-embryogenic callus, (ii) an off-white, compact, and nodular embryogenic callus (TYPE A), and (iii) a white and compact embryogenic callus (TYPE B). Repeated subculture (5-8 months) of the embryogenic calli gave rise to an aged, compact and nodular embryogenic callus (TYPE C) and an aged and friable embryogenic callus (TYPE D). During 1987, the TYPE C and TYPE D calli were obtained from 0.5% of the cultured embryos of FLA302. However, in 1988 these two types of calli were obtained in about 10% of the embryos of seven cultivars (Chris, Coker 983, FLA301, FLA302, Fremont, Hunter and Pavon). The aged calli (TYPE C and TYPE D) maintained their embryogenic capacity for over 19 months.

TYPE A and TYPE B calli of each of the eight cultivars used formed green shoots on regeneration media. Shoots were also obtained from the aged TYPE C and TYPE D calli. Rooted plants were grown to maturity in soil.

3.2. Plant Regeneration from Embryogenic Suspensions

One month old embryogenic calli (TYPE A and TYPE B) from any of the cultivars failed to dissociate in liquid media and did not give rise to suspension cultures. However, suspension cultures were established from both TYPE C and TYPE D calli of FLA302. An off-white, compact, organized and nodular embryogenic callus was formed upon plating of the regenerable suspension from TYPE C callus on MS medium containing 2% sucrose and various hormones. Transfer of these calli to regeneration media resulted in the formation of somatic embryos and green shoots. More than 200 rooted green plants were formed. No somatic embryos or plants were recovered from the suspension established from TYPE D calli. Later, stable cell suspension cultures were also obtained from aged but less compact and more friable embryogenic calli of Coker 983, Chris, FLA302, Fremont, Hunter and Pavon.

3.3. Plant Regeneration from Protoplasts

At the time of protoplast culture, the regenerable suspension of FLA302 was one year old, while the non-regenerable suspension was six months old.

High yields (ca 2.5 x 10^7/gm fresh weight) of healthy protoplasts were obtained from the regenerable suspension. Cell wall formation was completed within two days after culture, and two-, three-, and four-celled colonies were observed at the end of the first week. Within two weeks compact protocolonies (8-12 cells) were formed. Both MS and KM media supported the formation of protocolonies, although the regenerable suspension gave the highest plating efficiency (35-39%) on KM medium in 1.2% agarose. Most efficient plant regeneration took place when protocolonies were transferred to MS medium alone or with 1 mg/l IAA and 1 mg/l zeatin. More than 100 rooted green plants were produced, and several were transferred to soil in the greenhouse.

3.4. Genetic Transformation

The availability of regenerable suspension cultures and protoplasts of wheat has encouraged us to use these in a variety of experiments on genetic transformation (unpublished results). We have observed transient expression of various genes delivered into scutellar cells of immature embryos and plated cell suspension cultures by the Biolistics method, and into protoplasts by electroporation or polyethylene glycol (PEG) methods. We have also identified many independent putative stable transformants on selective media. Further molecular confirmation of the putative transformants is being carried out, along with the attempts to recover transgenic plants.

4. DISCUSSION

The results presented in this report once again highlight the importance of the identification and characterization of embryogenic calli, and their selective maintenance for long-term totipotent cultures. It has also demonstrated that in wheat totipotent cell suspension cultures can be obtained only from aged embryogenic calli.

Culture of protoplasts of cereals and grasses has proven to be very difficult. However, much progress has been made since the 1970's when sustained cell divisions were obtained in protoplasts isolated from cell suspension cultures (31). Then, in 1980 (15) it was shown that somatic embryos and plants can be obtained from protoplasts isolated from embryogenic suspension cultures. Although the establishment as well as the maintenance of embryogenic suspension cultures has been found to be extremely difficult, such cell lines have nevertheless been established in about ten species, including maize, rice, sugarcane and now wheat. All reported instances of plant regeneration from protoplasts in the Gramineae are based on the use of embryogenic protoplasts from these cell lines (1,3,14,16). Similarly, the recent recovery of somatic hybrids/cybrids of rice (32) and transgenic plants of maize (33), rice (34), and orchard grass (35) is also based on the use of embryogenic protoplasts. Finally, the totipotent suspension cultures are a very attractive target for the direct delivery of DNA into intact cells and tissues by the Biolistics technology (11).

Acknowledgements. This work was supported by the Monsanto Company, St. Louis, MO.

REFERENCES

1 Bright, S.W.J. and Jones, M.G.K., eds. (1985) Cereal Tissue and Cell Culture. Martinus Nijhoff/Dr W. Junk Publishers, Dordrecht

2 Vasil, I.K. and Vasil, V. (1986) in Cell Culture and Somatic Cell Genetics of Plants, Vol. 3, Plant Regeneration and Genetic Variability (Vasil, I.K., ed.), pp. 121-150, Academic Press, Orlando
3 Vasil, I.K. (1987) J. Pl. Physiol. 128, 193-218
4 Hanna, W.W., Lu, C. and Vasil, I.K. (1984) Theoret. Appl. Genet. 67, 155-159
5 Swedlund, B. and Vasil, I.K. (1985) Theoret. Appl. Genet. 67, 575-581
6 Cavallini, A., Lupi, M.C., Cremonini, R. and Bennici, A. (1987) Protoplasma 139, 66-70
7 Cavallini, A. and Natali, L. (1989) Pl. Sci. 62, 255-261
8 Vasil, V. and Vasil, I.K. (1982) Bot. Gaz. 143, 454-465
9 Jones, T.J. and Rost, T.L. (1989) Bot. Gaz. 150, 41-49
10 Fraley, R.T., Rogers, S.G. and Horsch, R.B. (1986) CRC Crit. Rev. Pl. Sci. 4, 1-46
11 Klein, T.M., Roth, B.A. and Fromm, M.E. (1980) in Genetic Engineering: Principles and Methods, Volume II (Setlow, J.K., ed.), pp. 13-31, Plenum Press, New York
12 Potrykus, I. (1989) TIBTECH 7, 269-273
13 Vasil, V., Hauptmann, R.M., Morrish, F.M. and Vasil, I.K. (1988) Pl. Cell Rep. 7, 499-503
14 Vasil, I.K. (1988) Bio/Technology 6, 397-402
15 Vasil, V. and Vasil, I.K. (1980) Theoret. Appl. Genet. 56, 97-99
16 Roest, S. and Gilissen, L.L.W. (1989) Acta. Bot. Neerl. 38, 1-23
17 Ozias-Akins, P. and Vasil, I.K. (1982) Protoplasma 110, 95-105
18 Ozias-Akins, P. and Vasil, I.K. (1983) Protoplasma 117, 40-44
19 Ozias-Akins, P. and Vasil, I.K. (1983) Amer. J. Bot. 70, 1092-1097
20 Harris, R., Wright, M., Byrne, M., Varnum, J., Brightwell, B. and Schubert, K. (1988) Pl. Cell Rep. 7, 337-340
21 Redway, F.A., Vasil, V., Lu, D. and Vasil, I.K. (1990) Theoret. Appl. Genet. 79, In Press
22 Redway, F.A., Vasil, V. and Vasil, I.K. (1990) Pl. Cell Rep. 9, In Press
23 Vasil, V., Redway, F.A. and Vasil, I.K. (1990) Bio/Technology 8, In Press
24 Murashige, T. and Skoog, F. (1962) Physiol. Plant. 15, 473-497
25 Muller, A.J. and Grafe, R. (1978) Molec. Gen. Genet. 161, 67-76
26 Dudits, D., Hadlaczky, G., Levi, E., Fejer, O., Haydu, Z. and Lazar, G. (1977) Theoret. Appl. Genet. 51, 127-132
27 Chu, C.C., Wang, C.C. and Sun, C.S. (1975) Sci. Sin. 18, 659-668
28 Chuang, C.C., Ouyang, J.W., Chia, H., Chou, S.M. and Ching, C.K. (1978) in Proceedings of Symposium on Plant Tissue Culture, pp. 52-56, Science Press, Beijing
29 Datta, S.K. and Wenzel, G. (1987) Pl. Sci. 48, 49-54
30 Kao, K.N. and Michayluk, M.R. (1975) Planta 126, 105-110
31 Vasil, I.K. (1983) Int. Rev. Cytol. Supp. 14, 79-88
32 Kyozuka, J., Kaneda, T. and Shimamoto, K. (1980) Bio/Technology 7, 1171-1174
33 Rhodes, C.A., Pierce, D.A., Mettler, I.J., Mascarenhas, D. and Detmer, J.J. (1988) Science 240, 204-207
34 Shimamoto, K., Terada, R., Izawa, T. and Fujimoto, H. (1989) Nature 338, 274-276
35 Horn, M.E., Shillito, R.D., Conger, B.V. and Harms, C.T. (1988) Pl. Cell Rep. 7, 469-472

TISSUE, CELL, AND PROTOPLAST CULTURE OF MAIZE (*Zea mays* L.)

Laudenir M. Prioli[1,2], William J. da Silva[1], and Maro R. Söndahl[3]

1. State University of Campinas (UNICAMP), Institute of Biology, Dept. of Genetics, 13081 Campinas, SP, Brazil; 2. Present address: North Carolina State University, Dept. of Genetics, Box 7614, Raleigh, NC 27695, USA; 3. DNA Plant Technology Corp., 2611 Branch Pike, Cinnaminson, NJ 08077, USA.

INTRODUCTION

In the past ten years considerable progress has been made on plant regeneration from tissues, cells, and protoplasts of various Gramineae species. In maize, plant regeneration was first described in 1975[1]. After the initial reports on somatic embryogenesis in maize in 1982[2,3], plant regeneration from embryo-derived callus was extended to numerous other genotypes[4,5,6]. Genetic variability for plant regeneration has been reported within maize germplasms adapted to temperate[7,8,9,10,11] and tropical[6] regions. Further studies on the genetic control of somatic embryogenesis are still needed to better understand and manipulate this trait. Considerable attention has also been directed at studying the genetic variability observed among plants derived from maize callus for use in genetics and breeding[12,13,14,15]. The possibility of regenerating plants from maize cell suspension and protoplast cultures has opened new perspectives for direct gene transfer. Nevertheless, reports on the establishment of maize cell suspension and protoplast cultures capable of regenerating plants remain limited to a few genotypes[16,17,18,19].

This report summarizes our results on evaluations of plant regeneration from embryo-derived callus, cell suspension, and protoplast cultures of maize inbred lines adapted to tropical regions. We have also evaluated the progenies of plants regenerated from callus for genetic variation.

PLANT REGENERATION FROM CALLUS

In initial studies[6], thirty-four maize homozygous inbred lines of high combining ability and adapted to tropical regions were evaluated for their ability to regenerate plants from callus. These inbred lines are from the Maize Genetics and Breeding Program of the Dept. of Genetics and Evolution, State University of Campinas (UNICAMP), Campinas, SP, Brazil. The inbreds were derived primarily from three gene pools: (1) Cateto, a flint autochthonous race adapted to the Southern Atlantic Coast of South America; (2) Tuxpeño, a dent race adapted to the coastal region of the Gulf of Mexico; and (3) an advanced generation population derived from Tuxpeño x teosinte (*Zea mays mexicana*, race Chalco) interspecific hybrid. Donor plants were grown in experimental field plots in Campinas (22°30' South Latitude) planted in November 1984, February 1985, and November 1985. In each growing season, callus cultures were initiated from 240 immature embryos of each genotype on N6 medium[20] containing 2,4-D (2.5 µM and 10 µM) and sucrose (3% and 6%). After 20-60 days, cultures were transferred to regeneration medium (RM-1) consisting of MS medium[21] without growth regulators. Regenerated plants were raised to maturity either in the greenhouse or field.

Twenty of the 34 maize inbred lines were capable of forming embryogenic callus cultures and regenerating plants under the culture conditions used in this study. Significant genetic differences for induction of somatic embryogenesis and regeneration were observed among the 20 inbreds responsive for plant regeneration. In addition, the conditions to achieve an optimal embryogenic response differed among genotypes as indicated by a significant interaction between environmental and genotypic effects. For example, several inbreds were more responsive in the fall when donor plants were not as vigorous as in the summer season. Despite the variation among growing seasons, some inbreds consistently tended to be highly responsive, while others always had poor responses. One Cateto inbred line designated Cat100-1 was always the most responsive for induction of somatic embryogenesis and plant regeneration.

GENETIC CONTROL OF SOMATIC EMBRYOGENESIS

Considering the genetic differences for somatic embryogenesis and plant regeneration demonstrated in the initial experiments, a few inbred lines were chosen to study the inheritance of the capacity to produce embryogenic callus from immature embryos. Two maize inbred lines (Cat100-1 and Cat100-8) capable of regenerating plants via somatic embryogenesis and four inbreds (Tux100-1, Tux100-5, Tux100-8, and Tux100-12) that did not develop embryogenic callus and did not regenerate plants were chosen for this study. Four families consisting of one parent capable of somatic embryogenesis, one parent not capable of somatic embryogenesis, F1 hybrids of reciprocal crosses, F2, and F1 hybrids backcrossed to each parent were evaluated. Donor plants were grown in Campinas during the summer of 1986/1987. Callus cultures were initiated on N6 medium supplemented with 2,4-D (7.5 μM) and sucrose (3% and 6%). A total of 240-400 immature embryos of each generation were plated on each culture medium. Cultures were evaluated every 6-8 days during five weeks using a stereomicroscope. Calli developing one or more somatic embryos were scored as embryogenic and calli not forming somatic embryos were considered non-embryogenic.

The segregation pattern for induction of embryogenic callus was similar among three families (Cat100-8 x Tux100-5, Cat100-8 x Tux100-12 and Cat100-8 x Tux100-1) and is represented in Figure 1A. Somatic embryogenesis segregated as a dominant trait. The segregation of embryogenic callus in all generations and especially in the backcrosses clearly indicates that only a few genes control the induction of somatic embryogenesis in these genotypes. The segregation pattern within these three families suggests that two major independent nuclear loci with different degree of dominance interact to control somatic embryogenesis. A third dominant nuclear locus seems to be a suppressor of somatic embryogenesis, as suggested by the segregation pattern observed when the inbred lines Cat100-1 and Tux100-8 were used as parents (Figure 1A). The capacity for somatic embryogenesis is recessive in this family and segregates close to the expected for a single Mendelian locus. High additive effect for regeneration in maize had been previously described by Tomes and Smith[8]. Studies of two families involving the maize inbred lines A188, Mo17, and B73[10,11] and studies in alfalfa[22] also suggested the existence of one or two primary locus conditioning regeneration as a dominant trait.

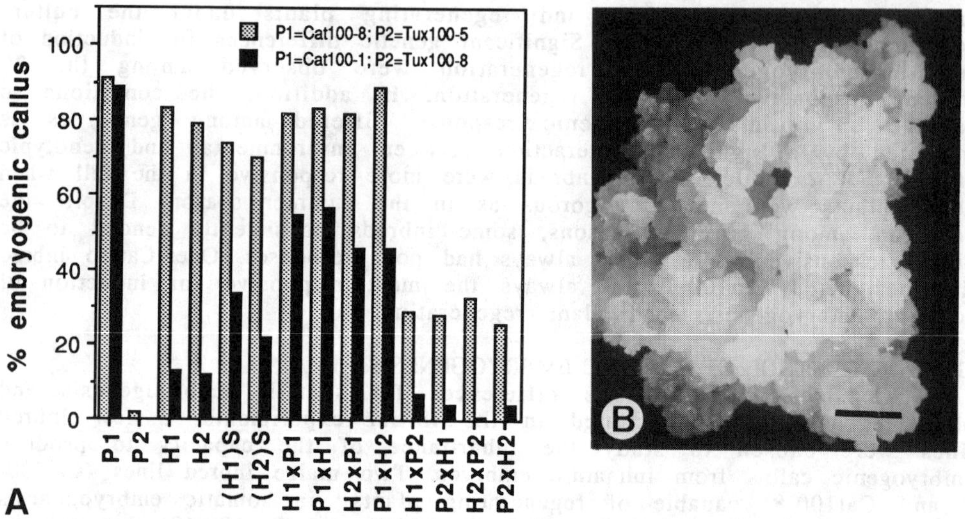

Figure 1. **A.** Inheritance of somatic embryogenesis in Type I maize callus of two families involving the inbred lines Cat100-8, Tux100-5, Cat100-1, and Tux100-8. A total of 240-400 immature embryos per generation were used for callus initiation. Cultures were evaluated every 6-8 days during five weeks. P=Parent; H1=P1xP2; H2=P2xP1; S=Self-pollination. **B.** Highly friable embryogenic callus of the inbred line Cat100-1 cultured on CM-1 medium. Bar = 2.8 mm.

SOMACLONAL VARIATION

Genetic variation was observed in approximately 10% of the 203 progenies of plants regenerated from callus of the inbred lines Cat100-1, Cat100-8, and Ast-1. Since the the donor inbreds were selfed for more than thirty generations before used in the tissue culture experiments, they were considered homozygous inbred lines. Regenerated plants (somaclones R0) were selfed and/or crossed with the donor genotype. Somaclone progenies (R1 plus several R2 and R3 generations) were grown in field plots at Campinas in the summers of 1985, 1986, and 1987. Variants included many mutations previously observed among maize somaclones[13,14,15], such as an apparent nuclear male sterility, alterations for plant morphology (brachytic, erect leaves, and wrinkled leaves), chlorophyll deficiency (albino and striped), and changes in the endosperm color (light yellow and lemon), in some cases associated with chlorophyll deficiency. In addition, alteration of one band of zein polypeptide as analyzed by isoelectric focusing and a cytoplasmic male sterile mutant were observed among Cat100-1 somaclones.

Heterosis was observed among several hybrids obtained in crosses between somaclones and the donor inbred line, suggesting the possibility of exploring somaclonal variation to obtain sister lines for hybrid production. Although the heterosis in these hybrids was obvious, further studies are necessary to analyze this phenomenon and its possible agronomic use.

CALLUS CULTURES

Most embryogenic calli initiated from immature embryos of the tropical inbred lines were similar to the compact Type I callus commonly described for maize cultures[2,4]. A large number of somatic embryos in various stages of development were observed on the surface of nodular, compact, and opaque tissues after 15-20 days of culture. Besides the Type I cultures, friable embryogenic calli (Type II) were initiated directly from the scutellum and/or as spontaneous sectors of Type I callus of four inbred lines (Cat100-1, Cat100-6, Cat100-8, and TuxMo-1). As previously described for other genotypes[3], the Type II calli contained numerous somatic embryos in the globular stage. In histologycal studies of Type II callus samples of the tropical inbreds, the globular embryos were observed both isolated and in groups of a few interconnected embryos distributed among non-organized and soft tissues of vacuolated cells. Selected friable calli were cultured on CM-1 medium[6,18] consisting of N6 salts, 30 μM glycine, 15 μM thiamine-HCl, 7.5 μM nicotinic acid, 7.5 μM pyridoxine, 550 μM inositol, 7.5 μM 2,4-D, 3% sucrose, and 2.3 g/l Gelrite.

Highly friable embryogenic (HFE) callus cultures (Figure 1B) were selected from spontaneous sectors of 10 to 20-month-old friable or semi-friable embryogenic callus cultures of the inbred line Cat100-1. As suggested by light and electron microscopy studies, the high friability of the HFE callus type seems to be due both to the low number or absence of embryos in the globular stage and to the presence of numerous individual aggregates of cytoplasmically dense cells distributed in non-organized and loose tissue of round-shaped vacuolated cells. In addition, the Cat100-1 HFE callus did not produce a mucilaginous substance frequently described in Type II maize callus. The HFE Cat100-1 calli doubled their cell mass every 5-6 days and frequently developed a few clumps of hard and globular embryogenic callus from highly friable areas. These calli were selectively subcultured every 10-15 days by transferring the most friable areas to fresh CM-1 medium. When HFE calli were either maintained on CM-1 medium for 30-40 days without subculturing or were transferred to RM-1 medium, numerous somatic embryos were formed in the callus mass. Plants were regenerated on RM-1 medium. The HFE callus lines were cultured for 2-3 years, most of the time as a population of one thousand or more callus samples (10-12 samples per Petri dish). During this period some callus samples were discarded because they lost the capability to form somatic embryos or only formed abnormal somatic embryos which did not germinate into plants.

CELL SUSPENSION AND PROTOPLAST CULTURES

Friable embryogenic callus (Type II) of three inbred lines (Cat100-1, Cat100-6, and TuxMo-1) were used to initiate cell suspension cultures. Callus samples (1.5-2 g) containing numerous somatic embryos in the globular stage were transferred to 125 ml Erlenmeyer flasks containing 50 ml of SM-1 medium[18]. Cultures were incubated on a gyratory shaker at 165 rpm in the dark at 26±1°C and subcultured every 3-4 days. In a second experiment[18], HFE callus cultures (Figure 1B) were used to initiate cell suspensions.

After several months, the suspension cultures initiated from Type II callus consisted mostly of globular somatic embryos, vacuolated round-shaped cells proliferating from the basal region of the somatic embryos, and a few large aggregates of cytoplasmically dense cells associated with round-shaped vacuolated cells. On the other hand, one-month-old suspensions initiated from HFE callus consisted of numerous proliferating large and small aggregates of cytoplasmically dense cells plus a few globular embryos, several small pieces of callus, and a relatively small number of vacuolated cells. After 20-30 days of culture, the fastest growing suspension cultures were selectively subcultured by sedimentation. Selective subculturing was periodically repeated to maintain a large number of small aggregates of embryogenic cells and to eliminate the large cell aggregates, which continuously form in these embryogenic cell suspension cultures. In addition, finely dispersed embryogenic cell suspensions, which were used for protoplast isolations, were filtered every 20-30 days through 520 μm stainless steel sieves. Embryogenic calli were obtained from cell suspension samples periodically plated and cultured on CM-1 medium. Fertile plants were recovered from suspension-derived callus transferred to regeneration medium (RM-1). Abnormalities, such as the absence of tassel and/or ear, terminal ear, development of female flowers in the tassel, and short internodes were also observed among regenerated plants.

Fast-growing, finely dispersed, and embryogenic cell suspensions initiated from Cat100-1 HFE callus cultures were used for protoplast isolations after 2, 5, and 9 months of culture[18]. Harvested cells were isolated in a solution of 0.45 M mannitol containing 2% cellulase Onozuka RS and 1-1.5% Pectolyase Y-23. Cell colonies and embryogenic calli were obtained at high frequencies when protoplasts were cultured in Petri dishes either in a thin layer of liquid medium or on a feeder layer of maize cells. Either a modified KM-8P[23] or a modified N6 media was used for protoplast cultures. Protoplast-derived calli were grown on CM-1 medium. Plants were regenerated on RM-1 medium and raised to maturity in the greenhouse. Fertile plants were recovered from protoplast-derived calli. Abnormalities were also observed among protoplast-derived plants, similar to those observed among plants regenerated from donor cell suspensions. The results with the Cat100-1 inbred line, and the results reported by Rhodes *et al.*[16], Shillito *et al.*[17] and Sun *et al.*[19] for other maize genotypes clearly demonstrate that plants can be regenerated from maize protoplasts. In addition, similar to the report on a B73-derivative inbred line by Shillito *et al.*[17], it was demonstrated that it is possible to recover fertile plants from maize protoplasts isolated from embryogenic cell suspensions capable of regenerating fertile plants.

CONCLUSIONS

Tropical maize inbred lines have significant genetic variation for somatic embryogenesis and plant regeneration. Our results suggest that somatic embryogenesis in cultures initiated from maize immature embryos is conditioned by only a few genes. Apparently there are two independent major nuclear loci conditioning somatic embryogenesis in the genotypes studied. A third dominant nuclear locus seems to be a suppressor of somatic embryogenesis. Further studies on determination of the exact number of genes and their interactions on the control of somatic embryogenesis and plant regeneration from in vitro cultures are necessary to better understand and manipulate this trait in maize. The genetic variation observed among progenies of regenerated plants substantiate the usefulness of somaclonal variation in rapidly providing small and stable changes in maize inbred lines. In addition, it was demonstrated that highly friable embryogenic callus are suitable for the establisment of finely dispersed and fast-growing embryogenic cell suspensions with plant regeneration capacity. Fertile plants with viable progenies can be recovered from maize protoplasts when embryogenic cell suspensions are used as donor cells.

Acknowledgments: The valuable discussions with Paulo Arruda and Alberto J. Prioli during the development of this work are gratefully acknowledged. The research on plant regeneration from callus, friable callus, somaclonal variation, and inheritance of somatic embryogenesis was made at State University of Campinas (UNICAMP), SP, Brazil, and was supported by FINEP-PADCT and CNPq-PIG, Brazil. The research on cell suspensions and protoplasts was made at DNA Plant Technology Corp., NJ, USA.

REFERENCES

1. Green, C.E. and Phillips, R.L. (1975) Crop Sci. 15,417-421
2. Lu, C., Vasil, I.K. and Ozias-Akins, P. (1982) Theor. Appl. Genet. 62,109-112
3. Green, C.E. (1982) in Plant Tissue Culture. Proc. Fifth IAPTC Congress (Fujiwara, A., ed.), pp. 107-108, Maruzen Publ., Tokyo
4. Lu, C., Vasil, V. and Vasil, I.K. (1983) Theor. Appl. Genet. 66,285-289
5. Duncan, D.R., Williams, M.E., Zehr, B.E. and Widholm, J.M. (1985) Planta 165,322-332
6. Prioli, L.M. and Silva, W.J. (1989) Rev. Bras. Genet. 12,553-566
7. Bechert, M. and Qing, C.M. (1984) Theor. Appl. Genet. 28,247-251
8. Tomes, D.T. and Smith, O.S. (1985) Theor. Appl. Genet. 70,505-509
9. Hodges, T.K., Kamo, K.K., Imbrie, C.W. and Becwar, M.R. (1986) Bio/Technology 4,219-226
10. Hodges, T.K., Kamo,K.K., Becwar, M.R. and Schroll, S. (1985) in Biotechnology in Plant Science (Zaitlin, M. *et al.* eds.) pp. 15-33, Academic Press, New York
11. Willman, M.R., Schroll, S. and Hodges, T.K. (1989) In Vitro Cell Develop. Biol. 25,95-100
12. Edallo, S., Zucchinali, C., Perezin, M. and Salamini, F. (1981) Maydica 26,39-56
13. Earle, E.D. and Gracen, V.E. (1985) in Tissue Culture in Forestry and Agriculture (Henke, R.R. *et al.*, eds) pp. 139-152, Plenum Press, New York
14. Zehr, B.E., Williams, M.E., Duncan, D.R. and Widholm, J.M. (1987) Can. J. Bot. 65:491-
15. Lee, M. and Phillips, R.L. (1987) Genome 29,834-838
16. Rhodes, C.A., Lowe, K.S. and Ruby, K.L. (1988) Bio/Technology 6,56-60
17. Shillito, R.D., Carswell, G.K., Johnson, C.M., DiMaio, J.J. and Harms, C.T. (1989) Bio/Technology 7,581-587
18. Prioli, L.M. and Sondahl, M.R. (1989) Bio/Technology 7,589-594
19. Sun,C.S., Prioli, L.M., and Sondahl, M.R. (1989) Plant Cell Rep. 8,313-316
20. Chu, C.C., Wang, C.C., Sun, C.S., Hsu, C., Yin, K.C., Chu, C.Y., Bi, F.Y. (1975) Sci. Scinica 18,659-668
21. Murashige, T. and Skoog, F. (1962) Physiol. Plant. 15,473-497
22. Reisch, B. and Bingham, E.T. (1980) Plant Sci. Lett. 20,71-77
23. Kao, K.N. and Michayluck, M.R. (1975) Planta 126,105-110

ANTIOXIDANTS INFLUENCE THE PLATING EFFICIENCY AND MICROCALLUS-GROWTH OF PROTOPLASTS IN LOLIUM PERENNE L.

J. CREEMERS-MOLENAAR* AND Y.VAN OORT**
* BARENBRUG HOLLAND BV, STATIONSSTRAAT 40, 6678 AC OOSTERHOUT
** CENTRE FOR PLANT BREEDING RESEARCH (CPO), P.O. BOX 16, 6700 AA WAGENINGEN. THE NETHERLANDS

1. INTRODUCTION

In cereals and grasses the possibilities for the application of genetic engineering are still largely dependant upon the availability of a reproducible procedure for plant regeneration from protoplasts. However, treatments such as protoplast fusion and direct DNA-uptake often unfavourably influence the vitality of protoplasts and reduce the plating efficiency, thus requiring that the basic culture conditions are as close to optimal as possible. Recently it has been shown that during protoplast isolation free radical species are generated (1) and that these radicals are involved in reducing the stability of the protoplast membrane by the process of lipid peroxidation (2).

According to the definition given by Halliwell (3) a free radical is "any species capable of independent existence that contains one or more unpaired electrons". The unpaired electron can make a molecule highly reactive against other molecules.

Ishii (1) has shown that during the enzymatic isolation of protoplasts from oat leaves the superoxide radical O^{2-} was generated and hydrogen peroxide accumulated in the reaction mixture. In the presence of metal-complexes, H_2O_2 will produce the reactive hydroperoxyl radical OH·. This radical is capable of initiating the process of lipid peroxidation by abstracting a hydrogen atom from a methyl group in a fatty acid located in e.g. the cell membrane (3). Under aerobic conditions the so-formed carbon radical will react with singlet molecular oxygen (O_2) to form a peroxy radical. Peroxy radicals are converted to lipid peroxides by abstracting hydrogen atoms from other lipid molecules, thus stimulating a chain reaction of lipid peroxidation. When cells are injured small amounts of lipid peroxides can also be formed enzymatically by the enzymes cyclo-oxygenase and lipoxygenase. In the presence of iron salts lipid peroxides will form reactive alkoxy and peroxy radicals, which further stimulate lipid peroxidation. The overall effects of lipid peroxidation in cell membranes are 1) decreased membrane fluidity 2)increased leakiness of the membrane (Ca^{2+}) and 3) inactivation of membrane-bound enzymes (3).

Several studies (4,5,6) demonstrated that the addition of natural or synthetic antioxidants to the protoplast isolation and culture medium significantly increased the protoplast stability (4), the frequency of colony formation (5) or the plant regeneration frequency (6).
In this study it was investigated whether the addition of antioxidants to the protoplast isolation and/or culture medium could improve the culture of suspension-derived protoplasts of Lolium perenne.

2. MATERIALS AND METHODS

2.1. Protoplast isolation: Protoplasts from non-regenerating suspension cultures of L. perenne were isolated and cultured as described previously (7).

2.2. Antioxidants: glutathione peroxidase (GSH-Px), phospholipase A2 (PLase A2) and superoxide dismutase (SOD) were purchased from Hoechst; n-propyl gallate (n-PG), glutahione (GSH) and catalase from Sigma; DL- tocopherol (acid ester disodium salt of vitamin E) from ICN and L-ascorbic acid (vitamin C) from BDH. Freshly prepared stock solutions in Milli Q water were added to the isolation and/or culture medium at concentrations as described in the "results and discussion" section. After addition, the pH and the osmolality of the media were adjusted to the appropriate values (7).

2.3. Procedure: The protoplast yield (number of protoplasts/g FW suspension callus) was determined after overnight enzymic digestion in the presence or absence of antioxidants. The plating efficiency (PE) was expressed as the percentage of plated protoplasts that had formed cell colonies after 3-4 weeks of culture. After 6 weeks microcalli were transferred onto solid medium lacking antioxidants. The results shown are representative for at least 2 experiments and all values are the means of duplicate measurements.

3. RESULTS AND DISCUSSION

3.1 Vitamin C. Vitamin C has been demonstrated to be an important antioxidant in plants (8). Addition of 10 mM vitamin C to the enzyme mixture increased the yield of protoplasts 2-5 times (Table 1). However, the PE of the protoplasts decreased when the vitamin C concentration of the enzyme mixture was higher than 4 mM.
Vitamin E. In biological membranes (e.g. chloroplasts) vitamin E functions as a protector against lipid peroxidation by reducing lipid peroxy and alkoxy radicals. Addition of vitamin E in a range of 0.1-2.0 mM to the enzyme mixture and/or protoplast culture medium did not influence protoplast yield or viability (data not shown). However, a minor increase of the PE (1.5X) at 0.5 mM vitamin E was observed in repeated experiments.

Vitamin C+E. Studies on herbicide-induced lipid peroxidation in seedlings of several plants showed that plants which contained a vitamin C to vitamin E ratio of between 10 and 15:1 were highly protected against membrane damage (9). A synergism between both vitamins in trapping free radicals has been suggested in which oxidized vitamin E is reduced again by vitamin C. Under our culture conditions a combination of vitamin C and vitamin E at a ratio of 10:1 (5 and 0.5 mM respectively) did not improve the PE, compared with vitamin E alone (results not shown).

Table 1. The effect of vitamin C on protoplast yield and PE in L perenne.

cell line	medium*	Y/PE**	vitamin C (mM)				
			0	1.0	2.0	4.0	10.0
22G	E	Y	3.0	8.4	8.6	8.3	16
		PE	0.4	0.3	0.3	0.3	0.1
22Mx	E+C	Y	1.5	1.4	1.4	2.5	3.5
		PE	0.6	0.5	0.6	0.6	0.2
22M	C	PE	0.2	0.06	0.09	0.02	0.02

* Vitamin C included in: E= enzyme mixture, C= culture medium. ** Y= protoplast yield (x 10^5/g FW), PE= plating efficiency (see also materials and methods).

3.2 The addition of catalase in a range of 0-5000 units/ml to the enzyme mixture decreased the protoplast yield in several experiments (results not shown). Concentrations of up to 1000 units catalase/ml in the enzyme mixture or the culture medium did not effect or either slightly improved the PE (Table 2). When SOD was added to the culture medium at a range of 0-1750 units/ml the protoplast PE decreased. However, when both SOD and catalase were added together to the enzyme mixture the PE of protoplasts improved twofold (Table 2).
In aerobic organisms SOD is a key antioxidant that catalyses the conversion of superoxide (O_2^-) to hydrogen peroxide. Hydrogen peroxide is not considered very reactive with organic molecules, but it probably acts as a precursor for very reactive oxidants such as the hydroxyl radical OH·. This OH· radical is formed by the metal-catalysed Haber-Weiss reaction (3):

$$O_2^- + H_2O_2 \xrightarrow[\text{catalyst}]{\text{metal}} O_2 + OH\cdot + OH^-$$

Hydrogen peroxide is removed by catalase and various peroxidases such as ascorbate peroxidase (chloroplasts) and glutathion peroxidase (mammalian cells). Thus the formation of OH· is inhibited by both SOD and catalase (10). It has previously been observed that the addition of SOD and catalase to the protoplast isolation medium of rice suspension-culture cells improved the PE (5). Although in our system the effect of the addition of catalase and SOD to the enzyme mixture was not very pronounced, an improvement in the PE was observed in repeated experiments. As stated by Ishii (5), differences in response between cell lines might be due to differences in the endogenous antioxidant concentrations. Further experiments are in progress to determine at which SOD/catalase concentrations the improvement of the PE is optimal.

Table 2. The effect of SOD and catalase on the PE of protoplasts from different suspension culture lines in L. perenne.

cell line	medium*	concentration catalase/SOD (units/ml)			
		0/0	500/0	1000/0	1000/150
22G	E	0.2	NT	0.2	NT
22M	C	0.1	0.2	0.3	NT
		0.3	0.3	0.3	NT
9A	E	0.05	NT	NT	0.1
	E	0.2	NT	NT	0.4
	E	0.2	NT	NT	0.5

*For addenda see Table 1. NT= not tested or not determined.

3.3 In leaf protoplasts of maize (4) and sugar beet (6) it has been shown that nPG has a pronounced membrane stabilizing effect. nPG is an inhibitor of lipoxygenase activity and is able to bind iron (3). The formation of reactive peroxy radicals is thus prevented and lipid peroxidation is inhibited. Addition of nPG to the protoplast isolation and/or culture medium at a range of 0-0.5 mM resulted in only variable and minor effects (results not shown).
These results indicate that the antioxidant properties of nPG are not exerted in our Lolium protoplast system.

3.4. Addition of GSH at a range of 0-5 mM to the enzyme mixture and/or the culture medium showed no consistent trend concerning the effect on the PE of Lolium protoplasts (results not shown).
When GSH-Px was added to the culture medium at a range of 0-7.5 units/ml no effect was observed on the PE of protoplasts (results not shown). However, when protoplasts were cultured at 7.5 units GSH-Px/ml, microcalli proliferated faster after the first subculture, as compared with the control.
Similarly, the PE of protoplasts was not effected by PLase A2 when added to the culture medium at a range of 0-100 units/ml. However, best microcallus growth after subculture was observed without PLase A2.
Remarkably, the PE of cultured protoplasts could be increased 2-5 times when a combination of GSH, GSH-Px and PLase A2 was added to the culture medium (Table 3). These results are in agreement with those found for suspension-derived protoplasts of rice (5).

Table 3. The effect of the inclusion of GSH (0.5 units/ml), GSH-Px (5 units/ml) and PLase A2 (50 units/ml) in the culture medium on the PE of protoplasts of <u>L. perenne</u>.

	cell line				
antioxidants	22D	B8	B8	B8	B8
none	0.001	0.1	0.2	0.1	0.1
GSH	NT	NT	NT	0.4	0.1
GSH-Px	NT	0.1	0.1	NT	NT
PLase A2	0.002	NT	NT	NT	NT
GSH+GSH-Px+Plase A2	0.003	0.4	0.4	0.5	0.2

In plant cells, apart from the action of catalase, the ascorbate-glutathione pathway serves to remove hydrogen peroxide (Fig. 1). In the cycle oxidized ascorbate is reduced by an enzyme-catalysed reaction with GSH. GSH also acts as a scavenger of O_2, O_2^- and OH·.
A comparable system works in mammalian cells; in the first step GSH is oxidized directly by GSH-Px, an enzyme that has not been demonstrated in plant tissue. GSH-Px can act on a range of peroxides, including lipid peroxides. It has been shown that only those peroxides that are not bound to the cell membrane can be reduced by GSH-Px (3). The enzyme PLase A2 cleaves out peroxidized fatty-acids, facilitating the repair of membranes damaged by lipid peroxidation (11).

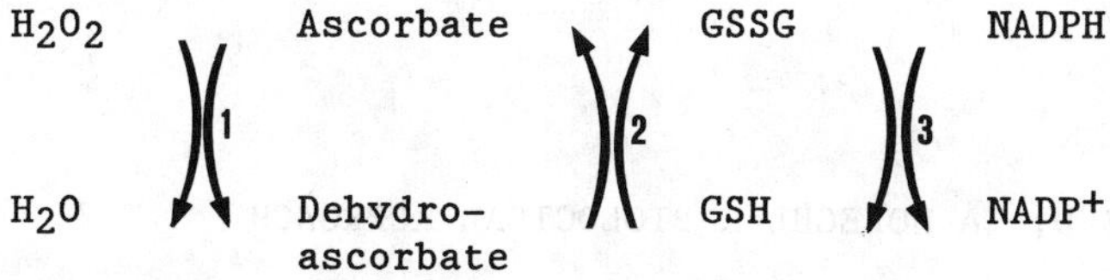

Fig. 1. Ascorbate-glutathion cycle in chloroplasts (2). 1, ascorbate peroxidase; 2, dehydroascorbate reductase; 3, glutathione reductase

GSH is found at very high (millimolar) concentrations in plant cells (9). Addition of GSH alone showed no marked effects, suggesting that under the described culture conditions the endogenous concentration of GSH was not restrictive. The addition of PLase A2 to the protoplast culture medium could make peroxidized fatty acids available for reduction by GSH-Px, using GSH as a substrate. Experiments are now in progress to determine the optimal concentrations of GSH, GSH-Px and PLase A2 when these compounds are combined in the protoplast culture medium.

In conclusion, our results indicate that the addition of enzymes and substrates that are involved in the reduction of peroxides, improve the PE of suspension-derived protoplasts of Lolium, when added to the enzyme mixture and culture medium. Peroxides are capable of generating reactive radicals which might in turn stimulate lipid peroxidation. Further research is underway to determine the effect of these antioxidants on plant regeneration from protoplasts.

REFERENCES

1 Ishii, S. (1987) In vitro cellular and developmental biology 23 (9), 653-658
2 Cutler, A.J., Saleem, M., Coffey, M.A. and Loewen, M.K. (1989) Plant Cell Tissue and Organ Culture 18, 113-127
3 Halliwell, B. and Gutteridge, J.M.C. (1989) in Free radicals in biology and medicine (Halliwell, B. and Gutteridge, J.M.C. eds.),Clarendon Press, Oxford
4 Saleem, M. and Cutler, A.J. (1987) J. Plant Physiol. 128, 479-484
5 Ishii, S. (1988) Plant Physiol. 88, 26-29
6 Krens, F.A., Jamar, D., Rouwendal, G.J.A. and Hall, R.D. (1990) Theor. Appl. Genet. 79, 390-396
7 Creemers-Molenaar,J., van der Valk, P., Loeffen, J.P.M. and Zaal, M.A.C.M. (1989) Plant Science 63, 167-176
8 Larson, R.A. (1988) Phytochemistry 27(4), 969-978
9 Finck, B.F. and Kunert, K.J. (1985) J. Agric. Food Chem. 33, 574-577
10 Elstner, E.F. (1982) Annu. Rev. Plant Physiol. 33, 73-96
11 van Kuijk, F.J.G.M., Sevanian, A., Handelman, G.J. and Dratz, E.A. (1987) TIBS 12, 31-34

ENHANCEMENT OF SUGAR YIELD: A MOLECULAR BIOLOGICAL APPROACH

J.S.GARTLAND, M.R.FOWLER, A.SLATER, N.W.SCOTT, K.M.A.GARTLAND & M.C.ELLIOTT
Plant Molecular Biology Group, Department of Applied Biology and Biotechnology, The David Attenborough Laboratories, Leicester Polytechnic, Scraptoft, Leicester, LE7 9SU, U.K.

1. INTRODUCTION

Sugar beet (Beta vulgaris L.) provides approximately 40% of the world's sucrose. Although selective breeding and improved agricultural practices have succeeded in increasing the fresh weight concentration of sucrose in beet from 4% to 18% (1), further progress by such methods seems unlikely because sugar concentration and root dry mass are strongly negatively correlated (2). The reason for this negative correlation lies in the anatomy of the sugar beet storage root. This organ is formed from a series of secondary cambial rings which are produced early in development. These cambial rings, once initiated, develop simultaneously. The width of the rings increases due to continued cell division and expansion of existing parenchyma tissue. Discrete vascular zones separate parenchyma zones to give the characteristic ringed structure. The outer cambial rings make no significant contribution to either the girth or sucrose storage capacity of the root (Fig. 1).

Milford (3) demonstrated that parenchyma storage cells with a volume of up to $15x10^{-8}$ cm^3 contain proportionately more sucrose than do larger cells. Both Wyse (4) and Milford (3) agreed that high sucrose yield would be achieved in storage roots with more active cambia. The ideal sugar beet can thus be envisaged as having a storage root with all cambial rings contributing to development, so that a larger number of vascular zones separate a corresponding number of parenchyma zones each made up of optimally sized cells with optimally short transport paths from the phloem to the vacuoles of the parenchymatous storage cells.

In order to modify storage organ development in a manner likely to increase sucrose content at harvest a genetic manipulation approach is being pursued. The strategy consists of two components: a) gaining an understanding of the controls acting over cell proliferation (both the controls in continually cycling cells and the controls acting during the transition from the non-dividing to the proliferative state) and b) identification of root specific mRNA species and their corresponding genes. Assessment of the regulatory regions involved in directing tissue specificity will allow the construction of chimaeric genes that will control cell division in specific areas of the storage organ.

FIGURE 1.

Transverse section through a mature sugar beet storage root.

2. MATERIALS AND METHODS

2.1 Plant Material: Plants were either field or greenhouse grown over a British summer. Cell cultures were maintained in MS (5) medium with a 17 day batch culture cycle.

2.2 Protein Extraction and Electrophoresis: A) Proteins from whole plant tissues were extracted according to Elliott et al. (2) with subsequent denaturing SDS-polyacrylamide gel electrophoresis, Coomassie staining and destaining according to Hames (6). B) Proteins from cultured cells were extracted by boiling in 5% SDS. Subsequent steps were as for A.

2.3 Sugar Beet DNA Isolation: High molecular weight DNA was isolated according to Hattori et al.(7)

2.4 Genomic Library Construction: Partially Sau 3a restricted genomic DNA was cloned into EMBL 3 according to Kaiser and Murray (8).

2.5 RNA Isolation: Sugar beet leaf and storage root RNA was isolated using the method of Slater (9). Cultured cell RNA was isolated according to Logemann et al. (10). In both cases poly A^+ RNA was isolated using messenger affinity paper (Amersham).

2.6 cDNA Synthesis and Cloning: Poly A^+ RNA from leaf and storage root was used as the template for cDNA synthesis using a Pharmacia kit (oligo-dT primer, 2nd strand synthesis using Klenow fragment), and Eco RI adaptors added. Fragments were cloned into lambda gt 10, packaged and titred on E. coli NM 514.

2.7 Biotin Labelling of Probe RNAs: Leaf and root poly A^+ RNA was labelled with photobiotin according to McInnes et al.(11).

2.8 Differential Screening: The root cDNA library was differentially screened with leaf and root poly A^+ RNA according to Perbal (13). DNA-RNA hybrids were detected chromogenically.

2.9 pUC18 Subcloning: Potential tissue specific sequences were excised from lambda gt 10 using Eco R1 and ligated into pUC 18. Plasmids containing inserts were purified using the cleared lysate method (12).

2.10 *In vitro* Translation of RNA: The method followed was similar to that of Elliott *et al*. (2)

2.11 *In vivo* Labelling of Cell Suspensions: Cells were radioactively labelled by adding ^{35}S-methionine to a concentrated sample of cells and incubated at 25°C for 2 hours.

2.12 IAA analysis was carried out according to Hall *et al*. (13)

3. RESULTS AND DISCUSSION

3.1 Division Related Gene Expression

Cell division studies have been concentrated on the changes in gene expression associated with the transition from the non-dividing to the proliferative state.

Extended periods in the stationary phase elongate the initial lag phase beyond the usual 2-3 days. Changes in gene expression when cells were transferred into fresh medium after varying times in the stationary phase were investigated.

Comparison of the most abundant proteins by SDS-polyacrylamide gel electrophoresis has revealed the presence of a group of proteins which may be divison related (200, 76, 45 and 25 KDa) (Fig. 2). These proteins can be detected in cells throughout the batch culture cycle provided that the cells are subcultured at the point of entry into the stationary phase. However, these proteins gradually decrease in abundance over several days when the cells are left in the stationary phase culture. The time of reappearance of the proteins after subculture is positively correlated with the duration of the stationary phase. A number of proteins specifically associated with the stationary phase have also been identified.

Radioactive labelling of cells *in vivo* indicates that the re-appearance of some growth related proteins results from *de novo* synthesis.

Analysis of mRNA populations by translation of RNA *in vitro* indicates that subculture does not cause any rapid, major changes in the translatable RNA population, although more detailed analysis may reveal a small number of changes. It seems certain that dividing and non-dividing cells express a small number of different mRNA species and it may be that their induction is a secondary response following subculture, resulting in a lag in the time of appearance.

One candidate for the role of causative agent of cell division related mRNA synthesis is IAA. Analyses of the levels of IAA in cells undergoing the transition from the non-dividing to the dividing state are therefore being undertaken. Again, the effect of duration of the stationary phase is being analysed. Levels of free IAA in the cells are comparatively low at the onset of stationary phase, and decrease with increasing duration of the stationary phase. Free IAA levels in cells subcultured normally increase from about day 6 of the batch culture cycle, whereas cells subcultured after 2 days in the stationary phase do not show increased levels until day 8. Increased levels of free IAA seem then to be associated with the late exponential and linear phases of the batch culture cycle.

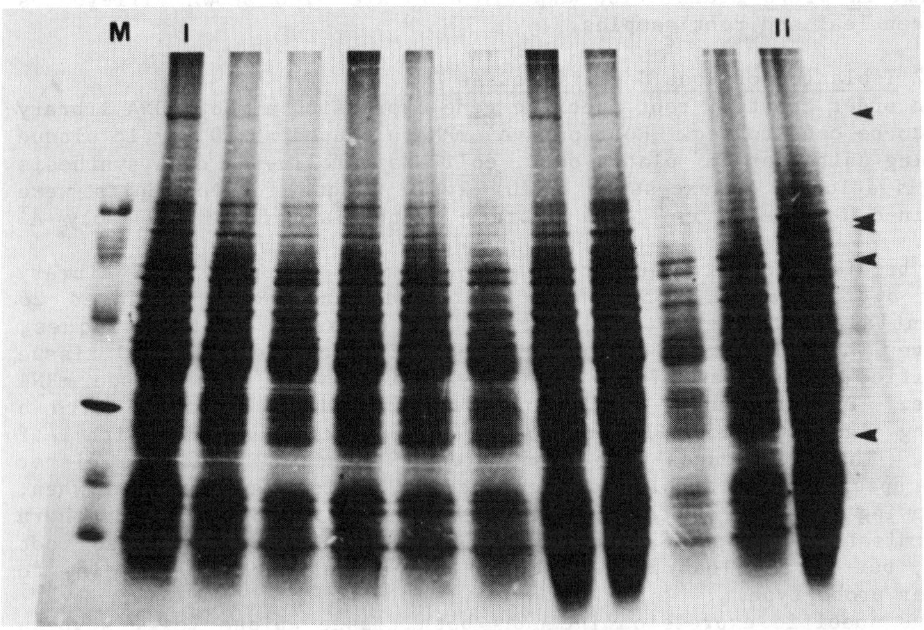

FIGURE 2. Growth Related Proteins. Molecular weight markers (track M) are 94, 67, 43, 30, 30 and 14 KDa. Other tracks are numbered from left to right. Tracks 1-5 are proteins from cells on day 17, 18, 19, 20 and 23 respectively, tracks 6-8 from day 1, 2 and 5 after subculture on day 17, tracks 9-11 from day 1, 2 and 5 after subculture on day 19.

Current work in this area is concentrating on identifying genes controlling the initiation of cell division, both by continuing present studies and by comparison with mammalian cells. Work is also underway to identify genes or gene products homologous to ones known to regulate cell division in continually cycling cells in other eukaryotes.

3.2 Comparison of Gene Expression in Root and Leaf

3.2.1 Protein gel electrophoresis

Plant gene expression is highly regulated, particular gene products being present in certain cell types or developmental stages, or following the application of an environmental stimulus.

Three major differences can be observed between proteins from a 200 g. fresh weight storage root and leaves from the same plant. Two proteins, of approximate molecular weights 15 and 23 KDa visible in the root extract track were not visible in the lane containing leaf protein. Additionally, a protein of approximately 30 KDa molecular weight was much more abundant in root compared to leaf samples. Poly A^+ RNA *in vitro* translation experiments have also shown differences between leaf and root samples.

3.2.2 Isolation of Root Specific cDNAs

In order to study root specific gene expression a root cDNA library had to be constructed. Root poly A^+ mRNA produced $>5x10^5$ lytic plaque forming units per cm^3 plated on *E. coli* NM514 following cDNA synthesis and ligation. In excess of $1x10^6$ lytic plaque forming units were obtained from each cm^3 cDNA library synthesised from leaf poly A^+ mRNA.

Hybridisation of a subset of plaques from the root cDNA library with biotinylated probes from root and leaf mRNAs produced 26 potential root specific plaques. The majority of the plaques, however, hybridised to both probe types. Each of the potential tissue specific plaques was picked and rehybridised with each tissue mRNA probe. Two particular cDNA clones were found to hybridise with a strong signal to root mRNA but not to any appreciable extent with leaf mRNA. These sequences were designated RS1 and RS2 (and are either root specific, or highly root enhanced). This was confirmed when, following cloning of the Eco RI cDNA inserts into pUC18, Southern hybridisations of pUCRS1 and pUCRS2 hybridised to biotinylated root mRNA, but not to leaf mRNA. pUC 18 DNA alone did not hybridise to either probe type.

Tissue specific expression includes both tissue unique (e.g. soybean seed storage protein (14), and tissue enhanced expression (15). The latter category is more fully represented and examples include; glutamine synthetase and patatin (16, 17). The majority of our studies to date have involved leaf and root mRNA probes. Detailed assessments with other organ probes will be performed.

3.2.3 Characterisation of Root Specific cDNA Clones

Agarose gel electrophoresis of Eco RI digested pUCRS1 and pUCRS2 revealed cDNA insert sizes of 550 and 750 base pairs respectively. Restriction endonuclease mapping of these fragments indicates that they contain different sequences, this is currently being confirmed by DNA sequencing. Southern blot analysis of sugar beet DNA using RS1 as probe indicates that this DNA sequence may be found on a 1,500 bp long Hind III fragment.

3.2.4 Genomic Library Construction

In order to examine the regulatory elements flanking a tissue specifically expressed gene a sugar beet genomic library has been constructed. Sugar beet DNA in the size range 15-20 Kb was ligated into EMBL 3 arms and packaged at a concentration of approximately

$1x10^6$ phage per microgram insert DNA. More than $7x10^5$ recombinant bacteriophage have been packaged. The genomic library is currently being screened to identify and analyse clones corresponding to RS1 and RS2. By analogy with observations made on many tissue specifically expressed genes it appears that the 5' regions are likely to contain regulatory elements essential for tissue specificity. The use of 5'-GUS transcriptional fusions will eventually allow gene expression studies in transgenic plants.

A knowledge of the molecular processes underlying control of the cell cycle and tissue specificity in sugar beet is central to our attempts to develop the "ideal" storage organ.

REFERENCES

1 Milford, G.F.J., Biscoe, P.V., Jaggard, K.W., Scott, R.K. and Draycott, A.P. (1980) in Opportunities for Increasing Crop Yields (Hurd, R.G., Biscoe, P.V. and Dennis, C. eds), pp.71-83, Pitman, London.
2 Elliott, M.C., Barker, R.D.J., Gartland, K.M.A., Grieve, T.M., Hall, J.F., Ryan, L.A., Scott, N.W and Slater, A. (1988) in Physiology and Biochemistry of Auxins (Kutacek, M., Bandurski, R.S. and Krekule, J. eds), pp.391-399, Academia, Prague.
3 Milford, G.F.J. (1973) Ann. Applied Biol. 75, 427-438
4 Wyse, R. (1979) J. Am. Soc. Sugar Beet Technol. 20, 268-385
5 Murashige, T. and Skoog, F. (1962) Physiol. Plant. 15, 473-497
6 Hames, D.B. (1981) in Gel Electrophoresis of Proteins (Hames, D.B. and Rickwood, D. eds), pp.1-86, IRL Press, Oxford
7 Hattori, J., Gottlob-McHugh, S.G. and Johnson, D.A. (1987) Anal. Biochem. 165, 70-74
8 Kaiser, K. and Murray, N.E. (1985) in DNA Cloning (Glover, D.M. ed), Vol 1, pp.1-47, IRL Press, Oxford
9 Slater, A. (1988) in Methods im Molecular Biology (Walker, J.M. ed), Vol 4, pp.437-446, Humana, New Jersey
10 Logemann, J. Schell, J. and Willmitzer, L. (1987) Anal. Biochem. 163, 16-20
11 McInnes, J.L., Dalton, S., Vize, P.D. and Robbins, A.J. (1987) Bio/Technology 5, 269-272
12 Perbal, B. (1988) in A Practical Guide to Molecular Cloning (Perbal, B. ed), Wiley, J and Sons Inc, U.S.A.
13 Hall, J.F.,Yen-Hung, Z., Fox, G.G., Ratcliffe, R.G. and Elliott, M.C. in preparation.
14 Walling, L., Drews, jG.N., Goldberg, R.B. (1986) Proc. Natl. Acad. Sci. USA 83, 2123-2127
15 Evans, I.M., Swinhoe, R., Gatehouse, L.N. Gatehouse, J.A. and Boulter, D. (1988) Mol. Gen. Genet. 214, 153-157
16 Tingey, S.V., Walker, E.L. and Coruzz, G.M. (1987) EMBO J. 6, 1-9
17 Twell, D. and Ooms, G. (1987) Plant Mol. Biol. 9, 345-375

APPLICATIONS OF THE PARTICLE GUN IN PLANT BIOLOGY

T. M. KLEIN*, S. A. GOFF, B. A. ROTH, AND M. E. FROMM
*E. I. Du Pont De Nemours & Co., Medical Products Dept., Glasgow Site, Newark, Delaware, 19714-6101; U.S. Department of Agriculture-ARS/UC Berkeley, Plant Gene Expression Center, Albany, California, 94710.

1. INTRODUCTION

The Biolistic™ technique represents a new approach that permits the direct delivery of DNA into intact cells and tissues (1,2). The process utilizes DNA-bearing particles that are accelerated to velocities that permit their penetration of cell walls and membranes. These microprojectiles are small enough so that penetration is not necessarily lethal and results in the direct deposition of DNA within the affected cell. Genes can be introduced into a wide range of tissues including suspension cultures, callus cultures, tissues isolated directly from plants or even tissues of whole seedlings. Table 1 lists the species and tissue types into which genes have been successfully introduced by microprojectiles.

There are three prime applications for the delivery of DNA by the Biolistic process: 1) Transient expression of genes introduced into intact tissues: This application is proving valuable for the study of how gene expression is controlled by environmental (8,6), tissue specific (4,10), and genetic factors (10). The Biolistic approach is an important complement to studying the expression of foreign genes in stable transformants since constructs can be tested within several days. The process is especially valuable for studying gene expression in plants for which efficient stable transformation procedures do not as yet exist. 2) The stable transformation of plants: Several plant species, including tobacco (3,5), soybean (12), and maize (17), have been stably transformed by microprojectiles. Recent results obtained at the USDA lab in Albany, CA have shown that stably transformed maize plants can be regenerated from embryogenic suspension cultures that have been bombarded with a gene that confers resistance to the herbicide chlorsulfurone. Progeny of the transformed plants also carried the foreign gene. Therefore, it appears that the process will be an important means for the modification of crop species by recombinant DNA techniques. 3) The delivery of DNA to organelles: Foreign DNA has been successfully introduced into the mitochondria of *Saccharomyces* (20,21)

Table 1. Species and tissues into which DNA has been delivered by microprojectiles.

Tissue Type and Species	Reporter Gene (Promoter/coding region)	Reference
Leaves:		
Tobacco	35S/GUS	3,4,5
Corn	35S/*Lc*	6
Seedlings (various tissues):		
Corn	35S/*Lc*	6
Eggplant	35S/luciferase (LUC)	7
Rice, Barley, Oats, Cucumber, Tobacco, Arabidopsis	35S/CAT, maize ubiquitin/LUC, or rice phytochrome/CAT.	8
Seed Tissues:		
Corn aleurone or scuttelum	Adh/GUS	9
	35S/GUS or *Lc*	6
	Adh/LUC or CAT	10
	Bz1 or *A1* /LUC	10
	A1 or Bz1 genomic clones	10
Barley embryos	35S/CAT	11
Epidermal tissues:		
Onion	35S/CAT	1
Corn	35S/*R*	6
Shoot apices:		
Soybean	35S/GUS	12
Maize, Wheat, Rice	35S/GUS	13
Tissue culture cells:		
Tobacco	35S/GUS	3,7
	35S/LUC	7
	35S/CAT	14
	chloroplast promoter/CAT	14
Soybean, Wheat, Rice	Adh/GUS or 35S/CAT	15
Corn	Adh/GUS	9,15
	35S/CAT	16
	Adh/GUS or 35S/NPTII	17
Barley	35S/CAT	11
	35S/GUS, NPTII	18
Rice	Adh/GUS	13
Soybean	35S/NPTII	19
Floral Tissues:		
Tobacco anthers or petals	35S/GUS	4
Pollen and Microspores:		
Tobacco	Pollen promoter or 35S/GUS	4

and the chloroplast of *Chlamydomonas* (22,23) and tobacco (14) by microprojectiles. No other method is currently available to deliver DNA directly to these organelles. It is clear that many new avenues to study the molecular biology of plants have been opened by the Biolistic process. This review will focus on the use of microprojectiles for the study of gene expression by transient assays in intact tissues.

2. STUDIES OF GENE REGULATION USING THE BIOLISTIC PROCESS

2.1 Transient assays for the analysis of plant genes.

An understanding of the factors that control gene expression requires effective means for analyzing the function and structure of cis and trans acting regulatory factors. One approach to assess the activity of promoters, 3' regions, and associated enhancer elements is to employ chimeric genes consisting of the regulatory regions of interest fused to a coding sequence specifying an enzyme that is easily assayed such as luciferase, ß-glucuronidase, or chloramphenicol acetyltransferase. A variety of mechanisms exist to introduce the resulting chimeric gene into protoplasts derived from tissues isolated from plants or from cells maintained in culture (24). The expression of the reporter gene can be analyzed within a short time (ie., several hours to several days depending on the strength of the promoter and sensitivity of the assay) after its introduction. Such transient expression assays have served to help define regulatory elements in a wide range of mammalian (25) and plant genes (26,27).

A prime requisite for using transient assays of gene expression is that the cell type of interest be competent to take up exogenous DNA. The procedures that are currently available for gene delivery, such as electroporation or PEG-mediated uptake, generally require protoplasts that are derived from cells maintained in culture (ie., callus or suspensions). Unfortunately, cells cultured *in vitro*, and the protoplasts derived from these cells, often do not exhibit proper regulation of either introduced or endogenous genes. Studies of genes that are light regulated have been particularly hampered by the lack of their faithful regulation and expression in protoplasts. The Biolistic process represents a potential means for circumventing the use of protoplasts in transient expression assays. Genes have been delivered to virtually every tissue type tested to date. For the most part, extensive optimization studies have not been necessary and substantial levels of expression have been detected in many tissue types using 'standard' bombardment conditions (9).

The majority of studies listed in Table 1 are directed at proving that the concept of microprojectile-mediated delivery of DNA to intact cells and tissues is valid. In addition to these optimization and proof of concept studies, several recent reports indicate that the Biolistic process can be applied as a tool in the study of the factors that control gene expression. These investigations have sought to determine if genes delivered by microprojectiles are faithfully regulated in a manner analogous to the corresponding native gene. For the system to be of value, the introduced genes must show proper responses to environmental, tissue specific, and genetic factors.

2.2 Regulation of Genes Introduced by Microprojectiles.

Genes responsible for the synthesis of anthocyanin pigments in maize were used to develop a model system to determine if microprojectile-introduced genes are properly regulated. The genetic and biochemical basis for the synthesis of these phenolic compounds has been extensively characterized (28). Classical genetic studies have identified the structural genes of the pathway and the complex, tissue-specific regulation of these genes by a set of regulatory loci. Most tissues of maize can produce the red to purple anthocyanin pigments if the plant has the functional structural genes that specify the enzymes responsible for each step of the pathway. These enzymes are coded for by the following genes: *C2*, *A1*, *A2*, *Bz1*, and *Bz2*. The expression of these genes is regulated by a set of loci that, by genetic and recent molecular analysis, appear to encode transcriptional activators (29-32). Depending on the set of regulatory genes present, the anthocyanin pathway can be activated in particular tissues at defined developmental stages. For example, anthocyanin production in the aleurone requires the presence of *C1* and *R*. Anthocyanin production in the plant body (ie., the leaves and stem) requires another set of regulatory genes, *B* and *Pl*. There are many alleles of these regulatory genes that are responsible for the complex developmental and tissue specific control of the pathway (28). For example, *Rscm2*, an allele of *R*, causes anthocyanin production in the scutellum of the embryo. *B-Peru*, an allele of *B*, conditions pigment production in the aleurone and embryo independent of *R*.

In addition to being well characterized genetically and biochemically, the anthocyanin pathway is being intensively investigated at the molecular level. Many of the structural and regulatory genes have been isolated and therefore are accessible for molecular characterization. For example, genomic clones of the *A1* (34) and *Bz1* (35) genes have been isolated by transposon tagging. Mutant plants lacking functional alleles of these structural genes cannot produce anthocyanins. A recent publication (10) described the microprojectile-mediated introduction of these genomic clones into tissues from plants that are blocked for pigment production because of a lesion in either the *Bz1* or *A1* gene.

To accomplish this, aleurone tissue of mature, hydrated seed was exposed by peeling back the pericarp. The entire seed was placed in a petri dish containing agarose-solidified medium. The tissue was then bombarded with plasmid DNA carrying the appropriate genomic clone. After 24 hours of incubation in the dark, a portion of the cells turned purple (Fig. 1). Therefore, delivery of these genomic clones resulted in the functional complementation of the mutant phenotype. Generally, these purple spots were comprised of several (3 to 5) cells with a central most cell being more intensely pigmented then the surrounding cells. Apparently, a precursor of the final product of the pathway can enter surrounding cells where the enzymes are present to convert it to the final purple anthocyanin pigment. Similar results were obtained using the *A1* clone. These experiments demonstrate that both the *A1* and *Bz1* genomic clones used encode functional proteins. Since a routine means for the genetic transformation of maize has not been developed, no other means currently exists to directly determine if such isolated genes are functional.

Figure 1. Delivery of the *Bz1* gene into aleurone tissue of plants lacking a wild type *Bz1* gene. The spots represent clusters of cells that contain anthocyanin pigment.

The introduction of either the *Bz1* or *A1* clones to aleurones carrying recessive alleles of either or both of the regulatory genes, *C1* or *R*, did not result in cells that expressed the anthocyanin pathway. This was expected since the presence of *C1* and *R* is required for pigmentation of the aleurone. The introduced genes could be expressed but it would not be possible to detect their expression phenotypically since the other endogenous structural genes must also be expressed for anthocyanin synthesis. To determine if expression of genes introduced into different regulatory backgrounds are properly controlled, chimeric genes consisting of the 5' and 3' regulatory regions of either the *Bz1* or *A1* gene were fused to a luciferase coding region. These chimeric genes were introduced into aleurones carrying various combinations of recessive or dominant alleles of the regulatory genes, *C1* and *R*. Following bombardment with microprojectiles carrying the appropriate chimeric gene, the tissue was incubated and luciferase activity was determined in tissue extracts after 24 hours. Levels of luciferase activity in permissive (*C1*, *R*) backgrounds were 30- to 200- fold greater than those detected in tissue carrying one or both of the recessive alleles of the regulatory genes. These results show that genes delivered to intact tissues by microprojectiles are regulated in a manner similar to the corresponding endogenous genes. This study helped form the basis for analyzing the regulation of the maize anthocyanin genes by molecular techinques.

2.3 Deletion Analysis of the *Bz1* Promoter.

A set of 5' deletions of the *Bz1* promoter has been constructed. These deletions were fused to a luciferase coding region and a *Bz1* 3' end was used as the polyadenylation signal. Deletions between -407 and -114 bp upstream of the start of transcription had minor effects on expression as compared to the full length (2.3 kb) promoter following their delivery into *C*,*R* aleurones. However, deletions leaving 80 or 60 bp upstream from the start of transcription had negligible levels of activity when compared to the full length promoter. Thus, these low relative activities indicate that sites important for the expression of *Bz1*, and perhaps the sites that interact with *C1* and *R*, are within 114 bp of the start of transcription. Linker scanner studies are currently under way to define the important sequences in this region of the *Bz1* promoter that are essential for expression. The results indicate that microprojectiles can be used to characterize and define important regulatory elements of promoter regions.

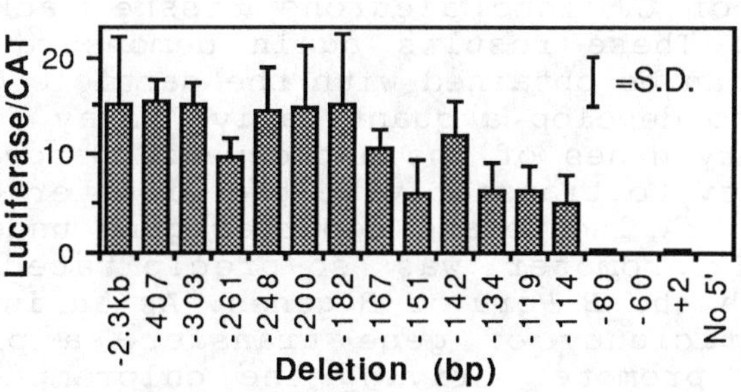

Figure 2. Deletion analysis of the *Bz1* promoter. Values are expressed as a ratio between luciferase activity from the *Bz1* constructs and CAT activity from an Adh1-CAT internal control plasmid that was co-delivered with each bombardment.

2.4 Regulatory Genes Introduced with Microprojectiles.

Microprojectiles have also been used to introduce regulatory genes of the anthocyanin pathway into intact tissues. A chimeric gene consisting of the coding region of *Lc* (a member of the *R* gene family) under the control of the 35S promoter from cauliflower mosaic virus was introduced into various tissues that lacked a functional *Lc* (or *R*) gene (6). Cell autonomous pigmentation was induced in aleurone, coleoptile, root and other tissues of germinating seed. Apparently the introduced gene activated the structural genes of the pathway. It is noteworthy that *Lc* does not normally pigment aleurone cells. This result led Ludwig and coworkers (6) to conclude that the *R* and *Lc* genes are functionally equivalent.

Another gene that appears to be functionally equivalent to *R* is *B.* The *B* gene is required for anthocyanin synthesis in many tissues of the plant body. An allele of *B*, *B-Peru*, conditions anthocyanin production in aleurone and embryo as well as the plant body. These genes have recently been cloned (32) and the particle gun was used to study their expression (33). Aleurone tissue lacking a functional *R* gene was bombarded with microprojectiles coated with plasmid containing the *B-Peru* genomic clone. Following incubation the bombarded tissue exhibited numerous cells containing the purple pigment (Fig. 3). Similar results were observed when aleurone tissue was bombarded with a cDNA of *B* or *B-Peru* under the control of the 35S promoter. These results demonstrate that the requirement for an *R* regulatory gene in aleurone tissue can be bypassed by the expression of an introduced *B* gene. Pigment formation in *r* tissue by the expression of the *B* cDNA demonstrates that the *B* gene product can functionally substitute for the *R* gene product. A similar set of experiments was performed with the *C1* regulatory gene. Aleurone cells turned purple following the delivery of a cDNA of *C1* into aleurone tissue lacking a functional *C1* gene. These results again demonstrate the rapid feedback that can be obtained with the particle gun.

In an attempt to develop a quantitative assay for the function of regulatory genes of the anthocyanin pathway, we examined their ability to transactivate the promoter of the *Bz1* structural gene. A luciferase reporter gene under the control of the *Bz1* promoter was co-precipitated onto microprojectiles with the *B-Peru* or *B* genes. As an internal control for the efficiency of gene transfer, a plasmid containing the *Adh1* promoter driving the chloramphenicol acetyltransferase (CAT) coding region was also precipitated onto these microprojectiles. This control is used to normalize the variation found between bombardments. The microprojectiles carrying the three plasmids were introduced into aleurone tissue of plants lacking a functional *B* or *R* gene (ie.,*C1*,*r*,*b*). The values of *B-Peru* or *B* transactivation

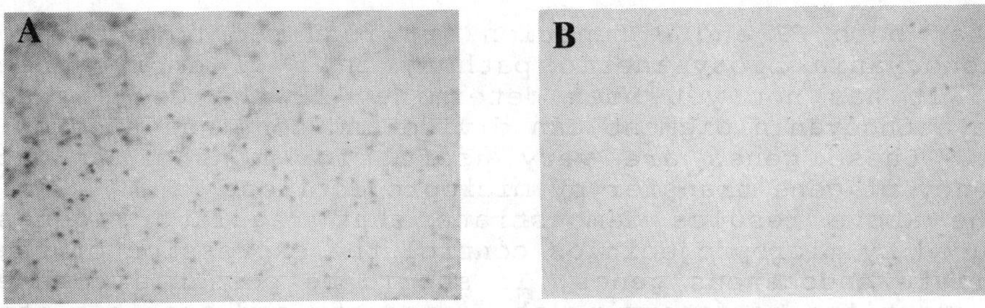

Figure 3. Introduction of regulatory genes into maize tissues. (a) Expression of *B-Peru* introduced into aleurones by microprojectiles. (b) Expression of cDNA's of *B* and *C1*, both under the control of the 35S promoter, in embryogenic maize callus. Each dark spot represents a single anthocyanin producing cell.

of the *Bz1* chimeric gene can therefore be expressed as a ratio of luciferase activity (from *Bz1*-luciferase) to CAT activity (from the *Adh1*-CAT control plasmid). Co-bombardment of the *Bz1*-luciferase reporter construct with the *B-Peru* gene, or an expressed *B-Peru* or *B-I* cDNA, resulted in an 80- to 130-fold increase in the ratio of luciferase to CAT activity relative to tissue receiving the reporter plasmid alone. The cDNA of *C1* also transactivated the *Bz1*-luciferase gene when both were introduced into *c1*,*R* aleurones. Co-bombardment of the *Bz1*-luciferase reporter construct with the *C1* gene or the cDNA of *C1* under the control of the 35S promoter resulted in an 85- to 300-fold induction of the *Bz1* promoter relative to tissue receiving the reporter plasmid alone.

Transactivation of the anthocyanin pathway was also examined in embryogenic callus cultures (33). These cells did not produce purple pigment following delivery of either the *B-Peru* genomic clone or an expressed *B-Peru* or *B* cDNA. Since this callus was derived from embryos lacking not only *R*, *B* or *Pl* function but also *C1* function, we determined the effect of delivering a construct carrying a *C1* cDNA under control of the 35S promoter with each of the *B* constructs. Purple cells were observed in callus receiving both the *C1* and *B* genes (Fig. 3). Transactivation of the *Bz1*-luciferase reporter gene was also observed in embryogenic callus when it was delivered together with both the 35S-*C1* plasmid and a plasmid carrying a gene coding for *B* function (either the *B-Peru* genomic or the cDNA of *B-I* or *B-Peru* under the control of the 35S promoter). The presence of both the *B* and *C1* genes resulted in a 30- to 190-fold induction of the *Bz1*-luciferase gene relative to its expression following delivery without either the *B* or *C1* genes. These results show that both *C1* and *B* function are required to activate the anthocyanin biosynthetic pathway in *r c1* embryogenic callus. It has not yet been determined if the cells that produce anthocyanin pigment can divide and continue to grow. However, these genes are very useful for monitoring the efficiency of gene transfer by microprojectiles.

The above results demonstrate that regulatory genes introduced by microprojecitles control the expression of the appropriate endogenous genes. It should be possible to use microprojectiles to determine the integrity and possibly the function of other cloned genes, especially if the expression of the introduced gene can be detected phenotypically. Transactivation of chimeric reporter genes by co-introduced regulatory genes will be useful to determine domains of regulatory proteins that are responsible for specific functions.

2.5 Tissue Specific and Environmentally Regulated Genes.

Other recent studies have shown that microprojectiles can be used to study promoter function in particular tissues. For example, microprojectiles have been used to deliver pollen specific genes directly into pollen (4). The promoter of a pollen specific gene from tomato, or the 35S promoter were linked as transcriptional fusions to coding DNA of the *Eschericia coli* ß-glucuronidase (GUS) gene. These chimeric genes were delivered into pollen on microprojectiles. Both genes showed expression in pollen although the tomato promoter was at least 1000 fold more active than the 35S promoter. Tobacco leaves bombarded with the 35S promoter construct showed GUS activity, but no significant activity was detected using the pollen-promoter fusion. The transient expression of the microprojectile-introduced genes was similar to that observed for these genes in stably transformed tobacco plants. Therefore, the particle gun will be useful for the study of cis-acting sequences important in pollen-specific gene expression.

Genes controlled by light can also be studied in transient expression assays with microprojecitles. The phytochrome gene in rice is subject to negative feedback by phytochrome. Bruce and co-workers (8) constructed a chimeric gene consisting of the 5' region of an oat phytochrome gene and a CAT coding region. This gene was introduced into seedlings and the response of the gene to red and far red light was determined. The gene was expressed following exposure to far red light while expression was repressed after exposure to red light. The endogenous phytochrome gene shows a similar pattern of response to light as the introduced gene. This system is being exploited to determine important elements of the phytochrome gene that are necessary for response to light related stimuli.

3. CONCLUSION

Microprojectile-mediated delivery is becoming an important element in studies of gene regulation and expression. The research that was outlined in this review indicates that the particle gun can deliver DNA into many types of tissues and that the expression of the introduced genes are faithfully regulated. This new technology will mature and develop over the coming years. It will be interesting to see the number of biological questions that will be made accessible to study with Biolistic technology.

4. ACKNOWLEDGMENTS. We thank Vicki Chandler and Karen Cone for the use of the *B* and *C1* clones, respectively. This work was supported by grants from the US Department of Agriculture/Agricultural Research Service.

5. REFERENCES

1 Klein, T.M., Wolf, E.D., Wu, R., and Sanford, J.C. (1987) Nature 327, 70-73
2 Sanford, J.C., Klein, T.M., Wolf, E.D., and Allen, N. (1987) Particlulate Sci. Technol. 5,27-37.
3 Klein, T.M., Harper, E.C., Svab, Z., Sanford, J.C., Fromm, M.E., and , P. (1988) Proc. Natl. Acad. Sci. (USA) 85,8502-8505.
4 Twell, D., Klein, T.M., Fromm, M.E., McCormick, S. (1989) Plant Physiol. 91, 1270-1274.
5 Tomes, D.T., Weissinger, A.K., Ross, M., Higgins, R., Drummond, B.J., Schaaf, S., Malone-Schoneberg, J., Staebell, M., Flynn, P., Anderson, J., and Howard, J. (1990) Plant Molec. Biol. 14,261-268.
6 Ludwig, S.E., Bowen, B., Beach, L., Wessler, S.R. (1990) Science 247, 449-450.
7 Morikawa, H., Iida, A., and Yamada, Y. (1989) Appl. Microbiol. Biotech. 31, 320-322.
8 Bruce, W.B., Christensen, A.H., Klein, T., Fromm, M., Quail, P. (1989) Proc. Natl. Acad Sci. (USA) 86, 9692f-9696.
9 Klein, T.M., Gradziel, T., Fromm, M.E., and Sanford, J.C. (1988) Bio/Technology 6, 559-563.
10 Klein, T.M., Roth, B.A., Fromm, M.E. (1989) Proc. Natl. Acad Sci. (USA) 86, 6681-6685.
11 Kartha, K.K., Chibbar, R.N., Georges, F., Leung, N., Caswell, K., Kendall, E., and Qureshi, J. (1990) Plant Cell Rep. 8, 429-432.
12 McCabe, D.E., Swain, W.F., Martinell, B.J., and Christou, P. (1988) Bio/Technology 6, 923-926.
13 Oard, J.H., Paige, D.F., Simmonds, J.A., Gradziel, T.M. (1990) Plant Physiol. 92, 334-339.
14 Daniell, H., Vivekananda, J., Nielsen, B.L., Ye, G.N., Tewari, K.K., Sanford, J.C. (1990) Proc. Natl. Acad Sci. (USA) 87-92.

15 Wang, Y.C., Klein, T.M., Fromm, M.E., Cao, J., Sanford, J.C. and Wu, R. (1988) Plant Molec. Biol. 11, 433-439.
16 Klein, T.M., Fromm, M.E., Weissinger, A., Tomes, D., Schaaf, S., Sletten, M., and Sanford, J.C. (1988) Proc. Natl. Acad. Sci. (USA) 85, 4305-4309.
17 Klein, T.M., Kornstein, L., Sanford, J.C., and Fromm, M.E. (1989) Plant Physiol. 91, 440-444.
18 Mendel, R.R., Muller, B., Schulze, J., Kolesnikov, V., Zelenin, A. (1989) Theor. Appl. Genet. 78, 31-34.
19 Christou, P., McCabe, D.E., and Swain, W.F. (1988) Plant Physiol. 87, 671-674.
20 Fox, T.D., Sanford, J.C., and McMullin, T.W. (1988) Proc. Natl. Acad. Sci. (USA) 85, 7288-7289.
21 Johnston, S.A., Anziano, P.Q., Shark, K.B., and Sanford, J.C. (1988) Science 240, 1538-1541.
22 Boynton, J.E., Gillham, N.W., Harris, E.H., Hoster, J.P., Johson, A.M., Jones, A.R., Randolf-Anderson, B.L., Robertson, D., Klein, T.M., Shark, K.B., and Sanford, J.C. (1988) Science 240, 1534-1538.
23 Blowers, A.D., Bogarad, L., Shark, K., and Sanford, J.C. (1989) The Plant Cell 1, 123-132.
24 Davey, M.R., Rech, E.L., and Mulligan, B.J. (1989) Plant Molec. Biol. 13, 273-285.
25 Maniatis, T., Goodbourn, S., and Fischer, J.A. (1987) Science 236, 1237-1245.
26 Kuhlemeier, C., Fluhr, R., Green, P.J., and Chua, N.-H. (1987) Ann. Rev. Plant Physiol. 38, 221-257.
27 Schell, J.S. (1987) Science 1176-1187.
28 Coe, E.H., Neuffer, M.G., and Hoisington, D.A. (1988) in Corn and Corn Improvement (Sprague, G.F., and Dudley, J.W.. eds.) pp.81-258 American Society of Agronomy, Madison, Wisconsin, USA.
29 Cone, K.C., Burr, F.A., and Burr, B. (1986) Proc. Natl. Acad. Sci. (USA) 83,9631-9635.
30 Paz-Ares, J., Ghosal, D., Wienand, U., Peterson, P.A., and Saedler, H. (1987) EMBO J. 6, 3553-3558.
31 Ludwig, S.R., Habera, L.F., Dellaporta, S.L., and Wessler, S.R. (1989) Proc. Natl. Acad. Sci. (USA) 86, 7092-7096.
32 Chandler, V.L., Radicella, J.P., Robbins, T.P., Chen, J., and Turks, D. (1989) Plant Cell 1,1175-1183.
33 Goff, S.A., Klein, T.M., Roth, B.A., Fromm, M.E., Cone, K.C., Radicella, P.J., and Chandler, V.L. (1990) EMBO J. In Press.
34 O'Reilly, C., Shepherd, N.S., Pereira, A., Schwarz-Sommer, Z., Bertram, I., Robertson, D.S., Peterson, P.A, and Saedler, H., (1985) EMBO J. 4, 877-882.
35 Fedoroff, N.V., Furtek, D.B., and Nelson, O.E. (1984) Proc. Natl. Acad. Sci. (USA) 81,3825-3829.

X-IRRADIATION AND CO-TRANSFORMATION OF PETUNIA "MITCHELL" PROTOPLASTS WITH TWO ANTIBIOTIC RESISTANCE GENES.

I. BENEDIKTSSON, F. KÖHLER AND O. SCHIEDER.
INSTITUT F. ANGEW. GENETIK, ALBRECHT-THAER-WEG 6, 1000 BERLIN 33, GERMANY.

1. INTRODUCTION.

Direct DNA transfer to protoplasts has become a widely applicated method for introducing new traits into plants that can be manipulated by means of cell and tissue culture (1,2,3,4). Due to the relatively low transformation rates in most plant species (Petunia: 10^{-4} - 10^{-3} relative transformation frequency) gene transfer experiments are usually done with selectable markers, like antibiotic resistance genes.

R.J. Schocher et al. (5) showed that non-selectable genes can be introduced into protoplasts by means of co-transformation. In this procedure two different plasmids are offered to the protoplasts at the same time and incorporated into the genome of the same recipient cells at a high frequency. This rate has been shown to be in the range of 30 - 80 % when testing colonies that express the one (selectable) marker for the precence of the other. Thus co-transformation is a powerful tool for introducing non-selectable genes into plants (6,7).

Due to the high applicability of these methods both in applied and basic research it is of great importance to understand the physiological procedures that steer the integration and expression of foreign genes in plants.

It has been previously shown in our laboratory that sublethal irradiation with X-rays highly increases the transformation rates to protoplasts of many plant species tested (8). For Petunia "Mitchell" protoplasts this enhance was in the range of sixfold over the non-irradiated controls. It was suggested that an activation of DNA-repair mechanisms in the irradiated cells was the cause for this raised transformation frequency.

In a set of experiments (9) it was shown that irradiation of cells before mixing with plasmid-DNA had the same effect as thereafter and that UV-rays showed similar results. By irradiating at different points after transformation it was possible to evaluate how long biologically active plasmid sequences were precent in the nucleus, i.e. how long after the transformation the irradiation had an effect on the transformation rates.

The incorporation of foreign DNA by direct gene transfer requiers many different steps. The DNA must pass the cell membrane, be transferred to the nucleus , pass the nuclear membrane, be integrated into the chromosomal DNA and finally expressed. It has not been answered what is the rate limiting step in this process. P. Meyer et al. (10) suggested that the passing of the nuclear membrane would be the hardest task for the genes while others concentrate on the expression (5,11).

In previous publications on co-transformation in plants (5,12,13) it has been postulated that the explanation for this phenomenon could be found in the tandem-like co-integration of both genes and that gene-tandems having the blessing of being integrated at an expression suspectible locus expressed both genes simultanously. This would require the integration into the genome to be a relatively common event and the expression to be the bottleneck for sucsessful transformation. If this is the case it is hard to explain the above mentioned effect of irradiation on the transformation frequencies.

Here we examined the possible effect of irradiation on co-transformation with two antibiotic resistance genes (hph and nptII) both in regard of expression as well as integration of plasmid DNA sequences.

2. MATERIALS AND METHODS.

2.1. Plant material.

The plants used were of a haploid _Petunia_ "Mitchell" line obtained by anther culture (14) and propagated on hormone free MS mediunm at 25 ^{0}C and 16 h light.

2.2. Plasmids.

The Kanamycin resistance conferring plasmid pHP23 has been described by Paszkowski et al. (6). The Hygromycin resistance conferring plasmid pGL2 by Shimamoto et al.(4) Both plasmids were kindly provided by J. Paszkowski (Eidgenössische Technische Hochschule, Zürich, Switzerland).

2.3. Protoplast isolation, transformation and culture.

Mesophyll protoplasts were isolated from sterile

shoot cultures as described by Köhler et al. (8) Transformation and irradiation has also been described (9). The culture of the protoplasts was as follows: The protoplasts were diluted to 3 x 10^4 cells pr. ml in V-47/1 medium (8) and kept in Erlenmeyer flasks for 7 days in the dark. Then the protoplasts were mixed 1:1 with a V-47/1 solution containing 1,2 % liquid Seaplaque agarose and transferred to 9 cm diameter Petri dishes, 12 ml pr. dish. The agar was allowed to solidify and subsequently 7 ml of a V-47/2 solution, with or without antibioticum, was added on top of the protoplast blocks before culturing in continous light.

This very simple method resulted in a high increse in the plating efficiency of Petunia protoplasts over the traditional bead-type culture system (15) (results not shown).

2.4. Callus culture.

Resistant colonies were transferred on NT-1 medium (8) for shoot regeneration. In co- transformation experiments with two antibiotic-resistance genes the calli were first selected on a medium containing one of the antibiotics (Kanamycin or Hygromycin B). When the calli had reached appr. 1 cm in size (diameter) they were halved and one part tested for sensitivity to the other antibioticum. The other half was cultured without selection pressure and used for further analysis.

2.5. DNA isolation and hybridization.

DNA was isolated from transformed calli by a method described by Paszkowski et al. (16). Other molecular genetic methods were performed from the protocols of Sambrook et al. (17).

3. RESULTS AND DISCUSSION.

Similar results were obtained when the co-transformed calli were first selected for hygromycin resistance and then for kanamycin resistance as vice versa (results not shown). For simplicity the calli in the following experiments were all first selected on Kanamycin containing medium and then tested for the precence of the Hygromycin resistance genes. When evaluating the results of the first selected resistance it was confirmed in all cases that X-irradiation raises the transformation frequency of direct gene transfer to protoplasts. By gradually increasing the dose of X-rays this phenomenon reached a maximum at about 8 - 10 Gy and stayed at this level when the dose was further increased.

Table I. Tranaformation factor: different doses of X-rays.

DOSE	EXPERIMENT A	B	C	D	E	MEDIAN
0 Gy	1,00	1,00	1,00	1,00	1,00	1,00
1 Gy	2,95	3,42	2,42	3,26	1,38	2,68
2 Gy	--	2,16	2,93	3,61	--	2,90
5 Gy	25,55	2,14	7,29	5,19	3,21	8,68
10 Gy	21,60	3,57	6,78	12,30	10,90	11,03
24 Gy	11,00	2,99	11,47	5,12	31,19	12,35

The transformation factor was calculated from relative transformation rates as described by Köhler et al. (9). The protoplasts were irradiated before transformation.

If repair enzymes are responsible for the elevation of transformation frequencies these results could indicate that at 10 Gy, in Petunia, the maximum level of activation of repair mechanisms has been reached.

In control experiments no escapes of colonies transformed with unspesific DNA were observed neither in the irradiated nor untreated plates. Colonies carrying only one of the antibiotic resistance genes newer showed "cross"-resistance to the other antibioticum.

In the co-transformation experiments a total of 1508 Kanamycin resistant calli were tested on Hygromycin medium whereof 748, or almost exactly 50% were Hygromycin resistant. This high rate is in accordance with other published results on co-transformation. In three independent experiments the rate of co-transformation after irradiation with different doses of X-rays was evaluated.

Table II. Co-transformation.

DOSE	CO-TRANSFORMATION FREQUENCY (%) A	B	C
0 Gy	48,3	54,5	34,8
1 Gy	45,9	53,4	25,5
2 Gy	44,2	59,6	38,9
5 Gy	68,6	66,4	34,7
10 Gy	40,7	65,2	36,4
24 Gy	43,8	73,5	45,5

The numbers indicate the percentage of tested Kanamycin resistant colonies that also showed resistance to Hygromycin.

Having in mind the drastic effect of X-rays when evaluating the transformation rate of the first selected gene it is plausible that irradiation does do not affect the co-transformation.
DNA of co-transformed calli expressing only the kanamycin resistance and sensitive for Hygromycin-B were tested in Southern and slot blots for the precence of sequences of the coding region of the hph gene. All tested colonies had incorporated the entire coding region in various quantities. Due to the high variability in the copy number of incorporated genes in transformed plants it is difficult to deduce the environmental effects on the transformed copy number. However no clear-cut differences could be observed between copy numbers or the electrophoretical properties of transformed DNA from irradiated and non-irradiated calli.

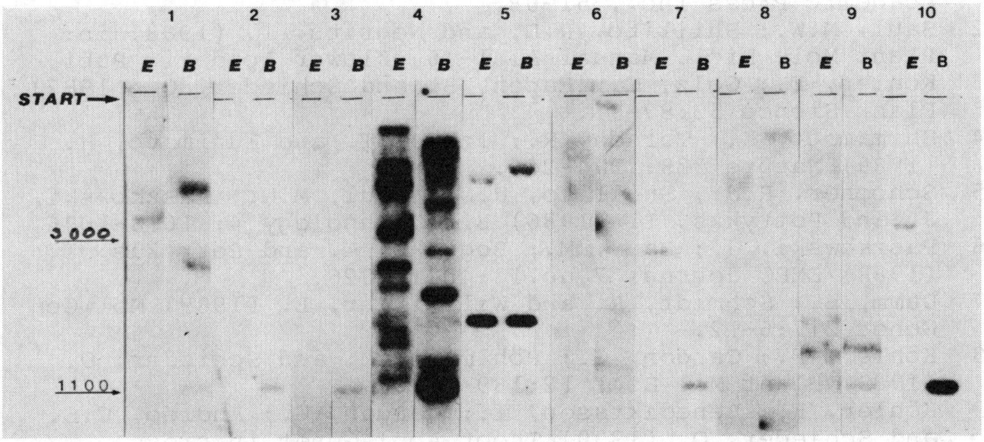

Figure I. Southern hybridization of DNA from ten Kanamycin resistant but Hygromycin B sensitive calli. No. 1-5 not irradiated; no. 6-10 irradiated with 10 Gy. E: cut with Eco RV, B: cut with Bam HI. Hybridized with the Bam HI fragment of the coding sequence of the hph gene from pGL2 (1100 bp).

As in these experiments the calli were not selected for Hygromycin resistance and the probes shown in the hybridizations above were from calli that did not express the hph-gene these results show the fate of a non-selected gene and can be used for comparing the events of integration with the occurance of expression. In these experiments (25 Hygromycin-negative calli tested) the ratio of expression to integration in co-transformed colonies was 50%, i. e. every second colony that integrated hph-sequences (in various copy numbers) also expressed this gene (only Kanamycin resistant colonies tested). Experiments testing the incorporation of silent genes into non-selected colonies after direct gene transfer are under way.

ACKNOWLEDGEMENTES.

This work was financed in part by a European Community grant no. CI 1*/0402-D for a cooperation between our institute and prof. C.S. Andreo, institut of CEFOBI, Rosario Argentina. The authors like to thank Dipl. Biol. Thomas Altman and Dr. Thomas Pickardt for skilful adwise.

REFERENCES.

1 Paszkowski, J.; Saul, M.W. and Potrykus, I. (1989) in: Cell culture and somatic cell genetics of plants, Vol.6, Academic Press Inc., 51-68.
2 Saul, M.W.; Shillito, R.D. and Negritu, I. (1988) in: Plant Mol. Biol. Manual A1 1-16, Kluwer Academic Publ.
3 Köhler, F.; Golz, C.; Eapen, S. and Schieder, O. (1987) Plant Sience 53:87-91.
4 Shimamoto, K.; Terada, R.; Izawa, T. and Fujimoto, H. (1989) Nature 338: 274-276.
5 Schocher, R.J.; Shillito, R.D.; Saul, M.W.; Paszkowski, J. and Potrykus, I. (1986) Bio/Technology 4: 1098-1096.
6 Paszkowski, J.; Baur, M.; Bogucki, A. and Potrykus, I. (1988) EMBO Journal 7,no.13:4021-4026.
7 Damm, B.; Schmidt, R. and Willmitzer, L. (1989) Mol Gen Genet 217:6-12.
8 Köhler, F.; Cardon, G.; Pöhlmann, M. and Schieder, O. (1989) Plant Mol Biol 12:189-199.
9 Köhler, F.; Benediktsson, I.; Cardon, G.; Andreo, C.S. and Schieder, O. (1990) Theor Appl Genet in press.
10 Meyer, P.; Walgenbach, E.; Bussmann, K.; Hombrecher, G. and Sädler, H. (1985) Mol Gen Genet 201 (3): 513-518.
11 Uchimia, H.; Hirochika ,H.; Hashimoto, H.; Hara, A.; Masuda, T.; Kasumimoto, T.; Harada, H.; Ikeda, J.E. and Yoshioka, M. (1986) Mol Gen Genet 205:1-8.
12 Czernilofsky, A.P.; Hain, R.; Baker, B. and Wirtz, U. (9186) DNA 5 (6): 473-482.
13 Tagu, D.; Berbounioux, C.; Cretin, C.; Perennes, C. and Gadal, P. (1988) Protoplasma 146: 101-105.
14 Ausubel, F.M.; Bahnsen, K.; Hanson, M.; Mitchell, A. and Smith, H.J. (1980) Plant Mol Biol Newsletter 1:26-32.
15 Shillito, R.D.; Saul, M.; Paszkowski, J. and Potrykus, I. (1985) Bio/Technology 3:1099-1103.
16 Paskowski, J.; Shillito, R.D.; Saul, M.; Mandak, V.; Hohn, T. and Hohn, B. (1984) EMBO Journal 3: 2717-2722.
17 Sambrook, J.; Fritsch, E.F. and Maniatis (1989) Molecular cloning, C.S.H. laboratory press.

GENE TRANSFER TO BARLEY

R.R. Mendel[1], E. Clauss[2], R. Hellmund[2], J. Schulze[2], H.H. Steinbiß[3], A. Tewes[1]

1. Central Institute of Genetics & Crop Plant Research, Academy of Sciences, GDR - 4325 Gatersleben
2. Institute of Breeding Research, Academy of Agricultural Sciences, GDR - 4300 Quedlinburg
3. Max-Planck-Institute of Breeding Research, FRG - 5000 Köln 30

1. INTRODUCTION

The stable transformation of crop plants belonging to the *Gramineae* family is of great importance for all further genetic manipulation. Since agrobacteria do not work (efficiently) with cereals the techniques of direct DNA transfer have been developed as an alternative (1). We tried different approaches of directly introducing DNA into cells and organs of barley in order to find a method for the stable integration of a foreign gene into the genome of barley with the aim to produce transgenic cell lines as well as transgenic fertile plants that inherit the new gene to their progeny.
The results of two cell genetic approaches (DNA transfer into protoplasts; DNA transfer into suspension culture cells by the particle gun) and of two plant genetic approaches (macroinjection into floral tillers; DNA transfer into growing pollen tubes) will be presented an discussed as preliminary data.. A further detailed analysis of the obtained putative transformants is in progress.

2. PROCEDURES, RESULTS AND DISCUSSION

2.1. Macroinjection into floral tillers

We used the technique of injecting DNA into the inflorescences of barley plants (*Hordeum vulgare*, cv. Erfa and Nebi). This technique was developed by De la Pena et al. and was successfully applied to rye (2). 2,055 plants between the 3-leaf- and 6-leaf state were injected with a syringe; about 0.2 - 0.4 µg of CsCl-purified circular plasmid DNA in a vol. of 5 - 10 µl were applied. Two constructs were used containing the neomycin phosphotransferase II (*npt*) gene as selectable marker thereby conferring resistance to kanamycin: pMLJ1glg, 12 kb, nos-promoter (3) and pGSGLUC1, 14.2 kb, TR1´ promoter, containing additionally TR2´-*gus* (4) (*gus* = gene for ß-glucuronidase). The 28,300 seeds obtained were selected on 150 mg/l kanamycin. 446 of them (= 1.6 %) grew in the presence of the antibiotic. Untreated control plants grown on kanamycin showed to a small degree (0.6 %) also resistance to kanamycin.
All 446 selected seedlings were tested for NPT activity, and 45 of them showed reproducibly low-activity levels corresponding to about 1 % of a stably transformed tobacco plant. These *F1* plants were selfed and the progeny was shown to segregate on

kanamycin, in some cases perfectly 3 : 1. However, in only very few F2 plants low NPT activity was detectable. It is under investigation whether the observed segregation is really linked to the introduced *npt* gene.
DNA isolated from F*1* and F2 plants was further analysed. PCR primers specific for an internal 412 bp fragment of the *npt* coding sequence allowed the amplification of the target sequence in selected F*1* and F2 plants also in those cases where no NPT activity was any longer detectable. When using the "left" PCR primer within the promoter region, pMLJ1glg-treated plants and pGSGLUC1-treated plants could be clearly differentiated. In pGSGLUC1-treated plants als a *gus*-specific 251 bp-fragment was PCR-amplifyable.
Southern DNA hybridization of genomic DNA (fig. 1) demonstrates that a hybridizing 1.7 kb *npt*-specific fragment is inherited from F*1* to F2 plants, however different plants show surprisingly the same fragment size even when cutting with *Pst*I inside the *npt*-coding sequence. Thus it appears that independent transformants show an identical hybridization pattern that is inherited from F*1* to F2. We see no difference in hybridization patterns when comparing young seedlings with adult plants.
To exclude that we have transformed endophytes instead of the barley plant, vigorous attempts were made to isolate microbes from F*1* and F2 leaves (with and without kanamycin in the medium: no microbes could be detected. Further, to exclude that we deal with extrachromosomal plasmid copies that have suddenly acquired the ability for self replication in plant cells, the isolated plant DNA was subjected to CsCl-gradient centrifugation: no plasmid bands could be observed. Furthermore, isolated plant DNAs were transformed into super-competent *E. coli* cells in order to detect free plasmid copies - with negative results.
A detailed comparative DNA analysis between different putative transformants on the F2 level is in progress in order to demonstrate unequivocally that DNA is integrated and to what extend vector sequences are included. The macroinjection approach was also used to introduce pGSGLUC1-DNA into wheat (cv. Chinese Spring). Both the *npt* and *gus* genes were inherited to the F2 where they were PCR-amplifyable and occurred non-rearranged on a 4 kb *Sal*I-fragment (M. Dittmann and R.R. Mendel, unpublished).

2.2 DNA transfer using growing pollen tubes

We used the approach developed by Picard et al. (5) to apply 3 µl CsCl-purified circular plasmid DNA (the same constructs as above) to the stigma of barley plants that were pollinated 5 - 20 min before DNA application. DNA was applied in a buffer mediating DNA stability for more than 1 hour: 1,058 plants treated; 11,200 seeds obtained and germinated on 150 mg/l kanamycin; 305 green plants selected (= 2.7 %); 25 out of 101 plants already tested show low NPT-signals; F*1* plants selfed to obtain the F2. Using PCR both the F*1* and F2 show the *npt*- and promoter-specific sequences. Again, genomic DNA hybridization shows the presence of the 1.7 kb *npt*-specific fragment in the F*1* and F2 plants tested (Fig. 1).
Comparing these preliminary data, we see no principal difference between the frequencies and results of the pollen tube approach and the macroinjection procedure. In both cases it appears that a foreign DNA sequence is inheritable from F*1* to F2 and in both cases the *npt* gene is only weakly expressed on the seedling level and loses its expression with plant growth to maturity and after inheritance to the next generation. This phenomenon of low expression is under further investigation. On the basis of NPT expression in the F*1*, a transformation frequency of about 10^{-3} to 10^{-4} could be deduced.

FIGURE 1. **Southern blot analysis of genomic DNA from plants DNA-treated by macroinjection and by using the pollen-tube approach.** The probe used was the 800 bp coding sequence of the *npt*-gene. Lane **1** - F*1* plant 1604; lane **2** - F*1* plant 1902; lane **3** - F2 plant 1902. 1 to 3 restricted with *Eco*RI cutting outside the *npt* coding sequence. Lane **4** - F2 plant 1902 not restricted. Note that the two independent transformants 1604 (obtained by macroinjection) and 1902 (obtained by the pollen-tube approach) give the same 1.7 kb fragment that does also occur in the F2 progeny of plant 1902 (lane 3). Fife other plants possess this 1.7 kb fragment as well (not shown here). Lane **5** - control plant, *Eco*RI-digested. Lane **6** - F2 plant 1604, *Pst*I-restricted. Lane **7** - same DNA as in lane 6, but not digested. Lane **8** - F2 plant 1341, *Pst*I-restricted. Lane **9** - same DNA as in lane 8, but not digested. Note that the two independent transformants 1604 and 1341 give the same 3.8 kb and 0.7 kb fragments although *Pst*I is cutting inside the *npt* coding sequence which should lead to different fragment sizes in independent transformants. Lanes **10** and **11** - 5 copy- and 1 copy-reconstructions, resp., of the 800 bp *npt* coding sequence. Note also some unspecific bands that are hybridizing to traces of the 2.8 kb vector sequences slightly contaminating the probe.

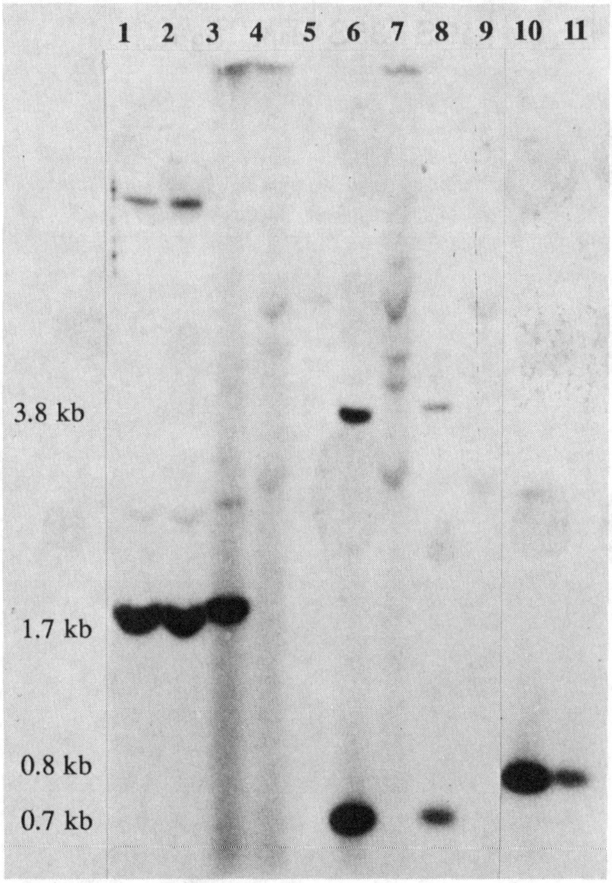

FIGURE 2. **Transient expression of chimaeric *npt* constructs in protoplasts of barley (mesophyll), sugarbeet (suspension culture) and *Nicotiana plumbaginifolia* (suspension culture).**

The following promoters were tested: mannopine synthase (TR1´) and nopaline synthase (nos) of *Agrobacterium tumefaciens*, 19S and 35S of cauliflower mosaic virus, pea Rubisco small subunit (SSU) and *Vicia faba* legumin (leg).

About 10^6 protoplasts taken up in a high calcium chloride/sodium chloride medium were heat-treated (5 min 45 C). DNA (10 µg circular plasmid) was added , followed by PEG to a final conc. of 16 %. After 2 - 3 days of cultivation in liquid medium, cells were tested for NPT activity. The spot of TR1´-*npt* comprises approx. 200,000 cpm.

TR1´= pGSGLUC1 (4); nos = pUCKN1 (4); CaMV19S = pABDI (6); CaMV35S = pHP23 (7); SSU=- pGSSTneo3 (4); leg = pGV180legP.FL (8).

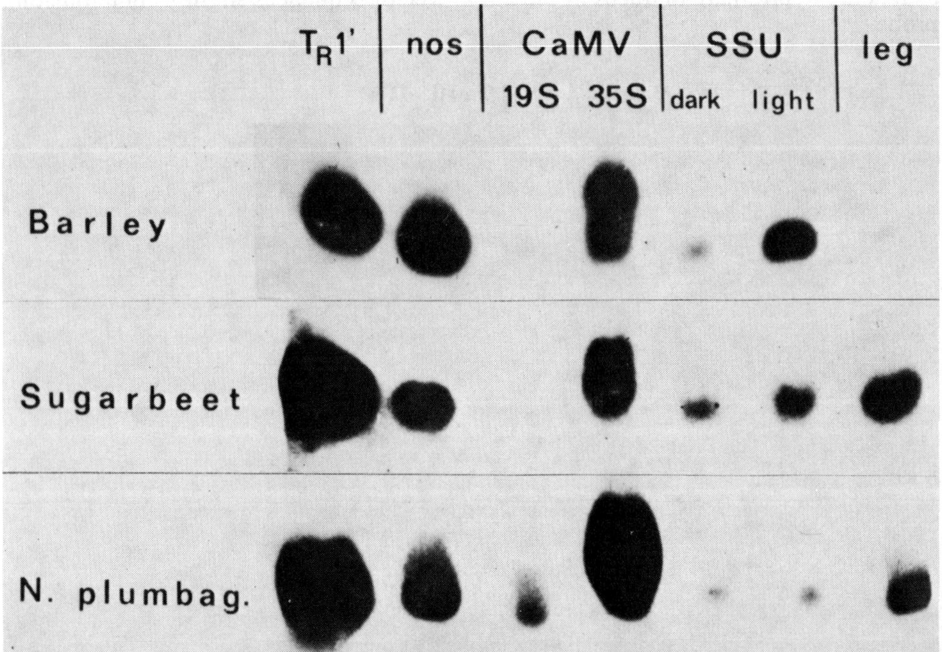

2.3 Transient gene expression in barley protoplasts

In order to show the principal expressibility in barley cells of the *npt* constrcuts used we determined at first their transient expression level (fig. 2). It turned out that TR1´-*npt* and nos-*npt* give high transient expression rates in barley mesophyll protoplasts. Furthermore, the comparison with the other promoters tested shows that a promoter mediating strong expression in tobacco and sugarbeet cells gives also strong expression in barley cells, and vice versa. Hence, the low *npt* expression in the above described DNA-treated plants is not due to a general lack of expressibility of these constructs in barley.
Concerning transient *gus* gene expression in mesophyll as well suspension culture protoplasts, only low activities could be detected (corresponding to 1 - 10 % of that level which was observed in tobacco mesophyll protoplasts when using the same constructs. Also pRT103GUS (9) (containing CaMV35S-*gus*) gave no higher activity.

2.4 DNA transfer into protoplasts

Suspension cultures were initiated from embryogenic calli derived from immature embryos, mesocotyle explants and inflorescences (cv. Borwina, Ilka, Golden Promise, Igri) and kept on MS-medium with 2 mg/l 2.4-D. Protoplasts isolated from these cultures could be readily regenrated to calli, in the case of Borwina with a plating efficienc of 25 %. These calli, however, did not regenerate.

About 10^6 protoplasts in a medium containing 75 mM calcium chloride and 60 mM sodium chloride were heat-treated (5 min 45 C) followed by the addition of 20 μg circular DNA (pHP23 (7), 4.4 kb, containing CaMV35S-*npt*) and PEG to a final conc. of 16 %. The protoplasts were imbedded into low-gelling agarose (1.5 %), selection on MS-medium containing 2 mg/l 2.4-D and 25 mg/l G418 started when the first microcolonies were visible (10 - 14 days after DNA-treatment). Colonies growing well after repeated subculturing on selective medium were tested for NPT activity. Also here, comparatively low NPT signals were observed. PCR analysis of isolated DNA shows the presence of the expected 412 bp *npt*-specific fragment. In genomic Southern blots, *npt*-signals differing from the patterns found in the DNA-treated plants were detectable. Further DNA analysis comparing the latter with the former patterns is in progress. The transformation frequency was calculated to be approximately 10^{-4}.

2.5 DNA-transfer into suspension cells by particle gun

Suspension cells (cv. Borwina) were bombarded with DNA-coated tungsten microprojectiles (0.6 μm diameter) using a particle gun according to Sanford (10). As plasmids, pHP23 (CaMV35S-*npt*), pGSGLUC1 (TR1´-*npt*, TR2´-*gus*) and pRT103GUS (CaMV35S-*gus*) were used and transient expression of these genes was reported previously by us (11). Bombarded cells were subsequently selected on MS-medium containing 0.5 mg/l 2.4-D and 5 mg/l G418. Three months after bombardment, the selected colonies showed only low NPT activities and no GUS activity (fluorimetric assay). Eleven months after bombardment, both the NPT and GUS activity were lost. DNA isolated at this time from the selected cultures and subjected to PCR analysis showed reproducibly the *npt*-specific 412 bp fragment and the 251 bp *gus*-specific fragment, respectively. DNA hybridizations are in progress. Experiments using embryogenic targets have been started.

3. CONCLUSIONS

The results of all four approaches show that the *npt* gene under control of three different promoters is only weakly expressed in barley cell cultures and plants compared to the expression level of the same constructs in tobacco. Hence a more suitable marker construct is needed for barley transformation. Nevertheless, the phenomenon of low expression is worth further investigation.

In a detailed further DNA analysis we will compare the hybridization patterns in blots of genomic DNAs isolated from the putative transformants that we obtained with the four approaches used. This will allow us to draw conclusions about the DNA integration when utilizing different techniques of direct DNA transfer.

The two cell genetic approaches offer the advantage of high numbers of target cells to be treated in a short time. However the calli obtained are not (yet) regenerable to plants. Thus the problem to obtain embryogenic calli derived from protoplasts has superior priority for the future. On the other hand, the two plant approaches give immediately fertile plants. The drawback with these approaches is , however, the high amount of manpower needed and the long time of the experiment. So the plant approaches represent principal solutions of the problem but are far away from any routine application.

4. REFERENCES

1. Klein, T.M., Bradley, A. and Fromm M.E. (1989) in Genetic Engineering, Principles and Methods (Setlow, J.K., ed.), Vol. 11, pp.13 - 31, Plenum Press, New York
2. De la Pena, A., Lörz, H. and Schell, J. (1987) Nature 325, 274-276
3. Saalbach, G. et al. (1988) Biochem. Physiol. Pflanzen 183, 211 - 218
4. kindly received from Marc Van Montagu, Rijksuniversiteit and Plant Genetic Systems Gent, Belgium
5. Picard, E., Jacquemin, J.M., Granier, F., Bobin, M. and Forgeois, P. (1988) Abstract Intl. Wheat Symposium, Cambridge UK
6. J. Paszkowski et al. (1984) EMBO J. 3, 2717 - 2722
7. kindly provided by J. Paszkowski, ETH Zürich, Switzerland
8. kindly provided by H. Bäumlein, Central Inst. Genetics, Gatersleben, GDR
9. Töpfer, R., Pröls, M., Schell, J. and Steinbiß, H.H. (1988) Plant Cell Rep. 7, 225-228
10. Klein, T.M., Wolf, E.D. and Sanford, J.C. (1987) Nature 327, 70 - 73
11. Mendel, R.R., Müller, B., Schulze, J., Kolesnikov, V. and Zelenin ,A. (1989) Theor. Appl. Genet. 78, 31 - 34

PROMOTERLESS REPORTER GENES AND THEIR USE IN PLANT GENE TRANSFORMATION

Klein B., R. Töpfer, A. Sohn, J. Schell & H.-H. Steinbiß.
Max-Planck-Institut für Züchtungsforschung, Abt. Genetische Grundlagen der Pflanzenzüchtung, Carl-von-Linné-Weg 10, D-5000 Köln 30

1. INTRODUCTION

Direct gene transfer into protoplasts is an alternative method to *Agrobacterium tumefaciens* mediated gene transfer in plants. As a result of this method two different phenomena can be observed, transient gene expression and stable integration of DNA, respectively. Transient gene expression monitored by reporter gene activity can be detected only for a limited time period after gene transfer (usually it last for one up to two weeks) and is considered to originate from non integrated DNA molecules. Such short term experiments are useful to elucidate conditions for DNA delivery or to test promoter constructions. However, conditions leading to stable integration of DNA can not be investigated by this phenomenon. Thus, to improve parameters for stable transformation long term experiments have to be performed leading to results at least 5 to 6 weeks post DNA transfer. Therefore, we decided to design experiments which would allow us to monitor DNA integration within one week after DNA transfer and thus enable us to improve parameters for direct protoplast transformation with a minimal time requirement.

2. PROCEDURE

Leaf mesophyll protoplast isolation from *Nicotiana tabacum* cultivar SR1 and DNA delivery were performed according to Negrutiu *et al.* (1) using 330000 protoplasts, 10 µg of plasmid DNA and 50 µg of calf thymus DNA. ß-Glucuronidase (GUS) activity was detected as described by Jefferson *et al.* (2).

Transformed plants were selected using 100 µg kanamycin per milliliter following the bead-type culture system of

Shillito *et al.* (3).

3. RESULTS AND DISCUSSION

The highest levels of transient gene expression are detected during the first few days after DNA delivery. In order to monitor gene expression resulting from integrated DNA, which at best is low in the beginning and is assumed to increase slowly within the first week, the phenomenon of transient gene expression has to be eliminated. Deletion mutants of a constitutive promoter construct leading to no significant transient gene expression might help in solving this problem. Therefore, based on the plasmid pRT104 (4) we created a series of deletion mutants of the cauliflower mosaic virus (CaMV) 35S RNA promoter (figure 1).

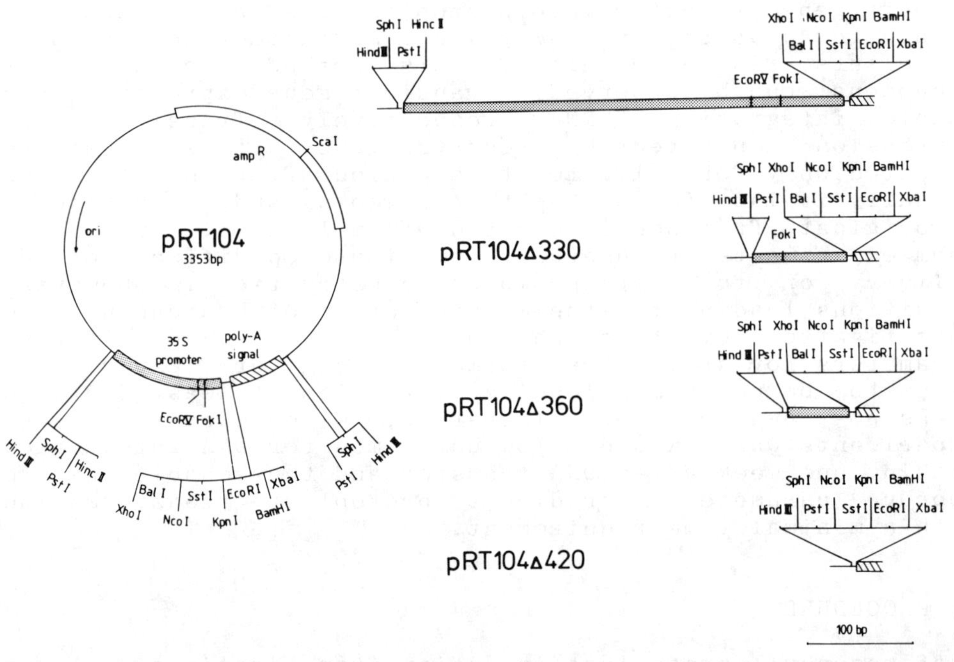

Fig. 1. The vector pRT104 (4) carrying the entire CaMV 35S RNA promoter was restricted with HincII and either EcoRV, eliminating the enhancer region, FokI, deleting both the enhancer and CAAT box, or BalI, removing the whole 35S RNA promoter, leading to the promoter deletion cassettes pRT104Δ330, pRT104Δ360, and pRT104Δ420, respectively. dotted area = promoter region; hatched area = poly-A signal.

Prior to expression and transformation studies, the reporter genes for ß-glucuronidase (GUS) and neomycinphosphotransferase II (NPT II) were cloned into the deletion cassettes. The GUS expression, which was measured in protoplasts 21 hours after DNA delivery (figure 2), showed a deletion dependent pattern with maximal activity in the case of pRT104gus decreasing to almost the control level with the promoterless pRT104gusΔ420. Plasmids pRT104gusΔ330 and pRT104gusΔ360 showed considerable expression above control level and thus were found not to be suitable for short term transformation studies.

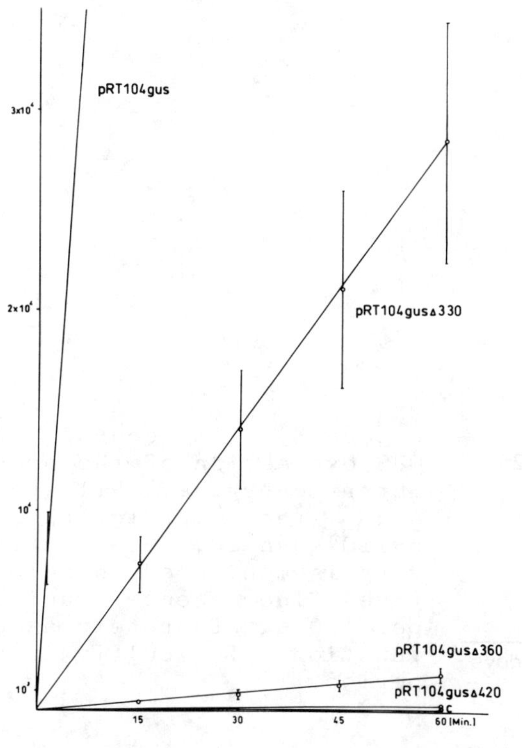

Figure 2. Transient GUS gene expression of the four deletion mutants was monitored after 21 hours of protoplast cultivation. The enzyme kinetics are given in relative fluorescence units per 100 µg of protein as a function of time.

The construct pRT104gusΔ420 did not show significant transient gene expression after 21 hours, however a time course over 7 days revealed an increase in GUS activity (figure 3). In contrast, pRT104gus gave a high expression level at day one decreasing slightly during the cultivation period. This seems to agree with the postulate that a construct showing no transient gene expression might result in an increasing expression level with prolonged cultivation period due to continous expression of integrated DNA. Pröls et al. (5) who investigated the time course of Chloramphenicolacetyltransferase (CAT) expression and the influence of transcriptional and translational inhibitors found transient gene expression to be a phenomenon of the early hours after DNA delivery. However, it can be monitored for up to 15 days due to protein stability. Consequently, instead of a rising GUS activity, found with pRT104gusΔ420 (figure 3), one would expect a decrease in GUS activity with continuing cultivation period where transient gene expression is concerned.

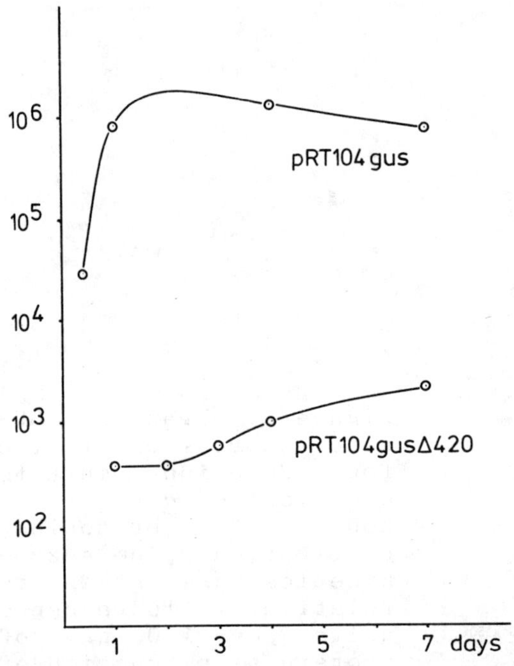

Figure 3. Time course of GUS expression of the promoter carrying pRT104gus and the promoterless pRT104gusΔ420. GUS activity is monitored in relative fluorescence units per 100 µg of protein as a function of cultivation period.

The increase in GUS activity of the promoterless vector might be an indication of expression due to integration of DNA. If this is the case one should find kanamycin resistant calli after application of the corresponding NPT II construct. On average, from five independant experiments we obtained 670 kanamycin resistant calli out of 330000 initially treated protoplasts after 6 weeks using pRT104neoΔ420 and approximately 2000 kanamycin resistant calli using pRT104neo. Plants were regenerated from such calli, self pollinated and seeds of the F1-generation were grown under selective conditions in order to test for the inheritance of the kanamycin resistance marker (figure 4). The inheritance of this phenotype after transfer of pRT104neoΔ420 gives evidence that in fact transformants can be obtained with a promoterless vector. However, the segregation of the F1-generation was not always found to follow the Mendelian rules.

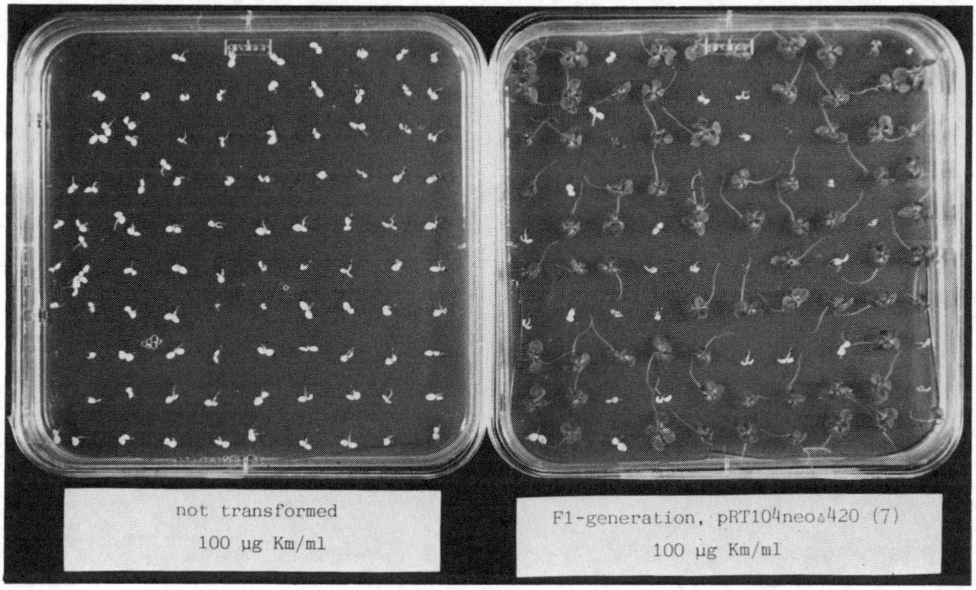

Figure 4. Inheritance of kanamycin resistance in the F1-generation of a plant regenerated after transformation with pRT104neoΔ420 (right hand side). The segregation is 71 (resistant) to 27 (sensitive) plants or 2.6 : 1. As a negative control untransformed SR1 tobacco seeds were tested (left hand side). Selection on 100 µg kanamycin per milliliter.

Since we found no transient gene expression but stably transformed plants using pRT104gusΔ420 and pRT104neoΔ420, respectively, the data indicate, that such promoterless constructs can be used to investigate conditions to improve parameters influencing transformation frequencies within one week post DNA transfer.

4. REFERENCES

1 Negrutiu, I., Shillito, R., Potrykus, I., Biasini, G., Sala, F. (1987) Plant Mol. Biol. 8: 363-373

2 Jefferson, R.A., Kavanagh, T.A., Bevan M.W. (1987) EMBO J. 6: 3901-3907

3 Shillito, R.D., Paszkowski, J., Potrykus I.(1983) Plant Cell Rep. 2: 244-247

4 Töpfer, R., Matzeit, V., Gronenborn, B., Schell, J., Steinbiß, H.-H. (1987) Nucl. Acids Res. 15: 5890

5 Pröls, M., Töpfer, R., Schell, J., Steinbiß, H.-H. (1988) Plant Cell Rep. 7: 221-224

TRANSFORMATION OF CUCUMBER VIA AGROBACTERIUM TUMEFACIENS

PAULA P. CHEE, MOLECULAR BIOLOGY DIVISION, THE UPJOHN CO., 301 HENRIETTA ST., KALAMAZOO, MI 49007, USA

1. INTRODUCTION

Studies on the development of tissue culture techniques and regeneration of species belonging to the family *Cucurbitaceae* have been reported (Jelaska, 1986; Chee, 1990). However, reports on regeneration of transformed cucurbits have not been well documented. Trulson et al. (1989) reported the transformation and regeneration of transformed cucumber plants which express the NPT gene. This transformation was achieved using *Agrobacterium rhizogenes* to transfer the T-DNA region of the binary plasmid pARC8, which contains a Nos-NPT II gene, by infecting cucumber leaf pieces. The resulting transformed roots were regenerated into whole plants (Trulson et al., 1986). The main limitation of this approach is its low frequency of plant regeneration from root tissues. Hence, only a few transformed plants were obtained. The present report describes the use of avirulent strains of *Agrobacterium tumefaciens* to mediate T-DNA transfers into cucumber cotyledon tissues and the use of somatic embryogenesis route to obtain a high number of morphologically normal transformed plants.

2. PROCEDURES

2.1 Materials and Methods

2.1.1 Plant materials: Seeds of cucumber, *Cucumis sativus* L. cv. Poinsett 76, (Asgrow Seed Co., Kalamazoo, MI) were soaked in tap water for approximately 15 minutes. Seed coats were removed manually. De-coated seeds were surface sterilized with 70% alcohol for 1 minute. A 25 minute treatment with 25% (v/v) solution of Clorox (commercial bleach, 5.25% sodium hypochlorite) followed. The seeds were then rinsed four times with sterile distilled water. Sterilized seeds were germinated at 28°C on 0.8% water agar (Difco Laboratories) in darkness. Unless otherwise stated, all media were supplemented with 3% sucrose and solidified with 0.8% phytagar (Gibco). The pH of all media was adjusted to 5.8 before being autoclaved. All media were autoclaved at 121°C for 20 minutes.

2.1.2 Transformation: Transformation of cucumber was performed by the method similar to that described by Horsch et al. (1985). Three to five day-old *in vitro* grown seedlings were used as donors of explants. The cotyledons were removed aseptically and cut into pieces of approximately 5 mm^2 in size. The pieces were submerged in an overnight culture of the avirulent *Agrobacterium* strain C58Z707 (Hepburn et al., 1985). The *Agrobacterium* strain contains the binary plasmid pGA482 (An et al., 1985; An, 1986). After gentle shaking to ensure that all edges were infected, the cotyledon pieces were blotted dry and cultured abaxial side down in a sterile petri dish containing 10 ml of initiation medium (MS basal medium + 2.0 mg/l 2,4-D + 0.5 mg/l KN) (Chee, 1990) followed by growth in the dark at 26°C. After 4 days, the *Agrobacterium*-infected cucumber cotyledon pieces were transferred to Petri plates containing the same medium supplemented with 500 mg/l Cb and 100 mg/l Km and cultured for five additional weeks in the dark at 26°C.

2.1.3 Plant regeneration: Regeneration of potentially transformed embryogenic callus tissue was done according to the procedure described by Chee (1990). Briefly, after five to six weeks on initiation medium, the cultures were transferred to MS medium + 1.0 mg/l NAA + 0.5 mg/l KN + 100 mg/l Km + 500 mg/l Cb for somatic embryo development. These cultures were incubated for an additional two weeks at 26°C under diffuse cool white fluorescent lamps (4 klx) with 16 hour photoperiod. The tissues were then transferred to hormone-free MS medium + 50 mg/l Km for conversion to plantlets. When plantlets developed an extensive root system on the latter medium, they were transplanted to pots containing planting medium and covered with Ziploc storage bags for hardening-off. Subsequently, the regenerated plants were potted in a mixture of (1:1, v/v) of soil and peatlite and grown in a greenhouse under standard conditions.

2.1.4 NPTII enzyme assay: NPT II enzyme activity was detected by the *in situ* gel assay procedure described by Reiss et al. (1984). Briefly, approximately 100 µg of cucumber callus or regenerated leaf tissue was mixed with 20 µl of extraction buffer in a 1.5 ml Eppendorf tube. The tissue was macerated with a Konte pestle and centrifuge for 15 minutes at 4°C to remove cell debris. From each sample tested, an aliquot of 35 µl of the supernatant solution was electrophoresed on a 10% non-denaturing polyacrylamide gel and the gel was then exposed to kanamycin sulfate and [^{32}P-γ]-ATP. Phosphorylated kanamycin, which is produced in those regions of the gel containing the modified bacterial Nos-NPT II enzyme activity is transferred by blotting onto phosphocellulose paper and detected by autoradiography.

2.1.5 Genomic DNA analysis: Total nucleic acids were extracted from cucumber leaves using the cetyltrimethylammonium bromide (CTAB) procedure described by Saghai-Maroof et al. (1984). The genomic DNA obtained was not subjected to any further purification steps as it was readily digestible by the restriction enzymes used for gene mapping.

2.1.6. Genomic blot hybridization: Genomic DNA from individual cucumber plant (between 5 to 10 μg)(Chee et al., 1989) was digested with five fold excess *Bam*HI and *Hin*dIII enzymes, and electrophoresed in a 0.7% agarose gel. The gel was then electroblotted onto nylon filter (Reed and Mann, 1985). Blotted filters were fixed by exposure for 3 min to a UV light source, then pre-hybridized in hybridization solution (6X hybridization solution contains 0.9 M NaCl, 0.09 M Na Citrate [pH7.2], and 0.02% BSA, PVP [40,000 mol wt. from Sigma, PVP-40], and Ficoll [400,000 mol wt. from Sigma, F-4375]) for at least 2 h, at 68°C. Filters were hybridized against a ^{32}P-labeled 600 bp *Bgl*II-*Nco*I fragment which contains mostly NPT II coding sequences. DNA labeling was done using a kit purchased from Bethesda Research Laboratories. Hybridizations were done using about 7.5 X 10^5 cpm of the labeled probe per ml of hybridization buffer containing 2% SDS and incubating at 68°C for at least 18 hours. Hybridized filters were washed in 1X hybridization solution for 1 hour at 68°C, dried and then exposed to film. Film exposures were generally for 12 to 24 hours.

3. RESULTS AND DISCUSSIONS

3.1. Cucumber cotyledon pieces were infected with *Agrobacterium* strain C58Z707 (Hepburn et al., 1985) which contained the binary *Agrobacterium*-derived plasmid pGA482 (An, 1986)(Fig. 1). Explants were subjected to antibiotic selection by transfering onto initiation medium which supplemented with 100 mg/l of kanamycin.

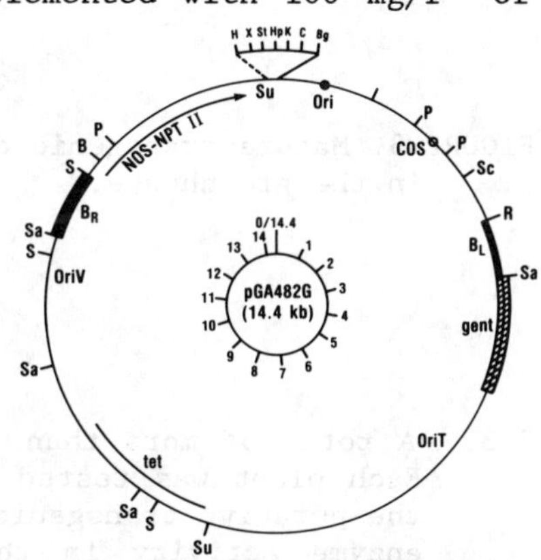

FIGURE 1. Structure of the *Agrobacterium* binary plasmid pGA482. Restriction enzyme sites are: C = *Cla*I; H = *Hin*dIII; Hp = *Hpa*; K = *Kpn*; St = *Stu*. Not all site locations are shown.

3.2 After about six weeks on initiation medium supplemented with 500 mg/l Cb and 100 mg/l Km, a characteristic gel-like callus was formed on the surface of the explants and at the site of tissue contact with the medium. The upper part of the gelatinous tissue contained small sectors of putative embryogenic tissue, which upon microscopic examination, consisted of cytoplasmically dense, multicellular aggregates, resembling the proembryonic masses. The sectors with proembryonic masses were selected for transfer to the secondary medium supplemented with 100 mg/l Km. Upon transfer, the putative transformed proembryonic structures developed into embryo-like structures. After two weeks, the embryos were transferred to hormone-free MS medium supplemented with 50 mg/l Km. Normal plantlet development followed. Transgenic plantlets were visually identifiable by their abiliy to form roots on MS medium supplemented with 50 mg/l Km (Fig. 2). In contrast, non-transgenic, control plantlets were inhibited from root development on medium supplemented with 50 mg/l Km. All transgenic plants were able to flower and set seeds normally (Fig. 3).

FIGURE 2. Regenerated cucumber plantlet cultured on MS medium supplemented with 50 mg/l kanamycin. Left: control plantlet, Right: transgenic plantlet containing NPTII gene.

FIGURE 3. Mature transgenic cucumber plants in the greenhouse.

3.3. A total of more than 100 transformed R_0 plants were obtained. Each plant was tested for NPT II enzyme activity. All of the the putative transgenic R_0 plants showed the presence of NPT II enzyme activity in their protein extracts which co-migrated with the bacterial-derived NPTII enzyme. Fig. 4 shows the analysis of NPT II enzyme activity in the protein extracts of some transformed R_0 plants. In contrast, the negative control plants, which resulted from co-cultivation with C58Z707 minus the binary plasmid showed no co-migrating NPT II enzyme activity in leaf extracts.

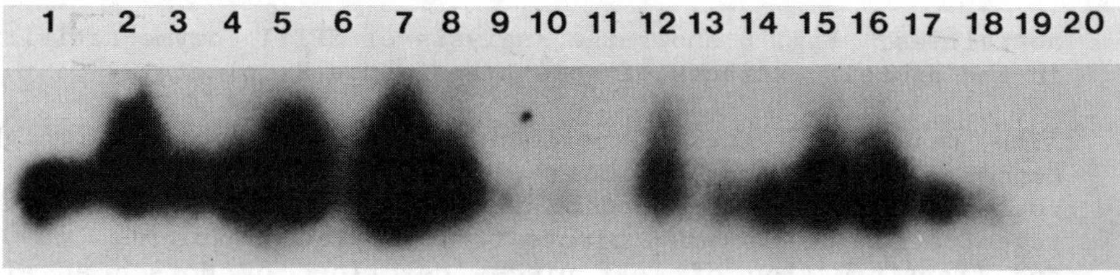

FIGURE 4. Analysis of NPT II activity in transgenic R_0 cucumber plants. Lane 1 contains the bacterial derived NPTII protein control; Lanes 11, 19 and 20 contain extracts from leaf material on non-transformed plants. Lanes 2 to 10 and 12 to 18 contain extract from leaf material of independent R_0 cucumber plants.

3.4. Total nucleic acids were extracted from leaves of the NPT II positive R_0 plants. This DNA was subjected to restriction endonuclease digestion with both *Bam*HI and *Hin*dIII. The restriction map of the Nos-NPT II fusion gene of pGA482 (An, 1986) predicts a fragment of about 2.0 kb in length with *Hin*dIII and *Bam*HI digest. Digested cucumber DNAs were attached to nylon filters which were hybridized against nick-translated ^{32}P-labeled *Bgl*II-*Nco*I 600 bp fragment isolated from the bacterial NPT II gene. This fragment contains only the bacterial NPT II gene portion of Tn5 (Beck et al., 1982) and was selected because of its low background hybridization against various plant genomic DNA (Chee et al., 1989). The predicted 2 kb Nos-NPT II hybridizing fragment was observed for DNAs isolated from all NPTII positive plants (Fig. 5).

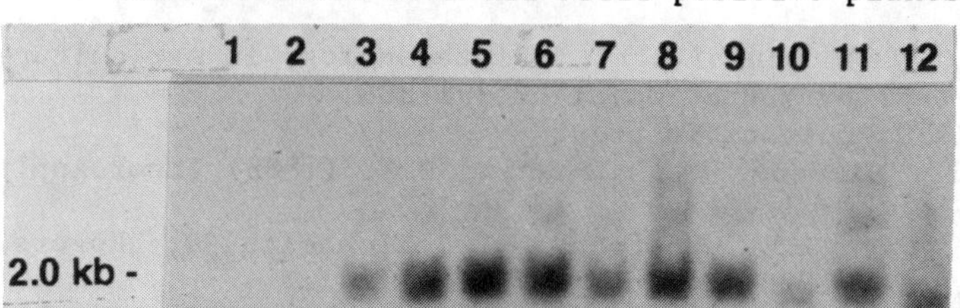

FIGURE 5. Genomic blot analysis of DNA. Lanes 1 and 2 contain a digest of DNA isolated from non-transformed cucumber plants. Lane 3 contains a digest of pGA482 DNA. Lanes 4 to 12 contain a digest of DNA isolated from independent transformed cucumber plants.

3.5. The most convincing evidence for the integration of the Nos-NPTII gene of pGA482 would be the transfer of the transgenic phenotype from the R_0 plants (self-pollinated) to their progenies. Assay for NPT II enzyme activity in forty progeny plants obtained from four randomly selected cucumber R_0 lines indicated that more than half of the progenies expressed NPT II activities. Fig. 6 shows the analysis of NPTII enzyme activity in the protein extracts of some transformed R_1 plants.

3.6. The results of these experiments clearly show that a high number of transformed cucumber plants can be obtained by using both the established procedure described by Chee (1990) for the regeneration of cucumber plants from cotyledon explants 1990) and transformation of leaf pieces described by Horsch et al. (Horsch et al., 1985). The work presented here also shows that the selection of transgenic Km^r cucumber tissues can be accomplished by supplementing culture media with 100 mg/l Km.

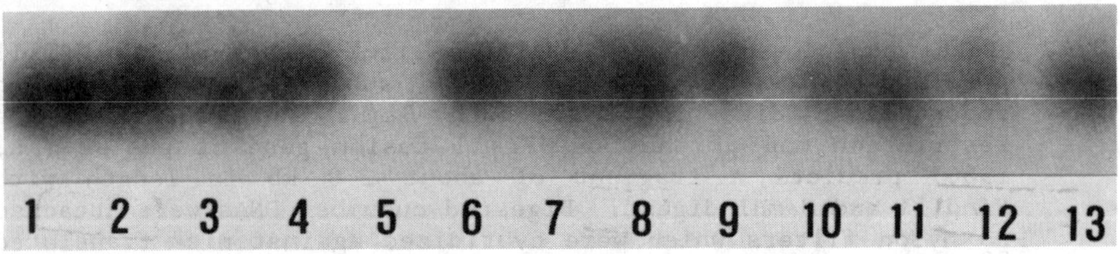

FIGURE 6. Detection of the NPT II gene in R_1 cucumber plants. Lane 1 contains the bacterial derived NPTII protein control. Lanes 5 contains extract from leaf material of non-transformed cucumber plants. Lanes 2 to 4 and 6 to 13 contain extract from leaf material of transgenic R_1 cucumber plants.

REFERENCES

1 Jelaska, S. (1986) Cucurbits, In: Biotechnology in Agriculture and Forestry (ed. by Y.P.S. Bajai) 2:371-386.

2 Chee, P.P. (1990) HortSci.in press

3 Trulson, A.J., Simpson, R.B., Shahin, E.A. (1986) Theor.Appl. Genet. 73:11-15.

4 Horsch, R.B., Fry, J.E., Hoffman, N.L., Eichholtz, D., Rogers, S.G., Fraley,R.T. (1985) Science 227:1229-1231.

5 Hepburn, A.G., White, J., Pearson, L., Maunders, M.J., Clarks, L.E., Prescott, A.G., Blundy, K.S. (1985) J. General Microbio. 131:2961-2969.

6 An, G. (1986) Plant Physiol. 81:86-91.

7 An, G., Watson B.D., Stachel, S., Gordon, M.P., Nester, E.W. (1985) EMBO J. 4:277-284.

8 Reiss, B., Sprengel, R., Will, H., Schaller, H. (1984) Gene 30:211-218.

9 Saghai-Maroof, M.A., Soliman, K.M., Jorgensen, R.A., Allard, R.W. (1984) Proc. Natl. Acad. Sci. USA 81:8014-8018.

10 Chee, P.P., Forber, K.A., Slightom, J.L. (1989) Plant Physiol. 91:1212-1218.

12 Beck, E., Ludwig, G., Auerswald, E.A., Reiss, B., Schaller, H. (1982) Gene 19:327-336.

REJUVENATION AND MICROPROPAGATION

R.L.M. PIERIK, DEPARTMENT OF HORTICULTURE, P.O. BOX 30, 6700 AA WAGENINGEN, THE NETHERLANDS.

1. GENERAL INTRODUCTION

Cloning in vivo and in vitro of adult or mature woody plants is adversely affected by characteristics accompanying maturation such as reduced growth rate, reduced or total lack of rooting ability or sometimes the unpleasant phenomenon of plagiotropy (1, 2). Maturation, a complex phenomenon, is the major problem preventing a wider application of tissue culture technology among woody spieces. Despite the problems mentioned, the possibility of multiplying mature trees by cloning and establishing field trials with micropropagated material, has been demonstrated e.g. for eucalypts (3), Tectona grandis (112) Sequoia sempervirens, Pinus pinaster and Pinus radiata (4, 5). Success with these trees has been achieved mainly by the use of special starting material e.g. from the base of the tree, by special pre-treatment in vivo and/or by in vitro culture (6). All of these tricks, which improve propagation are often described by the general term rejuvenation; a term which is difficult to appreciate since it usually does not say what it in fact entails. However, it is clear that rejuvenation is a prerequisite for possible cloning of adult trees and that the success in practice will mainly depend on the ability to rejuvenate them (7, 8, 9, 10).

2. MATURATION

Maturation or the transition from the juvenile to the adult phase (stage) involves an alteration during development from seedling to adult plant whereby one type of growth and development along an axis is succeeded either abruptly or gradually by another often sharply contrasting one (11, 12, 13, 14). Maturation can be defined as the developmental process inducing changes in morphological and physiological characteristics leading to the reproductive state. Old terminology for maturation such as phase change, ontogenetic ageing and meristem ageing should be avoided (14). However, the use of the term meristematic ageing, emphasizes that meristems play an important role in maturation. Increasing size of the shoot apex is one of the major factors underlying maturation (143). Juvenile meristems of Sequoia sempervirens are smaller and form more leaves than adult meristems (15). Juvenile apices of Hedera helix also have a smaller meristematic area, larger cells in the apex, and cell divisions occurring over the whole of the shoot for a longer period of time than adult apices (16, 17). Maturation in meristems or plants is never a permanent change in the genome, since it is the adult phase which produces the seeds which in turn give rise to seedlings with juvenile characters (12).

Maturation is strongly correlated with flowering. In early literature, the juvenile phase was often defined as the period from sowing, during which flowering does not occur and cannot be induced by the normal flower-inducing treatments (16, 17, 18). Generally the length of the juvenile phase increases in the following order: annual herbaceous plants, perennial herbaceous plants, shrubs, and trees (14). Some authors (17) propose that the ability to flower is a necessary condition to mark the end of the juvenile phase, although this cannot be used as the only criterion because: (1) conifers (19) can be made to flower in their juvenile phase, using gibberellins; (2) somatic embryos derived from root callus of mature Panax ginseng plants and from callus from sexual embryos of date palm can flower in the embryoid stage. Zimmerman et al. (17) therefore state that a plant has entered the adult phase: (1) If a plant can flower, no matter how flowering is induced; (2) If the said plant will then continue to flower in its natural environment without the application of any artificial stimulation that may have originally induced flowering.

Although in this definition the most general characteristic of true juvenility is the inability to flower, it is a narrow viewpoint and only considers one aspect of maturation. Maturation is certainly not a single on/off event restricted to flowering, but an accumulation of events in time viz. typical changes in: rooting ability and quality, growth rate, plagiotropism, leaf morphology, phyllotaxis, anthocyanin formation, hairiness, leaf retention, branch angle, thorniness and ability to flower (2, 13, 16, 18, 19, 23, 24, 25). That maturation is gradual is well illustrated by the following facts: (1) The ability to root Douglas fir and apple cuttings decreases long before flowering can occur (13, 20); (2) Juvenility (good rooting ability) can occur simultaneously with maturity (plagiotropic growth) in 'rejuvenated' shoots of Cunninghamia lanceolata (21, 22).

3. EARLY MATURATION IN JUVENILE MATERIAL

Unexpected features of early maturation have been observed in micropropagated plants derived from juvenile material (adventitious buds induced on cultured zygotic embryos) of Pinus radiata, Pinus pinaster, and Pinus taeda; the plants showing mature characteristics after 2 years in the field. Bonga et al. (31) have described 4 species (Thuja plicata, Sequoia sempervirens, Phoenix dactylifera, and Manihot esculenta) where very early flowering is seen in seedlings and meristems of vegetative shoot tips; flowering is also possible in explants regenerated from very young vegetative tobacco plants (32). The cause of the early apparent maturation is not known (26, 27).

It has been found in other species that plants rejuvenated completely in vitro mature more quickly than seedlings (28, 29). Accelerated maturation has also been shown in the early flowering of plants derived from tissue culture (30). This effect is especially paradoxical since it was often demonstrated that tissue culture of mature plants results in restoration of juvenile characteristics. The early flowering of tissue culture-derived plants can be explained only by assuming that the plants have been partially rejuvenated.

The unexpected results with micropropagated conifers showing advanced maturation may hypothetically be explained by the fact that the root systems of micropropagated plants are poorly developed as compared to those of seedlings in vivo. Since root systems of seedlings produce cytokinins, possibly causing rejuvenation or re-invigoration, the lack of roots (cytokinins) in micropropagated plants may induce early maturation (D.G. Thompson, personal communication).

4. LOCALIZATION OF MATURATION

An earlier theory (33) that maturation results from the increasing structural complexity and size of growing trees has recently been questioned (1, 9, 13, 34). Is maturation then determined at the level of the individual cell, the entire meristem, or is it due to correlative effects involving the whole plant (9, 14, 30)? Greenwood (30) proposes that there is evidence supporting all three possibilities.

The maintenance of characteristic differences in growth of juvenile and adult callus of Hedera helix (17, 30) and cherry rootstock (35) would indicate that phase change is cell-determined. However, the special role of the meristem is once again illustrated by the fact that rejuvenation is seen after regeneration of adult callus of Hedera helix (16). This has led a few authors (9, 36) to propose cautiously that changes of the cells in the meristems determine maturation.

Throughout the literature (1, 13, 143, 144) evidence is presented to support the view that maturation takes place in the growing meristems; correlative controls that exist between the different parts of a tree perhaps also play a role. It is believed that juvenility is an unstable state which exists in the meristems and which proceeds through a series of steps to a relatively stable state characteristic of the adult meristem. Changes occur in the buds with time, so that they acquire a certain individuality, persisting even after a bud is separated from the plant (1, 11). Arguments for an important role of meristems are: (1) When adult meristems are isolated and/or used in grafting experiments, they remain adult and are not easily changed (1, 6, 11, 36); (2) When an adult terminal meristem of e.g. Paliuris australis enlarges, lateral plagiotropy is seen with regard to the axillary buds (70); (3) Plants are continuously renewed only by meristematic growth (37); (4) When isolated meristems show an autonomy that is generally not so strongly expressed during the integrated development of the plant (38).

5. REJUVENATION VERSUS RE-INVIGORATION

True rejuvenation, also known as phase reversal (12) or return to the juvenile form (13), includes the complete reversal of maturation as a result of sexual reproduction or vegetative propagation via adventitious shoot formation or somatic embryogenesis (3, 7, 16, 39, 40). A disadvantage of using asexual reproduction methods to obtain true rejuvenation is that very little is known of the fundamental causes of shoot regeneration and somatic embryogenesis (41). There is not very good evidence to claim that all adventitious buds/shoots are juvenile (see also section 8). Particularly the work with adventitiously induced buds on conifer cultured embryos has shown that not all these buds are juvenile (D.G. Thompson, personal communication). Greenwood (30) concludes that adventitiously formed juvenile shoots probably arose from cells, which could be interpreted to mean that tissue culture conditions favoured the proliferation of residual cells rather than actual rejuvenation.

There is generally great uncertainty regarding the possibility of obtaining true rejuvenation, as shown by the loss of plagiotropy (12, 42, 43). Therefore the term 'apparent rejuvenation' should perhaps be replaced with other terms such as reinvigoration, reversal of ageing, reactivation or partial rejuvenation (3, 12, 20, 40). Re-invigoration is defined as the reversal of ageing, whereas ageing means the increase in disorders in time caused by the environment leading to a reduced vigour. For instance re-invigoration can be used when rooting-ability and vigour are increased as a result of e.g. pruning, hedging, repeated culturing, 6-benzylaminopurine (BA) treatment and grafting (44). In some cases of repeated micro-grafting true rejuvenation takes place (15, 21). Claims of complete rejuvenation of grape and Metrosideros spp after repeated subculture in vitro (28, 29) are incorrect, since the temporary modifications observed persist for a shorter time than the characteristics of normal seedlings. In fact the forcing of shoots, originating from buds in the juvenile zone of a tree, have incorrectly been described as leading to rejuvenation. Rejuvenation claims are based on appearance; in principle, the absence of markers other than morphological ones, strongly limits our ability to determine whether true rejuvenation or re-invigoration has occurred (2).

6. EXPLANATION OF REJUVENATION

Many authors (12, 45, 46, 144) have proposed that the modification in gene expression accompanying rejuvenation as well as maturation has an epigenetic basis and is not based on a change in the genome itself. It should not be ruled out, however, that plants with a so-called fixed juvenile phase are based on a mutation.

Explanations given for rejuvenation are often vague and speculative (47), since very little is known about the physiological and biochemical factors causing rejuvenation (16). Isolated meristems may rejuvenate as a result of wounding or as a result of the loss of correlative control (15); the reduction in the number of cells when a plant is formed via meristem culture is thought to be the cause of rejuvenation (48). The actual explanation for true rejuvenation is perhaps only possible if the changes are known which occur during sexual reproduction and/or adventitious bud formation or somatic embryogenesis.

7. BIOCHEMICAL MARKERS FOR MATURATION AND REJUVENATION

Maturation and rejuvenation cannot yet be characterized biochemically (6, 49), and very little is known about the molecular basis of the proces (50).

Maturation-related increase in chlorophyll content in Larix laricina was substantial from 1-4 years, whereas indications were obtained that the chlorophyll *a/b* binding protein gene is differentially expressed in juvenile and mature material (1, 51). In paper birch isozyme analysis was used to characterize rejuvenated and adult material (52), whereas peroxidase activity was used to quantify maturation in Cynara scolymus (53). Maturation in Juglans was characterized by an accelerated and asynchronous functioning of 2 enzyme systems, whereas rejuvenation was linked with high ratios of specific polyphenolic compounds (54). In juvenile and rejuvenated tissues of Sequoiadendron giganteum an unidentified membrane-associated protein was detected, which was absent in mature tissue (55). In Asparagus officinalis rejuvenation was linked with an increase in the amount of putrescine and spermidine (56). In Sequoia sempervirens the K/Ca ratio, the peroxidase/protein ratio, and the indole-3-acetic acid (IAA)/abscisic acid (ABA) ratio were all high in juvenile and low in adult material (3, 10).

Thus, there is a need to try and bring all of these observations together to attempt and form a working hypothesis to biochemically and physiologically explain and define maturation (D.G. Thompson, personal communication).

8. REJUVENATION BEFORE OR DURING SEXUAL REPRODUCTION

Rejuvenation occurring during sexual reproduction, is not directly attributable to either meiosis and/or fertilization as genetic processes (11, 13, 39, 57). The arguments for this statement are that plants arising through apogamy (apomictic or nucellar embryos) and somatic embryos are indistinguishable morphogenetically from those resulting from fertilization.

However, there is no doubt that the nucellar tissues of e.g. Citrus and Mangifera can be regarded as an island of juvenility in an otherwise adult organism (17, 143). At the time of meiosis a number of plant species are also capable of regenerating adventitious plants from reproductive structures (6), especially from anthers.

Although it was thought that a truly flowering meristem could not be formed adventitiously (16), this has been disproved. The first reports on regeneration of both vegetative and generative shoots from inflorescence explants were recorded around 1966 for tobacco (58) and Lunaria annua (57). It is remarkable that reversion from adult to juvenile in bulbous plants always occurs when isolated flower stem explants or parts of the inflorescence regenerate adventitious shoots (60, 61). Floral apices of banana (62), explants of female inflorescences of Casuarina equisetifolia (63), inflorescences of Cyperus fertilis (48), and floral tissues of cauliflower and broccoli (64) can reorganize and give rise to vegetative shoots. These results indicate that adventitious shoot formation does not necessarily induce rejuvenation and that two way expression of totipotency is possible (59).

It is generally thought that the developmental pathway of inflorescence explants, flower primordia, anthers, etc. has not been irreversibly determined and that floral tissues in principle can revert to a vegetative pattern of development and grow out into leafy shoots (45, 57, 59, 64).

9. GENERAL RULES FOR REJUVENATION

1. It is extremely difficult to separate reversal of ageing from actual rejuvenation (30). True rejuvenation is more difficult to achieve than reversal of ageing or re-invigoration (65).
2. The efficiency of rejuvenation decreases with ageing (7); the older the plant, the stronger is the treatment needed to reach rejuvenation. Juvenile characteristics can often only be recovered by repeated treatments (42).
3. Terminal and peripheral parts of a tree, chronologically the youngest ones, are ontogenetically the most mature ones (7, 14, 66). As the distance between shoot tip and root system increases with growth, the juvenile condition of shoot meristems gradually disappears (6, 46, 48), except at the base of the tree (the juvenile zone) (13, 16, 18, 67, 68). In principle, basal and interior parts may retain juvenile characteristics (16).
4. It is important to select young and vigorously growing buds to facilitate rejuvenation (3). Subsequent isolation in vivo or in vitro is an essential prerequisite; the smaller the isolated part the better the chances for rejuvenation (6).
5. Rejuvenation techniques (with the exception of adventitious organ formation and somatic embryogenesis) generally do not result in the loss of plagiotropy (6, 48, 69, 70).
6. Rooting ability, considered as an important parameter for rejuvenation, declines with the age of the plant (71); however a strict correlation between rooting and rejuvenation does not exist (34).
7. The cytokinin BA can be considered as an important anti-ageing substance and can be used both in vivo and in vitro to facilitate reinvigoration (30, 65). BA promotes the outgrowth of lateral and dormant buds. In a number of cases BA can induce adventitious bud formation, often leading to true rejuvenation.

10. RE-INVIGORATION AND REJUVENATION IN VIVO

Often rejuvenation is achieved by different methods being tried in sequence. In vivo methods are generally used before in vitro ones. For instance with Pinus taeda (72) first grafting on rootstocks, severe hedging, and BA treatment are carried out before subculturing in vitro starts. For Pinus pinaster (73) first in vivo cuttings are taken after which in vitro BA treatment is carried out on isolated brachyblasts. Since manipulation in vivo is often an important prerequisite for rejuvenation in vitro, the in vivo methods will be summarized below.

Methods often used for re-invigoration are pruning, hedging, and the use of stool beds. Severe pruning of trees results in the stimulation of lateral buds and shoots with enhanced rooting (7, 74, 75, 76); severe pruning has been used in apple, Pinus radiata (145) and in Eucalyptus spp. (71). Hedging involves the repeated pruning of stock plants of e.g. Pinus pinaster (47). Hedged plants provide a large number of cuttings with an apparently lower maturation rate than those from non-hedged plants (2, 7, 18, 77, 78). They are thought to be invigorated (71), since they flower relatively early. Hedging can also be used to keep plants in a juvenile condition by delaying (but not stopping) maturation (79) e.g. in Pinus radiata and Norway spruce (30, 145). Severe hedging may also be stimulating the growth of shoots from more juvenile regions of the stem. The use of stool beds, also called stool layering (repeated cutting back to soil level) results in newly developed vigorously growing shoots remaining juvenile and retaining good rooting ability (18); this method was introduced for root stock production of fruit trees (79).

Several tree species possess preformed, dormant buds which remain quiescent after their early initiation (9); the outgrowth of these buds often leads to the development of juvenile shoots (9, 76, 80), particularly when they arise in the so-called juvenile zone (the base of a tree: a stub or the trunk). However, they can also arise higher up in the tree. These basal shoots are called stump sprouts or epicormic shoots; in most cases they arise from suppressed accessary buds (additional buds), but they can also be of adventitious origin (Evers, personal communication). Preformed buds can be distinguished from adventitious buds by the presence of a vascular trace in the primary xylem (9). Epicormic shoots can be produced in coppice rotation plots, by making a notch at the base of the tree (47) or from felled trees. This phenomenon is seen in walnut, wild cherry (3), Tilia cordata, Hevea brasiliensis (6), Eucalyptus spp. (2, 47), Liquidambar styraciflua (142), Cunninghamia lanceolata, Pinus rigida (81), Quercus, Fagus (47); The most successful species used so far is Sequoia sempervirens (3, 47, 67, 81, 82). Another organ used as a source of juvenile material in an adult tree are lignotubers, which are swollen woody storage organs formed at the base of many Eucalyptus species during the first year of growth (46). Lignotubers contain juvenile meristems which persist throughout the life of a tree and they can give rise to juvenile shoots (68). Finally water shoots and reiterated shoots can also be used as a source of relatively young material. Water shoots arise from within the tree especially in fruit trees. They form upright shoots and are of most interest if they are formed in the juvenile zone (25). Reiterated shoots are, in principle, equivalent to epicormic shoots; they occur higher up in the tree and may retain or regain juvenile appearance (7, 11).

By taking cuttings of the ortet and particularly by serial cutting production (cuttings taken from rooted cuttings), maturation can be delayed (not reversed), resulting in an increased rooting potential (20, 34, 76, 78); often the combination of rooting cuttings with pruning or grafting on juvenile rootstocks were applied to obtain trees with juvenile characteristics (42, 76) in Douglas fir, Platanus acerifolia, Pinus pinaster, and Sequoia sempervirens.

Etiolation (induced by layering) has long been known to have a promotive effect on adventitious root formation (50, 79). Girdling and ring barking (50, 79) have a similar effect, probably due to blocking of the downward movement of carbohydrates and root promoting factors.

Pretreatment by special regulator treatments (sprays, pastes, or injection) of branches from mature trees is often necessary to activate parts of the tree or buds prior to explant selection (44, 50, 83). In most cases the cytokinin BA is applied which induces bud break (10, 16, 68, 82, 84). In a few plant species gibberellins are applied, which can substantially stimulate growth and sometimes (Hedera helix) can reverse some of the branches to the juvenile state (16, 74, 79, 83, 85). However, a cautionary statement about the general application of the rejuvenation results with the model plant Hedera to other species is necessary since a gibberellin treatment of Hedera, that results in rejuvenation, is doing exactly the opposite (stimulation of flowering) in species of the conifers.

In literature 3 types of results concerning the effect of grafting of adult scions on juvenile root stocks have been described: negative, intermediate and positive. Grafting of 12-year-old scions of Pinus taeda on 2-year-old root stocks did not result in any rejuvenating effect (36, 86). Reinvigoration was recorded in grafting experiments with Cupressus sempervirens (87), Larix laricina (51) and apple (43). In these cases grafting resulted in a temporary return to juvenility (7). True rejuvenation by grafting was first demonstrated in Hedera helix (88). Positive results were recorded when adult scions were grafted in succession on (very) juvenile vigorous root stocks of Eucalyptus camaldulensis, Cupressus dupreziana (76), Platanus acerifolia, Cupressus sempervirens, Sequoia sempervirens, Thuja plicata v. atrovirens, Pinus pinaster, Eucalyptus grandis, Sequoiadendron giganteum (3, 7, 21, 42, 47, 84) and Eucalyptus x trubutii (38). However, in these cases rejuvenation is gradual with more juvenile behaviour occurring after each graft (7, 89). The general conclusion from the grafting experiments is that the close proximity of the mature shoot apex to a juvenile root system may account for the reinvigoration or rejuvenation obtained (2, 6, 71). Prerequisites for success are repeated grafting on relatively young and vigorous root stocks (71). The rejuvenating influence of a juvenile rootstock may be explained hypothetically by the cytokinin production of the juvenile roots (see also section 3).

In a few plant species shoots arise from mature plants by forcing growth from sphaeroblasts, nodules of wood separated from the central cylinder. Sphaeroblasts or trace buds give rise to truly juvenile adventitious shoots (9, 79). They are uncommon in conifers (9), but occur in apple (44, 141), pear (9), plum, Ilex aquifolium (16), Platanus and Castanea (40). Sphaeroblasts are induced to develop by disbudding and heavily cutting back stock plants.

Root suckers are shoots arising from roots, spontaneously or by manipulation (wounding, root cuttings). They always originate adventitiously and are therefore often juvenile or rejuvenated if they originated from an adult tree (25, 79). Root suckers are most common in hardwood species (90) and are recorded in apple, pear, plum, birch, Platanus, Paulownia tomentosa, Rosa canina, Parthenocissus (25, 75), Populus tremula, Sequoia sempervirens, Pinus rigida, Cunninghamia lanceolata, Acacia, Robinia, Carya, Fraxinus, Liriodendron (47), Ulmus campestris, Sorbus spp (90), and Prunus avium (3, 47).

11. RE-INVIGORATION AND REJUVENATION IN VITRO

In vitro culture has become an important tool to propagate plants that may also have the added ability to re-invigorate and rejuvenate adult plants. It is essential that: (1) In vivo manipulation of stock plants (section 10) should often precede in vitro culture; (2) Relatively small structures, preferentially meristems, are isolated; (3) Repeated subculture takes place; (4) Correlations are broken (38, 84). The sterile isolation of adult material is often a large problem, and the induction of growth is also quite difficult (47). The conditions under which in vitro culture takes place can have an influence; it has been shown with grape that high carbon dioxide concentrations inhibit the transition from juvenile to adult (146), whereas ethylene accumulation stimulates this in Hemerocallis (147).

When stump sprouts or epicormic buds are available in adult trees, rejuvenation is not necessary; then it is a question of in vitro re-invigoration of the isolated material. Successful examples of micropropagation starting from stump sprouts or epicormic buds are: Acer pseudoplatanus (90), Sequoia sempervirens (10, 44), Castanea sativa (90, 91), Quercus robur (90, 92), Tilia cordata, Salix alba (90), and Eucalyptus grandis (93).

Rejuvenation can be obtained through micro-grafting, the grafting of adult meristems or shoot tips on juvenile rootstocks in vitro. The extent of rejuvenation is determined by the age and size of the scion, the genotype of the graft and rootstock, the composition of the nutrient medium and the number of graftings (94). Adult scions are more difficult to graft than juvenile ones (95, 96); small scions rejuvenate easier than large ones (94); several successive grafts (so-called cascade grafting) are often necessary for rejuvenation. Rejuvenation may not take place after micro-grafting as in the case of Citrus (46, 96, 97). Successful rejuvenation or re-invigoration by micro-grafting has been described for a few plant species: Sequoia giganteum (94), avocado (98), Hevea brasiliensis (99), and Thuja plicata (100, 101).

In two cases co-culture has had a remarkable effect. The reversion of adult Hedera helix to the juvenile phase was realized by growing shoots of the 2 phases together in the same vessel of nutrient solution (102). Shoot invigoration and rooting potential of hard-to-root Mazzard sweet cherry were progressively increased during 6 consecutive subcultures, only when co-cultured with an easy-to-root Prunus hybrid (35, 103).

In principle, meristem culture can be an efficient tool to obtain rejuvenation. A classical example is Cymbidium where an adult meristem is easily transformed into a protocorm, characteristic for the juvenile phase (38, 104). There are only a few examples of rejuvenation via meristem culture apart from orchids. These are: strawberry, grape (3, 38), Prunus avium (84), Pinus pinaster (105) and Sequoia sempervirens (15). The greatest problem with adult trees is, that true meristem culture is very difficult to achieve (3).

In quite a number of plant species repeated subculture of adult shoots results in re-invigoration or complete rejuvenation, by which shoot multiplication and rooting ability are strongly improved. Very often media for subculturing are rich in cytokinin (15) which promotes axillary branching. After a number of subcultures a so-called stabilization is reached in vitro (106). Rejuvenation is closely related to the number of subcultures (53, 65, 107, 108). The number of subcultures required strongly depends on the species and even cultivar: 4 in Asparagus officinalis and hardy deciduous azaleas, 9 in the apple cv. Jonathan, 13 in the apple rootstock M9, and 31 in the apple cv. Red Delicious (18, 56, 74, 109). In principle, the cause of rejuvenation or re-invigoration by subculturing is not known (110). Often rejuvenation is not complete because e.g. micropropagated plants flower earlier than seedlings (106, 111) or because the rooting ability of trees is not completely restored (65, 112, 113, 114, 115). Plagiotropic growth habit has never been overcome by repetitive subculturing (22, 65). In a number of plant species morphogenetic behaviour during subculturing changes so characteristically that it can be considered true rejuvenation: Solanum tuberosum (48), Asparagus officinalis (56), strawberry (116, 117), grape (29, 118, 119, 120), Syringa vulgaris (107), Vaccinium ssp (17, 121).

Adventitious shoot formation and somatic embryogenesis in vitro are in principle the most efficient methods of inducing rejuvenation. Rejuvenation also occurs after regeneration from anthers (48) and protoplasts; from at least 8 different species of adult trees and shrubs plants have been regenerated from protoplasts (122). It is proposed (39) that during adventitious shoot and somatic embryogenesis miniaturization of the meristem occurs, resulting in juvenility. The decrease in the number of cells of a newly formed meristem probably plays a role.

In a number of cases (57, 58) adventitious organ regeneration results in generative shoots, indicating that rejuvenation does not always occur (see section 8). In a few other plant genera (Passiflora, Panax, Betula) adventitiously formed shoots flower rapidly (14), but they loose the flowering ability afterwards, so that ultimately rejuvenation occurs. Bonga (123) demonstrated that adventitiously formed shoots from female cones of Larix decidua were only partly rejuvenated. Despite the exceptions given here adventitious shoot formation and somatic embryogenesis remain powerful tools to rejuvenate adult plants (Table 1). However, the induction of adventitious shoot formation and somatic embryogenesis remain a problem in woody species.

Table 1. Adventitious shoot (AS) or somatic embryo (SE) formation leading to rejuvenation in the following plant species. References in brackets.

Asparagus officinalis	AS + SE (56)	Musa spp.	AS (62,133)
Betula pendula	AS (23,90)	Myriophyllum heterophyllum	AS (134)
Brassica spp.	AS (46)	Phoenix dactylifera	AS+SE(118,135)
Cardamine pratensis	AS (124)	Populus ciliata	SE (136)
Citrus spp.	SE (125)	Prunus cerasus	SE (59)
Coffea arabica	SE (125,126,127,128,129)	Pyrus communis	AS (137)

Elaeis guineensis	SE (127,130)	Quercus petraea	SE (138)
Eucalyptus longan	SE (131)	Sequoia sempervirens	AS (47)
Hedera helix	AS (14)	Thuja plicata	AS (139)
Larix spp.	AS (140)	Vitis vinifera	SE (127)
Malus domestica	AS + SE (132)		

12. REFERENCES

1 Hutchinson, K.W. et al. (1988) in Molecular genetics of forest trees (Hällgren, J.E., ed.), Proc. Frans Kempe Symp., Umea, Sweden, pp. 101-114

2 Thompson, D.G. (1984) in Seedling physiology and reforestation success (Durya, M.L. and Brown, G.N., ed.), pp. 12-17, M. Nijhoff, Dordrecht

3 Boulay, M. (1985) in Proc. Int. Symp. on in vitro propagation of forest tree species (Anonymous, ed), pp. 51-81, A.R.F., Bologna

4 Ahuja, M.R. et al. (1988) in Somatic cell genetics of woody plants (Ahuja, M.R. ed.), pp. 127-135, Kluwer Acad. Publ., Dordrecht

5 Dunstan, D.I. (1988) Can. J. Forest Res. 18, 1497-1506

6 Bonga, J.M. (1987) in Cell and tissue culture in forestry (Bonga, J.M. and Durzan, D.J., ed.), Vol. 1, pp. 249-271, M. Nijhoff Publ., Dordrecht

7 Franclet, A. and Boulay, M. (1989) in Biomass production by fast growing trees (Pereira, J.S. and Landsberg, J.J., ed.), pp. 267-274, Kluwer Acad. Publ., Dordrecht

8 Franclet, A. (1983) Bull. Soc. Bot. France 130, 87-10

9 Bornman, C.H. (1984) in International symposium of recent advances in forest biotechnology (Anonymous, ed.), pp. 178-194, Michigan Techn. Univ., Michigan

10 Boulay, M. (1987) in Plant tissue and cell culture (Green, C.E. et al., ed.) pp. 367-382, Liss, New York

11 Brink, A. (1962) Quart. Rev. Biol. 37, 1-22

12 Wareing, P.F. (1987) in Improving vegetatively propagated crops (Abbott, A.J. and Atkin, R.K., ed.) pp. 263-270, Acad. Press, London

13 Schaffalitzky de Muckadell, M. (1959) Det Forstl. Forsogsv. Danmark 25, 309-455

14 Hackett, W.P. (1987) in Cell in tissue culture in forestry (Bonga, J.M. and Durzan, D.J., ed.), Vol. 1, pp. 216-231, M. Nijhoff Publ., Dordrecht

15 Walker, N. (1986) Annales Rech. Sylv. Afocel 1985, 25-47

16 Hackett, W.P. (1985) Hort. Rev. 7, 109-155

17 Zimmerman, R.H. et al. (1985) Encycl. Plant Physiol. NS 11, 79-115

18 Clark, J.R. (1982) Comb. Proc. Plant Prop. Soc. 1981, 31, 449-453

19 Metzger, J.D. (1988) in Plant hormones and their role in plant growth and development (Davies, P.J., ed.), pp. 431-462, Kluwer Acad. Publ., Dordrecht

20 Roberts, A.N. and Moeller, F.W. (1978) The Plant Propagator 24, 2-4

21 Franclet, A. (1981) Fruit Belge 49, 277-291

22 Engelmann, F. et al. (1987) Acta Hortic. 212, 479-482

23 Särkilahti, E. (1988) Tree Physiol. 4, 173-179

24 Nozeran, R. et al. (1982) Bull. Soc. Bot. France Lettres Bot. 129, 107-130

25 Passecker, F. (1973) Bodenkultur 24, 132-140

26 McKeand, S.E. (1985) J. Am. Soc. Hort. Sci. 110, 619-623

27 Mohammed, G.H. and Vidaver, W.E. (1988) Plant Cell Tissue Organ Culture 14, 137-160

28 Oliphant, J.L. (1988) Acta Hortic. 227, 482-484

29 Cancellier, S. and Cossio, F. (1987) Acta Hortic. 212, 391-399

30 Greenwood, M.S. (1987) Plant Growth Regul. 6, 1-12

31 Bonga, J.M. et al. (1988) in Genetic manipulation of woody plants (Hanover, J.W. and Keathley, D.E., ed.), pp. 57-77, Plenum Press, New York

32 McDaniel, C.N. et al. (1989) Devel. Biol. 134, 473-478

33 Borchert, R. (1976) Acta Hortic. 56, 21-36

34 Ununger, J. and Ekberg, E. (1987) Scand. J. For. Res. 2, 441-453

35 Ranjit, M. et al. (1983) HortScience 18 (4-2), 568

36 Greenwood, M.S. (1984) Physiol. Plant. 61, 518-522

37 Sax, K. (1962) Ann. Rev. Plant Physiol. 13, 489-506
38 Margara, J. (1982) in Bases de la multiplication végétative. Les Méristèmes et l'organogenèse (Margara, J., ed.) pp. 160-163, INRA, Paris
39 Nozeran, R. (1978) in Tropical trees as living systems (Tomlinson, P.B. and Zimmerman, M.H., ed.), pp. 424-443, Cambridge Univ. Press, Cambridge
40 Hartmann, W. (1984) Erwerbsobstbau 26, 135-140
41 Pierik, R.L.M. (1988) Acta Hortic. 230, 63-71
42 Franclet, A. (1981) Annales Rech. Sylv. Afocel 1980, 11-42
43 Dunberg, A. (1977) in Vegetative propagation of forest trees (Anonymous, ed.), pp. 55-64, Inst. Forest Improvement, Sweden
44 Boulay, M. (1987) in Cell and tissue culture in forestry (Bonga, J.M. and Durzan, D.J., ed.), Vol. 3, pp. 185-206, M. Nijhoff Publ., Dordrecht
45 Durzan, D.J. (1984) in Handbook of plant cell culture, crop species (Sharp, W.R. et al. ed.), Vol. 2, pp. 471-503, Macmillan, New York
46 George, E.F. and Sherrington, P.D. (1984) in Plant propagation by tissue culture (George, E.F. and Sherrington, P.D., ed.), pp. 160-166, Exegetics, Basingstoke
47 Franclet, A. (1979) Etudes Recherches Afocel 12, 3-18
48 Nozeran, R. and Rossignol-Bancilhon, L. (1977) Bull. Soc. Bot. France 124, 59-96
49 Montenuuis, O. et al. (1987) Acta Hortic. 212, 489-497
50 Davies, F.T. (1988) Acta Hortic. 227, 113-120
51 Greenwood, M.S. et al. (1989) Plant Physiol. 90, 406-412
52 Brand, M.H. and Lineberger, R.D. (1988) HortScience 23 (3), 787
53 Moncousin, C. (1982) in Plant tissue culture 1982 (Fujiwara, A., ed.), pp. 147-148, Proc. 5th Int. Congress Plant Tissue Cell Culture, Japan
54 Drouet, A. et al. (1989) Plant Physiol. Biochem. 27, 259-268
55 Bon, M.C. (1988) Tree Physiol. 4, 381-387
56 Doré, C. (1988) Agronomie 8, 843-850
57 Pierik, R.L.M. (1967) Meded. Landbouwh. Wageningen 67 (6), 1-71
58 Aghion-Prat, D. (1965) Physiol. Vég. 3, 229-303
59 Durzan, D.J. (1987) Plant Growth Reg. 6, 95-111
60 Pierik, R.L.M. and Steegmans, H.H.M. (1986) Neth. J. Agric. Sci. 34:217-223
61 Smits, P.C.H. et al. (1989) Acta Bot. Neerl. 38, 348
62 Cronauer, S.S. and Krikorian, A.D. (1988) in Forest and crop biotechnology. Progress and prospects (Valentine, F.A., ed.), pp. 45-57, Springer Verlag, New York
63 Duhoux, E. et al. (1986) Plant Cell Rep. 5, 161-164
64 Grout, B.W.W. (1988) Biotechn. Agric. Forestry 6, 211-225
65 Fouret, Y. et al. (1986) New Zeal. J. For. Sci. 16, 319-327
66 Barner, R.D. and Burley, J. (1982) in Improving vegetatively propagated crops (Abbott, A.J. and Atkin, R.K., ed.), pp. 211-227, Acad. Press, London
67 Ball, E.A. (1978) in Propagation of higher plants through tissue culture (Hughes, K.W. et al., ed.), pp. 259, Techn. Info Centre, Springfield, Virginia, U.S.A.
68 Mazalewsky, R.L. and Hackett, W.P. (1979) Comb. Proc. Int. Plant Prop. Soc. 29, 118-124
69 Gupta, P.K. and Durzan, D.J. (1987) J. Am. Soc. Hort. Sci. 112, 969-971
70 Roux, J. (1968) Annales Sci. Nat. Bot. Biol. Vég. 12, 109-256
71 Hackett, W.P. (1988) in Adventitious root formation in cuttings (Davis, T.D. et al., ed.), pp. 11-28, Dioscorides Press, Portland
72 Amerson, H.V. et al. (1988) in Genetic manipulation of woody plants (Hanover, J.W. and Keathley, D.E., ed.), pp. 117-137, Plenum Press, New York
73 Franclet, A. et al. (1980) C.R. Acad. Sci. Paris D 290, 927-929
74 Howard, B.H. et al. (1989) J. Hort. Sci. 64, 147-156
75 Ahuja, M.R. (1987) in Cell and tissue culture in forestry (Bonga, J.M. and Durzan, D.J., ed.), Vol. 3, pp. 207-223, M. Nijhoff, Dordrecht
76 Franclet, A. (1977) Etudes Recherches Afocel 12, 1-20
77 Bhojwani, S.S. et al. (1987) Acta Hortic. 212, 69-76

78 Clair, J.B.St. et al. (1985) Silvae Genetica 34, 42-48
79 Hartman, H.T. and Kester, D.E. (1959) in Plant propagation (Hartman, H.T. and Kester, D.E., ed.), pp. 184-204 Prentice Hall, New Jersey
80 Hanus, D. and Rohr, R. (1987) Acta Hortic. 212, 77-82
81 Bigot, C. and Engelmann, F. (1987) in Cell and tissue culture in forestry (Bonga, J.M. and Durzan, D.J., ed.), Vol. 3, pp. 114-127, M. Nijhoff Publ., Dordrecht
82 Thorpe, T.A. (1985) in Proc. International symposium on in vitro propagation of forest tree species (Anonymous, ed.), pp. 33-50, A.R.F., Bologna
83 Mascarenhas, A.F. et al. (1988) in Genetic manipulation of woody plants (Hanover, J.W. and Keathley, D.E., ed.), pp. 391-412, Plenum Press, New York
84 Franclet, A. et al. (1987) in Cell and tissue culture in forestry (Bonga, J.M. and Durzan, D.J., ed.), Vol. 1, 232-248
85 Robbins, W.R. (1957) Am. J. Bot. 44, 743-746
86 Greenwood, M.S. (1984) Physiol. Plant. 61, 518-522
87 Franclet, A. (1977) Unpublished results, stencil 1-15
88 Doorenbos, J. (1954) Proc. Kon. Acad. Wet. Amsterdam C 57, 99-102
89 Siniscalco, C. and Pavolettoni, R. (1988) Acta Hortic. 227, 98-100
90 Chalupa, V. (1987) in Cell and tissue culture in forestry (Bonga, J.M. and Durzan, D.J., ed.), Vol. 3, 224-246, M. Nijhoff, Dordrecht
91 Vieitez, A.M. et al. (1983) J. Hortic. Sci. 58, 457-463
92 Vieitez, A.M. et al. (1985) J. Hortic. Sci. 60, 99-106
93 Ikemori, Y.K. (1987) Commonwealth Forestry Rev. 66, 351-356
94 Montenuuis, O. (1987) Annales Rech. Sylv. Afocel 1986, 39-61
95 Deogratias, J.M. et al. (1986) Acta Hortic. 193, 139-145
96 Navarro, L. (1988) Acta Hortic. 227, 43-55
97 Starrantino, A. and Carusso, A. (1988) Acta Hortic. 227, 101-103
98 Pliego Alfaro, F. and Murashige, T. (1987) HortScience 22, 1321-1324
99 Muzik, T.J. and Cruzado, H.J. (1958) Nature 101, 1288
100 Misson, J.P. and Giot-Wirgot, P. (1985) Annales Rech. Sylv. Afocel 1984, 187-197
101 Misson, J.P. (1988) Can. J. Forest Res. 18, 473-477
102 Frank, H. and Renner, O. (1956) Planta 47, 105-114
103 Rangit, M. and Kester, D.E. (1988) J. Am. Soc. Hort. Sci. 113, 150-154
104 Champagnat, M. (1977) in La culture des tissus et des cellules des végétaux (Gautheret, R.J., ed.), pp. 238-253, Masson, Paris
105 David, H. et al. (1978) C.R. Acad. Sci. Paris D 287, 245-248
106 McCown, B.H. (1988) in Adventitions root formation in cuttings (Davis, T.D. et al., ed.), pp. 289-302, Dioscorides Press, Portland
107 Pierik, R.L.M. et al. (1988) Acta Hortic. 226, 195-204
108 McCown, B.H. (1989) Biotechn. Agric. Forestry 5, 324-341
109 Webster, C.A. and Jones, O.P. (1989) J. Hort. Science 64, 421-428
110 Lyrene, P.M. (1981) J. Am. Soc. Hort. Sci. 106, 396-398
111 Zimmerman, R.H. (1986) in Cell culture and somatic cell genetics of plants (Vasil, K., ed.), Vol. 3, pp. 243-258, Acad. Press, Orlando
112 Gupta, P.K. et al. (1980) Plant Sci. Letters 17, 259-268
113 Gupta, P.K. et al. (1981) Plant Sci. Letters 20, 195-201
114 Horgan, K. (1987) in Cell and tissue culture in forestry (Bonga, J.M. and Durzan, D.J., ed.), Vol. 3, 128-145, M. Nijhoff, Dordrecht
115 Franclet, A. et al. (1980) C.R. Acad. Sci. Paris D 290, 927-930
116 Jungnickel, F. (1988) Biotechn. Agric. Forestry 6, 38-103
117 Boxus, P. et al. (1977) in Plant cell, tissue, and organ culture (Reinert, J. and Bajaj, Y.P.S., ed.), pp. 130-143, Springer Verlag, Berlin
118 Mullins, M.G. et al. (1979) Annals Bot. 44, 623-627
119 Cancellier, S. and Cossio, F. (1988) Acta Hortic. 227, 508-513
120 Monette, P.L. (1988) Biotechn. Agric. Forestry 6, 3-37
121 Lyrene, P.M. and Perry, J.L. (1988) Biotechn. Agric. Forestry 6, 181-198

122 McCown, B.H. (1988) Scientia Hortic. 37, 257-265
123 Bonga, J.M. (1984) in International symposium of recent advances in forest biotechnology (Anonymous, ed.), pp. 64-68, Michigan Techn. Univ., Michigan
124 Pierik, R.L.M. (1967) Z. Pflanzenphysiol. 56, 141-152
125 Litz, R.E. (1987) in Plant tissue and cell culture (Green, C.E. et al., ed.), pp. 407-418, A.R. Liss, New York
126 Guimaraes, M.L. et al. (1987) Acta Hortic. 212, 315-321
127 Wann, S.R. (1988) Hort. Rev. 10, 153-181
128 Staritsky, G. (1970) Acta Bot. Neerl. 19, 509-514
129 Dublin, P. (1980) Café Cacao Thé 24, 121-130
130 Jones, L.H. (1989) Biotechn. Genet. Eng. Rev. 7, 281-296
131 Litz, R.E. (1988) J. Plant Physiol. 132, 190-193
132 Welander, M. (1988) J. Plant Physiol. 132, 738-744
133 Dore Swamy, R. and Sahijram, L. (1989) Scientia Hortic. 40, 181-188
134 Kane, M.E. and Albert, L.S. (1989) J. Aquat. Plant Management 27, 1-10
135 Drira, N. and Benbadis, A. (1985) J. Plant Physiol. 119, 227-235
136 Cheema, G.S. (1989) Plant Cell Rep. 8, 124-127
137 Chevreau, E. et al. (1989) Plant Cell Rep. 7, 688-691
138 Jörgensen, J. (1988) J. Plant Physiol. 132, 638-640
139 Coleman, W.K. and Thorpe, T.A. (1977) Bot. Gaz. 138, 298-304
140 Laliberte, S. and Lalonde, M. (1988) Am. J. Bot. 75, 767-777
141 Wellensiek, S.J. (1952) Proc. Kon. Ned. Acad. Wet. Amsterdam C 55, 567-573
142 Sommer, H.E. (1983) Bull. Soc. Bot. France Act. Bot. 130, 79-85
143 Romberger, J.A. (1976) Acta Hortic. 56, 301-317
144 Sussex, J. (1976) Acta Hortic. 56, 275-280
145 Libby, W.J. and Hood, J.V. (1976) Acta Hortic. 56, 91-98
146 Fournioux, J.C. and Bessis, R. (1986) Can. J. Bot. 64, 2608-2616
147 Smith, D.L. et al. (1989) Physiol. Plant. 76, 466-473

GENOTYPE-PHENOTYPE INTERPLAY IN MICROPROPAGATION

TIMO TÖRMÄLÄ, KEMIRA OY, ESPOO RESEARCH CENTRE, P.O. BOX 44, SF-02271 ESPOO, FINLAND

1. INTRODUCTION

The role of in vitro propagation (micropropagation) methods has increased steadily during the last two decades, especially with many ornamental species. In agronomic and forestry crops the low cost price of conventional propagules has limited the large scale application of micropropagation. Currently, several hundreds of millions of plants are produced through micropropagation primarily in Western Europe, United States and Far East.

Simplistically micropropagation can be described as a vegetative in vitro method of producing offspring identical to the stock plant. Micropropagation is in fact a relatively complex multistep process. Ideally, the aim of micropropagation is to produce with competitive costs plants, which perform as well or better than the original selected stock plant. The performance of the phenotype arising from the micropropagation process is affected by the environmental factors prior, during and after the in vitro phases as well as the genotype of the stock plant and possible genetic changes occurring during the process. This overview deals with the phenotype-genotype interplay in micropropagation. Data from ornamental (gerbera), food (strawberry) and forest (birch) crop species are used as examples.

2.MICROPROPAGATION PROCESS: CONCEPTUAL ISSUES

Micropropagation process (Table 1) is initiated by the selection of the stock or mother plant. The degree of the influence of environmental factors on the phenotype and selection criteria varies considerably from case to case. If the stock plants are selected from comparative clonal experiments the uncertainty in selection due to environmental factors is negligible. On the other hand, if the selection is based on one individual growing in the nature,the relative importance of the genotype and environment remains obscure.

In some species (e.g. strawberry and potato) after the

selection the stock plants are given a treatment (e.g. growing in an elevated temperature), which in combination with use of the meristem as an explant results in virus free culture. The virus eradication results into plants, which are phenotypically better than the original stock plant.

Table 1. Steps of the micropropagation process and some genotype-phenotype issues relevant to them.

Step of the micropropagation process	Issues
1. Selection of the stock plant	Relative importance of the genotype and environment on phenotypic performance of the stock plant
2. Pre-treatment of the stock plant	Phenotypic improvement e.g. through disease elimination.
	Phenotypic change in the in vitro performance.
3. Initiation	Genotypic and environmental variation in tissue culture performance.
4. Multiplication	Genetic and epigenetic variation.
	Role of tissue culture environment and the genotype in the induction of the variation.
5. In vitro rooting	Phenotypic variation e.g. due to media or way of cutting
6. Ex vitro growing	Genotypic and phenotypic fidelity.
	Stability of the epigenetic changes.
	Phenotypic performance.

The in vitro culture is commenced normally using meristem or bud as explant material. Within a species the success of initiation depens greatly on the genotype. Also the environmental factors affect the tissue culture potential of a particular phenotype. These factors include the conditions the stock plant has been grown prior to the taking of the explants as well as the physical and chemicals factors in the tissue culture environment (1).

There are three main biological systems for the actual multiplication in micropropagation: 1) Enhancement of axillary shoot formation, 2) Induction of adventitious shoots and 3) somatic embryogenesis or polyembryogenesis (in conifers). The adventitious system may be based on direct shoot formation on e.g. leaves or it may involve a callus stage. Sometimes the production is based on a combination of the axillary and adventitious system. The type of production system influences on the extent of genotypic variation created during the micropropagation process. The mechanisms of genetic variation (somaclonal variation) in micropropagation are poorly documented, but it is evident that the axillary shoot system genetically the most stable one. The incidence of variants tends to increase with the culture time. (2).The extent of the genetic change ranges from single gene mutations to ploidy level changes. There are indications that some of the mutations have already occurred in the stock plant and only revealed, when the plants has been micropropagated (3). Some of the lines with harmful mutations are already eliminated in multiplication phase on the basis of e.g. retarded growth.

The micropropagation process can induce phenotypic changes, which are not due to somaclonal variation. These epigenetic traits may be desirable (faster development, earlier flowering, improved growth habit) or undesirable (excessive branching and bushiness) (4). These phenotypic variations are generally attributed to the increased juvenility of tissue cultured plants and to carry over effects of growth regulators (cytokinins) used especially during the multiplication phase (5).

When appropriate amount of shoots has been produced, they are rooted either in vitro or directly ex vitro and grown subsequently in greenhouse, nursery or field.

The plants derived from tissue culture are judged on the basis of uniformity and phenotypic fidelity. Lack of uniformity reflects variations and inconsistency in the propagation and growing procedures or e.g. genetic changes (somaclonal variation). As discussed above there are many possible reasons, why the phenotype of the micropropagated plants may differ from the phenotype of the stock plant.

3.MATERIALS AND METHODS

Literature and data from laboratories of Kemira Oy (Espoo Research Centre, Hortus, Terra Nigra) were used for analysis of the genotype, environment and phenotype issues in gerbera, strawberry and birch. The data from the production laboratories do not allow the accurate calculation of e.g. frequency of variants as the feedback from clients is far from comprehensive.

Terra Nigra has produced annually several million gerberas (mainly cut) using micropropagation. The propagation is initiated using meristems as explants. Hortus has produced for the Finnish market a few million strawberry plants since late 1980' using standard micropropagation procedures and virus indexed stock plants. Kemira Oy, Enso-Gutzeit and Hortus have produced several hundreds of thousands of birch (Betula pendula) plantlets via micropropagation starting 1987. The propagation system is primarily based on axillary shoots. Over 100 genotypes have been tissue cultured by early 1990. The total number of initiations (buds) is about 5000.

4. RESULTS AND DISCUSSION

4.1. Gerbera

In gerbera the stock plant selection is based on clonal trials in standard greenhouse conditions. Thus the selection is based strongly on the genotype.

The easiest characteristics for the observation of variants are flower color and form. The frequency of altered color pattern in a variety has been between roughly 1:5000 to 1:500000. Some of the variants were not stable when re-micropropagated, but some altered flower types were stable in micropropagation. The nature of the variation has not been decisively shown, but there are indications that at least in some cases the change would be genetic. In 1990 a lot 5170 plants of a pot variety were grown until flowering. The flower color is normally red. In total 10 variants with yellow flowers were observed. The fact that they all were located next to each other indicates a common origin (same vessel) and the possibility of a mutation.

The frequency of real genetic changes is impossible to asses currently, because the nature of the variation is not clear. It is obvious that many of the reversible color changes are epigenetic.

Much of this variation seems, however, to be somehow genetically controlled, because the frequency is variety dependent.

In addition to somaclonal variation the phenotypic variation in the growth habit is a critical issue in gerbera micropropagation.

Multiple crowning is a problem, which is related to the effects of cytokinins used in the multiplication. The incidence of the phenomenon is dependent of the genotype. Some genotypes, which are excellent pot gerberas, when propagated with conventional vegetative methods, can not be micropropagated commercially due to unacceptable high portion of inferior phenotypes.

4.2. Strawberry

In strawberry the stock plants are normally established clonally propagated varieties. The main reason for micropropagation is the phenotypic improvement of the planting material through virus eradication.

Micropropagated strawberries deviate sometimes from conventionally propagated phenotypically (morphologically). In most cases plants originating from micropropagation have performed as well or better than the conventionally propagated ones (6,7). The are, however, also reports on the adverse effects of the phenotypic changes caused by micropropagation (8) especially, if the micropropagated plants are used directly in the berry production. Often, one will overcome these effect, if micropropagated plants are used for runner production (9). There is also evidence that micropropagation has altered the disease resistance of some strawberry varieties (10). This variation is most probably not genotypic.

Variation can be observed already during the in vitro culture. Plants with very slow growth or white leaves are usually eliminated from the production stock. Among the plants produced in Hortus and planted in the greenhouse there has not been an incidence of clearly aberrant phenotypes.

4.3. Birch

The concept of micropropagating forest tree species is based on the assumption that by selecting superior genotypes and mass cloning them the productivity of the forests can be elevated (11). Micropropagation allows the capture of non-additive genetic gains. It is obvious that the selection of the stock plant is a critical issue in the micropropagation of forest trees. If the selection is based on an single outstanding individual tree, the genotypic quality of that tree cannot be ascertained. Therefore it is much more advisable to select the stock trees from comparative (genotypic) trials.

In birch the suitability to micropropagation depends on the genotype, but there are also environmental factors (season), which have phenotypic effects on tissue culture reaction. The birch plants we have produced have been uniform and true to type (leaf morphology etc) with one interesting exception. A number of plants differed drastically in the nursery from the rest of the plants originating from the same stock plant. These plants grew horizontally on the ground and seemed to lack the apical dominance more or less totally. The growth was also slow and they wintering losses were higher than in the normal plants. When explants were taken from these plants and they were micropropagated for subsequent field testing they maintained these characteristics.

In spite of the slow growth rate in the nursery this

variant (mutant?) thrived in tissue culture (higher multiplication, growth rate, and biomass production).

5. Conclusions

In micropropagation the relative importance of the genotype and environment varies from case to case in a complicated way. Phenotypic (epigenetic) and genetic variation is often hard to distinguish decisively. There is even in the epigenetic tissue culture induced variation a genetic component involved as the frequency of variation strongly depends of the genotype. Especially in forest tree crops, which have extremely long rotation cycles, the stock plant selection should be based on solid evidence on the genotypic performance. The micropropagation system employed should be so generic that strong genotypic selection based on tissue culture suitability do not occur.

REFERENCES

1 Read, P.E. (1990) in Handbook of Plant Cell Culture (Ammirato, P.V., Evans, D.A., Sharp, W.R. and Bajaj, Y.P.S. eds.), Vol. 5, pp. 95-125, McGraw Hill, New York

2 Evans, D.A. and Bravo, J. E. (1986) in Tissue Culture as a Plant Production System for Horticultural Crops (Zimmerman, R.H., Griesbach, R.J., Hammerschlag, F.A. and Lawson, R.H., eds.), pp. 73-96, Matinus Nijhoff,Dordrecht

3 Preil, W. (1986) in Genetic Manipulation in Plant Breeding (Horn, W., Jensen, C.J., Odenbach, W. and Schieder, O., eds.), pp. 377-403, Walter de Gruyter & Co., Berlin

4 Chu, I.Y.E. and Kurtz, S.L. (1990) in Handbook of Plant Cell Culture (Ammirato et al., eds.), Vol 5, pp. 126-164, McGraw Hill, New York

5 Constantine, D.R. (1986) in Plant Tissue Culture and Its Agricultural Applications (Withers, L.A. and Alderson, P.G., eds), pp. 175-186, Butterworths, London

6 Damiano, C., Faedi, W. and Cobianchi, D. (1982) Ann. dell. Istituto Sperim. Fruttic. 13, 69-78

7 Sansavini, S. and Gherardi, G. (1980) Frutticoltura 42, 39-46

8 Anderson, H.M., Abbott, A.J. and Wiltshire, S. (1982) Scient. Hort. 16, 331-341

9 Beech, M.G., Crisp, C.M., Simpson, S.M. and Atkinson, D. (1988) J. Hort. Sci. 63, 77-81

10 Shoemaker, N.P., Swartz, H.J. and Galletta, G. J. (1985) Hortscience 20, 253-254

11 Timmis, R. (1985) in Crop Physiology of Forest Trees (Tigerstedt, P.M.A., Puttonen, P. and Koski, V. eds.), pp. 259-272, Helsinki University Press, Helsinki

EFFECT OF MINERAL NUTRITION ON THE GROWTH AND MULTIPLICATION OF IN VITRO CULTURED PLANTS

P.J. Lumsden[1], S. Pryce[1] and C. Leifert[2]
[1]**School of Applied Biology, Lancashire Polytechnic, Preston, PR1 2TQ, U.K.;** [2]**Neoplants Ltd., 197 Kirkham Road, Frickleton, Preston, PR4 1HU, U.K.**

INTRODUCTION

Cultivation of plants in vitro, where the culture conditions can be precisely controlled, is becoming increasingly widespread and important, e.g. commerical micropropagation of a wide range of plants; propagation of recalcitrant species; regeneration and propagation of novel materials produced by tissue culture methods.

The physiology and biochemistry of in vitro cultured plants is likely to be very different from that of plants growing in vivo. The relative growth rate in vivo is a function of net assimilation rate (E), which is usually related to photosynthetic rate, and leaf area ratio (F). In vitro, though, photosynthesis is negligible, and leaf area ratio irrelevant. The assimilation rate is therefore due solely to the uptake of nutrients, including carbohydrate, from the media. However, since plants are likely to differ in their response to nutrients when grown in vitro there is clearly a need to determine the effect of essential plant nutrients on plant growth in vitro. At a practical level an understanding of optimal nutrient concentrations and ratios could lead to increased growth rates and could enable plant morphogenesis in vitro to be directed more efficiently. At a more basic level, differences between the physiology and biochemistry of plants grown in vitro and of those grown in soil or in water culture needs to be determined.

Very little is known about the effects of mineral and carbohydrate nutrition on plant growth and morphogenesis in vitro. Most plant tissue culture media are still based on the nutrient composition described by Murashige and Skoog (1), a medium designed for Nicotiana cell cultures. As determined by sand and water culture experiments, the optimal concentrations of most nutrients, including NO_3^-, PO_4^{3-} and K^+, for in vivo cultured plants are much lower than the concentrations found in most tissue culture media. This would suggest that the concentration of nutrients in the media used for in vitro propagation is too high for optimal plant growth. However, other experiments have shown that in vitro the culture medium can be rapidly depleted of nutrients, including NO_3^-, PO_4^{3-}, K^+ and Mg^{2+}, while other nutrients such as Ca^{2+} are taken up less rapidly (2,3,4). Further, increasing the phosphate concentration of MS medium increased the growth rate of some species (5).

We have carried out experiments in which the concentration of individual nutrients remaining in the media has been determined after different periods of culture. This has been done using both liquid and agar medium. Plantlets of Dephinium, Hemerocallus and Iris have been used, and growth rates determined. Using this approach the uptake and possible depletion of individual nutrients, and the relationship between nutrient uptake and growth and multiplication rates has been determined.

METHODS

Liquid Culture. Single explants of either Hemerocallus or Delphinium were grown in universal bottles with 10.5 ml MS basal medium, supported on filter paper 'bridges'. The Hemerocallus media contained twice the normal concentration of phosphate and 4% sucrose. Cultures were maintained at 25°C under a 14h photoperiod with a light intensity of 14 Wm^{-2} provided by fluorescent tubes. Plantlets were weighed at the start of the culture period and after 28 days of culture, the end of the culture period. The liquid media was then sued for nutrient analysis.
Agar culture. The medium used was basal MS with 3% sucrose and 0.7% agar. Clear plastic containers were used as culture vessels, with about 50 ml media per tub. Different numbers of plants were placed into the tubs. Tubs were weighed before plants were introduced and after. At regular intervals some tubs were taken for analysis of the media.
Nutrient analysis. Liquid media were used directly for analysis. Agar media were frozen, thawed and centrifuged. The liquid supernatant was then used for analysis.

Nitrate (NO_3^-) was determined by a colourimetric assay based on the interference by nitrate with formation of the coloured complex between rhenium and phenyl pyridyl ketoxime.

Phosphate (PO_4^{3-}) was determined colourimetrically with ammonium molybdate and ammonium metavanadate.

Potassium (K^+) was determined by flame photometry.

Sugars (sucrose, glucose and fructose) were separated by HPLC using an anion exchange column and detected by refractive index detector.

RESULTS

1. Single plants in liquid culture

Single plantlets of Delphinium (10) or Hemerocallus (6) were cultured on liquid MS media. The Hemerocallus media contained twice the normal concentration of PO_4^{3-} and 4% sucrose. After 28 days plantlets were weighed and the concentrations of PO_4^{3-}, NO_3^- and K^+, in ppm, and of sugars, in g/l, were determined, firstly in bottles with no plants (controls) and then in the bottles with plants. A mean concentration for each nutrient was calculated for the controls. Subtracting the concentration of nutrient in the bottles with plants from the control value gave a value for the amount of nutrient taken up by each individual plantlet. Plotting uptake against plant fresh weight gave a linear relationship for all four nutrients. Only the results for nitrate are shown in Fig. 1.

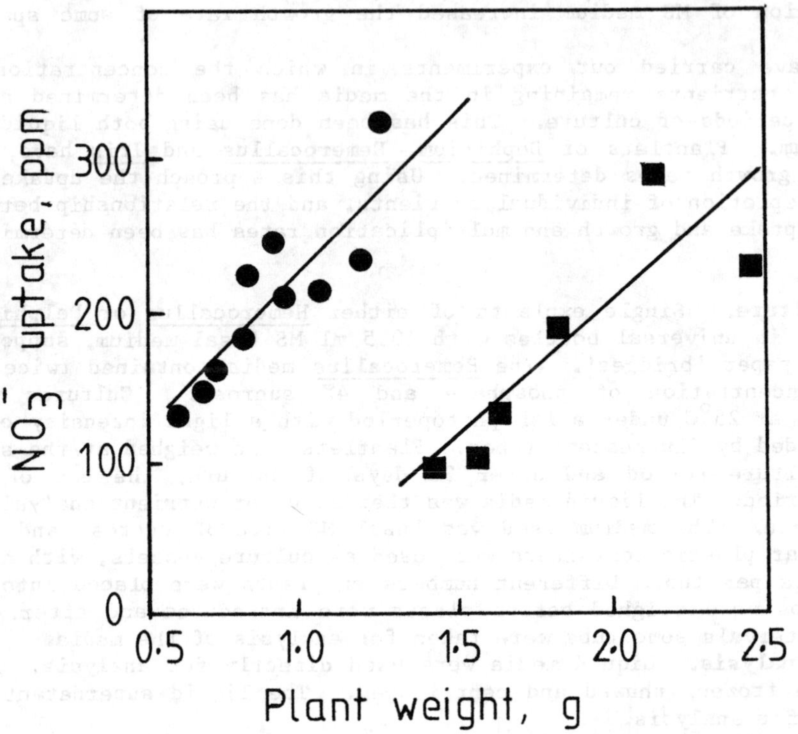

Fig. 1. Calculated uptake of NO_3^- from liquid MS media by single explants of Delphinium (●) and Hemerocallus (■).

The uptake for the other nutrients followed a similar pattern. Plantlets of Hemerocallus were heavier than those of Delphinium, but Delphinium appeared to take up more nutrient per g fresh weight. The uptake data for individual cultures was then combined to give a value for the mean uptake per plant, the mean uptake per g fresh weight, and the % uptake (Table 1).

The data in Table 1 shows that a significant amount of nutrients had been taken up after 28 days of culture; 30-40% NO_3^-, about 30% of the total sugars, and in the Delphinium, which had a normal level of phosphate, more than 60% of the PO_4^- had been taken up. In addition it was found that in the Hemerocallus media about 45% of the remaining sugar was sucrose, the rest being glucose and fructose, while in the Delphinium media only 5% of the remaining sugar was sucrose.

Table 1. Nutrient uptake from liquid MS media by single explants of Hemerocallus and Delphinium.

Nutrient	Mean uptake per explant	Uptake per g fresh weight	% uptake
Hemerocallus			
K^+	190+34.4 ppm	99.9 ppm	25.0
PO_4^{3-}	126+13.1 ppm	61.0 ppm	47.4
NO_3^{4-}	373+31.9 ppm	86.6 ppm	31.5
Sugar	47.3+1.6 g/l	7.6 g/l	29.8
Delphinium			
K^+	115+13.6 ppm	123.9 ppm	16.0
PO_4^{3-}	86.3+5.6 ppm	92.9 ppm	64.0
NO_3^{4-}	206+17.8 ppm	224.2 ppm	40.6
Sugar	9.1+0.9 g/l	9.8 g/l	29.4

These uptake data refer to the removal of nutrients from 10.5 ml liquid media by single plantlets. It is possible to extrapolate from these results and, assuming that the uptake had not become limiting, to predict the % of each nutrient that would be removed by different numbers of plantlets from 50 ml media, the standard volume in a normal culture tub. Such calculations clearly suggest that if each plant took up nutrient at the same rate as that of individual plantlets in the liquid media, then with more than 10 plants per 50 ml media PO_4^{3-} would be completely removed from the media of both Hemerocallus and Delphinium cultures, and NO_3^- and total sugar would be removed or significantly depleted.

2. Uptake from solid media by Delphinium

To investigate the actual effect of plant numbers on nutrient uptake, cultures of Delphinium were established on MS medium (approx. 50 ml), with either 1, 5, 10 or 20 plants per culture tub. After 28 days the media was extracted as described and analysed for PO_4^{3-}, NO_3^-, K^+ and total sugar. The uptake of individual nutrients was calculated as described above for the liquid cultures. The data are shown in Table 2.

These data indicate that the uptake of PO_4^{3-}, NO_3^- and sugar per plant decreased with increasing numbers of plants. The uptake of these nutrients may in fact have been saturated at more than 10 plants per tub. However, the uptake of K^+ did not seem to follow the same pattern; uptake continued to increase with increasing numbers of plantlets, and the uptake per plant remained more or less constant, indicating that there was no limitation to uptake by individual plants even with high numbers of plants per tub. The conclusion for this must be that K^+ is metabolised or utilised less rapidly than the other nutrients examined. With PO_4^{3-}, NO_3^- and sugar, the values for uptake by single plants were greater than for K^+. It is assumed that this more rapid uptake reflects greater utilisation. As a result of more rapid uptake, the amount of these nutrients must become limiting, so that with greater numbers of plants per tube individual plants are unable to take up as much as if they were growing alone.

Table 2. Uptake of nutrients from MS media by Delphinium plants, 1, 5, 10 or 20 plants per tub, after 28 days culture.

Nutrient	Plants per tub	Mean uptake	% uptake	Uptake per plant (ppm)
K^+	1 plant n=6	26.1 ppm	2.9	26.1
	5 plants n=7	156.7 ppm	16.7	30.6
	10 plants n=10	356.4 ppm	38.9	35.6
	20 plants n=6	491.6 ppm	53.6	24.8
PO_4^{3-}	1 plant n=6	43.2 ppm	27.3	43.2
	5 plants n=7	122.9 ppm	77.6	24.6
	10 plants n=10	141.2 ppm	89.2	14.1
	20 plants n=6	135.1 ppm	85.4	6.8
NO_3^-	1 plant n=6	64.3 ppm	10.4	64.3
	5 plants n=7	205.4 ppm	33.1	41.1
	10 plants n=10	372.3 ppm	60.0	37.2
	20 plants n=6	438.8 ppm	70.7	21.9
Sugar	1 plant n=6	2.5 g/l	6.8	2.5
	5 plants n=7	8.3 g/l	22.6	1.7
	10 plants n=10	21.5 g/l	58.5	2.2
	20 plants n=5	27.4 g/l	74.4	1.4

3. Uptake by Iris

In another experiment, cultures of Iris, 10 plants per tub, were established on MS media and the media was sampled at 5 day intervals to gain some information on the time course of uptake (data not shown).

The uptake data suggest that in Iris, nutrient uptake is slightly different from that in Hemerocallus and Delphinium. For example 77% of the K^+ had been taken up by day 15, whereas in Delphinium cultures with 10 plants per tub less than 40% of the K^+ had been taken up after 28 days (Table 2). However, the rate of uptake of the different nutrients clearly difered. PO_4^{3-} appeared to be taken up most rapidly, whereas NO_3^-, K^+ and sugar were taken up relatively slowly from day 0 to day 10, and much more rapidly between day 10 and 15. The uptake rate of all the nutrients would probably have decresed after day 15 due to their beocming depleted.

A further experiment was carried out with Iris in which the concentration of PO_4^{3-} in the media was varied, to give 0.5, 1.0 and 2.0 times the normal concentration. Growth rates were determined after culturing for 28 days. It was found that on media with 0.5 x PO_4^{3-} there was a slight reduction in growth rate, but that there was no significant diference in growth rates between plantlets on 1 x and on 2 x PO_4^{3-} (data not shown).

DISCUSSION

The data presented are limited in that only four major nutrients have been investigated, using only three plant species. However, a number of conclusions can be drawn.

Patterns of nutrient utilisation are clearly different between species. For example, with Hemerocallus just under 50% of the remaining sugar was sucrose while in Delphinium and Iris only 5% of the remaining sugar was sucrose.

The data from the liquid media indicated that nutrient uptake was proportional to growth rate. It is assumed that, since there is no transpiration, uptake will be due to diffusion, and that utilisation of the nutrients by the plants to bring about growth will create a diffusion gradient. The rates of uptake differed for the different nutrients; PO_4^{3-} appeared to be most rapidly taken up in all three species, but there was much greater uptake of K^+ by Iris than by the other species, suggesting differences in metabolism giving a greater diffusion gradient of K^+ in Iris. Given the observed rates of uptake then with the numbers of plants commonly placed in culture tubs depletion of PO_4^{3-} and in some cases NO_3^- would occur.

What is not shown by these data is whether depletion of individual nutrients would actually affect growth rates. Experiments carried out in vivo have shown that plants respond differently to the depletion of different nutrients. Lowering of external nitrate led to a rapid lowering of relative growth rate, thus maintaining a constant internal concentration of nitrate (6). However, adjustments to changes in other nutrients, e.g. K^+ and PO_4^{3-}, may be less rapid, and growth may continue despite a shortage of the nutrient until it actually becomes internally deficient (7). The evidence of our in vitro studies is that at least in Iris although the concentration of PO_4^{3-} in the media would appear to be severely depleted after 15 days, there was no significant effect of doubling the concentration of PO_4^{3-}, suggesting that as with in vivo studies there is not a rapid change in response to reduced PO_4^{3-}. Changing the concentration of other nutrients, such as NO_3^-, might have more significant effects on growth rates. There is evidence that the availability of some nutrient factor is important since reducing the number of explants per tub of Hemerocallus significantly increased mutliplication rates (C. Leifert, unpublished).

It should be noted that in all the experiments reported here the standard culture period was 28 days, at which point depletion may not have started to affect growth rates. However, if cultures were maintained for longer periods of time, reduced growth rates would be more likely to occur. The point at which this occurs is likely to differ between species.

REFERENCES

1. Murashige, T. and Skoog, F. (1962) Physiol. Plant. 15, 473-497.
2. Thom, C. (1981) Plant Cell Tissue Organ Culture 1, 3-14.
3. Geerts, O. (1987) Med. Fac. Landbouww. Rijksuniv. Gent 54, 1449-1456.
4. Taya, M. (1989) Enzyme Microb. Technol. 11, 170-176.
5. Meyer, M.M. (1976) Hort. Sci. 117, 891-892.
6. Ingestad, T. and Lund, A.B. (1979) Physiol. Plant. 45, 137-148.
7. Clarkson, D.T., Sanderson, J. and Scattergood, C.B. (1978) Planta 139, 47-53.

Rapid regeneration of sugarbeet (*Beta vulgaris* L.) plants from *in vitro* cultures

B.K. Konwar and R.H.A. Coutts
Department of Biology
Imperial College of Science, Technology & Medicine
Prince Consort Road
London SW7 2BB. U.K.

INTRODUCTION

Sugarbeet (*Beta vulgaris* L.) produces 35% of the annual global raw sugar and thus provides scope to plant breeders and seed producers for developing superior cultivars. Due to its allogamous nature seed production true to the type is difficult and laborious. Tissue culture has been considered as a useful tool for the improvement of sugarbeet (1). The development of genetically manipulated superior sugarbeet plants and its tissue culture is of vital importance. Several studies are being carried out world-wide on the *in vitro* regeneration of sugarbeet. Regeneration of shoots from callusing plant parts has been reported by a number of workers (2, 3, 4, 5, 6, 7, 8, 9, 10, 11). Regeneration of shoots from hormone autonomous callus has also been reported (12, 13). *In vitro* organogenesis and whole plant regeneration of sugarbeet is difficult due to the variations inherent in different genotypes, which can influence the hormonal response of callus (14). Up till now there is no reliably rapid method for whole plant regeneration of sugarbeet from *in vitro* cultures.

We report here an easy and quick approach for *in vitro* propagation, multiplication and finally transfer of rooted sugarbeet plants into the potting compost under greenhouse conditions.

MATERIALS AND METHODS

Seed of the sugarbeet cultivar 'Bella' was obtained from the School of Science, Trent Polytechnic, Clifton, Nottingham, U.K. In order to remove the pelleting material seeds were washed with 3 changes of 1% (v/v) 'Teepol' detergent solution for 15 min. For surface sterilization seeds were treated with 15% (v/v) 'Domestos' bleach (9.5% Sodium hypochlorite) for two 15 min periods followed by two rapid and four 15 min washes in sterile distilled water. Sterilized seeds were sown on sterile moist vermiculite in sterile germinating boxes. The boxes were sealed with parafilm, wrapped with black polythene sleeves and incubated at 25°C for seed germination which was observed three days later.

MS (15) medium containing 100 mg/l myo-inositol, 2mg/l glycine, 0.5 mg/l nicotinic acid, 0.5 mg/l pyridoxine-HCl, 0.1 mg/l thiamine-HCl, 3% (w/v) sucrose and 0.8% technical agar (Oxoid Agar No. 3) was used for all in vitro cultures. Prior to autoclaving the pH of the medium was adjusted to 5.8 with potassium hydroxide. On different occasions, the medium was supplemented with hormones at different concentrations, viz 0.25, 0.5, 1.0 mg/l 6-benzylaminopurine (6, BAP); 0.25, 0.5, 1.0 mg/l 2,4-dichlorophenoxyacetic acid (2,4-D); 0.5, 1.0 mg/l kinetin (Kin) and 0.5, 1.0 mg/l naphthalene acetic acid (NAA). All in vitro cultures were maintained in a growth room at 25 ± 3°C with constant illumination of 27 $\mu E\ m^{-2}\ sec^{-1}$ from 40 alternate glass surrounded fluorescent 65/80W GRO-LUX (Thorn) and 65-80W/35 warm white (Philips) tubes.

Shoot tips of 7 day old seedlings with 0.5 cm hypocotyl length were aseptically removed and cultured on MS basal medium twice for 8 days as a precautionary measure to detect microbial contamination. Aseptic growing tips from the second MS basal plates were cultured on different MS media supplemented with growth regulators. Leaf and petiole segments from 30-40 day old seedling tip-derived plants were cultured on MS media supplemented with BAP, 2,4-D and Kin.

Shoot cultures developed from petiole and leaf segments in MS+BAP after about 5 weeks were cultured in MS+Kin and MS+NAA media thereafter.

After 35 days shoot cultures in MS+0.5 mg/l Kin produced healthy plantlets with dense rooting. Five of these plantlets were transferred into potting compost under greenhouse conditions after carefully washing off the culture medium from roots. Plants were placed on plastic trays containing 5 cm water and covered with transparent polythene sheeting to avoid desiccation. Seed derived plants of the same cultivar were also grown in potting compost in the greenhouse for comparative purposes. In the green house the light intensity was 33.3 $\mu Em^{-2}\ sec^{-1}$ and the temperature was maintained at 25±3°C.

RESULTS AND DISCUSSION

Aseptic seed culture of the sugarbeet cultivar 'Bella' resulted in 86% successful germination. Seed treatment with 1% (v/v) 'Teepol' and 15% (v/v) 'Domestos' bleach for aseptic seed propagation on sterile moist vermiculite was found to be effective to control microbial contamination during germination. Ritchie et al (11) reported similar results but observed considerable cultivar variation in seed sterilization of sugarbeet, Hussey et al (16) experienced difficulty in producing aseptic cultures of sugarbeet from seeds and seedlings. Moreover, repeated seedling tip culture on MS basal plates before transferring into culture media offered us the chance to eliminate the weakly growing and/or contaminated material.

Seedling tips from the second MS basal plates were cultured on various MS media supplemented with BAP, 2,4-D and Kin. Cultures in MS+BAP initiated axillary shootlets within 12 days. After 35 days, vegetative growth of these cultures was found to be very dense with lateral shoots from callus and axillary shoots from leaf axils. Though the initial response of the cultures to three different concentrations of BAP was almost idential; the long term response after about 50 days growth was sectorial shoot blackening and death especially at the higher concentrations of the hormone. This sectorial shoot and leaf death may be due to overcrowded shoot development. Thus, out of the 3 BAP concentrations, 0.25 mg/l BAP may be the best one for successful <u>in vitro</u> shoot development of sugarbeet. Cultures in MS+2,4-D callused profusedly at the cut surfaces with limited shoot growth. Seedling tips cultured in MS+Kin produced roots within 15 days with rapid shoot elongation. Shoots on MS+BAP were light green with thick-succulent leaves and petioles whereas cultures on MS+Kin produced dark green leaves with a thin but firm structure.

Axillary shoots arising out of leaf axils can be cultured in a similar fashion to callus derived lateral shoots. Since genetic stability is best preserved in organized tissue (17), axillary shoots may be preferable to the lateral shoots where genetic variability may be caused by chromosomal irregularities.

Seedling tip derived <u>in-vitro</u> shoots were used as the parent material for adventitious shoot production. Leaf and petiole segments from these shoots were cultured in order to further reduce chance contamination and also to produce more shoots from subsequent cultures. Petiole cultures on MS+BAP medium produced shoots from the upper epidermis that mostly covered the region within 0.5 cm of the attached blade.

However, shootlet differentiation was also observed from the callus arising out of the cut surfaces of petiole and leaf segments. According to Ritchie <u>et al</u>. (11) development of adventitious shoots is generally from the upper sub-epidermal parenchyma of the petioles. Regeneration of shoots from petiole sub-epidermal parenchyma might be a good pointer for sugarbeet protoplast isolation and culture. Up until now there are no comprehensive reports on the successful culture of sugarbeet protoplasts although the culture of leaf mesophyll protoplasts with nurse suspension culture did produce some micro-colonies (18). The isolation of viable protoplasts from petiolar upper sub-epidermis and their subsequent culture has been successfully initiated (results not shown). Sectiorial death of adventitious and callus derived lateral shoots was also found to be common at higher concentration of 6,BAP. However, regeneration of lateral shoots was mostly from the white friable callus which was also observed by Saunders (19) and Ritchie *et al.* (11).

Petiole and leaf segments cultured on medium supplemented with 2,4-D callused profusely with or without little shootlet differentiation. The initial shootlets developed in low concentrations of 2,4-D were reconverted to callus after further culture. Thus, the use of 2,4-D for shoot induction in Bella sugarbeet appears to be counter productive. Cultures in MS+Kin produced some shoots with subsequent root growth after about 25 days of culture. However for development of shoots from all Bella plant parts 0.25 mg/l BAP was the best hormone concentration investigated.

Petiolar adventitious shoots cultured on MS+Kin showed root and shoot growth within 10 days of culture. Whereas the cultures on MS+NAA took more than 25 days for the induction of root and shoot elongation. Though both the concentrations of Kinetin were found to be equally effective for root and shoot growth of the cultured shoots, blackening of older roots was observed in 1 mg/l Kin after about 45 days of culture. Thus, for in vitro whole plant regeneration of sugarbeet 0.5 mg/l Kin is the most effective hormone concentration.

In vitro Bella plants developed from petiolar adventitious shoots in MS+0.5 mg/l Kin after 35 days of culture were transferred (Table 1) into potting compost under greenhouse conditions. These *in vitro* derived plants became established on potting compost immediately within 12 days of transfer; which was evident from their unfolding of first leaf.

Table 1. Characteristics of the *in vitro* plants transferred into potting compost.

Plant	Height (cm)	No. of leaves	Petiole length (cm)	Leaf size (cm)	No. of lateral shoots	Roots
1	12.5	19	6.8	4.2 x 3.9	7	Dense rooting; older roots reddish brown, whereas young roots at the centre are whitish.
2	9.0	15	4.2	3.6 x 3.5	3	
3	11.8	14	16.3	3.9 x 3.5	4	
4	12.8	19	7.2	4.6 x 3.6	6	
5	10.6	16	5.5	3.2 x 3.9	4	

In the greenhouse, the *in vitro* derived plants showed much faster vegetative growth than seed grown plants. The superior vegetative growth of the *in vitro* derived plants (Table 2) on potting compost may be due to better aeration and higher light intensity in the greenhouse. The *in vitro* derived plants produced more and larger leaves with longer petioles than seed derived plants. However, beet size was found to be much smaller in case of *in vitro* derived plants (Table 2).

Table 2. Performance of in vitro and seed derived plants on potting compost.

Plants	No. of leaves	Leaf size (cm)	Petiole length (cm)	Girth of beet (cm)	Length of beet (cm)
In vitro	16.4	24.7 x 13.4	15.6	16.2	15.8
Seed	12.2	20.6 x 12.7	14.1	26.8	14.2

The experimental procedure described here for obtaining fully mature sugarbeet plants from in vitro cultures is simple and rapid. The frequency and pattern of shoot development in in vitro cultures is greatly influenced by the growth regulators. However, the rooted in vitro plants under congenial environmental conditions perform better than seed derived plants at least with respect to vegetative growth.

REFERENCES

1. Coumans-Giles, F.M., Kevers, C.I., Coumans, M., Ceulemans, E., Gaspar, Th. (1981) Plant Cell, Tissue and Organ Culture 1:93-101.
2. Butenko, R.G., Atanassov, A., Upmantsera, W. (1972) Phytomorphology 22:140-143.
3. Hooker, M.P., Nabors, M.W. (1977) Z. Pflanzenphysiol 84:237-246.
4. Rogozinska, J., Goska, M. (1978) Bull de L. Acad Polonaise des Sciences 26:343-345.
5. Mohammad, A.M.S., Collin, H.A. (1979) New Phytol 82:293-299
6. Slavova, Y. (1981) Plant Physiol (Sofia) 7:71-77.
7. Hosemans, D., Bossoutrot, D. (1983) Z. Pflanzenzuchtg 91:74-77.
8. Szabados, L., Gaggero, C. (1985) Plant Cell Rep. 4:195-198.
9. Saunders, J.W., Shin, K. (1986) Crop Sci 26:1240-1245.
10. Yu, M.H. (1989) Crop Sci 29:205-209.
11. Ritchie, G.A., Short, K.C., Davey, M.R. (1989) J Exptl Bot 40:277-284
12. De Greef, W., Jacobs, M. (1979) Plant Sci Lett 17:55-61.
13. Saunders, J.W., Daub, M.E. (1984) Plant Sci Lett 34:219-223.
14. Jarl, C., Bornman, C.H. (1986) Hereditas 105:55-59.
15. Murashige, T., Skoog, F. (1962) Physiol Plant 15:473-497.
16. Hussey, G., Hepher, A. (1978) Ann Bot 42:477-479.
17. Binding, H., Krumbiegel-Schroeren, G. (1984) In: Vasil, I.K. (ed) Cell culture and somatic genetics of plants. Laboratory procedures and their applications, Vol. 2, Academic Press Inc., Florida, USA, pp43-48.
18. Bhat, S.R., Ford-Lloyd, B.V., Callow, J.A. (1986) Plant Physiol 124:419-423.
19. Saunders, J.W. (1982) Crop Sci 22:1102-1105.

MORPHOGENESIS OF GLADIOLUS BUDS IN BIOREACTORS - IMPLICATION FOR SCALED-UP PROPAGATION OF GEOPHYTES

MEIRA ZIV, THE HEBREW UNIVERSITY OF JERUSALEM, DEPARTMENT OF AGRICULTURAL BOTANY AND THE WARBURG CENTER FOR BIOTECHNOLOGY IN AGRICULTURE, REHOVOT 76100, ISRAEL

1. INTRODUCTION

Micropropagation which has been used successfully for several geophytes (corm and bulb plants) depends on both axillary and adventitious bud regeneration *in vitro*.

In vitro propagation of *Gladiolus* in agar cultures was reported and has been summarized recently by Ziv and Lilien-Kipnis (1). *In vitro* methods for mass propagation of pathogen-free geophytes are still slow and expensive since most tissue culture laboratories depend on manual separation and subculture, using agar cultures in small containers. The high price of labor which comprises 70-80% of the final cost of *in vitro* plants (2), makes the method uneconomical for geophytes. The use of large-scale liquid cultures in bioreactors (3), can decrease handling costs and increase the rate of bud proliferation (4). The problem of vitrification - abnormal leaf morphogenesis (5) in liquid culture, can be obviated by the reduction or prevention of leaf development, coupled with the induction of compact bud aggregates (6). *In vitro* storage organ development in *Gladiolus* (7, 8, 9, 10) and in other geophytes, can further reduce the problem of abnormal shoots and provide an efficient system for storage and international transfer.

Several growth retardants act as gibberellin antagonists by blocking their biosynthesis (11). Paclobutrazol (PAC) reduced stem extension in several ornamental species (12). In *Aechmea fasciata*, PAC enhanced massive shoot proliferation in planlets propagated *in vitro* (13) and in *Philodendron* cultured buds treated with growth retardants proliferated in liquid media (14). In *Gladiolus* growth retardants enhance protocorm development (9).

The use of liquid cultures in bioreactors for micropropagation of *Gladiolus* and other geophytes and the effect of growth retardants on enhanced shoot and cormlet development are reported.

2. MATERIALS AND METHODS

2.1 *Gladiolus* (cv Eurovision) - The apical buds were removed with a small, 3 mm long wedge, from *Gladiolus* corms, 5-6 mm in diameter and were cultured as reported previously (4, 9). The nutrient medium consisted of 1/2 strength MS minerals (15), full vitamins

and organic additives; with 0.25 µM naphthalene-acetic acid (NAA) and 10 µM benzyladenine (BA). Aseptic bud explants were transferred to 125 ml Erlenmeyer flasks containing 15 ml liquid medium (same as above, lacking agar), supplemented with ancymidol (ANC) (EL531, Elanco, France), or PAC (PP-333 ICI, England). The cultures were rotated on a gyratory shaker at 100 rpm under 32 $\mu Em^{-2}s^{-1}$ continuous fluorescent light at 25 ± 1°C.
Bud aggregates were subcultured to bubble bioreactors for further proliferation. The liquid medium included either ANC (2.0- 4.0 µM) or PAC (3.5-8.5 µM). The aeration rate was 0.5 vvm (volume air · 1^{-} 1 medium · min^{-2}) and as the plant biomass increased it was raised up to 1.0 vvm. After 28 days in liquid culture the proliferating clusters were separated into either small (3-4 mm) or large (6-12 mm) sections and subultured to agar medium. The MS medium consisted of 1/2 strength minerals with 3% or 6% sucrose, vitamins and inositol at full strength, 0.1mM ABA and polymer mixture 'K' 0.2% (developed by PBInd, Israel). The cultures were kept under continuous light at 25 ± 1°C. The number of shoots, protocorms, cormlet number per explant and their fresh weight (FW) were determined after 10-12 weeks in stage III. Corms allowed to dry *in vitro*, were planted 4-5 months from the time of subculture to stage III, in pots. The pots were transferred to the greenhouse under natural daylength at 25 ± 1°C and leaf emergence was determined after 3 weeks. Corms were harvested from *ex vitro* plants and their FW was determined.

2.2 *Nerine* - Explants were isolated from the inflorescence stalk, and cultured on agar medium to ensure asepsis. Young ovaries and 10-15 mm peduncle sections, from the branching zones of the pedicles responded best on MS minerals medium, supplemented with 4.5 µM 2,4-D, 13.2 µM BA, 0.1 g/L ascorbate, 0.15 g/L citrate and 1.0 g/L CH (Lilien-Kipnis, personal communication). Adventitious buds were subcultured to liquid media in flasks and cultured on a gyratory shaker as described for *Gladiolus* . Furt... proliferation in bubble bioreactor in 1/2 strength MS minerals medium supplemented with 10 µm BA, 0.25 µM NAA and 8.5 µM PAC was as described for *Gladiolus*. Proliferating clusters were subcultured to a hardening agar medium supplemented with or without 2.7 µM NAA and 6% sucrose for plantlets and bulblets development.

3. RESULTS AND DISCUSSION

Aseptic bud explants, in the presence of BA and NAA in the liquid medium, developed 4-5 shoots, in the absence of growth retardants, the leaves were succulent and continued to elongate. Leaf elongation and succulence were controlled by growth retardants, known to inhibit gibberellin biosynthesis (11). ANC and PAC inhibited leaf elon- gation and enhanced adventitious bud production in the liquid medium (Fig. 1). PAC induced the highest number of buds per explant. The buds developed into protocorm aggregates, the largest size protocorms were observed in ANC treated explants.

The rate of proliferation in a bubble bioreactor was higher than in the cultures agitated on a shaker, however, the aggregates in the bubble bioreactor were necrotic and high in anthocyanin as the aeration rate was increased (Fig. 2). This was probably due to vigorous agitation and the high rate of aeration required for circulation, as the plant aggregates increased in biomass.

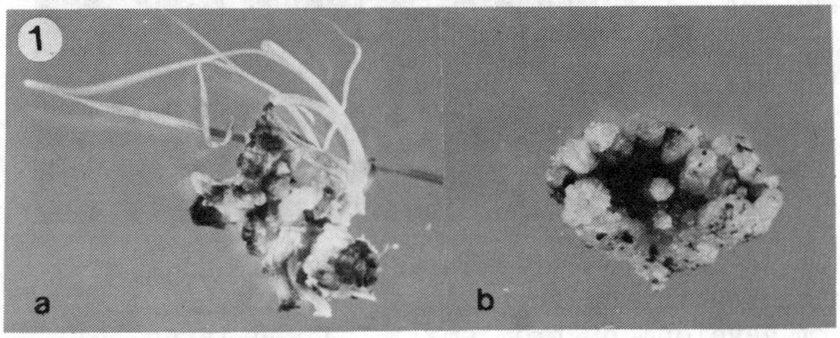

FIGURE 1 Bud aggregates from control (a) or PAC (b) treatments in bioreactor cultures.
FIGURE 2 Bud aggregates from liquid shaken (a) and bioreactor (b) cultures.

Renewal of the medium, once every 10 days, in a 'semi-continuous' culture, resulted in a higher growth value than in the 'batch' bioreactors in both the control and PAC treatments (Table 1). The size of

TABLE 1 - Growth of *Gladiolus* bud aggregates in 'batch' or 'semi-continuous bioreactor' cultures after 32 days, (Ziv 1990)

Treatment[a]	Growth value[b]	
	Batch	Semi continuous
Control	3.0 ± 0.2	5.0 ± 0.3
PAC 3.4 μM	2.2 ± 0.3	4.2 ± 0.6

[a]The rate of aeration was 0.7 vvm and the inoculum was 13 g FW/750 ml.
[b]Growth value $(FW_F-FW_I)/FW_I$ 'Batch'- no medium change, 'semi-continuous' - medium change once every 10 days; 4 bioreactors per treatment.

the inoculum affected growth, which was higher in bioreactors inoculated with a smaller inoculum in both the control and PAC treatments. The size of the inoculum did not effect bud proliferation, which was similar in the two inoculi in both the control and the PAC treatments. PAC treatment decreased the growth value but enhanced proliferation as compared to the control (Fig. 3). Growth retardants were reported to effect the level of cytokinins and ethylene (15) which can control bud proliferation and can also effect storage organ formation in geophytes.

3

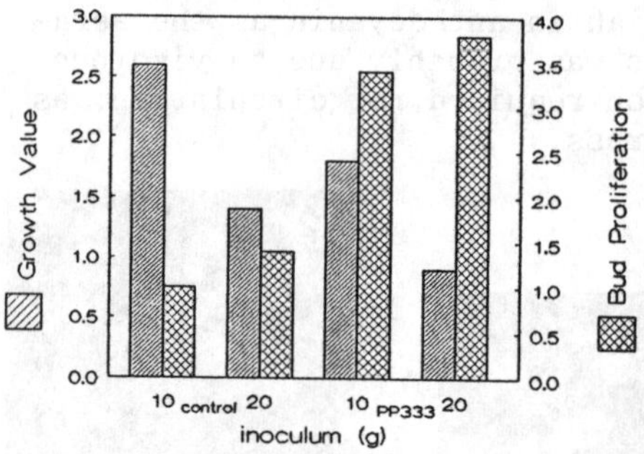

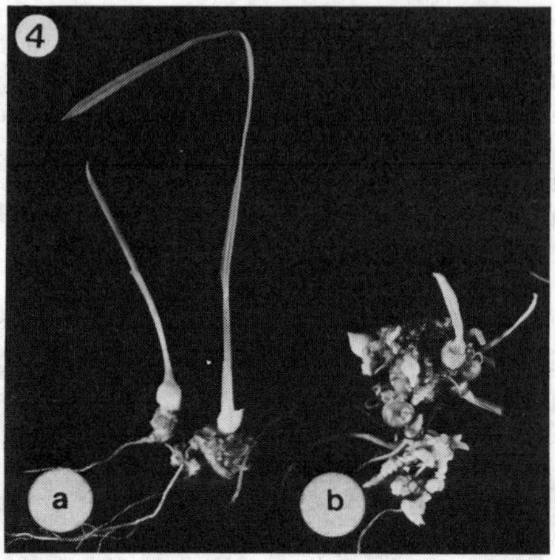

FIGURE 3 The effect of PAC and inoculum size on growth and bud proliferation in bioreactor cultures.

FIGURE 4 Corm and protocorm development from small (a) and large (b) clustres.

The proliferating bud aggregates were separated and transplanted to an agar hardening medium. The average size of the aggregates in the bioreactors, separated before inoculation to an agar hardening medium, was 10-15 mm. Bud aggregates and protocorms induced by growth retardants in the proliferation liquid medium developed into corms in the hardening medium. Two cluster sizes, 3-5 mm and 6-12 mm were compared for their proliferation potential. A higher number of shoots developed from the larger cluster but the FW of the developing corms was about one third of those developed from the small macerated clusters (Table 2; Fig. 4). The larger clusters were not in good contact with the medium and competed for nutrients which had to diffuse through the cluster to the developing buds. Separation of the clusters and embedding them in a double layered culture provided better contact and faster shoot growth. The growth retardants had a carry over effect and inhibited leaf elongation even 3-4 weeks after subculture of protocorms to a hardening medium.

TABLE 2 - The effect of *Gladiolus* aggregate size on shoot and corm development from bioreactor cluster subcultured to a hardening medium, (Ziv unpublished).

Aggregate size (mm)	Number of shoot	Number of corms	Corm FW (mg)
3-5	6.2 ± 0.7	3.1 ± 0.4	210 ± 3.1
6-12	19.1 ± 2.7	10.2 ± 1.9	68.5 ± 5.6

The cormlets were allowed to grow and eventually dry in culture 4-5 months from the time of subculture. The corms were transferred to pots or to plugs. All the cormlets germinated within 10-14 days after being transplanted to the greenhouse.

Nerine buds, developing from the initial explants, continued to proliferate in the liquid medium in the presence of ANC or PAC. The buds formed organogenic clusters (Fig. 5) and were subcultured to bioreactors. The rate of growth in the bioreactor was 5 to 6 folds higher than in the shaker cultures (Table 3). Unlike *Gladiolus* the size of the inoculum had no effect on the growth value which was similar in the small and large inoculi, in both shaker or bioreactor cultures.

FIGURE 5 *Nerine* protocorm and bud clusters from PAC treatment in bioreactor culture.

TABLE 3 - The effect of the inoculum size on the growth of *Nerine* cultures in shaken flasks or bubble bioreactor in the presence of 8.7 μM PAC after 28 days (Ziv, Lilien-Kipnis & Kahani, unpublished).

Culture type (ml medium)	Inoculum size (g)	Growth value[a]
Shaken flasks (80)	1.2 (1.7)[b]	0.4
	2.4 (3.6)	0.5
Bioreactor (750)	14.2 (54.8)	2.91
	28.6 (93.5)	2.26

[a]$(FW_F - FW_I)/FW_I$
[b]Figures denote final FW

The aggregates differentiated and developed plantlets. which upon subculture to a high sucrose (6%) medium, developed small bulblets.

A similar response which was observed in *Brodiaea* bud explants isolated from young corms and treated with 9.3 μM ANC and 10 μM BA will be presented.

The high propagation potential exhibited by *Gladiolus*, *Nerine* and *Brodiaea* buds in liquid cultures, in response to growth retardants treatments, resulted from both axillary and adventitious bud development. It is possible that growth retardants induced axillary bud development, by inhibiting the growth of the apical bud and leaves while addition of NAA and BA could have enhanced adventitious bud proliferation in the presence of the growth retardants, as was shown in *Aechmea* (13). Since leaf growth of the new developing buds was strongly inhibited, it permitted cormlet development as long as there was sucrose in the medium. Increased corm size by use of higher levels of sucrose in the medium, was reported by Dantu and Bhojwani (7), Steinitz and Lilien-Kipnis, (10)) as well as in the present work for all three geophytes studied.

The use of liquid cultures for scaling-up micropropagation of geophytes can greatly reduce manual handling. Storage organ production in response to growth retardants treatment has an advantage over plantlet development, which require acclimatization prior to transplanting *ex vitro*.

4. ACKNOWLEDGEMENT

The work on *Nerine* and *Brodiaea* was carried out by S. Kahani and A. Ilan, as part of their M.Sc. research studies, in cooperation with H. Lilien-Kipnis and A.H. Halevy respectively. The author would like to thank Ms Geula Gadasi for her excellent technical assistance and Ms Nili Ben-Yehezkel for the editing and typing of the manuscript.

5. REFERENCES

1. Ziv, M. and Lilien-Kipnis, H. (1990) In Handbook of Plant Cell Culture (Ammirato et al., eds.), Vol. 5, 461-478, McGraw Hill, N.Y.
2. Levin, R. and I.K. Vasil (1989) IAPTC Newslett. 59, 473-479.
3. Styer, D.J. (1985) In Tissue Culture for Forestry and Agriculture
4. Ziv, M. (1990) Acta Hort. (in press).
5. Ziv, M. (1986) In Plant Tissue Culture and Its Agricultural Applications (Withers, A.W., and Alderson, P.G. eds.) pp.187-196, Butterworth, London.
6. Ziv, M. and Hadar, A. (1990) Israel J. Bot. (in press).
7. Dantu, P.K., and Bhojwani, S.S. (1987) Gartenbauwiss 52, 90-93.
8. Ziv, M. (1979) Scientia Hort. 11, 257-260.
9. Ziv, M. (1989) Pl. Cell Tissue Organ Culture 17, 101-110.
10. Steinitz, B. and Lilien-Kipnis, H. (1989) J. Plant Physiol. 135, 495-500.
11. Graebe, J.E. (1987) Ann. Rev. Pl. Physiol. 38, 419-465.
12. Goulston, G.H. and Shearing, S.J. (1985) Acta Hort. 167, 339-348.
13. Ziv, M., Yogev, T., and Krebs, O. (1986) Israel J. Bot. 35, 175-182.
14. Ariel, T. (1987) The effect of culture condition on development and acclimatization of *Philodendron* 'Burgundy' and *Dianthus caryophyllus*. MSc thesis submitted to the Hebrew University of Jerusalem (in Hebrew with an English summary).
15. Murashige, T. and Skoog, F. (1962) Physiol. Plant. 15, 473-479.
16. Grossman, K. Kwiatkowski, J., Siebecker, H. and Junk, J. (1987) Plant Physiol. 84, 1018-1021.

DORMANCY DEVELOPMENT IN LILY BULBLETS PROPAGATED *IN VITRO*

G.J.M. DE KLERK[1], P. AGUETTAZ[1], A.M.G. PAFFEN[1], I. DELVALLEE[1], A.J.R. VAN DEN BERG[1,2], A.F. CROES[2], [1]Central Res. Lab. Tissue Cult. Hort. Crops, Postbus 85, 2160 AB Lisse, The Netherlands; [2]Dept. Exp. Plant Physiol., Univ. Nijmegen, Toernooiveld, 6525 ED Nijmegen, The Netherlands.

1. INTRODUCTION

Plants of many species develop dormancy to overcome adverse climatic conditions (1). Even when exposed to favourable conditions a dormant plant is unable to renew active growth and must first experience a specific dormancy breaking treatment such as chilling. The development of dormancy is important in tissue culture. In micropropagation of bulbs, non-sprouted (i.e. dormant) bulblets should be produced and not shoots or sprouted bulblets. Shoots of some woody crops, e.g. *Skimmia*, stop growing because of the development of dormancy (B. Kunneman, pers. comm.). Finally, somatic embryos generated in suspension cultures should be dormant to some extent to prevent precocious 'germination'.

Knowledge of the control of dormancy in vitro is limited. Therefore, we have initiated research on dormancy in bulblets generated on bulb-scale explants of lily (2). Here we report on the factors which affect the dormancy status and on the nature of dormancy in lily bulblets, which appears to be very similar to dormancy in woody species.

2. MATERIALS AND METHODS

2.1. Plant material and tissue culture procedure: Field-grown bulbs (circumference 18-20 cm) of *Lilium speciosum* Thunb. cv. 'Rubrum' nr. 10 were stored at 0.5°C until use. Scales were surface-sterilized for 30 min. in 1% (w/v) NaOCl. Explants of 7 x 7 mm were placed with the abaxial side on 15 ml medium in culture tubes (2.2 cm diameter). The medium was composed of MS macro- and microelements (3), 30 $g.l^{-1}$ sucrose, 0.4 $mg.l^{-1}$ thiamin, 100 $mg.l^{-1}$ myoinositol, 7 $g.l^{-1}$ agar (Difco-Bacto) and 0.05 $mg.l^{-1}$ NAA. The explants were cultured at 20°C at a light intensity of 30 $\mu E.m^{-2}.sec^{-1}$ (Philips TL 33) for 16 h per day. These standard conditions were changed as indicated. After 11 weeks of culture, the bulblets were harvested and the leaves were removed.

2.2. Determination of the dormancy status: The bulblets were treated for 20 min. with 0.2% (w/v) Benlate solution. Each sample was divided into two portions of 30 bulblets. One portion was planted immediately in soil and cultivated at 17°C. The second portion was planted in soil after 6 weeks at 2°C. Emergence was recorded after 15 weeks. The percentage of sprouting of cold-treated bulblets was determined to check viability. The percentage of non-cold-treated, viable bulblets that did not sprout and the chilling requirement were taken as measures of dormancy.

2.3. Analysis of newly synthesized proteins

Primordia, leaves and scales were excised from bulblets generated at 15°C and incubated for 24 h in 500 µl of the medium described before, but without agar. 0.75 MBq [^{35}S]methionine (23.1 TBq mmol^{-1}; Radiochemical Centre, Amersham) was added. Extraction of labeled protein and SDS-electrophoresis of individual parts were carried out as described previously (4).

3. RESULTS

3.1. Effect of physical conditions and medium composition. First we examined the effect of factors applied during the tissue culture (Tab. 1) and the effect of the length of culture (Fig. 1). Up to 6 weeks after the start of the culture, many viable non-cold-treated bulblets appeared to be non-dormant and sprouted after planting (Fig. 1). Thus dormancy only developed after 6 weeks. By consequence, a factor may affect the dormancy status by reducing the speed of development. Then dormancy develops only after prolonged culture but reaches a similar level. Alternatively, a factor may have an effect on the dormancy status independent of the duration of culture. Figure 2 shows that at 15°C dormancy did not develop after very long culture. Apparently, 15°C had a direct effect on the dormancy level. A low concentration of sucrose, though, had an effect mainly by reducing the speed of dormancy development (data not shown).

From Figure 1, it cannot be concluded whether dormancy continued to accumulate after 10 weeks, because the sprouting percentage then was always close to zero. The increase of the chilling requirement from 11 to 19 weeks (Fig. 3) shows that dormancy still accumulated during this period.

TABLE 1. The effect of various factors on the dormancy status of bulblets harvested 11 weeks after the start of the culture.

	cond. during culture	effect on dormancy status
light	0 or 30 µE.m^{-2}.sec^{-1}	no effect
day-length	8 h or 16 h per day	no effect
NAA/IAA	0.005 - 5 mg l^{-1}	no effect
BAP	0.05 - 0.5 mg l^{-1}	no effect
ABA	0.1 - 1 mg l^{-1}	no effect
MS-nutrients	0.1 - 2 times normal	no effect
temperature	15, 20 or 25°C	less dormancy at low temperature
sucrose	0 - 120 g l^{-1}	less dormancy at low sucrose conc.
GA_3	0.01 - 1 mg l^{-1}	less dormancy at high GA_3 conc.

3.2. <u>Nature of dormancy</u>. In bulblets generated at 20 or 25°C the number of scales continued to increase after the induction of dormancy (Fig. 4). The increase in scale number at 20 and 25°C parallelled the increase at 15°C, the temperature at which hardly any dormancy is induced.

In lily bulblets a primordium may develop either into a scale or into a leaf in which the base of the petiole swells to become a scale. In bulblets generated at 20°C, many of the third scales, but only few of the fourth and none of the fifth scales bore leaves (Tab. 2). The third scale was formed before and the fourth scale after 6 weeks (Fig. 4).

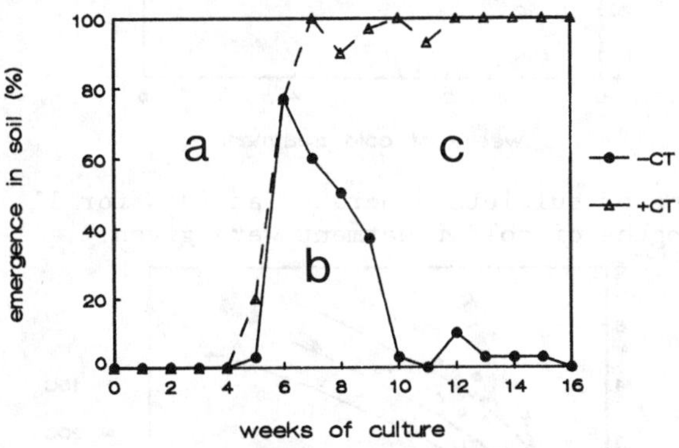

FIGURE 1. Percentage of sprouting of bulblets harvested at various times after the start of the culture. The bulblets were planted either immediately (-CT) or after a cold treatment (+ CT). Three phases can be distinguished: (a) non-viable (& non-dormant), (b) viable & non-dormant and (c) viable & dormant.

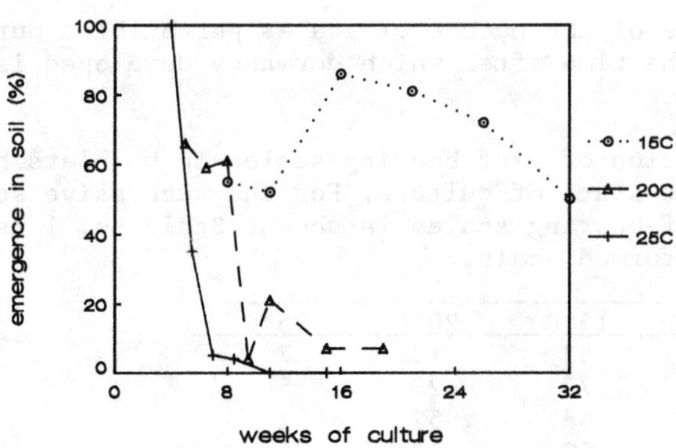

FIGURE 2. Development of dormancy during culture at various temperatures. The percentage of sprouting of viable bulblets is shown.

3.3. Protein synthesis. Newly synthesized proteins were fractionated by SDS-electrophoresis (Fig. 5). We analysed leaves, scales, primordia that were leaf-shaped (these primordia will develop into

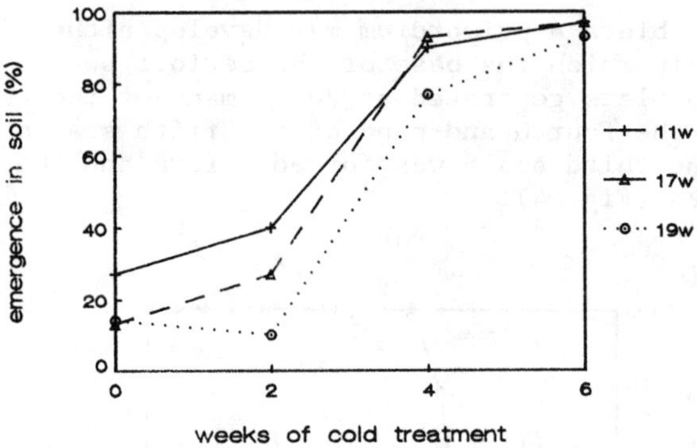

FIGURE 3. Sprouting of bulblets generated at 20°C for 11, 17 or 19 weeks. Various lengths of cold treatment were given.

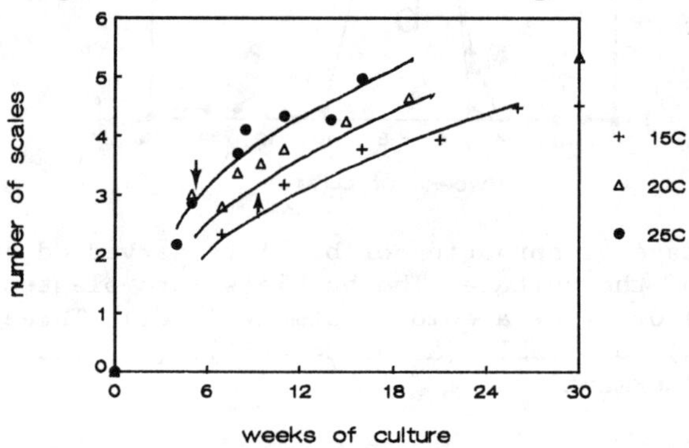

FIGURE 4. Increase of the number of scales per bulblet during culture at 15, 20 or 25°C. The time after which dormancy developed is indicated by arrows.

TABLE 2. Distribution of leaf-bearing scales in bulblets harvested ca. 20 weeks after the start of culture. For the successive scales, the percentage of leaf-bearing scales is shown. Scale no. 1 is the outermost, first formed scale.

	15°C	20°C	25°C
scale no. 1	53	13	7
scale no. 2	77	27	5
scale no. 3	68	52	2
scale no. 4	53	17	0
scale no. 5	64	0	0
scale no. 6	-	0	0

leaf-bearing scales) and primordia that were scale-shaped (these primordia will develop into scales). The overall patterns were similar. However, two proteins (indicated by arrows) were abundant in scales and leaf-shaped primordia, but occurred in (very) low amounts in leaves and scale-shaped primordia. Another protein (indicated by *) occurred in low amounts only in the scales.

4. DISCUSSION

With the exception of organogenesis and embryogenesis, developmental processes in vitro have been examined only infrequently. We have started a study on one such process, viz., the development of dormancy; we used lily bulblets generated on scale explants as a model system. We found that only few factors affected the dormancy status. Our results partly agree with data from ex vitro systems. The effect of temperature (Fig. 2) is well known in buds (5). Sucrose and GA_3 (Tab. 1) have the same effect on dormancy in developing seeds (6,7). However, day-length and ABA, which are known to affect the dormancy status in buds (8) and seeds (9), had no effect in lily bulblets (Tab. 1). With regard to ABA, it is interesting to note that added ABA suppresses leaf-formation, whereas in the presence of fluridon, an inhibitor of ABA synthesis, only leaves are formed (K.-S. Kim and G.J. de Klerk, unpubl. res.). This morphological effect of ABA correlates with dormancy (see below). Therefore, it may be that ABA influences dormancy anyhow. Finally, the various phases during bulblet formation, i.e., non-viable & non-dormant, viable & non-dormant, and viable & dormant, have also been reported for seeds (10). Together, these data indicate strong

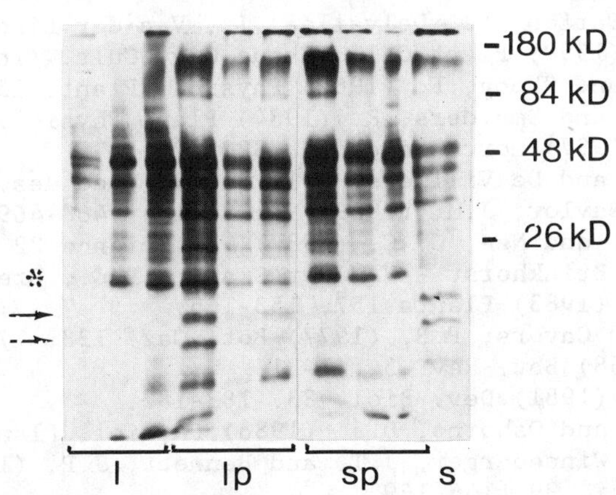

FIGURE 5. Fractionation by SDS-PAGE of proteins synthesized by leaves (l), leaf-shaped primordia (lp), scale-shaped primordia (sp) and scales (s) excised from bulblets generated for 10 weeks at 15°C.

parallels between dormancy development in bulbs, buds and seeds.

Dormancy connotes a state of inactivity (11). It has been found, however, that dormant seeds (4, 12) and buds (13) have high rates of metabolism. We have shown here that in lily bulblets the formation of new scales continued after the induction of dormancy (Fig. 4). The same has been reported for buds of pear (14) indicating that similar mechanisms are operating. We conclude from Figure 4 and Table 2 that the induction of dormancy involves a change in the developmental capability of the primordium. After the induction of dormancy, the primordium of a bulblet in vitro is only capable of developing into a scale. After planting the bulblet in soil, the primordium is still uncapable of leaf formation and requires a cold treatment to restore this capability.

The patterns of newly synthesized proteins were very similar in leaf-shaped primordia (that will grown into a leaf), in scale-shaped primordia (that will grow into a scale) and in scales themselves. There were few, but distinct differences with leaves. Therefore, we assume that all primordia were at first scale-like and then either continued this developmental programme or switched towards leaf development. When the primordia were excised for incorporation, this switch had already occurred and the primordia were clearly either leaf-shaped or scale-shaped. Nevertheless, all primordia were still similar to scales at the level of detectable proteins. It seems therefore that the developmental switch was based on only subtle changes in gene expression. Similar observations have been made with seeds (4, 15).

REFERENCES

1. Villiers, T.A. (1975) Dormancy and the survival of plants. Edward Arnold, London.
2. Aguettaz, P., Paffen, A., Delvallée, I., Van der Linde, P. and De Klerk, G.J. (1990) Plant Cell Tissue Org. Cult. (in press).
3. Murashige, T. and Skoog, F. (1962) Physiol. Plant. 15, 473-497.
4. De Klerk, G.J. and Smulders, R. (1984) Plant Physiol. 75, 929-935.
5. Dennis, F.G. (1987) HortScience 22, 820-824.
6. Cairns, A.L.P. and De Villiers, O.T. (1989) Weed Res. 29, 151-156.
7. Black, M. and Naylor, J.M. (1959) Nature 184, 468-469.
8. Fuchigami, L.H. and Nee, C.-C. (1987) HortScience 22, 836-845.
9. Karssen, C.M., Brinkhorst - Van der Swan, D.L.C., Breekland, A.E., Koorneef, M. (1983) Planta 157, 158-165.
10. Sidhu, S.S. and Cavers, P.B. (1977) Bot. Gaz. 138, 174-182.
11. Amen, R.D. (1968) Bot. Rev. 34, 1-31.
12. De Klerk, G.J. (1981) Dev. Biol. 83, 183-187.
13. McDonald, M.M. and Osborne, D.J. (1988) Physiol. Plant. 73, 392-400.
14. Young, L.C.T., Winneberger, J.T. and Bennett, J.P. (1974) J. Am. Soc. Hort. Sci. 99, 146-149.
15. De Klerk, G.J. and Linskens, H.F. (1979) Biochem. Physiol. Pflanz. 174, 537-545.

DO WE UNDERSTAND SOMACLONAL VARIATION?

R.L. PHILLIPS, S.M. KAEPPLER, AND V.M. PESCHKE[1], Department of Agronomy and Plant Genetics and Plant Molecular Genetics Institute, University of Minnesota, St. Paul, MN USA 55108

Tissue culture is clearly a mutagenic procedure. Although some progenies of regenerated plants may be highly productive and appear normal in all measured traits, most have detectable variation not present in the controls. The tissue culture process itself appears to be inducing these alterations as the cells proliferate in a relatively undifferentiated state. Probably all forms of mutational events that can occur in nature also happen in tissue cultured cells. However, certain mutational events appear to occur in exceedingly high frequencies in plant tissue cultures. These events may vary in frequency among species, genotypes, explant source, media, etc., but, in general, elevated mutational events appear to occur regardless of the specific cultured materials or conditions. The unusually frequent events include the following: 1) Single gene mutations; 2) Transposable element activation; 3) Quantitative trait variation; and 4) Chromosome breakage. In this paper, we will propose a hypothesis attempting to relate these four seemingly different kinds of heritable variation.

Hypothesis

The basic premise of the hypothesis advanced in this paper (Figure 1) is that the various mutational events are directly or indirectly related to the modification of DNA, specifically DNA hypo/hypermethylation.

---

[1]Current address: Department of Biology, Washington University, St. Louis, MO USA

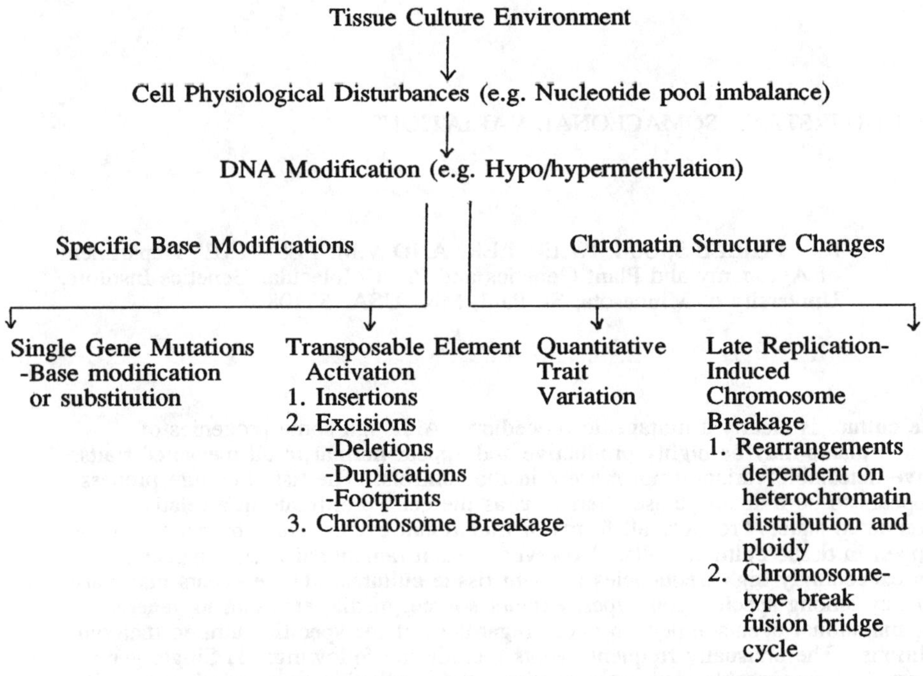

Figure 1. Hypothesis relating DNA modification to various mutational events elevated in frequency in tissue culture.

Qualitative Mutations

In studies at the University of Minnesota, analysis of first and second self-pollination generations of plants regenerated from maize tissue cultures revealed at least 50 different phenotypes segregating as if controlled by single, recessive genes. The frequency of variants per regenerated plant was 0.5 when the plants arose from 3- to 4-month-old cultures and 1.3 when from 8- to 9-month-old cultures (1). In many cases, the regenerated plants carried more than one mutation in a heterozygous state (10% for plants from 3-4 month cultures, 40% from 8-9 month cultures). Most studies with maize regenerants report an appreciable mutation frequency (2, 3, 4, 5, 6, 7). Zehr et al. (7) reported the absence of mutation for a few genotypes. Segregation of a new variant in the R_2 (second self-pollinated generation) and not in the R_1 occurs frequently and is thought to be the result of the regenerated plant being sectored (non-concordant) for the new mutation; i.e. if the tassel and ear of the R_0 plant are not both from the mutated sector, progeny from selfing the R_0 may be heterozygous in the R_1 and segregating only in the R_2. Of 160 R_0 maize plants identified as carrying a new recessive mutation, 83 (52%) were apparently sectored for the mutation (8).

Single recessive mutations can be the result of specific sequences modified by DNA methylation. The existence of 5-methylcytosine is common in plants (9) and Holiday (10) suggests that methylation may be involved in "controlling the cell-to-cell inheritance of gene-activity patterns". Increased methylation usually correlates with loss of gene activity and decreased methylation with activity (10, 11). Base substitutions also can occur when 5-methylcytosine is deaminated; the 5-methylcytosine converts to thymine. Two alcohol dehydrogenase mutations derived from tissue culture have been sequenced (12, 13); however, these did not represent changes to thymine. More examples of tissue culture-induced mutations need to be sequenced in an attempt to determine whether or not they are due to single base changes and, if so, the nature of the changes.

Transposable Elements and Somaclonal Variation

Barbara McClintock's (14, 15) finding that certain "shocks" to the genome caused the activation of transposable elements led to the idea that transposable element activation would be a likely outcome of tissue culture (16, 17). The high frequency of chromosome breakage observed in regenerated plants (18) as well as the variety of heritable genetic variants produced supported this possibility. We have detected both Activator (Ac) and Suppressor-mutator (Spm) genetic activity in the progeny of tissue culture-derived maize plants (19, Peschke and Phillips, submitted). In each case, the cell lines which produced plants containing transposable element activity also gave rise to a number of plants without such activity; this together with the lack of activity in plants from control seed indicate that activity was not present before culture. Evola et al. (20, 21) have also detected Ac and Spm activity in regenerated maize plants of strains with no history of transposable element activity.

In most cases, transposable element activity has not been obviously correlated with specific tissue culture-induced mutations or chromosome breakage events. However, transposable elements are capable of producing DNA sequence alterations ("footprints") at the site of their insertion and subsequent excision, so not all transposon-mediated events will be easily recognized as such (Peschke and Phillips, in preparation). A notable example of a tissue culture-induced mutant which does appear to be associated with transposable element activity is an unstable flower color mutant in alfalfa (22).

Methylation and transposable element activity

The activation of transposable elements by chromosome breakage has been extensively documented (23), but precisely how transposable elements become active at the DNA level is not known. It is generally believed that the newly-active elements originate from cryptic (i.e. stable, inactive) sequences normally present in the genome; such sequences have been detected by Southern hybridization in nearly every maize line tested (23). One hypothesis under current study is based on the numerous observations which correlate transposable element activity with lowered methylation at specific restriction sites in the element (24). For example, in two experiments involving Ac elements which cycle between activity and inactivity at a relatively high frequency, the elements were fully methylated when inactive but less methylated when they regained activity (25, 26). One of these cycling, currently inactive, Ac elements was put into tissue culture and became genetically active at 80X the normal rate for that allele (24). In addition, this activity was correlated with reduced methylation in the 5' end of the element. Some correlation between tissue culture-derived Ac genetic activity and reduced methylation within Ac-homologous sequences has also been detected in our materials (Peschke et al., in preparation). Although Ac-homologous sequences were often hypomethylated at internal HpaII and PvuII sites in the plants with Ac activity, this hypomethylation was not necessary for activity. In contrast, the Ac sequences in sibling plants without Ac activity were highly methylated (Peschke et al., in preparation). One possible explanation for the relatively high methylation we have observed even in plants with transposable element activity is that these elements have been inactive for many years (unlike the cycling elements) and were examined only two plant generations after activation. Kunze et al. (27) have noted that a newly-active element may become progressively less methylated over time, eventually reaching the methylation state of an element which has been active for many generations. Studies of our materials in later generations may reveal a similar trend.

Culley (28; personal communication) initiated endosperm cultures from genotypes without Ac but containing the Ac-responsive allele c-m2 to see whether transposable element activation could be detected in these cultures. He observed spotting on the cultures similar to the spotting produced in kernels containing both c-m2 and a genetically active Ac element. In addition, he observed that all of the Ac-homologous sequences in these cultures were highly unmethylated at internal PvuII sites, while leaf tissue of the same genotype showed the usual high degree of methylation. If this situation were also true for immature embryo-derived callus, it would indicate that activation of transposable elements in plants from such callus might result from extensive demethylation during culture followed by improper re-methylation at the time of regeneration and/or at meiosis or other phases of the sexual cycle. Our preliminary comparisons of A188 plant DNA and immature embryo-derived 4-month-old callus DNA did not show any significant difference between the two, when methylation within and around Ac-homologous sequences was examined (Peschke, unpublished observations).

Behavior of transposable elements in vitro

In our experiments and others mentioned above, the activation of transposable elements in tissue culture has been assayed using the genetic constitution of regenerated plants and their progeny. Two recent studies (29, 30) have examined the behavior of Mutator (Mu) transposable elements in vitro. Cultures were initiated from maize embryos containing either genetically active (unmethylated) or inactive (methylated) Mu elements, and samples of the cultures from different time points were examined using Southern hybridization. In general, their results indicated that the methylation of the elements in culture was similar to that of the progenitor materials, although some degree of increased methylation of initially active Mu elements was noted. In neither study did there appear to be activation of inactive Mu sequences; the reason for this is not known. An important observation of both laboratories was the appearance of novel restriction fragments in the cultures with active Mu elements, which is interpreted to represent transposition of Mu sequences in vitro. This finding indicates that a transposable element which has become active in tissue culture may transpose, causing additional genetic and possibly cytological rearrangements. Even if the element later becomes inactive (as did some of the Mu elements) it may have already had a permanent effect on the callus and (later) the regenerated plants.

Quantitative Trait Variation

Quantitative trait variation is prevalent among families derived from regenerated plants. Lee et al. (31) showed extensive variation among families derived from regenerated maize plants. Variation was seen for yield, maturity, moisture content at harvest, and plant and ear height. The trend was toward poorer agronomic performance as the length of time in culture prior to plant regeneration increased. However, the top performing lines in several of the trials were tissue culture-derived. Zehr et al. (7) found similar types of variation in studies of families derived from regenerants of seven inbred lines.

We have become interested in the testcross performance of inbred A188 regenerant-derived families, especially in testcrosses with the inbred that was the explant source. In most crosses of the explant source inbred with regenerant-derived families (which are up to 50% lower yielding than the uncultured sources), the yield is completely restored to normal levels (Kaeppler, unpublished). Fincher reported similar results with different maize strains (32). Should one interpret this to be the result of completely recessive gene mutations induced in the lower yielding regenerant-derived line? Or could this result be due to some mechanism in crosses that eliminates the undesirable gene effects? Evaluation of selfs of these crosses should provide further insight. The behavior of the maize transposable element Mu (Robertson's Mutator) indicates that methylation and expression of certain maize sequences can be modified in crosses. Loss of Mu excision correlates with an increase in Mu methylation (33, 34). All copies of Mu in the genome are modified when mutator activity is lost and this state is relatively stable. Crosses to other lines can modify the methylation state and the activity of Mu (35). If the restoration of normal yields in crosses of regenerant lines with the explant source inbred is the result of the masking of recessive genes, progenies from self-pollination should show extensive variation. On the other hand, if the restoration of normal yields in crosses is the result of erasure or reprogramming of epigenetic patterns, the self-pollination progeny may show relatively little variation. Preliminary evidence shows that heterosis over the better parent (usually the source inbred) for yield and maturity also occurs in some of the testcrosses (Kaeppler, unpublished; Woodman, personal communication; 32). These latter results indicate that new dominant or epistatic alleles are produced by the tissue culture process (possible mechanism is discussed below). If this heterosis is also manifested to some extent in testcrosses to unrelated inbreds, it would make tissue culture especially useful in regard to quantitative genetic variation in maize.

In oats (Avena sativa) Dahleen (36) evaluated R_4 and R_5 lines from the varieties Lodi and Tippecanoe. The R_0 plants giving rise to these lines were cytologically selected to be normal. Yet 82% of the Lodi lines and 91% of the Tippecanoe lines showed significant variation from controls in at least one of the eight traits measured.

Such variation in so many quantitative traits among regenerant families from inbred strains is unexpected. Each quantitative trait is controlled by many genes. Is it likely that the tissue culture process would mutate a sufficient number of genes or specific major genes to cause such variation in a trait? In addition, is it reasonable that so many quantitative traits would be modified in a single experiment? We suggest that such extensive changes in quantitative traits are difficult to imagine by classical mutations of structural or regulatory genes. Rather, the effect may be due to a more global effect on the genome. DNA methylation is correlated with the activity of chromosomes in certain cases. In mammalian females, one X chromosome is inactivated early in development and becomes late replicating; the DNA of this X chromosome is highly methylated (37). Several examples have been reported of proteins binding differentially to methylated DNA (38) and proteins exist that can bind to any methylated sequence. DNA methylation can thereby alter chromatin structure and indirectly affect the expression of many genes, perhaps making it more plausible that quantitative traits could be modified by passage through tissue culture. Chromatin structure changes associated with DNA methylation can affect regions as large as 40-50 kb or more (39).

Another puzzling question deals with how dominant mutations might be produced via tissue culture that would result in heterosis when a regenerant line is crossed with the source inbred. This question is especially puzzling since dominant qualitative mutations are relatively rare. Of course, dominant qualitative mutations such as those providing herbicide tolerance have been reported (40); the selection imposed allows the recovery of presumably rare dominant mutations. One possibility to consider relates to the observation that tissue culture may cause previously silent sequences to become active, such as the activation of transposable elements which act in a dominant fashion. Other genes that are normally silent or inactive could be activated by DNA demethylation. Such activated genes would be expected to behave as dominants or at least additively because the other homologous sequences would be inactive.

Late Replication-Induced Chromosome Breakage

Meiotic analysis of regenerated plants often reveals abnormal chromosomes that are the result of a breakage event. In oats, plants regenerated from 20-month-old cultures of the Tippecanoe and Lodi varieties carried a detectable structurally-altered chromosome 49% and 88% of the time, respectively (41). In maize, 1036 regenerated plants were analyzed meiotically (2, 42, 43, 44). Of these, 217 (21%) possessed detectable structurally-altered chromosomes. Only 16 plants (1.5%) were aneuploid. Polyploidy was even more rare except for one case in which several plants regenerated from an embryo-derived callus line were tetraploid. The breakpoints among the recovered interchanges and deletions/duplications appeared to be non-random. The breaks generally occurred in chromosome arms possessing heterochromatic (late replicating) knobs and were between the knob and the centromere (8). In oats, the breaks often occurred in the prominent pericentromeric heterochromatin. These observations led to the idea that the tissue culture environment somehow causes normally late replicating DNA to fail to complete replication before mitosis (41). Such an event would lead to anaphase bridges and subsequent breakage of the chromosomes. Various rearrangements are possible depending on the position of the breaks, the number of chromosomes broken, and the ploidy level of the cell (45). Our studies with oats and maize indicate that different types of aberrations occur in the two species, but the differences are largely predictable based on the known distribution of heterochromatin. Once a break occurs, the chromosome-type break fusion bridge cycle could be initiated and lead to further rearrangements (46). This type of cycle can continue in the sporophyte of maize and, therefore, may exist in cultured somatic cells. It should also be pointed out that aneuploidy conceivably can result from delayed replication of pericentromeric heterochromatin and subsequent nondisjunction.

What could be responsible for the delay in replication of late replicating DNA? As mentioned previously, DNA methylation and late replication can be correlated events. Riggs (47) discusses this possibility in regard to animal cells. Maize sequences repeated more than 1000 times normally are more than 90% methylated (24). Perhaps alterations in the amount or pattern of methylation can further affect the timing of the replication of these sequences.

DNA Methylation and Tissue Culture-Induced Variation

The high negative correlation of cytosine methylation and gene expression in eukaryotes has been reviewed by Holliday (48) and Cedar (11). Several observations suggest that a heritable epigenetic mechanism, such as DNA methylation, may be involved in tissue culture mutagenesis. The first observation is the tissue culture-induced activation of transposable elements discussed above. The activation of cryptic elements by tissue culture may be effected by DNA demethylation. These active elements may only be representative of a much broader array of genes altered in expression by DNA methylation or demethylation.

A second observation is the occurrence of mutants such as that described by Oono (49). One tissue culture-induced, dwarf mutant was stable upon eight generations of selfing but could not be recovered upon outcrossing to normal plants followed by selfing. Sano (50) induced mutants that acted in a similar manner by treatment of seeds with 5-azacytidine, a demethylating agent. He also found a correlation of DNA hypomethylation and dwarfism in maize (51).

A final observation is the extensive quantitative trait variation observed among regenerant-derived families of inbred lines. Although sequence alterations may be responsible for this variation, the frequency and magnitude of this variation is so great that we postulate that the more highly mutable, yet heritable, DNA methylation may be involved. We have detected heterosis for yield and plant height in testcrosses of regenerant-derived families to the source inbred (explant source). These observations suggest the activation of new dominant or epistatic alleles. This type of positive mutagenesis is not expected at a high frequency by random sequence alteration. Perhaps, as quiescent transposable elements are activated, so quiescent genes may also be reactivated by the tissue culture process. Dahleen (36) detected two new storage protein (avenin) bands in certain regenerated lines of oats. Nonfunctional avenin sequences do exist in the oat genome (52).

We have found, as has Brown (53) for inbred A188 maize and Muller et al. (54) for indica rice, extensive sequence-specific variation in methylation patterns in progenies of regenerated plants. In our study, DNA of R_1 progenies of A188 regenerants (R_0) was isolated and digested with the isoschizomers HpaII and MspI. Both HpaII and MspI recognize the sequence CCGG although only MspI will cleave if the internal C is methylated. The DNA was probed with several cDNA and random genomic clones. All regenerated plants were from callus derived from a single explanted embryo. The plants had no obvious qualitative trait mutants and appeared relatively normal phenotypically.

Figure 2 shows the sequence specific variation in twelve R_1 families. Each lane is pooled DNA of ten R_1 plants. The DNA was probed with a random genomic PstI clone (BNL 5.09). Only one of the families (lane 2) has the same HpaII pattern as the noncultured control (lane 13). In addition, the non-normal patterns vary among the regenerated families indicating that methylation changes occurred more than one time in the callus. In a similar analysis of fifteen non-cultured A188 control plants, no variation was seen for this probe or any of the others tested.

Figure 3 shows the variation when this same DNA was probed with a cDNA clone of alcohol dehydrogenase I. In this case four of the twelve families are obviously different from the non-cultured control. We have seen similar variation among R_3 families derived from inbred A634 and B73 regenerants.

In general, the results have shown a strong tendency for the changes in methylation to represent losses of methylation. In most cases the MspI patterns were constant, suggesting that base changes occur at a much lower frequency than methylation changes. Brown (53), in a similar study, also found a high frequency of sequence-specific methylation variation. In addition, Brown found base pair changes (as shown using methylation insensitive enzymes) and gene duplications. We have found only one case of a potential sequence change and have found no change in copy number using single copy probes.

Nucleotide pool imbalances influence several aspects of DNA metabolism including replication and repair (55). These pools could be sensitive to the stress imposed by tissue culture. Since newly replicated DNA is methylated, any disruption of replication or repair could alter methylation patterns.

(a)

(b)

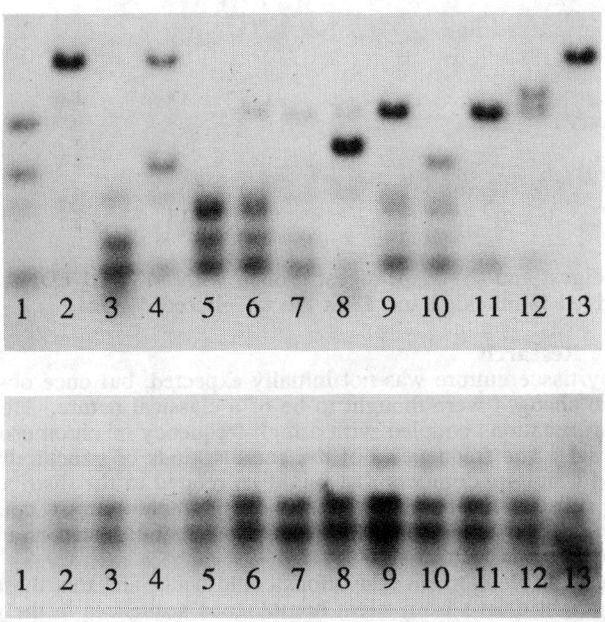

Figure 2. -- (a) HpaII digest and (b) MspI digest probed with a PstI genomic clone (BNL 5.09). Lanes 1-12 are R_1 derived families. Lane 13 is the uncultured control.

(a)

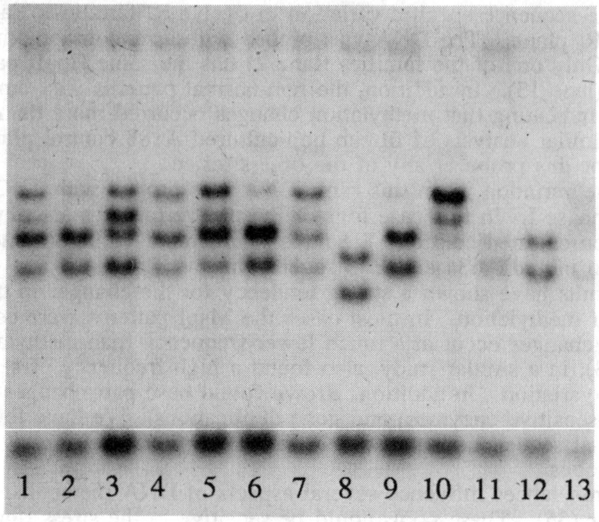

(b)

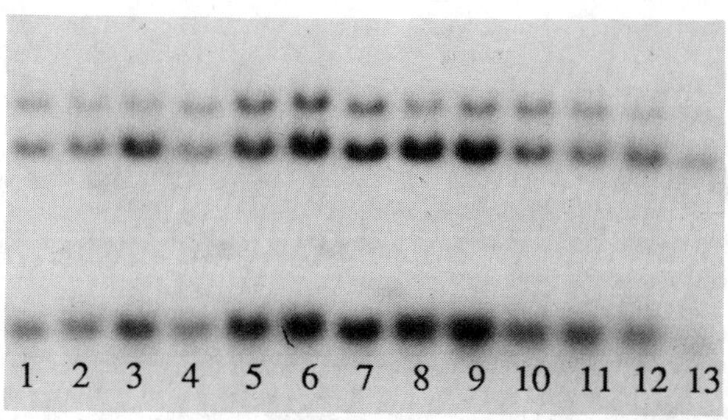

Figure 3 -- (a) HpaII digest and (b) MspI digest probed with an Adh1 cDNA clone. Lanes 1-12 are R_1 derived families. Lane 13 is the uncultured control.

Summary and Future Research

Variation induced by tissue culture was not initially expected, but once observed, the genetic and cytogenetic changes were thought to be of a classical nature. However, the high frequencies of mutations coupled with a high frequency of chromosome breakage seemed unusual. The frequencies of the various kinds of structurally altered chromosomes indicated a non-randomness that might be related to the distribution of heterochromatin. The subsequent observation that most quantitative traits could vary among lines derived via tissue culture of homozygous sources suggested an even more massive mutagenic action of tissue culture.

The finding that transposable element activations could occur and that these active transposable elements might derive from silent homologous sequences in the genome began to raise questions about DNA methylation being involved. Despite the strong correlations which have been observed between transposable element methylation and activity, both in tissue culture and in other systems, it is still unclear how these two events are related. It often appears that methylation changes may represent both a cause and a result of transposable element genetic activity. Several authors (56) have proposed that active transposons remain unmethylated due to the binding of their own gene product(s). In addition, the presence of one active transposable element appears to affect both the methylation and genetic activity of other transposable element-like sequences in the genome (57). Therefore, tissue culture-induced methylation changes may activate transposable elements which, in turn, may activate and/or alter the methylation of additional elements of the same family.

More recent analyses with random RFLP probes indicate that progenies of regenerated maize plants, even those from the same embryo-derived callus line, may differ in DNA methylation. The loss of methylation seems to be the general rule in our studies, although methylation increases can also occur.

Do we understand somaclonal variation? The answer is that we clearly do not. But it is exciting to consider the possibility that tissue culture may induce variation by an epigenetic mechanism that generates heritable variation. Many of the classical tissue culture-induced changes as described here could be due to an epigenetic mechanism such as DNA hypo/hypermethylation. In addition, some of the very unusual results of tissue culture such as the occurrence of a stable dwarf among rice regenerants which disappears upon crossing (49) could be DNA methylation effects.

Our goal is to understand the mechanism(s) underlying tissue culture-induced genetic change. Methods to avoid or control the occurrence of such variation would be highly desirable. If DNA methylation is a major factor in somaclonal variation, further studies employing agents that alter DNA methylation such as 5-azacytidine, ethionine, or ECP (2-amino-5-ethoxy-carbonyl-pyrimidine-4(3H)one) would be most interesting. Brown (53) found evidence for demethylation of DNA in plants grown at 5-azacytidine concentrations above 10 μM. Development of tissue culture/regeneration methods not employing an auxin also might be useful since auxins can dramatically increase methylation (58). More studies on the methylation state of the DNA of the explant tissue would provide useful insight to possible variances in behavior of different tissues in culture. Since the normal tissue-pattern of methylation may be seriously disrupted by tissue culture, tissue, genotype, and species differences in methylation would be valuable to know.

Another common feature of regenerated plants is their reduced stature and vigor. Yet the R_1 progeny usually appear nearly normal, although segregation of mutants may occur. Could the passage of abnormally modified cells of regenerants through meiosis erase or reprogram previous DNA modifications? In animals, the germ line reverses epigenetic modifications. Surani et al. (59) indicate that errors in the reversal process can occur and result in the transmission of epigenetic modifications to progeny.

Studies of methylation changes relative to culture time also would be important, especially since animal cells with a finite lifespan grown in vitro show a continual decline in methylation (48). In the case of the maize Ac element, Dennis and Brettell (24) have shown progressive demethylation with time spent in culture.

Finally, determination of nucleotide pools and how they may change in relation to tissue culture manipulation would be useful information, especially in regard to DNA replication/repair and methylation. A reduction in the level of the nucleotide-containing S-adenosyl methionine which serves as the methyl donor for the maintenance methylase reaction could result in altered enzyme efficiency.

The central conclusion from the discussion presented herein is not whether or not DNA methylation per se is the principal cause of tissue culture-induced variation. Rather, the discussion is intended to illustrate that the myriad of tissue culture effects, including single gene mutations, transposable element activation, quantitative trait variation, and changes in chromosome structure and number conceivably trace back to a single cause.

Acknowledgement: Research supported by the Minn. Agric. Exp. Sta., the Pillsbury Co., and a grant from the U.S. Department of Agriculture (USDA #88-37262-3919).

REFERENCES

1. Lee, M. and Phillips, R.L. (1987) Genome 29, 834-838
2. Armstrong, C.L. and Phillips, R.L. (1988) Crop Sci. 28, 363-369
3. Edallo, S., Zucchinali, C., Perenzin, M., and Salamini, F. (1981) Maydica 26, 39-56
4. McCoy, T.J. and Phillips, R.L. (1982) Can. J. Genet. Cytol. 24, 559-565
5. Rice, T.B. (1982) in Proc. 37th Annu. Corn and Sorghum Res. Conf. pp. 148-162, Chicago, IL
6. Woodman, J.C. and Kramer, D.A. (1986) in VI Intl. Congr. Plant Tissue Cell Culture, p. 215, Minneapolis, MN
7. Zehr, B.E., Williams, M.E., Duncan, D.R., and Widholm, J.M. (1987) Can. J. Bot. 65, 491-499
8. Phillips, R.L., Somers, D.A., and Hibberd, K.A. (1988) in Corn and Corn Improvement (Sprague, G.F. and Dudley, J., ed.) Monogr. 18, 3rd Ed. pp. 345-387, ASA-CSSA-SSSA, Madison
9. Gruenbaum, Y., Naveh-Mony, T., Cedar, H., and Razin, A. (1981) Nature 292, 860-862
10. Holliday, R. (1989) Sci. Am. 260, 60-73
11. Cedar, H. (1988) Cell 53, 3-4
12. Brettell, R.I.S., Dennis, E.S., Scowcroft, W.R., and Peacock, W.J. (1986) Mol. Gen. Genet. 202, 235-239
13. Dennis, E.S., Brettell, R.I.S., and Peacock, W.J. (1987) Mol. Gen. Genet. 210, 181-183
14. McClintock, B. (1978) Stadler Symp. 10, 25-47
15. McClintock, B. (1984) Science 226, 792-801
16. Larkin, P.J. and Scowcroft, W.R. (1981) Theor. Appl. Genet. 60, 197-214
17. Burr, B. and Burr, F. (1981) Stadler Symp 13, 115-128
18. Lee, M. and Phillips, R.L. (1988) Ann. Rev. Plant Physiol. Plant Mol. Biol. 39, 413-437
19. Peschke, V.M., Phillips, R.L., and Gengenbach, B.G. (1987) Science 238, 804-807
20. Evola, S.V., Burr, F.A., and Burr, B. (1984) Eleventh Annual Aharon Katzir-Katchalsky Conf., 8-13 January, Jerusalem. Abstract
21. Evola, S.V., Tuttle, A., Burr, F., Burr, B. (1985) First Int. Congr. on Plant Molecular Biology, Savannah, Georgia, Abstracts, p. 10
22. Groose, R.W. and Bingham, E.T. (1986) Plant Cell Rep. 5, 104-110
23. Burr, B. and Burr, F.A. (1988) in Plant Transposable Elements (Nelson, O., ed.) pp 317-323. Plenum Press, New York
24. Dennis, E.S. and Brettell, R.I.S. (1990) Phil. Trans. R. Soc. Lond. B 326, 217-229
25. Schwartz, D. and Dennis, E. (1986) Mol. Gen. Genet. 205:476-482
26. Chomet, P.S., Wessler, S., and Dellaporta, S.L. (1987) EMBO J. 6, 295-302
27. Kunze, R., Starlinger, P., and Schwartz, D. (1988) Mol. Gen. Genet. 214, 325-327
28. Culley, D.E. (1986) VI International Congress on Plant Tissue and Cell Culture, Minneapolis, MN, Abstracts, p. 220
29. James, M.G. and Stadler, J. (1989) Theor. Appl. Genet. 77, 383-393
30. Planckaert, F and Walbot, V. (1989) Genetics 123, 567-578
31. Lee, M.L., Geadelmann, J.L. and Phillips, R.L. (1988) Theor. Appl. Genet. 75, 841-849
32. Phillips, R.L. (1989) Genome 31:1119-1120
33. Chandler, V.L. and Walbot, V. (1986) Proc. Nat. Acad. Sci. USA 83, 1767-1771
34. Bennetzen, J.L., Brown, W.E., and Springer, P.S. (1988) in Plant Transposable Elements (Nelson, O., ed.) pp 237-250. Plenum Press, New York
35. Walbot, V., Britt, A.B., Luehrsen, K., McLaughlin, M., and Warren, C. (1988) in Plant Transposable Elements (Nelson O., ed.), pp 121-135, Plenum Press, New

York
36. Dahleen, L. (1989) Ph.D. Thesis, Univ. Minnesota
37. Monk, M. (1990) Phil. Trans. R. Soc. Lond. B 326, 299-312
38. Meehan, R., Antequera, F., Lewis, J., MacLeod, D., McKay, S., Kleiner, E., and Bird, A.P. (1990) Phil. Trans. R. Soc. Lond. B. 326, 199-205
39. Keshet, I., Lieman-Hurwitz, J., and Cedar, H. (1986) Cell 44, 535-543
40. Anderson, P.C. and Georgeson, M. (1989) Genome 31, 994-999.
41. McCoy, T.J., Phillips, R.L., and Rines, H.W. (1982) Can. J. Genet. Cytol. 24, 37-50.
42. Benzion, G. and Phillips, R.L. (1988) Genome 30, 318-325
43. Lee, M. and Phillips, R.L. (1987) Genome 29:122-128
44. Rhodes, C.A., Phillips, R.L., and Green, C.E. (1986) Can. J. Genet. Cytol. 28, 374-384
45. Johnson, S.S., Phillips, R.L., and Rines, H.W. (1987) Genome 29, 439-446
46. Stelly, D.M., Altman, D.W., Kohel, R.J., Rangan, T.S., and Commiskey, E. (1989) Genome 32, 762-770
47. Riggs, A.D. (1990) Phil. Trans. R. Soc. Lond. B 326, 285-297
48. Holliday, R. (1987) Science 238, 163-170
49. Oono, K. (1985) Mol. Gen. Genet. 198, 377-384
50. Sano, H., Youssefian, S., Kamada, I., and Itoh, M. (1990) Second International Workshop on Mol. Biol. of Rice, Jan 8-9, Kawasaki Medical School, Okayama, Japan
51. Sano, H., Kamada, I., Youssefian, S., and Wabiko, H. (1989) Biochem. Biophys. Acta. 1009, 35-38
52. Chestnut, R.S., Shotwell, M.A., Boyer, S.K., and Larkins, B.A. (1989) The Plant Cell 1, 913-924
53. Brown, P.T.H. (1989) Genome 31, 717-729
54. Muller, E., Brown, P.T.H., Hartke, S., and Lorz, H. (1990) Abstr. VIIth Int. Congr. Plant Tissue and Cell Culture
55. Haynes, R.H. (1985) in Genetic Consequences of Nucleotide Pool Imbalance (de Serres, F.J., ed.) Vol 31, pp 1-23, Plenum, New York
56. Gierl, A., Lutticke, S., and Saedler H (1988) EMBO J. 7, 4045-4053
57. Banks, J.A., Masson, P, and Fedoroff, N. (1988) Genes Dev. 2, 1364-1380
58. LoSchiavo, F., Pitto, L., Giuliano, G., Torti, G., Nuti-Ronchi, V., Marazziti, D., Vergara, R., Orselli, S., and Terzi, M. (1989) Theor. Appl. Genet. 77, 325-331
59. Surani, M.A., Allen, N.D., Barton, S.C., Fundele, R., Howlett, S.K., Norris, M.L., and Reik, W. (1990) Phil. Trans. R. Soc. Lond. B 326, 313-327

GENE EXPRESSION IN SALT TOLERANT ALFALFA CELL CULTURES AND THE SALT TOLERANT PLANTS REGENERATED FROM THESE CULTURES.

Ilga Winicov, Departments of Biochemistry and Microbiology, University of Nevada, Reno, Reno, NV 89557, USA.

INTRODUCTION

Food demand for human and animal consumption is increasing with the worlds growing population. At the same time it is becoming increasingly difficult to satisfy these needs with existing crops and limited land, much of which is being eroded by salinization. Increased salt tolerance in crop plants therefore would provide yield increases in marginal lands and protect plants exposed to transient osmotic stress.

The primary characteristics associated with salt tolerance to date are: 1) NaCl transport; 2) Na^+ distribution; and 3) components at the cellular level. Among those three characteristics, cellular components appear to be the most amenable to study and manipulation at the present time. Plant cell culture has been used selectively to isolate salt tolerant cell lines in alfalfa that do not exclude NaCl [1,2]; however, regeneration of salt tolerant whole plants from the selected cells has been difficult [3]. In order to develop alternates to lengthy breeding strategies, we have chosen to identify functional gene systems which are activated in cellular salt tolerance (Winicov, unpublished) with the aim of targeting those genes that can confer acquired salt tolerance to crop plants. The molecular approach at the cell level has been coupled to plant regeneration from the tolerant cell cultures in order to assess the function of the activated gene systems at the whole plant level, where tissue specific control repertoires contribute to overall regulation of gene expression.

We describe in this report enhanced expression of several genes as measured by mRNA accumulation in the salt tolerant cells grown in presence of salt. We have also regenerated vigorous plants from the salt tolerant lines with increased salt tolerance under greenhouse conditions, thus allowing us to compare gene regulation in selected systems between cellular and whole plant responses.

PROCEDURE

Materials and Methods.

Callus cultures were initiated from immature ovaries [3] of diploid or tetraploid Medicago sativa L and salt tolerant lines were selected as described previously [2]. Callus was subcultured at four week intervals on control medium or medium containing 0.5% (86 mM) or 1% (171 mM) NaCl in continuous light. Experiments testing for light dependence of the salt tolerant phenotype were carried out in a light tight enclosure. Plants were regenerated on Shenk and Hildebrandt medium [4] without hormone supplements.

RNA purification and hybridization on Northern blots was carried out as described previously [2] using 10 ug total RNA per lane. The alfalfa cDNA clones were isolated by differential screening from a lambda ZAP library constructed from poly A^+ RNA from salt tolerant alfalfa cells grown in presence of 1% NaCl. Pea chloroplast rbcL probe (Rubisco large subunit) was provided by Dr. J.D. Palmer [5].

RESULTS AND DISCUSSION

Experiments in my laboratory have shown that 1% NaCl induces high levels of mRNA accumulation of the chloroplast encoded psbA, psbD, and rbcL, as well as nuclear genes pcab4 and rbcS in the tolerant alfalfa line HG2-N1, relative to both stressed and unstressed salt sensitive HG2 cells (Winicov and Button, unpublished). Since plant cells in culture rarely show a significant level of photosynthetic competence, we wished to test the importance of this induction of photosynthesis specific gene expression for the salt tolerance phenotype in this cell line. We therefore tested the effect of light and Photosystem II inhibitors on the growth capacity of both sensitive and tolerant cell lines, with special emphasis on testing the effect of these inhibitors in salt stress growth conditions for the tolerant line. These results are shown in Table 1.

TABLE 1. Inhibition of growth by PSII electron transport inhibitors in the salt tolerant cells (HG2-N1) ± 1% NaCl but not salt sensitive HG2.

Inhibitor	HG2 - NaCl % wt.	HG2-N1 - NaCl % wt.	HG2-N1 + NaCl % wt.
Control	100	100	100
Cyanazine	92	83	69
DCMU (10 uM)	89	87	33

Growth of the salt sensitive cell line HG2 or salt tolerant HG2-N1 in absence of salt, but in the presence of either 10 uM DCMU or cyanazine, both specific PSII electron transport inhibitors [6], was only marginally decreased by the inhibitors. In contrast, growth of the tolerant line HG2-N1 was substantially inhibited by both cyanazine and DCMU on salt containing medium, indicating that photosynthesis may be important to the growth and/or salt tolerance capability of these cells under stress conditions.

These conclusions are strengthened by the results of measurements to determine the salt tolerant (HG2-N1) cell capacity to grow in presence of salt, but absence of light shown in Table 2. These results indicate a light requirement for induction and maintenance of the salt tolerant phenotype of HG2-N1.

TABLE 2. Effect of light on the salt tolerance phenotype by salt tolerant HG2-N1.

	Percent growth (wt.) +/- 1% NaCl	
Light/Dark	- NaCl	+ NaCl
4 wks. Light	100	100
4 wks. Dark	100	31

Our mRNA measurements indicated that salt stimulated accumulation of specific mRNAs in the tolerant HG2-N1 alfalfa cells correlated with the salt tolerant phenotype. In addition, inhibition of photosynthesis related gene expression appears to correlate with the salt tolerant phenotype as shown in data from Tables 1 and 2. Furthermore, these data are also consistent with the observation of light dependent accumulation of mRNA for the Rubisco holoenzyme in the salt tolerant line HG2-N1 shown for the transcripts of rbcL (Fig. 1) and similar data for rbcS (not shown).

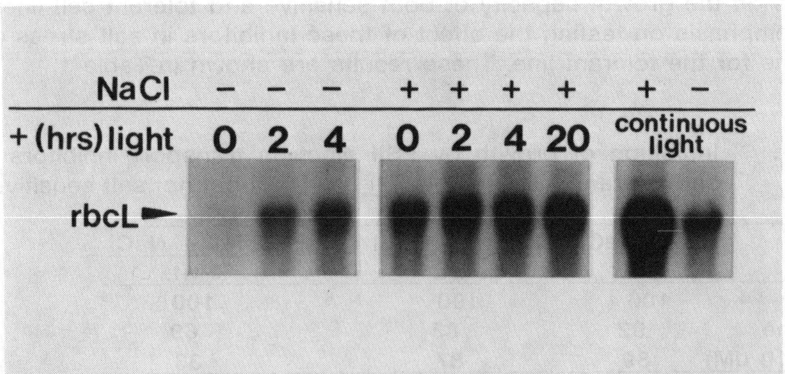

FIGURE 1. Northern blot analysis of rbcL mRNA concentrations in tolerant cells grown ± 1% NaCl and ± light for 4 wks and mRNA induction after transfer to light.

These results may be due to either transcriptional activation or enhanced mRNA stability [7,8], but the results in Fig. 1 indicate that NaCl may substitute to some extent for light in mRNA accumulation for these two gene transcripts.

In addition to photosynthesis related gene induction by salt in the tolerant line we have also isolated several cDNA clones from other mRNA species that appear to accumulate in the tolerant line in response to NaCl. Identification of the mRNA species for three of these clones is shown by Northern blot analysis in Figure 2. In each case we see the enhanced accumulation of a single mRNA species that is present after long term growth on 1% NaCl, but not on control medium in the tolerant HG2-N1 cells. These mRNA species do not appear to represent the expression of genes induced in acute stress, as shown by their absence in control lanes representing mRNA from the salt sensitive parent line after a 24 hr. salt stress. These results are consistent with our previous results from comparisons of protein analyses from salt regulated gene expression in tolerance and acute stress [2] and indicate that adaptive mechanisms of salt tolerance may differ in some aspects from those utilized in acute stress.

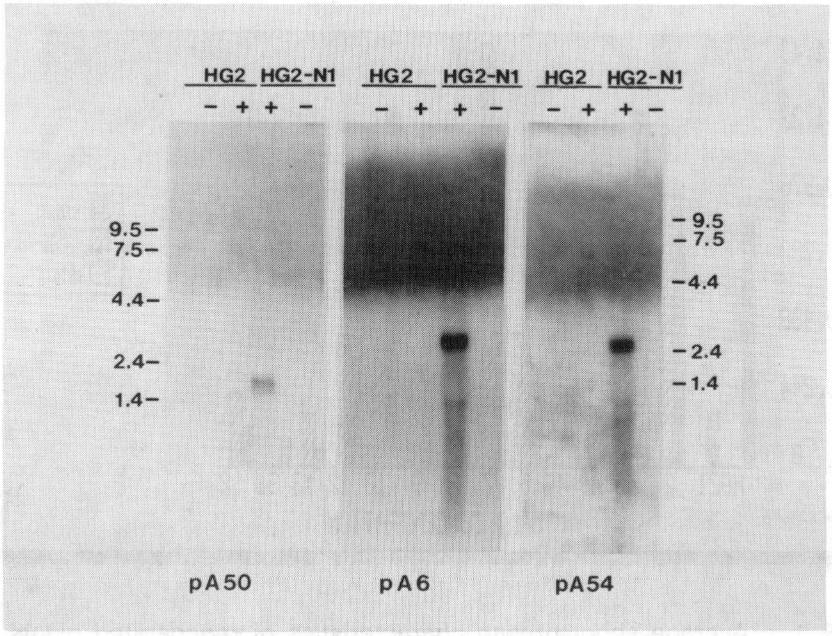

FIGURE 2. mRNA species accumulated in the tolerant line in salt as detected by three different cDNA clones.

We have identified several specific gene products that appear induced in the salt tolerant cells in response to salt. To extend these observations to the whole plant level and to assess the tissue specificity of this activation phenomenon, plants have been regenerated from several salt tolerant alfalfa lines. The regenerated plants have been tested in the greenhouse to determine their salt tolerance compared to parent diploid and tetraploid plants HG2 and #1 respectively as shown in Figure 3. Both HG2 and #1 plants did not survive growth for 4 weeks on 0.5 or 1% NaCl. The regenerated plants #2 through #32 showed different survival characteristics ranging from 0 to 80% at 1% NaCl and 20 to 100% survival at 0.5% NaCl. At least 5 rooted cuttings represent each plant and growth condition. The data in Figure 3 show the average growth characteristics of each plant at each salt concentration, indicating that the regenerated plants have acquired an improved capability to survive and grow in presence of NaCl as compared to the parent plants, from which they were derived.

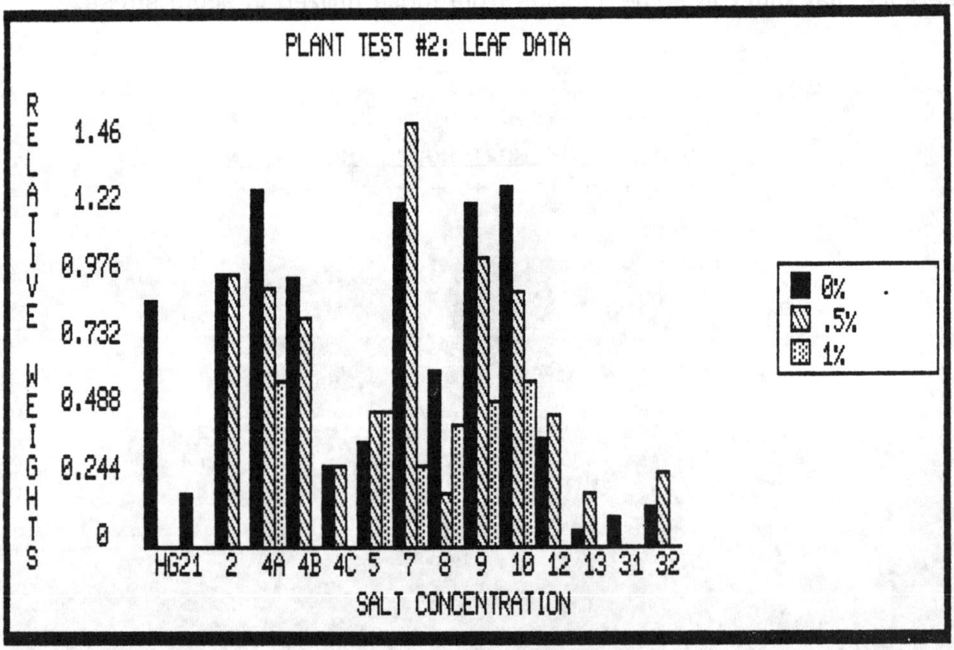

FIGURE 3. Average shoot growth characteristics of regenerated plants and parent plants after 4 weeks growth on the indicated salt concentration.

The majority of the regenerated 12 plants appear normal, have flowered, are self fertile and yield salt tolerant callus initiated from immature ovaries without additional selection on NaCl. These results indicate that the salt tolerance characteristic is stable in the plants and relates back to the cellular phenotype. Thus, activation of a subset of the physiological systems important in cellular metabolism may provide protection from stress and lead to incremental improvements in cellular and whole plant salt tolerance. Contributions of more than one physiological system to the tolerance phenotype could provide additive incremental improvements to tolerance. The regenerated plants will be instrumental in identification of cell and whole plant specificity of the genes activated in our salt tolerant cell lines.

REFERENCES

1 Stavarek, S.J. and Rains, D.W. (1984) in: Salinity Tolerance in Plants (Staples, R.C., Toenniessen, G.H. eds.) pp 321-334, Wiley-Interscience, N.Y.

2 Winicov, I., Waterborg, J.H., Harrington, R.E. and McCoy, T.J. (1989) Plant Cell Reports 8, 6-11

3 McCoy, T.J. (1987) Plant Cell Reports 6, 417-422

4 Shenk, R.U. and Hildebrandt, A.C. (1972) Can. J. Bot. 50, 199-204

5 Jensen, R.K. and Palmer, J.D. (1987) Current Genetics 11, 553-564

6 Izawa, S. (1980) Methods in Enzymol. 69, 413-434

7 Deng, X.W. and Gruissem, W. (1987) Cell 49, 379-387

8 Klein, R.R., Mason, H.S. and Mullet, J.E. (1988) J. Cell Biol. 106, 289-301

GLYPHOSATE RESISTANCE AND GENE AMPLIFICATION IN SELECTED *Daucus carota* SUSPENSION CULTURES

Y. Y. Shyr and J. M. Widholm, Department of Agronomy, University of Illinois, 1102 S. Goodwin, Urbana, IL 61801, USA

1. INTRODUCTION

There are many studies describing selected gene amplification with both animal cell cultures and microbial systems (1). However, in the case of plants, there are only two selection agents, phosphinothricin and glyphosate, both of which are nonselective herbicides, which have been shown to produce gene amplification when used in selection experiments.

Phosphinothricin is toxic to plants since it inhibits glutamine synthase activity. Stepwise selection of alfalfa cells with phosphinothricin produced cultures more resistant to the herbicide (10 to 100-fold) which also had elevated levels of the target enzyme glutamine synthase (3 to 7-fold), as well as an increased number of one of the glutamine synthase genes (4 to 11-fold) (2). The glutamine synthase mRNA levels were also elevated about 8-fold. The amplified alfalfa glutamine synthase gene has been cloned and sequenced (3). Southern hybridization results indicate that one glutamine synthase gene of the small family of such genes found in alfalfa is amplified, but is not changed noticeably in sequence. The minimum size of the amplified DNA unit is about 35 kb.

Glyphosate is a potent inhibitor of the shikimate pathway enzyme, 5-enolpyruvylshikimate-3-phosphate synthase (EPSPS) (4), which causes a very large buildup of shikimate (up to 10% of the dry weight) in plants and cultured cells (5). While the addition of the aromatic amino acids, phenylalanine, tyrosine and tryptophan can reverse the glyphosate toxicity to cultured cells, the decreases in the free levels of these amino acids would not seem to be sufficient to cause cell death (6). A more likely possibility for the actual cause of death is the draining of TCA cycle intermediates into shikimate since certain TCA cycle intermediates and related amino acids can reverse glyphosate toxicity (7).

The first selected resistance to glyphosate was accomplished with the bacterium, *Aerobacter aerogenes* and a higher plant suspension culture of *Corydalis sempervirens* (8). The bacteria were selected on 5mM glyphosate while the plant cells were selected stepwise with 0.5mM to 10mM glyphosate over a 9 month period. The EPSPS activities were increased about 10 and 33-fold in the resistant bacteria and C. sempervirens cells, respectively. Further studies showed that the normal herbicide-sensitive EPSPS form was increased in activity (9) and that the increase in synthesis was not due to increased EPSPS gene copy number but possibly due to increased gene transcription and mRNA stability (10).

We used stepwise selection, from 0.25 to 25 mM glyphosate, with the C1 Daucus carota cv. Danvers half long suspension culture to obtain a strain, denoted PR, which was 52-fold more tolerant to glyphosate than the C1 cells and which had 12-times the C1 level of EPSPS (11). Phosphatase, anthranilate synthase and shikimate dehydrogenase activities were not very different in the C1 and PR cells. The glyphosate inhibition curves of EPSPS from the two lines were identical indicating that the normal enzyme and not a different one composed the additional activity.

The PR glyphosate resistance was expressed in protoplast fusion hybrids with a Nicotiana plumbaginifolia universal hybridizer (12). Southern hybridization using a cloned portion of the petunia EPSPS gene showed that the PR cells and the fusion hybrids had more copies of the EPSPS gene sequences than the wild type. The PR cells were also used in protoplast fusion experiments with an amino acid analog resistant carrot strain and here again the glyphosate resistance was expressed in the selected fusion hybrids (13). The PR strain contained 5.5-times higher levels of EPSPS activity, 8.7-times higher levels of the EPSPS protein (determined by Western hybridization) and a corresponding increase in EPSPS gene copy number (determined by Southern hybridization) than the C1 wild type line.

A Petunia hybrida suspension culture was selected stepwise on glyphosate concentrations from 0.5 to 10mM over a 4 month period (14). The strain growing on 10mM glyphosate contained from 15 to 20-fold higher EPSPS levels. The amino terminus of the P. hybrida EPSPS protein was partially sequenced and olignucleoides corresponding to the codons were used to isolate cDNA clones (15). These cDNA clones were used in Northern and Southern hybridizations to show about 20-fold increases in both EPSPS mRNA and gene copies. A full length EPSPS cDNA clone driven by a CaMV 35S promoter was used to transform P. hybrida leaf discs. The transformed callus as well as regenerated plants exhibited glyphosate resistance.

Gamma-irradiated hybrid Lycopersicon esculentum X L. peruvianum suspension cultured cells were plated on 3mM glyphosate and one strain was obtained which was 100-times more resistant to glyphosate and which had 8 to 13-times the wild type EPSPS activity (16).

Single-step selection on solid medium containing 3mM glyphosate was also used to select Nicotiana tabacum cells with about 10-fold more resistance and 5 and 2-times the wild type 3-deoxy-D-arabino-heptulosonate-7-phosphate synthase and EPSPS activities, respectively

(17). Plants were regenerated which also expressed some glyphosate tolerance.

Several of the glyphosate resistance studies with plant cell cultures summarized above show that other shikimate pathway enzymes are not increased in activity, that glyphosate uptake is not decreased in the selected strains and the resistance is usually stable for at least one year.

2. PROCEDURES

Suspension cultures of <u>Daucus carota</u> L. var. Danvers half long, initiated in 1967 (denoted C1) and var. Takii, initiated in 1981 by A. H. Huang (denoted CAR) were selected stepwise for glyphosate resistance by inoculating 1 g fresh weight cells into 50 ml MS medium with 1.81 μM 2,4-D and partially inhibitory levels of glyphosate. The cultures were incubated on a gyratory shaker at 27 to 28°C.

The EPSPS activity was measured indirectly by measuring the quantity of the substrate PEP remaining after a 20 min. incubation at 37°C using a pyruvate kinase and lactate dehydrogenase coupled reaction (17). Protein was measured with the Bradford reagent and the specific activity calculated.

The carrot cell EPSPS gene copy number was estimated by Southern hybridization with EcoRI treated genomic DNA using ^{32}P oligo labeled pMON 9543 as probe. This probe is a 900 bp cDNA clone of petunia EPSPS in an EcoRI site of pUC9 (15). The same probe was also used to measure EPSPS mRNA content in total extracted RNA by Northern hybridization.

Four ml of cells (5 mg fresh weight ml^{-1}) in warm (40-45°C) 0.8% agar medium were plated on top of 16 ml of the same medium in a 100 x 15 mm Petri dish. Both medium layers contained glyphosate at various concentrations.

3. RESULTS AND DISCUSSION

Stepwise selection, beginning with 0.5 mM glyphosate, with the two carrot suspension cultures, produced strains growing on 35 mM levels after 10 to 13 steps. The increase in the estimated I_{50} concentrations were at least 70-fold. The EPSPS activity levels rose gradually during the selection process from about 70 nmol min^{-1} mg^{-1} protein in the wild type lines to about 1500 in the strains on 35 mM glyphosate. The EPSPS activity of the previously selected strain, PR, was near 760 nmol min^{-1} mg^{-1} protein by this assay. Thus the increases in EPSPS activity in the strains selected in this study were about 22-fold and that of the PR cells about half that.

Southern hybridization using a petunia EPSPS cDNA clone as probe showed that EcoRI digested genomic DNA from the selected strains contained hybridizing DNA fragments which increased in amount as the resistance increased. The maximal increases in comparison with the parental C1 and CAR cells were 26 and 34-fold, respectively, as estimated by densitometer analysis. The sizes of the amplified fragments in the Cl-derived strains were 10.5 and 5.2 kb and those in the CAR strains, 10.3, 6.6 and 4.8 kb. The sizes of the fragments in the selected strains are similar to those of the parental lines indicating that the EPSPS gene was not altered detectably during amplification.

Northern blot hybridization analysis of RNA from the selected strains also showed an increase in the amount of the EPSPS mRNA as the resistance increases. Thus there appears to be a close correspondence between EPSPS gene copy number, mRNA levels and EPSPS activity.

Glyphosate resistant cells were never obtained in liquid medium when selection was initiated with glyphosate concentrations of 10mM or higher. Likewise the wild type Cl and CAR cells did not form colonies within 45 d when plated in glyphosate concentrations of 1mM or higher while the PR and the selected strains from 35mM glyphosate grew quite well on 10mM glyphosate. A few colonies did form with 2mM glyphosate from Cl and CAR cells after several months and these putative resistant clones, when grown up without glyphosate and then retested, were clearly more resistant than the wild types but were less resistant than the 35mM strains. The 2mM selected clones also usually had increased EPSPS activity (2 to 4-fold) and increased EPSPS gene copy number. These results indicate that the wild type cell populations do not contain cells which are highly amplified but do have some that are amplified several-fold.

Selection in liquid medium with 2mM glyphosate showed that most of the cells died, but 30% remained viable by day 30 and then measureable cell growth began. This would indicate that a low % of the cell population carried resistance so were able to grow under these conditions. The proportion of the cells capable of growth was probably higher than that usually found for spontaneous mutations (about 10^{-6}) but it is difficult to calculate since the doubling times under the selective conditions are not known and there may well be continuous changes in the resistance occurring during the growth of the cells.

The glyphosate resistance is relatively stable as shown by the 2mM plating selection where the selected clones retain resistance, the increased gene copy number and the enzyme activity after being grown away from glyphosate for several months during the cell increase. Likewise a line denoted PRB, which is the PR strain grown without glyphosate for 2 years, showed nearly complete resistance and 65% of the PR EPSPS activity. Another strain denoted PRA, which was PR grown away from glyphosate for 5 years, had resistance and EPSPS levels similar to that of the wild type.

The reproducibility of the selection system was demonstrated by an additional selection experiment where Cl cells were inoculated into seven flasks of 0.3 or 0.6mM glyphosate and after eight serial transfers to higher glyphosate concentrations, all seven selected strains were growing on 35mM glyphosate. Enzyme and gene copy levels were not determined with these strains, however.

4. CONCLUSION

The carrot suspension cultures are reproducible systems for the study of gene amplification where glyphosate resistance, EPSPS gene copy number, mRNA level and enzyme activity increase coordinately. These increases are relatively stable in the absence of selection. Stepwise selection appears to select a small proportion of the wild type population which has a small increase (2 to 4-fold) in EPSPS activity. Additional selection with a slightly higher glyphosate concentration then selects for cells with higher amplification, and so on, until a 25-fold increase has occurred. The slow amplification rate and final increases found with the plant systems are quite different from that usually seen with animal cell systems indicating that the amplification mechanisms may be different.

REFERENCES

1. Stark, G.R. and Wahl, G.M. (1984) Ann. Rev. Biochem. 53, 447-491
2. Donn, G., Tischer, E., Smith, J.A. and Goodman, H.M. (1984) J.Mol. Appl. Genet. 2, 621-635
3. Tischer, E., DasSarma, S. and Goodman, H.M. (1986) Mol. Gen. Genet. 203, 221-229
4. Steinrucken, H.C. and Amrhein, N. (1980) Biochem. Biophys. Res. Commun. 94, 1207-1212
5. Amrhein, N., Deus, B., Gehrke, P. and Steinrucken, H.C. (1980) Plant Physiol. 66, 830-834
6. Haderlie, L.C., Widholm, J.M. and Slife, F.W. (1977) Plant Physiol. 60, 40-43
7. Killmer, J.L., Widholm, J.M. and Slife, F.W. (1981) Plant Physiol. 68, 1299-1302
8. Amrhein, N., Johanning, D., Schab, J. and Schulz, A. (1983) FEBS Let. 157, 191-196
9. Smart, C.C., Johanning, D., Müller, G. and Amrhein, N. (1985) J. Biol. Chem. 260, 16, 338-16, 346
10. Hollander-Czytko, H., Johanning, D., Meyer, H.E. and Amrhein, N. (1988) Plant Mol. Biol. 11, 215-220
11. Nafziger, E.D., Widholm, J.M., Steinrücken, H.C. and Killmer, J.L. (1984) Plant Physiol. 76, 571-574
12. Ye, J., Hauptmann, R.M., Smith, A.G. and Widholm, J.M. (1987) Mol. Gen. Genet. 208, 474-480
13. Hauptmann, R.M., della-Cioppa, G., Smith, A.G., Kishore, G.M. and Widholm, J.M. (1987) Mol. Gen. Genet. 211, 357-363
14. Steinrücken, H.C., Schulz, A., Amrhein, N., Porter, C.A. and Fraley, R.T. (1986) Arch. Biochem. Biophys. 244, 169-178
15. Shah, D.M. et al. (1986) Sci. 233, 478-481
16. Smith, C.M., Pratt, D. and Thompson, G.A. (1986) Plant Cell Rept. 5, 298-301
17. Dyer, W.E., Weller, S.C., Bressan, R.A. and Herrmann, K.M. (1988) Plant Physiol. 88, 661-666

RFLP-ANALYSIS OF RICE PLANTS REGENERATED FROM TISSUE CULTURES

E.MÜLLER, P.T.H.BROWN, S.HARTKE AND H.LÖRZ
UNIVERSITÄT HAMBURG
INSTITUT FÜR ALLGEMEINE BOTANIK
OHNHORSTSTR. 18
D-2000 HAMBURG 52, FRG

1. INTRODUCTION

The regeneration of plants from tissue cultures should result in individuals which are phenotypically and genetically identical to the explant material they are derived from. It is now known that in a proportion of tissue culture derived plants, variability can be found. This phenomenon has been called somaclonal variation (Larkin and Scowcroft, 1981).In rice, somaclonal variation has previously been decribed by Oono (1981), Sun et al. and by Ogura et al. (1987). They decribed phenotypic and cytological alterations amongst tissue culture derived rice plants, which were regenerated from both callus and protoplast cultures. To extent this phenotypic and cytological analysis.and to determine whether somaclonal variation could be detected at the molecular level and whether it could be related to specific aspects of the tissue culture process, an RFLP-analysis of regenerated Oryza sativa L. (ssp. Indica) plants was undertaken.

2. MATERIAL AND METHODS

Rice varieties were kindly provided by Dr. Kush, IRRI, Phillipines. Immature embryos were used as explants to establish tissue cultures according to the protocols as decribed in Hartke and Lörz, 1989. Methods for DNA isolation and RFLP-analysis were as described in Brown, 1989.

3. RESULTS AND DISCUSSION

In the first eperiments a number of rice lines were used to determine which restriction enzyme / gene probe combinations were most suitable for RFLP-analysis. Southern analysis based on Hind III digestion of genomic DNA and subsequent hybridization with a soybean actin gene (Shah et al. 1982) showed that this is the most suitable combination, with the ability to distinguish between different rice genotypes. Control plants were seed grown and were not subjected to tissue culture. Subsequent molecular analyses of genomic DNA from control plants revealed no heterogenity among them.

Experiments were started using direct regenerants (R0-generation) from the Indica type rice IR40. Preliminary results of the RFLP-analysis demonstrated that that these regenerants exibit DNA polymorphms regardless of whether the plants had a normal or a deviant phenotype. Variability was found to occur amongst plants from different calli as well as amongst plants which were regenerated from the same callus.

To determine the basis for this variation, we investigated plants from two different tissue culture experiments. First, plantlets were regenerated from a single immature embryo-derived callus via somatic embryogenesis after a callus induction phase of 28 days on 2.4 D containing media, followed by subculture on regeneration media (CCZI, Hartke and Lörz, 1989). In the second experiment, regenerants were obtained from two different calli after 67 days in callus induction phase and subsequently regenerated on CC-based media. A representative analyses (Figure 1) of regenerants from these two culture periods showed that molecular variability was mainly observed among rice plants from long term culture.

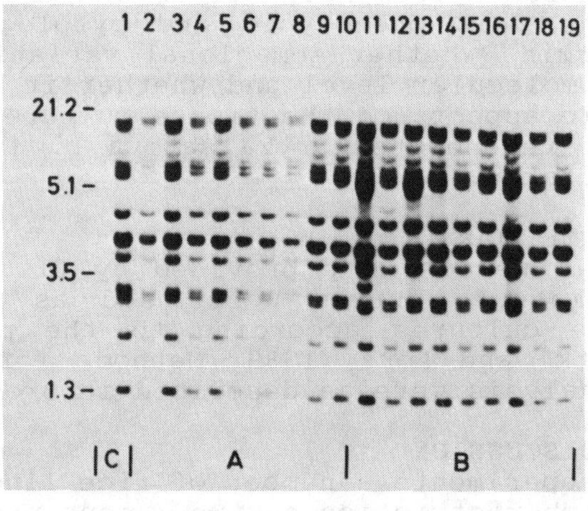

FIGURE 1
RFLP-analysis of regenerated IR40 rice plants. Comparison between regenerants after 28 days of callus induction(A) and 67 days of callus induction (B). Digestion of genomic DNA with Hind III; probe: actin gene from soybean. C represents control plant DNA.

The difference in genetic stability between regenerants from these two culture periods shown by RFLP-analysis suggests that the length of time may be one of the major factors for the induction of molecular and somaclonal variation. Of all the rice plants subjected to RFLP-analysis, approximately 19% showed DNA-variation, compared to less than 1% showing phenotypic variation such as dwarfism and albinism (data not shown). However, 23% of plants regenerated from long-term cultures showed RFLPs compared to only 6.3% of the plants regenerated from short induction cultures (Table 1). Results also confirm that no correlation can be made between those plants showing phenotypic alterations and the level of DNA polymorphisms detected with RFLP-analysis. The overall percentage of rice regenerants exibiting DNA alterations was considerably lower than that found in maize (Brown 1989).

TABLE

IR 40 regenerants: Time in culture on 2,4 D containing media and variation amongst regenerants.

Time in culture on callus induction media	Total number of regenerants	Number of molecular variants	No variation detectable
28 days	32 (100%)	2 (6.3%)	30 (93.7%)
67 days	135 (100%)	31 (23.0%)	104 (77.0%)

Detailed analysis of rice regenerated from long term callus cultures included the digestion of genomic DNA with different restriction enzymes and susbsequent hybridization with different gene probes (including the ATP/ADP-translocator and waxy gene from maize). The results showed that the level of polymorphism and the type of DNA changes resulting from this extended callus growth period were considerable. The form of polymorphism can vary extensively. Differences in hybridization pattern were found to occur not only between plants regenerated from different calli but also among regenerants which derived from a single callus.
In conclusion therefore, tissue culture and in particular the length of time in which the material is maintained in the stage of undifferentiated callus seems to be an important factor in the degree of DNA variation subsequently found in regenerated plants.

References

Brown, P.T.H. (1989)
DNA methylation in plants and its role in tissue culture.
Genome 31: 717-729.

Hartke, S.; Lörz, H. (1989)
Tissue culture of Indica rice (Oryza sativa L. ssp. Indica).
J. Genetics and Breeding, in press.

Larkin, P.J.; Scowcroft, W.R. (1981)
Somaclonal variation - a novel source of variability from cell cultures for plant improvement.
Theor. Appl. Genet. 60: 197-214

Müller, E.; Brown, P.T.H.; Hartke, S.; Lörz, H. (1990)
DNA variation in tissue culture-derived rice plants.
Theor. Appl. Genet., in press.

Ogura, H.; Kyozuka, J.; Hayashi, Y.; Koba, T.; Shimamoto, K. (1987)
Field performance and cytology of protoplast derived rice (Oryza sativa): high yield and low degree of variation in four Japonica cultivars.
Theor. Appl. Genet. 87: 670-676.

Shah, D.M.; Hightower, R.C.; Meager, R.B. (1982)
Complete nucleotide sequence of a soybean actin gene.
Proc. Natl. Acad. Sci. 79: 1022-1026.

Sun, Z.X.; Zhao, C.Z.; Zheng, K.L.; Qi, X.F.; Fu, Y.P. (1983)
Somaclonal genetics of rice.
Theor. Appl. Genet. 67: 67-73.

IN VIVO AND *IN VITRO* MUTAGENESIS OF A *BRASSICA NAPUS* ACETOLACTATE SYNTHASE GENE FOR THE PRODUCTION OF HERBICIDE RESISTANCE

Paul A. WIERSMA, D.L. DURDA, I. BABIC, W.L. CROSBY[1], M.M. MOLONEY Dept. Biological Sciences, University of Calgary, 2500 University Dr. N.W., Calgary, AB, Canada T2N 1N4 and 1) Molecular Genetics Section, National Research Council Canada, Plant Biotechnology Institute, 110 Gymnasium Road, Saskatoon, SK, Canada S7N 0W9

INTRODUCTION

Herbicide resistance in plants may occur as a result of point mutations in the gene encoding the herbicide target. Such mutations have been well characterized for resistance to herbicides which affect photosynthesis (1) and amino-acid metabolism (2,3). A better understanding of how these mutant enzymes interact with various herbicides will aid the development of new herbicide/plant systems. The enzyme acetolactate synthase (ALS (or AHAS)) has been the focus of several studies recently because it is the target for three classes of herbicides: sulfonylureas (4,5), imidazolinones (6) and triazolopyrimidines (7). Resistance to these herbicides has been produced, by mutagenesis of seeds or cultured tissure, for a number of plant species including: tobacco, *Arabidopsis thaliana*, oilseed rape, and sugar beet. Determining the structural basis of these mutations becomes more complex as crop species with multiple copies of ALS are used. For example, *Brassica napus* (oilseed rape) has 4 to 5 genes for ALS (8) and appears to regulate these sequences differentially. Characterization of new mutations requires isolation and sequencing of all ALS genes in that plant and then demonstration that the specific mutations found can cause a resistant phenotype upon re-introduction into plants. Since it is necessary to re-introduce these *in vitro* mutated genes into a suitable host to determine their phenotype, we sought to develop a bacterial expression system. This would enable us to take advantage of the convenience and simplicity of prokaryotic genetics and transformation. Expression of plant sequences in bacteria can be used in a number of ways, including: indentification of the function of unknown cDNA clones by complementing bacterial mutants; allowing production of plant enzymes and proteins which might be unstable or of very low abundance in the plant; and development of a system for mutagenesis where resultant clones can be rapidly analysed We describe here a method for efficiently expressing in bacteria a nuclear gene encoding an enzyme that is normally targetted to the plant chloroplast. The following steps describe a procedure for production of a herbicide resistant crop. It is first necessary to isolate a wild-type herbicide-sensitive ALS gene and modify it by truncation to allow expression in bacteria. Mutations are introduced into the gene by *in vitro* mutagenesis, making changes that are known from other genes to produce the desired phenotype, or by *in vivo* mutagenesis of bacteria that harbor the plant gene. The effects of these mutations can be evaluated initially in the bacterial host. The mutated ALS genes are re-introduced into the best cultivars of various species using native or enhanced plant promoters for expression of herbicide resistance.

RESULTS AND DISCUSSION

***Characteristics of a genomic ALS gene from* Brassica napus.** Isolation of a 3.3 kb clone (pPAW1) from *B. napus* has been described previously (8). An uninterrupted coding region (no introns) encodes 637 amino acids with a strong similarity along most of its length to published sequences from other species. The N-terminal sequence however shows no similarity to the others which is consistent with this being a chloroplast transit peptide as discussed below.

Truncation of a chloroplast transit peptide allows efficient ALS complementation in bacteria. *Salmonella typhimurium* strain DU2603 (9) was selected as the host strain for the experiments described here. This strain has mutations in both genes encoding the large subunits of ALS and consequently shows no ALS activity. We sought to render this strain prototrophic by the use of plant ALS sequences cloned into the bacterial expression plasmid pKK233-2 (10). This plasmid was chosen for its strong chimaeric promoter trc (trp-lac) and a NcoI site which not only contains the translational start codon (ATG) in the optimal context for bacterial expression, but also matches the restriction site enclosing the first codon on the *B. napus* ALS gene used in these studies.

The coding regions of all the plant genes sequenced to date have 60 to 90 amino acids at their N-termini which are not represented in the bacterial sequences. These probably encode a chloroplast transit peptide which would be cleaved from the polypeptide to produce the functional protein. We therefore tested the idea that a functional enzyme could be produced in bacteria by expressing the plant ALS coding region with part or all of this transit peptide already removed. To do this a plasmid containing the plant gene was linearized with Pst I and Hind III, digested with exonuclease III and mung bean nuclease and the plasmid recircularized with a NcoI linker. This produced a library of randomly sized truncations into the transit peptide sequence with a NcoI site to give a new start of translation and an efficient means of cloning into the bacterial expression vector. Members of this library were chosen at random and screened by DNA sequencing to identify a number of truncations which spanned the presumed processing site and which reproduced the correct reading frame for translation after the new ATG start codon. Several clones were selected for further complementation experiments and were assembled into the bacterial expression vector (Fig. 1).

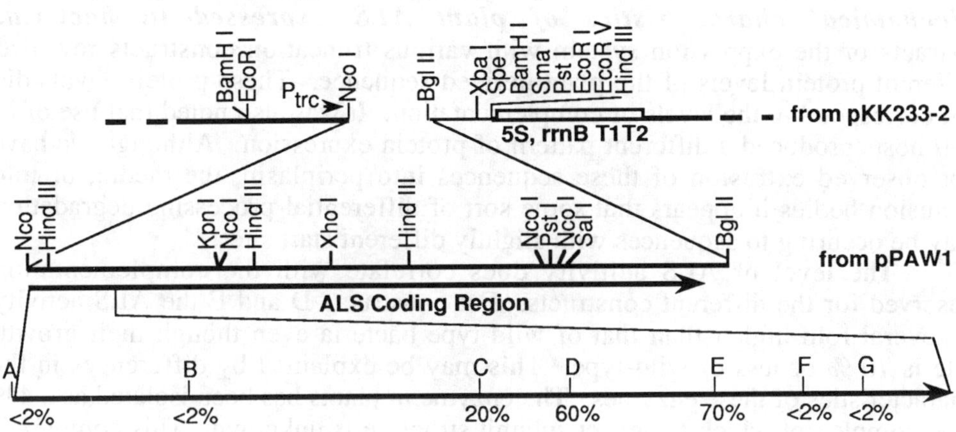

Figure 1 Construction of bacterial expression plasmids of the B. napus ALS coding region. The coding region fragment of ALS from pPAW1 was inserted as a transcriptional fusion behind the strong trc promoter of pKK233-2. The capital letters above the bottom line designate the relative start sites of the complete coding region (A) and a number of truncation mutants (B-G). The values beneath the line indicate the growth rate, relative to wild-type bacteria, of ALS-deficient bacteria containing the construct of that size.

The presence of the complete transit peptide is inhibitory to expression of plant ALS in bacteria. This may be due to normal interaction of this amphihelical sequence with membrane and in this case with the bacterial membrane. If trapped on the membrane the protein may be more susceptible to degradation or it may be unable to assemble into an active multimeric form as is normally found in both bacterial and plant ALS. The inhibition of assembly could also be caused by the presence of extra amino acids in the soluble form of the enzyme, although a precise N-terminus does not appear to be required, as a window of approximately 20-25 amino acids including the start sites of constructs D and E can give efficient complementation. Addition of further amino acids decreases the functioning of the protein without eliminating it completely until past the site of C. The longer N-terminus may also interfere with the normal folding of the monomer or may block interaction of the protein with substrates or coenzymes.

This method should be generally applicable to other proteins with transit or signal peptides for which the precise processing site is not known. It would be useful to know the specificity of the processing enzymes in plants so that an accurate prediction of the mature protein could be made for this type of heterologous expression experiment.

Biochemical characteristics of plant ALS expressed in bacteria. Extracts of the expression system with various truncation constructs revealed different protein levels of the plant derived sequence. These protein levels did not correspond to the levels of complementation. It was also noted that use of *E. coli* hosts produced a different pattern of protein expression. Although we have not observed extrusion of these sequences into periplasm, the media, or into inclusion bodies it appears that some sort of differential processing/degradation may be occuring to sequences with slightly different start sites.

The level of ALS acitivity does correlate with the complementation observed for the different constructs. For constructs D and E the ALS activity is several fold higher than that of wild-type bacteria even though their growth rate is 70 % or less of wild-type. This may be explained by differences in the characteristics of these enzymes. The enzyme in plants has been isolated as a 440 kDa complex of which the exact subunit structure is unknown. This complex is feedback regulated by valine. The active plant gene product expressed in bacteria migrates at approximately 120 to 140 kDa and is not inhibited by valine. This is similar in size to the bacterial enzyme which has two large subunits (equivalent to the plant gene used here) and two small subunits which are thought to be involved in feedback regulation (11). It is uncertain if the plant product interacts with the bacterial small subunit to form the observed active complex but genetic and biochemical evaluation of this possiblity are underway. Although large amounts of protein are produced from some constructs, it appears that only a fraction of this is active enzyme as the bulk of protein does not copurify with the activity. This is consistent with the requirement of interaction with a limiting bacterial component, or with a very low probability of assembly without additional plant factors, or with a bacterial processing event that either inactivates the bulk of protein or activates a very low level of the protein. Purification and microsequencing of the active versus inactive peptides can answer this question.

Herbicide resistance of plant ALS mutants expressed in bacteria. A well documented mutation which gives chlorsulfuron resistance in plant ALS genes (Pro to Ser at amino acid 173 in pPAW1) (8) was introduced into the truncated *B. napus* ALS gene by *in vitro* mutagenesis to test its efficacy in bacterially expressed plant ALS. As can be seen in the disc diffusion assay of Fig. 2 the bacteria expressing this construct ("173S") does not show the inhibition of the wild-type plant gene ("68", which is equivalent to construct E). It is also possible to detect cross-resistance to different classes of herbicides caused by the mutation as observed here for chlorsulfuron and two imidazolinones. A more quantitative evaluation of these data is seen in the

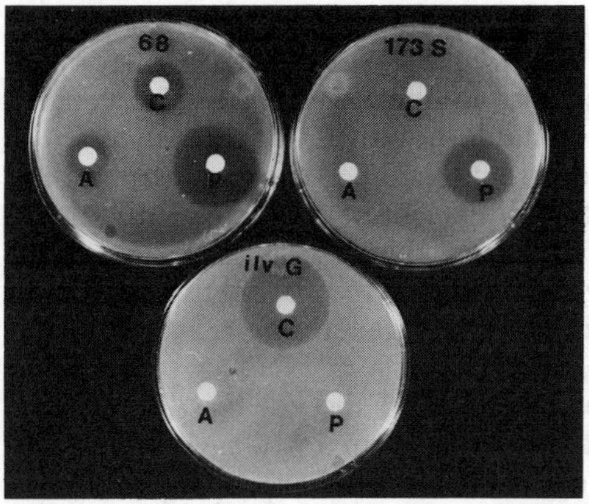

Figure 2. Disc diffusion assay for herbicide inhibition. 100 µl aliquots of overnight cultures of host bacteria containing the construct "E" (=68), the resistant mutant "173S", and the bacterial large subunit gene (*ilv* G) were poured in top agar onto minimal media plates containing pantothenate and ampicillin. Sterile 6 mm paper discs containing herbicides were applied to the surface and the plates incubated overnight. C, 40 µg chlorsulfuron; A, 2 mg Assert; P, 40 µg Pursuit.

Table 1. Inhibition curve analysis of ALS activity in enzyme extracts

	I_{50}, µM[a]		
Source	CS	Pursuit	Assert
E	20	30	400
173S	>1000	55	5500
LT2	0.08	2000	2000
leaves	0.01	1	40

[a]Extracts from bacteria containing construct E (Fig. 1); or E modified by a Pro to Ser mutation at amino acid 173 (173S); wild-type LT2; and *B. napus* leaves were assayed for ALS activity in the presence of increasing concentrations of herbicide. I_{50} designates the concentration of herbicide in which the enzyme activity is 50% of uninhibited. CS designates chlorsulfuon.

inhibition assays of Table 1. The presence of the mutation causes at least a hundred-fold increase in resistance to chlorsulfuron, a ten-fold increase in Assert resistance and little effect with Pursuit. This demonstates that the bacterial expression system can model the effect mutations in the plant gene can have on herbicide sensitity.

Mutagenesis is now being done on the plant gene using standard *in vivo* bacterial techniques. Also methods in which the plasmid is mutated before reintroduction into the bacterial host are being evaluated. This latter method has the advantage of not perturbing the host genes and also of ensuring that all copies of the plasmid contain the mutation which contributes to the observed phenotype. This also greatly facilitates accurate isolation and sequencing of the mutant plasmid.

***Herbicide resistance phenotypes of plant ALS transformed into* B. napus.** Cotyledonary petiole explants of *B. napus* cv Westar were transformed using *Agrobacterium* vectors as described by Moloney et al.(12). The vectors harbored within their T-DNA, constructs containing mutant ALS genes tested in the above bacterial expression system. Constructs included those containing the native promoter of plant ALS and also chimaeric genes using high level constitutive promoters such as CaMV 35S and its tandem repeat. Selections were performed both on kanamycin (15 mg L^{-1}) and directly on the herbicide chlorsulfuron (10 nM). In both cases we obtained regenerating plants showing resistance to the selection agents.

Analysis of the ALS activity in the presence and absence of sulfonylureas shows a 1 to 2 order of magnitude shift from the normal (10 nM) I_{50} of the enzyme in transformed plants, depending on event. The progeny of these transformants are currently being evaluated for their resistance at the whole plant level.

REFERENCES

1 Hirschberg J, Yehuda AB, Pecker I, Ohad N (1987) Plant Mol. Biol. 140, 357-366.
2 Comai L, Facciotti D, Hiatt WR, Thompson G, Rose RE, Stalker DM (1985) Nature 317, 741-744.
3 Yadav N, McDevitt RE, Benard S, Falco SC (1986) Proc. Natl. Acad. Sci. USA 83, 4418-4422.
4 Chaleff RS, Mauvais CJ (1984) Science 224, 1443-1444.
5 LaRossa RA, Schloss JV (1984) J. Biol. Chem. 259, 8753-8757.
6 Shaner DL, Anderson PC, Stidham MA (1984) Plant Physiol. 76, 545-546.
7 Kleschick WA, Gerwick BC (1989) Br. Crop Protection Council, Monograph 42; Copping LG, Dalziel J, Dodge AD, eds. p139-146.
8 Wiersma PA, Schmiemann MG, Condie JA, Crosby WL, Moloney MM (1989) Mol. Gen. Genet. 219, 413-420.
9 Weinberg RA and Burns RO (1984) J Bacteriol. 160, 833-841.
10 Amann E, Brosius J (1985) Gene 40, 183-190.
11 Lu M-F, Umbarger HE (1987) J. Bacteriol. 169, 600-604.
12 Moloney MM, Walker JM, Sharma KK (1989) Plant Cell Reports 8, 238-242.

CELL CULTURE OF ALIEN CHROMOSOME ADDITION LINES TO INDUCE SOMATIC RECOMBINATION AND GENE INTROGRESSION.

Larkin, P.J., Spindler, L.H., Banks, P.M.

CSIRO Division of Plant Industry
P.O.Box 1600 Canberra
A.C.T. 2601
Australia

1. INTRODUCTION

Crop breeding relies on the continual assessment and utilisation of new genes. Very often genes of particular utility can be identified in species related to the crop. Frequently, sexual crosses can be made to such related species, albeit with the application of embryo rescue techniques. However upon examination of subsequent generations from the initial hybrid the useful gene often continues to assort with the intact alien chromosome. In other words no meiotic recombination is occurring between the chromosomes of the crop and alien genomes. Almost without exception it is agronomically unacceptable to release a new variety carrying one or more whole alien chromosome. Thus, despite the prospect of successful sexual crosses to the alien species, in such a case the nuclear genes cannot be deployed for the crop's improvement.

To state this in other words: utilisation of alien genes is often precluded by the absence of meiotic recombination.

Wheat breeding and genetics has a long history of attempts to utilise alien germplasm (1,2). For a very limited number of alien genomes of the Triticeae there is little recombination barrier. For others such as the E genome certain genetic strategies have been employed successfully to obtain disease resistance genes. The most successful genetic strategy has been the use of mutants to the homoeologous pairing inhibitor locus, Ph1. In the absence of this locus or the presence of certain mutants there is a partial relaxation of the usual prevention of pairing between homoeologous chromosomes of different genomes. However there is a biological ceiling for homoeologous pairing which the present mutations appear to have met (3). Consequently, not all Triticeae genomes are able to recombine with the wheat genomes in the presence of the most effective mutation, ph1b. For example the rye genome is sufficiently unresponsive to the Ph1 mutation (ph1b), that it is unlikely to be of practical use in transferring genes from rye to wheat (4,5,6,7). This is attested by the fact that the only beneficial transfers to wheat from rye have tended to be by large translocations, probably whole arms (8). The barley genome does not respond at all to this approach.

We have promoted the potential of cell culture to facilitate gene transfer where meiotic recombination fails (9,10). Others have also suggested this application (11,12,13). The preliminary results reported here represent the first substantial confirmation of this concept and its application to two important breeding objectives.

The word *introgression* is used in population genetics to describe the transfer of genes from one species to another following hybridisation and backcrossing. The use of this word usually conveys a stabilisation and integration of the new genes in their new genetic environment. We are using the word *introgression* with this latter connotation and conveying that the introduced genes are linked on a fragment of donor chromatin smaller than a chromosome arm. Inheritance of such introgressed fragments would be expected to follow Mendelian patterns of segregation.

2. CELL CULTURE INDUCED TRANSLOCATIONS

We have previously reported that wheat plants regenerated from culture carry a high frequency of non-homologous translocations revealed by multivalent formation at meiosis (14). Our studies of the group 4 chromosomes of wheat indicate that 0.5% of regenerant families have a translocation, evident as multivalents, involving a group 4 chromosome. This extrapolates to 3.5% for the entire wheat genome. There are other examples of translocations following culture in many species. These have been recently summarized (10). More recent examples are *Zea* (15), *Hordeum* (16), *Festuca* (17) and cotton (18).

Cell culture-induced translocations (ie. chromosome interchanges) seem not to rely on homoeology and occur during mitotic cell cycles rather than meiotic cell cycles. We have observed non-homoeologous non-reciprocal interchanges in wheat plants following cell culture. For example translocations have been identified between chromosome 4A and 3B, 4A and 1B, 4A and 1B, and 4A and 6B (14,19). Similarly, following cell culture of wheat x rye hybrids, exchanges have been observed between 1R and 4D, and 3R and 2B (12,20).

In addition we have observed wheat somaclones forming multivalents at a low frequency in meiosis. The donor wheat did not show such multivalents. These we have tentatively supposed to be due to translocations involving smaller chromatin fragments than those reported above. Such translocations, if they were in fact occurring, would be most useful for recombination. The following sections describe projects aimed at putting this unsubstantiated concept to the test.

3. CEREAL CYST NEMATODE (CCN) RESISTANCE FROM RYE TO WHEAT

CCN is a serious economic constraint, infesting 1.2 million hectares of the Southern Australian wheatland. The integrated use of nematicides, crop rotation and manipulation of seeding dates can offer effective control against

CCN. However these methods are costly and the nematicides are a particular environmental hazard. The presently available wheat germplasm does not have a high level of resistance to CCN. Some level of field resistance has been described and released, but these genes offer only moderate resistance.

The source of the resistance used in this study was from the long arm of the group 6 rye chromosome from a particular triticale accession (21). This affords a significantly higher level of resistance than previously available in wheat and has been described as near-immunity. By backcrossing the triticale to wheat, an addition line carrying the whole 6R chromosome and the full degree of resistance has been produced. Such addition lines are not useful for cultivar production. This is firstly because they also carry many other genes from the rye chromosome which alter the agronomic characteristics of the wheat unfavourably. Secondly such chromosome addition lines are genetically unstable resulting in the loss of the rye chromosome and hence the CCN resistance from some progeny.

Meiotic recombination does not ordinarily occur in the case of crosses between wheat and rye. The 6R monosomic addition line was cultured for 1-2 months to induce exchange between the wheat chromosomes and the 6R. Progeny were screened by Jim Kollmorgen and Russell Eastwood at the Victorian Crops Research Institute. Individual plantlets were grown in thoroughly mixed nematode-infested soil in tubes placed in growth chambers. After 12 weeks at 15 C the roots were washed free of soil, and cysts (white females) counted on the roots to indicate the degree of resistance.

The initial criterion of selection for an introgression was the segregation pattern of CCN resistance amongst the progeny of the regenerants. Our research on transmission of monosomic 6R additions in wheat has shown that 41% of progeny from a selfed plant with a monosomic 6R addition will inherit at least one rye chromosome. Therefore, if no introgression has occurred 59% of progeny from a regenerant plant would be susceptible to CCN attack. If an introgression has occurred, Mendelian segregation of the resistance would result in only 25% of progeny being susceptible to CCN attack.

A number of isozyme markers have been found on the long arm of chromosome 6R. These are: esterase (EST_L) and glutamate oxaloacetate transaminase (GOT) which are isolated from leaves; 6-phosphogluconate dehydrogenase (6-PGD) and a second esterase (EST_G) which are isolated from grain; and alpha-amylase (alpha-AMY) isolated from germinating seed. The rye chromatin can also be recognised in the wheat background with a number of DNA probes homologous to nucleotide sequences repeated and dispersed throughout rye chromosomes but missing or rare in wheat chromosomes.

Plants were derived from cultured immature inflorescences or embryos which were monosomic for the added 6R. About 700 families have been screened to date. The most advanced investigations have been on lines initiated from immature inflorescences. After four generations, four of the advanced families are still exhibiting segregations for resistance consistent with the resistance gene being carried on a wheat chromosome.

Rye chromosome breakage was evident in these families by the separation of the resistance from all or some of the linked rye isozyme markers. Our results have allowed us to determine the order of these genes. Since the CCN resistance gene on the 6R is flanked on both sides by these markers, the introgressed segments are internal fragments of the $6R_L$ arm. Separation of resistance from all the other $6R_L$ markers suggests a small amount of introgressed chromatin. By contrast meiotic recombinants between disparate genomes tend to result from a single cross-over and therefore to carry the entire alien arm from the cross-over point to the telomere.

Two independent rye-specific dispersed repetitive probes are presently being used in Southern hybridisations to obtain further indications of the size of the resistance-bearing rye chromatin. With both probes there are consistent indications that some of the introgressions are small rye fragments. Recently a similar technique was used to assess the size of *Lophopyrum* chromatin segments in wheat (22).

4. BARLEY YELLOW DWARF VIRUS RESISTANCE FROM *THINOPYRUM* TO WHEAT

Barley yellow dwarf is the most important viral disease of small grain cereals. The disease is caused by a number of serologically distinct luteoviruses. No adequate resistance has been available in wheat. We have identified a number of Triticeae species to be sources of resistance. A number of these resistant species have been crossed and backcrossed to wheat and a limited number of addition lines were available to us. We have identified two resistant disomic addition lines to wheat, derived from independent sources of *Thinopyrum intermedium* (= *Agropyron intermedium*) (23,24).

Two methods were used to transfer the resistance gene to wheat chromosomes:
1. The use of the ph1b mutation to encourage homoeologous pairing and meiotic recombination;
2. The use of cell culture of one of the monosomic addition lines to induce somatic recombination. Resistance was assayed by feeding viruliferous aphids on individual seedlings and determining the virus content by ELISA after 3 weeks. A red coleoptile marker gene is associated with the short arm of the *T.intermedium* chromosome, while the resistance is on the long arm. This chromosome belongs to homoeologous group 7.

Families were selected in which the red coleoptile gene does not assort with the resistance. A family was identified from the ph1b-crossed material in which segregation of resistance has consistently conformed to Mendelian ratios. However the cytological data suggests that resistance assorts with a large *Thinopyrum* chromatin fragment.

The initial screen of over 2000 families derived from cell culture (20,000 individuals) is completed. Over 800 sub-families have been through a subsequent generation of screening. Families have been identified derived from cell culture which are missing the red coleoptile marker but carrying the resistance. Some resistant individuals of these families have 2n=42 chromosomes and meiosis in heterozygotes is normal.

Further evidence of the introgressions is being sought with DNA probes. Partially dispersed repetitive probes have been cloned from the N and E genomes. These have the capacity to recognise the presence of *T.intermedium* chromatin in the wheat background and to give some indication of the relative size of the introgressed fragment. A number of group 7 specific RFLP probes (25) are also effective markers of the resistance-bearing *T.intermedium* chromosome. The present preliminary indications in these studies is that cell culture has induced introgressions of the BYDV resistance to wheat chromosomes.

5. CONCLUSIONS

Cell culture of chromosome addition lines is a strategy for alien gene introgression which does not require the isolation of homoeologous pairing inhibitor mutants. Although such genes almost certainly exist in other crops, the discovery of mutations is time consuming and their use is restricting in the breeding context. The example of wheat indicates that such mutations are also not universally successful in overcoming recombination barriers. It is likely that greater relaxation of homoeologous pairing inhibition might be lethal. The cell culture approach is immediately applicable to crops where such mutants are not available and where there is very little background cytogenetic knowledge. The indications from our work are that some of the cell culture induced translocations are small interstitial fragments. This is in contrast to the ph mutant recombinants which are generally large exchanges requiring further induced recombinations before they can be used.

There is a further ramification of this research. Those involved in protoplast fusion for somatic hybridisation have argued for some time that the breeding benefits from their work will come from limited gene transfer rather than the recovery of very wide hybrids. This is undoubtedly true. These researchers can be encouraged from our demonstration that cell culture alone induces somatic exchange between unrelated chromosomes at high frequency.

REFERENCES

1 Sharma, H.C. and Gill, B.S. (1983) Euphytica 32, 17-31
2 Kimber, G. (1984) Z.Pflanzenzuchtg. 92, 15-21
3 Sears, E.R. (1984) in Gene Manipulation in Plant Improvement, 16th Stadler Genetics Symposium (Gustafson, J.P., ed.) pp.295-300, Plenum Press, New York
4 Koebner, R.M.D. and Shepherd, K.W. (1986) Theor. Appl. Genet. 73, 197-208
5 Jouve, N. and Giorgi, B. (1986) Can. J. Genet. Cytol. 28, 696-700
6 Naranjo, T., Roca, A., Giraldez, R. and Goicoechea, P.G. (1988) Genome 30, 639-646

7 Kimber, G. (1988) in First International Symposium on Chromosome Engineering in Plants. (Li,Z. and Swaminathan,M.S., eds.), Oct 1986 Xian, China, pp.12-13
8 Friebe, B., Hatchett, J.H., Sears, R.G. and Gill, B.S. (1990) Theor. Appl. Genet. 79, 385-389
9 Larkin, P.J. and Scowcroft, W.R. (1981) Theor. Appl. Genet. 60, 197-214
10 Larkin, P.J. (1987) Iowa State J. Res. 61, 393-434
11 Sharma, H.C. and Baezinger, P.S. (1986) Theor. Appl. Genet. 71, 750-756
12 Lapitan, N., Sears, R. and Gill, B. (1984) Theor. Appl. Genet. 68, 547-554
13 McCoy, T.J., Phillips, R.L. and Rines, H.W. (1982) Can. J. Genet. Cytol. 24, 37-50
14 Davies, P.A., Pallotta, M.A., Ryan, S.A., Scowcroft, W.R. and Larkin, P.J. (1986) Theor.Appl.Genet. 72, 644-653
15 Lee, M. and Phillips, R.L. (1987) Genome 29, 122-128
16 Jorgensen, R.B. and Andersen, B. (1989) Theor. Appl. Genet. 77, 343-351
17 Eizenga, G.C. (1989) Genome 32, 373-379
18 Stelly, D.M., Altman, D.W., Kohel, R.J., Rangan, T.S. and Commiskey, E. (1989) Genome 32, 762-770
19 Larkin, P.J., Banks, P.M., Bhati, R., Brettell, R.I.S., Davies, P.A., Ryan, S.A., Scowcroft, W.R., Spindler, L.H. and Tanner, G.J. (1989) Genome 31, 705-711
20 Lapitan, N., Sears, R., Rayburn, A. and Gill, B. (1986) J. Heredity 77, 415-419
21 Asiedu, R., Fisher, J.M. and Driscoll, C.J. (1990) Theor. Appl. Genet. 79, 331-336
22 Zhang, H.-B. and Dvorak, J. (1990) Genome (in press)
23 Xin, Z.-Y., Brettell, R.I.S., Cheng, Z.-M., Waterhouse, P.M., Appels, R., Banks, P.M., Zhou, G.-H., Chen, X. and Larkin, P.J. (1988) Genome 30, 250-257
24 Brettell, R.I.S., Banks, P.M., Cauderon, Y., Chen, X., Cheng, Z.-M., Larkin, P.J. and Waterhouse, P.M. (1988) Ann. Appl. Biol. 113, 599-603
25 Chao, S., Sharp, P.J., Worland, A.J., Warham, E.J., Koebner, R.M.D. and Gale, M.D. (1989) Theor. Appl. Genet. 78, 495-504

ANTIBIOTIC RESISTANT MUTANTS OF *SOLANUM NIGRUM*

P.J. DIX, C.P. Mc KINLEY AND P.F. Mc CABE[1]

Dept. of Biology, St. Patrick's College, Maynooth, Co. Kildare, Ireland ([1]Present address: Dept. of Horticulture, Michigan State University, East Lansing, MI. 48824, USA).

1. INTRODUCTION

Recently, an efficient system for expanding the number and range of plastome encoded antibiotic resistant mutants in Solanaceous species was reported (1,2). This involved mutagenesis of leaf strips, derived from axenic shoot cultures, with the plastome targetted mutagen, nitroso-methylurea, followed by incubation on medium normally associated with adventitious shoot initiation, but containing levels of antibiotics suppressing morphogenesis and causing bleaching of the explant. Resistant mutants appear as green nodules which develop into adventitious shoots. The previous report (1) concentrated on selection of streptomycin resistant mutants and demonstrated that while the technique could be used for several solanaceous species, *Solanum nigrum* was particularly responsive. Maternal inheritance of streptomycin resistance was demonstrated for three of the *S. nigrum* mutants and evidence was also found for occasional paternal inheritance in one of them. The present report extends this investigation to include the selection of additional antibiotic resistance markers both on the plastome (lincomycin and spectinomycin resistance) and nuclear genome (kanamycin resistance) using the leaf strip system for mutagenesis and *Agrobacterium* - mediated transformation respectively.

2. MATERIALS AND METHODS

2.1. Plant material

Shoot cultures of wild type and an atrazine resistant biotype (3) of *Solanum nigrum* were used, in addition to streptomycin resistant mutants already isolated using the wild type line (1).

2.2. *Agrobacterium tumefaciens* strain

A strain containing a cointegrate between modified Ti-plasmid, pTi B6535E and intermediate vector pMon 200 was kindly provided by T.A. Kavanagh. This vector confers kanamycin resistance and β-glucuronidase (GUS) activity.

2.3. Culture media

2.3.1 Shoot cultures: Nodal cuttings were maintained on RM medium, containing Murashige and Skoog, 1962 (4) salts and 30 gl^{-1} sucrose, pH 5.6, solidified with 7% Difco Bactoagar.

2.3.2 Shoot regeneration: Shoots were regenerated from explants placed on medium containing Murashige and Skoog, 1962 (4) salts, H-vitamins (5) and (per litre) 100 mg mesoinositol, 1 mg benzylaminopurine, and 30 g sucrose, solidified with 7 gl^{-1} Difco Bactoagar.

2.4. Mutagenesis procedure

Leaf strips were incubated for 90 min at 25°C in liquid RM medium containing 5 mM nitroso-methylurea (Sigma), washed three times in fresh medium and incubated on shoot regeneration medium supplemented with selective levels of antibiotics spectinomycin or lincomycin. Full details of the mutagenesis protocol, and safe handling of the mutagen have been reported elsewhere (1,2).

2.5. Leaf strip tests

Leaf strips from regenerated plants or their progeny were incubated on the same selective media used for mutagenesis, to confirm resistant phenotypes.

2.6. Transformation

Co-cultivation of leaf strips with the *A. tumefaciens* strain was carried out essentially according to established procedures (6). After 20 min incubation in 24-48 hour bacterial culture, resuspended in liquid RM medium, leaf strips were placed on shoot regeneration medium without antibiotics for 2-4 days, before being transferred to solid shoot regeneration medium supplemented with 100 mgl^{-1} kanamycin sulphate and 500 mgl^{-1} carbenicillin. In some cases resistant callus was transferred for a second passage on this medium to obtain shoot regeneration. β-glucuronidase (GUS) activity was assessed in transformed shoots, using published procedures (7).

3. RESULTS AND DISCUSSION

3.1. Spectinomycin resistance

After NMU mutagenesis spectinomycin resistant shoots appeared at a high frequency on medium containing 50 mgl^{-1} spectinomycin, and at lower frequencies on 75 or 100 mgl^{-1}. Results for one experiment, in which a previously isolated streptomycin resistant mutant (1) was used as starting material, are given in Table 1.

TABLE 1. Appearance of spectinomycin resistant shoots on mutagenised leaf strips of a streptomycin resistant mutant of *S. nigrum*.

Spectinomycin concentration (mgl^{-1})	Percentage explants producing one or more resistant shoots after 6 weeks
50	36
75	8
100	6

No resistant shoots were found among non-mutagenised controls. Resistance was confirmed in leaf strip tests for shoots obtained at all three selective levels.

The double mutant nature of the lines described on Table 1 was confirmed by chlorophyll retention and shoot regeneration of leaf explants on medium containing spectinomycin (selective level) and streptomycin sulphate (500 mgl^{-1}) supplied independently and together. Crosses were performed between one of these double mutants (StSp1) and wild type *S. nigrum* and maternal inheritance was demonstrated in progeny leaf strip tests. There was no evidence for paternal inheritance, as was found earlier for one of the streptomycin resistant mutants (1).

3.2. Lincomycin resistance

A high frequency (up to 28.2%) of NMU mutagenised leaf strips produced green shoots on medium containing 100 mgl^{-1} lincomycin hydrochloride. (see Fig.1). In leaf strip tests all lines were resistant to 100 mgl^{-1} lincomycin hydrochloride and some were resistant to much higher levels (>1000 mgl^{-1}) suggesting that different classes of mutant may have been obtained. By using an atrazine resistant biotype (3) of *S. nigrum* double ($atr^{r}lin^{r}$) mutants have been obtained. For one of these, (designated L17A1) maternal inheritance has been demonstrated (Table 2).

FIGURE 1. Development of a lincomycin resistant shoot, after mutagenesis with 5 mM NMU, on medium containing 100 mgl^{-1} lincomycin hydrochloride.

TABLE 2. Inheritance of lincomycin resistance in *S. nigrum* mutant L17A1. Resistance based on chlorophyll retention and shoot regeneration on 100 mgl^{-1} lincomycin hydrochloride. (WT = wild type)

	No. seedlings tested	No. resistant
WT (self)	10	0
L17A1 (self)	10	10
L17A1 x WT	15	15
WT x L17A1	15	0

3.3. Kanamycin resistance

The npt II gene, coding for neomycin phosphotransferase and conferring kanamycin resistant was successfully introduced into the streptomycin and spectinomycin resistant double mutant (StSp 1). *S. nigrum* leaf strips were sensitive to very high densities of *A. tumefaciens*, leaf strips browning in subsequent culture. Reduced bacterial densities however resulted in successful transformation (Table 3). Kanamycin resistant

TABLE 3. Production of kanamycin resistant shoots or calli after *Agrobacterium*-mediated transformation of StSp1 leaf strips (accumulated data from several experiments).

No. of explants tested	No of resistant: calli	shoots
671	14	5

shoots either developed directly on the explant, or after subculture of resistant callus (Fig. 2). None of the five resistant shoots, obtained directly, developed roots on transfer to RM medium containing 100 mgl^{-1} kanamycin sulphate, and they all lacked GUS activity suggesting they were not true transformants, but escapes from a leaky selection system. Shoots regenerated from two of the resistant calli were transferred to selective RM medium. Both rooted successfully and exhibited GUS activity, confirming their transformant nature.

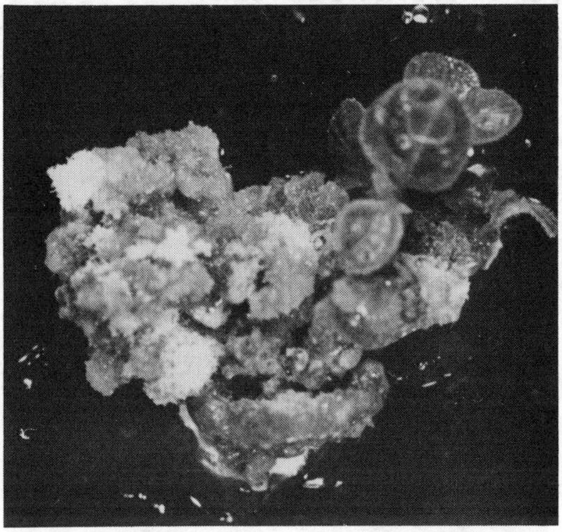

FIGURE 2. Recovery of a transformed shoot from kanamycin resistant callus, on medium containing 100 mgl^{-1} kanamycin sulphate, after *Agrobacterium* mediated transformation.

CONCLUDING REMARKS

Mutations to resistance to several different antibiotics have been successfully induced in *Solanum nigrum* by the plastome targetted mutagen nitroso-methylurea. These are maternally inherited and believed to reside in the chloroplast DNA. Consequently we now have double plastome mutants resistant to atrazine and lincomycin, or streptomycin and spectinomycin. The latter carry a third antibiotic resistance marker, to kanamycin, introduced by transformation and probably nuclear encoded. It is anticipated that these mutants will be useful in studies on interspecific cytoplasmic genetics involving *S. nigrum.*

Differing levels of resistance to antibiotics, manifested by some of the mutants, suggests different mutant classes, but the authors recognise that confirmation of this, and the wider use of these markers in genetic studies, depend on a more detailed molecular analysis of the mutants.

REFERENCES

1 McCabe, P.F., Timmons, A.M. and Dix, P.J. (1989) Mol. Gen. Genet. 216, 132-37
2 McCabe, P.F., Csésplö, A., Timmons, A.M. and Dix, P.J. (1990) in Methods in Molecular Biology, Vol. 6, Plant Tissue Culture, (Walker, J.M. and Pollard, J.W., eds.) Humana Press (in press)
3 Gasquez, J., Darmenzy, H. and Compoint, J.P. (1981) C.R. Acad. Sci. Paris. T. 292, 847-849
4 Murashige, T. and Skoog, F. (1962) Physiol. Plant. 15, 473-497

5 Nitsch, J.P. and Nitsch, C. (1969) Science 163, 85-87
6 Horsch, R.B., Fry, J.E., Hoffmann, N.L., Eichholtz, D., Rodgers, S.G. and Fraley, R.T. (1985) Science 227, 1229-1231
7 Jefferson, R.A., Kavanagh, T.A. and Bevan, M.W. (1987) EMBO J. 6, 3901-3907

ESTABLISHING A TOMATO TRANSPOSON TAGGING SYSTEM WITH HETEROLOGOUS TRANSPOSONS

J.D.G. JONES, G. BISHOP, K. HARRISON, D. JONES, S. SCOFIELD & C. M. THOMAS
Sainsbury Laboratory, John Innes Institute, Colney Lane, Norwich NR4 7UH, United Kingdom

1. INTRODUCTION

The study of maize transposons has proved to be not only interesting but also useful. Transposons, or transposable genetic elements, were first identified in maize (1), and were subsequently detected in all forms of life where they were looked for (2). Studies on transposon behaviour have given insight into cell lineages in development (3), into mechanisms which generate new variation on which natural selection can act (4,5) and into somatic processes which can modify gene expression during development (e.g. methylation- 6). Additionally, transposons have proved very useful for the isolation of new genes for which the biochemical properties of the gene product are too difficult or too poorly characterized to provide a route to gene cloning. This technique has been employed for the isolation of interesting genes from bacteria, *Drosophila*, maize and *Antirrhinum*. In particular, the maize regulatory genes *C1* (7), *R/Lc* (8) and *Vp1* (9) have been isolated with this technique, and also the developmental mutant *Knotted (10)*. In *Antirrhinum* the anthocyanin biosynthetic gene *pallida* (encoding dihydroflavanol reductase) was isolated using transposon tagging (11), as was *floricola* a developmental gene required for the initiation of floral development, (E Coen, pers. comm.).

Recently, with the advent of transgenic plant technology, it has become routine to introduce new DNA into a wide range of plant species and to generate fertile transgenic plants. This approach was used to demonstrate that the maize transposon *Activator* (*Ac*) is active after introduction into tobacco cells (12,13). Subsequent work has demonstrated the activity of *Ac* in tomato (14), *Arabidopsis* (15), carrot (15) and rice (16). This work opens up the possibility of carrying out gene tagging experiments in one species after introduction of transposons from another. It has more recently proved possible to monitor the somatic activity of *Ac* in cotyledons of tobacco using a chimeric streptomycin phosphotransferase (SPT) gene (17). This study demonstrated that *Ac* in tobacco, unlike in maize, exhibits a positive effect of increasing *Ac* dosage on excision frequency. The ability to generate an engineered variegated phenotype raises the possibility of propagating the introduced *Ac*-carrying T-DNA in the homozygous or heterozygous state as a variegated allele of a locus from which *Ac* has not yet departed.

It is the objective of our research program to use *Ac* for gene isolation in tomato (*Lycopersicon esculentum*). Specifically we are interested in isolating genes for resistance to the glasshouse disease tomato leaf mould (*Cladosporium fulvum* syn *Fulvia fulva*). We are seeking to optimize our use of the element by establishing a system for monitoring its behaviour, and in particular its somatic excision, in transgenic tomato. Since *Ac* in both maize (18) and tobacco (19) preferentially transposes to linked sites, we are also establishing a system to locate T-DNAs transferred to tomato chromomes on the genetic map using inverse PCR (IPCR) and RFLP blot analysis. This system will enable us to choose a specific stock that carries *Ac* at a known chromosomal location in order to increase our chances of tagging a specific gene.

2. GENERAL APPROACH

The optimal design of any transposon tagging experiment depends strongly on the specific objectives of the researcher.

1) Targeted tagging experiments, in which only a mutation in a specific gene is sought, will probably require a different approach from non-targeted tagging, in which transposon-induced mutations are sought at any location.

2) It is important to be able to easily identify a mutation in the target gene.

3) It will be useful to preselect individuals in which the *Ac* is known to have excised before screening for specific mutations, so that time and greenhouse space are not wasted in screening for mutations in plants in which *Ac* has not moved.

4) It would be very desirable to be able to propagate the T-DNA locus carrying *Ac* in the unexcised state. Setting up an engineered variegation is the easiest way to achieve this.

5) Since in transgenic tobacco *Ac* excision rates are higher in homozygotes than heterozygotes (unlike in maize- see ref 17), for targeted tagging we would wish to cross stocks which are homozygous for the T-DNA carrying *Ac* , in which this T-DNA is linked to the dominant allele of the targeted gene, to stocks carrying the recessive allele (Fig 1). F1 progeny would be screened for the mutation of interest. For example, to tag a disease resistance gene, we would start off with stocks homozygous for both the resistance gene and a linked *Ac,* and test cross to disease sensitive plants. Progeny individuals would be screened for disease sensitivity. Clearly, sensitivity to the disease must not be lethal.

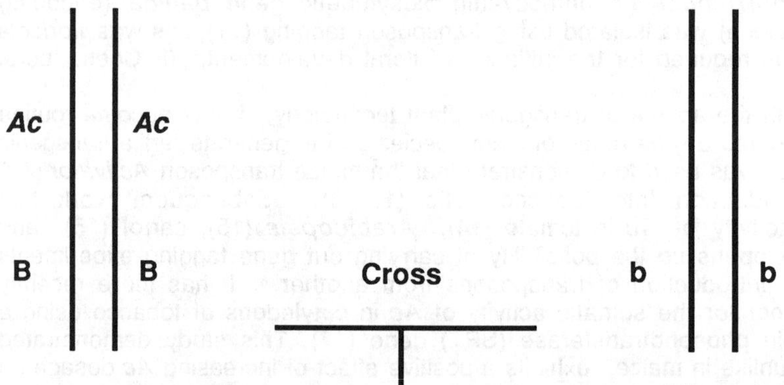

FIGURE 1. Idealized scheme for carrying out targeted transposon tagging of the dominant gene B. Homologous chromosomes are represented by vertical bars. Crosses will be carried out between individuals homozygous for both *Ac* and the dominant allele of the targeted gene, and individuals recessive for the targeted gene. The positive dosage effect should ensure high mutation frequencies in the *Ac*-containing parent and reduced somatic excision frequency in the F1 progeny. The F1 progeny would be screened for mutations in the B gene.

3. RESULTS

3.1 DNA CONSTRUCTIONS ASSAYED IN TOBACCO AND TOMATO

Using orthodox molecular biology techniques, we have constructed binary vector plasmids for tomato transformation based mostly on a chimeric 35S/neomycin phosphotransferase (NPT) gene confering kanamycin resistance to transformed plant cells. This was used in our more recent experiments because in our hands it is more effective than the nos/NPT constructs or 1'/NPT constructs that we worked with initially. Transformation experiments were carried out on young tomato cotyledons. Routinely, transgenic tomato plants were ready to go to the greenhouse 4 months after initiating the transformation experiment.

The DNA constructs which have been assayed at the time of writing are shown in Fig 2. All allow selection for transformed cells with kanamycin, except for the earliest used, SLJ11, which carries instead a hygromycin phosphotransferase (HPT) gene fused to the 1' promoter of the *Agrobacterium tumefaciens* T_R DNA. This construct confers hygromycin resistance, a marker which works more poorly than kanamycin resistance for tomato transformation.

SLJ11, SLJ581, SLJ601, SLJ721, SLJ871 and SLJ1515 are binary vector plasmids in which *Ac* was cloned into a chimeric gene in such a way that excision of the element should restore antibiotic resistance gene, or β-glucuronidase gene (SLJ721), function. Constructs SLJ11, SLJ581 and SLJ871 incorporate the *aadA* reading frame which when expressed in plants confers spectinomycin resistance (21), and SLJ601 incorporates a gene encoding the enzyme phosphinotricin acetyltransferase (BAR) which confers resistance to the herbicide Basta^{TM} (22).

In our studies on the tomato and tobacco plants derived from transformation with SLJ11 and SLJ581 (SPEC:*Ac*) and SLJ601 (BAR:*Ac*), no variegated phenotype was observed on plating progeny seed on spectinomycin-containing (100-500μg/ml) or Basta-containing (10 μg/ml) media respectively. However, a high proportion of the progeny were completely resistant to either spectinomycin or Basta respectively. Furthermore, in each family, the same proportion of the progeny are kanamycin resistant as were either spectinomycin resistant (SLJ11 and SLJ581) or Basta resistant (SLJ601). This observation is in marked contrast to the result in tobacco when *Ac* is cloned into the SPT gene (17). One possible explanation for this difference is that *Ac* always excises prezygotically from the SPEC and BAR genes. This interpretation was ruled out by both Southern blot and PCR experiments on DNA isolated from spectinomycin- or Basta- resistant progeny (data not shown). Another explanation would be that resistance to the compound is a non-cell-autonomous phenotype, in that sectors of resistant cells could detoxify enough of the antibiotic compound for the sensitive cells to survive. The streptomycin resistance confered by the SPT gene is in contrast manifestly cell-autonomous, since in variegated tobacco seedlings, green resistant sectors lie adjacent to white sensitive sectors. Unfortunately, we have so far failed to detect a variegation with tomato carrying an SPT: *Ac* gene in progeny of tomatoes transformed with SLJ1515, primarily because in untransformed seedlings it is difficult to reproducibly bleach cotyledons with streptomycin.

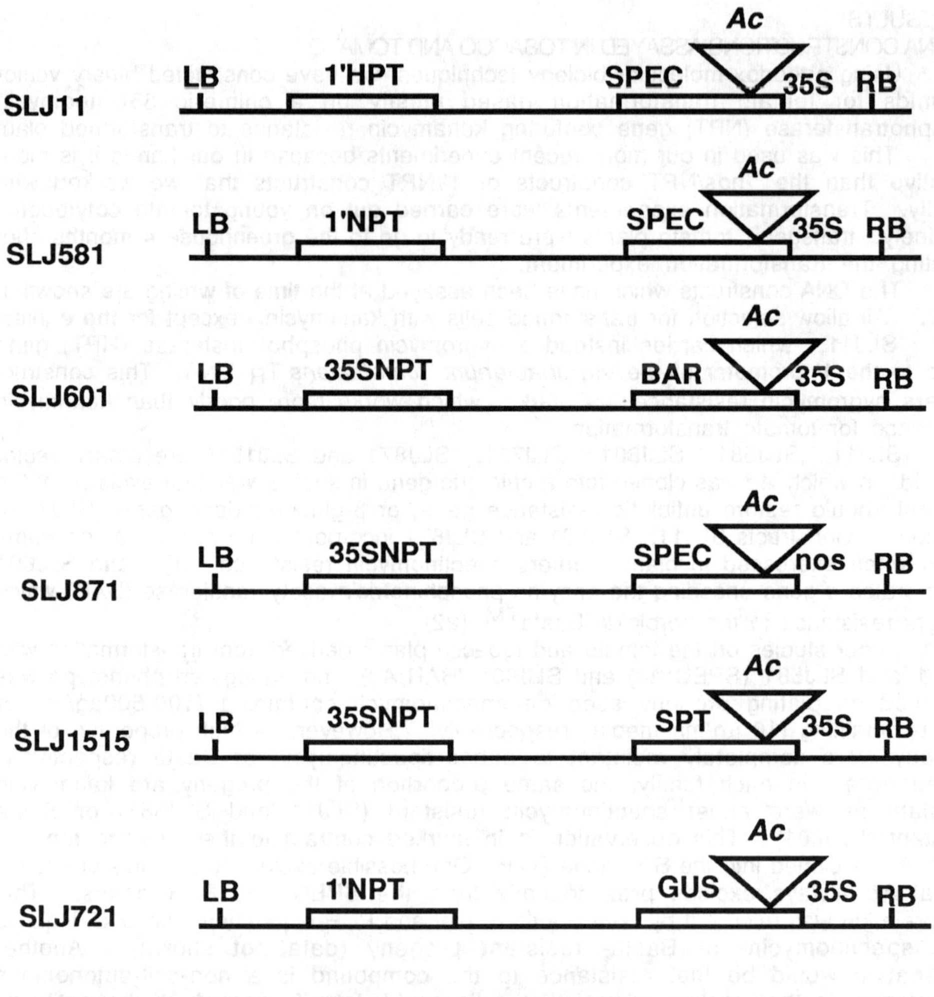

FIGURE 2 Constructs carrying *Ac* which were introduced into tomato and tobacco via *Agrobacterium* transformation. LB, left border of T-DNA; RB, right border of T-DNA; 1', promoter of T_R 1' gene; 35S, promoter of cauliflower mosaic virus 35S RNA; nos, promoter of nopaline synthase gene ; SPT, streptomycin phosphotransferase; SPEC, spectinomycin adenyltransferase; BAR, Basta^{TM} resistance (phosphinotricin adenyltransferase).

SLJ11, SLJ581, SLJ601, SLJ721, SLJ871 and SLJ1515 are binary vector plasmids in which *Ac* was cloned into a chimeric gene in such a way that excision of the element should restore antibiotic resistance or GUS gene (SLJ721) function.

This theory about the non cell-autonomous properties of the SPEC and BAR genes was tested genetically in tobacco. Similar tests are underway in tomato but are not yet complete. Individuals which carried a non-autonomous *Ds* element cloned into either the SPEC (SLJ1561) or BAR (SLJ1481) genes were either selfed or crossed to lines carrying an active source of *Ac* transposase. Our prediction was that if these resistance genes acted in a non-cell-autonomous way, self progeny of plants transformed with SLJ1561 or SLJ1481 would be sensitive to the relevant antibiotic, whereas progeny arising from crosses to stocks carrying a source of transposase would segregate for full resistance to the antibiotic. This prediction was fulfilled. The data are summarized in Table 1.

TABLE 1. Frequency of antibiotic resistance in progeny of SLJ1481 (BAR:*Ds*) and SLJ1561 (SPEC:*Ds*), either selfed or test crossed to a plant carrying 4 loci of an *Ac* transposase-providing T-DNA (s*Ac*). Numbers indicate ratio of resistant (R) to sensitive (S) seedlings. NA indicates not applicable.

	Kanamycin (R:S)	Basta (R:S)	Spectinomycin (R:S)
SLJ1481C SELF	3:1	0:100	NA
SLJ1481C X sAc	NA	58:54	NA
SLJ1561E self	3:1	NA	0:100
SLJ1561E X sAc	NA	NA	63:76

The results in Table 1 show that self progeny of individuals carrying a non-autonomous *Ds* element in a BAR or SPEC gene are all sensitive to the corresponding anitbiotic. However, on crossing to individuals carrying a source ot transposase that can act *in trans*, sufficient post-zygotic somatic excision takes place for full antibiotic resistance to be observed. Prezygotic excision of the *Ds* prior to formation of the zygote by gametic fusion with an *Ac*-active cell cannot take place, so all expression of the detoxifying enzyme must be postzygotic.

Our concern about the possibility that variegated seedlings could appear fully resistant due to detoxification of the antibiotic by a small proportion of the cells led us to make a construct, SLJ871 (Fig 2), in which *Ac* was cloned into a *nos*/SPEC fusion instead of a 35S SPEC fusion. We anticipated that since the *nos* promoter from the *Agrobacterium* nopaline synthase gene is much weaker than the 35S promoter, we stood more chance of seeing an intermediate spectinomycin resistance phenotype in individuals which genotypically showed somatic but not germinal excision. Preliminary results suggest that this expectation was well founded, and studies to confirm this are now in progress.

Our experiments with SLJ1515 transformants that carry an SPT:*Ac* fusion will be inconclusive until a way is found to reproducibly bleach tomato cotyledons by germinating seed on streptomycin- containing media.

We have carried out studies with SLJ721 transformants which indicate that GUS staining using the histochemical stain X GLUC may also prove to be a useful way of detecting transformants carrying an active element. However, it clearly does not permit a direct selection of individuals carrying germinal (prezygotic) excision events.

3.2 ACTIVITY OF *Ac* IN TRANSGENIC TOMATO

Our studies are consistent with those of Yoder (14) and colleagues, who observed a very high rate of *Ac* activity in tomato. We have carried out studies on two plants, SLJ11A (carrying SPEC:*Ac*) and SLJ581A (carrying BAR:*Ac*). In Southern blot and PCR studies on test cross progeny of the primary transformant SLJ11A, we found that only 5 out of 21 individuals retained a pre-excision form of the T-DNA. In the case of SLJ581A, at least 10 out of 15 test cross progeny showed a germinal excision event. This suggests a very high germinal excision rate in the primary transformant. We are still waiting to determine the excision rate at these loci from one seed generation to the next.

3.3 MAPPING T-DNAS IN TRANSGENIC TOMATO

Since *Ac* preferentially transposes to linked sites, we have been attempting to map introduced T-DNAs in order to identify transformants carrying an active *Ac* close to the genes of interest.

Since it would be unacceptably time-consuming to cross each transformant to a set of tester stocks and then cross again to look for linkage, we have adopted what should eventually be a faster strategy.

Dr Steve Tankesley of Cornell University has kindly availed us of about 50 F2 lines arising from an F1 cross between *L. esculentum* and the related *L. pennellii*. These F2 plants segregate restriction fragment length polymorphisms (RFLPs), and Dr Tankesley's lab has mapped the segregation of many different RFLP probes in these individuals (22). If plant DNA could be isolated adjacent to the T-DNA, and hybridized to RFLP blots from these same 50 individuals, we can establish which RFLP probes the T-DNA is linked to.

Our approach to isolating tomato DNA linked to T-DNA is to use the inverse polymerase chain reaction (IPCR) (23). In this technique, after first digesting total DNA from a transgenic plant and then recircularizing it in dilute ligations, DNA adjacent to known DNA is amplified by carrying out PCR amplification with primers which prime outwards from known DNA . The principle of this is illustrated in Figure 3. We have isolated DNA adjacent to 9 different T-DNAs, 2 of them in tobacco and 7 in tomato. Prior to carrying out such experiments , we routinely carry out orthodox PCR reactions to check for the integrity of the tranferred DNA, and in particular to look for bands which could only appear in the event of somatic excision of *Ac* .

To date we have scored two different IPCR-generated probes for segregation on RFLP blots, and we have established that one of our lines carrying an active *Ac* in a SPEC gene (SLJ11A) is on the long arm of chromosome 7.

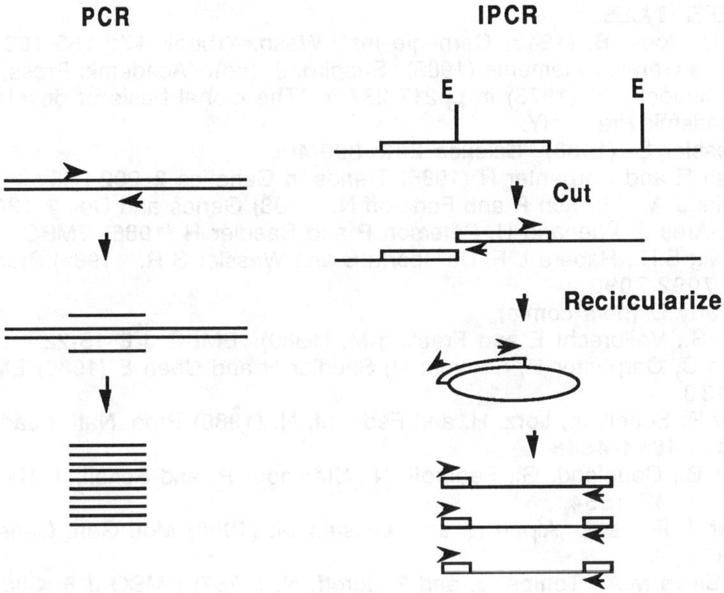

FIGURE 3. The polymerase chain reaction (PCR) contrasted with the inverse PCR (IPCR) reaction. The amplification step is the same, but in IPCR it is preceeded by a cut with an enzyme (E) followed by a recircularization reaction (23).

4. SUMMARY AND PROSPECTS

The work that we are carrying out to establish a transposon tagging system in tomato using the maize element *Ac* is a large undertaking. For this reason, since there are several European labs interested in developing and using such a system, we are conducting this work in collaboration with the groups of Prof. Peter Starlinger (Koln, FRG), Dr Jacques Hille (Amsterdam) and Dr Ian Taylor (Nottingham, UK). This will have the great benefit that these groups do not each independently have to invent the same wheel.

It seems reasonable to anticipate that within 12-18 months we will between us have a transposon mapped to most if not all of the chromosome arms of tomato, with the result that we can look forward to the ability to use transposon tagging to isolate and try and understand the functions of the many interesting and important tomato genes.

5. ACKNOWLEDGEMENTS

This work was supported by the Gatsby Charitable Foundation and also by a grant (PMB/523) from the Agricultural and Food Research Council of the UK. We thank Gunther Donn, of Hoechst, for supplying phosphinotricin, and also DNAP Inc. , of Oakland, CA, USA, for kindly permitting JJ to continue experimenting with the DNA constructs he made while he worked there.

REFERENCES

1) McClintock, B. (1948) Carnegie Inst. Wash. Yrbook **47**: 155-1697)
2) Mobile Genetic Elements (1983) Shapiro, J. (ed). Academic Press, New York
3) McClintock, B (1978) in pp217-237 in "The clonal basis of development" Academic Press, NY.
4) Wessler S. (1988) Science **242** 399-405
5) Coen E and Carpenter R (1986) Trends in Genetics **2** 292-296
6) Banks J A, Masson P and Federoff N. (1988) Genes and Dev **2** 1364-1380
7) Paz-Ares J, Wienand U, Peterson P and Saedler H (1986) EMBO J **5** 829-833
8) Ludwig S.R., Habera L F, Dellaporta S and Wessler S R. (1989) Proc Natl Acad Sci USA **86** 7092-7096
9) McCarty D (pers comm)
10) Hake S., Vollbrecht E and Freeling M. (1989) EMBO J **8** 15-22
11) Martin C, Carpenter R, Sommer H, Saedler H and Coen E (1985) EMBO J **4** 1625-1630
12) Baker B, Schell, J., Lorz, H. and Federoff, N. (1986) Proc. Natl. Acad. Sci. USA **83**: 4844-4848
13) Baker B., Coupland, G., Federoff, N., Starlinger P. and Schell, J. (1987) EMBO J **6**: 1547-1554
14) Yoder J, Palys J, Alpert K. amd Lassner M. (1988) Mol. Gen. Genet. **213**: 291-296
15) Van Sluys M.A., Tempe, J. and Federoff, N. (1987) EMBO J **6**: 3881-3889
16) K Shimamoto (1990) J. Cell Biochem Suppl 14E p258
17) Jones J.D.G., Carland F.C., Maliga, P. and Dooner H.(1989) Science **244**; 204-207
18) Greenblatt, I. M. (1984) Genetics **108**: 471-485
19) Jones J.D.G, Carland F.C. and Dooner H.D. (submitted)
20) De Block, M., et al (1987) EMBO J **6** 2513-2518
21) Svab Z., Harper E.C., Jones J.D.G. and Maliga P. (1990) Plant Mol. Biol. **14** 197-205
22) Bernatzky R and Tankesley S (1986) Genetics **112** 887-898
23) Triglia T, Peterson M and Kemp J (1988) Nucleic Acids Res. **16** 8186

THE USE OF THE POLYMERASE CHAIN REACTION TO DETECT SPECIFIC SEQUENCES IN TRANSFORMED PLANT TISSUES.

JOHN D. HAMILL, STEVEN ROUNSLEY, ANDREW SPENCER, GORDON TODD*, MICHAEL J.C. RHODES, PLANT BIOTECHNOLOGY GROUP, AFRC INSTITUTE OF FOOD RESEARCH (NORWICH LABORATORY), COLNEY LANE, NORWICH NR4 7UA, AND *DEPT. OF BIOLOGICAL SCIENCES, BIRMINGHAM UNIVERSITY, BIRMINGHAM B15 2TT, U.K.

1. INTRODUCTION

The polymerase chain reaction is a very powerful technique which is now used in many areas of biology. Two recent books discussing the merits and practicalities of PCR are recommended[1,2]. A simple and obvious use of PCR in plant transformation studies is to use it to confirm the presence and/or the expression of a foreign gene in transgenic plant tissues and the work reported here was intended to examine a number of parameters affecting the successful amplification of foreign genes from plant tissues. A recent report by Lassner et al.[3] also shows the potential of PCR for analysis of foreign genes in plant tissues.

2. MATERIALS AND METHODS

Axenic transformed root tissue of *Nicotiana rustica* and *N. tabacum* were generated by infection of plants with *A. rhizogenes* carrying pBI 121.1[4]. Axenic transformed root and plant tissue of *N. rustica* containing the yeast ornithine decarboxylase gene[5] and axenic transformed mint (*Mentha citrata*)[6] were also used as a source of plant DNA. DNA was extracted and purified on CsCl gradients as described previously[5] (designated pure DNA) or as crude DNA by previously published methods[3,7]. Primers used for amplification are shown in Table 1 and also the length of DNA of which they are expected to allow amplification. The standard buffer[13] was used and each reaction was carried out in 50μl containing 10mM Tris HCl (pH 8.3 at 25°C), 50mM KCl, 1.5mM $MgCl_2$, 0.01% w/v gelatin 200μM of dATP, dTTP, dCTP, 50μM dGTP, 150μM dC^7GTP, 0.1% v/v Triton X-100, DNA (usually 200ng), and approximately 20pmoles oligonucleotide primer (100ng each primer was used for 20-21mers, 150ng used for 30mers). A programmable thermal controller (MJ Research Inc.) with 1 min. at 92°C, 1 min. at Oligo hybridization temperature (usually 55°C) and 1.5 min. at 72°C was used to allow DNA replication to occur. Fragments were visualized on a 1% agarose gel with ethidium bromide at 0.3μg/ml. For Southern blotting, DNA was transferred to nylon membranes and hybridization to ^{32}P labelled GUS gene sequences from PBI 121 was carried out as described[5]. For gene expression studies c-DNA was synthesized using a commercially available kit (Pharmacia Ltd.) using 5μg poly A^+ mRNA and a small portion (2-10μl) used for PCR analysis. Quantification of PCR products was carried out using Hoechst 33258 (bisbenzimide) as recommended by Cetus Corporation[14].

3. RESULTS AND DISCUSSION

Initial experiments with both pure and crude DNA from N. rustica and N. tabacum transformed roots showed that the 1.2Kb GUS internal fragment could be amplified after 25 cycles using 55°C as hybridization temperature and 200ng (approximately for crude DNA) of DNA. We found that treatment of crude DNA preparations with ribonuclease (boiled, cooled and stored at -20°C) was not necessary and in fact was detrimental to the successful amplification of target DNA. Southern blotting confirmed the fragment to be from the GUS gene. Slightly diffuse bands of other sizes, both larger and smaller, were sometimes seen which did not hybridize to the probe (data not shown). Further experiments were done using 200ng of pure plant DNA with variations in the parameters as described in Figs. 1-6 and using the GUS primers. Unless stated to be otherwise, the hybridization temperature was 55°C.

The experiments indicate that the PCR buffer developed by Cetus is near optimal for amplification of the 1.2 Kb GUS internal band from transformed Nicotiana genomic DNA. A parallel series of experiments using the NPT II primers (Table 1) produced an almost identical set of results as described above. As a rule of thumb it would appear that an input level of genomic DNA the range of 20ng-2μg and 30 cycles should allow visible amplification of a foreign gene, on ethidium bromide agarose gels, from transformed plant tissues. The optimum temperature for annealing should be determined empirically but the approximate Tm (i.e. temperature at which 50% of oligonucleotides are bound to the target sequence) can be estimated for oligonucleotides up to 20 bases in length using the formula (Tm = 4(G+C) + 2(A+T)[15]. We determined the temperatures at which a number of primers of varying sizes would allow the amplification of a band of the expected size from transformed plant DNA. These results are summarized in Table 2. It is clear that the simple formula Tm = 4(G+C) + 2(A+T) is a reasonable guide to the actual annealling temperature when 20-21 mers are used. For 30 mers, the formula is not applicable.

We have used the standard buffer and temperature conditions as described above to amplify a number of fragments from c-DNA though 50-70 cycles may be necessary. In addition to being useful for monitoring expression of genes, the technique can also be useful to ensure that full length copies of genes are present, in a batch of c-DNA, before insertion into a vector and the subsequent construction and amplification of a c-DNA library, eg Rol B gene in c-DNA from transformed root tissue. A further use of PCR in plant transformation studies is to use it to quickly and efficiently check bacteria for the presence of a foreign gene (whether they be E. coli, A. rhizogenes or A. tumefaciens). A single bacterial colony is dispersed into 20μl sterile water; 5μl is removed, boiled for 2-3 minutes and added to a PCR reaction with the appropriate primers. After 25 cycles an aliquot is run on an EtBr-agarose gel and those colonies containing the foreign gene (and binary vector) can easily be identified. The remaining 15μl containing bacterial cells is then inoculated into nutrient broth or streaked onto plates.

A note of caution Due to the power of the PCR technique it is important that the proper controls are carried out. In plant transformation studies for example it is possible that a very low level of agrobacterial contamination can persist in cultures which may, using the approaches as described in this report, give a false positive. In addition to checking for the absence of contaminating bacteria by growth in nutrient media it is possible to carry out additional PCR reactions with primers designed to amplify regions outside the T-DNA, eg from the virulence region of the Ti or Ri plasmid which are not present in axenic tissues.

4. CONCLUSION

In conclusion, we have found the PCR technique to be very useful in the identification and analysis of transformed plant tissues. In general, the technique is flexible, requires DNA of moderate quality and is capable of specifically amplifying foreign genes from plant genomic DNA within a few hours. The standard buffer as developed by members of the Cetus Corporation[1,2,13] is near optimal using genomic or c-DNA from a number of plant species and a regime of 30 cycles of amplification should allow visualization of bands on etbr agarose gels using 20ng-2μg genomic DNA. In agreement with others we have found that it is advisable to use as high an annealling temperature as possible to avoid artifact bands being amplified. In practice oligonucleotides of 21 bases in length work well with TM calculated by the formula 4 (G+C) + 2(A+T) being a good guide as to the annealling temperature. However, it is advisable to determine the precise annealling temperature for any pair of oligonucleotides by a process of trial and error.

REFERENCES

1) Erlich, H.A. (Ed.) (1989) PCR Technology: Principles and applications for DNA amplifications. Stockton Press, New York, London, Tokyo, Melbourne, Hong Kong. 2) Innis, M.A., Gelfard, D.H., Sninsky, J.J., White, T.J. (1990) PCR protocols: A guide to methods and applications. Academic Press Inc., San Diego, New York, London, Sydney, Tokyo, Toronto. 3) Lassner, M.W., Peterson, P. Yoder, J.I. (1989) Plant Mol. Biol. Reporter 7, 116-127. 4) Jefferson, R.A., Kavanagh, T.A., Bevan, M.W. (1987) EMBO J 6: 3901-3907. 5) Hamill, J.D., Robins, R.J., Parr, A.J., Evans, D.M., Furze, J.M., Rhodes, M.J.C. (1990) Plant Mol. Biol. (in press). 6) Spencer, A., Hamill, J.D., Rhodes, M.J.C. (1990) Plant Cell Rep. 8: 601-604. 7) Dellaporta, S.L. Plant Mol. Biol. Rep. 1: 19-21. 8) Jefferson, R.A., Burgess, J.M., Hirsch, D. (1986) Proc. Natl. Acad. Sci. (USA) 83: 8447-8451. 9) Beck, E., Ludwig, G., Averswald, E.A., Reiss, B., Schaller, H. (1982) Gene 19: 327-336. 10) Furner, I.J., Huffman, G.A., Amasino, R.M. Garfinkel, D.J., Gordon, M.P., Nester, E.W. (1986). Nature 319: 422-427. 11) Fonzi, W.A. and Sypherd, P.S. (1987) J. Biol. Chem. 262: 10127-10135. 12) Depicker, A., Stachel, S., Dhaese, P., Zambryski, P., Goodman, H.M. (1982) J. Mol. Appl. Genet. 1: 561-574. 13) Saiki, R.K. and Gelfard, D.H. (1989). Perkin Elmer Cetus Amplifications 1, 4-5. 14) Haff, L.A. and Mezei, L.M. (1989) Perkin Elmer Cetus Amplifications 1, 8-10. 15) Thein, S.L. and Wallace, R.B. (1986). In Human gentic disorders: a practical approach (ed. K.E. Davis), p33-50 IRL Press, Herndon, Virginia.

Table 1

Oligonucleotides used as primers	Source of DNA	Expected size of fragment
GGTGGGAAAGCGCGTTACAAG GTTTACGCGTTGCTTCCGCCA positions 400→420 and 1599→1579 from the GUS gene[8]	*N. rustica*, *N. tabacum* transformed with pBI 121[4]	1.2Kb
GAGGCTATTCGGCTATGACTG ATCGGGAGCGGCGATACCGTA positions 201→22 and 900→879 in NPT II gene[9]	*N. rustica*, *N. tabacum* transformed with pBI 121	0.70Kb
ATGGATCCCAAATTGCTATTCCTTCCACGA TTAGGCTTCTTTCTTCAGGTTTACTGCAGC positions 1861→1891 and 2637→2607 in T_L T-DNA (entire Rol B gene)[10]	*N. rustica*, *N. tabacum* transformed with *A. rhizogenes* LBA9402[5]	0.78Kb
GAGCCTGGCAGATACTTTGT CCATCACATGTGGGTCCCCA positions 1526→46 and 1809→1790[11] in yeast ODC gene	*N. rustica* transformed with yeast ODC gene[5]	0.28Kb
ATGGCAATTACCTTATCCGCAACTTCTTTA CGAAAGACGCTTTCATTTCCTTGATGTTGA positions 585→615 and 1110→1080 in Nopaline synthase gene[12]	*M. citrata* transformed with *A. tumefaciens* T37[6]	0.53Kb

Table 2.

Oligonucleo-tude primers (Table 1)	Predicted Tm Simple Formula (4(G+C)+2(A+T))	Expected Size of Fragment	Annealling Temperature 55°C	60°C	62°C	64°C	67°C	70°C	72°C
GUS	66	1.2Kb	+	+	+	+	+	+	-
NPT II	64	0.7Kb	+	+	+	+	?	-	-
Rol B	86	0.78Kb	+	+	+	+	+	-	-
Yeast ODC	60	0.28Kb	+	-	-	-	-	-	-
NOS	82	0.53Kb	+	+	+	-	-	-	-

+ signifies a single clear band of expected size seen on a 1% EtBr agarose gel.
? signifies a weak band.
- signifies the lack of any bands on the gel.

Figure I. The effect of varying Magnesium Chloride concentration. Experiment was conducted using 200ng of input N. rustica DNA, 55°C annealling temperature and GUS primers. Magnesium levels were varied from 0, 0.5, 1.0, 1.5, 2.0, 2.5, 5.0 and 10.0mM. Amplification of the 1.2Kb GUS internal fragment is shown on an EtBr-agarose gel (A) and the corresponding Southern blot (B). Tracks 1→8 represent 0→10mM Mg^{++} respectively. 1mM is optimal.

Figure II. The effects of altering the annealling temperature. Experiment was conducted using 200ng of N. rustica DNA in standard buffer and GUS primers. Amplification of the 1.2 Kb internal GUS fragment is shown (A) on an EtBr-agarose gel and (B) on the corresponding Southern Blot. Lanes 1-11 represent annealling temperatures of 45, 48, 50, 53, 55, 58, 61, 64, and 67°C respectively. Artefacts disappear above 61°C.

Figure III. The effect of varying the template concentration. Experiment was conducted using variable levels of DNA from transformed N. rustica, with GUS primers, standard buffer and 55°C as annealling temperature. Input DNA levels were varied from 0, 1, 5, 10, 25, 50, 100, 200, 500, 1000 and 2000ng of genomic DNA. Amplification of the 1.2Kb internal GUS fragment is shown (A) on EtBr-agarose gel and (B) on the corresponding Southern Blot. Tracks 0→11 represent 1→2000ng DNA respectively.

Figure IV. The effects of varying the input level of genomic DNA and cycle number. Experiment was conducted using standard buffer, with 55°C annealling temperature and GUS primers. Amplification of the 1.2Kb internal GUS fragment is shown on an agarose gel. Lanes 1-7 represent 20, 25, 30, 35, 40, 45 and 50 cycles of amplification using; A 20ng input DNA, B 200ng of input DNA and C 2μg of input DNA. Quantification of DNA yield using Hoechst 33258 showed that DNA synthesis was no longer exponential after 20 cycles and stops after about 35 cycles for Reaction B i.e. 200ng of input DNA.

Figure V. The effects of varying the enzyme concentration. Experiment was conducted using 200ng of N. rustica DNA, standard buffer with 55°C annealling temperature and GUS primers. Lanes 1-6 represent 0, 0.25, 0.50, 1.0, 2.0 and 4 units of Taq polymerase/50μl reaction respectively. Optimal levels are 0.25 units/50μl to avoid smearing.

Figure VI. The effect of varying Potassium Chloride concentration. Experiment was conducted using 200ng of input N. rustica DNA, 55°C annealling temperature and GUS primers. Potassium levels were varied from 0, 25, 50, to 100, and 200mM. Amplification of the 1.2Kb internal GUS fragment is shown on an EtBr-agarose gel. Tracks 1→5 represent 0→200mM respectively. 25-50mM is optimal.

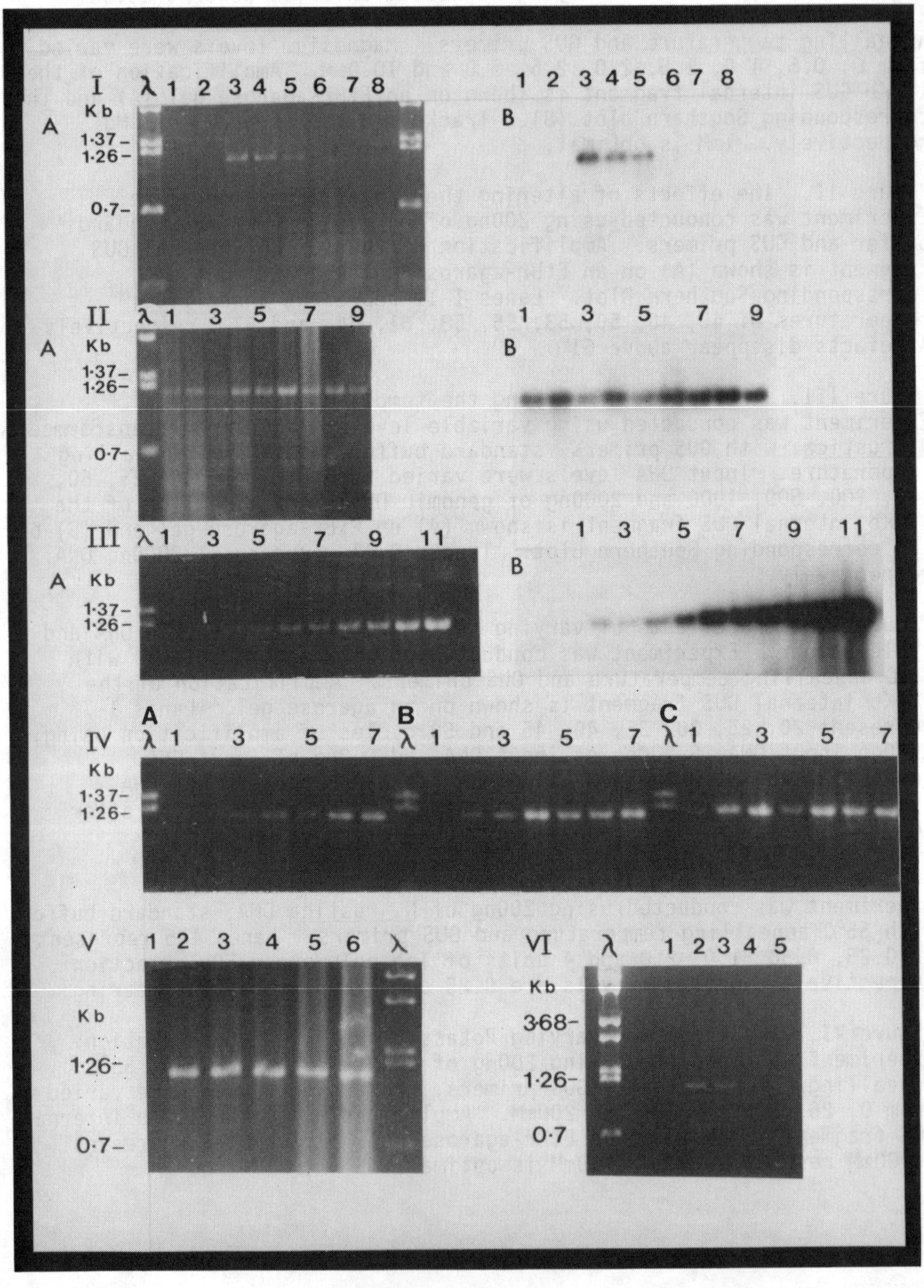

I λ 1 2 3 4 5 6 7 8
Kb
A
1·37–
1·26–
0·7–
1 2 3 4 5 6 7 8
B
II λ 1 3 5 7 9
A
Kb
1·37–
1·26–
0·7–
1 3 5 7 9
B
III λ 1 3 5 7 9 11
A
Kb
1·37–
1·26–
1 3 5 7 9 11
B
IV
A λ 1 3 5 7
B λ 1 3 5 7
C λ 1 3 5 7
Kb
1·37–
1·26–
V 1 2 3 4 5 6 λ
Kb
1·26–
0·7–
VI λ 1 2 3 4 5
Kb
3·68–
1·26–
0·7

DIFFERENTIAL EXPRESSION OF AGROPINE SYNTHASE GENE DURING MORPHOGENESIS IN VITRO

M. INOGUCHI[1], H. KAMADA AND H. HARADA
INSTITUTE OF BIOLOGICAL SCIENCES, UNIVERSITY OF TSUKUBA
1-1-1 TEN-NOH-DAI, 305 TSUKUBA, JAPAN
[1]PRESENT ADDRESS:DEPARTMENT OF BIOCHEMISTRY, OKAYAMA UNIVERSITY OF SCIENCE, 1-1 RIDAI-CHO, 700 OKAYAMA, JAPAN

1. INTRODUCTION

The agropine synthase gene (*ags*) is a kind of opine synthase genes found in T-DNA regions of Ti and Ri plasmids. Opine synthase genes and oncogenic (*onc*) genes are important for genetic colonization of host plants by *Agrobacteria* (1). Historically these genes were assumed to be constitutive in plants and the 5' control region of the octopine synthase gene (*ocs*), the nopaline synthase gene (*nos*) and the 1' and 2' genes in TR-DNA of the Ti plasmid for mannopine synthesis have been widely used as promoters for foreign genes in plants. However, it was recently found that the TR-DNA 1'2' dual promoter and the TL-DNA gene 5 promoter are influenced by phytohormone levels (1, 2). Furthermore a recent study of the *nos* promoter revealed that it's expression is organ-specific and developmentally regulated in transgenic tobacco plants (3).

During our plant transformation experiments, we have isolated a transformant of tobacco (*Nicotiana tabacum* cv. Wisconsin 38) that possesses only a TR-DNA of an octopine type Ti plasmid pTiA66. Since the TR-DNA of this type of plasmids does not encode *onc* genes but agropine synthesis functions, this transformant line (A66-20) was not hormone autonomous but showed agropine production. We found that its agropine content varied with morphological stages of the *in vitro* culture.

We report here that agropine synthetic activity in the transgenic tobacco line is consistently higher in callus than in leaf tissues and that the *ags* promoter fused to the GUS reporter gene (Pags-GUS) is indeed more active in callus than in leaf tissues, though it is expressed in apical meristems of regenerated plants.

2. MATERIALS AND METHODS

2.1 Plant material: Axenic shoots of tobacco (*N. tabacum* cv. Wisconsin 38) were used in this study. They were micropropagated monthly on MS basal medium (4) and grown under a 16 h photoperiod (ca. 3000 lux) at 25°C.

2.2 Bacterial strains: Bacterial strains and plasmids used in this study are shown in Table 1.

Table 1. Bacterial strains and plasmids.

			Ref.
BACTERIAL STRAINS			
Agrobacterium tumefaciens			
A66	(pTiA66)	*tms*⁻ mutant of an octopine type A6	(5, 6)
LBA4404	(pAL4404)	helper strain for binary vector system	(7)
PLASMIDS			
pTTR1		pUC19 carring an entire TR–DNA(10.5 kb *Hap* I–*Kpn* I fragment) of an octopine type pTiB6806	this study
pTTRN3		pUC19 carring 2.4 kb *Nco* I fragment of pTTR1, which contains the *ags* gene	this study
pBI101		a binary vector carring a GUS gene cassette	(8)
pBI121		pBI101 carring CaMV 35S promoter	(8)
pPags2601		pBI101 carring *ags* promoter	this study

2.3 Plant tissue culture: Leaf or stem segments of the transformed plants were cultured on MS medium supplemented with 1 mg/L of naphthaleneacetic acid (NAA) and 1 mg/L of 6-benzylaminopurine (BA) to induce calli. Induced calli were subcultured on fresh medium with the same composition every month. Shoots were regenerated from calli by reducing the concentration of NAA to 0.1 mg/L.

2.4 Agropine synthesis in a transformant A66-20

2.4.1 Isolation of A66-transformants: Decapitated tobacco shoots were infected by directly applying *A. tumefaciens* A66 to the apical cut-surface. Adventitious buds that formed around the sites of inoculation were isolated and transferred to fresh MS medium containing 1 g/L of carbenicillin to eliminate the bacteria. Axenic shoot lines established after sequential subculturing on the same medium were subsequently maintained on MS medium without the antibiotic.

2.4.2 Opine test: Opines were detected by high voltage paper electrophoresis (HVPE) system of Petit et al. (9). Agropine was detected with an alkaline silver nitrate reagent.

2.4.3 Quantification of agropine synthesis: Fifty mg (fresh weight) of tissue were incubated for 2 days at 25°C in 1 mL of liquid MS medium containing 2 µCi of D-[U-^{14}C]-glucose (specific activity, 270 mCi/mmol, Amersham International plc.). Tissues were blotted with paper towelling and an opine test was performed. Each HVPE lane was cut transversely into 5-mm segments. The radioactivity of each segment was counted by a liquid scintillation counter (Beckman LS-8100, U.S.A.) using toluene. To determine the mobility of agropine and mannopine, cold samples of authentic agropine and mannopine had been added to each lane and the segments were stained with alkaline-silver nitrate after the counting.

2.5 Agropine synthase promoter

2.5.1 Plasmid construction: pTTR1, which possessed an entire TR-DNA copy of octopine Ti plasmid pTiB6806, was made by cloning the 10.5 kb *HapI*-*KpnI* fragment of pTiB6806 into the poly-linker of pUC19. A 2.4 kb *NcoI* fragment of pTTR1, which contains the full length *ags* gene and the right border of the TR-DNA copy, was subcloned into pUC19 to make up pTTRN3. pPags2601 was made by inserting the 260 bp *AccI* fragment of pTTRN3, which carries the 5' flanking region of the *ags* gene (Pags) upstream of the GUS gene of a binary vector pBI101 (Table 1).

2.5.2 Plant transformation: The binary vector pPags2601 in *E. coli* JM109 was mobilized into *A. tumefaciens* LBA4404 by triparental conjugation described previously (10). Leaf discs of *N. tabacum* cv. Wisconsin 38 were transformed with *A. tumefaciens* LBA4404, which harbored the pPags2601 or pBI121 as reported previously (11). Transformed plants were selected on MS medium containing 100 mg/L of kanamycin. Plants were maintained in axenic culture on hormone-free MS medium under a 16 h photoperiod (ca. 3000 lux) at 25°C.

2.5.4 GUS Assay: Histochemical and fluorometric GUS assays (8) were performed with X-Gluc (Sigma) and MUG (Sigma) as substrates, respectively. Five µL of extract obtained from ca. 50 mg fresh weight of tissue were used per fluorometric assay, which was carried out for 30 min at 37°C. Reaction rates were in a linear range during incubation. Total protein content in extracts was determined by dye-binding (12) with bovine serum albumin as the standard.

3. RESULTS & DISCUSSION

3.1 Agropine synthesis in a transformant A66-20

3.1.1 Isolation of A66-20: Among over 120 shoots induced by the infection of *A. tumefaciens* A66 on tobacco, only one line (A66-20) was not hormone autonomous and produced agropine. This phenotype implies that A66-20 is transformed with TR-DNA, but without a functional TL-DNA copy. Southern analysis of A66-20 DNA confirmed this. Non-transformed cells around transformed cells may respond to phytohormones released from the transformed cells to form non-transformed shoots (13) and this might be the mechanism through which A66-20 was formed. By an unaccountable reason, however, shoots from this line remained very small even after a prolonged culture. Normal growth occurred after grafting it to a *N. glauca* rootstock.

3.1.2 Agropine content: The presence of agropine in the A66-20 line was examined in shoot and callus tissues by paper electrophoresis followed by alkaline-silver nitrate staining. Opine tests could not be conducted with the original A66-20 shoot due to its small size. However, the elongated A66-20 shoot from the *N. glauca* graft tested positive for agropine. Detection of agropine was remarkably more distinct when A66-20 callus was tested. Calli were induced from A66-20 shoots prior to and after grafting. Adventitious buds were induced from these calli and the regenerated shoots were of a similar phenotype to the original A66-20 shoot. These secondary A66-20 shoots elongated in a manner similar to the original A66-20 shoot when grafted on *N. glauca*.

A comparison of the size and intensity of stained spots indicates that agropine accumulates to higher concentrations in callus than in shoot tissues (contents in root tissue was similar to that of leaf tissue). Agropine content in nongrafted secondary shoots (obtained from the callus of grafted A66-20 shoots) was higher than in grafted shoots but lower than in callus. This relationship was observed during two additional cycles of callus induction and plantlet regeneration.

3.1.3 Synthetic activity: The rate of label incorporation into the agropine moiety after HVPE of transformed tissues was compared with that of control (non-transformed) tissues. Whereas the callus tissue showed distinct label incorporation that correspond to agropine and mannopine, no label incorporation was detected for agropine or mannopine when shoot tissues were used. This HVPE data were confirmed by HPLC analysis.

This suggests that agropine synthetic activity itself varies between shoot and callus tissues.

3.2 Agropine synthase promoter

3.2.1 Expression of the Pags-GUS in leaf and callus tissues: GUS activity in extracts prepared from leaves and calli was assayed with several independently isolated Pags-GUS transformants and a CaMV-GUS transformant. GUS activity in Pags-GUS transformants was consistently higher in callus than in leaf tissue, although levels of activity varied depending on the transformant used (Table 2).

Table 2. GUS activity in callus and leaf tissue of several transformed lines.

Tissue lines	4-MU released			
	pmol/h/100 ng fr wt		pmol/h/μg protein	
	leaf	callus	leaf	callus
Pags-GUS				
#5	< 0.1	0.6	< 0.1	1.2
#7	< 0.1	36.2	< 0.1	149.9
#12	0.3	14.3	0.5	24.5
#23	1.9	59.5	3.7	149.5
35S-GUS	23.7	18.5	38.2	67.7
non-transformant	< 0.1	< 0.1	0.7	3.7

GUS activity in leaf extracts was usually undetectable. GUS activity in the CaMV-GUS transformant was high in both leaf and callus tissues (Table 2). This is strong evidence that expression of the *ags* gene promoter of the Ti plasmid is tissue specific, which was the cause of variation in agropine synthetic activity mentioned above.

3.2.2 Spatial distribution of the Pags-GUS expression in a whole plant: Fluorometric assays of GUS activity were positive for extracts prepared from basal parts of stems and roots of the Pags-GUS transformed plant but were negative for leaves taken from any position on the plant (Fig. 1).

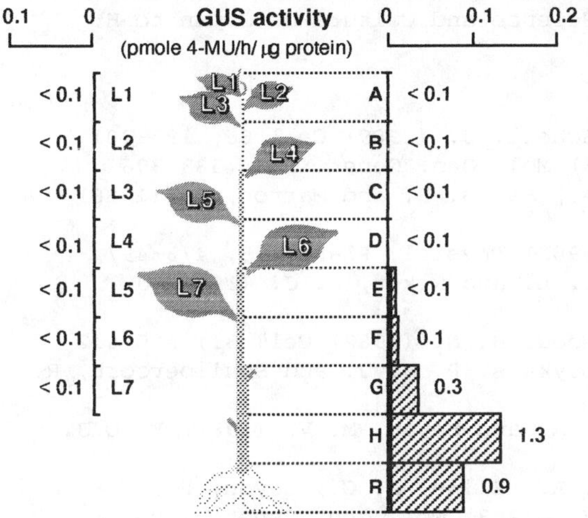

Fig. 1. Positional distribution of GUS activity in a Pags-GUS transformant. L1 to L7, completely expanded leaves from apical to basal end; A to H, stem segments; A, shoot apex and several nodes with unexpanded leaves; B to H, nodes at apical ends; R, root system.

GUS activity of Pags-GUS transformants was histochemically localized to meristematic tissues of shoot apices and root tips. Although the staining was clear and consistent at root tips, it varied shoot apical meristems. Staining could not be observed in either leaf blades or root caps. Cambial tissue usually did not stain, but sometimes, in the lower part of the stem, staining appeared in root primordia. In contrast, a CaMV-GUS transformant showed staining in both mesophyll and veins of leaf blades, along vascular systems in stems and roots, and, most intensely, at apical meristems.

This suggests that activity of the *ags* promoter might be specific to meristematic tissues of whole plants and callus. The relative absence of staining from stem segments seems to be contradictory to rather high GUS activity detected by fluorometric assay. There might be unlocalized GUS activity perhaps in cambial meristems, that could be detected by fluorometric assay but not by histochemical detection. The reasons for variability of staining at shoot apical meristems are not clear, so it remains unclear in which state of meristematic tissues the *ags* expression is specific. Thus we are currently investigating the relationship between the *ags* promoter activity and stages of cell division.

The *ags* gene is obviously not constitutive but appears to be specifically expressed during certain stages of meristematic activity of callus and whole-plant meristems.

4. ACKNOWLEDGEMENT

Agropine was kindly provided by Professor J. Tempé of Université de Paris-Sud.

This work was supported in part by a Grant-in-Aid for Scientific Research on Priority Areas and a Grant-in-Aid for Special Projects from the Ministry of Education, Science and Culture of Japan to H. Kamada and H. Harada.

REFERENCES

1 Zambryski, P., Tempé, J. and Schell, J. (1989) Cell 56, 193-201
2 Koncz, C. and Schell, J. (1986) Mol. Gen. Genet. 204, 383-396
3 An, G., Costa, M. A., Mitra, A., Ha, S.-B. and Marton, L. (1988) Plant Physiol. 88, 547-552
4 Murashige, T. and Skoog, F. (1962) Physiol. Plant. 15, 473-497
5 Hendrickson, A. A., Baldwin, I. L. and Riker, A. J. (1934) J. Bacteriol. 28, 597-618
6 Binns, A. N., Sciaky, D. and Wood, H. N. (1982) Cell 31, 605-612
7 Hoekema, A., Hirsch, P. R., Hooykaas, P. J. J. and Schilperoort, R. A. (1983) Nature 303, 179-181
8 Jefferson, R. A., Kavanaph, T. A. and Bevan, M. V. (1987) EMBO J. 6, 3901-3907
9 Petit, A., David, C., Dahl, G. A., Ellis, J. G., Guyon, P., Casse-Delbart, F. and Tempé, J. (1983) Mol. Gen. Genet. 190, 204-214
10 Ditta, G., Stanfield, S., Corbin, D. and Helinski, D. R. (1980) Proc. Natl. Acad. Sci. 77, 7347-7373
11 Horsch, R. B., Fraley, R. T., Rogers, S. G., Sanders, P. R., Lloyd, A. and Hoffmann, N. (1984) Science 223, 496-498
12 Bradford, M. M. (1976) Anal. Biochem. 72, 248-254
13 van Slogteren, G. M. S., Hoge, J. H. C., Hooykaas, P. J. J. and Schilperoort, R. A. (1983) Plant Mol. Biol. 2, 321-333

INTRINSIC GUS-LIKE ACTIVITY

Ching-yeh Hu, Paula P. Chee, Robert H. Chesney, James H. Zhou, Paul D. Miller and W. Timothy O'Brien
Biology Department, Wm. Paterson College, Wayne, NJ 074370 USA (C-y.H., R.H.C., W.T.O.); Molecular Biology Division, The Upjohn Co., Kalamazoo, MI 49001 (P.P.C.); Department of Biology, OML, Yale Univ., New Haven, CT 06511 (J.H.Z.); and DNA Plant Technology Corp., Cinnaminson, NJ 08077 (P.M.)

Introduction

One of the major reasons leading to the popularity of the *beta*-glucuronidase (GUS)-fusion system (1) was the general assumption that there was no detectable intrinsic GUS activity in higher plants (2). Nevertheless, more and more investigators were troubled by the so-called "false positive" results or "background" activities when GUS assays were performed. We decided to look into this phenomenon with a large variety of selected seed plants. Positive GUS tests were observed in certain parts of some Gymnosperm and every key group of Angiosperm species. In fact, the presence of GUS activities in seed plants were documented in a handful of early reports (reviewed in 3) and some recent reports (4,5).

The objective of this report is to document and alert workers in the field about this wide-spread phenomenon of intrinsic GUS-like activity in seed plants.

Materials and Methods

Plant Materials

Survey of intrinsic GUS-like activity in seed plants -- The following 52 species were chosen and assayed for the presence of intrinsic GUS-like activities: amaranth *(Amaranthus gangeticus* L.), apple (*Pyrus Malus* L), aster (*Aster* L.), avocado (*Persea americana* Mill.), beet (*Beta vulgaris* L), blue grass (*Poa pratensis* L.), cabbage (*Brassica oleracea* L. var. capitata, L.), carrot (*Daucus carota* L. var. sativa, DC.), celery (*Apium graveolens* L. var. dulce, Pers.), *Chrysanthemum coronarium* L., Corn (*Zea mays* L.), cucumber (*Cucumis sativus* L.), daffodil (*Narcissus Pseudo-Narcissus* L.), ginkgo (*Ginkgo biloba* L.), *Glycine soja* Sieb. & Zucc., grapefruit (*Citus paradisi* Macf.), hazelnut (*Corylus americana* Marsh.), holly (*Ilex verticillata* Gray), iris (*Iris* L.), kohlrabi (*Brassica caulorapa* Pasq.), lettuce (*Lactuca sativa* L.), lotus (*Nelumbo nucifera* Gaertn.), maple (*Acer saccharinum* L; *A. palmatum* Thunb.), muskmelon (*Cucumis melo* L.), pak-choi (*Brassica chinensis* L.), parsley (*Petroselinum crispum* Nym.), pea (*Pisum sativum* L.), pear (*Pyrus communis* L.), pecan (*Carya illinoensis* Koch), pine (*Pinus* L.), poppy (*Papaver orientale* L.), pumpkin (*Cucubita pepo* L.), radish (*Raphanus sativus* L.), red cedar (*Juniperus virginiana* L.), rice (*Oryza sativa* L.), rose (*Rosa multiflora* Thunb.), scallion (*Allium fistulosum* L.), snake plant (*Sansevieria trifasciata* Prain), soybean [*Glycine max* (L.) Merr. cv. Burpee's Tenderpod], spider flower (*Cleome spinosa* L.), spiderwort (*Tradescantia virginiana* L.), spinach (*Spinacia oleracea* L.), string bean (*Phaseolus vulgaris* L. cv. Burpee's Tenderpod), sweet pepper (*Capsicum frutescens* L. var. grossum, Bailey), tobacco (*Nicotiana Tabacum* L.), tomato (*Lycopersicon esculentum* Mill.), tree-of-heaven (*Ailanthus altissima* Swingle), turnip (*Brassica Rapa* L.), watermelon (*Citrullus vulgaris* Schrad.), wheat (*Triticum aestivum* L. cv. Durum and Hard Red Spring), yew (*Taxus cuspidata* Sieb. & Zucc.), zucchini (*Cucurbita Pepo* L. var. Melopepo, Alef.). The roots of holly, soybean, string bean and tobacco from *in vitro* cultured plants were also tested.

Tissues from five different plants of a given species (or variety) were surface disinfacted and tested under aseptic conditions with either histochemical or qualitative fluorometric or both procedures.

Changes in the pattern of intrinsic GUS-like activities during seed germination -- Sterile seeds of soybean and string bean were germinated under aseptic conditions. Histochemical GUS tests were performed on the seed sections or germinated seedlings at the 0, 2, 4, 6, 8, 10, 12, and 14th day of germination. The roots, hypocotyls, primary leaves, trifoliate leaves, and flowers of mature string bean plants were also tested.

Changes in intrinsic GUS-like activities of immature embryos during *in vitro* culture -- Immature pods of soybean and string bean were harvested and surface disinfected. Embryos, ranging from heart to early cotyledon stages, were excised and cultured in modified B5 liquid medium (6). Histochemical GUS tests were performed at the 0, 1, 2, 3, 5 and 7th day of culturing.

Histochemical GUS Assay

A modified Jefferson's procedure (7) was followed. The testing solution contained 2 mM substrate, X-Glu, in a pH 7.0 phosphate buffer (29.3 mM K_2HPO_4 + 20.7 mM KH_2PO_4) in which 10 mM EDTA and oxidation catalysts of 0.5 mM potassium ferricyanide and 0.5 mM potassium ferrocyanide were added. The sterile tissues were incubated with filter sterilized testing solution overnight in a 37°C dark incubator. The chlorophylls, if present, were cleared by soaking the tissues in 95% ethanol before data taken.

Qualitative Fluorometric GUS Assay

The mature dry seeds and seedling leaves of the following 12 species were tested with a qualitative fluorometric assay procedure described by Jefferson (7): carrot, cucumber, lettuce, pea, radish, spinach, squash, string bean, sweet corn, tomato, turnip and watermelon. Soybean seeds at various developmental stages (0.5, 1, 2, 3, 4, 5, and 6 mm in length plus mature dry seeds) were tested along with leaves of corn (negative controls) and E. coli GUS gene containing transgenic tobacco (positive controls).

Tissue extract in MUG containing lysis buffer was incubated at 37°C for 0 and 24 h for the first 12 species and for 0, 6, 12 and 24 h for soybean embryos. The blue fluorescence was observed visually with a long-wave UV light box.

ELISA GUS Assay

Standard ELISA procedure was carried out to assay the presents of E. coli GUS in plant tissue with a dilution series of purified E. coli GUS as the positive control. Immature soybean seeds of 3, 5, 7 and 12 mm in length were sampled. Other tissues sampled were young corn and rice leaves, mature rice leaves, and embryogenic oil palm callus. The detection level of the assay was as low as 7.2 ng.

Results

Histochemical Assays

Survey of intrinsic GUS-like activity in vegetative organs -- The leaves of 32 tested species expressed negative and 11 species expressed positive intrinsic GUS-like reactions (Table I). Only scallion had positive staining reactions on the entire leaf of every tested specimen. In rest of the positive species, either only certain portions of the leaf (mostly the major veins near the base of the blade and the petioles) stained or some of the tested specimens failed to express stain. The intensities of the staining reactions on leaves, except for scallion, were quite low. Roots of all four tested species expressed negative reactions.

Survey of intrinsic GUS-like activity in reproductive organs -- The intrinsic GUS-like activities in fruits and/or seeds of 46 species are listed in Table I. Table II shows that species expressing positive activities are distributed in every key group of Angiosperms (classification is based on Porter (8)) as well as in some of the Gymnosperms. All except five tested species showed positive reactions in one or more parts of fruit wall, seed coat, endosperm/gametophyte and/or embryo. The immature fruits of soybean, for example, showed strong staining reactions in all the above four tested parts. Usually, the vascular bundles expressed stronger staining activities than the surrounding tissues. Like its leaves, a high intensity staining reaction was expressed by scallion endosperms. In the immature seeds of cucumber, the only positively stained tissues were the vascular bundles connecting the seeds to the placenta. The five negative species were avocado, maple, lettuce, tobacco and yew.

The intensity of histochemical staining varies from species to species as well as from tissue to tissue. Among the 45 species with their immature (i) and/or mature (m) embryos tested the following 10 species expressed strong positive embryonic staining reactions: apple (i), cabbage (m), hazelnut (i, m), kohlrabi (m), pear (i), string bean (i, m), soybean (i, m), spider flower (m), tomato (i). The following 14 species expressed moderate embryonic staining reactions: *Chrysanthemum* (m), corn (m), cucumber (i), *Glycine soja* (m), holly (i), muskmelon (m), parsley (m), pecan (m), pok-choi (m), radish (m), rose (m), scallion (m), spinach (m), tree-of-heaven (m). The following 12 species expressed weak embryonic staining reactions: amaranth (m), aster (m), beet (m), carrot (m), grapefruit (i), iris (i), lotus (m), pumpkin (m), rice (m), snake plant (i), sweet pepper (i), wheat (m). Only 11 species had negative embryonic staining reactions: avocado (i), blue grass (m), celery (m), maple (i), lettuce (m), pine (m), rice (i), spiderwort (m), tobacco (m), yew (i, m) zucchini (m).

Changes in the pattern of intrinsic GUS-like activities during seed (Table III) -- The intrinsic GUS-like activities existed in every part of mature dry seeds of soybean and string bean with the exception of the primary leaves of string bean. The intensities of staining increased during the early period of the germination process (even the primary leaves of string bean expressed positive staining) then dropped rapidly on or after the fourth day of germination. Practically no detectable positive staining reactions were found on any part of the seedlings by the tenth day of germination. Negative results were detected on all the six tested organs of the mature string bean plants.

Changes in intrinsic GUS activities of immature embryos during *in vitro* culture -- In both species immature embryos expressed strong GUS-like staining activities at dissection. All the staining activities of smaller embryos and most of the staining activities of larger embryos disappeared after one to two days of *in vitro* culture. By the fifth and seventh days, only small blue spots infrequently appeared on certain larger embryos.

Qualitative Fluorometric Assay

Among the 12 species with seedling leaves tested, only tomato expressed positive fluorescence. Among the seeds tested, pea and sweet corn were clearly positive. Soybean seeds expressed positive reactions at all tested developmental stages, but the intensities were not as strong as those of transgenic tobacco leaves (the positive control). The negative controls, the sweet corn leaves, showed no activity (Fig. 1).

ELISA GUS Assay

Anti-GUS activity was detected in all of the positive control wells. No anti-GUS activity was detected in any of the test samples assayed.

Discussion

This work revealed that, contrary to the general belief, intrinsic GUS-like activity is widely distributed in the seed plants. Such activities are mainly expressed during fruit and seed development and infrequently expressed in the vegetative organs of seedlings and mature plants. On the basis of the ELISA tests, it appears that the protein(s) responsible for the intrinsic GUS-like activity is antigenically dissimilar to the *E. coli* GUS enzyme.

The intrinsic GUS-like activities in the seeds diminished during the germination process. It is possible that this activities may not completely disappear even at the mature phase of the plant life. Wenzler et al. (4) detected low levels of endogenous GUS activities in the leaves and roots of the potato and tobacco plants with the highly sensitive quantitative fluorometric assays. High level of activities reappear at the on set of reproductive phase, as demonstrated during anther (5), fruit and seed developments.

In vitro culture conditions quickly turned off most of the intrinsic GUS-like activities in excised immature soybean and string bean embryos. But our work in culturing various types of soybean tissues indicated that *in vitro* culture conditions did not always turn off such activities (unpublished data). These data suggested that the microenvironment around the cell played an important role in the expression of gene(s) for such activities.

References

1. Jefferson RA, Burgess SM, Hirsh D (1986) Natl Acad Sci USA 83: 8447-8451
2. Jefferson RA, Kavanagh TA, Bevan MW (1987) EMBO J 6: 3901-3908
3. Fishman WH (1955) Adv In Enzymology 16: 361-409
4. Wenzler H, Mignery G, Fisher L, Park W (1989a) Plant Mol Biol 13: 347-354
5. Plegt L, Bino RJ (1989) Mol Gen Genet 216: 321-327
6. Hu CY, Sussex IM (1986) VI Int'l Congr Pl Cell & Tissue Cult (Abs# 28) Univ Minnesota
7. Jefferson RA (1987) Plant Mol Biol Rep 5: 387-405
8. Porter CL (1967) Taxonomy of Flowering Plants. Freeman and Co, San Francisco

Figure 1. Qualitative fluorometric assay of intrinsic GUS-like activities in soybean seeds at various developmental stages. Well 1-7: extracts from immature seeds of 0.5, 1, 2, 3, 4, 5, and 6 mm in lengths, respectively. Wells 9-13: negative controls; leaf extracts from 2-week old corn plants. Wells 14-16: positive controls; leaf extracts from transgenic tobacco plants.

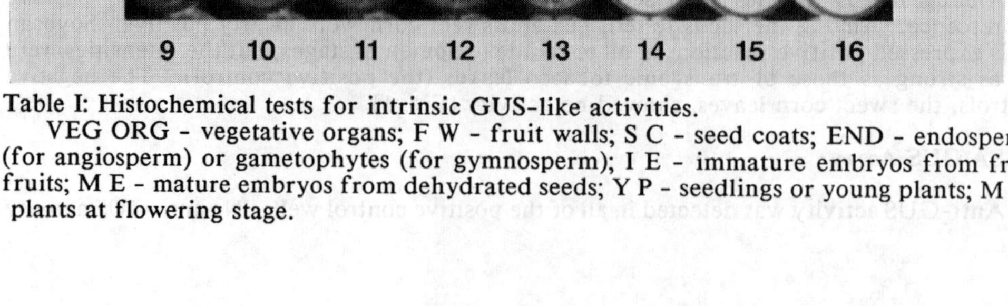

Table I: Histochemical tests for intrinsic GUS-like activities.

VEG ORG - vegetative organs; F W - fruit walls; S C - seed coats; END - endosperms (for angiosperm) or gametophytes (for gymnosperm); I E - immature embryos from fresh fruits; M E - mature embryos from dehydrated seeds; Y P - seedlings or young plants; M P - plants at flowering stage.

	REPRODUCTIVE ORGANS					VEG ORG	
	F W	S C	END	I E	M E	Y P	M P
Amaranth		−	−		+	−	
Apple	−	+			+		−
Aster		+	+		+		
Avocado	−	−			−		
Bean, string				+	+	−	−
Beet	+	−	+		+	−	
Blue grass	+	+	−		−		−
Cabbage		−			+	−	
Carrot		+	+		+	+	
Cedar, Red			+				−
Celery		+	+		−	+	
Chrysanthemum		−			+		−
Corn	+	+	−		+	−	
Cucumber	−	+		−	+	−	
Daffodil							−
Ginkgo							+
Glycine soja					+	−	
Grapefruit		−		+			
Hamlock		+	+		+	−	
Hazelnut	−	+		+	+	−	
Holly	−	−	+	+		−	−
Iris	−	−	+	−			−
Kohlrabi		−			+	−	
Lettuce		−			−	−	
Lotus			−		+		
Maple	−	−		−			−
Muskmelon		−			+		−
Pak-choi		−			+	+	
Parsley		+			+	+	
Pear	−	+		+			−
Pecan	−	−			+		
Pepper, sweet	+	+	+	+		−	+
Pine		−	+		−		−
Poppy							−
Pumpkin		−			+		
Radish		−			+	−	
Rice	+	+	−	−	+	+	+
Rose	−	+			+		−
Scallion		−	+		+		+
Snake plant	−		+	+			−
Soybean	+	+	+	+	+	−	−
Spider flower	+	−	+	−	+	+	
Spiderwort	+	+		−			−
Spinach		+	−		−	−	
Tobacco		−	−		−	−	
Tomato	+	+	+	+	+	−	+
Tree-of-Heaven	−	−	+		+	+	
Wheat	+	+	−		+	−	
Yew		−	−	−			−
Zucchini		+	+		−		−

Table II. A survey of seed plants on the presence of intrinsic GUS-like activities in various parts of fruits and/or seeds.

Classification of angiosperms is based on Porter (1969). Non-underlined species expressed positive GUS-like reactions in one or more tested parts via histochemical test. Underlined species expressed negative reactions in all tested parts. See Table I for specific fruit/seed parts which were tested in each species.

GYMNOSPERM: hamlock, pine, red cedar, yew
ANGIOSPERM
I. CLASS MONOCOTYLEDONEAE
A. SUBCLASS CALYCIFERAE: spiderwort
B. SUBCLASS COROLLIFERAE: iris, scallion, snake plant
C. SUBCLASS GLUMIFLORAE: blue grass, corn, rice, wheat
II. CLASS DICOTYLEDONEAE
A. THE APETALAE
1) THE AMEMTIFERAE: hazelnut, pecan
2) THE FLORIFERAE: amaranth, beet, spinach
B. THE POLYPETALAE
1) THE HYPOGYNOUS
a) THE APOCARPOUS: avocado, lotus
b) THE SYNCARPOUS: cabbage, grapefruit, holly, kohlrabi, maples, pak-choi, radish, spider flower, tree-of-heaven
2) THE PERIGYNOUS: apple, pear, rose
3) THE EPIGYNOUS: carrot, celery, *Glycine soja*, parsley, soybean, string bean
C. THE SYMPETALAE
1) THE HYPOGYNOUS: sweet pepper, tobacco, tomato
2) THE EPIGYNOUS: aster, *Chrysanthemum*, cucumber, lettuce, muskmelon, pumpkin, zucchini

Table III. Change in intrinsic GUS-like activities via histochemical test during seed germination.

PLUM - plumule; COTY - cotyledon; HYPO - hypocotyle; RADI - radical or tap root.

SPECIES	DAYS UNDER GERM COND	PL. PARTS PLUM	COTY	HYPO	RADI
Soybean	0	++	++	++	++
	2	++	++	++	++
	4	+++	+++	+++	+++
	6	–	++	+	–
	8	–	+	–	–
	10	–	–	–	–
	12	–	–	–	–
String Bean	0	+	++	+	++
	2	+	+++	++	++
	4	+	++	+++	–
	6	–	++	++	–
	8	–	+	+	–
	10	–	–	–	–
	12	–	–	–	–

FIELD TESTING OF CUCUMBER PLANTS WHICH EXPRESS THE CMV COAT PROTEIN GENE: FIELD PLOT DESIGN TO TEST NATURAL INFECTION PRESSURES

Jerry L. Slightom[1], Paula P. Chee[1], & Dennis Gonsalves[2]
[1]Molecular Biology-Unit 7242, The Upjohn Company, Kalamazoo, MI 49007, USA; [2]Department of Plant Pathology, Cornell University, New York State Agriculture Experiment Station, Geneva, NY 14456, USA

1. Introduction

The development of plants which are tolerant to virus infections has been shown by the expression of the tobacco mosaic virus (TMV) coat protein gene in transgenic tobacco plants (1). However, TMV is not the most prevalent or severe virus TMV. A more important virus pest, with respect to crop losses, is the cucumber mosaic virus (CMV) which is known to have a very broad host range including both monocotyledonous and dicotyledonous plants, as well as being widely distributed throughout the world (2).

The genome of CMV consists of three positive-stranded (+) RNAs and a fourth RNA, a subgenomic RNA derived from RNA3 (3). Both RNA3 and RNA4 encode the coat protein gene. The CMV family contains many different strains which fall into two basic subgroups, subgroup I (includes strains C, D, Fny, and Y) and subgroup II (includes strains Q and WL) (4,5,6).

The CMV strain C coat protein gene (6) was engineered for expression using the cauliflower mosaic virus (CaMV) 35S transcript regulatory sequences, cloned into an *Agrobacterium*-derived binary vector (7), and transformed into the tobacco genome to produce CMV tolerant plants (H. Quemada, D. Gonsalves, and J. Slightom, unpublished). Tobacco plants resistant to CMV infection have also been produced by expressing the coat protein gene of CMV strain D (8). However, the development of CMV tolerant plants belonging to the family cucurbitaceae is much more important because many members of this family are extremely susceptible to CMV infections. Towards this goal, we used the *Agrobacterium*-mediated transfer mechanism and the cucumber plant regeneration system described by Chee (9) to transfer the engineered CMV coat protein gene into the CMV susceptible cucumber cultivar, Poinsett 76. We report here the results of the preliminary test of these transgenic cucumber plants in a field environment with natural CMV infection pressure.

2. PROCEDURES

2.1 Materials and methods

2.1.1 Plant materials: Cucumber seeds (*Cucumis sativus* L. Poinsett 76, Asgrow Seed Co.) were decontaminated and germinated as described by Chee (9).

2.1.2 <u>Transformation</u>: Transformation of cucumber was performed using a method similar to that described by Horsch et al. (10). The cotyledons, from 3 to 5 day old seedlings, were removed aseptically and cut into pieces of approximately 5 mm^2. These pieces were then submerged into an overnight culture of the avirulent *Agrobacterium* strain C58Z707 (11) which contained the binary plasmid pGA482GGcpCMV19. After exposure to the bacteria, the cotyledon pieces were blotted dry and cultured abaxial side down in a sterile petri dish containing 10 ml of initiation medium [MS basal medium +2.0 mg/l 2,4-dichlorophenoxyacetic acid (2,4-D) + 0.5 mg/l kinetin (KN)] (9) followed by growth in the dark at 26°C. After 4 days the cotyledon pieces were transferred to the same medium supplemented with 500 mg/l carbenicillin (Cb) and 100 mg/l kanamycin (Km) and cultured for five weeks in the dark at 26°C.

2.1.3 <u>Plant regeneration</u>: Regeneration of potentially transformed embryogenic callus tissue was done according to the procedure described by Chee (9). Briefly, after six weeks on initiation medium, the cultures were transferred to MS medium + 1.0 mg/l naphthaleneacetic acid (NAA) + 0.5 mg/l KN + 100 mg/l Km + 500 mg/l Cb for the development of somatic embryos. These cultures were incubated for an additional two weeks followed by transferred to hormone-free MS medium + 50 mg/l Km for conversion to plantlets. After the plantlets developed an extensive root system they were transplanted to pots as described by Chee (9).

2.1.4 <u>NPTII, GUS, and CMV coat protein assays</u>: The transformation status of the regenerated cucumber plants was determined by NPT II enzyme activity using the *in situ* gel assay procedure described by Riess et al. (12) and GUS enzyme activity using the procedure described by Jefferson et al. (13). However, to ensure that NPT II and GUS positive plants expressed the CMV coat protein gene, leaf protein extracts were assayed by ELISA using antibodies raised against the whole CMV-C virion as described by Edwards and Gonsalves (14).

2.1.5 <u>Genomic DNA analysis</u>: Total nucleic acids were extracted using the cetyltrimethylammonium bromide (CTAB) procedure described by Saghai-Maroof et al. (15). These DNAs (between 5 to 10 μg) were digested with 5 fold excess *Bam*HI and/or *Hin*dIII enzymes, electrophoresed in a 0.7% agarose gel, followed by transfer onto nylon filters. These filters were hybridized against ^{32}P-labeled NPT II and/or GUS gene-containing fragments as described previously by Chee et al. (16).

3. RESULTS AND DISCUSSION

3.1. Cucumber cotyledon pieces were infected with the avirulent *Agrobacterium* strain C58Z707 which contained the binary vector pGA482GGcpCMV19 (Fig. 1). Putative transformed cucumber tissues were subjected to selection by transferring onto

Figure 1. Structure of the *Agrobacterium* binary plasmid pGA482GG/cpCMV19. Restriction enzyme sites are: Bg/B=fusion between *Bgl*II and *Bam*HI sites; C=*Cla*I; H=*Hind*III; Hp=*Hpa*I; K=*Kpn*I; St=*Stu*I; X=*Xho*I. Not all site locations are shown.

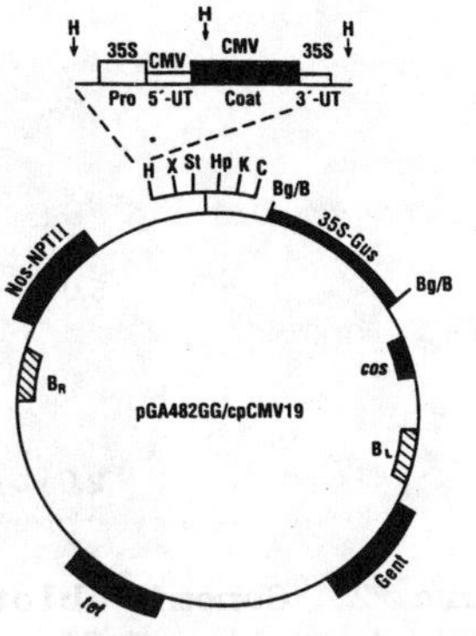

initiation medium supplemented with 100 mg/l Km; this level of kanamycin was found to be sufficient for the selection of transformed tissues (P. Chee, unpublished). A total of 100 putative transformed R_0 plants were obtained and all of these plants were tested for the presence of NPT II enzyme activity. All of the putative transgenic R_0 plants showed NPT II enzyme activity (data not shown) indicating that at least the Nos-NPT II gene of the binary vector (Fig. 1) had been transferred. To determine if the T-DNA region spanning the CMV coat protein gene had been transferred these tissues were next checked for GUS expression (13). GUS activity was found in each transgenic cucumber plant indicating that the CMV coat protein gene should also be present (see Fig. 1).

3.2. Transfer and integration of the Nos-NPT II to GUS gene region of the vector pGA482GGcpCMV19 (Fig. 1) was also analyzed by genomic blot hybridizations. Total genomic DNA was extracted from the leaves of 57 transformed R_0 plants and digested with *Bam*HI and *Hind*III restriction endonucleases, electrophoresed in a 0.7% agarose gels followed by transfer onto nylon filters. These filters were hybridized against ^{32}P-labeled DNA fragments isolated from the NPT II gene, and all R_0 plants were positive (data not shown). Fig. 2 shows an example of results obtained for several R_1 progeny obtained from transformed plant T-36. The ^{32}P-labeled NPT II fragment hybridized to the predicted 2 kb Nos-NPT II *Bam*HI fragment (7) and the ^{32}P-labeled GUS fragment hybridized to a fragment of about 5.0 kb.

3.3 Six of the transformed R_0 plants were then tested for expression of the engineered CMV coat protein gene using the ELISA technique described by Edwards and Gonsalves (14). All six transgenic R_0 plants showed the presence of CMV coat protein; however, their expression levels varied considerably. These six R_0 plant lines were grown in the greenhouse, self-pollinated, and about 200 R_1 seeds were collected for each plant line. [In addition to self-pollination, the T-47 line was also crossed with the CMV susceptible cucumber cultivar

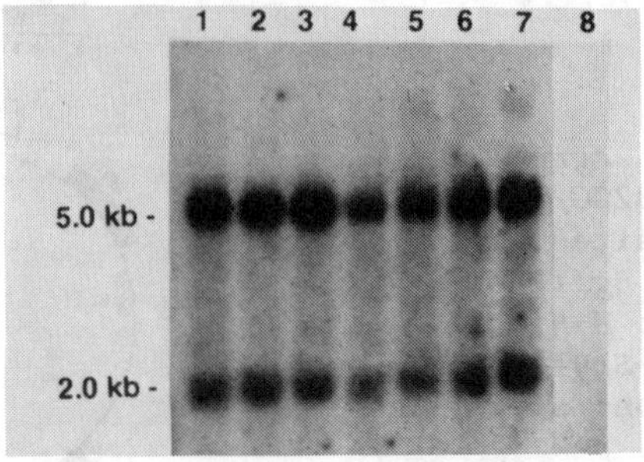

Figure 2. Genomic blot hybridization of R_1 progeny from cucumber plant line T-36. DNAs were digested with *Hin*dIII and *Bam*HI and hybridized against ^{32}P-labeled NPT II and GUS gene fragments. Lanes 1 to 7 contain DNAs from the respective T-36 progeny plants and lane 8 contains DNA isolated from a non-transformed cucumber plant.

Straight 8] The ability of these plant lines to tolerate CMV infection was first tested in the greenhouse by planting 20 R_1 seeds from each of the six lines, and after about two weeks of growth they were inoculated with CMV-C. Two leaves from each plant were infected with CMV-C (inoculum strengths were in the range of 100 to 400 virus lesions per inoculum) as described by Edwards and Gonsalves (14) and the plants were checked for visual symptoms and for virus growth using ELISA assay procedures described by Edwards and Gonsalves (14). All six transgenic plant lines showed delay in the onset of CMV infection as compared to non-transformed control plants (data not shown). Transformed cucumber lines, T-47 and T-60, were selected for testing in a field environment.

3.4. The field test site was an isolated plot which was surrounded by a hardwood forest on two sides and apple and plum orchards on the remaining borders. There were no cucurbits, cultivated or wild, growing in the area of the test. The USDA field release permit allowed for the late season planting of the transgenic cucumber plants, but they were not allowed to set seed, so only data on virus spread was collected. The field plot was designed for heavy disease pressure by incorporating the use of CMV infected susceptible cucumber plants. Plants were planted 4 feet along rows which were separated by 6 feet, and every fifth plant was a CMV infected plant. The infected non-transformed Poinsett 76 plants were mechanically infected with CMV-C five days prior to planting (ID #1, Table 1). None of the transgenic plants (ID #47 and 60, Table 1) were inoculated with CMV as virus spread relied only on natural transmission by the indigenous aphids which were plentiful at the test near the end of August. Virus spread was monitored by the present of a second set of non-transformed Poinsett 76 plants (ID #2, Table 1). The plots also contained F1 hybrid plants obtained from a cross of R_0 T-47 with Straight 8 (ID #4, Table 1) and a CMV tolerant plant line, Marketmore (ID #5, Table 1). Thus, this field test was quite different from that previously reported (17) in that transgenic plants were not directly inoculated. Virus spread to the transgenic plants relied on the natural aphid vector, and this pressure was monitored with the susceptible Poinsett 76 plants (ID # 2).

TABLE 1. Plant identifications used in the field test

Plant ID #	Genotype	Treatment	Number
1	Non-transformed Poinsett 76	CMV-C infected	39
2	Non-transformed Poinsett 76	none	92
47	Transformed Poinsett 76 T-47	none	49
60	Transformed Poinsett 76 T-60	none	42
4	F1 hybrid T-47 X Straight 8	none	91
5	Marketmore (CMV tolerant)	none	94

3.5. The field plot was planted on August 21, 1989 and the plants were scored visually and by ELISA (14) twice during the remaining growing season. The first collection occurred about three weeks after planting (September 13, Table 2) and the second collection occurred about 5 weeks after planting (September 25, Table 2). These field test results were very encouraging because the transgenic lines, T-47 and T-60, were only about 23% infected, while the non-transformed Poinsett-76 plants were 70% infected (Table 2). This indicates that the field experienced a strong virus infection pressure. The number of infected plants found in these transgenic R_1 plant lines may be expected because they should be heterozygous for the coat protein gene. If this is true then an extrapolation could be made that the level of tolerance found for the transgenic plants is similar to that found for the CMV

TABLE 2 Cucumber field trial CMV-ELISA results

Plant identification (from Table 1)	September 13 #infected/#tested	September 25 #infected/#tested
1	39/39 (100% infected)	39/39 (100% infected)
2	16/41 (39% ")	29/41 (70% ")
47 (R_1)	6/22 (27% ")	5/22 (23% ")
60 (R_1)	2/17 (12% ")	4/17 (24% ")
4	11/38 (29% ")	18/47 (47% ")
5	5/39 (13% ")	4/39 (10% ")

tolerant cucumber cultivar Marketmore (Table 2). In addition, seedlings resulting from the cross between R_0 T-47 and CMV susceptible Straight 8 (Tables 1 and 2) showed the expected higher infection rate (near 50%, in Table 2).

3.6. The conclusions which we draw from this preliminary field test are as follows: (1) field tests designed using natural disease pressure do yield meaningful results; (2) the results from using these R_1 transgenic cucumber plants suggest that CMV coat protein gene expression does increase their ability to tolerate CMV virus infection. These results have encouraged us to conduct another test in 1990 using R_2 transformed cucumber plants, and the test will be more informative because it will span a complete growing season.

REFERENCES

1 Powell-Abel, P., Nelson, R.S., De, B., Hoffmann, N., Rogers, S.G., Fraley, R.T., and Beachy, R.N. (1986) Science 232, 738-743
2 Kaper, J.M. and Waterworth, H.E. (1981) in Handbook of Plant Virus Infections and Comparative Diagnosis (Kurstak, E., ed.) pp. 257-332, Elsevier/North-Holland Biomedical Press
3 Peden, K.W.C. and Symons, R.H. (1973) Virology 53, 487-492.
4 Piazzolla, P., Diaz-Ruiz, J.R., and Kaper, J.M. (1979) J. gen. Virol. 45, 361-369
5 Rizzo, T.M. and Palukaitis, P. (1988) J. gen. Virol. 69, 1777-1787
6 Quemada, H., Kearney, C., Gonsalves, D., and Slightom, J.L. (1989) J. gen. Virol. 70, 1065-1073
7 An, G. (1986) Plant Physiol. 81, 86-91
8 Cuozzo, M., O'Connell, K.M., Kaniewski, W., Fang, R.-X., Chua, N.- H., and Tumer, N.E. (1988) Bio/Technol. 6, 549-557
9 Chee, P.P. (1990) Hort. Sci. in press
10 Horsch, R.B., Fry, J.E., Hoffmann, N.L., Eichholtz, D., Rogers, S.G., and Fraley, R.T. (1985) Science 227, 1229-1231
11 Hepburn, A.G., White, J., Pearson, L., Maunders, M.J., Clarke, L.E., Prescott, A.G., and Blundy, K.S. (1985) J. gen. Microbio. 131, 2961-2969
12 Reiss, B., Sprengel, R., Will, H., and Schaller, H. (1984) Gene 30, 211-218
13 Jefferson, R.A., Kavanagh, T.A., and Bevan, M.W. (1987) EMBO J. 6, 3901-3907
14 Edwards, M.C., and Gonsalves, D. (1983) Phytopath. 73, 1117-1120
15 Saghai-Maroof, M.A., Soliman, K.M., Jorgensen, R.A., and Allard, R.W. (1984) Proc. Natl. Acad. Sci. USA 81, 8014-8018
16 Chee, P.P., Fober, K.A., and Slightom, J.L. (1989) Plant Physiol. 91, 1212-1218
17 Nelson, R.S., McCormick, S.M., Delannay, X., Dube, P., Layton, J., Anderson, E.J., Kaniewska, M., Proksch, R.K., Horsch, R.B., Rogers, S.G., Fraley, R.T., and Beachy, R.N. (1988) Bio/Tech. 6, 403-409

TWO YEARS OF TRANSGENIC FLAX FIELD TESTS: WHAT DO THEY TELL US?

ALAN McHUGHEN MARK JORDAN AND SANDRA McSHEFFREY

Crop Development Centre, University of Saskatchewan,
Saskatoon, Sask. S7N 0W0 CANADA

Introduction

Crop improvement is often cited as the ultimate goal of gene transfer technology research. Before this goal can be achieved, however, we need to know what effect the presence of foreign genes (or DNA) will have, not only on the crop plant but also on the environment. It is easy to presume that the physical addition of a few hundred extra base pairs of DNA to a plant genome will have no effect of normal plant growth and development, or that the expression of two or three foreign genes with highly specific functions will not interfere with unrelated plant activities. What experimental evidence is there to support these presumptions? As responsible people and scientists we need to protect our environment and health. In order to do this we need to know how transgenic organisms will affect the natural, agricultural and human environment. These points raise many questions, but the first one to be addressed while transgenic plants are in field testing stage is how to minimize the risk of escape of a transgenic organism (or gamete) from the test site. In order to provide data to answer some of these questions, 14 transgenic lines of commercial cultivars of linseed flax (Linum usitatissimum) were generated and field tested in 1988 and 1989. Two of the four tests were designed to determine outcrossing frequency (and therefore risk of transgenic gamete escape), and two were tests of agronomic performance of lines carrying introduced chimeric herbicide resistance genes.

Materials and Methods

Transgenic plants:

Linseed flax (Linum usitatissimum) commercial cultivars NorLin, McGregor and Vimy were transformed with disarmed Agrobacterium tumefaciens vectors carrying chimeric genes for kanamycin resistance (NPT-II) alone or with genes conferring resistance to the herbicides glyphosate (modified EPSP synthase gene provided by Monsanto, Co., St. Louis, USA) or sulfonylurea (see Jordan and McHughen, 1988; McHughen, 1989 for details of transformation procedure).

Outcrossing tests:

A 2.5m row of transgenic McGregor flax plants was seeded parallel to normal McGregor flax plants; the rows were spaced 30cm apart, so that at anthesis, the inflorescences would be in physical contact with adjacent rows. Ten rows of control McGregor plants were seeded on each side of the transgenic row for a total of 20 rows of McGregor and one of transgenic McGregor. A sample of seed from each row was germinated on petri dishes containing Murashige and Skoog basal medium plus 100mg/l kanamycin. In this medium, normal (non-transgenic) flax will germinate but the root is stunted and does not branch, while the true leaves are bleached. Heterozygous transgenic (NPT-II containing) seedlings appear normal with vigorous root growth and branching and healthy green leaves, so this is a quick assay for detecting transfer of the NPT-II carrying gametes from the homozygous transgenic plants to controls at different distances. Although this test was conducted in 1988, useful data were limited by the ravages of drought, so it was repeated in 1989. Observations were made at least weekly through the growing season to ensure, among other things, that the time of flowering coincided between transgenic and non-transgenic lines. In accordance with federal government regulations, this and the other transgenic tests were isolated from other flax by at least 100m.

Sulfonylurea resistance test:

Eight lines of McGregor and NorLin transformed to contain a mutant sulfonylurea resistant ALS originally isolated from Arabidopsis (Haughn et al., 1988) were tested in soil incorporated with metsulfuron methyl (active ingredient of Ally (DuPont)) at zero, 2.25g/ha and 4.5g/ha, being the control, half and full recommended rate for broadleaf weed control in cereal crops. Non- transgenic McGregor and NorLin lines served as the check, sown alternately with parallel 2.5m rows of the transgenic lines. There were two replicates of each treatment, and one row of each transgenic line per rep, for a total of six plots with 17 rows per plot. Observations were made on such agronomic traits as germination, days to flowering, days to maturity, fertility and seed set, as well as the effects of the herbicide.

Glyphosate resistance test:

Four transgenic lines carrying a modified EPSP synthase gene conferring resistance to the herbicide glyphosate (active ingredient in Roundup (Monsanto)) were grown in a replicated test consisting of alternating rows of transgenic NorLin or Vimy with one of non-transgenic NorLin, Vimy or McGregor. The treatments were zero, 0.28 and 0.56 Kg/ha, being the zero, 1x and 2x recommended rate for grassy weed control. There were three reps with three treatments randomized within each; placement of the specific transgenic lines within each plot was randomized, although always bordered by non- transformed controls on either side. The herbicide was applied when the seedlings were about 10cm tall, corresponding to a usual herbicide treatment application time. The plants were observed throughout the season, recording information on agronomic traits as well as effects of the herbicide treatments, along with records of weed species present. At the end of the season, plants were harvested by hand and threshed.

Results and Discussion

Outcrossing tests:

Although small, the tests provided useful data for the determination of safe isolation distances for use with transgenic flax containing genes less benign than NPT-II. At a sowing distance of 30cm, such that inflorescences are touching at anthesis, pollen transfer occurs at a frequency of less than 1% (see table 1). No transgenic seed was found from a plant sown more than 30cm away from the homozygous NPT- II carrier plants. Therefore, maintaining the standard isolation distance for commercial select seed of flax (as determined by the Can. Seed Growers Assoc., and sanctioned by federal laws) of 100m will almost certainly preclude escape of genetic material via pollen.

Table 1. Frequency of outcrossing of NPT-II carrying pollen from transgenic plants to non-transgenic plants. Data combined from results of 1988 and 1989 tests

Distance from transgenic row	No. seeds tested	No. positive	% outcrossed
30cm west	304	2	0.66
30cm east	330	1	0.30
>60cm, east and west	931	0	0

This isolation distance for flax is useful in the early field testing stages when determining the utility and safety of a novel trait, but of course becomes meaningless if the transgenic lines are commercially released. Others factors must also be considered. Seed dispersal mechanisms, vegetative reproduction, regional prevalence of related species (potential outcrossing host), etc. also need to be addressed before risk of escape can be determined. For these reasons, the consideration of the host plant species is as important as the transferred trait.

The outcrossing tests were also useful for looking at some agronomic effects of the presence of T-DNA. Of critical importance to farmers (and also to an outcrossing experiment) is the timing of flowering and time to maturity. Transgenic and non-transgenic plants reached flowering at the same time (ca. 48d after seeding), produced flowers for the same duration (ca. 35d), and matured (ready for harvest) simultaneously. Oilseed flax is grown because of its high quality industrial drying oil, so any alterations to the oil quality would be of major concern. Samples of seed from the transgenic line and from control plants grown in the 1989 test were analyzed by gas chromatography for oil profile. Table 2 shows that T-DNA does not disrupt the normal oil profile. The differences in absolute values are well within the variation expected for this type of test.

Table 2. Fatty acid profile for transgenic and non-transgenic McGregor flax. Mean percentages for each sample of 10 seeds.

line	palmitic	stearic	oleic	linoleic	linolenic
Control	6.1	5.25	21.5	18.25	48.4
Transgenic	5.9	4.8	18.7	17.1	53

Sulfonylurea resistance test:

In the zero treatment plots (ie. those with no herbicide application), there were no apparent differences between the transgenic lines and the non-transgenic lines of the corresponding cultivar for any measured parameter. Germination, days to anthesis, fertility, days to mature and apparent yield were the same, whether the rows were of transgenic plants or controls. Normal inter-varietal differences were retained however, in that, for example, the NorLin plants (transgenic as well as non-transgenic) were earlier to flower and mature than the McGregor derivatives. Actual yield estimates could not be taken from this test because yield is a function of productivity per unit of area; single plants or single rows cannot be used as reliable indicators of yield in a true field situation. Proper replicated, multi-site yield tests of this material are in progress, however.

In the herbicide treated plots, the plants appeared either normal or completely susceptible. Control (non-transgenic) plants in the herbicide plots germinated, but growth was stunted, with the leaves small and clasping to the stem. Most of these plants grew only about 10- 15cm and did not flower, even if they did survive.

Table 3. Mean numbers of plants harvested from each row in each treatment. All transgenic lines included here are homozygous for the mutant ALS enzyme. Line 12103 is derived from McGregor, the others from NorLin.

Line	Control	2.25g/ha	4.5g/ha
McGregor	10.5	7	1.5
12103	13	13.5	15
Norlin	18	11	5.6
12115	12	11	11
12116	20	12.5	11.5
12087	11.5	17	13
12140	13	12.5	15
12174	17.5	15	11

Although the data in table 3 are complicated by the uneven herbicide application (permitting some control plants to thrive in a "treated" plot) and some unrelated premature plant death due to wind and/or insect damage, significantly fewer control plants were harvested from the treated plots than from the untreated, while the transgenic lines as a group did not show the same regression. However, from these and other collected data (not shown) the transgenic lines did vary in their degree of resistance expressed. One of the transgenic lines (12174) was noticeably weak at expression of the herbicide resistance, but the others showed no significant differences between treatments for any of the measured parameters, including seed yield. This indicates that flax can express this mutant ALS and have it confer sulfonylurea resistance while not interfering with normal agronomic functioning of the plant.

Glyphosate resistance test (in collaboration with Monsanto):

Four transgenic lines (one NorLin, three Vimy derivatives) were grown in three reps of three treatments (0, 1x and 2x rates of Roundup) for a total of nine plots. In the control treatments, several different weed species soon infiltrated the plots. These included lamb's quarters, prostate pigweed, stinkweed, kochia, Russian thistle, wild buckwheat and wild tomato, all commonly occurring weeds in this area. Because flax is a poor biological competitor, the weeds soon outgrew the flax and severely inhibited vigor. Nevertheless, no observable differences were recorded between control and transgenic plants in the untreated plots, indicating again that the presence of T-DNA does not alter normal functioning of the plant.

In the sprayed plots, all non-transgenic control plants and all weeds were killed within a week of application with the exception of a few wild buckwheat plants. Damage to the transgenic plants did not appear for several more days, and again differences in degree of damage were noted between transgenic lines. Table 4 shows stand reduction scores (in terms of % damage to individual plants). Although the transgenic plants

Table 4. Stand reduction scores following Roundup treatment of transgenic and non-transgenic flax; figures are percentage damage compared to untreated plants of the same line, mean of three reps.

line	no herbicide	0.28 Kg/ha	0.56Kg/ha
NorLin	0	100	100
McGregor	0	100	100
Vimy	0	100	100
15	0	15	42.5
12363	0	63	72
12370	0	37	75
12372	0	62	63

did not die from the herbicide treatment, they did cease growth and the apical meristem eventually died. Growth resumed from axillary buds and they flowered and set seed normally, but the recovery period caused a delay in flowering and ultimate maturity.

Delayed maturity and consequent yield loss of these transgenic flax lines precludes direct use of them as commercial cultivars using Roundup as a weed control agent. However, variation in the degree of resistance expressed, and the small number of transgenic lines tested, indicate that perserverence will provide a transgenic line showing minimal damage from field-level applications of Roundup.

Herbicide resistant crops will be among the first commercial products of gene transfer technology for crop improvement. Field trials of transgenic crop species are beginning, with impressive results (DeGreef et al., 1989). A conventionally produced cultivar of canola with resistance to triazine herbicides has considerable market share in Canada in spite of a 30% yield deficit. Flax is very susceptible to yield reduction due to weeds, and farmers are anxious to obtain herbicide resistant cultivars. It should be possible to obtain through transformation technology flax lines fully resistant to more efficacious, environmentally friendlier herbicides than are available now.

Conclusion

Flax is a very infrequent outcrosser, has no local "weedy" relatives, is a poor biological competitor, does not disperse seed, and does not spontaneously propagate vegetatively, making it a model system for testing of transgenic traits where containment is important. The test results show that the presence of foreign DNA does not change the character of the plant, as in every test, the transgenic and non-transgenic plants were indistinguishable in every character in the untreated control plots.

The results of the field tests of herbicide resistant transgenic flax lines show that herbicide resistance is a feasible goal of transformation technology. The availability of better, more environmentally friendly weed control mechanisms will be beneficial to farmers in particular and society in general. Development of such transgenic crops will lead to greater rotational options for farmers, higher yields and better quality of product due to better weed control, and less net chemical pollution of our environment.

References:

DeGreef, W., R. Delon, M DeBlock, J. Leemans and J. Botterman. 1989. Bio/Technology 7:61-64.

Haughn, G., J. Smith, B. Mazur and C. Somerville. 1988. Mol. Gen. Genet. 211:266-271.

Jordan, M, and A. McHughen. 1988. Plant Cell Rep. 7:611-614.

McHughen, A. 1989. Plant Cell Rep. 8:445-449.

FROM MICROSPORE TO EMBRYOID: FACES OF THE ANGIOSPERM POLLEN GRAIN

V. RAGHAVAN, Department of Botany, The Ohio State University, Columbus, Ohio 43210, USA

1. INTRODUCTION

In the context of sexual reproduction in angiosperms, the pollen grain represents the beginning of a short-lived male gametophytic phase and provides the vehicle for the delivery of sperm to the vicinity of the egg. Under certain experimental conditions, pollen grains of some angiosperms follow a different developmental program leading to the production of embryo-like structures or embryoids and calluses. Thus, pollen development in angiosperms is like two faces of the same coin: during its normal life the pollen grain is programmed for terminal differentiation; under experimental conditions it embarks on a pathway leading to immortality. It is a testament to the pollen grain as a model developmental system that it is able to display two dissimilar phenotypic forms.

The objective of this chapter is to reflect on the early cytological and biochemical events that precede the normal gametophytic and induced sporophytic pathways of the angiosperm pollen grain. Ontogeny of the male gametophyte will be dealt with first, then embryogenic type of development, and finally the differences that emerge in the pattern of macromolecule synthesis associated with normal and embryogenic types of growth.

2. RESULTS AND DISCUSSION

2.1. Ontogeny of the pollen

The pollen producing organ of the flower is the anther, which in its primordial stage is a cylindrical organ harboring a homogeneous mass of cells surrounded by a layer of epidermis. Later, as the anther becomes four-lobed, one or more rows of hypodermal cells differentiate through the entire length of each anther lobe. These cells designated as archesporial initials, divide periclinally to give rise to an inner primary sporogenous cell and an outer primary parietal cell. The former functions directly as the microsporocyte or divides mitotically to yield microsporocytes. The microsporocytes undergo meiosis, each microsporocyte yielding four haploid microspores held together in a tetrad. The primary parietal cell divides anticlinally and periclinally to form a multilayered anther wall.

Eventually, the microspores are liberated into the anther locule by dissolution of the callose wall of the tetrad. This is followed by a cytoplasmically asymmetric division of the microspore (first haploid mitosis) to generate a large vegetative cell and a small generative cell. Frequently this precedes chemical differentiation which is the transcription of genetic information and its translation into proteins. Typically, the young male gametophyte at the two-celled stage is known as the pollen grain. In many plants, pollen grains are shed at the two-celled stage by the dehiscence of the anther wall.

Another phase of male gametogenesis is the division of the generative cell to give rise to two sperm or male gametes. Especially in plants belonging to the Gramineae, this division is completed in the pollen grain before its release from the anther; in others it occurs after the pollen grain germinates on the receptive surface of the stigma of the same or of a different flower. During germination, the pollen grain extrudes a tube which grows through the tissues of the stigma and style towards the embryo sac carrying within it the nuclear package of the generative cell and vegetative cell or the latter along with two sperm. As is well-known, one of the sperm fuses with the egg and the other with the diploid polar fusion nucleus in the embryo sac. The fate of the nucleus of the vegetative cell is a moot question; it is believed that it either disintegrates or survives as a vestigial organelle in the embryo sac or in the pollen tube. What is badly needed to resolve this issue are better methods of observing the vegetative cell nucleus in the pollen tube cytoplasm prior to or at the time of fertilization.

2.2. Embryogenic development of the pollen

During the past quarter century, the great majority of studies on the male gametophyte of angiosperms have been concerned much more with the induction of embryogenic type of growth of pollen grains than with their normal ontogeny. These studies were spawned by the discovery of Guha and Maheshwari (1) who showed that when anthers of Datura innoxia (Solanaceae) were cultured at the pollen grain stage in a mineral salt medium supplemented with casein hydrolyzate, indoleacetic acid (IAA) and kinetin or with coconut milk, grape juice or plum juice, embryo-like structures surfaced from the sides of the anther in about 6 to 7 weeks. Although it was initially suspected that embryoids originated from the somatic tissues of the anther, later work confirmed their origin from pollen grains and hence their haploid nature. In a brief ontogenetic study it was also found that in anthers cultured in a medium containing coconut milk, a sizable number of pollen grains began to enlarge and divide repeatedly forming multicellular units. Later, the multicellular pollen forced its way out of the exine into the anther locule and organized into a typical bipolar embryo-like structure (2). After the publication of these results, Niizeki and Oono (3) showed that when anthers of rice (Oryza sativa) were cultured at the mature pollen grain stage in a medium containing IAA, 2,4-dichlorophenoxyacetic acid (2,4-D) and kinetin, calluses of pollen grain origin appeared from within the anther in about 4 to 8 weeks. Transfer of the callus to a medium containing IAA and kinetin was found to elicit the formation of plantlets by organogenesis. Later studies showed that plantlet formation by embryogenesis or by organogenesis from pollen grains of cultured anthers is a secure generalization applicable to plants distributed within a broad spectrum of families. Using anthers of different stages of development, it was also shown that the potentialities of the pollen grain at the uninucleate stage are not yet fixed, but by the binucleate stage they are irrevocably determined in the gametophytic pathway.

2.3. Embryogenic division patterns of pollen grains

By bringing together cytological observations of embryogenic development of pollen grains of different species, it has been possible to look closely at the involvement of the generative cell and vegetative cell in this event (4). Beginning with a uninucleate pollen grain, the first division in the embryonic pathway generates a facsimile of the typical gametophytic pollen with a large vegetative cell and a small generative cell (or nuclei) or a pollen with two identical cells or nuclei. From this point onwards, depending upon the contributions of the two cells to the final product, five different cytological pathways of embryogenesis have been recognized. Perhaps the most widespread pathway is one in which the embryoid is formed by repeated divisions of the vegetative cell, as was first shown in Nicotiana tabacum (5). As the vegetative cell begins to divide, the generative cell either disintegrates or undergoes but a few divisions without contributing to the formation of the embryoid. An independent role for the generative cell in embryogenesis was slow to be recognized, but it is the prevalent pathway in cultured anthers of Hyoscyamus niger (6). Here the vegetative cell does not divide or undergoes but a few divisions; in both cases, the vegetative cell or its division products function as a suspensor-like structure on the organogenetic part of the embryoid formed by derivatives of the generative cell. In this plant, repeated divisions of both the generative and vegetative cells also give rise to embryoids, undoubtedly of chimeric nature. Somewhat earlier to the publication of the work on H. niger, the role of both pollen cells in embryoid formation had acquired a new dimension as a result of observations on pollen embryogenesis in cultured anthers of Datura innoxia (7). Here the generative cell and vegetative cell, instead of dividing independently, fuse with one another and repeated divisions of the fusion product give rise to the embryoid. One of the more curious and as yet unexplained aspects of fusion is the unpredictable way it occurs, such as between one or two haploid vegetative cell nuclei and a haploid or endopolyploid generative cell nucleus. The formation of embryoids with different ploidy levels from the same anther accords with the idea that pollen nuclei with nonhaploid complement of chromosomes are formed in cultured anthers.

Some investigators agree in reporting that when two symmetrical cells or nuclei result from the first haploid pollen mitosis in cultured anthers, both cells or nuclei contribute derivatives to the formation of the embryoid (8). While nuclear divisions in the pollen are generally accompanied by cytokinesis, this is not always the case. As shown in Triticum aestivum, the two identical nuclei divide repeatedly to form a bunch of free nuclei, whose fate has not been clarified (9).

It appears from the above that the generative and vegetative cells of the angiosperm pollen grain display a range of developmental capabilities when they are primed for embryogenic divisions. Further work is necessary for an understanding of the factors that control the relative extent of contributions of derivatives of the two cells to embryoid formation.

2.4. Biochemical and molecular cytology of pollen developmental strategies

Experimental assaults on the biochemical and molecular changes associated with pollen developmental transformations have been rare and the available information is largely based on histochemical and autoradiographic studies of nucleic acid and protein synthesis. As is well-known, the synthesis and accumulation of these macromolecules provide the operational criteria for gene transcription during developmental episodes. For a complete review of these investigations, the reader is referred to Raghavan (4, 10). In the following account, references will be made mostly to work done in my laboratory on Hyoscyamus niger for the past 15 years and on rice during recent years. Anthers of H. niger possess many advantages to study pollen embryogenesis. The anther is relatively small in size which makes histological approaches to follow pollen development feasible. Embryogenic divisions are initiated in pollen grains within 8 to 12 hours after culture of anthers and multicellular pollen grains are visible as early as 72 hours after culture. Stains such as acetocarmine for squash preparations and azure B for sections embedded in glycol methacrylate color the embryogenic pollen grains and embryoids, while the large majority of nonembryogenic pollen grains do not take up the stain. Finally, embryogenic divisions are initiated not only by culture of whole anthers, but also by culture of segments (one-fourth) of the anther; limited success has also been achieved in inducing embryogenic divisions in isolated pollen grains in a single step method (11, 12). Although rice anthers do not offer these advantages, they were chosen because of the availability of a cloned rice gene has made it possible to follow the expression of a specific gene during pollen development and embryogenesis.

2.4.1. RNA synthesis. Reynolds and Raghavan (13) followed autoradiographically the timing and pattern of RNA synthesis during maturation and germination of pollen grains of H. niger using ^{3}H-uridine as a label. This study showed that incorporation of the radioactive precursor during pollen maturation was periodic with peak RNA synthesis occurring in the uninucleate, vacuolate pollen grains and in the vegetative cell of bicellular pollen grains. During the early stages of germination, the vegetative cell actively synthesized RNA, although with the appearance of the pollen tube and its elongation, RNA synthetic activity of this cell decreased and came to a complete halt. The generative cell synthesized little or no RNA during its life span; the sperm formed from this cell did not also synthesize any RNA. With minor variations, the pattern and timing of RNA synthesis during maturation of pollen grains of H. niger are similar to those described in pollen grains of Tradescantia paludosa (14), Lilium longiflorum (15, 16) and Paeonia tenuifolia (17, 18). In accord with the above-mentioned studies and those of La Cour (19), Woodard (20) and Jalouzot (21, 22), the vegetative cell of the pollen of H. niger appears to synthesize RNA more actively than the generative cell.

A strikingly different pattern of RNA synthesis was observed in embryogenic pollen grains of cultured anthers. It was mentioned earlier that in H. niger pollen grains begin to divide in the embryogenic pathway within a few hours after culture of anthers. This consideration suggested that pollen grains might become embryogenically competent during the early period of culture of anthers and macromolecule synthesis necessary for embryogenesis initiated simultaneously. To verify this, anthers were pulsed in a medium containing ^{3}H-uridine immediately after excision and label incorporation into embryogenic pollen grains was localized autoradiographically in embryoids formed during a chase in the unlabeled medium. This experiment established that a small number of pollen grains become embryogenically determined as early as the first hour of culture of the anther and that this is associated with the synthesis of RNA (23). Incubation of anthers in actinomycin D inhibited embryogenic development of pollen grains as well as ^{3}H-uridine incorporation. To the extent that actinomycin D inhibits the synthesis of mRNA or rRNA, these findings are in harmony with the view that pollen embryo-genesis in H. niger is associated with the synthesis of one or both of these RNA fractions.

In cultured anthers of H. niger only a small percentage of embryogenically determined uninucleate pollen grains divide in the embryogenic pathway to form vegetative and generative cells. The subsequent fate of these bicellular pollen grains depends upon RNA synthetic activity of the two cells. Thus, only those pollen grains in which continued RNA synthesis occurs in the generative cell or in both generative and vegetative cells become embryogenic, while those in which RNA synthesis occurs exclusively in the vegetative cell become starch-filled and nonembryogenic (24). Autoradiography also revealed a suggestive correlation between the initial level of incorporation of ^{3}H-uridine into the generative and vegetative cells and their propensity for division and in RNA synthetic activity of the cells derived from them. Thus, the generative cell which synthesizes RNA actively gives rise to new cells incorporating ^{3}H-uridine, while the metabolically inactive vegetative cell forms no new cells or very few new cells which also synthesize little or no RNA.

Since in our experiments using ^{3}H-uridine no chromosomal labeling (where mRNA synthesis would occur) was observed before nucleolar labeling, the assumption is made here that most of the labeled RNA is rRNA. The obsertion that uninucleate pollen grains of cultured anthers begin to synthesize RNA at about the same stage of development as pollen in the intact anther (13) suggests that no new genes for rRNA synthesis are activated for embryogenic determination. On the other hand, that transcription of rRNA in the generative cell is an important prerequisite for embryogenic divisions is indicated by the absence of appreciable incorporation of the isotope into this cell during normal pollen development and its accelerated incorporation into the generative cell of potentially embryogenic pollen grains.

The above studies have served primarily to focus attention on the problem of embryogenic development of pollen grains of cultured anthers at the genic level.

2.4.2. Changes in poly(A)+RNA accumulation. In situ hybridization technique using ^{3}H-polyuridylic acid [^{3}H-poly(U)] as a probe has been extensively employed to localize polyadenylic acid-containing RNA [poly(A)+RNA] or mRNA in several animal systems. Comparable experiments on normal and embryogenic pollen grains of H. niger have given some insight into how they respond to different developmental signals (25). For example, from the time of their release from the tetrad to the bicellular stage, pollen grains do not anneal the probe; this reinforces the earlier observation that synthesis of rRNA is the main chore of uninucleate pollen grains. There was a transient binding of ^{3}H-poly(U) in the generative and vegetative cells of mature pollen grains, although only the generative cell was active in this role during pollen germination.

Because of the significance of mRNA synthesis in pollen embryogenesis, we looked for poly(A)+RNA changes in uninucleate pollen grains which are most prone to embryogenic induction in cultured anthers. When sections of anthers cultured for different periods were hybridized with ^{3}H-poly(U), a small number of embryogenically determined pollen grains were found to anneal the probe within the first hour and with further periods of culture, there was an increase in the number of pollen grains binding the probe and in the number of binding sites. ^{3}H-poly(U) binding in the embryogenically determined pollen grains was blocked by actinomycin D, indicating that it represents newly synthesized mRNA. After the first haploid mitosis, only those pollen grains in which the generative cell nucleus alone or along with the vegetative cell nucleus accumulated poly(A)+RNA in the surrounding cytoplasm were found to divide in the embryogenic pathway, while accumulation of poly(A)+RNA by the vegetative cell alone perpetuated the gametophytic program resulting in starch-filled pollen. These data suggest that in contrast to normal pollen development, embryogenic development is due to transcription of new mRNA in the uninucleate pollen grains and subsequently by the generative cell or by both generative and vegetative cells. The trauma of excision and culture of the anther is apparently the trigger than initiates the new rounds of transcriptional activity. A complete and unequivocal account of mRNA changes during normal and induced embryogenic pollen development in H. niger must now await a careful analysis by a combination of biochemical and molecular techniques. Because of the small number of pollen grains that become embryogenic in a given anther, this is easier said than done.

Although specific genes that are activated during pollen embryogenesis in cultured anthers of H. niger have not been characterized, we have recently monitored the temporal and spatial expression of a rice histone H3 gene during developmental transformations in the pollen grains of rice. Our general strategy to follow histone gene expression was by in situ hybridization of sections of normal and cultured anthers with a ^{3}H-labeled RNA probe made from the gene. For details see, Raghavan (26, 27). Briefly, during the whole spectrum of pollen development, expression of the gene occurred only at the starch-filled bicellular stage and it continued at a reduced rate in the mature stage. In contrast, embryogenically determined pollen grains which were identified in anthers cultured from 8 to 10 days gave strong hybridization signals. After the first haploid mitosis, potentially embryogenic pollen grains revealed label in the vegetative cell, but not in the generative cell. This is a fair testimony to the function of the vegetative cell in initiating embryogenic divisions in pollen grains of cultured rice anthers. The expression of histone H3 gene in the bicellular pollen grain during its transformation into an embryoid has not been examined; however, sections of multicellular pollen grains have been hybridized in situ with the labeled probe.

As a result we know that all newly formed haploid cells express the histone gene. This picture of gene expression during pollen developmental changes in rice, admittedly incomplete, simply indicates that of the battery genes that may be activated during pollen embryogenesis, histone H3 gene has an important role.

2.4.3. Protein synthesis. The impressive correlation noted earlier between RNA synthesis and poly(A)+RNA accumulation and embryogenic division of pollen grains in cultured anthers of *Hyoscyamus niger*, brings into focus the regulation of protein synthesis at different stages of transformation of the pollen grain into an embryoid. As shown in a recent work involving autoradiography of incorporation of ^{3}H-arginine, ^{3}H-leucine, ^{3}H-lysine and ^{3}H-tryptophan (28), the synthesis of proteins by the uninucleate pollen grains and subsequently by the generative cell of bicellular pollen grains is linked to embryoid development in cultured anthers of *H. niger*. This is not really surprising because enzyme proteins should certainly be involved in modulating embryogenic divisions.

A distinction between the synthesis of certain general and basic proteins and between the pattern of protein synthesis during normal pollen development and during induced embryogenesis may be useful to explain the results. One of the major attributes of the pollen grain was incorporation of ^{3}H-arginine and ^{3}H-lysine before the onset of vacuolation at the uninucleate stage and at the bicellular stage on the threshold of starch accumulation. In the bicellular pollen grains, labeled protein precursors were incorporated mostly into the vegetative cell. Uninucleate pollen grains did not incorporate ^{3}H-leucine and ^{3}H-tryptophan at any stage of their development, but the vegetative cell of the bicellular pollen again actively incorporated these precursors.

That during embryogenic transformation the pollen grain is charged with a different pattern of protein synthesis was evident from the initiation of incorporation of ^{3}H-leucine and ^{3}H-tryptophan as well as by an accelerated incorporation of ^{3}H-arginine and ^{3}H-lysine into uninucleate embryogenically determined pollen grains. Subsequent division of uninucleate pollen grains was associated with the incorporation of ^{3}H-arginine, ^{3}H-leucine and ^{3}H-lysine into the generative cell or into both generative and vegetative cells. The course of development pursued by the bicellular pollen grain is markedly affected by protein synthetic activity in its constituent cells. Pollen grains which incorporate ^{3}H-arginine, ^{3}H-leucine and ^{3}H-lysine in the vegetative cell fail to become embryogenic and continue the gametophytic program. However, when the generative cell or both generative and vegetative cells incorporate these labeled precursors, the pollen grains continue embryogenic divisions. In the bicellular pollen grains, incorporation of ^{3}H-tryptophan was sparse and was found only in the vegetative cell.

Summarily, these results demonstrate that an intense period of protein synthesis is associated with deflection of the gametophytic program of the pollen grain of *H. niger* in the embryogenic pathway. The hope that genome transcription to support these early events of pollen embryogenesis should also occur during this period is fulfilled to some extent by the results of experiments reviewed earlier. Thus, changes in macromolecule synthesis might not only represent an outcome of embryogenic transformation of the pollen grain, but may also play an important role in this transformation.

3. CONCLUDING COMMENTS

Embryogenic development of pollen grains of cultured anthers provides an attractive system for studying certain fundamental problems of dedifferentiation of plant cells. As a developmental phenomenon, pollen embryogenesis serves as yet another demonstration of the concept of totipotency of plant cells. From all considerations it seems that what is observed is a masking of the gametophytic program and unfolding of the sporophytic program in but a few pollen grains. Further studies are needed to determine the types of proteins that are synthesized during the face change of the angiosperm pollen grain. Our ultimate goal is to decipher and describe at all levels of consideration how an embryoid arises from a single progenitor cell of the pollen grain.

4. ACKNOWLEDGEMENTS

Parts of the work described in this paper were supported by grants from the Rockefeller Foundation and the National Science Foundation.

REFERENCES

1 Guha, S. and Maheshwari, S.C. (1964) Nature 204, 497.
2 Guha, S. and Maheshwari, S.C. (1966) Nature 212, 97-98.
3 Niizeki, H. and Oono, K. (1968) Proc. Jap. Acad. 44, 554-557.
4 Raghavan, V. (1986 Embryogenesis in Angiosperms. A Developmental and Experimental Study. Cambridge, New York.
5 Sunderland, N. and Wicks, F.M. (1971) J. Expt. Bot. 22, 213-226.
6 Raghavan, V. (1976) Science 191, 388-389.
7 Sunderland, N., Collins, G.B. and Dunwell, J.M. (1974) Planta 117, 227-241.
8 Zhou, C. and Yang, H. (1980) Acta Bot. Sinica 22, 211-215.
9 Zhu, Z., Sun, J. and Wang, J. (1978) Acta Bot. Sinica 20, 6-12.
10 Raghavan, V. (1987) Cell Diffn. 21, 213-226.
11 Raghavan, V. (1978) Amer. J. Bot. 65, 984-1002.
12 Nagmani, R. and Raghavan, V. (1983) Z. Pflanzenphysiol. 109, 87-90.
13 Reynolds, T.L. and Raghavan, V. (1982) Protoplasma 111, 177-188.
14 Taylor, J.H. (1953) Expt. Cell Res. 4, 164-173.
15 Taylor, J.H. (1958) Amer. J. Bot. 45, 123-131.
16 Steffensen, D.M. (1966) Expt. Cell Res. 44, 1-12.
17 Sauter, J.J. (1969) Z. Pflanzenphysiol. 61, 1-19.
18 Sauter, J.J. and Marquardt, H. (1969) Z. Pflanzenphysiol. 58, 126-137.
19 La Cour, L.F. (1949) Heredity 3, 319-337.
20 Woodard, J.W. (1958) J. Biophys. Biochem. Cytol. 4, 383-389.
21 Jalouzot, J.-F. (1969) Expt. Cell Res. 55, 1-8.
22 Jalouzot, M.-F. (1969) Rev. Cytol. Biol. Veg. 32, 115-120.
23 Raghavan, V. (1979) Amer. J. Bot. 66, 36-39.
24 Raghavan, V. (1979) Amer. J. Bot. 66, 784-795.
25 Raghavan, V. (1981) J. Cell Biol. 89, 593-606.
26 Raghavan, V. (1989) J. Cell Sci. 92, 217-229.

27 Raghavan, V. (1990) in the Impact of Biotechnology in Agriculture (Sangwan, R.S. and Sangwan, B.S., ed.), Kluwer, Dordrecht.
28 Raghavan, V. (1984) Canad. J. Bot. 62, 2493-2513.

TWO PATHWAYS IN POLLEN PROTOPLAST CULTURE: CELL DIVISIONS AND TUBE GROWTH

C. ZHOU and Y. WU
DEPARTMENT OF BIOLOGY, WUHAN UNIVERSITY, WUHAN 430072,CHINA

1. INTRODUCTION

So far the manipulation of male reproductive cells in higher plants has developed from anther culture to pollen culture. The experimental control on alteration of developmental pathways has already been realized in tobacco pollen culture (1). Logically, pollen protoplast culture should inevitably be a further advance from pollen culture. The pollen protoplast combines the merits of both pollen and protoplast: In comparison with pollen, it presents a protoplast system; in comparison with somatic protoplast which is currently the main source of protoplasts, it provides a new, sexual and haploid system. Obviously, to obtain haploid protoplasts directly from pollen is more meaningful than indirectly via haploid plants from anther or pollen culture.

Although the study on pollen protoplasts has begun early in 1970s (2-5), due to technical difficulties the breakthrough in isolation and purification of pollen protoplasts in large quantities and subsequently their culture has not appeared until recent years (6-11). The present report deals with the two developmental pathways of pollen protoplasts in vitro, i.e. along a gametophytic pathway to continue their in vivo program toward germination and growth of pollen tubes, or along a sporophytic pathway switching toward cell divisions which might result in final regeneration of plants. For convenience, the term 'pollenplast' is suggested to simplify 'pollen protoplast' in the following text.

2. ISOLATION

Isolation of large number protoplasts from pollen is the indispensable technical prerequisite to their culture. The pollen grain is surrounded by an exine and an intine. The exine consists of sporopollenin characterized by resistance to biodegradation; the intine is composed of cellulose and pectin easy of enzymatic digestion (12). Therefore, the exine has been believed as the main barrier to pollenplast isolation. Recently, two approaches have been developed: One way was to release pollenplasts by chemical degradation of the pollen wall by 4-methyl N-oxide morpholine monohydrate (MMNO) at 75° C (6). Owing to the damage to the pollenplasts during chemical degradation at high temperature, the technique has been improved by combining chemical and enzymatic treatments and lowering the temperature (7). The other way was to isolate pollenplasts by enzymatic method. The principle of this technique does not lie in digesting the exine, but in causing its dehiscence and exposing the intine to direct enzymatic action. The advantage of this method is easy to obtain viable pollenplasts. Tanaka et al. isolated about 80% protoplasts from nearly matured pollen grains of _Lilium longiflorum_ (8). Our method was reported previously (10) and has been adopted since then for much more materials with minor technical modifications. Usually the pollen grains are incubated in an enzymatic solution containing 0.5-2% cellulase, 0.5-2% pectinase, 0.5% potassium dextran sulfate, 1% calf serum, and mannitol and/or sucrose dissolved in K3 medium. The kinds and concentrations of osmotica vary with plant species and pollen stages. So far protoplasts have been isolated from young pollen at late uninucleate to early binucleate stage in four species, i.e. _Hemerocallis fulva_, _H. minor_, _Gladiolus gandavensis_ and _Iris tectorium_, and from mature pollen in six monocot species, including _Zephyranthes grandiflora_ and _Hippeastrum vittatum_ in addition to the above-mentioned four species. More recently, in two dicots, _Brassica campestris_ var. _purpurea_ and _B. napus_, pollenplasts have been isolated in our lab as well (S.Q. Li and H.Y. Yang, unpublished). Although this method has so far been limited to only a part of taxa, it did provide a basis for further culture studies. We have used the mature pollenplasts to maintain their gametophytic development and the young pollenplasts to induce their sporophytic development under in vitro conditions.

3. GAMETOPHYTIC PATHWAY IN MATURE POLLENPLAST CULTURE

Soon after success in isolation of pollenplasts, their culture in _L. longiflorum_ (8) and _H. fulva_ (10) were tried respectively and similar results were obtained such as wall regeneration in most pollenplasts, formation of a few tube-like and other abnormal structures, and limited nuclear divisions. Further experiments in our lab were aimed at production of large number pollen tubes from cultured pollenplasts. Mature pollenplasts of _G. gandavensis_ were inoculated in K3 medium supplemented with 32% sucrose, 0.1 mg/l 2,4-D, 1 mg/l NAA and 0.2 mg/l BA. After 18 h of culture, pollen tubes were yielded with a frequency up to 59.3% for pollenplasts isolated from flowers one day before anthesis, and 47.7% for pollenplasts at the day of anthesis. Usually, only one tube emerged from a pollenplast, but in some cases two or even three tubes were produced (Tab. 1.). The pollen tubes produced from pollenplasts looked quite normal in morphology and growth behavior (Fig. 1-2).

Fluorecence localization of actin filament distribution with FITC-phalloidin as a microprobe demonstrated that the pollenplasts just after isolation usually contained an irregular network of actin filaments. When they underwent germination, the actin filaments became polarized, converging into the tube, and distributed along the tube just like in normal pollen germination (Fig. 3.).

Ultrastructurally, the pollenplasts just isolated showed typical protoplast features: global outlook, naked plasma membrane and intact organelles (Fig. 4.). A vegetative nucleus and a generative cell were located inside (Fig. 5.). After 18 h of culture, cell wall regeneration of the pollenplasts proceeded intensively and exhibited multilayered figure quite different from the original pollen wall structure (Fig. 6.), while the tubes were covered with a thin layer of wall (Fig. 7.).

All the above observations indicate that under suitable condition pollenplasts have the capability of continuing their gametophytic pathway to form normal pollen tubes.

TABLE 1. Tube growth in pollenplast culture of G. gandavensis

Pollenplasts	Number observed	Frequency of germinated pollenplasts (%)		
		1 tube	2-3 tubes	Total
One day before anthesis	403	56.3	3.0	59.3
The day of anthesis	224	44.6	3.1	47.7

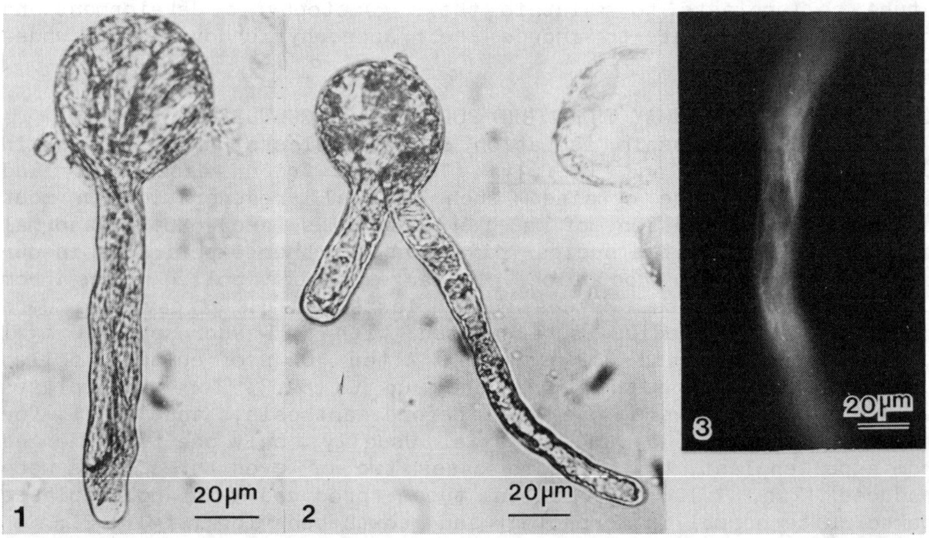

FIGURES 1-3. Germination and growth of one (Fig. 1.) or two (Fig.2) tubes from pollenplasts in G. gandavensis and longitudinal arrangement of actin filaments along the tube (Fig. 3).

4. SPOROPHYTIC PATHWAY IN YOUNG POLLENPLAST CULTURE

The isolation of protoplasts from young pollen at uninucleate or early binucleate stage has faced technical difficulty all along, so that it was impossible to carry out culture experiments (8). Recent

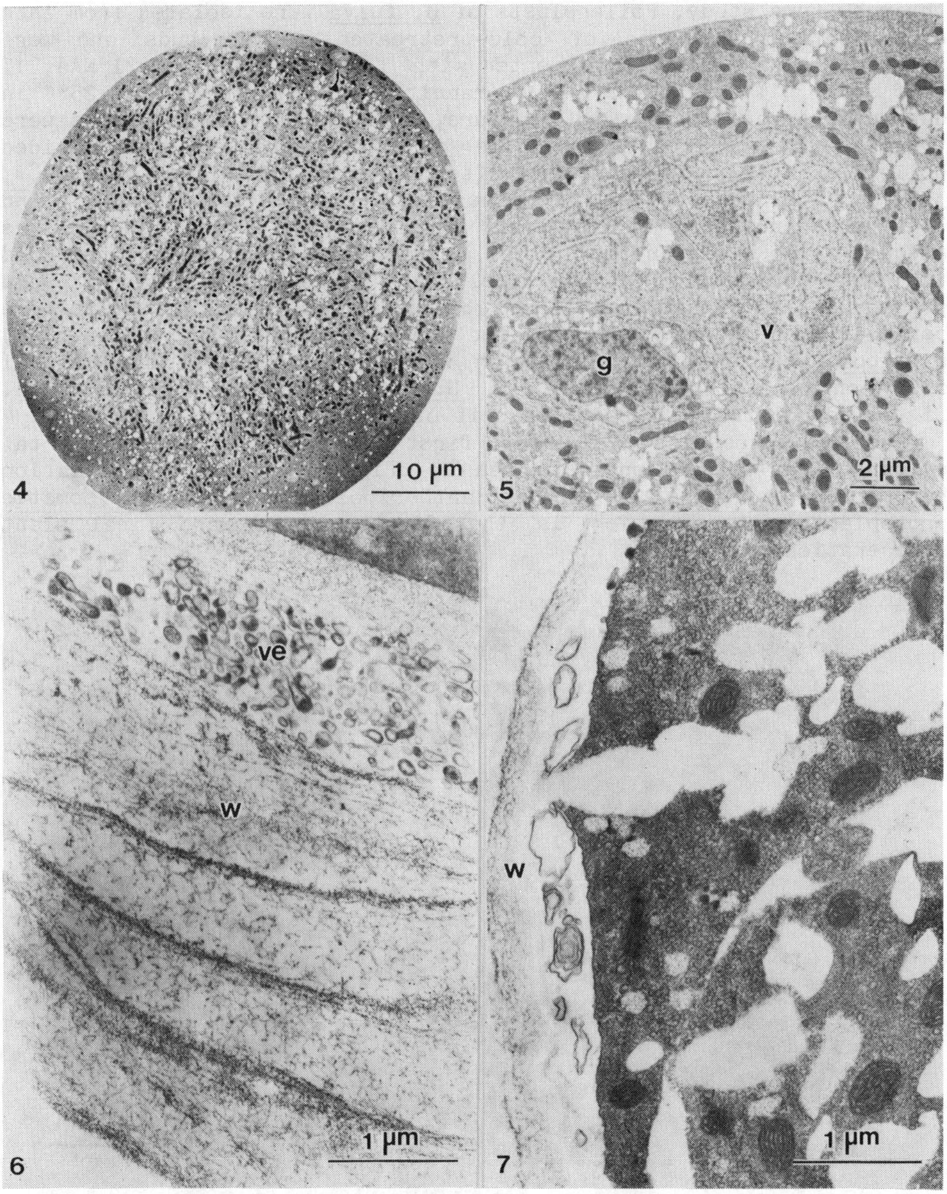

FIGURES 4-7. ultrastructural features in mature pollenplast culture of G. gandavensis. Fig.4. Whole view of a pollenplast. Fig.5. Part of a pollenplast showing its lobed vegetative nucleus (v), generative cell (g) and organelles. Fig.6. Involvement of numerous vesicles (ve) in pollenplast wall (w) regeneration. Fig.7. Transection of a pollenplast-derived pollen tube showing its wall (w) formation.

success in isolating young pollenplasts has offered us a basis for their culture study. Pollenplasts of H. fulva were isolated from late uninucleate microspores of cold-pretreated flower buds and were cultured in K3 medium supplemented with 1 mg/l 2,4-D, 2 mg/l NAA, 1% calf serum, 1% Polybuffer 74, 9% mannitol and 17% sucrose at 25°C in darkness. After 4-5 days of culture, some of the pollenplasts were triggered to cell division. The induction frequency of divided pollenplasts and cell clusters after 15 days of culture was 0.8%, 3.5% and 7.2% respectively from the materials pretreated for 0, 4 and 7 days. Microscopic observations revealed that equal as well as unequal first division took place, and subsequent divisions resulted in the formation of multicellular clusters including some proembryo-like structures (Figs. 8-11.) (11). Similar results were obtained in a second year repeated experiment. With another species, H. minor, the first cell division was also induced. Ultrastructural study on pollenplast culture of H. fulva has been carrying on. Fig.12. demonstrates a 2-celled unit observed.

The above results, for the first time, provide experimental evidence that the pollenplasts can be triggered to dedifferentiation resembling to those in either anther/pollen culture or somatic protoplast culture, which is the first step toward final plant regeneration.

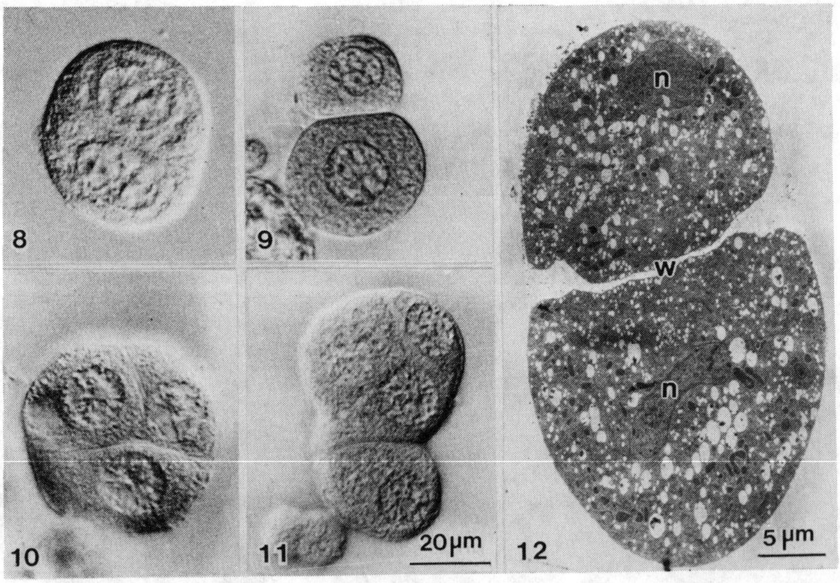

FIGURES 8-12. Cell divisions in young pollenplast culture of H. fulva showing equally (Fig.8.) and unequally (Fig.9.) divided 2-celled units, a 4-celled unit (Fig.10.) and a proembryo-like structure (Fig.11.). Ultrastructure of a 2-celled unit separated by a wall (w) is shown in Fig.12. (n: nucleus)

5. PROSPECTS

Pollenplast as a new experimental system will be of great significance in plant cell engineering. It can be utilized in respect of either gametophytic or sporophytic pathways. Along its gametophytic pathway, the next task should be realization of pollination and fertilization with pollenplasts leading to seed formation, thus might provide a promising vector to introduce foreign genes into the egg cell. Theoretically, the efficiency of transformation via pollenplast pollination, provided be successful, will be higher than via pollen grains. Along the sporophytic pathway, the further goal will be regeneration of plants from pollenplast culture. This might offer an ideal system for cell fusion as well as gene manipulation owing to its haploid nature. Moreover, pollenplast may present a new means for studying pollen cell biology, in particular the mechanisms of wall formation and germ aperture destination, pollen membrane structure and function, cytoskeletal organization, and male gametogenesis.

REFFERENCES

1 Benito Moreno, B.M., Macke, F., Hauser, M.T., Alwen, A. and Heberle-Bors, E. (1988) in Sexual Reproduction in Higher Plants (Cresti, M., Gori, P. and Pacini, E., eds.), pp. 137-142, Springer-Verlag, Berlin Heidelberg New York London Paris Tokyo

2 Bhojwani, S.S. and Cocking, E.C. (1972) Nature, New Biol. 239, 29-30

3 Power, J.B. (1973) in Plant Tissue and Cell Culture (Street, H.D., ed.), pp. 118-119, Blackwell, Oxford

4 Bajaj, Y.P.S. and Cocking, E.C. (1973) In Vitro, 8, 420

5 Bajaj, Y.P.S. and Davey, M.R. (1974) in Fertilization in Higher Plants (Linskens, H.F., ed.), pp. 73-84, North-Holland Publishing Company, Amsterdam

6 Loewus, F.A., Baldi, B.G., Franceschi, V.R., Meinert, L.D. and McCollum, J.J. (1985) Plant Physiol. 78, 652-654

7 Baldi, B.G., Franceschi, V.R. and Loewus, F.A. (1987) Protoplasma, 141, 47-55

8 Tanaka, I., Kitazuma, C. and Ito, M. (1987) Plant Sci. 50, 205-211

9 Miki-Hirôsige, H., Hakamura, S. and Tanaka, I. (1988) Sexual Plant Reprod. 1, 36-45

10 Zhou, C. (1988) Plant Sci. 59, 101-109

11 Zhou, C. (1989) Plant Sci. 62, 229-235

12 Knox, L.B. (1984) in Embryology of Angiosperms (Johri, B.M., ed.), pp. 197-272, Springer-Verlag, Berlin Heidelberg New York Tokyo

BIOCHEMICAL AND CYTOLOGICAL CHANGES IN YOUNG TOBACCO POLLEN DURING *in vitro* STARVATION IN RELATION TO POLLEN EMBRYOGENESIS

V. ŽÁRSKÝ, L. ŘÍHOVÁ and J. TUPÝ
Department of Genetics, Institute of Experimental Botany, Czechoslovak Academy of Sciences, Vltavská 17, CS-15000 Praha 5, Czechoslovakia

INTRODUCTION

The potentiality of microspores or young pollen grains to deviate from the normal gametophytic development to embryogenic pathway and sporophyte formation raises the possibility to use the pollen system as an unique tool to study fundamental questions of cytodifferentiation and developmental biology. From this aspect, isolated pollen culture presents a number of advantages over anther culture as it eliminates uncontrolled effects of anther wall and enables directly affect pollen grains by changes of culture conditions.

An important progress in the study of pollen embryogenesis has been achieved with the model system tobacco by demonstration that the same pollen population isolated from immature anthers can *in vitro* develop either into normal mature pollen or to be induced to form early embryos following a starvation treatment (1,2). In the experiments with *Nicotiana tabacum* L. (2) a highly homogenous population of midbinucleate pollen grains with no central vacuole and no or a few small starch grains has been prepared using Percoll density gradient centrifugation. When cultured as suspension in a mineral medium with glutamine and mannitol the pollen grains attained the mature form and germinated. A 2 d culture of the same initial sample in the absence of glutamine resulted in a regression of cytoplasm and starch disappearance and most pollen grains showed a characteristic central organization of cytoplasm. When transferred to a medium with sucrose and glutamine up to 74 % of these grains produced multicellular structures. Similar starvation treatment has been successfully applied to induce the formation of large quantities of proembryos from unfractionated pollen and an efficient regeneration of these proembryos into plantlets has been achieved (3). Starvation can also be considered as a principal factor causing the formation of embryogenic pollen grains in the anthers cultured *in vitro* and also during their development *in situ* (see 4). It thus appears that the initial events leading to a switch from a gametophytic to a sporophytic development should be studied in the course of changes taking place in starved pollen grains.

MATERIALS AND METHODS

Anthers were excised just before pollen isolation from *Nicotiana tabacum* L. cv. Samsun plants grown in a greenhouse in free soil and under natural light conditions. Pollen for starvation treatment was isolated at midbinucleate stage at the beginning of starch deposition corresponding to bud length of 25 mm. The developmental stage was checked in each flower bud and usually anthers from only one bud were used per experiment to obtain a relatively homogenous pollen population. The starvation treatment was either according to Kyo and Harada (2) for 2 d in a medium containing 20 mM KCl, 1 mM $MgSO_4$, 1 mM $CaCl_2$, 1 mM KH_2PO_4 and 300 mM mannitol (medium KH) or according to Benito Moreno et al. (3) in a medium with Miller macrosalts and 400 mM mannitol (medium BM). The length of starvation in the latter medium was up to 6 d which was found in preliminary experiments as optimal for the subsequent proembryo formation in a medium with sugar. Similarly, Benito Moreno et al. (3) applied 1 week starvation in this medium which differs from the starvation medium of Kyo and Harada (2) essentially by the presence of inorganic nitrogen. The density of pollen grains was adjusted to 5×10^4 grains per 1 ml.

Starch, protein and RNA were assayed according to the methods described in refs. 5,6 and 7, respectively. Soluble sugars were determined by the anthrone method and expressed as sucrose.

The size of nucleolus was measured using differential interference contrast. Membrane bound Ca^{2+} was localized by fluorescence of complexes with chlortetracycline (CTC, 8), callose was visualized with decolorized aniline blue (DAB) and nuclear DNA was stained by 4,6-diamidino-2-phenylindole (DAPI). DNA content was estimated by cytophotometry of nuclei stained by Feulgen method on a Leitz microspectrophotometer MPV 3.

RESULTS AND DISCUSSION

Starvation induced decrease of some pollen components

Early development of tobacco pollen after microspore mitosis is characterized by an intense synthesis of starch, RNA and proteins and can be characterized as a differentiation phase of male gametophyte formation (9). During this phase RNA increases from 34 to 230 pg (10) and the protein content from about 400 to 1400 pg (11) per pollen grain. At the midbinucleate stage one pollen grain contained 49 pg of RNA and 740 pg of proteins and these levels decreased by about 16 and 64 %, respectively, during 2 d of starvation (Table 1).

TABLE 1. Levels of sucrose, starch, proteins and RNA in midbinucleate pollen before and after starvation treatment

	Sucrose	Starch	Proteins	RNA
	(ng per pollen grain)			
Starting population	0.41	0.24	0.74	0.049
2 d culture in KH	0.10	0.06	0.27	0.041
2 d culture in KH + 0.01 mM IAA	0.10	0.02	0.23	0.028

The presence of IAA in the starvation medium significantly enhanced the decrease of RNA. A much more marked regression was registered in soluble sugars and starch which fell to about 25 % of their initial levels. Similar changes, in addition to mitochondria degradation, have been described by Journet et al. (12) in starved cells of *Acer pseudoplatanus* in suspension culture and a decrease in RNA in association with carbon starvation is reported for various plant systems (13, 14,15). The data presented indicate that the starvation results in cessation of the normal differentiation processes of gametophytic cytoplasm formation and that the energy and carbon requirements in starved pollen grains are covered by successive degradation of sugars, proteins and RNA.

Cytological changes

It could be confirmed that the starvation treatment results in formation of pollen grains with central ball structure of cytoplasm (C-type) giving rise to proembryos after transfer to an embryogenic medium and of randomly vacuolized grains (V-type) which mostly die during the second step of culture or follow their normal gametophytic development. The proportion of the latter type increases when 0.01 mM IAA is present in the starvation medium. In the C-type the cytoplasm is organized around the vegetative nucleus and is connected by thin cytoplasmic strands with pollen grain wall (Fig. 1b).

The activation of RNA synthesis during pollen differentiation from microspore is associated with an increase of the volume of nucleolus and the observed decrease in its size during the culture in a medium lacking carbon sources (Table 2, Fig. 1) presents an additional evidence for the cessation of gametophytic ribosome formation as a result of starvation. This together with the decrease of protein and RNA content (Table 1) demonstrates a regression of gametophytic information-carrying molecules in starved pollen grains.

TABLE 2. Changes in the size of nucleolus in pollen grains developing *in situ* and starved for 6 d in medium BM

Developmental stage		Volume of nucleolus (μm^3)*
After microspore mitosis		34
Midbinucleate		65
Advanced differentiation (maximal size)		125
Starved midbinucleate	C-type	14
	V-type	48

*Calculation for presumed spherical shape of nucleolus

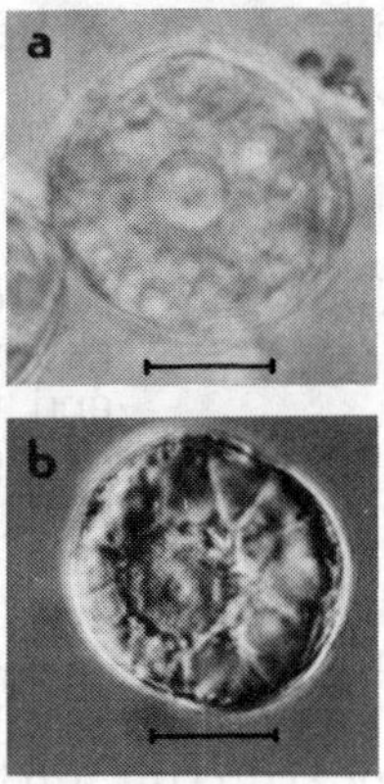

FIGURE 1. Midbinucleate pollen grain before (a) and after 3 d starvation in medium BM (b); bar = 10 µm.

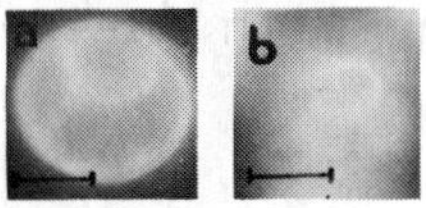

FIGURE 2. Fluorescence of calcium with CTC round about the first pollen mitosis spot in the microspore (a) and callose fluorescence (after staining with DAB, b) on the same place in midbinucleate pollen grains (bar = 10 µm).

During early stages of normal pollen development membrane bound calcium visualized by CTC fluorescence is concentrated around the region of first pollen mitosis (Fig. 2a). In the same region a transient callose wall is formed which confirms the already well established association between a selective accumulation of Ca^{2+} ions and callose synthesis (16). After breakdown of the callose wall, there remains a callose ring localized on the surface of intine at the pole of first pollen mitosis. Fig. 2b shows this ring in the midbinucleate stage used for pollen culture. During the starvation treatment, both the CTC-calcium and DAB-callose fluorescence disappear. It is suggested that the polarized calcium distribution in young pollen grain may be an important factor of the differential pathway of the vegetative and generative cell development and that changes in the calcium localization could take part in embryogenesis induction (17).

In pollen grains cultured for 2 d in an embryogenic medium after the starvation treatment, mitotic division of both generative and vegetative nuclei was observed (Fig. 3). The products of generative nucleus mitosis are condensed like in sperm cells while the vegetative nuclei remain diffuse and undergo further division. The origin of proembryos from vegetative cells in tobacco pollen has already been well demonstrated (18).

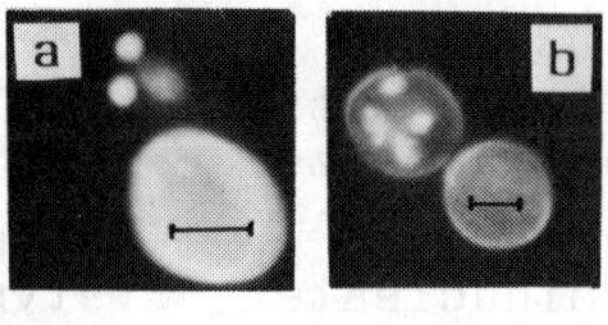

FIGURE 3. Pollen grains after division of generative (a) and of generative and vegetative (b) nuclei in embryogenic medium. Staining with DAPI, bar = 10 µm.

DNA cytophotometry in starved pollen

Cytophotometric measurements of DNA content in normally developing pollen grains of tobacco (data not given) confirmed the findings of Chu et al. (19) that in the vegetative nucleus DNA remains at 1C level. During the starvation treatment of midbinucleate pollen an increase of DNA was registered in the generative nucleus, as during *in situ* pollen maturation, and also in the vegetative nucleus in nearly 50 % of pollen grains (Fig. 4). In young tobacco pollen cultured in distilled water for 7 d, 2 % of grains showed incorporated radioactivity in the vegetative nucleus after labelling with ^{3}H-thymidine, indicating DNA replication (20).

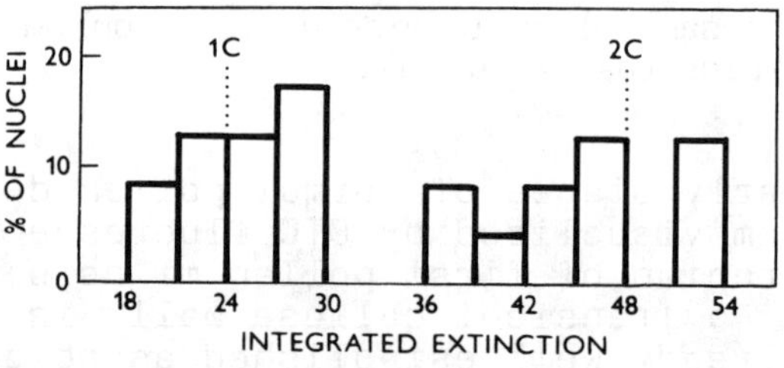

FIGURE 4. Cytophotometric estimation of DNA content in vegetative nuclei of pollen grains starved in BM medium for 6 d.

The starvation, probably through gametophytic cytoplasm regression, calcium redistribution, structural changes and vegetative cell dedifferentiation, thus results in derepression of the arrest of the vegetative cell cycle at G1 and in its progression to S phase required for embryogenic pathway of pollen development.

REFERENCES

1 Kyo, M. and Harada, H. (1985) Plant Physiol. 79, 90-94
2 Kyo, M. and Harada, H. (1986) Planta 168, 427-432
3 Benito Moreno, R.M., Macke, F., Hauser, M.-T., Alwen, A. and Heberle-Bors, E. (1988) in Sexual Reproduction in Higher Plants (Cresti, M., Gori, P. and Pacini, E., eds), pp. 137-142, Springer, New York
4 Heberle-Bors, E. (1989) Sex. Plant Reprod. 2, 1-10
5 Luštinec, J., Hadačová, V., Kamínek, M. and Procházka, Ž. (1983) Anal. Biochem. 132, 265-271
6 Schaffner, W. and Weissmann, C. (1973) Anal. Biochem. 56, 502-514
7 Tupý, J., Süss, J. and Říhová, L. (1986) J. Plant Physiol. 123, 467-476
8 Reiss, H.-D. and Herth, W. (1978) Protoplasma 97, 373-377
9 Tupý, J., Süss, J., Hrabětová, E. and Říhová, L. (1983) Biol. Plant. 25, 231-237
10 Tupý, J. (1982) Biol. Plant. 24, 331-340
11 Žárský, V., Čapková, V., Hrabětová, E. and Tupý, J. (1985) Biol. Plant. 27, 438-444
12 Journet, E.-P., Bligny, R. and Douce, R. (1986) J. Biol. Chem. 261, 3193-3199
13 Rosner, A., Gressel, J. and Jacob, K.M. (1977) Biochim. Biophys. Acta 479, 386-397
14 Trewavas, A. (1970) Plant Physiol. 45, 742-751
15 Tupý, J. (1988) Plant Sci. 55, 137-144
16 Tretyn, A., Bednarska, E. and Górska-Brylass, A. (1987) Cell Biol. Int. Rep. 11, 157-164
17 Trewavas, A.J. (1985) in The Cell Division Cycle in Plants (Bryant, J.A. and Francis, D., eds), pp. 133-156, Cambridge Univ. Press, Cambridge
18 Sunderland, N. and Wicks, F.M. (1971) J. Exp. Bot. 22, 213-226
19 Chu, C., Liu, H. and Du, R. (1982) Acta Bot. Sin. 24, 1-9
20 Aruga, K., Nakajima, T. and Yamamoto, K. (1985) Japan. J. Breed. 35, 50-58

MOLECULAR AND PHYSIOLOGICAL DIFFERENTIATION OF POLLEN IN-VITRO

VAN HERPEN, M.M.A., SCHRAUWEN, J.A.M., WETERINGS, K.A.P., DE GROOT, P.F.M., REIJNEN, W.H., VAN DEN HEUVEL, K.J.P.T. and WULLEMS, G.J., DEPARTMENT OF EXPERIMENTAL BOTANY, MOLECULAR PLANT PHYSIOLOGY RESEARCH GROUP, UNIVERSITY OF NIJMEGEN, NL-6525 ED NIJMEGEN, THE NETHERLANDS

1. INTRODUCTION

It is unknown how the programs underlying plant developmental processes manifest themselves in terms of ordered gene expression (1). An important developmental process in plants is pollen grain development (gametogenesis).
The pollen grain is the progenitor of the sperm cells and the vehicle for their delivery to the embryo sac (2). The pollen grains are produced in the anther which consists of four chambers or pollen sacs. In a young anther, the pollen sac encloses a homogeneous group of diploid cells (microspore mother cells). These microspore mother cells undergo meiosis resulting in clusters of four haploid microspores (tetrad). Each microspore passes through the first haploid mitosis resulting into a pollen grain composed of a large vegetative and a small generative cell, the latter situated in the middle of this grain (the mid-binucleate stage). The further development of this young and infertile pollen into fertile pollen i.e. the development of the male gametophyte, is a process which can be followed in-vitro. The in-vitro development of young tobacco pollen into mature and fertile tobacco pollen proceeds in a rather simple medium with glutamine and sucrose (3). What we observe in the in-vitro situation is that the wall surrounding the pollen becomes thickened and that the pollen starts with the accumulation of starch immediately after initiation of the in-vitro culture. Within three days of in-vitro culture at a temperature of 28 ºC the gametophytes have accumulated an abundant amount of starch granules. At any moment during pollen development in-vitro the pollen can be tested for in-situ pollination, in-vitro germination (formation of a pollen tube), and subjected to labeling experiments without the interfering presence of the anther tissue. We wanted to study gene expression underlying the developmental process of gametogenesis in-vitro. Therefore, the system was first compared with in-vivo development.

2. COMPARISON OF IN-VITRO AND IN-VIVO POLLEN DEVELOPMENT

2.1. Germination, seed set and time of pollen development

Pollen development in-vivo finally results in fertile pollen i.e. pollen capable of fertilization. The pollen matured in our in-vitro system at least got to have that potential. So first we compared pollen developed in-vivo and in-vitro for their ability to germinate (to form a pollen tube) and to set seed. There is a peak in the percentage of grains germinating in-vitro after three days of in-vitro maturation (see Table 1.). This percentage is comparable with the germination percentage of in-vivo developed pollen. Moreover, three days of in-vitro culture produce pollen which shows an optimum in the number of produced seeds after pollination of emasculated flowers. Preliminary results indicate that the number of seeds from pollen developed for three days in-vitro equals that of pollen developed in-vivo. Furthermore, there is seed set from pollen developed either for one or two days in-vitro (see Table 1.). However, this pollen not only produces reduced numbers of seeds but the time necessary for seed set is extended (M.M.A. van Herpen, in preparation). The period of time necessary for the in-vitro development of fertile pollen equals that of in-vivo development i.e. 3 days. However, pollen developed in-vitro for four days is no longer able to germinate or to set seed indicating that the process of gametogenesis in-vitro can not be fully compared with the process in-vivo in which the pollen remains fertile over a long period of time.

Table 1. Germination and seed set ability of pollen during development in-vitro starting from the mid-binucleate stage (0 days). Germination and seed set ability of pollen developed in-vivo serve as a control.

Development	0d	1d	2d	3d	4d	In-vivo
Germination in-vitro (%)	0	0	1	±75	1	±80
Seed set	-	±	±	+	-	+

2.2. Gene expression

Further similarities between pollen development in-vivo and in-vitro became apparent when gene expression was studied with the help of a granule-bound starch synthetase probe (M.M.A. van Herpen, in preparation) and two other pollen cDNA clones prepared from mRNA from mature tobacco pollen developed in-vivo (K.A.P. Weterings, in preparation). No differences in expression of these genes could be detected between pollen development in-vitro compared to pollen development in-vivo. The cDNA clones λ202 and λ303 (K.A.P. Weterings, in preparation) were analysed by Northern blot hybridization which show that the corresponding genes are only expressed in pollen. The clones are characteristic for pollen since no signal appeared on blots containing RNA from either different parts of the plant or diffrent parts of the flower (K.A.P. Weterings, in preparation). The clones were tested in the in-vitro system and the accumulation of the corresponding mRNAs was found to proceed in exactly the same way as in-vivo. The expression of the corresponding genes starts after pollen mitosis and the mRNA level continues to rise until the mature stage. This fact, together with the presence of the transcripts in pollen from a broad range of plant species suggests an important role for both mRNAs during pollen development and/or germination. Preliminary results from Northern blot analysis of RNA from germinating pollen show that one of the mRNAs is still present during pollen tube growth (K.A.P. Weterings, in preparation). Hybridization experiments with the granule-bound starch synthetase cDNA showed no differences in gene expression between pollen development in-vitro compared to pollen development in-vivo. Accumulation of the corresponding mRNAs in-vitro is regulated in approximately the same way as in-vivo (M.M.A. van Herpen, in preparation).

2.3. Protein pattern

We also looked at the patterns of proteins present in pollen developed in-vitro and in-vivo. Proteins were extracted from pollen grains and analyzed by one-dimensional gel electrophoresis. The pattern of proteins from pollen after 3 days of in-vitro culture is not the same as the protein pattern of pollen developed in-vivo. However, the protein pattern of pollen developed in-vitro for four days is more or less the same as that of pollen developed in-vivo (see Figure 1.). Although the pollen development in vitro leads to the same end-product as in-vivo i.e. fertile pollen, the accumulation of proteins in vitro compared to in-vivo is not the same. The similarities and differences can be used as a first screening to relate the presence of individual proteins to the process of pollen development. Proteins in the two samples that do not band at identical positions on the gel are not supposed to play a role in pollen maturation.

FIGURE 1. Fluorogram of proteins synthesized in tobacco pollen, incubated with 555 kBq [^{35}S] methionine during 2 hours. Pollen developed in-vitro (IVP) after 0,1,2,3 and 4 days in culture. Pollen developed in-vivo (ISP)

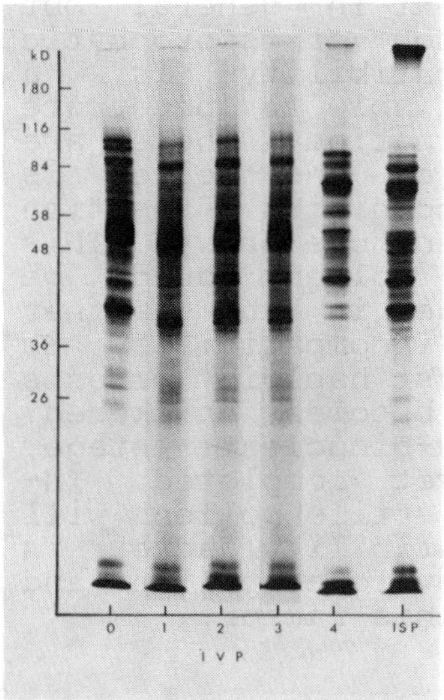

3. Advantages of pollen development in-vitro

So it looks as if pollen development is an autonomous process: mid-binucleate pollen can be separated from the interfering anther tissue and leads in a simple environment, via an autonomous process of sequential induction (coordination in time) to mature and fertile pollen with an abundant amount of starch granules. We are able to disturb this process by omitting glutamine and sucrose from the culture medium. The cells react to carbon and nitrogen deprivation by progressive dedifferentiation and the result is a steady decrease in pollen fertility upon prolonged culturing (3). Transferring the pollen after a period of 7 days into the medium used for pollen development causes the cells to develop into cell clumps and finally into embryoids. So, differentiated tissue (mid-binucleate pollen) is also capable of expressing a different developmental route (embryogenic pathway) after dedifferentiation when suitable environmental signals are provided. This switch in development in-vitro offers unique possibilities to study either gene expression or protein accumulation that underlies the process of gametogenesis. So future research will be focused on isolating proteins associated with germination and/or seed set ability. The primary structure obtained after sequencing of individual proteins could be compared with that of proteins in a database. The synthesis of synthetic oligonucleotides on basis of the amino acid sequence and the isolation of the corresponding genes would be a promising approach.

The advantages of in-vitro pollen development are enormous,

not only when studying pollen development in general but also when dealing with the phenomenon of sporophytic incompatibility. Sporophytic incompatibility is a phenomenon in which mature pollen is not able to penetrate the stigma of a gynaecium after pollination. There are indications that a factor in the wall of the pollen is recognized by the stigma. This recognition reaction finally leads to the inhibition of pollen tube growth. This factor in the wall surrounding the pollen would be synthesized by the sporophytic tissue in the anther surrounding the pollen. Sporophytic incompatibility is established after completion of the first haploid mitosis when the wall surrounding the pollen becomes thickened. When we isolate the young pollen in mid-binucleate stage, wall synthesis has started but is not yet completed. In-vitro development of this pollen into fertile pollen will give information about when the incompatibility factor is incorporated into the wall of in-vivo developed pollen and may provide the opportunity to isolate this factor.

REFERENCES

1 Goldberg, R.B., Barker, S.J. and Perez-Grau, L. (1989) Cell 56, 149-160
2 Raghavan, V. (1987) Cell Differentiation 21, 213-226
3 Benito Moreno, R.M., Macke, F., Hauser, M.T., Alwen, A. and Heberle-Bors E. (1988) in Sexual Reproduction in Higher Plants (Cresti, M., Gori, P. and Pacini, E., eds.), pp.137-142, Springer-Verlag, Heidelberg

SEX DETERMINATION IN THE DIOECIOUS *MELANDRIUM* : THE *IN VITRO* CULTURE APPROACH.

Y. Wu(1), D. Ye(1), P. Installé(2), M. Jacobs(1), I. Negrutiu(1,2)
(1) Plantengenetica, Vrije Universiteit Brussel, 65 Paardenstraat B-1640 Belgium
(2) Phytotec, 77 ch Romaine, B-5800 Gembloux, Belgium

1. INTRODUCTION

Among dioecious species, *Melandrium album* (2n = 24) (Caryophyllaceae, subfam Dianthoideae) shows a clear-cut correlation between the phenotypic sex and the presence of heteromorphic sex chromosomes. The genetic control is by far the most strict among such plant species due to a relatively well characterized male (Y, X) and female (X, X) chromosome system. Thus in normal, 2A XY plants, and also in tetraploids (4A XXXY), the male "determiners" on a single Y chromosome are able to suppress the female influence of the X chromosome(s) alone or in combination with suspected autosomal loci (Westergaard, 1958; van Nigtevecht, 1966).

As a matter of fact, it has not been possible to induce flowering or to reverse sex in *Melandrium*, neither by exogenous application of growth regulators, nor by grafting (Lyndon, 1985). This contrasts with a series of monoecious (for ex. cucumber) or dioecious (for ex. *Mercurialis*) species where sex reversal was achieved through hormone applications (Durand and Durand 1984).

However cell culture techniques have not been used within such species so far as an experimental tool to challenge the sex determination and/or differentiation programme.

We present here two distinct approaches, namely (A) the comparative induction of plant regeneration via callus culture versus protoplast culture in male and female plants and (B) production of haploid plants via androgenesis on a medium containing either an auxin or a cytokinin.

2. PLANT REGENERATION FROM TISSUE EXPLANTS IN MALE VERSUS FEMALE PLANTS.

For several tissue explants analysed (leaves, hypocotyls, roots, germinated seeds), callus induction and root formation were observed regularly, while shoot regeneration varied from 2% for leaf derived calli and 19 % for calli originating from germinated seeds.

Leaf explants from sexed plants were cultured for callus and root induction : higher rates (on average of two fold) were registered in explants from male plants. The results correlated with higher auxin levels measured in leaves from male plants as compared with female ones.

The general high tendency towards root differentiation (under a variety of hormonal regimes) raises important difficulties in modulating the culture system towards shoot regeneration.

3. SCREENING FOR SHOOT REGENERATION POTENTIAL

A search was initiated among regenerated plants derived from germinated seed explants. Among them, line T37 exhibited a particularly high capacity for embryogenesis (Fig 1) The regeneration could reach about 70% and 100 % from the leaf explants on hormone-free and low concentration of hormone additive media respectively.

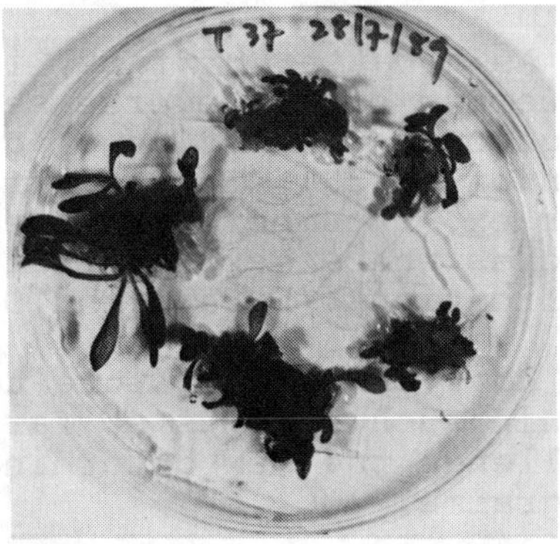

FIGURE 1. T37 plantlet culture

4. PLANT REGENERATION THROUGH PROTOPLAST CULTURE.

From leaf protoplasts (fig 2a)isolated under standard conditions and cultured in an agarose bead system, first divisions were observed after about 5 days in culture (fig 2b).

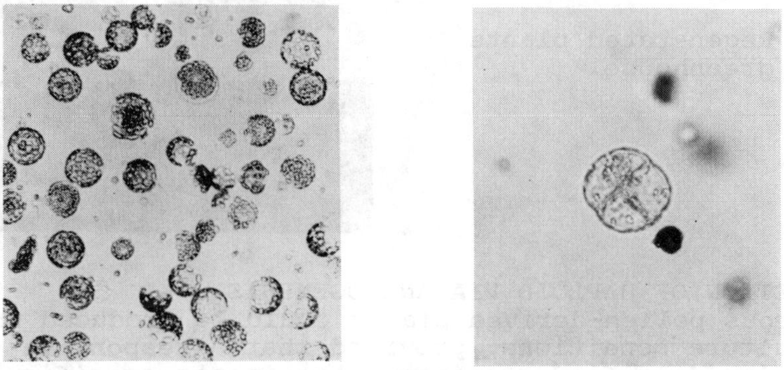

FIGURE 2 a)Leaf protoplasts. b) The first protoplast division after 5 days culture.

Protoplasts isolated from different "isolates" showed a remarkable variability in division rates (ranging from 1 to 40 %). In general, the isolation of protoplasts from male plants was more difficult and division rates and colony formation were also lower as compared to females.

Plant regeneration has been obtained from protoplast isolated from seedling leaves either via organogenesis or embryogenesis (regeneration frequencies around 10 %). Somatic embryogenesis was clearly induced in 2,4D containing media, while cytokinin was promoting shoot formation.

A high aptitude for embryogenesis is routinely being observed in the selected T37 line. Embryos are produced within 7 weeks at 75 % rates.

Such regenerants are being grown under greenhouse conditions (fig 3) and analysed for genetic sex versus phenotypic sex expression.

FIGURE 3 Regenerated plants grown in greenhouse.

5. PRODUCTION OF HAPLOID VIA ANDROGENESIS.

Numerous pollen-derived plants could be produced under simple culture conditions, provided that a responsive genotype is identified. Optimized treatments specifically contributed to increase the number of responsive anthers.

The fact that a similar yield and quality (in terms of survival, plant recovery and phenotype etc) of androgenic plants was obtained on a medium containing distinct hormone molecules (IAA versus BAP), suggested to us that the hormonal supplement :

- improved the overall condition or responsiveness of the anthers;
- did not exert a specific and long-lasting effect on sex determination and differentiation.

The results were indicative of non-specific parameters acting to select in favour of "competent" pollen grains that could divert genuine fate and embark on the embryogenic pathway.

Phenotypic (plant size and leaf shape), cytological (chromosome counts) and genetic (seed set) evidence indicated that the great majority of the plants were haploids and maintained the haploid state to maturity.

The particular thing about obtaining androgenic haploids in a dioecious plant with heteromorphic sex chromosomes is that a complete separation of single X and Y chromosomes can be achieved, and this in a background of a haploid set of autosomes. Our results showed that independent of genotype and hormonal supplement in the culture medium, all the plants investigated so far were phenotypic female. Whenever a cytological analysis was performed the phenotypic sex corresponded to the genetic sex. The data suggested to us that absence of X chromosomes in *Melandrium* is lethal, at least at zygotic to torpedo embryonic stages. An alternative explanation is that under the described experimental conditions, Y-bearing microspores are not "competent" for embryonic development. If so, X contains embryo-promoting genes. Since beyond that stage the observed lethality was negligible (<20%) we concluded that no corresponding vital functions are to be expected on the X chromosome exclusively. In the absence of Y and in the haploid condition (A+W), we could not identify thus far autosomal loci that are epistatic to X-borne loci in promoting male organogenesis

Sex determination is an important biological problem : it representing a challenging matter of plant development (and here is almost nothing known about plant development at cellular and molecular level at present). In addition, it is the basis of breeding concepts and practice. Our purpose was to open experimentally this field for new approaches. Our results indicate that *Melandrium* behaviour in culture and regeneration is influenced by both genotype and sex. Gene transfer has become a powerful tool of analysis and experimentation. We therefore intend to apply on *Melandrium* direct gene transfer techniques, and check the possibility of monitoring sex expression

REFERENCES

1 Durand R and Durand B. (1984) Phisiol Plant. 60, 267-270
2 Lyndon RF. (1985) in Handbook of flowering plants", Vol IV, pp 313-319 CRC Press
3 Van Nigtevecht (1966), Genetica 37, 307-344.
4 Westergaard (1958) Adv. Genet. 9, 217-281
5 D. Ye, P. Installé, D. Ciupercescu, J Veuskens, Y Wu, M. Jacobs, I Negrutiu (1990) Sexual Plant Reproduction (in press).

TRANSFORMATION OF POLLEN

ERWIN HEBERLE-BORS, ROSA MARIA BENITO MORENO, ANNA ALWEN, EVA STÖGER, OSCAR VICENTE, INSTITUTE OF MICROBIOLOGY AND GENETICS, UNIVERSITY OF VIENNA, ALTHANSTR. 14, 1090 VIENNA, AUSTRIA

WHY POLLEN AS A VECTOR FOR GENE TRANSFER?

In most transformation systems today, in vitro regeneration of plants is an essential step and very often the main limiting factor for yield and success. Circumvention of in vitro regeneration would therefore be highly advantageous. The simplest way to do this would be to transfer DNA into the gametes and, after gamete fusion, to exploit seed embryogenesis for plant formation. If feasable, such an approach would be technically simpler and faster than methods based on in vitro regeneration. It should also be universal in its application. In addition, one would avoid the problem of somaclonal variation resulting from in vitro regeneration. Somaclonal variation in transgenic commercial lines of crop plants can spoil the advantage obtained by transferring a transgene. The problem of chimaera that is intrinsic to in vitro regeneration, would also be avoided.

Gametes in the strict sense are not easily accessible in plants although progress has been achieved in isolating both sperm and egg cells (1). Both, male as well as female gametes, can be used for gene transfer. The main advantage of the egg cell might be the efficient formation of a whole plant either in situ or after isolation and in vitro culture. The latter remains to be shown, however. The male gamete in higher plants is not a free, mobile cell as in lower plants or in animals, but is part of a larger structure, the pollen. And only the pollen as such is able to perform fertilization in situ, while the isolated sperm cell would have to be fused in vitro with the egg cell. Attempts to use male gametes for gene transfer have therefore concentrated on the transfer through pollen.

POLLEN DEVELOPMENT AND POLLEN FUNCTION

The pollen grain in angiosperms is a comparatively simple structure, consisting of either two or three cells (one generative plus one vegetative, or two sperm cells plus one

vegetative).
Pollen grains represent the male gametophytes in the alternation of generations in higher plants. They develop from microspores that are themselves the result of meiosis. Microsporogenesis and pollen development take place in the male sex organ of the sporophyte, the anther. Sporophyte and gametophyte interact in the regulation of pollen development and pollen function, with the sporophyte exerting control through the activity of particular tissues in the anther, mainly the tapetum.
The mature pollen is released after dehydration and is transferred to the stigma of a pistil. Transfer can be to the stigma of the same flower (self-pollination) or to stigmas of flowers of other plants of the species (cross-pollination). Various mechanisms and modes of pollen transfer have evolved in the plant kingdom to ensure either inbreeding or outbreeding.
After passing the style and the ovarian locule(s), the pollen tube reaches the micropyle of an ovule and penetrates it. It is lead towards one of the synergids of the embryo sac. There it bursts and discharges its cytoplasm and the two sperm cells into the usually degenerated or degenerating synergid. One of the sperm cells fuses with the egg cell to release the sperm nucleus into the egg cytoplasm that then fuses with the egg nucleus to form the zygote. The other sperm cell fuses with the central cell to release the second sperm nucleus that fuses with the two polar nuclei to form the triploid endosperm nucleus. A coordinated development of embryo, endosperm and sporophytic tissues of the ovule and ovary then leads to the formation of seeds and fruit.

THE DIFFERENT TECHNIQUES OF POLLEN TRANSFORMATION

The use of pollen as a vector for DNA preceded the rise of plant molecular biology. Since the mid seventies DNA transfer by pollen was considered a possibility to improve breeding by providing a means of transferring specific traits from one variety to another or from one species to another. The earliest and simplest technique of pollen transformation was intra- or interspecific pollination with irradiated pollen (2,3).
A more refined technique was to incubate mature pollen with exogenous genomic DNA extracted from other organisms or a genetic variant of the plant to be transformed. The genomic DNA contained genetic markers that were used to screen the offspring population (4,5,6,7,8).
After this classical period of pollen transformation, the eighties brought the break-through in plant transformation with the discovery of the genetic colonization of plants through _Agrobacterium tumefaciens_ and the use of disarmed agrobacteria as vectors. Other gene transfer systems soon

followed, based mainly on regeneration from protoplast (PEG-mediated, electroporation, microinjection). Recently, biolistics (use of DNA-gun) joined the arsenal of weaponry to deliver DNA into living tissues (9). Still, however, there exist plant species that cannot be transformed by these methods, among them most of the economically interesting cereals. Interest in the old method of pollen transformation reappeared, now complemented by recombinant DNA technology and the modern DNA delivery devices.

Recombinant DNA technology potentially removes one of the drawbacks of the classical pollen transformation experiments. It allows the transfer of clearly defined DNA pieces that can be detected unequivocally in the offspring, and that often allow selection of transformed seeds.

The first technique of modern pollen transformation was the classical incubation of dry pollen in a solution of cloned DNA before pollination (10,11).

A variation of this technique is the application of DNA and pollen on the stigma surface (12,13, E. Picard, personel communication). In a third variation, rice styles are cut off after pollination and pollen tube growth and DNA is applied on the cut surface (14,15).

In another set of techniques, DNA is delivered into pollen grains with particular devices. DNA has been injected into maize pollen grains (A. Crossway, personel communication), and electroporation has been used in tobacco (16). Agrobacteria have also been applied to transfer DNA into tobacco pollen. Mature pollen has been germinated in the presence of <u>Agrobacterium</u> <u>tumefaciens</u> or Ti-plasmid DNA, and the germinated pollen was then used to pollinate tobacco flowers (6, 10). The most recent attempt is to use the particle gun to deliver DNA into maize pollen grains (17).

In the case of macroinjection into tillers of cereals (18), it is not clear yet which of the sexes is the primary recipient of the DNA. Therefore, macroinjection is not considered in detail in this context.

OBSTACLES ON THE WAY

At present, it is difficult to conceive how DNA enters the mature pollen without being degraded and how it is transported to the egg cell or integrated into the sperm DNA. Only one report has been published on the expression of DNA transferred into pollen grains. In this study, the DNA was introduced by a DNA gun and expression was achieved only when the marker gene used had been cloned behind a pollen-specific promoter (19).

It has been proposed that foreign DNA can enter pollen grains through imbibition (6). In fact, mature, released pollen grains have a plasma membrane that does not form an osmotic barrier (20). They take up water through imbibition. However, there is still the barrier of the pollen wall (exine and intine) that has to be penetrated by the DNA macromolecules. Fluorescence-labelled proteins derived from the stigma gave no evidence for transfer into the pollen grains (21). In some plants, including the cereals, pollen rehydrates in a controlled way on a dry stigma, and rapid rehydration - by incubating pollen in an aqueous solution - in these plants destroys pollen function. In addition, pollen grains excrete nucleases (22,23,16) after imbibition that degrade exogenously supplied DNA.

Nucleases may well be present on the stigma also and in the stylar channel. This organ has not only the function of allowing the pollen tube to enter the ovary, but also of protecting the ovary from alien agents (pathogens) and to nourish the pollen tube during its passage through the style. The presence of intercellular hydrolytic enzymes in the style has been demonstrated (24). The nourishing function implies an intimate interaction of pollen tube and style. Such intimacy is also implied by the finding that the style may be involved in the pollen tube movement (25). Even if DNA enters the pollen grains, DNA passage inside the pollen tube is conceivable only if the DNA has been transferred into the pollen grain or tube during early pollen tube growth. DNA introduced into older tubes (after cutting the pollinated styles) cannot reach the embryo sac because of the callose plugs that are formed in intervalls along the tube and that restrict the living protoplast of the pollen tube to the tip. If the cut is performed within the tip, the tube cell will most likely die and fertilization would be prevented.

Another barrier is the plasma membranes of the generative and then sperm cells that persist until their fusion with egg and polar cell. Only during imbibition of dry pollen with a DNA solution or through the action of a DNA gun is it conceivable that DNA can enter the generative or sperm cells.

A further obstacle for DNA flowing through the stylar channel may be the micropyle and the filiform apparatus. Usually the nucellus forms a tight channel through which the pollen penetrates. This channel should be blocked even more so once the pollen tube has passed the micropyle.

Assuming that foreign DNA can pass all the barriers and enter the sperm nuclei, it still has to be integrated into the nuclear DNA. Although the general mechanism of integration of foreign DNA into genomic DNA is not understood, there is circumstantial evidence that enzymes participating in DNA replication may be involved. However, no DNA synthesis takes place during pollen tube growth of angiosperms. On the other hand, pollen has a very active DNA repair system (26), and it is conceivable that DNA that has entered the nuclei of the generative or sperm cells is integrated into their genome by the action of the repair system.

After reaching one of the synergids, the pollen tube contents are discharged into it, but only the sperm nuclei with the cytoplasm of the sperm cells are transferred into the egg cell and the central cell. The tube cytoplasm is left behind in the synergid. This is indicated by electron microscopical studies that show that the paternal plastids stay in the synergid (27). The fusion of the sperm cell with the egg cell should effectively prevent the penetration of tube cytoplasm into the egg cell. It is hard to imagine how foreign DNA being present in the tube cytoplasm can be transferred into the egg cytoplasm given the efforts that have to be made to transfer DNA into in vitro cultured protoplasts.

In some species, including most cereals, the distance from stigma to ovule and the time between pollination and fertilization is very short (less than an hour in barley, 27). DNA flowing in large amounts alongside the pollen tube in the stylar channel or being present inside the tip of the pollen tube may in these cases override the barriers and reach the synergid and eventually the egg nucleus. Experiments will have to find out if this is possible without interferring with pollen function.

MICROSPORES AND IMMATURE POLLEN GRAINS AS TARGETS FOR GENE TRANSFER

We have developed an in vitro system for tobacco that allows effective simulation of in vivo maturation conditions for microspores and young pollen grains (28). The in vitro matured pollen is used to pollinate emasculated flowers in situ. The flowers set seed and the seeds germinate normally. Selfings and reciprocal backcrosses involving in vitro matured pollen from wild-type and transgenic plants containing marker genes (KAN^R, NOS) show Mendelian segregation, proving that it was the in vitro matured pollen which performed fertilization. In vivo seed set after pollination with in vitro matured pollen is now also possible in wheat (Staufer et al., see poster).

During the period of in vitro maturation, the pollen can be used as a target for gene transfer. Attempts to use electroporation for gene transfer into tobacco microspores have so far been unsuccessful in our hands, as were attempts with microinjection. Co-cultivation of microspores with <u>Agrobacterium</u> <u>tumefaciens</u> turned out to be very promising as indicated by the expression of the ß-glucuronidase (GUS) gene in the in vitro matured, germinating pollen. This GUS activity has been obtained at pH 7, the pH that excludes false positives due to cleavage of glucuronides by a plant endogenus GUS (29). Potential transgenic plants have been produced after pollination with this pollen and are presently investigated further. The first transformant produced in this way has been selected as a kanamycin-resistant seedling, is positive in the Southern, and the NPT II gene segregates in the two subsequent generations in a Mendelian way.

Transfer of DNA into microspores or immature pollen, followed by in vitro maturation and in vivo pollination circumvents some of the problems mentioned above on the use of mature pollen. The foreign DNA enters the single celled microspore or the generative cell of the young pollen grain which are the cells of the plant male germ line. Integration into the genome should be possible during S-phase of the generative cell. After second pollen mitosis, the foreign DNA should therefore be present in the two sperm cells.

In cereals, regeneration from protoplast was and still is a very tedious and inefficient process that is, in addition, prone to somaclonal variation. Pollen embryogenesis is an alternative and often efficient in vitro regeneration process in these species. Microinjection has been used to transfer DNA into early pollen embryos which then have been regenerated into haploid and then diploid and homozygous plants. First demonstrated in rape seed (30), it is at present being attempted in several cereals (31, I. Potrykus, personel communication). Cocultivation of microspores and young pollen grains of rape seed with agrobacteria and subsequent pollen culture has also lead to putative transformed haploid plants (32).

WHAT IS A TRANSGENIC PLANT?

From the work on Agrobacterium-mediated and direct gene transfer, a number of criteria have developed that have to be fulfilled in order to verify the transgenic nature of a plant.

First, the marker gene that is used to select or identify transgenic plants amongst the non-transformed ones has to be assayed biochemically. Second, the marker gene (and the gene of interest) have to be physically detected in the genome of the plant by Southern blotting. In addition, it has to be shown that the foreign DNA is covalently linked to the nuclear DNA of the recipient. Fourth, the gene has to be shown to pass through meiosis by demonstrating Mendelian segregation in the offspring of the primary transformant.

There is a growing awareness that these tests are crucial in order to accept a plant as being transgenic. Particularly in cereals it appears that foreign DNA can be detected in the primary transformants, but fails to be detected in the next generations (G. Feix, personel

communication, I. Potrykus, personel communication).
A primary criterium for successful transformation is reproducibility. Unfortunately, the history of pollen transformation is darkened with negative claims. For instance, the early work on the use of irradiated pollen has later been seriously challenged (33,34,35). The work on incubation of pollen with DNA has equally been challenged (35,36,11). Other attempts like electroporation (16) or microinjection (A. Crossway, personel communication) of preincubated pollen did not result in the formation of transgenic plants.
It is interesting to note that an approach analogous to pollen transformation, i.e. the creation of transgenic animals through fertilization with DNA-treated sperms (37), faces the same problem of non-reproducibility (38).
The future will show which of the techniques of pollen transformation will survive. Pollen transformation is still an open issue.

REFERENCES

1 Sexual Reproduction in Higher Plants (Cresti M, Gori P, Pacini E, eds) Springer, Berlin Heidelberg New York
2 Pandey KK (1975) Nature 256:310-313
3 Pandey KK (1980) Heredity 45:15-29
4 Hess D (1969) Z Pflanzenphysiol 60:348-358
5 Hess D (1980) Z. Pflanzenphysiol 98:321-337
6 Hess D (1987) Int Rev Cytol 107:169-190
7 De Wet JMJ, Bergquist RR, Harlan JR, Brink DE, Cohen CE, Newell CA, de Wet AE (1985) in Experimental manipulation of ovule tissue (Chapman GP, Mantell SH, Daniels W, eds), pp 197-209, Longman, London
8 Ohta Y (1986) Proc Natl Acad Sci USA 83:715-719
9 Klein TM, Wolf ED, Wu R, Sanford JC (1987) Nature 327:70-73
10 Sanford JC, Skubik KA (1986) in Biotechnology and ecology of pollen (Mulcahy DL, Bergamini, Mulcahy G, Ottaviano E, eds), pp 71-76, Springer, New York Berlin Heidelberg Tokyo
11 Booy G, Krens FA, Huizing HJ (1989) J Plant Physiol 135:319-324
12 DNA Plant Technology Corporation and Du Pont European Patent Application Nr. 88100267.9 from 11.01.1988)
13 Picard E (1988) Genome 30 (Suppl), abstract Nr. 25.8.2
14 Zhou GY, Weng J, Gong ZZ, Zhen YS, Yang WX, Shen WF, Wang ZF, Tao QZ, Huang JG, Qian SY, Lin GL, Ying MC, Xue DY, Hong AH, Xu YJ, Chen SB, Duan XL (1988) Scientia Agricultura Sinica 21:1-6
15 Luo ZX, Wu R (1989) Plant Molec Biol Rep 7:69-77
16 Negrutiu I, Heberle-Bors E, Potrykus I (1985) in Biotechnology and ecology of pollen (Mulcahy DL, Bergamini, Mulcahy G, Ottaviano E, eds), pp 65-70,

Springer, New York Berlin Heidelberg Tokyo
17 AGRACETUS European Patent Application Nr. 87310612.4 from 02.12.1987
18 Pena de la A, Lörz H, Schell J (1987) Nature 325:274-276
19 Twell D, Klein TM, Fromm ME, McCormick S (1989) Plant Physiol 91:1270-1274
20 Heslop-Harrison J (1979) Am J Bot 66:737-743
21 Vithanage HIMV, Heslop-Harrison J (1979) Ann Bot 43:113-114
22 Matousek J, Tupy J (1984) Biol Plant (Praha) 26:62-73
23 Matousek J, Tupy J (1985) J Plant Physiol 119:169-178
24 Herrero M, Dickinson HG (1979) J Cell Sci 36:1-18
25 Sanders LC, Lord EM (1989) Science 243:1606
26 Jackson JF (1988) in Sexual Reproduction in Higher Plants (Cresti M, Gori P, Pacini E, eds), pp 81-86, Springer, New York Berlin Heidelberg Tokyo
27 Engell K (1988) in Sexual Reproduction in Higher Plants (Cresti M, Gori P, Pacini E, eds), pp 383-388, Springer, New York Berlin Heidelberg Tokyo
28 Bennito Moreno RM, Macke F, Alwen A, Heberle-Bors E (1988a) Planta 176:145-148
29 Alwen A, Vicente O, Heberle-Bors E (1989) in "Molecular communication in higher plants", pp 75, Abstracts of EMBO-EMBL Symposium 1989 on Molecular Communication in Higher Plants
30 Neuhaus G, Spangenberg G, Mittelsten Scheid O, Schweiger H-G (1987) Theor Appl Genet 75:30-36
31 Potrykus I, Datta SK, Neuhaus G, Spangenberg G (1988) in Abstracts of Second International Congress on Plant Molecular Biology, Nr. 57
32 Pechan PM (1989) Plant Cell Reports 8:387-390
33 Engvild KV (1985) Theor Appl Genet 69:457-461
34 Cornish MA, Werner CP (1985) Heredity 55:321-326
35 Sanford JC, Skubik KA, Reisch BI (1985) Theor Appl Genet 69:571-574
36 Sanford JC (1988) in Forest and crop biotechnology (Valentine HA, ed), pp 163-173, Springer, New York Berlin Heidelberg Tokyo
37 Lavitrano M, Camaioni A, Fazio VM, Dolci S, Farace MG, Spadafora C (1989) Cell 57:717-723
38 Brinster RL, Sandgren EP, Behringer RR, Palmiter RD (1989) Cell 59:239-241

IN VITRO FERTILIZATION OF SINGLE, ISOLATED GAMETES, TRANSMISSION OF CYTOPLASMIC ORGANELLES AND CELL RECONSTITUTION OF MAIZE (<u>ZEA</u> <u>MAYS</u> L.)

E.KRANZ*, J.BAUTOR, H. LÖRZ*
MAX-PLANCK-INSTITUT FÜR ZÜCHTUNGSFORSCHUNG,
D-5000 KÖLN 30, FRG.
*PRESENT ADDRESS:
UNIVERSITÄT HAMBURG, INSTITUT FÜR ALLGEMEINE BOTANIK
OHNHORSTSTR. 18, D-2000 HAMBURG 52, FRG.

1. INTRODUCTION

In vitro fertilization in higher plants has been accomplished in various plant species using excised, embryo sac-containing tissue and pollen. Little is known, however, about the fertilization process at the cellular level for which methods of manipulation of single gametes have to be employed. Recently sperm cells of various plant species have been isolated, and viable female gametes have been isolated from the embryo sacs of <u>Nicotiana</u> (1), <u>Torenia</u> (2), <u>Zea</u> (3,4) and <u>Plumbago</u> (5). To attempt artificial fertilization, sperm cells were injected into the embryo sac of <u>Torenia</u> (6).

In vitro fertilization performed with single gametes will also represent an effective method for studies of effects of non-gametic cells on the development of artifically-produced zygotes.

So far the techniques of transfer of cytoplasmic characters via protoplast fusion (7) as well as the fusion of cytoplast and protoplast or karyoplast, respectively, to perform cell reconstitution (8, 9, 10, 11, 12) have been applied to somatic cells only. Using gametes, the possibilities for combining genetic material will be extended.

We now report on the isolation of the gametes of maize, the culture of egg cells, the electrofusion-mediated in vitro fertilization of single sperm and egg cells as well as the development of the fusion products in culture. This is to our knowledge the first report on in vitro fertilization of higher plants using isolated single gametes.

Additionally, single sperm cells were fused with non-gametic cells of the embryo sac and with individually selected cytoplasts of maize. Furthermore, using cytoplasts, cytoplasmic organelles have been transferred through the fertilization process.

2. PROCEDURE

2.1 Materials and methods

2.1.1. Plant material

Pollen and ears from maize plants (Zea mays L.) inbred lines A 188 from A.Pryor, C.S.I.R.O., Canberra, Australia, pollen from inbred lines Ac 718, 1262, 1605, 1635 from P.A.Peterson, Department of Agronomy, Iowa State University, USA, pollen from inbred Ac line Zm651 and ears from inbred 2360 (Origin: Maize Genetics Cooparation) from U. Wienand, Max-Planck-Institut für Züchtungsforschung, Köln, FRG, as well as inbred line 1206 (Dr. Schmidt, KWS, Einbeck, FRG) for the establishment of a feeder-cell suspension, were used. All plants were grown in the greenhouse.

2.1.2 Pollen and ear collection

Pollen was collected as described elsewhere (13, 14). Ears were bagged before silk emergence, and the outer leaves of the ears were surface sterilized with ethanol (70%). Ovules were selected from the middle part of the ears under a dissecting microscope.

2.1.3 Isolation and selection of sperm cells, egg cells and cytoplasts

Sperm cells were released from the pollen grains after rupture by osmotic shock (15,16,17), in the fusion medium and individually selected by microcapillaries under microscopic observation as described elsewhere (14).

About 20 pieces of ovular tissue, containing embryo sacs, were collected in mannitol solution (570 mOsm/kg H_2O), following digestion in 3-cm diameter plastic dishes with 1.5 ml of enzyme solution (0.75% pectinase, Serva; 0.25% pectolyase Y23, Seishin; 0.5% hemicellulase, Sigma and 0.5% cellulase "Onozuka" RS (Yakult Honsha), pH 5.0, adjusted to 570 mOsmol/kg H_2O with mannitol. The mixture was agitated on a rotary shaker at 140 rpm and 24 ± 0.5°C for 50 min ± 10 min, followed by mechanical isolation of the embryo sacs and the egg cells (14).

Cytoplasts were prepared from green or etiolated leaf tissue of 12-14 day old seedlings of maize. The tissue was enzymatically digested in 3-cm diameter plastic dishes with 1% cellulase "Onozuka" RS (Yakult Honsha) and 0.5% macerozyme R 10 (Serva), adjusted to 570 mOsmol/kg H_2O mannitol, pH 5.6 and incubated at 40 rpm and 27°C for 1.5 h. The digest was filtered through a sieve (mesh size 50 µm) followed by centrifugation at 150 g for 3 min and washing of the pellet in 0.5 M mannitol solution. After centrifugation for another 3 min, the pellet was resuspended in 3-cm plastic dishes in mannitol solution (0.5 M) for selection.

Embryo sacs, egg cells, synergids, central cells and cytoplasts were individually selected and transferred by microcapillaries. These capillaries were connected via teflon tubing, which was filled with mineral oil, to a computer-controlled dispenser/dilutor (18).

2.1.4 Electrofusion

Fusion was performed as described previously (19, 20, 21) and adapted to gametes (14). Ten fusion droplets of 1000 nl, consisting of 0.50 M and 0.55 M mannitol, respectively, were overlayered by mineral oil on a coverslip and selected single gametes were transferred to these droplets. Fusion was performed with a pair of electrodes fixed to an electrode support mounted under the condensor of the microscope. Single or multiple (2-3) negative DC-pulses (50 µs; 0.7 - 1.0 kV x cm^{-1}) were applied on one of the electrodes after dielectrophoretic alignment of defined gametes and cytoplasts (1MHz; 48-71 V) with an electrofusion apparatus (CFA 400, Krüss, FRG) (Fig.1).

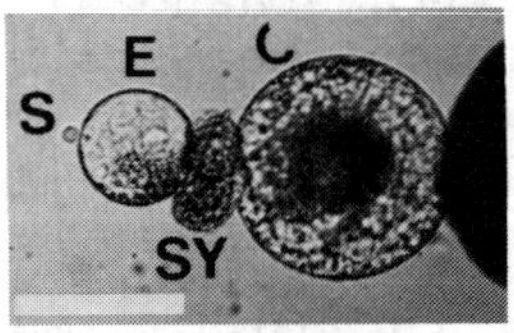

FIGURE 1. Alignment of sperm cell (S) and unit of egg cell (E), synergids (SY) and central cell (C) of maize on one electrode for fusion of sperm and egg cell. Bar = 100 µm.

2.1.5 Culture procedures

Fusion products were cultured on the transparent, semipermeable membrane of "Millicell"-dishes (diameter 12 mm, Millipore). These dishes were inserted in the middle of 3-cm plastic dishes previously filled with 1.5 ml of a maize feeder-suspension. Fusion products as well as feeder-cells were cultivated in a modified MS medium (22) with 1.0 mg/l 2.4 D and 0.02 mg/l kinetin, adjusted to 600 mOsmol/kg H_2O with glucose, pH 5.7. Culture conditions: 24 ± 0.5°C; light/dark cycle of 16/8 h and approximately 50 $\mu E^{-2} s^{-1}$.

3. RESULTS AND DISCUSSION

The timing (50 min ± 10 min) of the exposure of the embryo sac containing tissue to the enzyme solution was important to obtain sufficient cells of the embryo sacs of maize, as also found by Wagner et al. (4). The yield was determined finally by the manual step of isolation (mean frequency 25%).

Sperm cells fused very efficiently with other sperm cells (mean frequency of one to one-fusion 85%). This is a further indication that sperm cells in a mature pollen grain of maize exist as protoplasts (16, 17, 23). The fusion processes, involving sperm cells, exceeded not more than one second in most experiments. The mean frequency of fusion of a sperm cell with an egg cell was 79%.

Isolated egg cells showed protoplasmic streaming during 22 days of culture, but no cell division could be observed.

However, under comparable culture conditions, the fusion products of a sperm with an egg cell started to divide within 2.5-3 days. Multicellular structures developed with a mean frequency of 83%. (Fig.2)

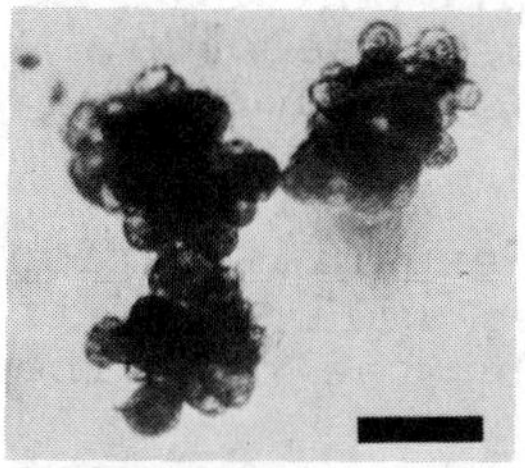

FIGURE 2. Development of three fusion products of sperm cell and egg cell of maize 12 days after fusion on a semipermeable membrane in a "Millicell"-dish. Bar = 250 µm.

Additionally, the following fusion combinations could be accomplished: Sperm with synergid (mean 75%) and sperm with central cell (mean 55%); sperm with synergid (with adherent egg cell); sperm with egg cell (with adherent synergids and partially with adherent central cell) and cytoplast with egg cell (with adherent synergids) with sperm (Fig.3).

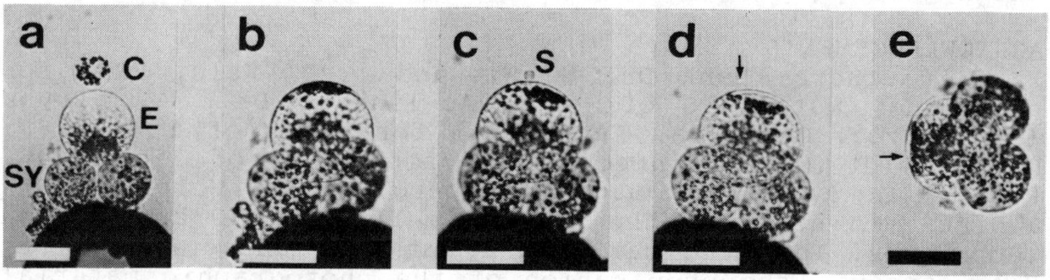

FIGURE 3. Electrofusion of cytoplast (C), egg cell (E) and sperm cell (S) of maize. Synergids (SY) adhered to the egg cell. After application of the DC-pulses the time interval from a-b was 4 seconds, from c-e 2 seconds. Arrows indicate fusion site of sperm cell. Bars = 50 µm.

The fusion of sperm cells with non-gametic cells of the embryo sac may facilitate studies of the function of such cells. Additionally, the transfer of cytoplasmic organelles directly through the fertilization process will promote studies of cytoplasmic inheritance.

The high fusion ability of the sperm cells allowed a high yielding cell reconstruction (mean 73%) (Fig.4).

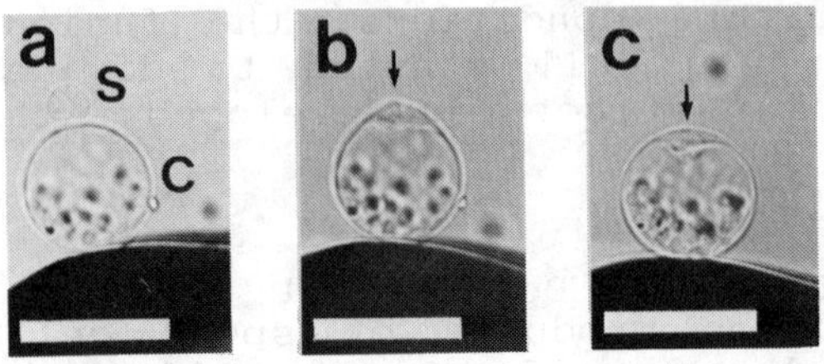

FIGURE 4. Cell reconstitution by fusion of one sperm cell (S) with one cytoplast (C) of maize. The time interval from a-c was 3 seconds. Arrows indicate fusion site. Bars = 25 µm.

Sperm cells are an attractive tool for experimentation with the haploid genome and further studies will show if they are useful in creating haploid lines by the described methods.

Recently fusion products of the gametes have been used for transformation studies. After microinjection of plasmid DNA high survival rates (mean 83%) and division rates (mean 67%) were observed (E.Kranz, unpublished). The creation of fusion products consisting of cells with "natural" competence for division and regeneration may facilitate studies of the development of artificial produced zygotes as well as studies of transformation through the sexual route or of the zygote.

ACKNOWLEDGEMENTS

The authors thank Dr. U. Wienand (MPI, Köln, FRG) for providing material of the Ac-lines, Dr. A. Pryor (C.S.I.R.O., Canberra, Australia) for seeds of line A188, Dr. Schmidt (KWS, Einbeck, FRG) for seeds of line 1206, Prof. H.U. Koop, Dr. G. Spangenberg, Mr D. Wolff and Mr J.Dirk for help on the single cell culture and fusion techniques. The critical reading of the manuscript by Dr. P.A. Lazzeri, the preparation of the photographic material by Mrs C. Adami, the typing of the manuscript by Mrs M. Modersohn and the financial support by the BMFT-Bonn (grant BCT 0390-2) are gratefully acknowledged.

REFERENCES

1 Hu, S.Y., Li, L.G. and Zhu, C. (1985) Acta Bot.Sin.27, 337-344
2 Mol, R. (1986) Plant Cell Reports 3, 202-206
3 Wagner, V.T., Song, Y., Matthys-Rochon, E. and Dumas, C. (1988) in Sexual reproduction in higher plants (Cresti, M., Gori, P. and Pacini, E., eds.), pp. 125-130, Springer, Berlin, Heidelberg
4 Wagner, V.T., Song, Y.C., Matthys-Rochon, E. and Dumas, C. (1989) Plant Sci. 59, 127-132
5 Huang, B.Q. and Russel, S.D. (1989) Plant Physiol. 90, 9-12
6 Keijzer, C.J., Reinders, M.C. and Leferink-ten Klooster, H.B. (1988) in Sexual reproduction in higher plants (Cresti, M., Gori, P. and Pacini, E., eds.) pp. 119-124,

Springer, Berlin, Heidelberg

7 Pelletier, G., Vedel, F. and Belliard, G. (1985) Hereditas, Suppl.3, 49-56

8 Binding, H., (1979) in Plant cell and tissue culture, principles and applications (Sharp, W.R., Larsen, P.O., Paddock, E.F. and Raghavan, V. eds.), pp. 789-850, Ohio State University Press, Columbus, Ohio

9 Maliga, P., Lörz, H., Lazar, G. and Nagy, F. (1982) Mol.Gen.Genet. 185, 211-215

10 Wallin, A., Glimelius, K. and Eriksson, T. (1979) Z. Pflphys. 91, 89-94

11 Spangenberg, G. and Schweiger, H.G. (1986) Eur.J.Cell Biol. 41, 51-56

12 Spangenberg, G., Neuhaus, G. and Schweiger, H.G. (1986) Eur.J.Cell Biol. 42, 236-238

13 Kranz, E. and Lörz, H. (1990) Plant Sex. Reprod. 3 (in press)

14 Kranz, E., Bautor, J., Lörz, H. (1990) In vitro fertilization of maize by electrofusion of isolated gametes. Submitted

15 Russel, S.D. (1986) Plant Physiol. 81, 317-319

16 Matthys-Rochon, E., Vergne, P., Detchepare, S., and Dumas, C. (1987) Plant Physiol. 83, 464-466

17 Cass, D.D. and Fabi, G.C. (1988) Can. J. Bot.66, 819-825

18 Koop, H.U. and Schweiger, H.G., (1985) J.Plant Physiol. 121, 245-257

19 Koop, H.U., Dirk, J., Wolff, D. and Schweiger, H.G. (1983) Cell. Biol. Int. Rep. 7, 1123-1128

20 Koop, H.U. and Schweiger, H.G. (1985) Eur.J.Cell Biol. 39, 46-49

21 Schweiger, H.G., Dirk, J., Koop, H.U., Kranz, E., Neuhaus, G., Spangenberg, G. and Wolff, D. (1987) Theor. Appl. Genet. 73, 769-783

22 Murashige, T. and Skoog, F. (1962) Phys. Plant. 15, 473-497

23 Dupuis, I., Roeckel, P., Matthys-Rochon, E. and Dumas, C. (1987) Plant Physiol. 85, 876-878

ANTHER CULTURE SELECTION TO ENHANCE SNOW MOLD DISEASE RESISTANCE IN WINTER WHEAT

J. H. MCBEATH, F. MEHDIZADEGEN AND H. LOCKWOOD. AGRICULTURAL AND FORESTRY EXPERIMENT STATION, UNIVERSITY OF ALASKA FAIRBANKS, FAIRBANKS, AK 99775-0080, USA.

1. INTRODUCTION

Snow mold is one of the most serious diseases of winter wheat (*Triticum aestivum*) and grasses in Alaska and other high latitute regions (1-9). Snow mold disease can be caused by one or more of several low temperature parasitic fungi (7,9). In Alaska, *Sclerotinia borealis* (*=Myriosclerotinia borealis*) and a sclerotial low temperature basidiomycete (sLTB) (identified as *Coprinus sp.* by Traquair in 1980 [10]) are the most prevalent pathogens (5,9). These fungi infect host plants in late fall or winter when the soil is not yet frozen. During long winters, under a thick snow layer, these fungi proliferate and spread in host tissues because of the dark, humid conditions. Death of the plant is due to depletion of carbohydrate reserves (11,12) and maceration of plant tissues (13,14) (McBeath and Adelman, unpublished data).

Controlling snow mold disease can be quite difficult. Disease control through application of fungicides is costly and often ineffective. Because of the fragility of northern environments, chemical use also elicits great concern. Breeding resistance to snow mold in winter wheat has yet to yield significant results due to the complexity of this disease. Furthermore, the short growing season and long harsh winter of the north further tax the reserves of most winter wheat cultivars which have been bred for more moderate climates.

Plant cell cultures provide a rich source of genetic variation (15,16). For winter wheat and other self-pollinated crops, somaclonal or gametoclonal variation could be especially useful (16, 17). Crop improvement through gametoaclonal variation induced from anther culture has aroused considerable interest in recent years as methods of creating doubled haploids have become more efficient (17) and our understanding of the nature and origin of somaclonal and gametoclonal variation also has progressed (16, 18).

We report here on a system being developed to enhance resistance to snow mold disease in early maturing, cold hardy, hard red, winter wheat by selecting calli and androgenic plantlets resistant to the extracellular enzymes of snow mold fungi.

2. PROCEDURE

2.1 Material and methods

2.1.1 Anther culture : Seeds of cold-hardy, early-maturing, hard red winter wheat cultivars Capitan, Froid, Norstar and Roughrider were germinated and vernalized, and seedlings were transplanted and grown in a greenhouse. After spikes were harvested and cold-shock treated, anthers were excised and cultured on a boiled potato extract medium (19) for 4 weeks at 25 C, in the dark. Embryoids were then transferred to tissue culture increase medium containing Murashige-skoog basal salts (20), 1 mg 2,4 Dichlorophenoxy acetic acid (2,4-D), 1 mg Naphthalene acetic acid (NAA) and 0.1 mg cytokinin. Embryoids were kept under constant light, at 23 C for callus proliferation and differentiation (21,22).

2.1.2 Calli propagation : Only embryogenic calli with green centers (shoot primordia) were propagated and tested for their resistance to snow mold extracellular enzymes. Nonembryogenic calli, calli producing albino shoots or those producing only roots were discarded. Calli were subcultured every 12-14 days by subdividing calli into pieces (1-3 mm dia.). Characteristics of these calli and their passages were recorded.

2.1.3 Androgenic plantlets : To promote differentiation of roots and shoots, calli with shoot primordia were transferred to the above medium without 2,4-D and NAA but containing 1 mg indole-3-acetic acid (IAA) and 1 mg kinetin. Green androgenic plantlets with roots were transferred to a 250-ml Erlenmeyer flask containing 50 ml liquid medium with Murashige-Skoog basal salts, 5 mg IAA and 0.1 mg kinetin on a shaker. Androgenic plantlets with well developed root systems were planted in perlite and moved to the greenhouse after a transition period in the growth chamber.

2.2.1 Snow mold cultures : Alaska isolates of *S. borealis* and sLTB were grown in 250-ml Erlenmeyer flasks containing 100 ml of minimal basal salt medium (23) amended with 1.0% straw at 10 C for 6 and 4 weeks, respectively. Culture filtrates were collected by filtering the cultures through Whatman #1 filter papers.

2.2.2 Extracellular Enzyme preparations: Extracellular emzymes of *S. borealis* and sLTB were precipitated at 4 C by adding an equal volume of cold acetone to culture filtrates and centrifuging at 20,000 g for 20 min. Pellets were resuspended in a 0.01 M potassium phosphate buffer (pH 6.5). Aliquots were dialyzed, lyophilized and rehydrated. Cellulolytic and pectolytic enzyme activities were determined by measuring the increase in reducing group following incubation with substrates (24). Carboxymethyl cellulose and citrus pectin were used as substrates for cellulase and polygalacturonase (PG), respectively. PG activity was also determined by viscometric assay (25). The enzymes in the 0.01 M potassium phosphate buffer were adjusted to obtain a final concentration of 25,000, 2,500 and 250 units/ml cellulase activities and 12,000, 1,200 and 120 units/ml PG activities. Unless specified, the pH regimes for enzymes of *S. borealis* and sLTB were 6.0 and 6.8, respectively.

2.3.1 Calli treatment : Calli which were uniform in size and similar in shoot development were treated with 20 μl of various concentrations of enzymes. Equal volumes of 0.01 M potassium phosphate buffer were applied to equal numbers of calli which served as controls. After incubation at 10 C, calli were evaluated weekly for 2-6 weeks. The extent of their reactions was recorded using a numerical rating: 0=no visible change, 1=0% to 25% tissue discoloration, 2=25% to 50%, 3=more than 50% but less than 75% and 4=more than 75% to total discoloration (death).

2.4.1 Plantlet treatment : Approximately 2.0 g of fresh leaf tissue was randomly excised from each plantlet and washed in 15 ml sterile distilled water for 15 minutes. Leaf tissues were then cut into 2.5 cm segments, washed in 10 ml sterile distilled water for 15 min. and blot dried. Four leaf segments were randomly selected, weighed, and placed in each test tube containing 1 ml enzyme preparation and 5 ml of 0.01 M potassium phosphate buffer. No enzymes were added to buffers in the controls. Following 7 days incubation at 10 C, leaf tissues were homogenized in 5 ml of 80% (v/v) acetone. After incubation at 4 C overnight, chlorophyll extracts were filtered, centrifuged at 2,000 g for 5 min. Pellets were resuspended in fresh acetone and centrifuged. Supernatants from both centrifugations were combined and the total volume was brought to 20 ml with acetone. Absorbance values of chlorophyll were determined spectrophotometically at 562 nm. The total chlorophyll contents were calculated by using an equation developed by Hipkins and Baker (26).

3. RESULTS AND DISCUSSION

3.1. Genotype of donor plants is one of the most important factors affecting androgenesis in cereals (27). Great differences in the morphogenic potential and regenerating abilities have been observed on the four cold-hardy, early-maturing, hard red winter wheat cultivars. Among them, anthers of Roughrider have the highest calli initiation abilities (1.8%). These calli multiplied rapidly and aged slowly (could sustain at least 17 subcultures). Calli derived from anthers of Roughrider have relatively high differentiation abilities; approximately 12% of the calli produced green shoots. The callus initiation ability of Froid was 1.2%, and its green shoot regeneration rate was 6.1%. Both Capitan and Norstar have poor morphogenic potentials. High frequencies of albino plant and green root regeneration were found on calli derived from Capitan.

3.2. Enzyme preparations containing 2,500 units/ml of cellulase and 1,200 units of pectinase were found to elicit the widest range of responses of calli and were used throughout the study. Temperature also seemed to play a significant role in the activities of the snow mold extracellular enzymes and their abilities to affect calli cells. The optimal temperature for enzyme treatment appeared to be around 10 C.

3.3. Over 3,000 calli derived from anthers of Roughrider were treated with the extracellular enzymes of *S. borealis* and sLTB. Responses of clones rangied from no change (rated 0) in calli appearance to rapid deterioration of green centers and death (rated 4).

3.4. Responses of regenerated plantlets rangied from little loss of chlorophyll to severe chlorosis as measured by chlorophyll loss in leaf segments.

3.5 Extracellular enzymes from *S. borealis* and sLTB elicited different responses in clones. However, responses of clones to a given enzyme preparation were found consistent. Enhancements in snow mold resistance were observed in clones derived from Roughrider.

REFERENCES

1. Tupenevich, S.M. and Shirko, V.N. (1939) Elen. Plant Prot. Leningrad. 18, 85-99
2. Ekstrand, H. (1955) Stat. Vaxiskyddsanst. Madd. 67,1-125
3. Groves, J.W. and Bowerman, C.A. (1955) Can. J. Bot. 33,591-594
4. Tomiyama, K. (1955) Rept. Hokkaido Agr. Exp. Sta. 47, 1-123
5. Lebeau, J.B. and Logsdon, C.E. (1958) Phytopathol. 48,148-150
6. Bruehl, G.W., Sprague, R., Fischer, R., Nagamitsu, Nelson, W.L. and Vogel, O.A. (1966) Wash. Agr. Exp. Stn. Bull. 677,1-21
7. Årsvoll, K. (1973) Norw. Plant Prot. Inst. Div. Plant Pathol. Rept. 56,1-21
8. Jamalainen, E.A. (1974) Ann. Rev. Phytopathol. 12,281-302
9. McBeath, J.H. (1985) Plant Dis. 69,722-23
10. Traquair, J.A. (1980) Can. J. Plant Pathol. 2,105-115.
11. Kiyomoto, R.K. and Bruehl, G.W. (1977) Phytopathol. 67,206-211
12. Amano, Y. and Osanai, S.I. (1983) Bull Hokkaido Prefect. Agric. Expt. Sta. 50,83-97
13. McBeath, J.H. and Wenko, L. (1986) Phytopathol. 76,1143
14. Mehizadegan, F. and McBeath, J.H. (1988) Phytopathol. 78, 1592
15. Chaleff, R.S. (1983) in Genetic Engineering of Plants (Kosuge, T., Meredith, P.C. and Hollaender, A., ed.) pp.257-270, Pleunum, New York and London
16. Reisch, B.I. (1987) in Plant Cell Biotechnology (Pais, M.S.S., Mavituna, F. and Novais, J.M., ed.) pp.87-95, Springer-Verlag, Berlin Heidelberg
17. Baenziger, P.S., Kudirka, D.T., Schaeffer, G.W. and Lazar, M.D. (1984) in Gene Manipulation in Plant Improvement (Gustafson, J.P., ed.) pp.385-414, Pleunum, New York and London
18. Larkin, P.J. and Scowcroft, W.R. (1983) in Genetic Engineering of Plants (Kosuge, T., Meredith, P.C. and Hollaender, A., ed.) pp.289-314, Pleunum, New York and London
19. Anonymous, Research Group 301, (1976) Acta Genet. Sin. 3,30-31

20.Murashige, T., and Skoog, F. (1962) Physiol. Plant. 15, 473-497
21.Schaeffer, G.W., Baenziger, P.S. and Worley, J. (1979) Crop Sci. 19,697-702
22.McBeath, J.H. and Schaeffer, G.W. (1986) Phytopathol. 76,1146
23.Marsh, P.B., Bollenbacher, K., Butler, M.L. and Raper, and Raper, K.B. (1949) Textile Res. Jour. 19,462-484
24.Miller, G.L. (1959) Anal. Chem. 31,426-428
25.Almin, K.E. and Erickson, K.E. (1967) Biochem. et Biophys. Acta. 139,238-247
26.Hipkins, M.F. and Baker, N.R. (1987) Photosynthesis Energy Tranduction: A Practical Approach. 197p. IRL Press Oxford, Washington D.C.
27.Lazar, M.D., P.S. Baenziger, and Schaeffer, G.W. (1984) Theor. Appl. Gennt. 68:131-134.

HAPLOID SUSPENSION AND PROTOPLAST CULTURE FROM ISOLATED MICROSPORES OF MAIZE

J.C. MITCHELL AND J.F. PETOLINO
DOWELANCO/UNITED AGRISEEDS, INC., P.O. BOX 4011, CHAMPAIGN, IL, 61824, USA

1. INTRODUCTION

Procedures for the routine regeneration of large numbers of plants from isolated microspore cultures of maize have been recently developed (1,2). Haploid microspores are competent to undergo abnormal cell divisions culminating in the formation of embryo-like structures from which plants can be recovered. These procedures have now been adapted for the production of suspension and protoplast cultures capable of plant regeneration. The present paper describes the establishment of regenerable, haploid cell suspensions directly from cultured maize microspores. The recovery of plants from protoplasts isolated from these microspore-derived suspension cultures is also discussed.

2. PROCEDURE

2.1 Plant Material

The genotype used was a highly anther culturable S_2 line (139/39-02) developed by inter-mating microspore-derived doubled haploids descended from a three-way cross, (H99xFR16)xPa91. Donor plants were field-grown during April-August 1988 and 1989 in Champaign, IL.

2.2 Anther Culture and Microspore Isolation

Anthers containing predominantly early binucleate microspores, as determined by mithramycin-fluorescent staining, were excised and placed in 20x60 mm Petri dishes (60 anthers per dish) containing 10 mL of induction medium. The induction medium consisted of YP major salts, N6 minor salts, 60 g/L sucrose, 27.9 mg/L $FeSO_4.7H_2O$, and 37.3 mg/L Na_2EDTA adjusted to pH 5.8. Dishes were sealed with Parafilm and placed in plastic boxes covered with aluminum foil. Cultures were maintained in the dark at 28°C for 10 days after which microspores were isolated. Anthers were placed onto a 380 micron mesh and gently squashed using a glass stirring rod. Microspores passed through the screen while the majority of the anther wall debris was left behind.

2.3 Suspension Initiation and Maintenance
The entire contents of 4 dishes (induction medium, released microspores from 240 anthers, and some anther wall debris) were placed into a 125 mL Erlenmeyer flask along with 10 mL of initiation medium for a final volume of 50 mL. The initiation medium consisted of MS salts and vitamins, 30 g/L sucrose, 100 mg/L myo-inositol, and 2 mg/L 2,4-D adjusted to pH 5.8. The cultures were then placed onto an orbital shaker (150 rpm) and maintained in the dark at 28°C. After approximately 10-14 days, the cultures were allowed to settle for a few seconds and 15 mL of the supernatant was removed and replaced with fresh initiation medium. After approximately 4-8 weeks of this replenishment procedure, the cultures became thick and mucilaginous. At this point, they were centrifuged (AccuSpin, Beckman Instruments, Inc., Palo Alto, CA) at 2500 rpm for 3 min and the upper portion of the sediment along with some of the supernatant was transferred to a maintenance medium. The maintenance medium consisted of LM salts (3), MS vitamins, 60 g/L sucrose, 0.7 g/L L-proline, 8.0 mg/L dicamba, and 0.8 mg/L 2,4-D adjusted to pH 5.8. Once established, sub-culturing was performed every 7-10 days by allowing the culture to settle, selecting 20 mL from the sediment-supernatant interface, and placing it into 30 mL of fresh maintenance medium.

2.4 Protoplast Isolation and Culture
Three days after subculture, 1 mL of suspension was transferred to a 60x15 mm Petri dish along with 2 mL of BH3 medium (4) and 2 mL of filter sterilized enzyme solution. The enzyme solution contained 128 g/L mannitol, 3.6 g/L $CaCl_2$, 1.2 g/L MES buffer, 0.1 g/L NaH_2PO_4, 1% (w/v) Onozuka RS cellulase, 1% (w/v) macerozyme R10, and 0.2% (w/v) pectolyase Y-23 adjusted to pH 5.6. All enzymes were obtained from Karlan Chemical Corporation (Torrance, CA). After 16 hours, the partially digested preparation was passed through a 45 micron stainless steel screen to remove the debris and large cell clusters and then centrifuged at 700 rpm for 5 minutes. The supernatant was removed and the pellet was re-suspended in 5 mL of 25% sucrose solution containing 250 mg/L $MgSO_4$, 150 mg/L $MgSO_4$, 100 mg/L KNO_3, 27.2 mg/L KH_2PO_4, 2.5 mg/L $Fe(SO_4)_3 \cdot 6H_2O$, 0.16 mg/L KI, and 0.0003 mg/L $CuSO_4$. Before centrifuging at 700 rpm for 5 minutes, 2 mL of a 13% mannitol solution was carefully over-layered. Protoplasts were removed from the sucrose-mannitol interface and washed twice in BH3 medium (4) before diluting to a final density of approximately 1 x 10^5

protoplasts/mL. Yields ranged from 0.1 x 10^4 to 5 x 10^5 protoplasts per mL of suspension culture.

Protoplast samples of 100 uL (containing approximately 1 x 10^4 protoplasts) were placed onto 47 mm-diameter nylon membrane filters having a 0.2 micron pore size (Nalge Company, Rochester, NY). The filters were placed onto a feeder layer of maize cells from a Black Mexican Sweet (BMS) suspension culture growing in maintenance medium. The feeder layer was made by thinly spreading 9 mL of 1.2% Sea Plaque agarose (FMC BioProducts, Rockland, ME), cooled to 37°C, and 9 mL of a BMS suspension culture (cells plus old maintenance medium) in exponential growth phase over an agar-solidified callus induction medium (see below). After sealing the dishes with Parafilm, the cultures were incubated in the dark at 28°C. After 3-4 weeks, small callus colonies (1-2 mm) were transferred to fresh callus induction medium (see below).

2.5 Callus Induction and Plant Regeneration

Samples of suspension culture (0.5-1.0 mL) or protoplast-derived callus colonies were transferred to callus induction medium. The callus induction medium consisted of N6 major salts and vitamins, B5 minor salts, 30 g/L sucrose, 2.9 g/L L-proline, 100 mg/L myo-inositol, 100 mg/L casein hydrolysate, 2.5 mg/L dicamba, 0.1 mg/L 2,4-D, and 8 g/L agar (Hazelton Research Products, Inc., Lenexa, KS). After 2-3 weeks, callus displaying a compact, nodular appearance was transferred to a regeneration medium and maintained in a 16/8 hour photoperiod (60 umol/m^2/sec). The regeneration medium consisted of SH basal salts, 10 g/L sucrose, and 3.5 mg/L BAP. After 5 days, callus was transferred to fresh regeneration medium without BAP. After approximately 4 weeks, plantlets were placed into 100x25 mm culture tubes containing fresh regeneration medium before transfer to soil.

2.6 Analytical Procedures

Chromosome counts of suspension cells were made after fixation in 50% formic acid using Feulgen staining. Root tip cytology of regenerated plants was performed after fixation in Farmer's solution. Growth rates of the suspension cultures (exressed as increase in cell number over time) were determined after chromium trioxide incubation. Protoplast densities were measured using a Speirs-Levy Eosinophil hemocytometer (C. A. Hausser and Son, Philadelphia, PA).

3. RESULTS AND DISCUSSION

Ten days after the onset of anther culture (the time of microspore isolation and suspension culture initiation), approximately 4% of the microspores displayed abnormal nuclear divisions. Over the course of several weeks, these androgenic proembryos proliferated into a thick, mucilaginous culture consisting primarily of elongated, vacuolated cells along with a few densely cytoplasmic cell clusters. After several cycles of selecting for cells at the sediment-supernatant interface, enrichment for clusters of densely cytoplasmic cells occurred. Upon examination, over 87% (218 out of 250) of the cells were observed to be haploid 6 months after culture initiation.

Over time, variation in overall appearance became apparent among different flasks of suspension derived from the same original microspore culture. Differences in color, texture, and relative viscosity were used to classify the cultures into various categories. The suspension cultures also differed with respect to their relative growth rates with doubling times ranging from 6 to 12 days. Cultures generally started out yellowish-brown in color, somewhat coarse textured, thick, and slow growing. These cultures, upon re-plating onto callus induction medium, consistently produced embryogenic callus from which plants could be recovered. From a sample of 16 regenerated plants, all were haploid as determined by root tip cytology, however, one was self-fertile (probably due to spontaneous chromosome doubling during development). Over several sub-culture passages, some of the cultures became dull yellow, less thick, and more rapidly growing. Plants could be regenerated from such cultures, but not as readily. Similarly, over time, some of the suspensions became bright yellow, very fine, and rapidly growing. Plants could not be recovered from these cultures under the conditions used.

With timely sub-culturing, involving judicious selection for cells at the sediment-supernatant interface, most of the cultures could be maintained as yellowish, moderately thick suspensions capable of regenerating plants. This type of culture consisted of a combination of single cells and cell clusters averaging 0.4 mm in diameter with the majority falling between 0.2 and 0.6 mm. Upon transfer to callus induction medium, typically 5-10 regenerable callus pieces were obtained per 0.5 ml of suspension. Cultures have been maintained, as such, for as long as 18 months.

Enzymatic digestion of these suspensions has resulted in preparations containing single cells and protoplasts which, after density gradient centrifugation, have yielded protoplasts with a minimal amount of cellular contamination (0.1-0.6%). Although viable protoplasts could be isolated from all of the suspensions, the thin, rapidly growing cultures gave the best yields. After three weeks of culture on feeder layers, macroscopic callus colonies were observed. On average, from 50 to 250 colonies per 10^4 protoplasts cultured were obtained. Upon transfer to regeneration medium, haploid plants were recovered. On average, 20% of the macroscopic callus colonies regenerated plants.

The addition of hormones to anther culture medium is not typically required for maize androgenesis to occur. In the present study, microspores were induced to initiate the process of androgenesis (i.e., proembryo development) in the absence of added hormones. After 10 days, exposure to elevated auxin (2 mg/L 2,4,-D) resulted in the proliferation of androgenic proembryos, such that mucilaginous cultures formed. Enrichment for densely cytoplasmic cell clusters resulted in the establishment of regenerable, haploid suspension cultures. The origin of these haploid suspension cultures differed from that previously reported (5) in that they were established <u>directly</u> from cultured microspores without an intervening callus phase, thus, circumventing the requirement for the appearance of callus with a particular morphology (i.e., Type II).

As observed in other studies, suspension cultures derived from the same explant source over time differ in appearance. Three general types of cultures were identified in the present study. Thicker, slower growing suspensions tended to regenerate plants more efficiently than the finer-textured, faster-growing cultures. However, the later suspension cultures yielded considerably more protoplasts under the same enzymatic digestion conditions. Cultures could be maintained, however, which possessed characteristics (i.e., less thick and more rapidly growing) which allowed them to yield adequate amounts of protoplasts while maintaining acceptable levels of regeneration. Moreover, further modification of the enzymatic digestion procedure itself should result in more efficient recovery of protoplasts from the thicker cultures.

As reported in other systems (6,7), once regenerable suspension cultures are available, totipotent protoplasts can be readily obtained. The callus colony recovery frequencies measured in the present study are in the range of those previously reported for maize (8). Although most of the plants regenerated from these cultures were sterile (as expected since they were haploid), colchicine-induced chromosome doubling of the callus should result in more efficient seed recovery.

The microspore can now be added to the limited list of explants of maize capable of being used as a source of embryogenic suspension cultures and totipotent protoplasts. This not only expands the germplasm base now available for protoplast manipulation to include those genotypes capable of undergoing androgenesis but also makes haploid protoplasts available for genetic and developmental studies. Haploid protoplasts would represent unique recipients for genetic transformation since their use would result, upon chromosome doubling, in the gene of interest being incorporated into a completely homozygous genotype.

REFERENCES

1 Pescitelli, S.M., Mitchell, J.C., Jones, A.M, Pareddy, D.R., and Petolino, J.F. (1989) Plant Cell Reports 7,673-676

2 Pescitelli, S.M., Johnson, C.D., and Petolino, J.F. (1990) Plant Cell Reports 8,628-631

3 Litvay, J.D., Verma, D.C., and Johnson, M.A. (1985) Plant Cell Reports 4:325-328 (1985).

4 Grosser, J.W. and Chandler, J.L. (1987) Scientia Horticulturae 31,253-257

5 Sun, C.S., Prioli, L.M., and Sondahl, M.R. (1989) Plant Cell Reports 8,313-316

6 Prioli, L.M. and Sondahl, M.R. (1989) Biotechnology 7,589-594

7 Shillito, R.D., Carswell, G.K., Johnson, C.M, DiMaio, J., and Harms, C.T. (1989) Biotechnology 7,581-587

8 Carswell, G.K., Johnson, C.M., Shillito, R.D., and Harms, C.T. (1989) Plant Cell Reports 8,282-284

EMBRYO RESCUE OF INTERSPECIFIC HYBRIDS OF BRASSICA SPINESCENS X BRASSICA CAMPESTRIS AND DNA ANALYSIS

ABHA AGNIHOTRI, MALATHI LAKSHMIKUMARAN,
K.R. SHIVANNA*, V. JAGANNATHAN

TATA ENERGY RESEARCH INSTITUTE, 90 JOR BAGH, NEW DELHI 110003, INDIA
*BOTANY DEPARTMENT, UNIVERSITY OF DELHI, DELHI 110007, INDIA

1. INTRODUCTION

The members of the genus Brassica, B. juncea and B. campestris are the main crucifers grown in India for edible oil. Substantial losses in yield occur due to susceptibility of cultivated varieties to diseases, such as white rust caused by Albugo candida and alternaria blight caused by Alternaria brassiceae, aphids and drought. Wide hybridization can be used to transfer desirable agronomical traits from wild allies of Brassicas into the cultivated varieties. However, due to crossability barriers the hybrid embryos abort during early stages of development. We have recently obtained a fertile, amphidiploid intergeneric hybrid between Eruca sativa, which is tolerant to drought, white rust and aphids, and B. campestris by embryo rescue (1). The studies on wide hybridization has now been extended to interspecific hybridization between B. spinescens (2n=16), a dwarf plant with thick dark green leaves and tolerant to white rust and alternaria blight and B. campestris (2n=20).

2. MATERIALS AND METHODS

Seeds of B. spinescens and B. campestris var. Pusa Kalyani were obtained from Indian Agricultural Research Institute, New Delhi. Sequential embryo rescue technique and other experimental methods were the same as those used for obtaining the Eruca-Brassica hybrid (1). The embryos which emerged from cultured ovules were transferred to MS medium (2) containing 0.05 mg/l kinetin, 0.005 mg/l naphthalene acetic acid, 0.5 mg/l gibberellic acid and 50 mg/l casein hydrolysate (CH) and grown to healthy plantlets.

The plantlets were screened by DNA probes and those showing the presence of DNA from the male parent were used for further micropropagation on the same medium without gibberellic acid (medium A). Healthy plantlets were transferred to MS basal medium (without phytohormones and CH) containing 1% sucrose and 0.7% agar (Medium B) for further growth. New shoots derived from axillary buds in 3-4 weeks were cultured on medium B supplemented with 0.5 mg/l indole acetic acid or indole butyric acid for 2 weeks for induction of rooting prior to being transferred to medium B for further growth.

Leaf discs of B. spinescens x B. campestris hybrid were cultured on MS medium used for ovary culture for inducing callus and somatic embryogenesis. Embryos obtained from leaf discs within 4-5 weeks were transferred to MS basal medium containing 3% sucrose (without agar and phytohormones) and maintained on a rotating shaker (20 rpm) for 2-3 days before transferring to medium B for further growth and development. When the plantlets were 8-10 cm in height, they were transferred to autoclaved vermiculite for 8-10 days and subsequently planted in pots containing soil, peat moss and farm yard manure (2:1:1) and grown to maturity under field conditions.

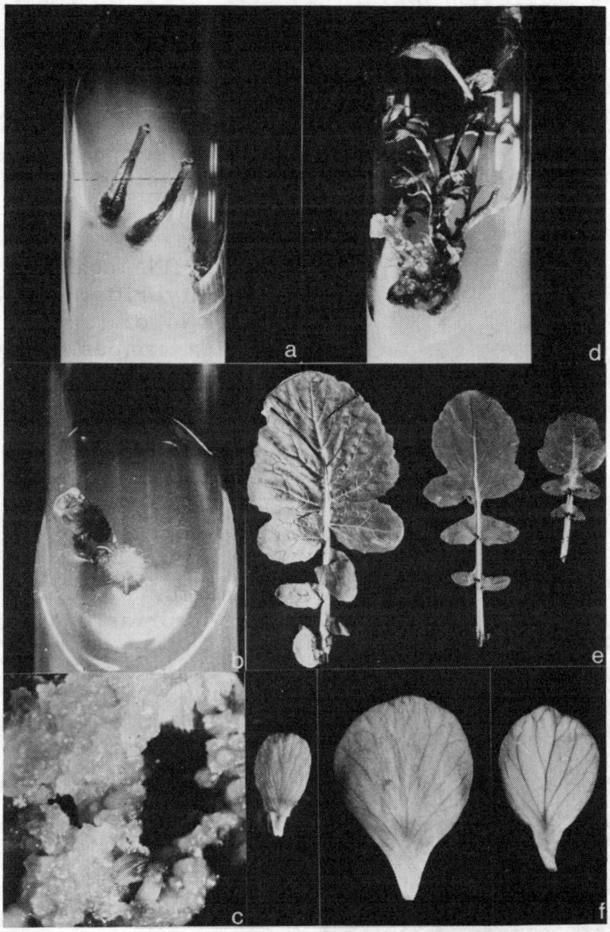

Fig. 1. a-f. Production of hybrid plantlets, its multiplication and morphology. a) Ovaries after 2 weeks of culture. b) rescued embryo from a cultured ovule. c) embryogenic callus showing differentiation. d) multiple shoots from nodal segment. e) leaves of B. campestris (left) B. spinescens (right) and the hybrid (centre). f) petals of B. campestris (left), B. spinescens (right) and the hybrid (centre).

3. RESULTS AND DISCUSSION

To the best of our knowledge, this is the first report of the hybrids of B. spinescens with cultivated Brassica. Ovaries of B. spinescens pollinated with B. campestris pollen started withering within a week when left on the plant. The pistils showed good pollen germination and growth of a considerable number of pollen tubes into the ovary indicating that pre-fertilization barriers are not the major constraints for obtaining hybrids (Fig. 3.a.) However, the hybrid embryos failed to develop further on the plants. Hybrid plantlets could, however, be raised through sequential culturing of ovaries and ovules (Fig. 1.a,b.)

Wide hybridization in Brassica has been reported to produce matromorphic plants (3,4). In the present investigation also, out of the three plants raised following interspecific pollination, one was found to have no DNA derived from the male parent and was probably a matromorph.

Nitrocellulose membranes containing total DNA of both the parents and different hybrids restricted with Alu I were analysed using a 177 bp tandem repeat DNA fragment isolated from B. campestris (5),

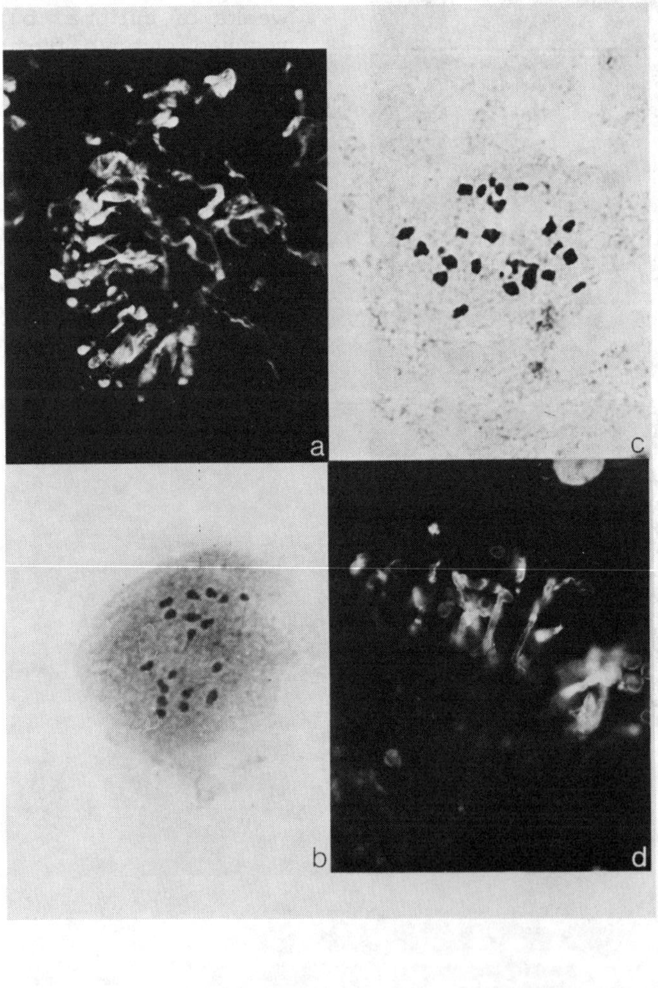

Fig. 2. DNA analysis of the hybrids. Total DNA of Brassica campestris a) B. spinescens b) and hybrids c-e) digested with Alu I and hybridized with B. campestris tandem repeat probe pA-7 showing the presence of B. campestris genome in two of the three hybrids.

after restriction with Hind III. DNA of the plants obtained from two out of the three successful embryos showed strong signals with Brassica specific probe pA-7, thus showing the presence of DNA from the male parent (Fig. 2.) They were selected for further multiplication and growth.

Multiplication of hybrids was achieved through production of multiple shoots from nodal explants as well as through somatic embryogenesis (Fig. 1. c,d.) One of the interesting observations has been the need for the somatic embryos to pass through the liquid medium for normal growth. The reason for this is not clear. The hybrid plants showed characteristics of both the parental species. Leaves were intermediate to both the parents in size and shape but clasping nature characteristic of B. campestris was absent. The flowers were dark yellow and resembled those of B. campestris, while petal venation resembled that of B. spinescens. (Fig. 1.e,f.) Flowering period was also intermediate to that of both the parents.

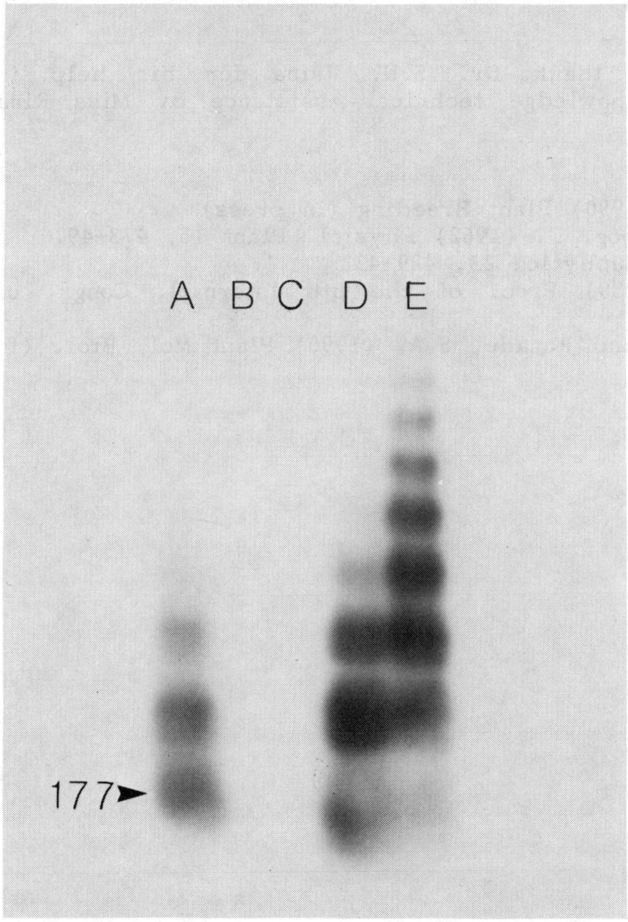

Fig. 3. a) B. spinescens stigma pollinated with B. campestris showing profuse pollen germination and pollen tube growth. b) Squash preparation of meiocytes of the FI hybrid showing 2n=18. c) Squash preparation of meiocytes of the amphidiploid showing 2n=36. d) Stigma of FI hybrid pollinated with B. campestris showing good pollen germination and pollen tube growth.

The male meiosis of the hybrids (2n=18) revealed the occurrence of a majority of univalents and very few bivalents at metaphase-I. Anaphase I and subsequent stages were highly irregular. Hybrids had rudimentary anthers and were completely pollen sterile. Although attempts to induce colchiploids in FI hybrids were not successful, all hybrid plants obtained through somatic embryogenesis turned out to be amphidiploids (2n=36) and showed about 20% pollen fertility, probably by doubling of chromosome at the callus stage (Fig. 3.b,c.)

Pistils of the F1 hybrids pollinated with pollen of *B. campestris* and *B. juncea* showed good pollen germination and pollen tube growth on the stigmatic surface (Fig. 3.d.). However, very few pollen tubes could be seen growing into the stylar region and the ovary. Both F1 and amphidiploids have been back crossed with *B. campestris* and *B. juncea* and progenies will be screened under field conditions for pollen and seed fertility, yield and resistance to fungal diseases.

ACKNOWLEDGEMENTS: We thank Dr. S.N. Raina for his help in cytology. We also acknowledge technical assistance by Miss Rina Agarwal.

REFERENCES

1. Agnihotri, A. *et al* (1990) Plant Breeding (in press)
2. Murashige, T. and Skoog. F. (1962) Physiol. Plant 15, 473-497
3. Eenink, A.H. (1974) Euphytica 23, 429-433
4. Batra, V. *et al* (1989) Proc. of the 6th Internatl. Congr. of SABRAO, 443-445
5. Lakshmikumaran, M. and Ranade, S.A. (1990) Plant Mol. Biol. (In press)

HAPLOIDS FROM UNPOLLINATED OVARIES OF ALLIUM CEPA - SINGLE PLANT SCREENING, HAPLOID DETERMINATION, AND LONG TERM STORAGE

JOACHIM KELLER

Zentralinstitut für Genetik und Kulturpflanzenforschung Gatersleben, Corrensstraße 3, 4325 Gatersleben, German Democratic Republic

1. INTRODUCTION

In vitro induced haploidy is a prerequisite for obtaining homozygous lines for plant breeding purposes on a large scale. Onion, *Allium cepa* L., is a species in which the frequently used method of anther culture has been unsuccessful till now. We were also not able to obtain haploids from 98,000 anthers cultivated on 25 modifications of commonly used adequate media (1). Because of the successful attempts to obtain haploids via gynogenesis reported on sugar beet (2) and other plant species, we cultivated isolated ovules, ovaries, or whole flower buds (3) on medium BDS (4).

2. MATERIAL AND METHODS

2.1. *Haploid induction:* Organs were incubated on BDS + IBA, BAP, glutamine (concentrations according to 2), and 10 % sucrose during 1 month in dark at 27 °C. After transfer to BDS with reduced contents of BAP and sucrose the plantlets emerged and were further cultivated in 16 h day at 25 °C (light) and 20 °C (dark), respectively.

2.2. *Microbulblet induction and micropropagation:* Bulblet induction was achieved by addition of 12.5 mg l^{-1} BAP and increasing the sucrose contents up to 15 %. The basal part of the microbulblets was cut into two or four pieces. Developing adventitious plantlets were isolated and further cultivated on media without hormones. In long term cultures of plantlets on media with BAP adventitious shoot formation will be induced also without cutting.

2.3. *Ploidy determination:* Chromosome counts were performed on metaphases in squash preparates of root tips. In material from *in vitro* culture chromosome counting is difficult because of rare metaphases and badly preparable root tips. Therefore, the clones with exactly counted chromosomes served as standard material for following series of measurements. Light extinction of FEULGEN stained nuclei was measured by means of a VICKERS microdensitometer. Length and width of stomata guard cell nuclei and of the whole stomata were recorded with the help of computer aided image processing developed in the Institute of Plant Breeding Research in Quedlinburg (R. Ahne, personel communication).

3. RESULTS AND DISCUSSION

3.1. Genotype screening on haploid induction: As onion is an outbreeding species genotype effects have to be considered. A screening programme has been performed comprising 50,641 ovaries from 273 individual plants of 50 varieties and breeding lines (185 ovaries per plant on the average). Single plants of the varieties 'Holska', 'Calbenser Gerlinde', 'Zerti', and some of 'Stuttgarter Riesen' revealed to be the most productives. In the latter variety, the individual differences were reproducible in a two year comparison, plants being cultivated in phytotron (5). In the variety 'Stuttgarter Riesen', plants grown in phytotron were significantly more reactive than such grown in the greenhouse. Table 1 present an extract of the entire screening.

3.2. Bulblet induction, storage, and micropropagation: Bulblet induction could not be achieved on basic medium with 3 % sucrose. It could, however, be favoured by 12.5 mg l^{-1} BAP and 15 % sucrose. The percentage of bulb forming plantlets increased up to 38.9 %. Bulblet storage of one year was proved to be possible in dark at 4 °C. On the bulb scale bases, rows of adventitious buds developed after cutting. This was followed by outgrowth of 'fans' of new plantlets. In short term tests a medium with NAA, BAP, inositol, and proline had the greatest effect. In longer cultivation periods it was surpassed by a medium with only NAA and BAP (Table 2). A further possibility of multiplication is a long term culture on medium with BAP up to 6 months (several subcultures). After a period of bulblet swelling, the adventitious plantlets emerge as bunches breaking through the outer bulb scales. In all cases, plantlets could be isolated from each other without difficulties after one month of culture on BDS with 8 % sucrose.

TABLE 1. Genotype screening

Variety	Plants tested	Plants reacting	Ovaries of reacting plants	No. of regenerants abs.	%
Stuttg. Riesen (greenhouse)	65	27	5783	44	0.76 abc
(phytotron)	8	6	6665	82	1.23 ade
Karatalski	11	3	406	5	1.23
Zittauer Gelbe	8	5	939	12	1.28
Ambros	3	2	263	5	1.90
Zerti	7	2	209	5	2.39 bf
Calbens. Gerlinde (greenhouse)	3	1	41	1	2.44
(phytotron)	1	1	792	22	2.78 dgi
Holska	2	1	175	5	2.86 chj
all other lines	165	36	7596	57	0.75 efghk
sum/ average	273	84	22869	238	1.04 ijk

Letters a...k indicate pairs of figures with statistically significant differences.

TABLE 2. Multiplication factors in cuttings of bulblets

Medium supplements (mg l^{-1})	Multiplication factor after 1 month	3 months
1.8 IAA + 2.2 KIN	4.9	9.4
1.8 IAA + 21.5 KIN	7.1	30.3
1.8 IAA + 1.2 BAP	8.3	18.3
5.9 NAA + 1.2 BAP	9.3	14.2
5.9 NAA + 12.5 BAP	10.3	37.7
4 NAA + 4 BAP + 100 Inositol + 100 Proline	11.7	30.0

TABLE 3. Examples of length and width measurements in stomata (measures in um)

Plant material	Number	Length $\bar{x}$	Length s	Width $\bar{x}$	Width s	Ploidy
Donor plants						
'Stuttg. R.'pl.1	200	36.8	2.5	29.6	1.3	2n
'Stuttg. R.'pl.2	200	39.6	2.0	30.6	2.5	2n
Regenerants						
'Stuttg. R.' r.1	95	29.0	2.8	23.6	2.4	n
'Stuttg. R.' r.2	107	28.4	2.1	24.2	1.7	n
'Stuttg. R.' r.3	100	30.4	1.9	26.0	1.6	n
'Carmen'	116	32.4	3.5	25.6	2.8	2n
'Ambros'	101	28.8	2.3	22.0	1.9	n
line 76/232	112	31.0	2.3	22.2	2.1	n
'Hylight'	86	34.8	2.1	25.6	2.6	2n
'Zittauer G.'	78	39.4	3.2	27.6	2.0	2n

TABLE 4. Examples of length and width measurements in stomata guard cell nuclei of the same plants as in table 3 (measures in um)

Plant material	Number	Length $\bar{x}$	Length s	Width $\bar{x}$	Width s	Ploidy
Donor plants						
'Stuttg. R.'pl.1	200	11.9	1.4	6.4	0.8	2n
'Stuttg. R.'pl.2	200	14.0	1.7	5.7	1.1	2n
Regenerants						
'Stuttg. R.' r.1	104	8.2	0.9	6.8	0.8	n
'Stuttg. R.' r.2	99	8.9	1.1	6.8	0.8	n
'Stuttg. R.' r.3	104	8.9	1.0	7.2	0.9	n
'Carmen'	113	11.7	1.4	8.7	1.0	2n
'Ambros'	107	8.3	0.9	6.9	0.7	n
line 76/232	111	9.2	1.0	7.0	0.8	n
'Hylight'	105	11.3	1.3	8.3	1.0	2n
'Zittauer G.'	84	13.1	1.3	8.8	1.1	2n

Differences between the haploid and the diploid length measures are statistically significant.
$\bar{x}$: average value, s: standard deviation.

3.3. Ploidy determination: Haploid chromosome number - 8 - could be exactly counted. Microdensitometry revealed that, in haploid nuclei, the degree of light extinction amounts to about 0.4 ... 0.5 times of that in diploid ones. Among 164 of the 457 obtained regenerants 62.2 % were haploid, 31.7 % diploid, and 6.1 % mixoploid. Length measures of guard cell nuclei and of the whole stomata were clearly different. Haploid plants had generally smaller and more circular stomata and nuclei than diploids (Tables 3,4).
A low percentage of the regenerants degenerated in early stages of development. Degeneration was accompanied by formation of giant nuclei, callus formation and vitrification. Mixoploids and cytologically aberrant degenerating plantlets are possibly caused by instable phases in the first steps of gynogenesis. Further investigations are necessary to clear up this and also to elucidate the origin of diploid regenerants.

4. ABBREVIATIONS

BAP: Benzylaminopurine. BDS: medium after Dunstan and Short (4), IAA: Indole-3-acetic acid, IBA: Indole-3-butyric acid, KIN: Kinetin, NAA: Naphthalene-1-acetic acid

REFERENCES

1 Keller, J. (1990) Arch. Züchtungsforsch. 20, 211-219
2 Hosemans, D. and Bossoutrot, D. (1983) Z. Pflanzenzücht. 91, 74-77
3 Keller, J. (1990) Euphytica, in press
4 Dunstan, D.I. and Short, K.C. (1977). Physiol. Plantar. 41, 70-72
5 Keller, J. (1990) Wiss. Z. Ernst-Moritz-Arndt-Univ. Greifswald, in press

PARTIAL GENOME TRANSFER IN INTERSPECIFIC TOMATO HYBRIDS

J. WIJBRANDI* AND M. KOORNNEEF, DEPARTMENT OF GENETICS, AGRICULTURAL UNIVERSITY WAGENINGEN, DREIJENLAAN 2, 6703 HA WAGENINGEN, THE NETHERLANDS (*PRESENT ADDRESS: KEYGENE, P.O. BOX 216, 6700 AE WAGENINGEN, THE NETHERLANDS)

1. INTRODUCTION

The transfer of desirable traits from wild into cultivated species is a method to improve crops. For this purpose, interspecific sexual or somatic hybrids may be constructed. However, these species hybrids contain many unwanted traits of the wild species besides the desired ones, and are often sterile. Fertility is required to perform several backcrosses with the cultivated species to remove unwanted characters of the wild species. These problems may be circumvented by asymmetric somatic hybridisation. By this procedure protoplasts of one species, the recipient, are fused with inactivated (mostly by Röntgen- or gamma-irradiation) protoplasts of another species, the donor. The donor genome will be fragmented and the asymmetric hybrids will contain the complete genome of the recipient species and a small part of the donor genome. Several asymmetric hybrids were obtained already: the fraction of donor genome that was transferred varied from a few traits (1) to many chromosomes (2).

We were interested in the transfer of traits of the wild tomato species *Lycopersicon peruvianum* to the cultivated tomato, *Lycopersicon esculentum*. *L. peruvianum* has many desirable characters for tomato improvement (3), but only a few of these have been introduced into the cultivated tomato, because sexual hybrids of both species are very difficult to obtain. In the present experiments we obtained asymmetric somatic hybrids by fusing *L. esculentum* protoplasts with irradiated *L. peruvianum* protoplasts. The hybrids were selected for the dominant regeneration capacity character of *L. peruvianum* (4). The regenerated plants were analysed for chromosome number, morphology, fertility and the retention of *L. peruvianum* specific morphological, isozyme and RFLP (= restriction fragment length polymorphism) markers.

2. MATERIAL AND METHODS

2.1. Plant materials

As recipient (*Lycopersicon esculentum*) the Dutch hybrid cultivar Bellina (Rijk Zwaan Seed Company, de Lier, The Netherlands) and the genotypes LA291, LA1164, LA1166, LA1182, LA1189, LA1444 and LA1665 (Prof. Rick, Tomato Genetics Stock Center, Davis, USA) were used. Plants from the *Lycopersicon peruvianum* accession PI128650 were used as donor.

2.2. Cell culture

The isolation, fusion and culture of protoplasts were described by Wijbrandi et al. (5). The selection of hybrids was based on regeneration capacity when *L. peruvianum* was irradiated. When no irradiation was applied, the hybrids were selected on basis of kanamycin resistance (present in the *L. esculentum* parent) and regeneration capacity (6).

2.3. Irradiation treatment

Prior to fusion, leaf protoplasts of *L. peruvianum* were irradiated with gamma rays from a ^{60}Co source at a dose rate of approximately 2000 Gray/hour. Three different doses were used: 50, 300 or 1000 Gy (= 5, 30 or 100 kRad). Putative hybrids that resulted from these fusions, are indicated as 5H, 30H and 100H, respectively. Symmetric hybrids, which resulted from fusions of unirradiated protoplasts, are indicated as 0H-hybrids.

2.4. Plant characterisation

The determination of chromosome numbers in root tip cells and the assessment of several morphological characters distinguishing both species as well as fertility were described earlier (6).
The *L. esculentum* LA-genotypes that were used for the fusions, were homozygous for two to four recessive morphological markers (7). A total of 18 different marker genes, located on 10 of the 12 tomato chromosomes, were available. The asymmetric hybrids were assayed for the phenotype of the relevant markers to determine the retention of the dominant *L. peruvianum* wild type alleles.
The procedures for isozyme analysis and RFLP analysis were described earlier (6 and 8, respectively). The RFLP markers were tomato single copy clones, obtained from Dr. Tanksley (Cornell University, Ithaca, USA).

3. RESULTS

3.1. Cell culture and plant regeneration

Irradiated *L. peruvianum* protoplasts did not result in microcalli. The fusion cultures yielded many calli of which several formed shoots. Within each experiment there was a large variation in callus as well as shoot morphology. The rate of shoot regeneration as well as the fraction of calli that could form shoots decreased with the irradiation dose.
The characteristics of the putative hybrid calli that regenerated shoots, with respect to root formation, establishment in soil, flowering and seed set, are shown in Table 1 for each irradiation dose.

3.2. Chromosome number

The chromosome numbers of the asymmetric somatic hybrids varied between 29 and 85. The majority of the 5H-hybrids had chromosome numbers below the tetraploid level ($2n < 48$), whereas the 30H- and the 100H-hybrids usually were above the tetraploid level. On the average, the 5H-hybrids had 16 additional chromosomes above an euploid number (range 5-22), while the 30H-hybrids had 13

TABLE 1. Characteristics of different hybrid calli, that regenerated shoots: root formation of shoots (*in vitro*), establishment in soil in the greenhouse (at least one month), flowering, fruit and seed set.

Hybrids	Regenerating calli	Root formation	Greenhouse plants	Flowers	Fruits	Seeds
0H	ND	40	32*	31	28	21
5H	53	38	27	21	13	0
30H	139	63	32	15	5	0
100H	13	9	3	1	1	0

ND not determined
* not all rooted hybrids were transferred to soil

additional chromosomes (range 2-22). In several asymmetric hybrids small centric chromosome fragments were observed.

3.3. Morphology and fertility

The asymmetric hybrid plants were less vigorous than both parental species and the symmetric hybrids. Especially the 30H- and 100H-hybrids were very retarded in growth and often died after transfer to soil. The leaves of the asymmetric hybrids differed much in colour, shape and size.

A majority of the greenhouse-grown 5H-hybrids and a minority of the greenhouse-grown 30H- and 100H-hybrids did flower (Table 1). The asymmetric hybrids flowered less abundantly than the parents and the symmetric hybrids. The corolla of the 5H-hybrid flowers were in general dark yellow, like that of *L. peruvianum* and the symmetric hybrids, whereas most 30H-hybrids and the only flowering 100H-hybrid had pale yellow flowers, such as *L. esculentum* ones. The flowers of the asymmetric somatic hybrids produced no or small amounts of pollen which were not viable. Several asymmetric hybrids set small fruits (Table 1). Most fruits were orange, which is intermediate between the fruit colour of the symmetric hybrids (yellow) and of the cultivated tomato (red).

No seed set was obtained in the asymmetric hybrids after selfing or backcrossing with the parental species (Table 1); the symmetric hybrids, on the contrary, did set seeds. In some fruits of five 5H-hybrids and one 30H-hybrid very small, abortive seeds were observed.

3.4. Marker genes

On the average, 73% of the *L. peruvianum* specific alleles (one or both) of the morphological markers were detected in the relevant 30H-hybrids. Several of the *L. esculentum* genotypes used were homozygous recessive for two morphological markers on the same chromosome. In 43% of the asymmetric hybrids derived from these parents, only one of the markers was complemented by the *L. peruvianum* allele. This is an indication for frequent breakage of the *L. peruvianum* chromosomes.

The isozyme analysis revealed that *L. peruvianum* specific alleles of acid phosphatase and glutamate oxaloacetate transaminase were present in at least 70% of the hybrids; the retention of donor alleles was lower in 30H- than in 5H-hybrids.

3.5. RFLP analysis

The genome composition of eight 5H-hybrids and seven 30H-hybrids was characterised by Southern blot analysis using 29 single copy clones as markers. Each tomato chromosome contained at least two such RFLP markers. The hybridisation patterns of three RFLP markers are shown in Figure 1. The 5H-hybrids contained more alleles, loci and complete chromosomes from *L. peruvianum* than the 30H-hybrids, whereas the 30H-hybrids contained more incomplete chromosomes (Table 2). In most hybrids also some *L. esculentum* alleles were lost.

TABLE 2. The amount of *L. peruvianum* genome (number of loci, alleles and chromosomes) in the 5H- and 30H-hybrids, as determined by autoradiographic analysis and densitometric scanning of Southern blots hybridised with 29 single copy clones (29 RFLPs = 29 loci = 58 alleles). A chromosome was considered complete when for each locus of the chromosome the same number of alleles was found.

	5H-hybrids (n=8)	30H-hybrids (n=7)
Loci (n=29) *	27.3 ± 3.4 (94%)	20.4 ± 4.3 (70%)
Alleles (n=58)	38.6 ± 7.8 (67%)	30.6 ±11.2 (53%)
Complete chromosomes *	13.4 ± 2.8	7.1 ± 4.1
Incomplete chromosomes *	4.9 ± 1.6	11.4 ± 3.2
Lost chromosomes	5.8 ± 3.6	6.9 ± 3.9

* significant difference between the two groups of hybrids, as tested with Student's t test ($P < 0.01$).

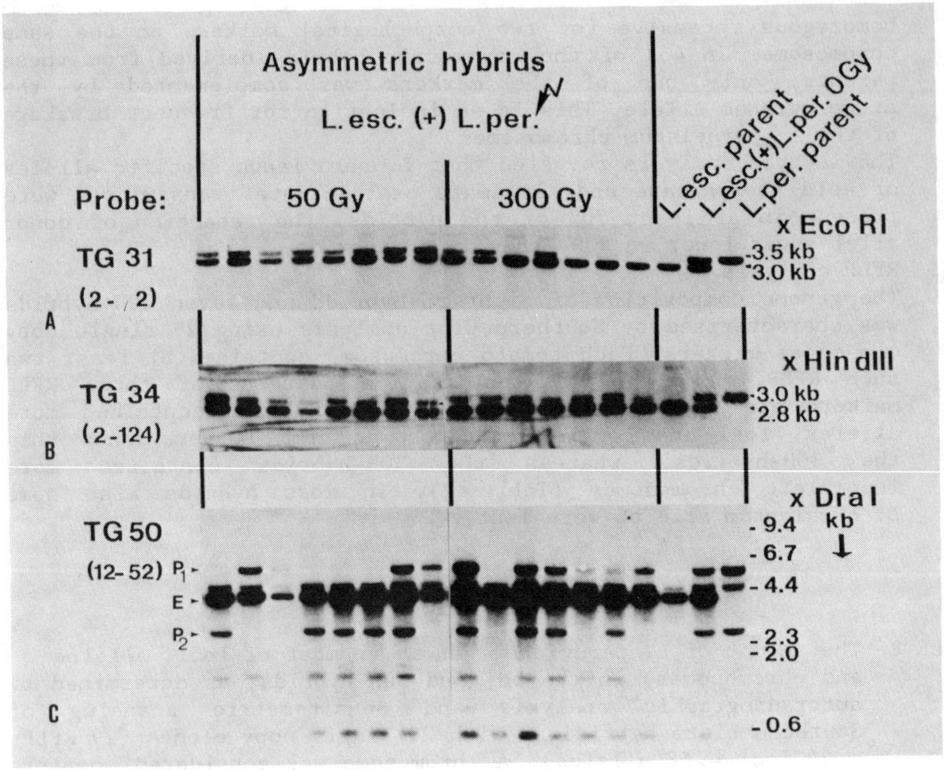

FIGURE 1. DNA from eight 5H-hybrids (lane 1-8), seven 30H-hybrids (lane 9-15), *L. esculentum* cv. Bellina (lane 16), one symmetric hybrid (lane 17) and *L. peruvianum* PI128650 (lane 18) was digested with the restriction enzymes indicated, electrophoresed, blotted and hybridised with the clone indicated. a) Hybridisation with clone TG31. A single fragment was detected in the respective parents. The size of the fragments is indicated at the right. b) Hybridisation with clone TG34. A single fragment was detected in the respective parents. The size of the fragments is indicated at the right. c) Hybridisation with clone TG50. P_1 and P_2 at the left indicate the two major *L. peruvianum* specific bands and E the *L. esculentum* specific band. The position of the *Hin* dIII fragments of phage lambda DNA is indicated at the right.

4. DISCUSSION

The present experiments show that asymmetric somatic hybrid plants can be obtained after fusion of *L. esculentum* protoplasts with irradiated *L. peruvianum* protoplasts. Selection of these hybrids was achieved on the basis of the dominant regeneration capacity derived from the irradiated parent. All asymmetric hybrids were aneuploid. The asymmetric hybrid nature could be confirmed by several morphological characteristics, which were intermediate between those of the symmetric somatic hybrids and those of *L. esculentum*, and by the analysis of specific markers. However, the irradiation of *L. peruvianum* before fusion did not have the desired effect, namely elimination of the donor genome to such an extent, that only a small fraction was present in the asymmetric hybrids.

Many of the asymmetric 3OH-hybrids contained incomplete chromosomes of *L. peruvianum*, as indicated by the analysis of morphological and RFLP markers. Apparently, the high dose (300 Gy) of gamma irradiation induced many breaks in the *L. peruvianum* chromosomes. This was confirmed by the presence of chromosome fragments in the cells.

The limited elimination in the tomato asymmetric hybrids can be a consequence of the unintended selection for good callus growth, because this trait is superior in *L. peruvianum* as compared with *L. esculentum*, is multi-genic and not linked to regeneration capacity (4). Another explanation can be the selection against aneuploid chromosome numbers that deviate too much from the triploid or pentaploid level in this species combination.

ACKNOWLEDGEMENTS

This research was supported by the Foundation for Fundamental Biological Research (BION), which is subsidised by the Netherlands Organisation for Scientific Research (NWO).

REFERENCES

1 Dudits, D. et al. (1987) Proc. Natl. Acad. Sci. USA 84, 8434-8438
2 Gleba, Y.Y. et al. (1988) Theor. Appl. Genet. 76, 760-766
3 Rick, C.M. (1982) in Plant Improvement and Somatic Cell Genetics (Scowcroft, W.R. and Frey, K.J., eds.), pp. 1-28, Academic Press, New York
4 Koornneef, M. et al. (1987) Theor. Appl. Genet. 74, 633-641
5 Wijbrandi, J. et al. (1990) Theor. Appl. Genet., in press
6 Wijbrandi, J. et al. (1990) Plant Science, in press
7 Rick, C.M. (1982) Rep. Tomato Genet. Coop. 32, 3-10
8 Wijbrandi, J. et al. (1990) Mol. Gen. Genet., in press

TRANSFER OF RESISTANCE TO PLRV, PVX AND PVY FROM S.brevidens TO POTATO BY SOMATIC HYBRIDISATION: CHARACTERISATION AND FIELD EVALUATION

M G K JONES, R DUNCKLEY, S STEELE, A KARP, R GIBSON, [1]N FISH, [2]J VALKONEN, [2]T POUTALA AND [2]E PEHU

AFRC Institute of Arable Crops Research, Depts of Biochemistry and Physiology, Plant Pathology, Rothamsted Experimental Station, Harpenden, Herts AL5 2JQ, UK; present address [1]Shell Research Ltd, Sittingbourne, Kent; [2]Dept of Crop Husbandry, University of Helsinki, Finland.

1. INTRODUCTION

To understand and apply the technology of protoplast fusion, it is useful to study populations of somatic hybrid plants produced rather than just those hybrids which have the desired properties. This enables a more complete picture of the processes involved to be built up, and any limitations to the approach to be assessed realistically.

For a number of years there has been a programme of somatic hybridisation at Rothamsted, in particular we have studied somatic hybrids produced by fusing protoplasts of the South American diploid wild potato species S brevidens and dihaploid potato (S tuberosum). S brevidens was initially identified as resistant to Potato Leaf Roll Virus (1), but it is not possible to make a direct sexual cross with potato to transfer this property. The desirability of introducing these resistance genes into potato germplasm is reflected by a number of studies in which somatic hybrids between these two species have been produced (eg 2,3,4,5). In our programme on Solanum, we have studied protoplast isolation and culture (6), culture induced variation (7), fusion methodology (8,9,10,11,12), production and identification of somatic hybrids both by chemical and electrofusion (10,11), their characterisation cytologically, mophologically, at the molecular level (13) and in the field (14 and unpublished). In addition, we have found that S brevidens is also resistant to the other major virus diseases of potato, Potato Virus X (PVX) and Potato Virus Y (PVY) (15,16), that the resistances are expressed in somatic hybrids, and that these resistances are closely linked (17). The mechanism appears to be inhibition of cell-to cell spread (18). One disadvantage of S brevidens is that it is non-tuber bearing, so the present thrust of the work is to apply techniques of asymmetric hybridisation to transfer small blocks of genetic material encoding the desired properties, using species-specific repetitive DNA probes to identify asymmetric hybrids (19).

2. PROCEDURES

2.1 Materials and Methods

2.1.1 Protoplast isolation and culture: see Foulger and Jones (6). Protoplasts were isolated from shoot cultures of diploid *S brevidens* (24 chromosomes CPC2451) and dihaploid potato PDH40 (24 chromosomes, kindly provided by M J de Maine, and derived from cv Pentland Crown).

2.1.2 Electrofusion: see Fish et al (11)

2.1.3 Isoenzyme analysis: see Fish et al (10)

2.1.4 Cytological analysis: see Pehu et al (13)

2.1.5 Molecular analysis: see Pehu et al (13)

2.1.6 Species specific probes: see Pehu et al (19)

2.1.7 Morphological and field analysis: see Fish et al (14)

3. RESULTS AND DISCUSSION

3.1 Production of somatic hybrids

58 somatic hybrids between *S brevidens* and the dihaploid breeding line of potato PDH40 have been studied in detail. These were produced both by chemical fusion (11) and by electrofusion (47), followed by mass culture. Hybrid plants were initially identified by isoenzyme analysis. Without selection, 2.6% of shoots were hybrid after chemical fusion, and 12.3% after electrofusion.

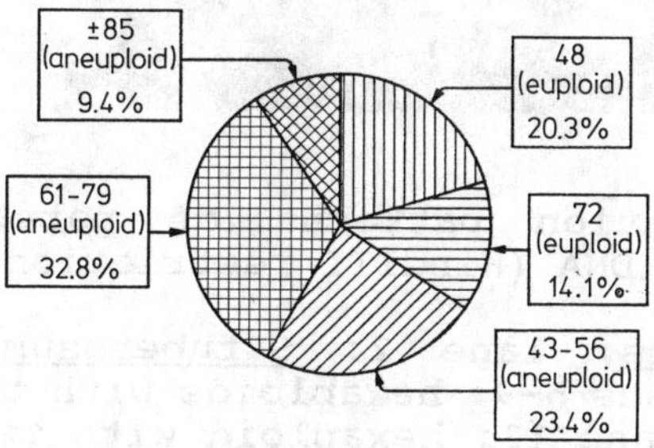

FIGURE 1. Percentages of different chromosome number classes of somatic hybrids

3.2 Cytology

The cytology of these plants was examined by root tip squashes, Fig 1. Hexaploids are derived from 3 protoplasts fusing together. The potentially useful tetraploid and hexaploid hybrids were 34% of the total, the remainder were aneuploid at the tetra-, hexa- or octaploid levels.

3.3 Chloroplast origins

Chloroplast DNA (cpDNA) was extracted from the hybrids and from the parents, restricted and the banding patterns of DNA fragments obtained after gel electrophoresis. Of the 58 hybrids analysed, 55% had cpDNA of S brevidens and 45% that of potato. Segregation of chloroplasts appeared to be random, it was not correlated to genome dosage in hexaploids (see below), and there was no evidence for recombination.

3.4 RFLP analysis

Molecular (RFLP) analysis of the plants, using either the patatin sequence probe pGMO1, or SS12 (a random cDNA probe), verified the nuclear hybridity of the plants (Fig 2). In addition, from the intensity of the bands of the autoradiographs, for hexaploid hybrids it was possible to identify the parental genome doses, ie 2 protoplasts of S brevidens to 1 of potato, or vice-versa (Fig 2). All 6 possible combinations of chloroplasts and parental nuclear genomes were found for tetraploid and hexaploid hybrids.

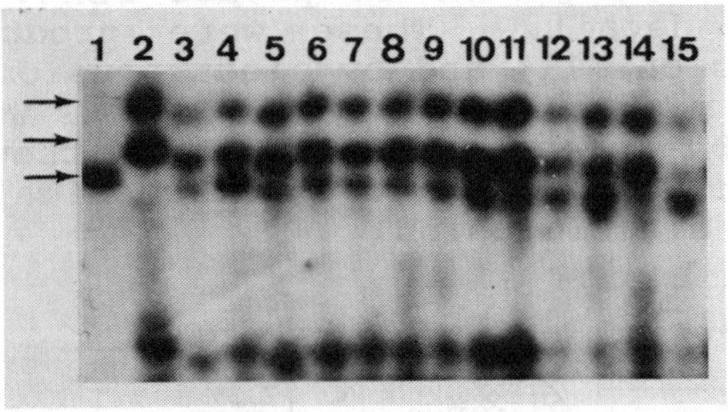

FIGURE 2 Hybridisation patterns of patatin probe to parental and hybrid DNA (HindIII restriction).

Lane 1: S.brevidens; lane 2: S.tuberosum; lanes 3-15 somatic hybrids; lanes 5-9: hexaploids with two doses of S. tuberosum genome; lane 13: hexaploid with two doses of S. brevidens genome.

3.5 Virus resistance

S brevidens was chosen for hybridisation because of its resistance to PLRV (1,2,15). We have since found, that S brevidens CPC2451 is also resistant to the other two major virus diseases of potato, PVX and PVY (15,16). In contrast, the potato line PDH40, is susceptible to all three viruses. When the tetraploid hybrids were analysed for expression of these virus resistances, expression of resistance ranged between that of the fusion parents. Genotypes with a morphology more resembling _S brevidens_ tended to be more resistant to virus infection. In contrast, hexaploid hybrids could be divided into high and low resistance groups, resistance being associated with two doses of the _S brevidens_ nuclear genome. Similarly, the genotypes more closely resembling _S brevidens_ were more resistant (17).

3.6 Mechanism of virus resistance

Having resistance to all three viruses in one plant is attractive to breeders, both because it may simplify selection and it is known that PLRV resistance can break down in plants infected with PVY or PVX (20). The fact that all three viruses have unusually low titres in _S brevidens_ suggested that there may be a common mechanism of resistance. To test this possibility, a group of hypo-aneuploid hybrids, that is those which had lost one to three chromosomes, were infected with the viruses, to find out if loss of one to three chromosomes led to loss of only one virus resistance, or to all three. Aneuploid hybrids lacking parental chromosomes were mostly either resistant to all three viruses or susceptible to all of them (17). This suggests that resistance is associated with the loss of a single chromosome of _S brevidens_. Close linkage, or a common resistance mechanism, would make this character easier to exploit.

3.7 Resistance is associated with slow cell-to-cell spread

The low virus titre in _S brevidens_ suggested that either virus replication or movement was inhibited. To distinguish between these possibilities, protoplasts were prepared from leaves of _S brevidens_ and PDH40, infected with PVY or PVX, and virus replication was monitored by immunofluorescence and by ELISA (18). The results showed that PVX and PVY could multiply equally well in protoplasts of _S brevidens_ and PDH40. In contrast, protoplasts of PDH40 made from virus infected plants showed 45-100 times more infection for PVX and 1000 times more infection for PVY than similarly infected plants of _S brevidens_. This result indicates that virus resistance in _S brevidens_ is associated with slow cell-to-cell spread rather than with reduced virus replication.

3.8 Field evaluation of hybrids

In the UK, somatic hybrids between sexually incompatible species are regarded as "genetically manipulated organisms", and can only be grown under guidelines laid down by the "Advisory Committee for Genetic Manipulation" (ACGM). As the first somatic hybrid plants grown in the field in the UK, the hybrids between S brevidens and PDH40 were a test case for the ACGM. Growth of the hybrids in the field at Rothamsted Experimental Farm was permitted in 1987 and 1988 with the conditions given in Table 1.

Table 1 ACGM restrictions for field testing hybrids

1. Inform local environmental health officer and press
2. Limited access to the plot
3. All flowerbuds to be removed during the growing season
4. Harvest all tubers, incinerate all remaining plant material
5. Examine plot for "ground-keepers" during the following year, spray with total herbicide
6. No potatoes to be grown on the plot for the following 4 years

In 1989 and 1990, permission to grow flowering hybrids was given, on the condition that the aim of the experiment was to monitor cross-pollination from hybrids, to assess any possible risk of transfer of hybrid pollen to surrounding trap potatoes and the wild Solanum species S nigrum and S dulcamara.

The majority of somatic hybrids grew vigorously in the field. They typically had much broader terminal and lateral leaflets than either parent, but anthocyanin stem pigmentation intermediate between that of the 2 parents (PDH40 has no anthocyanin stem pigment). About half the tetraploid hybrids produced tubers. Their morphology ranged from swollen stolons to reasonably formed tubers. Of the hexaploid hybrids, those with 2 genomic doses of potato gave the highest yields, in one case outyielding PDH40 nearly threefold.

In 1989, there was virtually no berry set because of the unusually hot, dry growing season, and so the risk assessment is being repeated in 1990. However, the main concern expressed by ACGM, that the hybrids might exhibit some form of "weedy potential", does not appear to be a problem, and in fact the hybrids are less persistant than normal potatoes, because they show less dormancy, and tubers do not survive so well in cold storage.

3.9 Production of species-specific probes

As an aid to rapid identification of complete or asymmetric hybrid plants, a series of highly repetitive DNA probes were developed which hybridised specifically to DNA of S brevidens or potato, (19). There are a number of useful applications of such probes, including verification of putative hybrids and the determination of parental genome dosage.

3.10 Asymmetric hybridisation

When complete genomes are combined together in hybrids, particularly if one partner is a wild species, then unwanted traits (eg non tuberisation in S brevidens) can also be transferred. By irradiation or otherwise fragmenting the donor genome, it is possible to transfer blocks of genes encoding the useful characters, thus avoiding transfer of unwanted genes, with the production of asymmetric hybrids. This approach is now being investigated, using S brevidens specific probes to identify asymmetric hybrid calli. Identified asymmetric hybrids will be screened for presence of virus resistance, and the linkage group encoding the resistance can be identified with the same chromosome specific probes developed for tomato (21).

4. CONCLUSIONS

Somatic hybridisation can be applied for production of complete hybrids, particularly if the parental material has been selected for good agronomic characters. Asymmetric hybridisation can usefully be used to reduce transfer of unwanted genes, and such application can be facilitated by using molecular techniques to identify hybrids and linkage groups encoding useful properties.

5. ACKNOWLEDGEMENTS

We thank the following for financial support: The Potato Marketing Board, The Biotechnology Action Programme of the EEC (No BAP.0101.UK(H)M G K Jones), The Finish Academy of Sciences and The University of Helsinki.

6. REFERENCES

1 Jones, R.A.C. (1979). Potato Research 22, 137-144.
2 Austin, S., Baer, M.A. and Helgeson, J.P. (1985). Plant Sci. 39, 74-82.
3 Austin, S., Ehlenfeldt, M.K., Baer, M.A. and Helgeson, J.P. (1986). Theor Appl Genet 71, 628-690.
4 Barsby, T.L., Shepard, J.F., Kemble, R.J. and Wong, R. (1984). Plant Cell Rep 3, 165-167.
5 Helgeson, J.P., Hunt, G.T., Haberlach, G.T. and Austin, A. (1986). Plant Cell Rep. 3, 212-214.

6 Foulger, D. and Jones M.G.K. (1986). Plant Cell Rep. 5, 72-76.
7 Karp, A., Jones, M.G.K., Foulger, D., Fish, N. and Bright, S.W.J. (1989). Amer. Potato J. 66, 669-684.
8 Tempelaar, M.J. and Jones, M.G.K. (1985a). Planta 165, 205-216.
9 Tempelaar, M.J. and Jones, M.G.K. (1985b). Plant Cell Rep. 4, 92-95.
10 Fish, N., Karp, A. and Jones M.G.K. (1987). In Vitro 23, 575-580.
11 Fish, N., Karp, A. and Jones M.G.K. (1988a). Theor. Appl. Genet. 76, 260-266.
12 Jones, M.G.K. (1988). Trends Biotechnol. 6, 153-158.
13 Pehu, E., Karp, A., Moore, K., Steele, S., Dunckley, R. and Jones M.G.K. (1989). Theor. Appl. Genet. 78, 696-704.
14 Fish, N., Steele, S.H. and Jones M.G.K. (1988b). Theor. Appl. Genet. 76, 800-806.
15 Gibson, R.W., Jones M.G.K. and Fish, N. (1988). Theor. Appl. Genet. 76, 113-117.
16 Gibson, R.W., Pehu, E., Woods, R.D. and Jones M.G.K. (1990). Annals Appl. Biol., in press.
17 Pehu, E., Gibson, R.W., Jones, M.G.K. and Karp, A. (1990a). Plant Sci., in press.
18 Valkonen, J.P.T., Pehu, E., Jones, M.G.K. and Gibson, R.W. (1990). J. Gen. Virol., submitted.
19 Pehu, E., Thomas, M., Poutala, T., Karp, A. and Jones M.G.K. (1990b). Theor. Appl. Genet., in press.
20 Jayasinghe, C., Chuquillanqui, C. and Salazar, L.F. (1987). Amer. Potato J. 66, 134-144.
21 Bonierbale, M.W., Plaisted, R.L. and Tanksley, S.D. (1988). Genetics 20, 1095-1103.

TRANSFER OF TOBACCO MOSAIC VIRUS RESISTANCE BY ASYMMETRIC PROTOPLAST FUSION

George W. Bates
Department of Biological Science, Florida State University,
Tallahassee, FL. 32306, USA

1. INTRODUCTION

Protoplast fusion has been used successfully for organelle transfers (1), species reconstruction experiments (2) and genetic complementation studies (3). However, application of protoplast fusion to the interspecific transfer of nuclear genes for plant improvement, has been hampered by morphological abnormalities and sterility in the hybrids. Recent studies have shown that donor-recipient fusions can be used to form genetically asymmetric hybrids in both interspecific and intergeneric species combinations (4-8). These hybrids sometimes have improved fertility when compared with symmetric hybrids between the same species.

In donor-recipient fusions, protoplasts from a donor species are irradiated and fused with recipient protoplasts. Irradiation blocks division of the donor protoplasts and results in elimination of donor chromosomes from the hybrids after fusion. However, selection for a nuclear-encoded trait of the donor can result in recovery of asymmetric hybrids. In 1987, Bates et al. (4) fused *N. tabacum* protoplasts with irradiated protoplasts of kanamycin-resistant (Km^r) *N. plumbaginifolia*. Selection on kanamycin led to the regeneration of *N. tabacum*-like plants that retained the Km^r trait of *N. plumbaginifolia*. Although male sterile, these plants could be backcrossed with *N. tabacum* and Km^r was inherited by the progeny. Recently, Piastuch and Bates (9), using *in situ* hybridization of a species-specific repetitive DNA probe, showed these hybrids retain one to several *N. plumbaginifolia* chromosomes.

This paper describes results of donor-recipient fusions between *N. tabacum* and Km^r protoplasts of *N. repanda* and *N. glutinosa*. The goal was to transfer the N gene, which encodes the hypersensitive response to TMV infection, to tobacco by asymmetric protoplast fusion. The N gene was originally transferred to tobacco from *N. glutinosa* by a sexual cross (10). Somatic hybridization between *N. tabacum* and *N. repanda* would recreate this gene transfer. *N. repanda* also gives a hypersensitive response to TMV inoculation; indeed *N. repanda* is resistant to a large number of diseases of tobacco. Sexual crosses between *N. tabacum* and *N. repanda* have never been successful (11). Thus, *N. tabacum* (+) *N. repanda* hybrids may be of value.

2. METHODS

2.1 Plant Material

Plants of *N. repanda* and *N. glutinosa* were transformed to kanamycin resistance (12) using *Agrobacterium* carrying pMON200 (the generous gift of Dr. Stephen

Rogers, Monsanto). Km^r calli that grew from leaf discs of *N. repanda* were regenerated and rooted on media containing 100 mg/l kanamycin. In the case of *N. glutinosa* regeneration failed; thus, this species was maintained as a transformed suspension culture line on liquid media containing kanamycin. The recipient in all fusions was *N. tabacum* var. Xanthi (genotype nn, kanamycin sensitive, nopaline negative).

2.2 Protoplast Isolation, Irradiation, Fusion, and Culture

Mesophyll protoplasts were isolated from *N. tabacum* and *N. repanda* as described previously (4). Protoplasts were isolated from *N. glutinosa* suspension-culture cells by digestion with 1% Cellulysin + 0.2% Pectolyase Y23, followed by flotation on sucrose. Protoplasts were irradiated by exposure to ^{137}Cs for 30 min (total gamma dose, 120 Gy).

The protoplasts were washed twice with 0.5 M mannitol, resuspended in 0.5 M mannitol + 0.5 mM $CaCl_2$ at a density of 4 X 10^5 protoplasts/ml, and mixed together to give a five-fold excess of *N. tabacum* protoplasts. Electrofusion was carried out as described previously (4). Following fusion, the protoplasts were transferred to petri dishes and diluted with protoplast culture medium.

2.3 Selection and Regeneration

Km^r calli were recovered after fusion by the agarose-bead method (13) using 100 mg/l kanamycin for selection. After 4 to 6 weeks, Km^r calli were transferred to solid medium + 100 mg/l kanamycin for further growth. Shoot regeneration and rooting were induced on media containing 100 mg/l kanamycin.

2.4 Esterase and Nopaline Analyses

Esterase isozymes were visualized on polyacrylamide gels (4). Nopaline assays were performed by paper electrophoresis and phenanthrenequinone staining of leaf extracts (12).

2.5 Crossing and Seedling Tests for Kanamycin Resistance

Hybrid plants were used as the female parent in backcrosses with *N. tabacum* var. Xanthi. Inheritance of Km^r was determined by surface sterilizing the seeds and sowing them on rooting medium containing 1% sucrose and 100 mg/l kanamycin. Resistant seedlings grew normally, whereas sensitive seedlings germinated but never produced true leaves or secondary roots and bleached after 3-4 weeks.

2.6 Tests for TMV Resistance

Leaves of plants to be tested were rubbed with a cotton swab dipped in a slurry consisting of TMV particles diluted into 0.1 M phosphate buffer (pH 7) plus 10% w/v Carborundum powder. The presence of local lesions was scored five days later.

3. RESULTS

3.1 *N. tabacum* (+) *N. repanda* Fusions

The following controls were included in each fusion between *N. tabacum* and irradiated *N. repanda* protoplasts: *N. tabacum* protoplasts cultured alone, irradiated *N. repanda* protoplasts cultured alone, and *N. tabacum* protoplasts mixed with irradiated *N. repanda* protoplasts and cultured without being fused. No Km^r calli were recovered from any of these controls. A fourth control consisted of a mixture of *N. tabacum* and unirradiated *N. repanda* protoplasts that were fused, cultured, and selected for Km^r. A plating efficiency of 10% (fraction of *N. repanda* protoplasts in the initial sample recovered as calli after agarose-bead culture) was observed for this control. When fused mixtures of *N. tabacum* + irradiated *N. repanda* protoplasts were subjected to selection on kanamycin the plating efficiency dropped to 1.5%, suggesting that about 15% of the *N. repanda* protoplasts were recovered as Km^r hybrid clones.

Rooted plants were established from 19 independent, Km^r calli. Twelve of these clones regenerated vigorously and gave rise to numerous plants that survived transfer to soil.

3.1.1. Morphology of regenerated plants: All the regenerants had the upright growth habit of *N. tabacum*. Although similar in appearance to *N. tabacum*, the regenerants were shorter and displayed less apical dominance. Considerable variation in stature, degree of branching, and leaf morphology existed between plants from different clones. The leaves ranged from ones similar to *N. tabacum* to elongate, strap-like leaves, irregularly-shaped leaves and thick succulent leaves.

All twelve of the vigorously regenerating clones gave rise to plants that flowered. Floral morphology was much closer to that of *N. tabacum* than *N. repanda*. Plants regenerated from two clones produced about 5% viable pollen; plants from the remaining clones were completely male sterile.

3.1.2. Biochemical analyses: The *N. repanda* fusion partner carried a functional nopaline synthase gene linked to the neo gene, which encodes Km^r. Nopaline tests were used as a rapid screen for the presence of *N. repanda* genetic material in the hybrids. Nopaline was detected in leaf extracts from all the hybrids.

PAGE gels of *N. tabacum* leaf extracts, stained for esterase isozymes, show one major band and two minor bands; *N. repanda* has two major bands and two minor bands of esterase staining. The two species had no esterase bands in common. Esterase isozyme profiles were examined for plants from all twelve Km^r clones. Seven clones had all the *N. tabacum* esterase bands plus one *N. repanda* band. Plants from four clones had the three *N. tabacum* bands plus two *N. repanda* bands. One clone displayed only the *N. tabacum* esterase bands.

3.1.3. Chromosome numbers and inheritance of kanamycin resistance: Chromosome counts for the hybrids ranged from 51 to 90 (Table 1). Most of the clones regenerated plants having from 56 to 62 chromosomes, three clones yielded plants with more than 70 chromosomes. Five plants examined were clearly mixoploid, having variable chromosome numbers in the same plant. Two other plants (988 and 985) may also be mixoploid.

Plants from six clones formed a small number of viable seeds when backcrossed with *N. tabacum* as the pollen parent. Three clones (9, 11, and 12) yielded only sterile plants; plants regenerated from clones 3 and 4 were lost to fungal infections before seeds were collected, and plant 1089 (clone 8) has set pods but has not yet been analyzed for seed viability.

The ratio of kanamycin-resistant to sensitive (Km^r:Km^s) seedlings ranged from 4:1 to 1:2 for the different regenerants (Table 1). No Km^r progeny were recovered from plants 983 and 1076.

The inheritance of Km^r has been followed for additional backcross generations for some of the hybrids. Some of the backcross data obtained for plant 992 is shown in Fig. 1. This plant was mixoploid, having from 53 to 58 chromosomes and gave a 2:1 ratio of Km^r:Km^s seedlings. All of its BC_1, Km^r progeny had lower chromosome numbers, greatly improved seed sets, and were male sterile. These Km^r, BC_1 progeny tested positive for nopaline and had only *N. tabacum* esterases (except plant 1011, which also had one *N. repanda* band).

TABLE 1. Chromosome numbers, fertility, and inheritance of kanamycin resistance among *N. tabacum* (+) *N. repanda* asymmetric hybrids. Fertility and inheritance were assessed by backcrossing the hybrids with N. tabacum. Asterisks indicate plants with mixoploid chromosome numbers. ND, not determined.

Clone	Plant	Chromosome Number	Viable Seeds/pod	Inheritance of Km^r Km^r:Km^s
1	992	53-58*	25	17:8
2	1079	52,56*	16	12:3
3	1031	59	ND	ND
4	988	51-52	ND	ND
5	984	70,90*	215	20:39
6	983	60	14	0:14
7	985	50-51	41	9:17
8	1089	80	ND	ND
9	1086	53,60*	0	-
10	1076	69-75*	90	0:79
11	1023	62	0	-
12	995	62	0	-

Four of the Km^r, BC_1 plants were backcrossed with *N. tabacum*. The ratio of Km^r: Km^s, BC_2 progeny was lower than in the previous generation. The Km^r, BC_2 plants examined had either 48 or 49 chromosomes (data not shown). Two of these BC_2 plants have reached maturity and been backcrossed. Although male sterile, they show further improvement in seed sets and continued sexual transmission of Km^r.

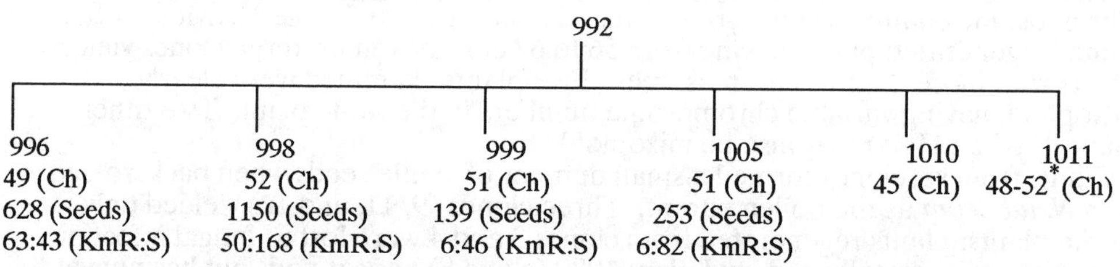

FIGURE 1. Inheritance of kanamycin resistance following backcrossing of *N. tabacum* (+) *N. repanda* hybrid 992 with *N. tabacum*. Ch, chromosome number; Seeds, number of viable seeds per capsule; KmR:S, segregation of kanamycin resistant and kanamycin sensitive progeny following the backcross.

3.1.4. Analysis of the hybrids for TMV resistance: *N. repanda* gives a hypersensitive response when inoculated with TMV. The *N. tabacum* line used in these fusions lacks the N gene and is systemically infected when inoculated with TMV. Of the hybrids, only plants 1076 and 1086 gave a hypersensitive response when inoculated with TMV. The other hybrids eventually developed systemic TMV infections.

3.2 *N. tabacum* (+) *N. glutinosa* Fusions

Kanamycin-resistant calli were recovered after fusion of *N. tabacum* (+) *N. glutinosa* at frequencies similar to those obtained for fusion of *N. tabacum* (+) *N. repanda* protoplasts. From approximately 150 Km^r calli, 13 fast growing, regenerable lines were obtained. Plants regenerated from these lines have not yet been thoroughly analyzed. However, preliminary observations indicate that they are all hybrids and retain a significant amount of genetic material from the donor, *N. glutinosa*.

The *N. tabacum* (+) *N. glutinosa* regenerants are morphologically intermediate between the two parental species. The leaf bases of many of the plants were slightly truncate, reminiscent of *N. glutinosa*. Also, the corollas of those plants that have flowered are intermediate in shape and size between the two species. Those plants that have been analyzed are positive for nopaline and have all the esterase isozyme bands of *N. tabacum* plus two or three esterase bands of *N. glutinosa*. Chromosome numbers have been determined for six of the *N. tabacum* (+) *N. glutinosa* hybrid lines. Three of the lines have from 65 to 68 chromosomes, which is only 5 to 7 less than would be expected for a balanced hybrid between *N. tabacum* and *N. glutinosa*. To other lines have 85 to 90 chromosomes and the sixth line has about 130. These plants are partially male and female fertile. Seeds have been recovered from backcrosses with both *N. tabacum* and with *N. glutinosa*, as well as by self fertilization, for all of these plants except the one with 130 chromosomes. Progeny from these crosses have not yet been tested for viability or Km^r.

3.2.1. <u>TMV resistance of the *N. tabacum* (+) *N. glutinosa* hybrids</u>: All of the hybrid plants gave a hypersensitive response when inoculated with TMV.

4. DISCUSSION

The experiments described here indicate that the extent of radiation-induced chromosome elimination varies greatly. In the *N. tabacum* (+) *N. repanda* fusions, most of the regenerants had from 56 to 62 chromosomes, although some had as few as 51 and some as many as 90. If all the chromosomes above 48 are from *N. repanda*, then these hybrids retain from 4 to 88% of the *N. repanda* genome, with most having 23%. In the fusions involving *N. glutinosa*, much less chromosome elimination was observed. This may be due to the use of *N. glutinosa* suspension-culture derived protoplasts in these fusions, rather than mesophyll protoplasts, or it may reflect a greater nuclear compatibility between the *N. glutinosa* and *N. tabacum* than between *N. repanda* and *N. tabacum*. Work in other laboratories on asymmetric hybridization has revealed similar variability in the degree of chromosome elimination, with some studies reporting extensive elimination (4,5) and others less elimination (6-8,14). At present the factors controlling chromosome elimination are poorly understood. Interestingly, two studies have reported that the degree of chromosome elimination is not strongly affected by radiation dose (6,8).

Both the fusions described in this paper resulted in fertile hybrids. In the *N. tabacum* (+) *N. repanda* fusions, the hybrids were female fertile but male sterile. Fertile hybrids have not been reported before for this species combination, whether formed by sexual crosses or somatic hybridization (11). Thus, asymmetric hybridization may provide a way to mobilize useful traits from *N. repanda* to *N. tabacum*. Alloplasmic type male sterility has been observed before when the *N. tabacum* nucleus is combined with *N. repanda* cytoplasm. When the *N. tabacum* (+) *N. repanda* hybrids were backcrossed with *N. tabacum*, Km^r was passed on to the progeny in many cases. The BC_1 and BC_2 plants displayed improved seed sets and reduced numbers of excess chromosomes. Thus, backcrosses may allow stable introgression of *N. repanda* traits into *N. tabacum*. By contrast, the *N. tabacum* (+) *N. glutinosa* hybrids are both male and female fertile as are sexual hybrids between these species. The inheritance of Km^r has not yet been evaluated for these hybrids.

Only two of the *N. tabacum* (+) *N. repanda* hybrids gave a hypersensitive response when inoculated with TMV. One of these hybrids proved to be sterile and the other set seed when backcrossed with *N. tabacum* but none of the progeny were Km^r. These progeny are presently being evaluated for TMV resistance. By contrast, all the *N. tabacum* (+) *N. glutinosa* hybrids were TMV resistant. Together these observations indicate that it is possible to transfer specific, unselected traits by asymmetric hybridization. However, the effectiveness of N gene transfer decreased with increasing chromosome elimination. Thus, effective gene transfer by asymmetric hybridization may require examining large numbers of hybrids, direct selection for the trait of interest, or establishment of linkage between the trait to be transferred and a selectable marker such as the Km^r gene.

REFERENCES

1. Galun, E. and Aviv, D. (1983) in The Hnadbook of Plant Cell Culture (Evans, D., ed.), Vol. 1, pp.358-392, Macmillian, New York.
2. Sunberg, E., Landgren, M. and Glimelius, K. (1987) Theor. Appl. Genet. 75, 96-104.
3. Harms, C.T., Potrykus, I. and Widholm, J.M. (1981) Z. Pflanzenphysiol. 101, 377-390.
4. Bates, G.W., Hasenkampf, C.A., Contolini, C.L. and Piastuch, W.C. (1987) Theor. Appl. Genet. 74, 718-726.
5. Dudits, D., Maroy E., Praznovszky T., Olah, Z., Gyorgyey, J. and Cella R. (1987) Proc. Natl. Acad. Sci. 84, 8434-8438.
6. Gleba, Y.Y., Hinnisdaels, S., Sidorov, V.A., Kaleda, V.A., Parokonny, A.S., Boryshuk, N.V., Cherep, N.N., Negrutiu, I. and Jacobs, M. (1988) Theor. Appl. Genet 76, 760-766.
7. Famelaer, I., Gleba, Y.Y., Sidorov, V.A., Kaleda, V.A., Parokonny, A.S., Boryshuk, N.V., Cherep, N.N., Negrutiu, I. and Jacobs, M. (1989) Plant Sci. 61, 105-117.
8. Yamashita, Y., Terada, R., Nishibayashi, S. and Shimamota, K. (1989) Theor. Appl. Genet. 77, 189-194.
9. Piastuch, W.C. and Bates, G.W. (1990) Mol. Gen. Genet. (in press).
10. Holmes, F.O. (1938) Phytopathology 28, 553-561.
11. Shintaku, Y., Yamamoto, K. and Nakajima, T. (1988) Theor. Appl. Genet. 76, 293-298.
12. Rogers, S.G., Horsch, R.B. and Fraley, R.T. (1986) Meth. Enzymol. 118, 627-640.
13. Shillito, R.D, Paszkowski, J. and Potrykus, I. (1983) Plant Cell Rep. 2, 244-247.
14. O'Connel, M.A. and Hanson, M.R. (1987) Theor. Appl. Genet. 75, 83-89.

IMPROVEMENTS OF THE BRASSICA CROPS BY TRANSFER OF GENES FROM ALIEN SPECIES VIA SOMATIC HYBRIDIZATION

K. Glimelius, J. Fahlesson, M. Landgren, C. Sjödin* & E. Sundberg.

DEPARTMENTS OF PLANT BREEDING AND GENETICS*, UNIVERSITY OF AGRICULTURAL SCIENCES, BOX 7003, S-750 07 UPPSALA, SWEDEN.

1. INTRODUCTION

The oilseed *Brassica* crops have during the last decade increased in importance. For production of vegetable oils it is the third most important oil source after soybean and palm (1). The oil from the different Brassica crops is used both for consumption and for technical purposes. Although the *Brassica* crops belong to some of the earliest domesticated plants they have a rather narrow genetic background, due to breeding from a few ancestors to meet the demands of the quality requirements for the oil and the meal. A widening of the genepool within these crops would, thus, be of great practical importance. The improvements needed are several different traditional breeding goals, such as cold and drought tolerance as well as disease resistance. Furthermore, large efforts in trying to modify the oilcomposition in these crops have also recently started (2). The fatty acids produced by the oil seed crops have potentials of several different kinds and can be used as fuels, lubricants, water repellents, plasticizers, waxes and surface active agents. With such a wide range of different traits to improve there is a need to broaden the genetic variation in the crops and thus genetic engineering, including somatic hybridization, has been suggested as potential tools.

With the aim to investigate the use of somatic hybridization for transfer of genes from alien species into rapeseed a research program was started at our department. The investigation was made with the purpose to evaluate the protoplast fusion techniques, but also the limitations regarding production of functional hybrids depending on taxonomic relations between the different species combined.

2. RESULTS AND DISCUSSION

An overview of the different somatic hybridization experiments performed and in progress in our laboratory is illustrated in Fig 1. where the newly produced hybrids are viualized by the hatched circles. In all combinations a large number of hybrid calli were obtained. Usually, from each combination, more than 20 plants were regenerated from separate calli. Below follows a more detailed presentation and discussion of the hybrids produced.

2.1 Interspecific somatic hybrids

To start with several different cultivars of *B campestris* were fused with *B. oleracea* (3). Thus, a production of somatic hybrids with a combination of two of the *Brassica* genomes was obtained, resulting in resynthesized rapeseed. The question was then whether all the three *Brassica* genomes, A, B and C could be combined. Fusions of protoplasts from *B. napus* and *B. nigra* succeded in combining all the three genoms into a novel species, never reported before (4). Of the trigenomic *B. naponigra* hybrids, about 70% had the expected chromosome number of 54 (Table 1). According to isozyme analysis the pattern typical for both species

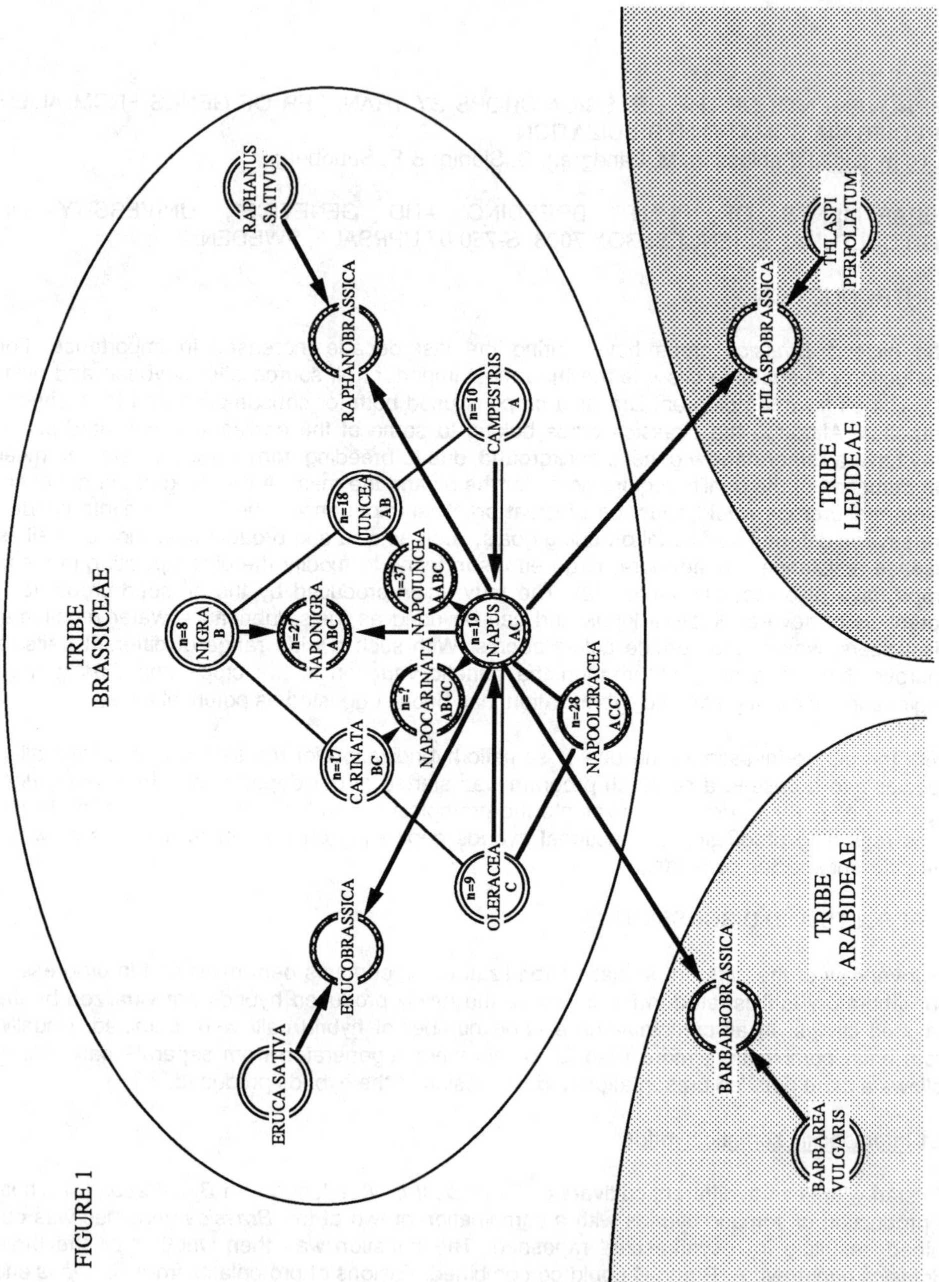

TRIBE BRASSICEAE
RAPHANUS SATIVUS
RAPHANOBRASSICA
n=10 CAMPESTRIS A
n=18 JUNCEA AB
n=37 NAPOJUNCEA AABC
n=8 NIGRA B
n=27 NAPONIGRA ABC
n=19 NAPUS AC
n=? NAPOCARINATA ABCC
n=17 CARINATA BC
n=28 NAPOOLERACEA ACC
n=9 OLERACEA C
ERUCOBRASSICA
ERUCA SATIVA
BARBAREOBRASSICA
BARBAREA VULGARIS
TRIBE ARABIDEAE
THLASPOBRASSICA
THLASPI PERFOLIATUM
TRIBE LEPIDEAE

FIGURE 1

was obtained, revealing production of true hybrids. The hybrids had different degrees of fertility, but most of them were able to set seeds both after selfing as well as after backcrossing to rapeseed.

Other experiments with production of allopolyploids, where several of the *Brassica* genomes are combined have also been made. Somatic hybrids have been produced between *B. napus* and *B. oleracea* and between *B. napus* and *B. juncea* (Sundberg, unpublished) as well as between *B. napus* and *B. carinata* (6). These hybrids contain three or four diploid *Brassica* genomes resulting in the amphitriploid *B. napooleracea* (AACCCC) and the amphitetraploids *B. napojuncea* (AAAABBCC) and *B. napocarinata* (AABBCCCC). The combinations where several of the *Brassica* genomes were combined into somatic hybrids proved to be very successful in culture. The calli obtained were easy to differentiate into shoots, with a regeneration frequency of 18% for *B napooleracea* and 8% for *B. napojuncea*. Furthermore, a large number of the hybrid plants had the expected chromosome number and a relatively high fertility (Table 1).

TABLE 1 Chromosome number, chloroplast segregation and fertility in somatic hybrids produced between different species within the tribe Brassiceae.

HYBRID COMBINATION (NO OF HYBRID PLANTS) A B	CHROMOSOME NUMBER	% OF HYBRID PLANTS	% OF PLANTS WITH CHLOROPLASTS FROM PARENT A	B	SELF FERTILITY % OF RAPE (AVERAGE)
B. ole (+) B. cam	38	30			17
	<38	9			-
(23)	>38	52	54	46	0.2
	CHIMERIC	9			-
B. nap (+) B. nig	54	67			4.6
	<54	23			6.4
(30)	>54	7	88	12	0.4
	CHIMERIC	3			-
B. nap (+) B. ole	56	44			50
	<56	0			-
(18)	>56	50	68	32	2.0
	CHIMERIC	6			-
B. nap (+) B. junc	74	88			10
	<74	0			-
(8)	>74	12	100	0	10
	CHIMERIC	0			-
B. nap (+) E. sat	60	8			2
	<60	59			7
(24)	>60	29	85	15	1
	CHIMERIC	4			-

Even though sexual crosses can be performed between several of the different *Brassica* species there are limitations. For example, in crosses between some of the combinations only one of the parents can be used as the female parent (5), and some combinations such as *B. naponigra* could not be obtained at all (7). Since the somatic hybridization can be performed independent of gametes, restrictions regarding the cytoplasm is not present. Instead new unique combinations of organelles can be obtained using protoplast fusion (see further under 2.3). Furthermore, by utilizing fusions of protoplasts isolated from the diploid species, production of complete allopolyploid hybrids will be obtained directly. Thus, the chromosome doubling necessary to perform after sexual hybridization, to obtain fertile hybrids, is not required.

2.2 Intergeneric somatic hybrids

The next level to investigate was, whether species from different genera could be combined via protoplast fusion. A selection of hybridization material was made with regard to chromosome homology, but also with regard to agronomic value. Both *Eruca sativa* and *Raphanus sativus* belong to the tribe *Brassiceae*, as the different *Brassica* species do. Both species have chromosome homologies with *B. campestris* (8). Traits of value for introgression into rapeseed would be drought tolerance and insect resistance from *Eruca* and CMS as well as restorers to the Ogu cytoplasm from *Raphanus*.

In both combinations somatic hybrids have been obtained and the hybrids between *Eruca* and *Brassica* have been investigated in detail (9). From the analysis of chromosome number a large variation was found with only a few plants containing the sum of the chromosome number. According to the isozyme pattern a preferential elimination of nuclear genetic material was recorded for the *Erucobrassica* hybrids. About 50% of the plants were partial hybrids, even among those having the expected chromosome number. Even though most of the *Erucobrassica* plants contained less chromosomes, than the expected, these had a higher fertility than those containing the sum of the parental chromosome numbers (Table 1). Investigations are in progress of the more recently synthesized *Raphanobrassica* hybrids to determine chromosome number, isozyme pattern and fertility of those plants.

The reasons for the variation in chromosome numbers could be several. In the cases where a higher chromosome number than the expected was found multiple fusions could be the reason. In mass fusion systems using e.g. PEG it is difficult to regulate the number of protoplasts involved in each fusion event. However, abnormalities induced by the combination of a mitotic active cell (hypocotyl) with a cell in interphase (mesophyll) could also result in an abnormal ploidy level. Furthermore, culture of cells and calli *in vitro* can result in changes in the ploidy level. The most pronounced deviation from the expected chromosome number was found in the intergeneric combination between *Eruca* and *Brassica*. The most probable reason to this might be the taxonomic distances between the species, causing irregularities during mitosis and sorting out of chromosomes from one of the fusion partners. Irregularities like sticking of chromosomes, chain formation and multiconstrictional chromosomes have been registered especially in intergeneric somatic hybrids (10).

2.3. Chloroplast and mitochondrial segregation in the somatic hybrids.

Chloroplast DNA has been characterized in most of the produced hybrids. In all hybrids a chloroplast type identical to either one of the parental types was found. Regarding the segregation of chloroplasts random distribution was found in the resynthesized rapeseed, while biased segregation was found in all the other combinations (Table 1). However, even though biased segregation was obtained, all different chloroplast types could be established with the hybrid nucleus, except in the case of the hybrids between *B. napus* and B. *juncea*. Analysis of the biased segregation revealed, that protoplast type did not affect the segregation. The segregation favouring the *B. napus* chloroplasts in most of the combinations could more be due to unequal input of organelles when fusing an aphidiploid species with a diploid. According to Butterfass (11) more nuclear DNA always involves more plastids per cell. However, other possibilities for the uneven distribution of chloroplasts could reflect the different phylogenetic distances between the species combined in the fusions, which might result in differences in compatibility or replication rates between the species specific chloroplasts (12).

Investigations of the mitochondrial DNA, mt-DNA, in the hybrids have been performed analysing the restriction patterns of the mt-DNA as well as using probes of different mitochondrial genes. From the investigations made so far, a much more complex result than the one obtained for the chloroplasts was found. Overall, two main patterns were found. Some of the hybrids had a mt-DNA identical to either one of the parental types. Another pattern, however, revealed that a rearranged mt-DNA with presence of fragments from both parents, although not the sum of the two, was obtained. Moreover, new unique fragments were discovered. From a more detailed investigation using hybridizations to different mt-genes the rearrangements were localized to specific regions in the mt-genome (Landgren unpublished). The results also indicate that the rearrangements involved intergenomic recombinations between the two mitochondrial genomes.

2.4. Intertribal somatic hybrids.

Since one of the aims in the project was to evaluate the limitations caused by taxonomic differences between the species combined to somatic hybrids, fusions between species belonging to different tribes in *Brassicaceae* were also performed.

The species chosen in this case were *Barbarea vulgaris* and *Thlaspi perfoliatum*. The two species belong to different tribes of *Brassiceae* as well as to each other (Fig. 1), and attempts were made to combine them with rapeseed. Both species have valuable traits of interest to transfer to rapeseed. *Barbarea* is known for its cold tolerance and *Thlaspi* for a rather high concentration (16%) of the fatty acid nervonic acid, with a carbon chain of 24 C.

In both hybridization experiments a large number of calli as well as shoots have been obtained. Isozyme investigations of the shoots revealed that partial hybrids were obtained in the combination between rapeseed and *Thlaspi*, while more markers are needed for evaluating the putative hybrids between rapeseed and *Barbarea*. Analysis have started with different DNA probes to reveal hybridity between these two species. Even though a development into shoots was easily obtained in both cases, there were problems to root the hybrid plants and transfer them to the greenhouse (Fahleson, unpublished).

3. CONCLUSIONS

The presented examples of somatic hybrids produced between different species within the family of *Brassicaceae* prove that fertile hybrids can be obtained. Interspecific, intergeneric and even intertribal combinations can result in hybrid plants. It remains to be seen, however, whether the intertribal combinations will be fertile and of practical use. Since the importance of the allopolyploids produced primarily will be for introgression of specific traits into our crops, and not as novel hybrid species, the investigations have now to be focused on if stable transfer of alien genes is obtained in the receptor genome.

Even though the above presented somatic hybrids were mainly produced to investigate the outcome of nuclear and organellar composition in the hybrid plants, the potential agronomic value of the artificial hybrids has also been investigated. A good example of that is the transfer of genes for resistance to *Phoma lingam* present in some accession of *B. nigra, B. juncea* and *B. carinata,* which all have been hybridized with *B. napus* and resulted in a transfer of the resistance trait (6). Other traits that are under evaluation are cytoplasmic male sterility (CMS) from *B. juncea* containing the Anand cytoplasm and a potential new male sterility from the *Eruca* material. Furthermore, in field tests of the somatic hybrids backcrossed once to rapeseed a search for resistance to different pathogens and insects are made.

REFERENCES

1. Downey, R.K. and Röbbelen, G. (1989) in Oil crops of the world. (Röbbelen, G. Downey, R.K., Ashri, A. eds.) pp. 339-362, McGraw-hill publishing company, New York.
2. Knauf, V.C. (1987) The application of genetic engineering to oilseed crops. Trends in Biotechn., 5, 40-46.
3. Sundberg, E., Landgren, M. and Glimelius, K. (1987) Fertility and chromosome stability in *Brassica napus* resynthesised by protoplast fusion. Theor. Appl. Genet., 75:96-104.
4. Sjödin, C. and Glimelius, K. (1989) *Brassica naponigra*, a somatic hybrid resistant to *Phoma lingam*. Theor. Appl. Genet. 77:651-656.
5. Olsson, G. and Ellerström, S. (1980) in Brassica crops and wild allies, biology and breeding. (Tsunoda, S., Hinata, K., Gómez-campo, C. eds.) Japan Scientific Soc. Press, Tokyo, pp. 167-187.
6. Sjödin, C. and Glimelius, K. (1989) Transfer of resistance against *Phoma lingam* to *Brassica napus* by asymmetric somatic hybridization combined with toxin selection. Theor. Appl. Genet., 78:513-520.
7. Olsson, G. (1963) In Recent Plant Breeding Research, Svalöv 1946-1961. John Wiley & Sons, New York and London, pp. 179-192.
8. Mizushima, U. (1980) in Brassica crops and wild allies, biology and breeding. (Tsunoda, S., Hinata, K., Gómez-campo, C. eds.) Japan Scientific Soc. Press, Tokyo, pp. 89-105.
9. Fahleson, J., Råhlén, L. and Glimelius, K. (1988) Analysis of plants regenerated from protoplast fusions between *Brassica napus* and *Eruca sativa*. Theor. Appl. Genet. 76:507-512.
10. Nehls, R., Krumbiegel-Schroeren, G., Binding, H. (1986) In differentiation of protoplasts and of transformed plant cells. (Reinert, J. and Binding, H. eds.) Springer-Verlag, Berlin, pp. 67-108.
11. Butterfass, T.H. (1989) Nuclear control of plastid division. In: Division and segregation of organelles. Society for experimental biology seminar series, volume 35. (Boffey SA, LLoyd D. eds) Cambrigde University Press, Cambridge, pp 21-38.
12. Landgren, M. and Glimelius, K. (1990). Analysis of chloroplast and mitochondrial segregation in three different combinations of somatic hybrids produced within Brassicaceae. Theor. Appl. Genet. (submitted).

Section B
Morphogenesis and Metabolism

MECHANISMS OF SOMATIC EMBRYOGENESIS IN CELL CULTURES——PHYSIOLOGY, BIOCHEMISTRY AND MOLECULAR BIOLOGY

ATSUSHI KOMAMINE, MASAMI MATSUMOTO, MASAYOSHI TSUKAHARA ARIHITO FUJIWARA, RYOICHI KAWAHARA, MASAKI ITO, JANE SMITH KOJI NOMURA AND TATSUHITO FUJIMURA

BIOLOGICAL INSTITUTE, TOHOKU UNIVERSITY, SENDAI, 980 JAPAN

1.INTRODUCTION

Somatic embryogenesis is an ideal system for investigation of the whole process of differentiation of plants as well as the mechanisms of expression of totipotency in plant cells. The intention of the first attempt of Haberlandt to establish plant tissue culture systems was to provide evidence for totipotency of plant cells. Thus, the mechanism of somatic embryogenesis is one of the most fundamental problems in plant physiology. Recently, somatic embryogenesis has attracted attention in plant biotechnology, because it can provide useful systems to produce transgenic clone plants as well as to obtain materials of artificial seeds.

In this paper, physiological, biochemical and molecular biological aspects of somatic embryogenesis will be presented mainly on the basis of works on carrot suspension cultures in my laboratory.

2. Establishment of high frequency and synchronous somatic embryogenesis systems.

The first and epoch-making reports on somatic embryogenesis were published in 1958 by Steward et al.(1) and Reinert (2). In the following twenty years, however, little progress was made in understanding the mechanisms of somatic embryogenesis, because somatic embryogenesis occurred in vitro only at low frequency and asychronously in the system used at that time. In such systems, biochemical and molecular events specific for embryogenesis should be diluted by the activities of cells not engaged in embryogenesis. Furthermore, only average values for biochemical parameters related to various stages of embryogenesis could be determined when asynchronous embryogenesis systems are used. Thus, high frequency and synchronous embryogenesis systems were required for investigation of mechanisms of somatic embryogenesis at molecular level.

We established suitable systems for this purpose using carrot suspension cultures (3). Embryogenic cell clusters were selected by sieving with nylon screen and desity gradient centrifugation in Ficoll solution and then transferred to media lacking auxin and containing zeatin at 10^{-7} M. Synchronous embryogenesis occurred from cell clusters at about 90% frequency in this system. This system is useful for investigating the process of embryogenesis from embryogenic cell clusters, which are designated State 1 cell clusters.

However, since these clusters can differentiate to embryos in auxin-free media without any trigger, embryogenesis can be considered to have been already determined in embryogenic cell clusters, State 1. Thus, the process of formation of State 1 cell clusters from single cells is also important for analyzing the process of embryogenesis, and a system is required in which high frequency embryogenesis occurred from single cells. We attempted to establish such a system (4).

The competent single cells, which were small, round and cytoplasm rich and were designated State 0 cells, were collected by sieving with nylon screen, density gradient centrifugation and manual picking up. When State 0 cells were pretreated with auxin (2,4-D at $5X10^{-8}$ M) for 6 days and then transferred to the auxin-free medium, embryos were formed at high frequency (85-90%). Auxin pretreatment was essential and zeatin (10^{-6} M), mannitol (10^{-3} M) and a high concentration of oxygen (40%) were promotive.

This system provides a useful system to investigate the whole process of somatic embryogenesis from single cells to plants.

Furthermore, when State 0 cells were cultured directly in auxin-free media, cells were elongated and they could not differentiate to embryos even if they were transferred to the medium containing auxin. Therefore the process, in which State 0 cells were cultured in auxin-free media, can be regarded as the process of losing totipotency, while the process, in which State 0 cells were cultured in media containing auxin and differentiated to embryos at high frequency, can be regarded as the process of expression of totipotency. These two processes are useful to investigate what events occur during expression or losing of totipotency.

3. Phases in somatic embryogenesis

Detailed morphological observations revealed that four phases, 0, 1, 2 and 3 Phase, were recognized in the early process of embryogenesis in the system mentioned above(5).

In Phase 0, competent single cells (State 0) form embryogenic cell clusters (State 1) in the presence of auxin. During this phase, the cell clusters formed from single cells gain the ability to develop to embryos when auxin is removed from the medium, giving rise to State 1 cell clusters. The subsequent phase, Phase 1 is induced by transfer of State 1 cell clusters to the auxin-free medium. During Phase 1, cell clusters proliferate slowly and apparently undifferentiatedly. After Phase 1, rapid cell division occurs in certain parts of cell clusters, leading to the formation of globular embryos. This phase is designated Phase 2. In the following phase, Phase 3, plantlets develop from globular embryos via heart shaped and torpedo shaped embryos (Fig.1).

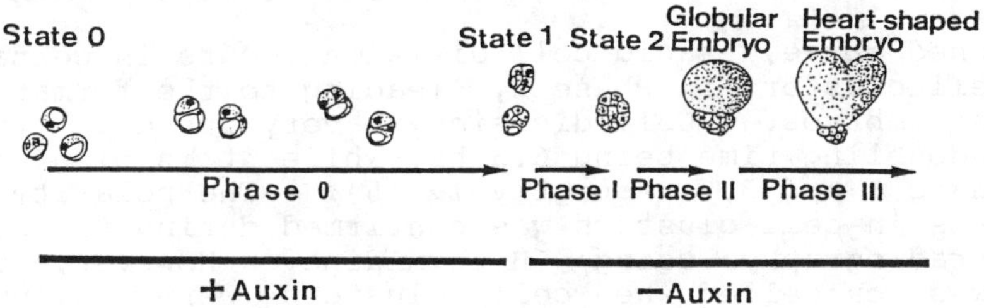

Fig.1. Developmental phases of somatic embryogenesis in a carrot suspension culture.

4. Physiological factors affecting somatic embryogenesis.

Auxin is the most important factor for regulation of induction and development of embryogenesis and it has different effects in different phases of embryogenesis. The presence of 2,4-D or other auxins is requested for the formation of embryogenic cell clusters (State 1) from single cells (State 0). This indicate that auxin is an essential factor for induction of embryogenesis (Phase 0). In other words, auxin is necessary for "competent" cells (State 0) to express totipotency. However, auxin is inhibitory on embryogenesis in Phase 1 and in following phases. Auxin is most inhibitory in Phase 1. Since the original single cells cannot differentiate directly to form embryos in auxin-free media, there are at least two stages in somatic embryogenesis: Stages requiring auxin and inhibited by auxin. Antiauxins, 2,4,6-trichlorophenoxyacetic acid and p-chlorophenoxyisobutyric acid (PCIB) inhibited embryogenesis after Phase 1 (6). These findings suggest that auxin is required for induction of embryogenesis (the process from competent cells to embryogenic cell clusters), but is inhibitory for development of embryogenesis (the process .pa from embryogenic cell clusters to plantlets).

A cytokinin, zeatin, shows promotive effect on embryogenesis in every phase. It is most promotive in Phase 2, in which active cell division occurs. Action of zeatin may be involved in promotion of cell division.

Other phytohormones, gibberellin and abscisic acid inhibit embryogenesis from cell clusters (7).

Besides phytohormones, cell-to-cell interaction is another important factor in somatic embryogenesis. A rather high cell density (10^5 cells/ml) is required for the formation of embryogenic cell clusters from single cells (4), whereas a lower cell density ($2X10^4$ cells/ml) favours the development of embryos from embryogenic cells (3).

5. Expression of polarities in early stages of somatic embryogenesis.

As mentioned above, rapid cell division occurs in certain parts of cell clusters in Phase 2, leading to the formation of globular embryos. Cell division is very rapid in this phase, the doubling time being 6.3 hr, while it is 51 hr and 36 hr in Phase 1 and 3, respectively (5). The polarity of DNA synthesis in cell clusters was confirmed during Phase 1-2 by autoradiography using ^{3}H-thymidine. However, the polarity was cancelled when cell clusters were cultured under the non-embryogenic condition, i.e., in the presence of auxin. The polarized rapid cell division or DNA synthesis is, therefore, considered specific to embryogenesis. It is important to investigate the mechanisms of expression of polarity of active DNA synthesis and rapid cell division to understand embryogenesis.

We analysed the polarity of DNA synthesis during Phase 1-2 by three dimensional reconstruction of serial sections of pulse-chase labelled cells with ^{3}H-thymidine using a computer. The initial material was State 1 cell clusters. It was confirmed that DNA synthesis occurred randomly in the first 3 days of culture in the absence of auxin (Phase 1), but polarized DNA synthesis was clearly observed during 3-4 days of culture (Phase 2), giving rise to globular embryos, in which high activity of DNA synthesis observed in procambial and proepidermal cells, while no DNA synthetic activity was detected in the suspensor like structure.

In State 2 cell clusters, polarized localization of actively DNA synthesizing cells was observed. We attempted to separate actively DNA synthesizing cells (ADS cells) from non-actively DNA synthesizing ones (NDS cells) by density gradient centrifugation in Percoll solution after maceration to protoplasts to investigate the difference at molecular level between ADS and NDS cells. Protein patterns were compared in both cells using SDS-PAGE and ^{35}S-methionine. Three proteins could be detected in ADS cells, but not in NDS cells. These proteins can be candidates of markers of polarity of DNA synthesis specific for embryogenesis.

Polarized localization of poly(A)$^+$RNA was also detected in cell clusters at the end of Phase 0. This polarity also disappeared when cell clusters were transferrted to non-embryogenic conditions. Moreover, localization of free calcium during embryogenesis was observed with a fluorescent calcium indicator.

6. Molecular aspects of somatic embryogenesis.

6.1 Molecular markers of somatic embryogenesis.

Many efforts have been made to find specific molecular markers for somatic embryos (8, 9, 10).

We found several molecular markers in our carrot system mentioned above. As the results of two-dimensional gel electrophoresis, three proteins, a,b and c could be detected throughout the process of expression of totipotency (Phase 0-3), but disappeared during the process of losing totipotency. Two mRNAs also showed the same pattern as protein a,b and c. Additionally, protein d appeared during Phase 0. In Phase 1-2, the pattern of in vitro translated products of mRNA extracted from embryogenic cultures coincided exactly with that of mRNA from a non-embryogenic culture (in the presence of auxin), with exception of four proteins; two appeared and two disappeared during Phase 1-2. These results indicate that only a few proteins may play important roles during embryogenesis and that changes in protein patterns are regulated at mRNA level.

Smith et al.(11) reported a nuclear protein associated with cell division, antigen against 21D7 monoclonal antibody. We applied 21D7 to our system and analyzed by Western blotting and immocytochemical method to examine whether antigen 21D7 (21D7 protein) can be a candidate for molecular marker of totipotency. 21D7 protein could be detected throughout the process of expression of totipotency, while it disappeared within 48 hr during the process of losing totipotency, that is, when State 0 single cells were cultured in the absence of auxin. Furthermore, when State 0 single cells were microinjected with 21D7, they elongated and no longer divided nor differentiated even if they were cultured in the presence of auxin. These results indicate that expression of 21D7 protein may be essential for expression of totipotency.

6.2 Gene expression during somatic embryogenesis.

The most attractive approach to elucidate the mechanisms of somatic embryogenesis is to isolate genes which are expressed specifically during embryogenesis and to reveal their function. Thus, many attempts have been made for this purpose (12,13,14,15). However, functions of the genes involved in embryogenesis have been still unclear.

We constructed λgt11cDNA libraries from poly(A)$^+$RNA of hypocotyls and roots of carrot seedlings. We screened the cDNA libraries differentially to isolate the hypocotyl- or root-specific cDNAs. Finally we isolated 2 cDNAs, CAR 3 and CAR 4, which were specifically expressed in hypocotyls, and other 2 cDNAs, CAR 5 and CAR 6 specifically expressed in roots. Expression of these 4 cDNAs was investigated in Phase 1-3 by Northern hybridization.

Relative expression of CAR 4 and 5 increased after globular embryos and that of CAR 6 increased a little later (after heart-shaped embryos), while CAR 3 was expressed earlier (before globular embryos), that is, in Phase 2. Expression of these cDNAs was at very low level in cells which were cultured in the medium containing 2,4-D, and was strongly supressed when 2,4-D was added to heart shaped embryos. In situ hybridization analyasis revealed that CAR 4 was expressed in epidermis and in the regions surrounding tracheary elements in torpedo shaped embryos.

The nucleotide sequence of CAR 4 cDNA was determined by dideoxy-chain termination method. The deduced CAR 4 protein was rich in proline (N-terminal region) and in leucine (C-terminal region). The characteristic repeated motif was found in the proline-rich region, which resembled repeated sequences found in proline-rich cell wall proteins, such as p33 (carrot) (16) or PRP(soybean) (17).

7. Perspectives

The final goal of studies on somatic embryogenesis is complete understanding of its mechanisms at molecular level. Establishment of high frequency and synchronous embryogenesis system promoted the elucidation of mechanisms of somatic embryogenesis.

However, it is still a far way. Genes involved in embryogenesis have been and will be revealed more and more. Regulation of expression of these genes will be also elucidated by molecular biological techniques. It can be investigated whether or not a gene isolated as a specific one for embryogenesis play a critical role in embryogenesis by introduction of anti-sense RNA or others at specific stages using microinjection or other techniques and by observation of the following fate of treated cells or cell clusters. Mutants concerning with various developmental stages of embryogenesis provide useful tools for genetic analysis of embryogenesis (18). But it will be still difficult to know the biochemical function of a gene if it is a novel one. Furthermore, several or rather many genes can be involved in induction and development of embryogenesis, which may be regulated by the balance of .pa expression of the genes. It will make the analysis of the mechanisms of embryogenesis at molecular level more complicated.

Thus, we have many problems to solve on the mechanisms of somatic embryogenesis. However, the most fundametal and characteristic function of plant cells, totipotency,should be elucidated using high frequency and synchronous somatic embryogenesis in cell cultures.

REFERENCES

1 Steward, F.C., Mapes, M. and Mears, K. (1958) Amer. J. Bot. 45, 705-708

2 Reinert, J. (1958) Ber. dtsch. Bot. Ges. 71, 15

3 Fujimura, T. and Komamine, A. (1979) Plant Physiol. 64, 162-164

4 Nomura, K. and Komamine, A. (1985) Plant Physiol. 79, 988-991

5 Fujimura, T. and Komamine, A. (1980) New Phytol. 86, 213-218

6 Fujimura, T. and Komamine, A. (1979) Z. Pflanzenphysiol. 95, 13-19

7 Fujimura, T. and Komamine, A. (1975) Plant Sci. Lett. 5, 359-364

8 Sung, Z. R. and Okimoto, R. (1981) Proc. Natl. Acad. Sci. USA, 78, 3683-3687

9 Sung, Z. R. and Okimoto, R. (1983) Proc. Natl. Acad. Sci. USA, 89, 2661-2665

10 Chibbar, R.N., Polowick, P.L., Newsted, W.J., Shyluk, J. and Georges, F. (1989) Plant Cell Tissue and Organ Culture 18, 47-53

11 Smith, J. A., Krauss M. R., Borkird, C. and Sung, Z.R. (1988) Planta 174, 462-472

12 Choi, J. H., Liu, L. S., Borkird, C. and Sung, Z. R. (1987) Proc. Natl. Acad. Sci. USA, 84, 1906-1910

13 Borkird, C., Choi, J. H., Jin, Z. H., Franz, G., Hatzopoulos, P., Chorneau, R., Bonas, J., Pelegri, F. and Sung, Z. R. Proc. Natl. Acad. Sci. USA, 85, 6399-6403

14 Franz, G., Hatzopoulos, P., Jones, T. J., Krauss, M. and Sung, Z. R. (1989) Mol. Gen. Genet. 218, 143-151

15 Wilde, D., Nelson, W.S., Booij, H., de viries, S. C. Thomas, T. L. (1988) Planta 176, 205-211

16 Hong, J. C., Nagao, R. T. and Key, J. L. (1987) J. Biol. Chem. 262,, 8367-8376

17 Chen, J. and Varner, J. E. (1985) Proc. Natl. Acad. Sci. USA, 82, 4399-4403

18 Lo Schiavo, F. L., Giuliano, G. and Sung, Z. R. (1988) Plant Sci. 54, 157-164

LIPID SYNTHESIS AND TRANSFER

Jean- Claude KADER, and Pascal LAURENT

Laboratoire de Physiologie Cellulaire et Moléculaire, URA CNRS 1180, Université Pierre et Marie Curie, Tour 53, E4, 4 place Jussieu, 75252 PARIS Cedex 05, France.

1. INTRODUCTION

The high interest of the studies of plant lipids is justified by their various roles in plant growth and development : i/ lipids play important structural roles since they are major constituents of cell membranes ; ii/ lipids can control, in some extent, the activity of membrane-bound enzymes ; iii/ various membrane proteins are esterified at their N-terminal end by fatty acids ; iv/ some lipids (phosphatidylinositol, PI) play a role in the transduction of signals by their phosphorylated derivatives ; v/ storage lipids are used as fuels to supply energy during the first steps of the germination of seeds.

The study of the regulation of lipid biosynthesis is of major interest since this synthesis is both a housekeeping function, by participating to membrane biogenesis and renewal in all plant tissues and a developmentally regulated, organ-specific function (1). For the housekeeping function, the genes for major lipid constituents must be expressed constitutively in all organs. In contrast, the develoment of organs like seeds, and especially in oil-forming ones, will need an expression at a high level of particular genes in precise phases of formation and maturation of seeds.

It is thus essential to progress in our knowledge of the enzymes and genes involved in the various steps of lipid metabolism. However, only few of these genes have been characterized up to now. In this review, we will summarize recent data in this domain with special emphasis to the enzymes and genes which can be of particular interest in the control of lipid synthesis. Additional details can be found in several reviews recently devoted to plant lipid metabolism (2-6).

2. FATTY ACID SYNTHESIS WITHIN PLASTIDS

Plastids play major role in fatty acid synthesis. It is accepted that photosynthesis provides the necessary carbon for fatty acid synthesis *via* pyruvate dehydrogenase in chloroplasts whereas glycolysis is involved in non green plastids (3).

Various enzymes and proteins are involved in fatty acid synthesis (Fig. 1).

2.1. Acetyl-CoA carboxylase

This multifunctional protein, which catalyzes the formation of malonyl-CoA from acetyl-CoA , has been purified from various plants, including rapeseed . Its molecular mass varies from 220 to 240 kDa . It is to be noted that some herbicides inhibit this enzyme.

2.2 Acyl carrier proteins (ACP)

ACP, almost exclusively associated to plastids, is one of the components of the fatty acid synthetase which , in contrast with the multienzymatic complex from animals, is composed of separated proteins. ACP is the most extensively studied protein of plant lipid metabolism (1). The primary structure of ACPs has been determined for several plants including spinach, barley and rapeseed. High homology was found between barley and spinach ACPs whereas a significant homology was observed with E. coli ACP. In addition, isoforms of ACP (two to three) have been separated, differring in their amino terminal end by several residues. These isoforms could be involved in the regulation of the glycerolipid metabolism. The *in vitro* synthesis of ACP was studied and revealed that ACPs were synthesized as precursor proteins with a molecular mass 6 kDa greater than the mature protein. These observations are consistent with the suggestion that ACP is a nuclear-coded protein imported into plastid after removal of the transit peptide.

A synthetic gene coding for spinach ACP has been constructed, cloned and expressed in E. coli (7). Site-directed mutagenesis, implying modifications in the site of interaction with phosphopantetheine, has been recently conducted (8).

All these experiments will give important informations for the regulation of fatty acid synthesis, especially in the phases of formation of oil seeds. These results open the possibility to modify the intensity of fatty acid synthesis by acting on the levels of the expression of ACP genes in transgenic plants.

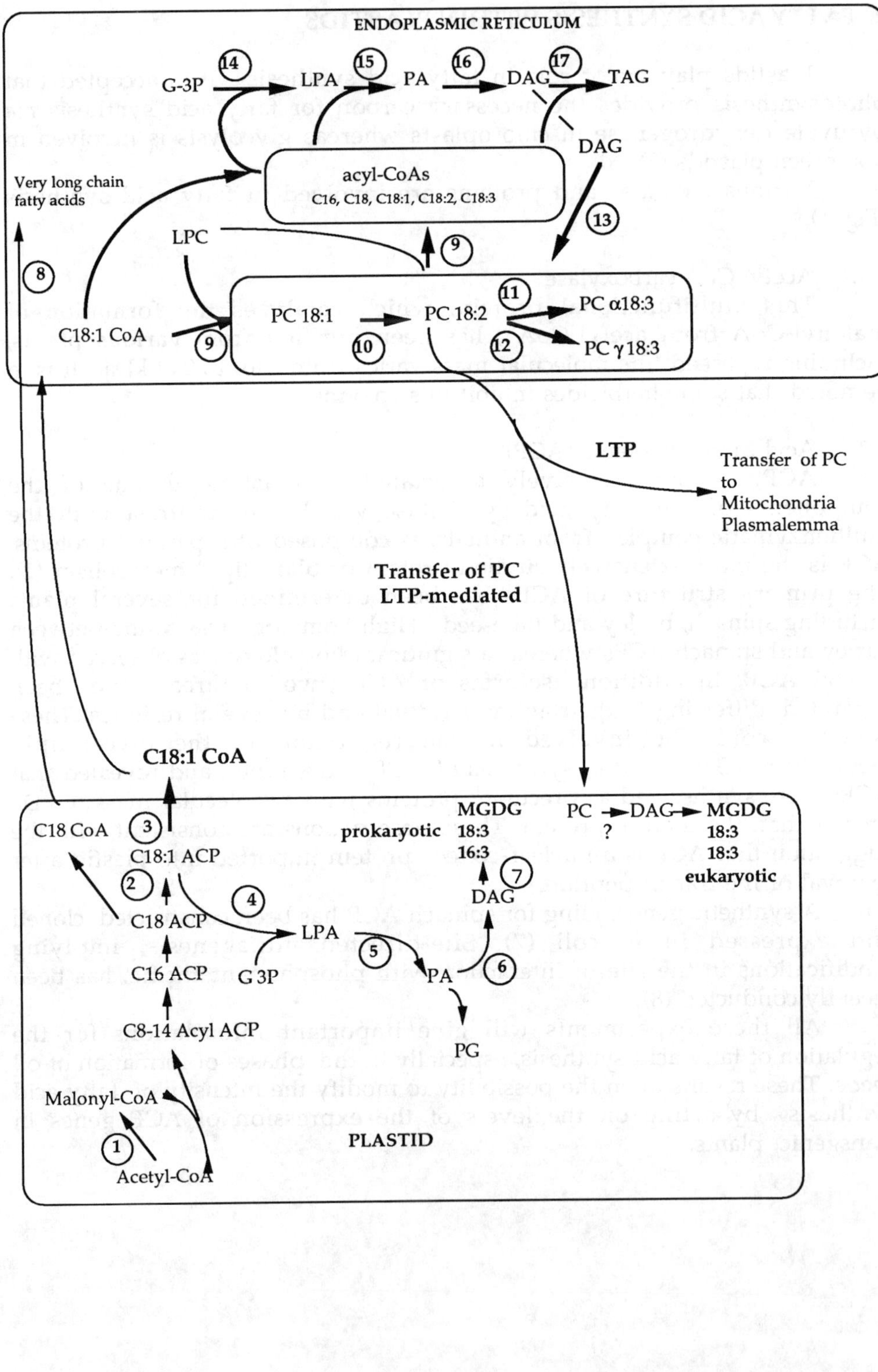
ENDOPLASMIC RETICULUM
G-3P
LPA
PA
DAG
TAG
DAG
14
15
16
17
acyl-CoAs
C16, C18, C18:1, C18:2, C18:3
Very long chain
fatty acids
13
LPC
9
8
11
PC 18:1
PC 18:2
PC α18:3
C18:1 CoA
9
10
12
PC γ18:3
LTP
Transfer of PC
to
Mitochondria
Plasmalemma
Transfer of PC
LTP-mediated
C18:1 CoA
MGDG
18:3
16:3
PC
DAG
MGDG
?
18:3
18:3
C18 CoA
3
prokaryotic
C18:1 ACP
7
eukaryotic
2
4
DAG
C18 ACP
LPA
5
C16 ACP
G 3P
PA
6
C8-14 Acyl ACP
PG
Malonyl-CoA
1
PLASTID
Acetyl-CoA

2.3 Other enzymes of fatty acid synthetase

Fatty acid synthesis is mainly concentrated within platids. The enzymes catalyzing the six individual partial reactions (acyltransferases, reductases, synthetase and dehydrases) have been partially or totally purified mainly from spinach or barley (4, 9). Among these enzymes, acetyl-CoA-ACP transacylase could be a rate-limiting for overall fatty acid synthesis, leading to chain length lower that the 16 or 18 carbons usually obtained. Also the condensing enzymes (keto acyl ACP synthetases) are of particular interest since the isoforms of these enzymes are differently sensitive to inhibitors (cerulenin and arsenite) and have a different molecular specificity towards acyl-ACP (3).

2.4. Oleate synthesis

A soluble plastid desaturase, with a molecular mass of 68 kDa, has been purified from developing safflower seeds. Since this desaturase is specific for stearoyl-ACP, this explains why the major monounsaturated fatty acid which is synthesized is oleate (10).

3. OLEATE METABOLISM WITHIN PLASTIDS

Once oleate formed, two major pathways can be followed.

According to **the first pathway,** the oleate is preferentially transferred to a glycerol-3-phosphate (at position *sn*-1) by a soluble plastidacyltransferase which leads to the formation of lysophosphatidic acid . This acyltransferase has been isolated from two plants , squash and pea, (11, 12) and a cDNA coding for the squash enzyme has been characterized (13). A second acyltransferase (lysophosphatidic acid-acyl ACP-acyltransferase), bound to plastid envelope membranes, is then operative, preferentially with palmitoyl -ACP at position *sn*- 2. The phosphatidate formed is then dephosphorylated to diacylglycerol (DAG) which allows the synthesis of a

FIGURE 1. ***Simplified scheme of lipid synthesis and transfer in plant cells.***

The main enzymes are indicated here : 1 : acetyl-CoA carboxylase ; 2: stearoyl-ACP desaturase ; 3: acyl-ACP thioesterase and acyl-CoA synthetase ; 4 : glycerol-3-phosphate acyl ACP glycerol 3 phosphate acyltransferase ; 5 : lysophosphatidate acyl-ACP acyltransferase ; 6 : phosphatidate phosphatase ; 7 : diacylglycerol galactosyltransferase : 8 : acyl-CoA elongase ; 9 : lyso-PC acyltransferase ; 1O : PC-oleate desaturase ; 11 : linoleate Δ 15 desaturase ; linoleate Δ6-desaturase; 13 : diacy lglycerol choline phosphotransferase ; 14 : gllycerol-3-phosphate acyltransferase ; 15 : 1-acyl glycerol 3 phosphate acyltransferase ; 16 : phosphatidate phosphatase ; 17 : diacylglycerol acyltransferase. ***The abbreviations*** *are : PC : phosphatidylcholine ; PG : phosphatidylglycerol ; PA : posphatidate ; G3P : glycerol-3 phosphate ; LPC: lysophosphatidate ; LPC : lysophosphatidylcholine ; DAG : diacylglycerol ; TAG : triacylglycerol ; MGDG : monogalactosyldiacylglycerol ; LTP :lipid transfer protein ; C16 : palmitate ; C18 : stearate ; C18:1 : oleate ; C18:2 : linoleate ; C18:3 α linolenate,* major plastid lipid, a galactolipid, monoacyl- diacylglycerol (MGDG). A sulfolipid (sulfoquinovosyl- diacylglycerol) is also formed. It is important to note that the major molecular species of MGDG formed contain polyunsaturated fatty acids, like α linolenate, exclusively esterified to the position *sn*-1, whereas position 2 is esterified by a C16 carbon fatty acid, hexadecatrienoate, synthesized by desaturation of palmitate. Phosphatidate can also be used to form an important phospholipid of thylakoids, phosphatidylglycerol. This phospholipid is assumed to be involved in chilling-sensitivity of plants (14) since it can contain saturated C16-carbon fatty acid in both positions. All this pathway, found in plants like spinach, called "**prokaryotic**", by reference to cyanobacteria, is characterized by the fact that the molecular species of lipids, like MGDG, have C18 acids at position 1, and C16 acids at position 2, essentially triunsaturated. This also led to the use of the name of **C16:3** plants.

By contrast, the **second pathway,** called **"eukaryotic"** or **C18:3 type ,** will lead to MGDG species containing α-linolenate in both positions. This pathway needs a remarkable intracellular cooperation between plastids and endoplasmic reticulum. According to this pathway, acyl-ACPs are transformed to acyl-CoAs by an acyl-ACP thioesterase and acyl-CoA synthetase. Acyl-CoAs are then exported to endoplasmic reticulum membranes. The other steps of this eukaryotic pathway will be described in the next chapters.

5. POLYUNSATURATED FATTY ACID SYNTHESIS

Oleate exported from plastids is used to form diunsaturated fatty acid (linoleate) thanks to a membrane-bound desaturase which has not been isolated up to now (10). This desaturase is assumed to be a part of a protein complex comprising a cytochrome b5 , which is a 16-kDa protein purified from potato tuber (15) and a flavoprotein (NADH cytochrome b5) , which is a 43-kDa protein isolated from the same material (16). It is to be noted that a specific polyclonal antibody has been recently raised against this flavoprotein (P. Laurent, unpublished). The oleate-desaturase needs as a substrate a phospholipid like phosphatidylcholine (PC) . An acyltransferase (lyso-PC acyltransferase) is then important to esterify oleate into PC molecules for further desaturation to linoleate (17). However, it is not clear up to now if PC can be also a substrate for α-linolenate synthesis. Of major interest is the use of mutants for the studies of desaturases. Several mutants of Arabidopsis defective in acyl-desaturases have been isolated (18) whereas algae defective in some desaturases have been used to isolate genes involved in the desaturase process (Murata, personal communication). It has also to be remembered that acyl-elongase, active with palmitoyl-CoA, has been recently solubilized and partially purified from leek microsomes(19).

6. PHOSPHOLIPID SYNTHESIS

The endoplasmic reticulum contains not only the desaturases for polyunsaturated acyl chain synthesis but also for the formation of the polar heads of phospholipids. The enzymes responsible for the synthesis of PC or phosphatidylethanolamine (PE) need cytidine diphospho-choline or cytidine diphospho-ethanolamine whereas the enzyme responsible for synthesis of phosphatidylinositol (PI) requires cytidine diphosphate diacylglycerol (20). However, neither of these enzymes have been purified up to now.

7. TRANSFER OF LIPIDS BETWEEN MEMBRANES

The fact that a phospholipid like PC is present in all plant organelles whereas the enzymes necessary for its synthesis are exclusively located in endoplasmic reticulum raised the idea that intracellular movements of phospholipids can occur. Movements of very-long-chain fatty acids have been studied *in vivo* between various membranes (21). Since spontaneous movements of phospholipids which are weakly soluble in aqueous phases are unlikely, the other mechanisms could imply either transfer of vesicles or protein-carrier mechanisms. The discovery of soluble proteins, able to facilitate intermembrane movements of phospholipids *in vitro*, reinforced the latter hypothesis. These proteins, called lipid-transfer proteins (LTP) , have been purified to homogeneity from various plant tissues (spinach leaves, castor bean seedlings, maize sedlings) (22-24). Their primary structure has ben established and revealed that these basic, 9-kDa proteins (91 to 93 residues) have eight cysteines at conserved positions (25). A full-length cDNA has been isolated and characterized (25). From this cDNA was deduced the presence of a signal peptide of 27 residues at the N-terminal end. A similar result was found by the team of Somerville with spinach and Arabidopsis (unpublished) and by Mundy and Rogers (26) for a barley protein which was recently identified as a LTP (27). These LTPs are active with various membranes, including mitochondria (23), plasmalemma (23) or chloroplasts (28). This could explain the origin of phospholipids which are not formed in these organelles. It is especially the case with PC which is located in the outer leaflet of the outer membrane of chloroplasts (29)

8. GALACTOLIPID SYNTHESIS

As emphasized by Joyard and Douce (30), the chloroplast envelope plays a major role in the synthesis of the two major galactolipids , MGDG and DGDG (digalactosyl-diacylglycerol) . A diacylglycerol galactosyl transferase has been solubilized and partially purified from spinach envelopes and was demonstrated to be lipid-dependent (31). Concerning the DGDG synthesis, it is assumed that this galactolipid is formed either by successive galactosylations of DAG molecules or by a galactolipid:galactolipid -galactosyltransferase (32).

One of the problems concerns the synthesis of α linolenate esterifying both positions of "eukaryotic " MGDG (3O, 33). Although the desaturase has not been purified, MGDG is assumed to be the major substrate for desaturation of linoleate to linolenate (30,34). The linoleate could originate from PC which is transferred by LTP from the endoplasmic reticulum (30, 32-36). However, no clear demonstration of a release of DAG from PC in the chloroplast envelope has been done. An acyltransferase activity might be involved in the exchange of the acyl chains from PC to MGDG.

9.TRIACYLGLYCEROL SYNTHESIS

This synthesis is particularly active in oil-forming seeds. The developing seeds possess acyltransferases, able to esterify acyl-CoAs, successively to glycerol-3-phosphate and lyso-phosphatidic acid to produce phosphatidic acid (2,37,38). These acyl-CoAs are either imported from plastids, or desaturated on PC (or at a lower extent PE) molecules. These membrane-bound acyltransferases remain to be purified. After hydrolysis of phosphatidic acid to DAG, two different routes are possible. The first one, driven by a DAG acyl transferase, acting at position sn-3, leads to the ultimate step of the formation of triacylglycerols (TAG). This enzyme has been purified to homogeneity from soybean seedlings (39). Alternatively, DAG can be used to synthesize PC by a CDP-choline phosphotransferase. Of interest, PC molecules can be, as already indicated, used as substrates for desaturation of fatty acids. Acyl-CoA can be released by lyso-PC acyltransferases and then reincorporated to the DAG-TAG chain. This allows a renewal of acyl composition of TAG and determines a balance between mono- and poly-unsaturated chains. It is important to note that the specificity of the acyltransferases for the esterifications in the various positions of the molecules play a major role. Improvements of the oil quality imply an extended knowledge of these acyltransferases.

In addition to oleate, linoleate and α-linolenate-containing TAG, other unusual fatty acids are observed : γ-linolenate in *Borago* or *Oenothera*, obtaining through a Δ6 desaturation, erucic acid (docosenoic acid), obtained by elongation of oleoyl-CoA, ricinoleic acid, whis is an oxygenated fatty acid, and medium-chain fatty acids (10 to 12 carbons) in *Cuphea* (37).

9. FUNCTIONAL ROLES OF LIPIDS IN PLANT CELL PHYSIOLOGY

Lipids play various important roles in plant cells, especially at the membrane level. It is known that the lipid environment controls several membrane-bound proteins ; some examples can be cited : cytochrome oxidase (role of diphosphatidyl glycerol), diacylglycerol galactosyl galactosyltransferase (role of PG)(31), thylakoid organization (role of PG)(40). The interactions between various stress and membrane lipids (41) were also studied as well as the relation of PG with chilling-sensitivity (14).

A growing interest is now devoted to PI and its phosphorylated derivatives. By comparing with the mechanisms described for animal cells, it was assumed that PI derivatives act as secondary messengers for transduction of external signals (hormones , light or stress). Several arguments have been recently obtained in favor of the functionning of such process in plant cells : i/PI is phosphorylated to phosphatidyinositol phosphate (PIP) and phosphatidylinositol bisphosphate (PIP2) by kinases (42,43) ; ii/polyphosphoiniositides are degraded to inositol trisphosphate (IP3) and DAG by phospholipase C(44-46). ; iii/ IP3 is able to release calcium from endoplasmic reticulum or vacuoles (47,48) . However, DAG was found to be ineffective on protein kinase C in contrast with the obsrvations made in animal cells (49) whereas lysolipids were demonstrated to be effective on the phosphorylation of proteins (50) or ATP-dependent proton accumulation(51). Moreover auxin provoked the activation of phospholipase A2 responsible for the release of these lysophospholipids (50). These observations led the group of Scherer (50) to develop a theory about the major role of these lysocompounds in the signal transduction.

10. CONCLUSIONS

Although the roles of lipids are various and important in plant cell physiology , in addition to the interest of storage lipids in oil seeds, knowledge about enzymes and genes involved in lipid transfer and synthesis is still limited. Only few genes or cDNA are available : ACP, LTP, glycerol-3-phosphate acyltransferase. No membrane-bound desaturase has been isolated from higher plants. The study of lipid metabolism is thus a considerable field open to investigation . The development of the methods of molecular biology, leading to the use of transgenic plants, will lead to exciting possibilities of the study of the regulation of lipid metabolism.

REFERENCES

1 Ohlrogge, J. (1987) in The Biochemistry of Plants (Stumpf, P.K., ed.), vol.9, pp. 137-157, Academic Press, New York.
2 Frentzen , M. (1986) J. Plant Physiol., 124, 193-209.

3 Harwood, J.L. (1988) Ann. Rev. Plant Physiol. Plant Mol. Biol. 39, 101-138.
4 Harwood, J.L. , Walsh, M.C. and Walker, K.A. (1990) Methods Plant Biochem. 3, 193-217.
5 Moore, J.C. (1990) Methods Plant Biochem. 3 , 229-239.
6 Huang, A.C. (1990) Methods Plant Biochem. 3, 219-227.
7 Guerra, D.J. and Browse, J. (1989) Arch. Biochem. Biophys., 271, 246-253.
8 Jaworski, J. (1989) Eur. J. Biochem. 184, 603-609.
9 Stumpf, P.K. (1987) in The Biochemistry of Plants (Stumpf, P.K., ed.), vol.9, pp. 121-136, Academic Press, New York.
10 Jaworski , J.C. (1987) in The Biochemistry of Plants (Stumpf, P.K., ed.), vol.9, pp. 159-174 , Academic Press, New York.
11 Nishida, I., Frentzen, M., Ishizaki, O. and Murata, N. (1987) Plant Cell Physiol. 28, 1071-1079.
12 Douady, D. and Dubacq, J.P. (1987) Biochim. Biophys. Acta 921, 615-619.
13 Ishizaki,O., Nishida, I., Agaka,K., Eguchi, O. and Murata, N. (1988) FEBS Lett. 238, 424-430.
14 Murata, N. (1983) Plant Cell Physiol. 24, 81-86.
15 Bonnerot, C., Galle, A.M., Jolliot, A. and Kader, J.C. (1985) Biochem. J. 226, 331-334.
16 Galle A.M. and Kader, J.C. (1986) J. Chromatog. 366, 422-426.
17 Browse J., Kunst,L., Anderson, S., Hugly, S. and Somerville, C. (1989) Plant Physiol. 90, 522-529.
18 Demandre,C., Trémolières, A., Justin, A.M. and Mazliak P. (1986) Biochim. Biophys. Acta 877, 380-386.
19 Bessoule, J.J., Lessire , R. and Cassagne C. (1989) Arch. Biochem. Biophys. 268, 475-484.
20 Justin, A.M., Demandre, C. and Mazliak, P. (1989) Biochim. Biophys. Acta 1005, 51-55.
21 Moreau,P., Bertho,P., Juguelin, H. and Lessire R. (1988) Plant Physiol. Biochem. 26, 173-178.
22 Kader, J.C. (1985) Chem. Phys. Lipids 38, 51-62.
23 Kader, J.C. (1990) Subcell. Biochem. 16, 69-111.
24 Watanabe, S.I. and Yamada, M. (1986) Biochim. Biophys. Acta, 876, 416-423.
25 Tchang F., This P., Stiefel V., Arondel V., Morch M.D., Pagès M., Puigdoménech P., Grellet F., Delseny M., Bouillon P., Huet J.C., Guerbette F., Beauvais-Canté F., Duranton H., Pernollet J.C. and Kader J.C., 1988 - J. Biol. Chem., 263, 16849-16855.
26 Mundy, J. and Rogers, J.C. (1986) Planta 169, 51-63
27 Breu, V., Guerbette, F., Kader, J.C., Kannangara, C., Svensson, B. and von Wettstein-Knowles, P. (1989) Carlsberg Res. Commun. 54, 81-84.

28 Miquel, M., Block, M.A., Joyard, J., Dorne, A.J., Dubacq, J.P., Kader, J.C. and Douce, R. (1987) Biochim. Biophys. Acta 937, 219-228.
29 Dorne, A.J., Joyard, J., Block, M.A. and Douce, R. (1985) J. Cell. Biol. 100, 1690- 1697.
30 Joyard, J. and Douce, R. (1987) in The Biochemistry of Plants (Stumpf, P.K., ed.), vol.9, pp. 215-274 , Academic Press, New York.
31 Covès, J., Joyard, J. and Douce , R. (1988) Proc. Natl. Acad. Sci. USA 85, 4966- 4970.
32 Heemskerk, J.W.M. and Wintermans, J.F.G.M. (1987) Physiol. Plantarum 70, 558-568.
33 Roughan, G. and Slack, R. (1984) Trends Biochem. Sci (1984) 9, 383-386.
34 Andrews,J., Schmidt, H. and Heinz, E. (1989) Arch. Biochem. Biophys. 270, 611-622.
35 Dubacq, J.P., Drapier, D., Trémolières, A. and Mazliak, P. (1984) Plant Cell Physiol. 25, 1197-1200.
36 Wang, X.M. and Hidebrand, D.F. (1988) Plant Physiol. Biochem. 26, 777-792.
37 Stymne, S. and Stobart, A.K. (1987) in The Biochemistry of Plants (Stumpf, P.K., ed.), vol.9, pp. 175-214, Academic Press, New York.
38 Battey,J.F., Schmid, K.M. and Ohlrogge, J.B. (1989) Tibtech., 7, 122-125.
39 Kwanyuen P. and Wilson R.F. (1986) Biochim. Biophys. Acta 877, 23-245.
40 Rémy, R., Trémolières, A., Duval, J.C., Ambard-Breteville F. and Dubacq, J.P. (1982) FEBS Lett. 137, 271-275.
41 Kuiper, P.J.C. (1984) in Structure, Funstions and Metabolism of Plant Lipids (Siegenthaler, P.A. and Eichenberger, W.A, ed.) , pp. 525-530, Elsevier, Amsterdam.
42 Sommarin, M. and Sandelius, A.S. (1988) Biochim. Biophys. Acta 958, 268-278.
43 Drobak, B.K., Ferguson, I.B., Dawson, A.P. and Irvine, R.F. (1988) Plant Physiol. 87, 217-222.
44 Tate, B.F., Schaller, G.E., Sussman, M.R. and Crain , R.C. (1989) Plant. Physiol, 91, 1275-1279
45 Einspahr, K.J., Peeler, T.C. and Thompson, G.A. (1988) Plant Physiol, 90, 1115-1120.
46 Memon, A.R., Rincon, M. and Boss, W.F. (1989) Plant Physiol. 91, 477-480.
47 Ranjeva, R., Carrasco, A. and Boudet, A.M. (1988) FEBS Lett 230, 137-141.
48 Schuhmaker, K.S. and Sze, H. (1987) J. Biol. Chem. 262, 3944-3946.
49 Morré, D.J., Pfaffmann, H., Drobes, B., Wilkinson, F.E. and Hartmann, E. (1989) Plant Physiol. 90, 275-279. DAG
50 Martiny-Baron, G. and Scherer, G.F.E. (1989) J. Biol. Chem. 264, 18052-18059.
51 Palmgren, M.G. and Sommarin, M. (1989) Plant Physiol. 90, 1009-1014.

BIOSYNTHESIS OF TRIACYLGLYCEROLS IN PLANT CELL AND EMBRYO CULTURES
THEIR SIGNIFICANCE IN THE BREEDING OF OIL PLANTS

NIKOLAUS WEBER* and DAVID C. TAYLOR+

*BUNDESANSTALT FÜR FETTFORSCHUNG, INSTITUT FÜR BIOCHEMIE UND TECHNOLOGIE, H. P. KAUFMANN-INSTITUT, D-4400 MÜNSTER, FRG; +PLANT BIOTECHNOLOGY INSTITUTE, NATIONAL RESEARCH COUNCIL, SASKATOON, SASK. S7N OW9, CANADA

1. INTRODUCTION

Vegetable oils containing a mixture of triacylglycerols are of high value in human nutrition. Economically valuable fats and oils for the preparation of oleochemicals, however, require a more uniform composition of acyl moieties, preferentially triacylglycerols with only one type of acyl moiety. Therefore, many efforts are being made at developing modified oil plants that produce triacylglycerols with an altered composition of acyl moieties (1). The benefits of the application of plant cell and embryo cultures in research on the biochemistry and physiology of triacylglycerol formation in such plants are discussed.

2. BIOSYNTHESIS OF TRIACYLGLYCEROLS

Fatty acids such as palmitic, stearic, and oleic acids are formed by fatty acid synthetase in the plastids of plant cells. They are translocated as acyl-CoAs from these organelles to the endoplasmic reticulum (ER) and partly modified, for example, by elongation, desaturation, and

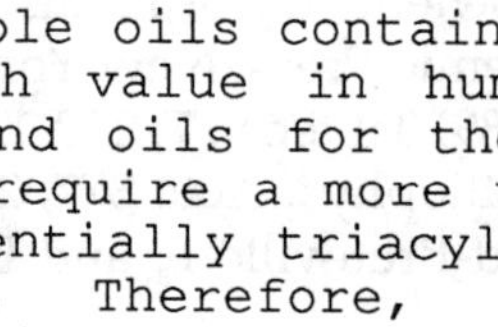

I

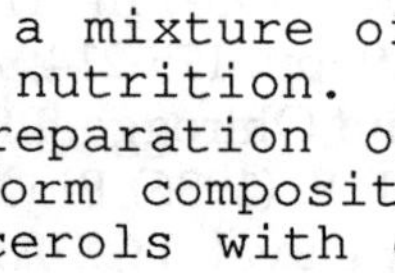

II

Scheme: Chemical structure of triacylglycerols (I), i. e. the typical reserve lipids of oil seeds, and of ionic and non-ionic polar lipids (II), such as glycerophospholipids as well as glycerogalactolipids other reactions. Triacylglycerols are assembled by enzymes of the Kennedy pathway which are also located in the ER; the following enzymatic reactions are involved (2)

1. *sn*-glycero-3-phosphates + acyl-CoAs --(AT 1)--> 1-acyl-2-lyso-*sn*-glycero-3-phosphates

2. 1-acyl-2-lyso-*sn*-glycero-3-phosphates + acyl-CoAs --(AT 2)--> 1,2-diacyl-*sn*-glycero-3-phosphates

3. 1,2-diacyl-*sn*-glycero-3-phosphates --($-P_i$)--> 1,2-diacyl-*sn*-glycerols

4. 1,2-diacyl-*sn*-glycerols --(AT 3)--> triacylglycerols

The acyl pattern of the final mixture of triacylglycerols is characteristic for each seed oil. The specific acyl chain targeting depends on both the selectivity of the various acyltransferases (AT 1-3) of the Kennedy pathway and the compartmentation of triacylglycerol biosynthesis and/or acyl chain modification (3,4).

Large proportions of triacylglycerols occur in the seeds of oil plants, e.g. rapeseed (*Brassica napus*) and soybean (*Glycine max*). In many plants, they are formed in the embryo during maturation and stored in subcellular organelles, i.e. oil bodies or oleosomes (5). Triacylglycerols are cleaved by seed lipases which are activated during germination. The resulting fatty acids serve as sources of energy for the growth of the embryo during germination and provide biochemical intermediates for the early metabolism, for example the biosynthesis of membrane lipids.

3. TRIACYLGLYCEROLS IN PLANT CELL CULTURES

Seed tissues of oil plants such as rape and soya contain generally high proportions of triacylglycerols (I). Plant cells in culture, however, are rapidly growing organisms which synthesize predominantly ionic and non-ionic polar lipids (II) for the assembly of new membranes.

The contents of triacylglycerols in photosynthetic and non-photosynthetic plant cell cultures (Table 1) range from 5-10 % of total lipids (6,7). Only a few examples of cultured plant cells with higher proportions of triacylglycerols are described in the literature. Suspension cultures of carrot (*Daucus carota*) cells, for instance, contain up to 30 % of triacylglycerols in total lipids (8). Cultured cells of anise (*Pimpinella anisum*) store even larger proportions (around 100 mg/g dry weight) of these reserve lipids (9,10). It is of particular interest that callus cultures of cocoa (*Theobroma cacao*) are able to double the content of triacylglycerols by incorporating exogenous stearic acid predominantly into these lipids (11).

Table 1: Proportions of triacylglycerols (TAGs) in total lipids of various plant cell cultures

Plant species	Proportions (%) of TAGs	Reference
Tropaeolum majus	8	(12)
Brassica napus	8	(13)
Glycine max	9;13	(14,15)
Petroselinum crispum	8-13	(16)
Theobroma cacao	17	(17)
Daucus carota	16;30	(8,18)
Pimpinella anisum	*)	(9,10)

*) About 100 mg TAGs/g dry weight, according to Refs 9 and 10. This level is 10-15 times higher than levels of TAGs found in most other plant cell cultures.

The constituent fatty acids of the various lipid classes occurring in cultured plant cells are similar to those found in the original plants. An exception to this rule are various seed tissues that contain triacylglycerols bearing certain unusual acyl moities, such as medium-chain and very long-chain or cyclic acyl moieties (19). It is known from a number of plant cell cultures that some of those unusual fatty acids, which are constituents of seed lipids, are strongly excluded from the lipids of the cultured cells. For example, fatty acids containing hydroxy groups, e.g. ricinoleic acid in seeds of *Ricinus communis* (20), or certain positional isomers of unsaturated fatty acids, e.g. petroselinic acid in seeds of *P. anisum* (9,10) and *D. carota* (21) are not detected in the lipids of cell cultures of these plants, whereas lipids, including triacylglycerols, containing cyclic fatty acids, e.g. cyclopropenyl fatty acids in seeds of *Malva* ssp. (22) and cyclopentenyl fatty acids in seeds of *Hydnocarpus anthelminthica* (23) or very long-chain fatty acids, e.g. erucic acid in *B. napus* seeds (24), occur in small proportions in the lipids of cell cultures of these plants. In this context it is of interest that typical storage compounds of the seeds of *Euonymus europaea*, tri*acetyl*glycerols, are not synthesized by cell cultures of this plant (25).

Radioactively labeled precursors, e.g. acetate (26) and fatty acids of various chain-lengths and degree of unsaturation (13,27,28) as well as glycerol (8,18), were used for studying the biosynthesis and metabolism of glycerolipids, including triacylglycerols, in plant cell cultures.

Figure 1 shows the incorporation of different radioactive fatty acids into the triacylglycerols of rape and soya cells in suspension. It is worth noting that erucic acid was incorporated predominantly into the neutral lipid fraction of rape cells, though the lipids of these cells, that had been derived from a low-erucic rapeseed variety, were obviously free of this very long-chain fatty acid (13). -- In addition to these labeled precursors, 1-*O*-alkyl-*sn*-glycerols and 2-*O*-alkyl-*sn*-glycerols may be used as non-radioactive precursors for studying the metabolism of both neutral and ionic ether glycerolipids in plant cell cultures (29,30).

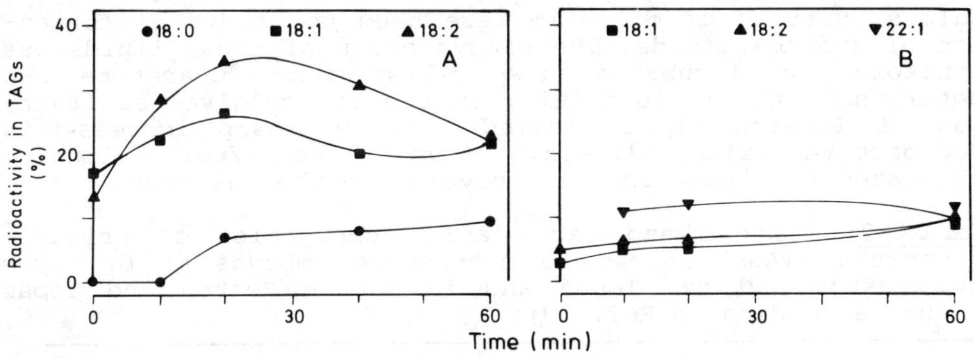

Figure 1: Time course of the incorporation of various ^{14}C-labeled fatty acids into the triacylglycerols of soya cells (A) and rape cells (B) in suspension culture (according to Refs 13 and 27)

It is obvious from the low levels of triacylglycerols and the relatively high proportions of ionic and other polar lipids that plant cells in culture are quite different from cells of oil storage tissues. Their ultrastructure which is rich in plasma membranes is characteristic of undifferentiated cells. Most likely, the induction of genes involved in early differentiation of the cells is necessary in order to express the various enzymatic reactions of triacylglycerol assembly, as is evident in undifferentiated anise cells (9,10). Such biosynthetically active cells may be of particular interest in future studies of storage lipid assembly.

4. TRIACYLGLYCEROLS IN EMBRYO CULTURES

Both somatic and microspore-derived embryos of oil plants are valuable tools in applied lipid biochemistry. It is well known from the literature that cultured embryos of various plants, such as *B. napus*, *G. max*, *D. carota*, *T. cacao*, *Elaeis guineensis*, and *Papaver orientale* accumulate triacylglycerols during maturation (8,31-37). Moreover, cell cultures which have started embroygenesis are found to synthesize intensively unusual fatty acids found in the mother plants. For example, the biosynthesis of erucic and petroselinic acids, is stimulated in developing embryos of *B. campestris* (38) and *P. anisum* (9,10), respectively. The formation of reserve lipids in *B. napus* embryos is paralleled by the biosynthesis of storage proteins, e.g. cruciferin and napin (33,39). As a consequence, the cellular structures of developing embryos derived from maturing seeds are closely similar to those of cultured embryos both containing high amounts of oil bodies and storage proteins.

Callus cultures of oil palm were used to study the formation of storage lipids. The biosynthesis of these lipids was monitored by incubating the cells with ^{14}C-acetate and determining the radioactivity in the triacylglycerol fraction of labeled lipids formed (36). A sharp increase in radioactive triacylglycerols five weeks after culturing indicated the induction of embryoids in the cultures.

Table 2: Content and fatty acid composition of triacylglycerols (TAGs) in mature microspore embryos of *B. napus* cv. Reston and cv. Topas and in mature Reston and Topas seeds (according to Ref. 33)

B. napus cv.	TAGs (mg/g FW)	Fatty acid composition (mol %)*) 16:0	18:1	18:2	18:3	20:1	22:1
Reston							
Cultured embryos	141	4	31	10	9	11	31
Seed embryos	408	4	21	13	7	11	40
Topas							
Cultured embryos	-	5	63	18	7	-	2+)
Seed embryos	-	4	64	19	8	-	1+)

*) Minor proportions of various other fatty acids are not shown. +) (20:1 + 22:1)

Recently, microspore-derived embryos of high-erucic (cv. Reston) and low-erucic (cv. Topas) rapeseed were used for studying the accumulation of reserve lipids and the formation very long-chain fatty acids as well as biochemical aspects of triacylglycerol formation (33). In Table 2, the quantities of storage lipids and their acyl composition in mature embryos are compared with those of mature seeds. It is obvious from these values that content and acyl patterns of triacylglycerols of the two varieties of rapeseed closely resemble those of the corresponding embryos.

The activities of various enzymes involved in the biosynthesis of triacylglycerols were determined in mature microspore-derived Reston embryos (33). It is striking that their specific activities, e.g. of diacylglycerol:erucoyl-CoA acyltransferase, are much higher than those observed in zygotic embryos of rape (40) or crambé (*Crambe abyssinica*) (41) (Figure 2). -- It is evident from these findings that cultured microspore embryos are excellent model systems for studying alteration and biosynthesis of storage lipids in genetically modified oil seeds at an early stage.

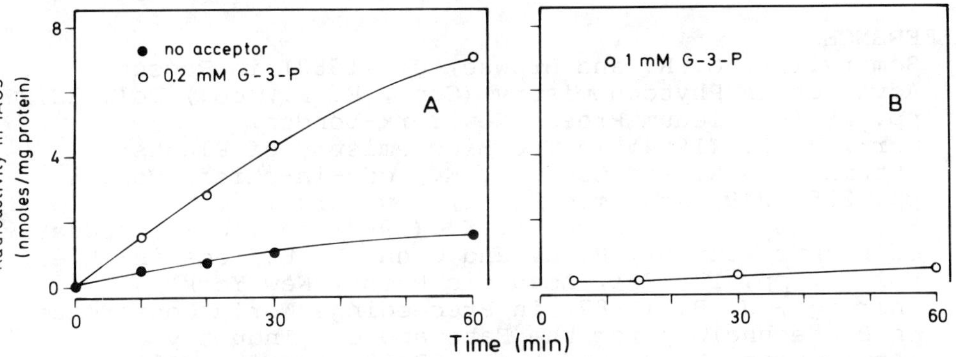

Figure 2: Incorporation of ^{14}C-erucoyl-CoA into triacylglycerols in homogenates of cultured microspore embryos of *B. napus* (A) and in homogenates of developing seeds of *C. abyssinica* (B) with and without glycerol-3-phosphate (G-3-P) supplied as acceptor; according to Refs 33 and 41

5. CONCLUSION

In plant breeding, cell cultures are eminently useful tools for selecting and propagating clones. In addition, studies aimed at developing improved oil crops may be carried out more efficiently by the use of the biosynthetic capacity of cultured plant cells. The pathway leading to the formation of triacylglycerols in plant cell cultures is apparently not different from that prevalent in developing seeds. Biosynthesis of appreciable proportions of triacylglycerols and various unusual fatty acids, however, seems to be correlated with embryogenesis in many plants. Obviously, the genetic control of the biosynthesis of triacylglycerols and unusual fatty acids in plant cell cultures comprises two different developmental signals which are of particular interest for utilization of such embryogenic cells as models in plant breeding.

Recent studies on developing somatic and microspore-derived embryos of *B. napus* and other oil crops show that embryo formation in culture parallels embryogenesis in developing seeds in many respects, e.g. storage lipid assembly and fatty acid elongation. Moreover, specific activities of the various enzymes of the Kennedy pathway are found to be much higher in cultured somatic and microspore embryos than in zygotic embryos. Cultured embryos may, therefore, be exploited as models to facilitate studies on the assembly of storage lipids and the regulation of the corresponding genes in oilseeds. Thus, genetically modified embryos in culture may be useful for the prognosis of alterations in the fatty acid composition of seed oils at a very early stage.

REFERENCES

1 Somerville, C. R. and Browse, J. (1988) in Recent Advances in Phytochemistry (Conn, E. E., ed.) Vol. 22, pp. 19-44, Plenum Press, New York-London
2 Gurr, M. I. (1984) in The Biochemistry of Plants (Stumpf, P. K. and Conn, E. E., eds-in-chief) Vol. 4, pp. 205 -249, Academic Press, New York
3 Stymne, S. and Stobart, A. K. (1987) in The Biochemistry of Plants (Stumpf, P. K. and Conn, E. E., eds-in-chief) Vol. 9, pp. 175-214, Academic Press, New York
4 Ohlrogge, J. B. (1987) in Proceedings World Conference on Biotechnology for the Fats and Oil Industry (Applewhite, T. H., ed.) pp. 87-92, American Oil Chemists' Society, Champaign, IL.
5 Roughan, P. G. and Slack, C. R. (1982) Annu. Rev. Plant. Physiol. 33, 97-132
6 Radwan, S. S. and Mangold, H. K. (1976) Adv. Lipid Res. 14, 171-211
7 Radwan, S. S. and Mangold, H. K. (1980) in Advances in Biochemical Engineering (Fiechter, A., ed.) Vol. 16, pp. 109-133, Springer-Verlag, Berlin-Heidelberg
8 Kleinig, H. and Kopp, C. (1978) Planta 139, 61-65
9 Theimer, R. R. and Schöpf, U. F. M. (1989) Fat Sci. Technol. 91, 434
10 Schöpf, U. F. M. and Theimer, R. R. (1989) Biol. Chem. Hoppe Seyler 370, 798-799
11 Tsai, C. H., Wen, M. C. and Kinsella, J. E. (1982) J. Food Sci. 47, 768-773
12 Radwan, S. S. (1976) Phytochemistry 15, 1727-1729
13 Schneiders G. (1981) Thesis, University of Münster
14 Radwan, S. S. (1975) Fette.Seifen.Anstrichm. 77, 181-184
15 Wilson, A. C., Kates, M. and de la Roche, A. I. (1978) Lipids 13, 504-510
16 Ellenbracht, F., Barz, W. and Mangold, H. K. (1980) Planta 150, 114-119
17 Leathers, R. R. and Scragg, A. H. (1989) Plant Sci. 62, 217-227
18 Gregor, H.-D. (1977) Chem. Phys. Lipids 20, 77-85
19 Mangold, H. K. (1977) in Plant Tissue Culture and its

Bio-technological Application (Barz, W., Reinhard, E. and Zenk, M. H., eds) pp. 55-65, Springer-Verlag, Berlin-Heidelberg
20 Gemmrich, A. R. (1982) Plant Cell Rep. 1, 233-235
21 Dutta, P. C. and Appelqvist, L.-Å. (1989) Plant Sci. 64, 167-177
22 Yano, I., Nichols, B. W., Morris, L. J. and James, A. T. (1972) Lipids 7, 30-34
23 Spener, F., Staba, E. J. and Mangold, H. K. (1974) Chem. Phys. Lipids 12, 344-350
24 Radwan, S. S., Grosse-Oetringhaus, S. and Mangold, H. K. (1978) Chem. Phys. Lipids 22, 177-184
25 Gemmrich, A. R. and Schraudolf, H. (1980) Chem. Phys. Lipids 26, 259-264
26 Kates, M., Wilson, A. C. and de la Roche, A. J. (1979) in Advances in the Biochemistry and Physiology of Plant Lipids (Appelqvist, L.-Å. and Liljenberg, C., eds) pp. 329-342, Elsevier/North-Holland Biomedical Press, Amsterdam
27 Stumpf, P. K. and Weber, N. (1977) Lipids 12, 120-124
28 Weber, N., Richter, I., Mangold, H. K. and Mukherjee, K. D. (1979) Planta 145, 479-485
29 Weber, N. and Benning, H. (1985) Eur. J. Biochem. 146, 323-329
30 Weber, N. and Mangold, H. K. (1988) in Cell Culture and Somatic Cell Genetics of Plants (Constabel, F. and Vasil, I. K., eds) Vol. 5, pp. 509-534
31 Schuchmann, R. and Wellmann, E. (1983) Plant Cell Rep. 2, 88-91
32 Hara, S., Falk, H. and Kleinig, H. (1985) Planta 164, 303-307
33 Taylor, D. C., Weber, N., Underhill, E. W., Pomeroy, M. K., Keller, W. A., Moloney, M. M., Wilen, R. W., Scow croft, W. R. and Holbrook, L. A. (1990) Planta, in press
34 Shoemaker, R. C. and Hammond, E. G. (1988) In Vitro Cell. Develop. Biol. 24, 829-832
35 Wright, D. C., Park, W. D., Leopold, N. R., Hasegawa, P. M. and Janick, J. (1982) J. Am. Oil Chem. Soc. 59, 475-479
36 Turnham, E. and Northcote, D. H. (1984) Phytochemistry 23, 35-39
37 Warren, G. S. and Fowler, M. W. (1979) Planta 144, 451 -454
38 De la Roche, A. J. and Keller, W. A. (1977) Z. Pflanzenzüchtg. 78, 319-326
39 Crouch, M. L. (1982) Planta 156, 520-524
40 Sun, C., Cao, Y.-Z. and Huang, A. H. C. (1988) Plant Physiol. 88,56-60
41 Battey, J. F. and Ohlrogge, J. B. (1989) Plant Physiol. 90, 835-840

LIPID COMPOSITION OF PLASMA MEMBRANE FROM CROWN GALL TUMOUR CELLS IN VITRO

BANIBRATA PANDEY
Tissue Culture Section, Bose Institute, Centernary Campus, Calcutta -700054 INDIA (present address: Department of Biology, University of Saskatchewan, Saskatoon, SK., S7N 0W0, CANADA)

1. INTRODUCTION

Plasma membrane is the fundamental unit of transport and govern all permeability process. The ubiquity of high proportion of acyl lipids in plasma membrane is some indication of the essentiality of these compounds for the physiology of plant cells, and the wide variety of lipid structure involved suggest that thier roles may be many and varied. Analysis of a variety of plasma membranes have demonstrated striking differences in the chemical nature and quantitative distribution of lipid classes which may be related to differences in the specific properties and function of the inerfaces, such as permeability, fluidity etc. This relationship is underlined by the fact that for a given membrane the composition is normally very constant and specific (1). Earlier investigations have exhibited that tumourisation involed alteration in the properties of membrane between normal and crown gall tumour cells and that relatively high level of phytohormones play an important role in perpetuating the persisting neoplastic state of the tumour tissues (2). Crown gall tumour tissue contain higher level of auxin and several cytokinins (2), and some of the genes of the core DNA sites of integrated T-DNA of transformed cells have been identified to be responsible for IAA and zeatin level of the tumour tissues (3-6). The information obtained in recent past have projected the idea that phytohormones do influence lipid profile of normal plant cells (7). Based upon the above findings, present study is intended to provide a meaningful comparison between lipid profile of crown gall tumour and that of normal cell to resolve:

a) Whether or not the alteration of membrane properties in crown gall tumour exhibit any correlation with its lipid profile.

b) in case there is a correlation whether this is due to a direct/primary result of the transformation process

c) or a secondary/indirect effect of the endogenous phytohormone imbalance in tumour cells after T-DNA integration.

2. PROCEDURE

Normal cell line and <u>Agrobacterium</u> <u>tumefaciens</u> (strain C_{58}) induced tumour cell line of <u>Tagetes</u> <u>erecta</u> L. (2n=24)were isolated from single cell clones and maintained in suspension culture, in identical cultural conditions in Murashige and Skoog's (8) basal medium but for normal cell line the medium was supplimented with IAA (1.0 mg/l) and zeatin (0.05 mg/l). The method followed for the induction of tumour, isolation of tumour and normal tissues, development of suspension culture and single cell clones isolation was described elsewhere (9-12). Protoplasts were isolated (9) and purified (13) from suspension culture cells at mid-exponential growth phase. Purified protoplasts were used for the isolation of plasma-membrane enriched fraction (PMEF) (9) followed by its purification using an aqueous two polymer phase system (9, 14) consisting of 5.6% (w/w) each of dextran T-500 and polyethylene glycol 3350; 0.4 M mannitol, 25 mM Tris, 30 mM NaCl, pH 7.3. ATPase activity was measured as described by Hodges <u>et</u> <u>al</u>. (15). The stoping and developing reagent for the ATPase assay have been used as described by Ames (16). The membrane protein was determined by the method of Markwell <u>et</u> <u>al</u>. (17). Cytochrome c oxidase (CCO) and IDPase activity were measured following reported procedure (18, 19). The total lipid extraction, fractionation of phospholipids, free sterol and sterol esters from total lipids was done using established procedure (9, 20, 21). Phospholipid classes were separated on unidimensional TLC using solvesnt system of Skipski <u>et</u> <u>al</u>. (22, 23). The quantitative determination of phospholipids was done by colorimetric method (24). The fatty acid methyl esters were prepared by methanol-sulphuric acid (25). The fatty acid methyl esters were analysed by gas chromatography (7). The trimethyl silyl (TMS) derivatives of the sterols were prepared by the method of Sweeley <u>et</u> <u>al</u>. (26), and gas chromatography of the TMS derivatives of sterols followed by their identification and determination were done as that described by Knights (27).

3. RESULTS AND DISCUSSION

The distribution of plasma membrane-marker enzyme K^+ stimulated, Mg^{2+} dependent ATPase activities among particulate fractions after differential centrifugation is presented in Table-1. ATPase (pH 6.5) in PMEF of tumour cell line exhibited high specific activity compared to that in the normal tissue. The activity of CCO and IDPase were negligible in PMEF of both the cell lines (data not presented). The lipid composition of PMEF of both the cell lines is presented in Table-2. Esterfied sterols and glycolipids were not detected in PMEF indicating that they were free of contamination by plastid membranes. Free sterols were more in normal cell line, and the tumour cells had more of phospholipids. The ratio of total sterol: phospholipids was higher in normal cell line. The data on the composition of individual classes of phospholipids of PMEF are presented in Table-3. In normal cell line higher amount of PC and PI and in tumour cell line, PE was detected. Table-4 records the data on GC analysis of the total phospholipids of PMEF in tumour and normal cell lines. Both the cell lines exhibited close agreement qualitatively but showed certain quantitative differences. Palmitic and linolenic acids were more in normal cell line, but linoleic acid was more in tumour cell line. Tumour cell line exhibited more of total unsaturated fatty acid than normal cell line.

It is nearly three decades that Wood and Braun (28, 29) pointed out altered membrane behaviour of crown gall tumour cells. The possibility that the lipids of plasma membrane of crown gall tumour is responsible for such an altered behaviour has also been assumed by many authors (30, 31). However, attempts has not been made to isolate and analyse the plasma membrane of crown gall tumour cells. The present study describes an in-depth comparative study of plasma membrane lipids of crown gall tumour cells. Tumour cell line shows higher specific activity of ATPase. As plant cell's growth and development of polarity depends on the activity of the ATPase (32), this study thus indicates a possible relation between a change in ATPase activity and alteration in the membrane properties of crown gall tumour cells and calls for further investigation of the specific role of ATPase in developing and perpetuating neoplastic process. The ratio of free sterols: bound sterols, affect the physical state and functional properties of membrane (33, 34). Thus as first approximation it would be rational to assume that the observed alteration of membrane permeability in crown gall tumour cells is linked with the lower amount of free sterols. Phospholipid composition for a given membrane is normally very constant and specific (1). The observed change in phosphopholipid composition in crown gall tumour cells may suggest altered membrane properties in the cells. It seems that this finding can also be useful in extending the study of biogenesis and regulation of phospholipids in crown gall tumours plasma membrane and their specific role in alteration of the membrane fluidity. Van Deenen and De Gier (35) reported that permeability of membrane depends primarily on the degree of unsaturation and length of the hydrocarbon chain of the constituent lipids. Thus it can be stated that the elevated lavel of unsaturation in the fatty acids of crown gall tumour cells plasma membrane may have an important role in its changed membrane proterties. The crown gall tumour cells in culture can be very useful for elucidating the mechanism of fatty acid elongation and desaturation, particularly the formation of polyunsaturated fatty acids in plant cells. Findings of present study indicate that the modification of lipids of plasma membrane in crown gall tumour is associated with altered plasma membrane properties, and most likely a transformation induced event and possibly it is not assocaited with the phytohormone autonomy of the tumour cells.

Table-1: Cellular Distribution of ATPase (pH 6.5) Activities and Proteins of Particulate Fractions in T. erecta tumour and Normal Cell lines (The data given are average of triplicate determination)

Fraction	Protein (mg)	ATPase (pH 6.5) Specific activity*	Total activity**
1,000 g pellet			
Tt	11.57	5.25	60.74
Tn	10.11	4.96	50.14
30,000 g pellet			
Tt	26.43	7.40	195.58
Tn	25.38	6.69	169.79
110,000 g pellet			
Tt	5.50	11.92	65.56
Tn	5.08	8.92	45.31

* μ mol Pi/mg protein/h; ** μ mol Pi/h (equivalent to 10 g of fresh tissue)
Tt = T. erecta tumour cell line; Tn = T. erecta, normal cell line

Table-2: Phospholipids, sterol and Sterylglycoside Content of Plasma-membrane Enriched Fraction (PMEF) in T. erecta Tumour and Normal Cell Lines (The data given are average of triplicate determination. μ mol*/mg protein)

Tissue	Phospholipid	Free sterol	Sterylglycoside**	Total sterol: phospholipids (mol:mol)
Tt	1.33	1.29	0.08	1.03
Tn	1.25	1.37	0.06	1.14

* Average molecular weights are assumed as 750, 387 and 806 respectively for phospholipid, sterol and sterylglycosides.
** Acylated sterylglycosides are not determined.

Table-3: Phospholipid composition of Plasma-membrane Enriched Fraction (PMEF) in T. erecta, Tumour and Normal cell Lines. (The data given are average of triplicate determination.

Phospholipids / Tissue	PC	PE	PI	PS	PG	PA
Tt	44.6	39.6	5.3	1.6	5.4	1.9
Tn	46.9	35.6	7.7	1.4	5.3	1.7

Table-4: GC Analysis of Fatty Acid Composition of Phospholipids of Plasma-membrane Enriched Fraction (PMEF) in T. erecta, Tumour and Normal Cell Lines. (The data given are average of triplicate determination percent w/w)

Fatty acids Tissue	14:0	14:1	16:0	16:1	18:0	18:1	18:2	18:3	20:0	Total Unsat.	Total Sat.
Tt	0.8	tr*	20.6	tr	3.7	7.4	51.2	15.4	0.5	74.0	25.6
Tn	0.9	tr	28.4	tr	5.8	7.1	40.8	16.3	0.5	64.2	35.6

* tr=trace (less than 0.1%)

REFERENCES

1. Hitchcock, C. and Nicholas, B.W. (1971) In plant Lipid Biochemistry pp. 1-387, Academic Press, London.
2. Bruan, A.C. (1978). Biochem. Biophys. Acta 516: 167-191.
3. Akiyoshi, D.E., Morris, R.O., Hinz, R., Mishke, B.S., Kosuge, T., Garfinkeel, D.J., Gordon, M.P., and Nester, E.W. (1983) Proc. Natl. Acad. Sci. U.S.A., 80: 407-411.
4. Schell, J., van Montagu, M. Zamboyski, P., Schorder, J., Herrera-Estrella, L., DeBlock, M., Iuze, D. (1984). In Genetics: New Frontiers (Chopra, V.L. et al, ed.) vol-1, pp. 475-485., Oxford-IBH publishing Co. New Delhi.
5. Kemper, E., Waffenschmidt, S., Weiler, E.W., Rousch, T. and Schorder, J. (1985) Planta 163: 257-262.
6. Beaty, J.S., Powell, G.K., Lica, L., Regier, D.A., MacDonald, E.M.S., Hommes, N.G. and Morris, R.O. (1986) Mol. Gen. Genet. 203: 274-280.
7. Pandey, B. and Gadgil, V.N. (1984) Phytochem. 23: 51-53.
8. Murashige, T. and Skoog, F. (1962) Physiol. Plant 15: 473-497.
9. Pandey, B. (1987) In Lipids of Crown Gall Tumour and Normal Callus Tissue. Ph.D. Thesis, pp. 1-146, Univ. of Calcutta, India.
10. Gadgil, V.N. (1960) In Abnormal plant Growth: Physiological studies on Crown Gall Tumour Tissue of Hollyhock (Althea rosea), Ph.D. Thesis, pp.1-175, Univ. of Calcutta, India.
11. Biswas, M and Gadgil, V.N. (1974) In Proc. 3rd Intern. Cong. of Plant Tissue and Cell Culture (Leicester, U.K.) Abst. Concurrent paper no. 98.
12. Pandey, B. and Gadgil, D.R. (1987) In Recent Advances in Plant Cell and Tissue Culture of Economically Important Plants (Reddy, G.M. ed) pp. 129-133, Osmania University, Hyderabad, India.
13. Slabas, A.R., Powell, A.J. and Lloyd, C.W. (1980) Planta 147: 283-286.
14. Uemura, M. and Yoshida, S. (1983) Plant Physiol. 73: 586-597.
15. Hodges, T.K. and Leonard R.T. (1974) Methods Enzymol. 32: 392-406.

16. Ames, B.N. (1966) Methods Enzymol. 8: 115-118.
17. Markwell, M.A.K., Hass, S.M., Bieber, L.L. and Tolbert, N.E. (1978) Anal. Biochem. 87: 206-210.
18. Widell, S. and Larsson, C. (1983) Physiol. Plant 57: 196-202.
19. Lundborg, T., Widell, S. and Larsson, C (1981) Physiol. Plant 52: 89-95.
20. Lundborg, T., Sandelim, A.S., Widell, S., Larsson, C. and Kylin, A. (1983) In Biochemistry and Metabolism of Plant Lipids (Wintermans, J.F.G.M. and Kuiper, P.J.C. ed) pp. 183-186, Elsevier, Amsterdam.
21. Rochester, C.P., Kjellbom, P. and Larsson, C (1987) Physiol. Plant 71: 257-263.
22. Skipski, V.P., Peterson, R.F. and Barclay, M (1964) Biochem. J. 90: 374-378.
23. Skipski, V.P., Barclay, M., Reichman, E.C. and Good, J.J. (1967) Biochem. Biophys. Acta 137: 80-89.
24. Raheja, R.K., Kaur, C., Singh, A. and Bhatia, I.S. (1973) J. Lipid Res. 14: 695-697.
25. Chalvardjian, A., Morris, L.J. and Holman, R.T. (1964) Biochem. J. 90: 518-521.
26. Sweeley, C.C., Bently, R., Makita, M. and Wells, W.W. (1962) J. Am. Chem. Soc. 39: 381.
27. Knights, B.A. (1966) J. Gas Chromatog. 4: 329-331.
28. Wood, H.N. and Braun, A.C. (1961) Proc. Natl. Acad. Sci. USA 47: 1907-1913.
29. Wood, H.N. and Braun, A.C. (1965) Proc. Natl. Acad. Sci. USA 54: 1532-1538.
30. Butcher, D.N., Philips, R., Powell, R.G. and Sogeke, A. (1974) In Proc. 3rd. Intern. Cong. of Plant Tissue and Cell Culture (Leicester, U.K.) Abst. Concurrent paper no. 170.
31. Cockerham, L.E. and Lundeen, C.V. (1979) Plant Physiol. 64: 543-545.
32. Serrano, R. (1989) Annu. Rev. Plant Physiol. Plant Mol. Biol. 40: 61-94.
33. Grunwald, C. (1971) Plant Physiol 48: 653-655.
34. Grunwald, C. (1980) In Secondary plant products (Bell, E.A. and Charlwood, B.V. ed.) pp. 221-256, Springer-Verleg, Berlin.
35. van Deenen, L.L.M. and DeGier, J. (1964) In The Red Cells (Bishop, C. and Surgenor, D.M. ed) p. 243, Academic Press, London.

INTRACELLULAR Pi-COMPARTMENTATION DURING PHOSPHATE STARVATION-TRIGGERED INDUCTION OF AN EXTRACELLULAR RIBONUCLEASE IN TOMATO CELL CULTURE

K. Glund, T. Nürnberger, S. Abel, W. Jost, J. Preisser* and E. Komor*
Martin-Luther-Universität, Sektion Biowissenschaften, Neuwerk 1, DDR-4020 Halle, and *Universität Bayreuth, Lehrstuhl für Pflanzenphysiologie, Universitätsstrasse 30, D-8580 Bayreuth.

INTRODUCTION

Depletion of phosphate (Pi) in plants causes metabolic effects that modify plant growth and development at the organ, tissue and cellular level (1). With respect to its intracellular distribution, *in vivo* ^{31}P-NMR studies using different plant cells have shown that much of the Pi is stored in the vacuole and the remainder is present in the cytosol (2,3). If plants are grown on low Pi nutrient, the concentration of Pi in the cytoplasm is maintained relatively constant, while the level in the vacuole decreases (4,5,6). Thus, the vacuole operates as a storage pool for surplus Pi.

A current model (7) for plants proposes that exogenous depletion of Pi induces an emergency rescue system aimed to overcome this stress situation. Using cultured tomato (*Lycopersicon esculentum*) cells, the phosphate starvation-induced synthesis and secretion of a number of proteins into the culture medium was shown (8,9,10). Among them, an acid phosphatase (8) and a ribonuclease (RNase)(10) were identified.

Using ^{31}P-NMR, we present data on cytoplasmic and vacuolar compartmentation of Pi during induction of an extracellular ribonuclease.

MATERIALS and METHODS

Plant Material

Cell suspension cultures of tomato (*Lycopersicon esculentum* Mill cv. Lukullus) were propagated as described (11). In experiments carried out under conditions of phosphate starvation ($-P_i$), KH_2PO_4 was omitted from the modified MURASHIGE-SKOOG medium (11). For experiments carried out under conditions of nearly constant P_i supply ($+P_i$), the extracellular P_i concentration was determined every 12 hours and adjusted to the starting concentration (3.5 mM) by adding 0.2 M KH_2PO_4-K_2HPO_4, pH 6.0. Growth under $+P_i$ or $-P_i$ conditions was initiated by transferring 2 x 10^7 cells grown for 3 days under $+P_i$ conditions into 100 ml of $+P_i$ or $-P_i$ medium after twofold washing the cells with the respective medium.

Assays

RNase activity was assayed as described (12). Extracellular P_i determination was carried out as in (10). Cellular phosphate was obtained by washing cells twice with phosphate-free medium, sonifying (3 x 30 sec) in 10 % trichloroacetic acid, centrifuging for 5 min at 5,000 x g and using the supernatant for determination.

Electrophoresis, Preparation of Antibodies, Blotting, Labeling

Disc gel electrophoresis was carried out as described (10). Polyclonal anti-RNase antibodies were raised by intramuscular inoculation of rabbits with 100 ug purified tomato extracellular RNase and antibodies were purified according to (13). Protein blotting was done as in (14). For protein labeling cells were incubated with 100 uCi of a L-[U-^{14}C]-amino acid mix for 3 h. Preparation of extracellular proteins and SDS-PAGE (10) followed by fluorography (14) were performed as described.

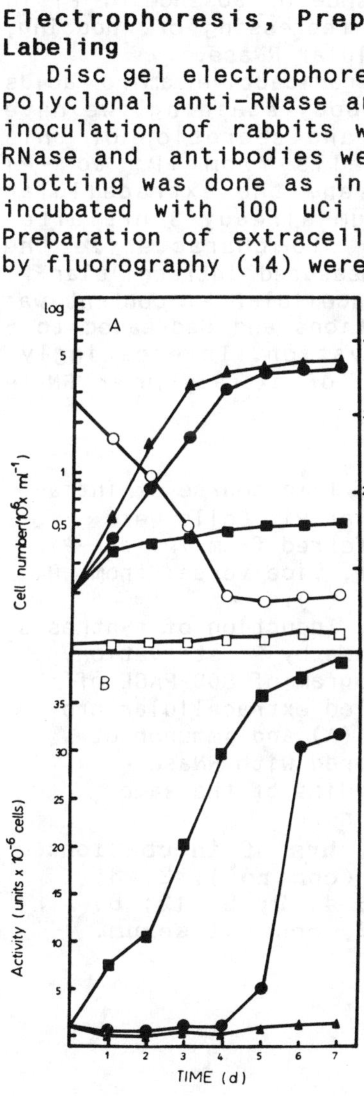

Fig.1: Extracellular RNA-degrading activity during growth of cultured tomato cells. Number of cells (A, closed symbols), extracellular Pi concentration (A, open symbols) and extracellular RNA-degrading enzyme activity (B) during culture growth under +Pi (●,○), -Pi (■,□), and conditions of constant Pi supply (▲).

^{31}P-NMR Measurements

A AC 300 Bruker NMR spectrometer operating in the pulsed-Fourier transform mode at 121.5 MHz was used. In vivo spectra were recorded after 400 accumulations with a repitition time of 5 s. In order to get a better signal to noise ratio, ^{31}P-NMR spectra were recorded with compressed cells within the detection coil in a medium devoid of Fe^{3+} and Mn^{2+} according to (15).

RESULTS AND DISCUSSION

Time Kinetics of Induction and Intracellular Pi Concentration

From Fig.1 it becomes evident that prsence or absence of Pi in the culture medium is a prerequisite for repressing or inducing, respectively, the activity of the extracellular RNase.
By labeling extracellular proteins with radioactive amino acids followed by fluorography as well as antibody staining, we have studied the time kinetics of synthesis and secretion of this protein after transferring tomato cells from +Pi to -Pi conditions (Fig.2, Inset). As can be seen, the extracellular RNase is detectable in the culture medium already 3 hrs after transferring cells from +Pi to -Pi. To characterize the process of induction in more detail, we measured intracellular Pi content. As can be seen (Fig.2), the intracellular Pi content was about 33 nmoles/10^6 cells under +Pi conditions and decreased to 6 nmoles/10^6 cells after 3 days of Pi starvation. Interestingly, during the time of induction of synthesis of extracellular RNase

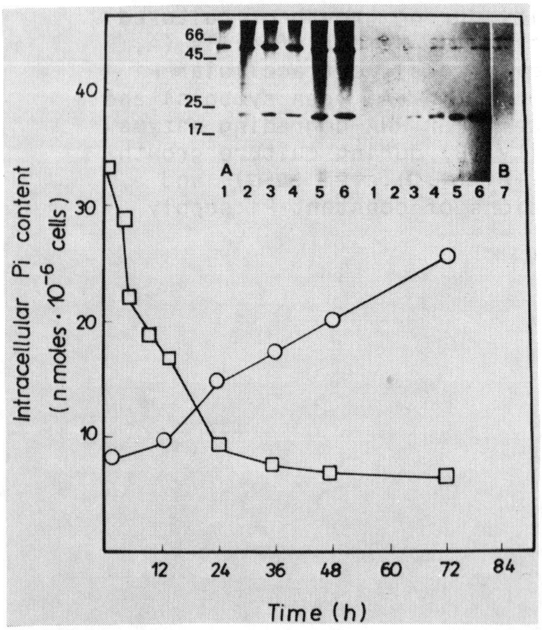

Fig.2: Time course of intracellular Pi. Cells were transferred from +Pi to -Pi ( ) or, vice versa, from -Pi to +Pi ( ).
Inset: Induction of synthesis of RNase by Pi starvation. Fluorogram of SDS-PAGE of labelled extracellular proteins (A) and immunoblot developed with RNase antibodies of the same proteins.
Lane, hrs of incubation: 1, 0 (control); 2, 3; 3, 6; 4, 9; 5, 12; 6, 48; 7, control serum.

(3 hrs, see above), the cellular Pi content decreased only to 28 nmoles/10^6 cells, i.e. to 85 % of the starting amount. These data reveal that the induction process under -Pi conditions occurs though the intracellular content of Pi is very high.

Fate of Intracellular Pi Compartmentation in Tomato Cells

Fig.3 shows ^{31}P-NMR spectra of Pi-containing medium and of intact tomato cells which were either supplied or not supplied with Pi. Two signals were identified at -14.69 ppm and -15.89 ppm for cells with Pi (peaks 2 and 3 in B) corresponding to pH 5.5 and pH 7.4, respectively, (calibration not shown). These values reflect presence of a vacuolar Pi pool (peak 3) at the acidic pH and a cytoplasmic Pi pool (Pi pool) at the slightly alkaline pH (6). Medium alone gave exactly the same chemical shift as the vacuolar Pi (cf. peak 3 in A and B). We have measured the NMR-detected extracellular volume under these conditions by measuring the distribution of ^3H_2O and ^{14}C-dextrane and found that it represents 40.6 % of the whole volume in the NMR-tube (data not shown). Therefore, 40.6 % of the Pi content, determined by NMR for +Pi-medium had to be substracted from the vacuolar peak obtained for +Pi cells. By doing this, a vacuolar to cytoplasmic Pi distribution of 30.8 % to 69.2 %, respectively, becomes apparent for +Pi cells. Interestingly, after pumping medium without Pi through the NMR-tube for about 3 hrs, the Pi signal at -15.89 ppm nearly completely disappeared, whereas the cytoplasmic one remained constant during this process.

Summarizing, our results demonstrate: (i) Tomato cells which were incubated in -Pi medium induce rapidly (within 3 hrs)

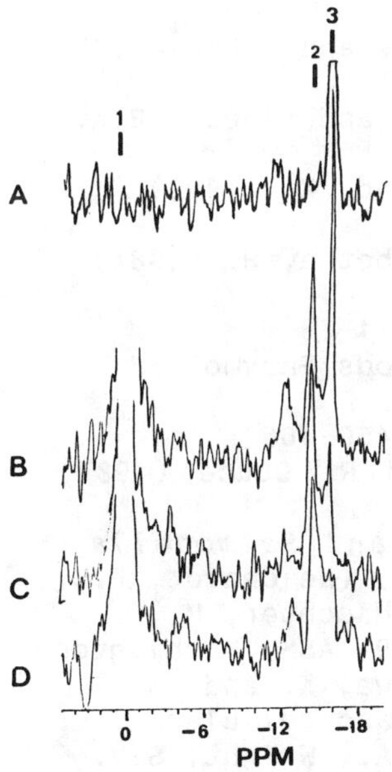

Fig.3: ^{31}P-NMR spectra of +Pi medium and tomato cells grown in +Pi (B) and, subsequently, washed with -Pi medium for 1 (C) and 3 (D) hrs. Spectra are referenced to methylene diphosphonic acid (peak 1) as internal standard. Peak assignments were made in reference to published data as cytoplasmic Pi pool (peak 2) and medium or vacuolar Pi pool (peak 3).

synthesis and secretion of a ribonuclease. (ii) Induction occurs under high intracellular Pi content. (iii) The vacuolar Pi pool disappeared after transferring cells from +Pi to -Pi conditions whereas the cytoplasmic one remained unchanged. Assuming transcriptional regulation of the induction process [which is true for many gene products of the pho-regulon in bacteria (16) and yeast (17) but has to be shown for higher plants], our data argue against changes in the cytoplasmic Pi pool as part of the intracellular signal transduction pathway as suggested (1). We rather propose a mechanism of signal transduction through the plasma membrane.

REFERENCES

1 Bieleski, R.L. and Ferguson, J.B. (1983) in Inorganic Plant Nutrition (Läuchli, A. and Bieleski, R.L., eds) Encyclopedia of Plant Physiology, New Series , Vol. 15A, pp. 422-449, Springer, Berlin
2 Roberts, J.K.M., Ray,P.M., Warde-Jardetzky, N. and Jardetzky, O. (1980) Nature 283, 870-872
3 Waterton, J.C., Bridges, J.A. and Irving, M.P. (1983) Biochim. Biophys. Acta 763, 315-320
4 Lauer, M.J., Blevins,D.G. and Sierzputowska-Gracz, H. (1989) Plant Physiol. 89, 1331-1336
5 Roby R.J. and Ratcliffe, R.G. (1984) Plant Cell Rep. 3, 234-236
6 Rebeille, F., Bligny,R., Martin, J.D. and Douce, R. (1983) Arch. Biochem. Biophys. 225, 143-148
7 Goldstein, A.H., Baertlein, D.A. and Danon, A. (1989) Plant Mol. Biol. Rep. 7, 7-16
8 Goldstein, A.H., Danon, A., Baertlein, D.A. and McDaniel, R.G. (1988) Plant Physiol. 87, 716-720
9 Goldstein, A.H., Mayfield, S.P., Danon, A. and Tibbot, B.K. (1989) Plant Physiol. 91, 175-182
10 Nürnberger, T., Abel, S., Jost, W. and Glund, K. (1990) Plant Physiol, in press
11 Tewes, A., Glund, K., Walther, R. and Reinbothe, H. (1984) Z. Pflanzenphysiol. 113, 141-150
12 Abel, S. and Glund, K. (1987) Planta 172, 71-78
13 Hurn, B.A.L.and Chantler, S.M. (1980) Methods Enzymol. 70, 104-142
14 Bonner, W.M. (1984) Methods Enzymol. 104, 460-465
15 Roby, C., Martin, J.-P., Blighny, R. and R. Douce (1987) J. Biol. Chem. 262, 5000-5007
16 Wanner,B.L. (1987) in *Escherichia coli* and *Salmonella typhimurium*: Cellular and molecular biology (Neidhardt, F.C., Ingram, J.L., Low, K.B., Magasanik, B., Schaechter, M., Umberger, H.E., eds.) Vol II, pp. 1326-1339, ASM, Washington
17 Yoshida, K., Kuromitsu, Z., Ogawa, N., Ogawa, K. and Oshima, Y. (1987) in Phosphate metabolism and cellular regulation (Torriani-Gorini, A., Rothman, A., Wright, S.V. and Yagil, E., eds.), pp. 49-55, ASM, Washington

AEROBIC HEXOSE METABOLISM IN EMBRYOGENIC CELL LINES OF *DAUCUS CAROTA* STUDIED BY ^{13}C NMR USING ^{13}C-ENRICHED SUBSTRATES.

C.DIJKEMA[1], H.BOOIJ[2], P.A. DE JAGER[1], S.C. DE VRIES[2], T.J. SCHAAFSMA[1] & A. VAN KAMMEN[2].
Departments of Molecular Physics[1] and Molecular Biology[2], Agricultural University Wageningen, Dreyenlaan 3, NL 6703 HA Wageningen. The Netherlands.

1. INTRODUCTION

High resolution (h.r.) ^{13}C NMR is increasingly applied to studies of carbohydrate metabolism in plant cells and tissues, as an analytical technique allowing non-invasive observation of endogenous intermediates and end products in intact biomass (1). One of the major advantages of h.r. ^{13}C NMR apart from being non-destructive is its high spectral resolution (2), as it shares a large spectral width (about 250 ppm of the applied magnetic field) with relatively narrow linewidths of the individual resonances(10-15 Hz in spectra of cell suspensions). High resolution ^{13}C NMR therefore allows the separate observation of the resonances of all chemically distinct carbon atoms present within mobile, low-molecular weight compounds, as cellular intermediates and endproducts generally are. We have recently exploited some of the advantages offered by h.r. ^{13}C NMR in a study of the long-term utilization of sucrose by embryogenic cell suspensions of *Daucus carota* (cv Flakkese) growing in the presence of the phytohormone 2,4-D and by somatic embryos derived from these cultures in the absence of 2,4-D (3).

The high intracellular sucrose, glucose and fructose concentrations allowed us to monitor the utilization of medium sucrose by natural abundance (n.a.) h.r. ^{13}C NMR, observing only the 1.1% n.a.^{13}C carbon nuclei. Fig.1A and B show separately the changes in sugar composition of both media and cells during the 14 days subculture period of an embryogenic cell suspension of *D. carota* Flakkese in the presence of 2,4-D. The media spectra (Fig.1A) reflect most clearly the high degree of resolution that can be obtained for low-molecular weight, highly mobile compounds in low-viscous solutions, as all 36 resonances belonging to sucrose(12), glucose(12) and fructose(12) are clearly distinguished. Also the cell spectra, being less resolved compared to the media, still show sufficiently resolved resonances of all the endogenous sugars in their intracelluar environment, whether this environment is the cytoplasm or the vacuole, as shown in Fig.1B. Fig.1 shows that the plant cells first hydrolyze the medium sucrose within one day into the constitutive hexoses. Thereafter glucose is taken up preferentially over fructose within seven days when 2,4D is in the medium, followed by fructose uptake between d 7 and d 9 of the 14 days subculture period. Both hexoses are stored intracellularly predominantly as glucose and sucrose, which are subsequently utilized between d9 and d14 when the medium hexoses are exhausted.

This paper presents a ^{13}C NMR study of aerobic hexose metabolism carried out at increased sensitivity by using 99% ^{13}C-enriched glucose and fructose as substrates, specifically labelled at the C1 position. Hexose metabolism and its short-term kinetics was studied in batch culture as involved in biweekly subculturing, and compared with metabolism as taking place in a 20 mm o.d. NMR tube for the carrot cell line Flakkese used in (3). This cell line has relatively low numbers of proembryogenic masses (about

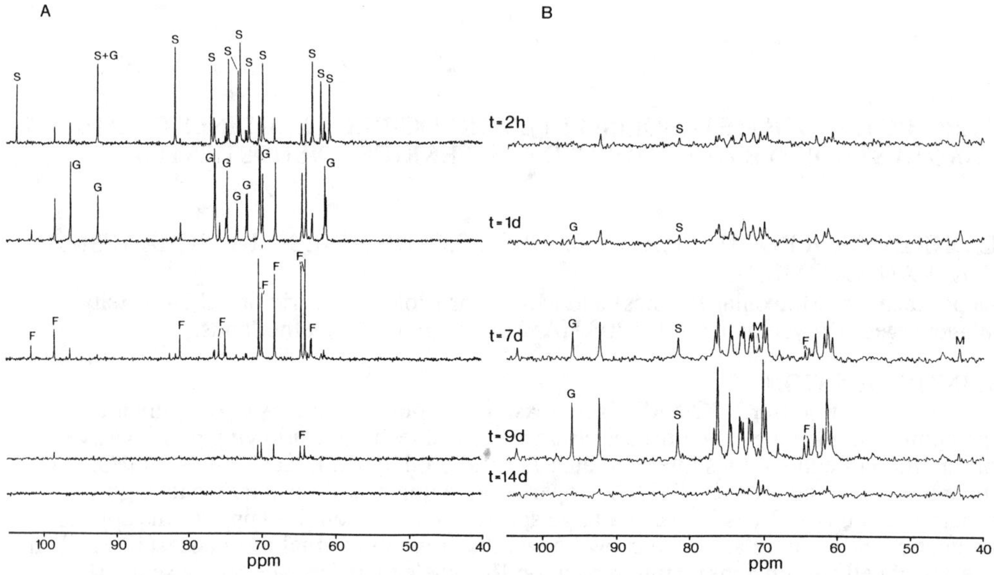

FIGURE 1. Natural abundance ^{13}C NMR spectra obtained from 3 mL of medium (A) and 2 mL packed cells (B) during the 14 d culture cycle of an embryogenic suspension culture. A, For reasons of clarity, only the 12 sucrose resonanances are indicated by S for t = 2 h, for t = 1 d only the 12 glucose resonances are indicated by G; for t = 7 d only the 12 fructose resonances are indicated by F. B, Sucrose (S) and glucose (G) are indicated each by only one single characteristic resonance; two fructose resonances are indicated by F. M indicates the malate resonances.

20%) present in suspension culture. In addition, hexose metabolism of this cell line was compared with that occurring in the A$^+$ cell line of *D. carota*, differing from Flakkese by the high amounts of proembryogenic masses present (> 90%). It is shown that ^{13}C NMR not only reflects which intracellular compounds are being labelled upon incubation, but also which carbons within these compounds are labelled, thus being of help in elucidating the metabolic routes involved in hexose conversion.

2 MATERIALS AND METHODS

Hexose metabolism in batch culture was studied by incubating whole (50 mL) cultures, after washing with fresh medium without carbon source, with 5 mM ^{13}C-enriched substrate. Samples corresponding to 2 mL packed cells were withdrawn from these batch cultures at the times indicated, cells were separated from the media by centrifugatiation and subsequently washed three times with ice-cold medium without carbon source to remove labelled substrate. Cells and media were thereafter measured as described in (3). Hexose metabolism in the NMR tube (20 mm o.d.) was studied by resuspending washed cells into fresh medium at a final density of 40% v/v packed cell volume in order to obtain a reliable spectral S/N ratio. To 12 mL of these suspensions ^{13}C-enriched substrate (5 mM final conc.).was added. Aeration in the NMR tube (C. Dijkema, to be published) was effected by leading a dense stream of small gas bubbles (95% O2/5% CO2, 1L/min.) via a sintered glass stone down from the bottom through the cell suspension, thus establishing a constant O2 tension in the suspension of about 80%

of its maximal value in solution at room temperature. A 20 mm $^{13}C/^{31}P$ dual-probehead was used for the study of hexose metabolism in situ in a 20 mm o.d. NMR tube. The kinetics of substrate uptake and conversion in the 20 mm NMR tube was monitored by sequential recording and storing on disk of appropriate time blocks of FIDs (20,000 Hz spectral width), yielding a reliable spectral S/N ratio. They were accumulated using 8k data points, applying a 60° pulse angle (12 µsec pulse duration) and a pulse delay of 0.5 sec between the pulses. Proton decoupling was applied throughout the experiments, using a composite WALTZ pulse scheme. Intensities of labelled compounds were determined by integration of the spectral resonances.

3. RESULTS AND DISCUSSION

Fig.2 shows the time course of ^{13}C label incorporation into the cells of a batch culture of the Flakkese cell line of day 3, by taking samples at the times indicated after incubation with $^{13}C1$-glucose (5mM). The spectrum recorded at t=0h is the n.a. ^{13}C NMR spectrum of intracellular sugars that were already taken up by or formed in the cells during the three days before incubation with $^{13}C1$-enriched glucose (see Fig.1). Maximum label incorporation was observed between 5 and 6 h after the onset of incubation. This was most clearly reflected by the spectral amplitudes observed for the C1 carbon resonances of both hexose moieties constituting the sucrose molecule: the C1 of the α D-gluco-pyranose anomer (93.0 ppm) and the C1 of the β D-fructo-furanose anomer (62.3 ppm). In addition, also the C6 carbons of both hexose moieties in sucrose, resonating at 61.0 and 63.4 ppm, were equally labelled although to less extend than both C1 carbons. In addition, accumulation of intracellular C1-labelled glucose was observed as the C1 resonances of its anomers at 96.8 (b D-gluco-pyranose C1) and 93.0 ppm (α D-gluco-pyranose C1) increased upon incubation. The latter resonance at 93.0 ppm coincides with that of the C1 of the α D-gluco-pyranose unit in sucrose. When the contribution of this sucrose resonance was subtracted from the total area of the 93.0 ppm resonance, assuming equal intensities for both C1 sucrose resonances just as observed for both C6 resonances, the relative amplitudes of both α and β glucose C1 resonances reflected anomeric equilibrium to exist between both glucose anomers. Finally, the C1 carbon of β D-fructo-pyranose (64.8 ppm) was labelled to only minor extend. After the time of maximum labelling, the resonance amplitudes of all labelled compounds decreased gradually on a much slower timescale than observed during labelling.

Since the batch cultures contained a certain amount of biomass, only a limited number of samples could be withdrawn, thus seriously restricting the time resolution. To enhance time resolution, a comparable experiment was carried out in a 20 mm NMR tube using a somewhat higher cell density (40% v/v) in order to guarantee a sufficient S/N ratio at a desired time resolution. After the ^{13}C-enriched substrate was added, aeration was started immediately and appropriate time blocks of FIDs were sequentially recorded and stored on disk. Spectra continuously collected as a function of time now showed the total of labelled medium and intracellular glucose, contrary to the spectra of sole cells as in Fig.2, in addition to the other labelled compounds. The spectra thus recorded in situ in the NMR tube for the time points indicated in Fig.2, were simular to those shown in Fig.2 for a batch culture after being corrected for the medium glucose. It was thus concluded that metabolism taking place under both conditions was comparable. Since throughout all the experiments anomeric equilibrium was observed for both medium and intracellular glucose anomers, the total of glucose could be determined from the isolated β glucose C1 resonance at 96.8 ppm Alternatively, this could be done by subtracting the contribution of the β D-fructo-pyranose C1 resonance of sucrose at 62.3 ppm from the total area of the resonance at 93.0 ppm in the other way. Both determinations lead to equal values for the

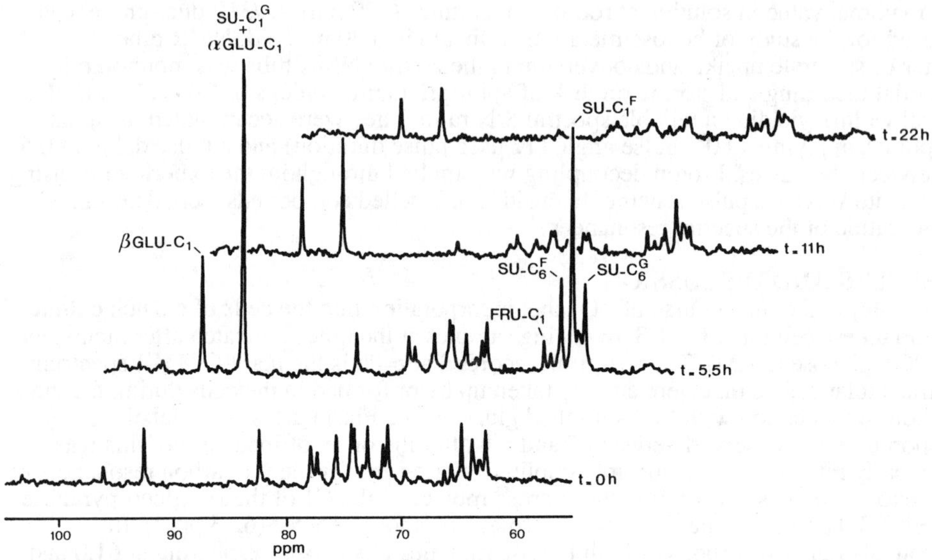

FIGURE 2. ^{13}C NMR spectra of cells of a batch culture of the Flakkese cell line of d 3, recorded at various times after incubation with 5 mM ^{13}C1-enriched glucose.

amount of total glucose present. Fig.3A shows the time course of all observed labelled compounds, by plotting their intensities (areas) as function of time.

Total labelled glucose disappeared at two different rates, rapid at first (about 1 mM/h) representing cellular uptake and partial conversion into labelled sucrose, followed by slower breakdown of intracellular glucose (0.1 mM/h). During the rapid phase of glucose disappearance, sucrose became increasingly labelled at the C1 and C6 carbons in sucrose, labelling reaching a maximum after about 5-6 hrs, just as observed in the case of batch cultures in Fig.2, C1 sucrose being about 2-3-fold more labelled as C6 sucrose. When maximum sucrose labelling was observed, medium glucose became exhausted and the endogeneous glucose remained solely visible in the spectra together with the intracellular sucrose, reaching low levels only after about 15hrs. The level of endogenous labelled fructose remained very low compared to the sucrose and glucose levels, just as can be inferred from Fig.2.

The labelling of the C6 carbons in both hexose moieties constituting the sucrose molecule, reflects scrambling of ^{13}C label in going from the C1 position in glucose/fructose towards the C6 position. This necessarily implies scrambling of ^{13}C label at C3 level in glycolysis from the labelled C1 carbon in dihydroxy-acetone-3P (DHAP), originating from C1 labelled fructose-1,6P2 (Fru-1,6P2), towards the C1 carbon in glyceraldehyde-3P (GAP) involving triose-phosphate-isomerase (TPI) activity. Subsequently, glucose /fructose become labelled at C6 when they become subsequently resynthesized via aldolase and fructose-1,6-diphosphate-phosphatase (Fru-1,6P2ase) activity. The fact that equal amplitudes were observed for both C1 and C6 resonances in the labelled sucrose molecules implies fast isomerization taking place between the Fru-6P and Glu-6P pools.

When cells of the A$^+$ cell line were incubated with ^{13}C1-enriched glucose, kinetics of glucose breakdown was comparable to that observed for the Flakkese cell line, as shown

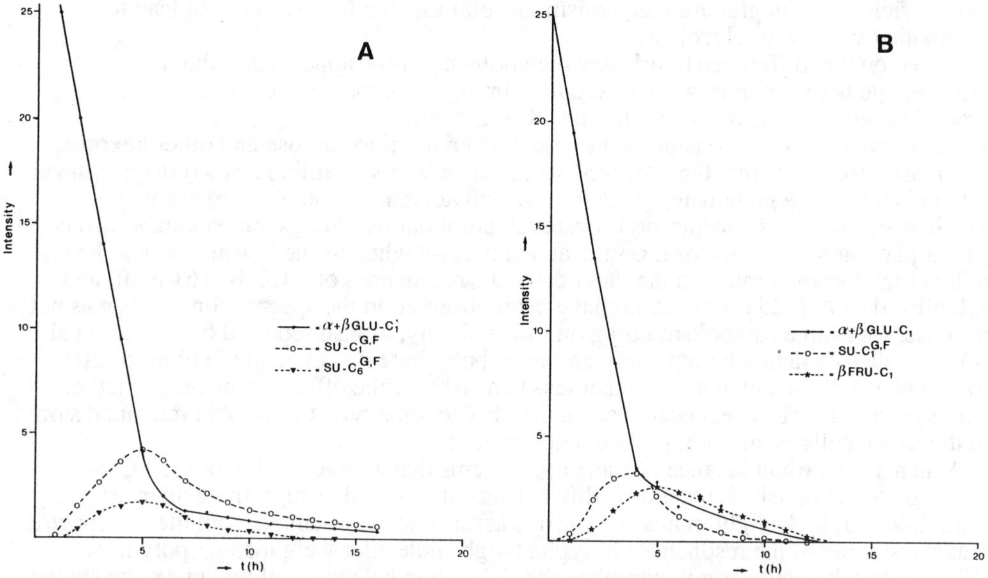

FIGURE 3. Time course of glucose breakdown and conversion into sucrose, labelled at the C1 (su-C1G,F) and C6 (su-C6G,F) carbons in both constituting hexose moieties glucose (G) and fructose (F) for the Flakkese cell line (A) and then A+ cell line (B); β fru-C1 indicating fructose being labelled at the C1 carbon.

in Fig.3B. However in this case, C1 labelled sucrose was the only sucrose species being formed, in addition to considerable synthesis of C1 labelled fructose, being formed with some lag time after sucrose synthesis. When the medium glucose became exhausted, the intracellular glucose and fructose levels became comparable. The simultaneous absence of C6 labelled sucrose and presence of C1 labelled fructose, strongly suggests that cells of the A+ cell line have a lower phospho-fructo-kinase (Pfk) activity compared to cells of the Flakkese cell line, thus tending to accumulate Fru-6P. To overcome this apparent metabolic blockade, excess Fru-6P in glycolysis is extruded from the glycolytic pathway by fructose formation involving phosphatase activity.
It thus seems that both cell lines, differing largely w.r.t. the number of proembryogenic masses present in suspension culture, behave quite different w.r.t. regulation of their metabolism in the top of glycolysis at the level of Fru-6P.

On the other hand, when cells of the A+ cell line were incubated with ^{13}C1-enriched fructose instead of glucose under further identical conditions, the kinetics of fructose breakdown was comparable to that shown in Fig.3B for glucose feeding, revealing the same biphasic behavior. Again, C1 labelled sucrose was the only sucrose species formed, no C6 labelling being observed, whereas in addition virtually no C1 labelling of glucose occurred. The site of hexose entry into the glycolytic pathway is thus clearly of influence on the mode of hexose conversion in the top of glycolysis. Glucose feeding, besides resulting into regular sucrose synthesis, thus also leads to overflow downwards in the glycolytic pathway at the level of Fru-6P, resulting in fructose formation. Fructose feeding on the other hand only leads to regular sucrose formation in adidition to normal fructose accumulation, no overflow taking place at the level of Glu-6P in view of the absence of free glucose. These observations suggest that fructose metabolism is regulated more efficiently than glucose metabolism, in sofar that the former does not lead to metabolic overflow in glycolysis.

Based on this difference in primary metabolism, it now appears possible to discriminate between lines with different embryogenic capacity, i.e. the presence of proembryogenic masses, by employing ^{13}C h.r. NMR.

Sofar, we have only considered hexose conversion into sucrose and other hexoses. The reason for this is that the observed steady state levels of amino acids reflecting sugar catabolism (as e.g. glutamate, glutamine, succinate, malate and alanine) are very low. The low steady state concentrations observed point out two things: either catabolism is taking place at a low level, or it occurs at a high level whereas the low amino acid levels reflect high turnover rates. In the latter case, the resonances of HCO3- (161 ppm) and solubilized CO2 (125 ppm) should have been observed in the spectra. Since this was not the case, we asume catabolism going on rather slowly, as also reported by Bender et al (4). In their 14C radiolabeling study on the carbohydrate metabolism of cultured carrot root explants, these authers claim that less than 10% of the offered substrate, whether being glucose or fructose, became catabolized. The remainder of the label remained stored in the carrot cells as sucrose, glucose and fructose.

When a ^{13}C carbon balance is made up, it seems that a large portion of the ^{13}C label becomes NMR invisible, most probably because it is stored as high-molecular weight, immobile starch. A slight indication for this assumption is obtained from the h.r. spectra, that show some of the resonances of typical high-molecular weight sugar polymers. Whether starch synthesis will complete the ^{13}C carbon balance, still remains to be shown by the application of ^{13}C MAS NMR, a technique able to visualize also the resonances of these h.r. NMR invisible sugar polymers.

REFERENCES

1 Ashworth, D.J., Lee, R.Y. and Adams, D.O. (1987) Plant Physiol. 85, 463-468

2 Breitmaier, E. and Voelter, W. (1987) Carbon-13 NMR Spectroscopy, High-resolution Methods and Applications in Organic Chemistry and biochemistry, third edition, VCH Verlagsgesellschaft mbH, D-6940 Weinheim (Federal repuplic of Germany)

3 Dijkema, C., De Vries, S.C., Booij, H., Schaafsma, T.J. and Van Kammen, A. (1988) Plant Physiol. 88, 1332-1337

4 Bender, L., Pauler, B. and Neuman, K.-H. (1987) Plant Cell Tissue Organ Cult 8, 135-146

ARE TWO LOCALIZATION SITES OF NITRATE REDUCTASE RESPONSIBLE FOR THE DIFFERENTIAL EXPRESSION IN TOBACCO CELLS DURING THE GROWTH CYCLE ?

A. NATO, D. LAVERGNE, V. FLIPO and J. HOARAU
Laboratoire Structure et Métabolisme des Plantes, associé au CNRS (URA 1128), Bât. 430, Université Paris-XI, 91405 ORSAY Cédex, France

1. INTRODUCTION

For higher plants, nitrate is the major form of nitrogen source. Nitrate uptake, an active process requiring energy and the presence of special carrier proteins is considered to be the initial limiting step allowing plants to assimilate nitrate. The assimilatory pathway consists of Nitrate Reductase (E.C.1.6.6.1-NR) which converts nitrate to nitrite. This NAD(P)H dependent enzyme is generally thought to be soluble and localized in the cytoplasm. Nitrite reductase, localized in the plastids then reduces nitrite to ammonia (1).

The NO_3^- assimilation pathway is affected by a variety of environmental (light, temperature, pH), nutritional factors (nitrate, ammonia, amino-acids) and growth regulators. The expression of the nitrate reductase activity (NRA) is highly regulated (2, 3). Substrate level, quality and adequate illumination and plant age are NRA key modulators (4, 5).

Plant cell suspension cultures offer a useful tool to study the aspects of the regulation of NO_3^- pathway. They allow to avoid problems inherent to a whole plant system. Some reports have described the induction, synthesis and turnover of NR in relation to cell growth cycle (6, 7, 8, 9, 10). In our previous report (11), physiological aspects of NRA of tobacco cells, in relation to cellular reduction power were analyzed under different growth conditions. Recently (12), a comparison of *in vivo* and *in vitro* NRA outlined the discrepancy between the two types of measurements in relation to growth cycle.

The aim of the present work is to analyze the characteristics of the two *in vivo* NRA peaks during a 3 week growth cycle of tobacco cells. The early peak is detected, immediately after subculture (0 to 7 hours) in a new medium. The second peak appears later (3 to 8 days) in batch culture conditions. A double recessive albino mutant α *Nicotiana tabacum*, totally lacking chlorophyll (13, 14) was used in order to evaluate the light control on NR expression without the interference by photosynthetic apparatus development (12).

The main objective of this paper is to present evidence that a microsomal form of NR protein exists in tobacco cells and to characterize it along the growth cycle.

2. PROCEDURE

2.1. Materials and methods

The selection and the culture of the double recessive albino plant mutant have already been described (13, 15). The cell suspensions derived from mesophyll friable callis are grown in 500 ml bottles on a tube roller (16) in a Murashige and Skoog modified medium : macroelements with only KNO_3 (20 mM) as the nitrogen source, microelements, vitamins, NAA (1 mg/l), kinetin (0.1 mg/l). Light is provided (photoperiod 16 h) by fluorescent tubes at 40 W m^{-2}. Experiments with phenylglyoxal (1 mM) or cycloheximide (2 µM) are conducted by adding aseptically the compounds to the culture medium.

In vivo NRA is done in the dark, under anaerobiosis by bubbling nitrogen gas in the culture medium containing KNO_3 at different concentrations (0 to 65 mM).

In vitro NRA is measured classicaly with NADH (0.8 mM) as the electron donor and KNO_3 (10 mM). The extraction of NR from tobacco cells is done according to Hoarau and *al.* (12).

For microsomal fraction isolation, tobacco cells (10 g) are ground at 4°C in 25 ml of an extraction buffer containing 100 mM K-phosphate pH 7.6, 3 mM EDTA, 2.5 mM DTT, 3 mM PMSF, 250 mM sucrose in a chilled mortar and pestle. The homogenate is filtered through 4 layers of miracloth and centrifuged at 2,500 g for 10 min. The 2,500 g supernatant is centrifuged at 120,000 g for 20 min to obtain a microsome pellet and a soluble fraction. The microsomal pellet is washed twice by resuspension in extraction buffer containing 0.75 M NaCl (in order to eliminate simply ionic association of soluble NR to membranes). NRA in microsomal fraction is tested after solubilization of membrane proteins by CHAPS (3-cholamidopropyl-dimethylammonial-1 propane sulfonate) 1 % during 45 min at ambient temperature. Inactivation of the enzyme activity is performed by incubation with an anti-NR serum spinach leaf polyclonal antibody. Proteins are measured by using the protein dye binding procedure (17) and the estimation of the intracellular content of NO_3^- is performed according to Cataldo (18) method.

3. RESULTS AND DISCUSSION

3.1. *In vivo* NRA measurements

a. Characteristics of the first early peak: after subculture in new culture medium (20 mM KNO_3), a rapid increase (8 fold) of NRA reaching a maximum at 7 hours is noted. Both light-grown and dark-grown cells exhibit the same magnitude of NRA during this period (Fig. 1). The early burst can be evidenced only in the presence of NO_3^- addition. During this period, no significant nitrate accumulation increase is detected in cells (results not shown). *In vivo* NRA is highly stimulated by increasing the concentration nitrate (up to 60 mM) in culture medium tests (Fig. 2). This suggests that a control of the intracellular concentration of NO_3^- took place, keeping it at a low level.

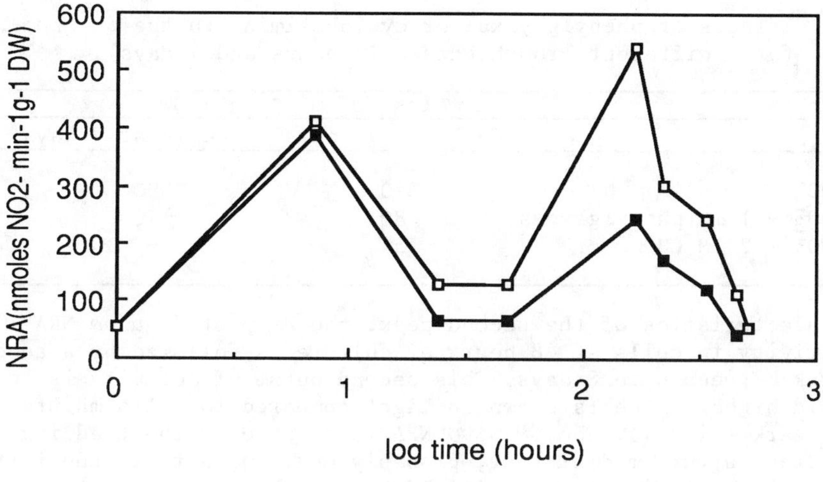

FIGURE 1. *In vivo* NRA during the cell growth cycle (light □-□ or dark ■-■). Activities were measured with cells (1 g) in 5 ml of culture medium) in anaerobic conditions.

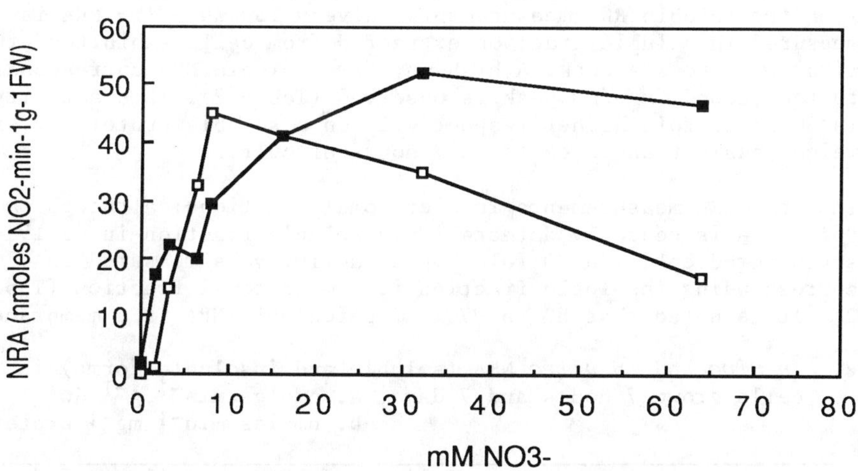

FIGURE 2. *In vitro* NRA as a function of NO_3^- concentration in culture medium tests. ■-■ : cells after 7 hours of subculture ; □-□ : cells after 7 days of culture.

In addition, the more efficient system for NO_3^- uptake seems to be inducible as demonstrated by applying phenylglyoxal or cycloheximide (Table 1).

TABLE 1. Effects of phenylglyoxal or cycloheximide in the *in vivo* NRA for 2 different growth phases (7 hours and 7 days).

	n moles NO_2^- min^{-1} g^{-1} DW	
	Cells at 7 hours	Cells at 7 days
20 mM KNO_3^-	370	590
20 mM KNO_3^- + 1 mM phenylglyoxal	80	76
20 mM KNO_3^- + 2 μM CH	54	-

b. Characteristics of the second peak: the drop of *in vivo* NRA activity in cells at 48 hours of culture is followed by a second peak between 6 to 8 days. This second burst of activity is two-fold higher in cells grown in light compared to cells maintained in dark (Fig. 1). The *in vivo* NRA is detected without adding NO_3^- to the culture medium test, probably due, in part, to the intracellular nitrate content, 5 fold higher than in early stage (results not shown). In addition, higher concentration of NO_3^- in medium tests inhibits the second phase increase of NRA (Fig. 2). The presence of 1 mM phenylglyoxal in culture medium greatly reduced growth rate and the second peak does not appear (Table 1).

3.2. *In vitro* soluble RNA measurements: a very low *in vitro* NRA is measured in soluble fraction extracted from cells exhibiting the first *in vivo* NRA peak. A high *in vitro* soluble NRA corresponding to the second *in vivo* peak is observed (Table 2). This activity is 30 or 80 fold higher respectively on a soluble protein or dry weight basis than in cells at 7 hours of culture.

3.3. *In vitro* NRA measurements in microsomal fractions: although *in vitro* NRA is scarcely detectable in soluble fraction in early subcultured cells, a 30 fold higher activity is measured in corresponding insoluble fraction i.e. microsomal fraction (Table 2). It is noted that 80 to 97 % of detectable NRA is a membrane

TABLE 2. *In vivo* and *in vitro* NRA (soluble and insoluble forms) in cells grown 7 hours and 7 days. a. nmoles min^{-1} g^{-1} dw
b. nmoles min^{-1} mg^{-1} protein

		Cells at 7 hours	Cells at 7 days
In vivo NRA	a	470	530
	b	22.3	8.1
In vitro NRA	a	1.5 (12 %)	125 (80 %)
Cytosolic fraction (soluble)	b	0.07 (3 %)	1.9 (24 %)
In vitro NRA	a	5.3 (78 %)	31 (20 %)
Microsomal fraction (insoluble)	b	2.5 (97 %)	6.0 (76 %)

bound form for cells exhibiting the first *in vivo* peak. In the second phase of NRA appearance, its expression per dry weight unit shows that 80 % of the activity remains in the soluble fraction. The expression per protein unit shows a 3 fold higher specific activity of the insoluble enzyme form compared to soluble fraction activity (Table 2). In addition, microsomal NRA is by least 50 % inactived by anti-NR polyclonal antibodies while the soluble fraction is completely inhibited (results not shown).

In conclusion, the results suggest two differently expressed forms of NRA in tobacco cells. Our previous results pointed out the same observations with green tobacco cells (19). A stimulating working hypothesis is the existence of an entirely different NR, acting as a membrane bound form in tobacco cells. Evidence for this possibility was conforted by recent reports on insoluble form of NR protein in higher plants and algae (20, 21, 22).

REFERENCES

1 Guerrero, M.G., Vega, J.M. and Losada, M. (1981) Annu. Rev. Plant Physiol. 32, 169-204
2 Jackson, W.A., Flescher, D. and Hageman, R.H. (1973) Plant Physiol. 51, 120-127
3 Remmler, J.L. and Campbell, W.H. (1986) Plant Physiol. 80, 442-447
4 Rufty, T.W. Jr, Israel, D.W. and Volk, R.J. (1984) Plant Physiol. 76, 769-775
5 Duke, S.H. and Duke, S.O. (1984) Physiol. Plant. 62, 485-493
6 Zielke, H.R. and Filner, P. (1971) J. Biol. Chem. 246, 1772-1779
7 Nelson, R.S., Horn, M.E., Harper, J.E. and Widholm, J. (1984) Plant Sci. Lett. 34, 145-152
8 Maki, H., Yamagishi, K., Sato, T., Ogura, N. and Nakagawa, H. (1986) Plant Physiol. 82, 739-741
9 Zink, M.W. (1982) Can. J. Bot. 60, 386-396
10 Privalle, L.S., Lahners, K.N., Mullins, M.A. and Rothstein, S. (1989) Plant Physiol. 90, 962-967
11 Nato, A., Hoarau J., Bhatnagar, R., Dulieu, H. (1984) In Advances in Photosynthesis Research (Sybesma, C. ed) Vol. III pp. 799-802, Martinus Nijhoff/Dr W Junk Publishers
12 Hoarau J., Nato, A., Lavergne, D., Ouhmidou, B. and Cherel, I. (1989) In Advanced Course on Inorganic Nitrogen Metabolism. June 4-10, 1989, Napoli
13 Dulieu, H. (1974) Mutation Res. 25, 289-304
14 Nguyen-Quoc, B., Prioul, J.L., Nato, A. and Dulieu, H. (1989) Plant Physiol. Biochem. 27, 811-816
15 Nato, A., Brangeon, J., and Dulieu, H. (1981) In Photosynthesis, V. Chloroplast development (Akoyunoglou, G. ed) Balaban Intern. Services, Philadelphie, pp. 821-830
16 Nato, A., Bazetoux, S. and Mathieu, Y. (1977) Physiol. Plant. 41, 116-123
17 Sedmak, J.J. and Grossberg, S.E. (1977) Anal. Biochim. 79, 544-552

18 Cataldo, D.A., Haroon, N., Schrader, L.E. and Young, V.L. (1975) Commun Soil Sci. and Plant Analysis 6, 71-80
19 Hoarau, J., Lavergne, D., Nato, A., Flipo, V., 1990 (submitted)
20 Ward, M.R., Tischner, R. and Huffaker, R.C. (1988) Plant Physiol. 88, 1141-1145
21 Ward, M.R., Grimes, H.D. and Huffaker, R.C. (1989) Planta 177, 470-475
22 Tischner, R., Ward, M.R. and Huffaker, R.C. (1989) Planta 178, 19-24

PHYTOCHROME IN PLANT CELL CULTURES

E. SCHÄFER, B. BRUNS, H. FROHNMEYER, K. HAHLBROCK, K. HARTER, T. MERKLE and S. OHL, INSTITUT FÜR BIOLOGIE II DER UNIVERSITÄT FREIBURG, SCHÄNZLESTR. 1, 7800 FREIBURG, FRG

1. INTRODUCTION

At least three different photoreceptor classes (phytochromes, blue UV-A- and UV-B receptors) control the transition from skotomorphogenesis to photomorphogenesis (1). The control of elongation rate, photoperiodism as well as mRNA accumulation after partial reetiolation by phytochromes and blue UV-A receptors is also documented (2, 3).

In nature plants are exposed to sun light which acts on all possible photoreceptors pigments simultaneously. This may lead to complex interactions among them. Detailed studies of these interactions have been carried out during the last 10 years. However, before describing these apparante interactions it must be established that each photorecpetor system is, in fact, involved in a particular response.

Unfortunately, phytochrome, which is most responsive in red and far-red light, is the only photoreceptor analysed in molecular terms so far. Phytochromes have been purified from several plants and cDNA's and/or genomic clones encoding for phytochrome polypeptides have been sequenced for several monocots and dicots (cf. 4). These data proved that not only two different phytochromes - as predicted from spectroscopical and physiological data (5, 6) - but at least three different phytochromes are present in plants (4) and probably also control photoregulation.

The participation of phytochrome in a given situation can be demonstrated under induction (light pulses of 5 min) conditions by testing for repeatable red/far-red reversibility. A reversion of a red light-pulse after a prolonged dark period by a far-red light pulse is generally taken as indirect evidence for the control by the so-called stable (type II) phytochrome (7). In continuous light (high irradiance response, HIR) dichromatic irradiations are required as a decisive test. Dependence of the HIR on preirradiations with red light pulses can be taken as indirect evidence that this HIR is under the control of type I or labile phytochrome (8). The role of phytochrome under continuous irradiation in green plants is not understood up to now (2).

The demonstration that the blue UV-A and/or UV-B photoreceptor systems are active in those systems where also the phytochrome system is active is much more difficult. This is due to the fact that both phytochromes and blue UV-A receptors absorb in blue and UV-A light and that all photoreceptors absorb in UV-B light (Fig. 1). Thus, any response to light may be due to the absorption by a single photoreceptor or to several which may either act independently to each other or may interact.

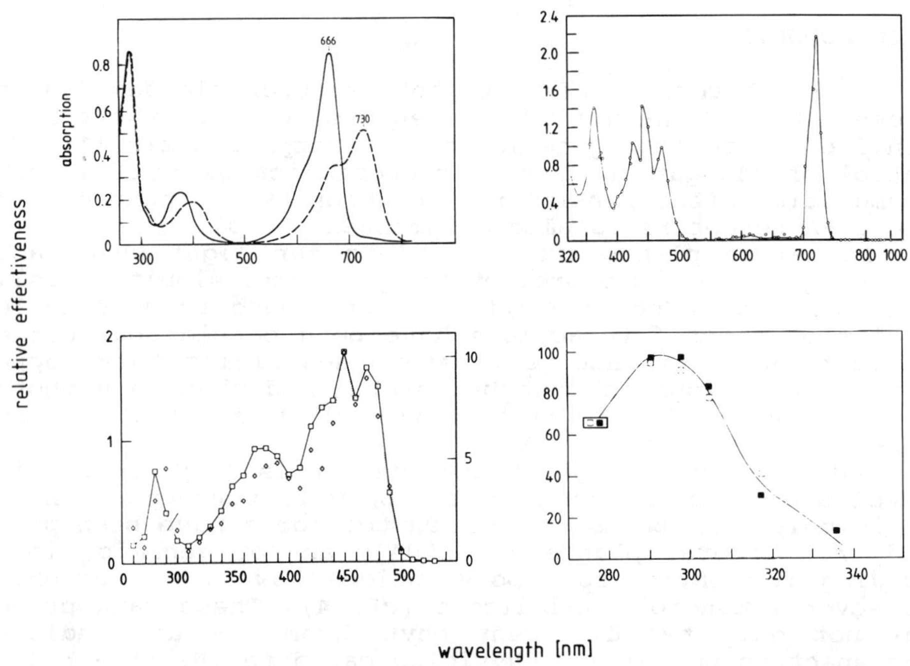

FIGURE 1. Comparison of absorption spectra of purified phytochrome (P_r and P_{fr}) (a), action spectra for light-controlled inhibition of hypocotyl elongation in lettuce seedlings (HIR, 27) (b), action spectra for phototropism of oat coleoptiles and of alfalfa hypocotyls (blue light receptor, 28) (c) and action spectra for UV induced synthesis of flavonoid glycosides in cell cultures of parsley and anthocyanin in maize coleoptiles (UV-B receptor, 14).

2. PHYTOREGULATION IN HIGHER PLANTS

Interactions of photoreceptors in higher plants have been extensively studied for the photocontrol of anthocyanin (9 - 10).

At the mRNA level the response of the multigen family encoding for the chlorophyll a,b binding protein has been studied for young tobacco and tomato seedlings. The photon fluence rate response curves and the responsiveness to P_{fr} in the dark after a continuous light pretreatment demonstrate clearly that besides phytochrome induction (11) a HIR and action of blue UV-A and UV-B receptor is observed. Based on these data differential photocontrol by different members of a multigen family cannot be excluded. Recent studies of the expression of chalcone synthase (CHS) in mustard shows for an individual gene of this small multigen family that a complex spatial pattern of gene expression is observed (12). In addition a photoregulation by induction responses due to labile and stable phytochrome, an HIR and blue light receptor could be demonstrated (13). The molecular basis of these complex regulation patterns is still unknown.

3. PHOTOREGULATION IN CELL CULTURES

3.1 Photoreceptors

Only for two cell suspension cultures it was possible so far to demonstrate a photocontrol of gene expression by phytochrome (14, 15). Generally speaking gene expression in cell cultures and calli is predominantly under the control of blue UV-A receptors (see chapter by G. Richter) but for cell suspension cultures of dill and parsley the prominent effect of UV-B receptor was demonstrated (14).

3.2 Detection of phytochrome in cell suspension cultures

The phytochrome system has been studied only in a few tissue cultures: callus tissue cultures from *Nicotiana tabacum* (16), some Cruciferae (17) and cell suspension cultures from Daucus carota (18, 19), Glycine max (19, 20) and Petroselinum hortense (19). The phytochrome content - measured spectrophotometrically - was changing with the age of the culture as demonstrated for *Daucus carota* and *Nicotiana tabacum* (16, 18).

Figure 2a shows the time courses of accumulation of spectrophotometrically detectable phytochrome in three different cell suspension cultures. These kinetics of dark accumulation are very similar. As demonstrated for parsley cell suspension culture the kinetics followed the changes in cell number (Fig. 2b). This is in contrast to phytochrome accumulation kinetics in etiolated tissues.

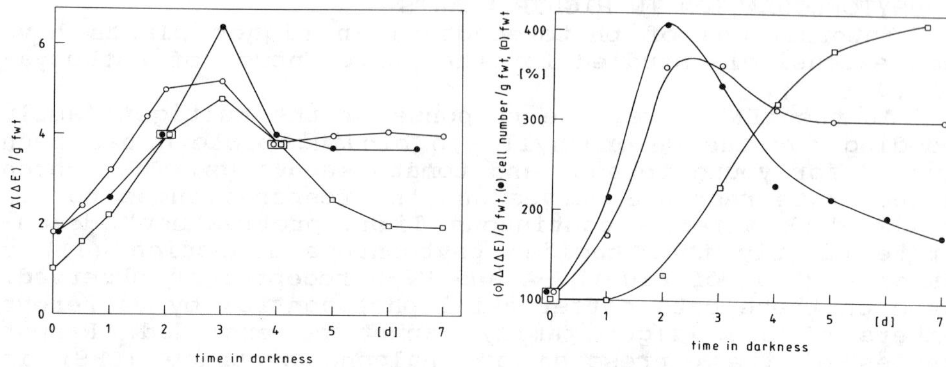

FIGURE 2.a) Kinetics of phytochrome accumulation (ΔE)/g fwt in the dark in cell suspension cultures after transferring the cells to fresh medium. ○ Glycine max, 0 Petroselinum hortense, □ Daucus carota.

2.b) Time course in the dark of changes of phytochrome content per g fwt in cell suspension cultures of Petroselinum hortense (0), changes of cell number pro g fwt (○) and changes of fwt (□) after transferring the cells to fresh medium. Data normalized to the initial level measured at time of inoculation.

Also the kinetics of phytochrome destruction (Fig. 3a) in parsley cell cultures are clearly distinct from those observed in etiolated tissue. Significant is a strong P_{fr} P_r dark reversion and a slow destruction rate. This kinetic behaviour mainly reflects the properties of type II or stable phytochrome (5) and has been reported also for other cell cultures (19).

Using a series of antibodies against phytochrome it was observed that phytochrome in parsley cell cultures was only recognized by an antibody which binds to an epitope which is strongly conserved in all types of phytochrome. A polyclonal antibody against type I - phytochrome shows no reaction with phytochrome from parsley cell cultures but with phytochrome from leaves of reetiolated parsley plants (Fig. 3b). This indicates that in this cell suspension culture the type II (stable) phytochrome is dominating. This is supported further by the observation that in these cell cultures only a weak red light induced pelletability can be observed, whereas this seems to be a common characteristic of phytochrome in etiolated seedlings (21).

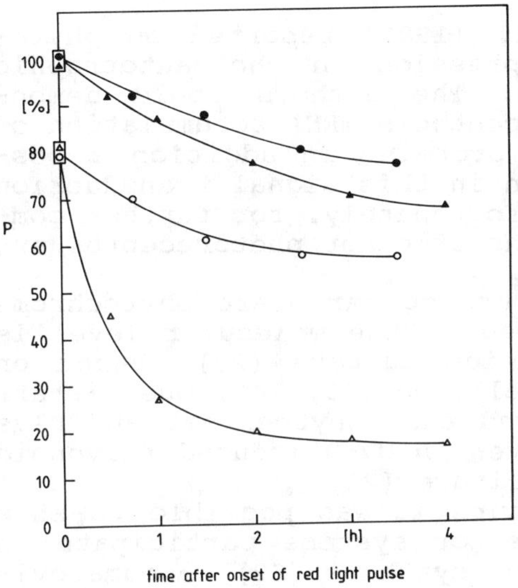

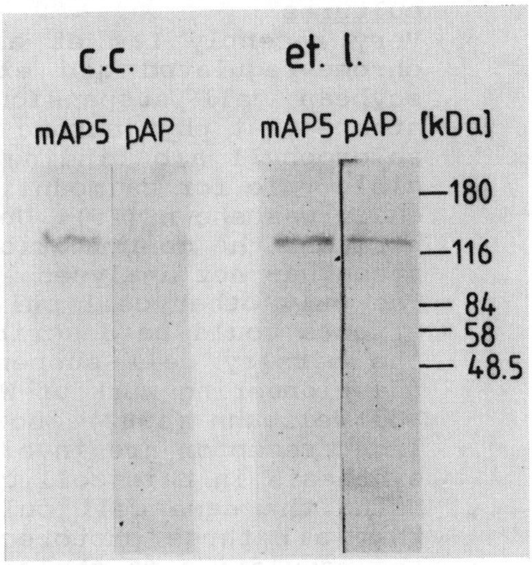

FIGURE 3.a) Effect of red light on the change in phytochrome content in two identically prepared parsley (Petroselinum hortense) cell suspension cultures. Cells were grown in continuous darkness given a 5 min red light pulse and then returned to darkness for various times until harvesting for in vivo phytochrome measurements.
P_{tot} (●), P_{fr} (▲) in cell suspension culture 1
P_{tot} (○), P_{fr} (△) in cell suspension culture 2

3.b) Affinity of antibodies mAP5 and pAP to phytochrome from dark grown parsley cell cultures (c.c.; cell culture 2 of Fig. 3a)) and from etiolated leaves of parsley plants grown in darkness for 7 days (et.l.)
Identical amounts of phytochrome $\Delta(\Delta E)$ were applied to each lane.
mAP5: Monoclonal antibody with high affinity to the conserved, chromophor attachment site of type I and type II phytochrome originally

produced against type I phytochrome of pea (Pisum sativum)
pAP: Polyclonal antibody against type I phytochrome of pea.

3.3. <u>Photoregulation of gene expression in cell suspension cultures</u>

Very recently Lam et al. (1989) reported on phytochrome-regulated gen expression in photoautotrophic soybean cell suspension. The authors could demonstrate that phytochrome controls mRNA accumulation of chlorophyll a/b binding protein. In addition a possible role for calmodulin in this signal transduction chain was shown (15). Unfortunately, for further comparisons the contributions of other photoreceptor systems was not analysed.
The only other cell culture so far where phytochrome effects could be described at the molecular level is the parsley cell suspension culture (22). Based on the pioneering work of Wellmann (23, 24), Duell-Pfaff and Wellmann (1982) showed that phytochrome and blue light receptor are involved in UV-B induced flavonoid synthesis in this cell culture (25).
Using the same cell culture it was possible to show that all three photoreceptor systems participate in the regulation of chalcone synthase (22) 4-coumaroyl-CoA ligase mRNA accumulation and another (LF 53) mRNA encoding for a still not identified transcript (Fig. 4). In these studies it was concluded that the absorption by the UV-B receptor is essential for the induction of mRNA accumulation, whereas the blue light receptor and the phytochrome systems modify the amount of mRNA accumulation. Using a newly established parsley cell suspension culture a strong contribution of the phytochrome system could no longer be demonstrated. Only under prolonged red light irradiation a weak modulating effect was observed (26). With this culture also a UV-B independent effect of blue light and red light on the chalcone synthase mRNA accumulation was measurable. The accumulation in blue light was only 10%, in red light approx. 1% of that observed after UV-B light treatment (26).

For all three gene families tested the effect of the blue light and the UV-B receptor was detectable when the transcription rate was measured in run off experiments (Fig. 5).

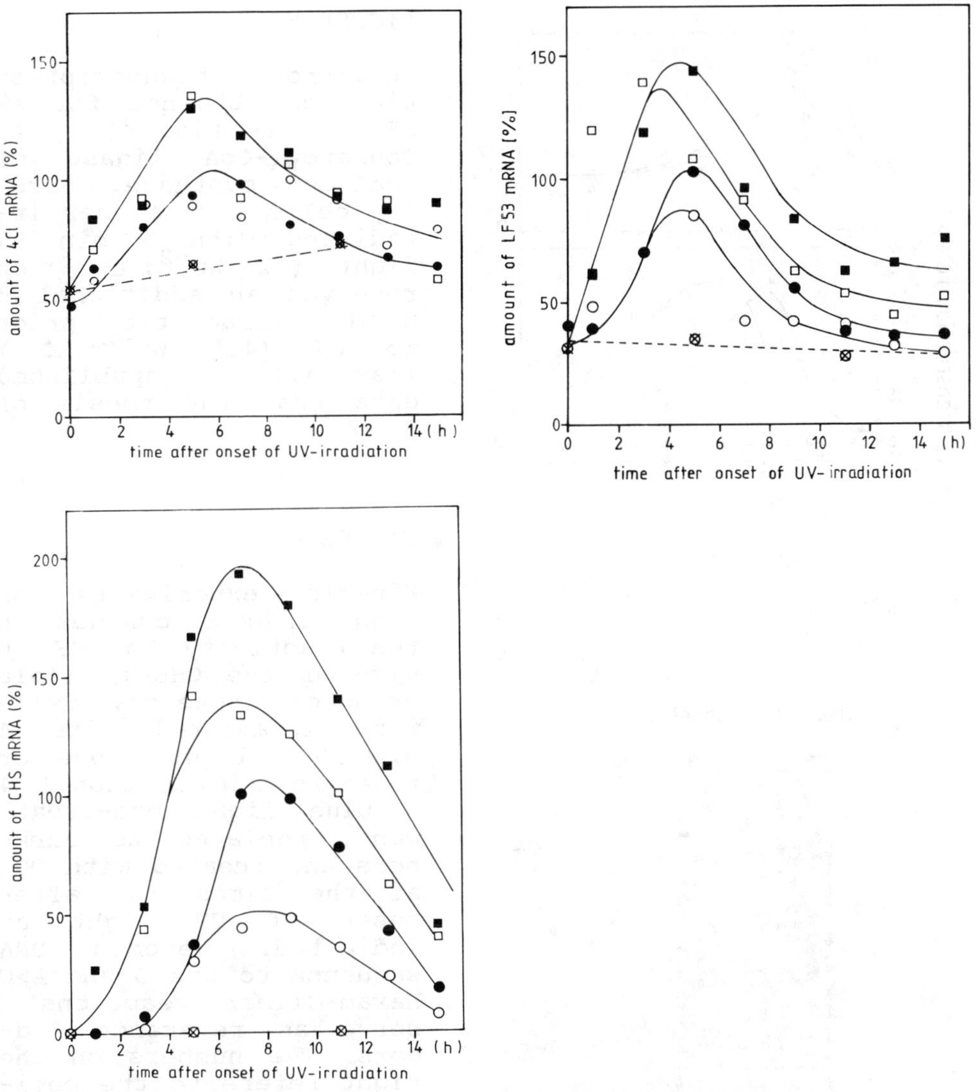

FIGURE 4. Time courses of the transient accumulation of mRNAs in cultured parsley cells in response to irradiation with light of different wavelengths: (⊗) dark controls, (●) 10 min UV, (○) 10 min UV - 10 min RG 9 light, (■) 6 h blue light + 10 min UV, (□) 6 h blue light + 10 min UV + 10 min RG 9 light.

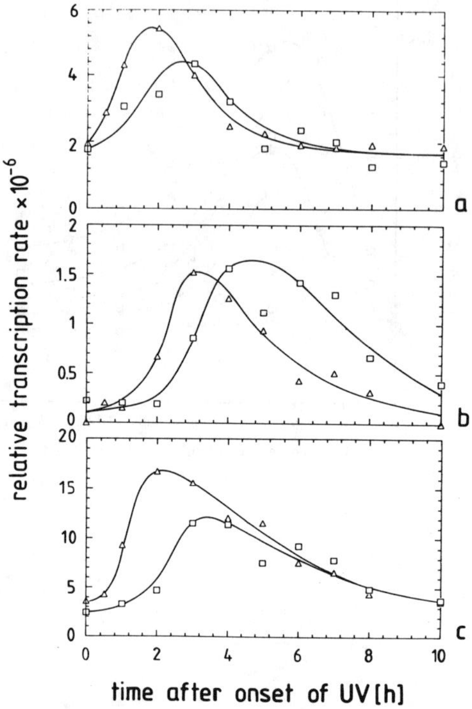

FIGURE 5.

In-vitro transcription kinetics obtained for a) LF 53 (see Fig. 4) b) 4-Coumaroyl-CoA ligase c) chalcone synthase. Parsley cells were either irradiated with 5 min UV light (32 Wm^{-2}; □) or received an additional 5 h blue irradiation prior to UV (4.8 Wm^{-2}; △) (partially unpublished data from the thesis of S. Ohl).

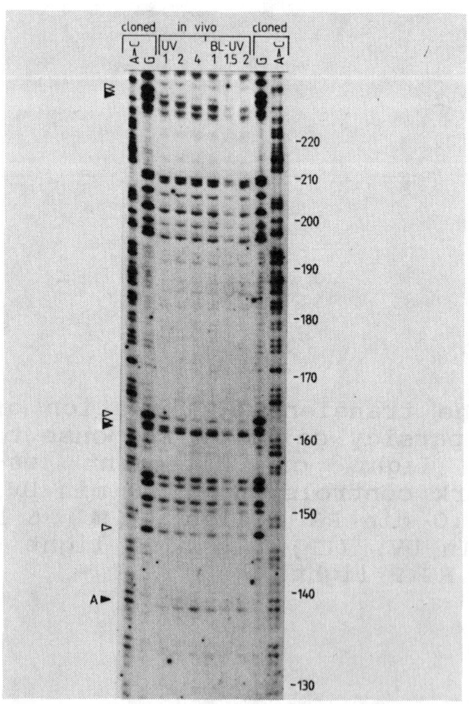

FIGURE 6.

Kinetic experiments of light induced changes in the reactivity to DMS *in vivo* of the CHS-1 allele promoter sequence. Cells were irradiated with 4 min UV-B light alone or received an additional 4 h blue light pre-treatment, replaced to darkness and treated with DMS at the times [h] after onset of UV light as indicated. Genomic DNA sequence to the G and A>C Maxam-Gilbert-reactions serve as reference ladders. The numbers on the right refere to the position relative to the transcription start site. Protections from and enhancements of DMS methylation *in vivo* are shwon as open and closed triangles, respectively (unpublished data by T. Merkle).

Therefore the question arised whether the same cis acting elements in the promotors of the genes are the targets for the signal transduction chains of the different photoreceptor systems. This problem has been approached by analysing DNA protein interaction using in vivo foot printing technique on the CHS promotor. The observed footprints were identical after UV-B and Blue light UV-B treatments (Fig. 6). Only the time course of appearence of the footprints was changed by a blue light pretreatment. Another approach was to use the promotor of the chalcone synthase gene fused to GUS as a reporter in a transient gene expression assay with protoplasts of parsley suspension culture. In these experiments even the minimal promotor could transfer responsiveness to blue and UV-B light (Fig. 7).

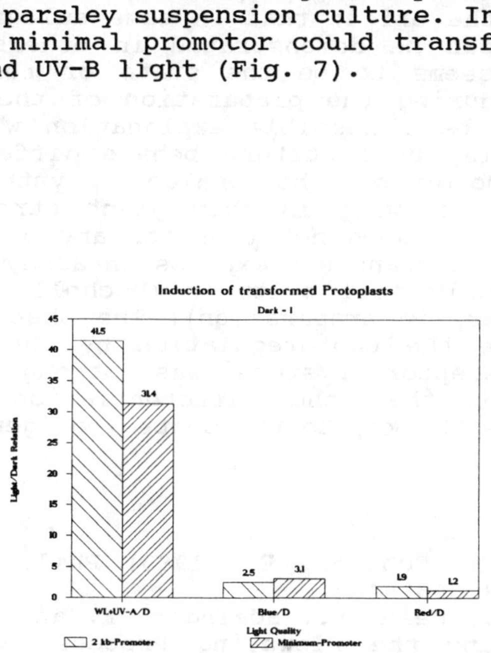

FIGURE 7. Photocontrol of parsley CHS promoter GUS fusion constructs tested in a transient gene expression assay in transformed parsley protoplasts (29). The transformed protoplast were irradiated for 9 h and GUS was extracted after the end of the irradiation period. The dark levels were 43 or 30 [pmol MU/mg Prot/min] for the 2kb or minimal promoter. The fluence rates were 4.2, 3.7 and 5 Wm^{-2} for WL + UVA, Blue and Red, respectively

3.4 Conclusions

Our knowledge about the phytochrome system and photoregulation in cell cultures is still very limited. The phytochrome species predominantly expressed in cell cultures seems to be the stable (type II) phytochrome. Photoregulation even in dark adapted cell cultures therefore probably reflect the photoregulation of light grown seedlings. Unfortunately, we must keep in mind that phytochrome content, kinetic properties of phytochrome are extremely variable in more or less identically prepared cell cultures. This can lead to the effect that in some cultures phytochrome control can be demonstrated in others not. A major problem seems to be the state of redifferentiation reached during the preparation of the cell culture. This may be a possible explanation why the old and new parsley cell culture behave differently and why the promotor of the chalcone synthase gene from mustard - showing in this plant strong phytochrome regulation - does not transfer any phytochrome regulation in a transient expression assay in protoplasts from parsely cell cultures (Rocholl, Frohnmeyer and Batschauer, in preparation). The observation that in this case the photoregulation by the UV-B and blue light receptor systems was strong point to the conclusion that the differentiation state of the cells are the key to the control of gene expression.

REFERENCES

1 Mohr, H. and Schäfer, E. (1983) Phil. Trans. R. Soc. Lond. B 303, 461-467

2 Schäfer, E., Heim, B., Mösinger, E. and Otto, V. (1984) in Light and the Flowering Process (Vince-Prue, D., Thomas, B. and Cockshull, K.E., eds.) pp. 17-32, Academic Press, London

3 Kuhlemeier, C., Green,P.J. and Chua, N.-H. (1987) Annu. Rev. Plant Physiol. 38, 221-257

4 Sharrock, R.A. and Quail, P.H. (1989) Genes and Development 3, 1745-1757

5 Brockmann, J. and Schäfer, E. (1982) Photochem. Photobiol. 35, 555-558.

6 Jabben, M. and Holmes, M.G. (1983) in Photomorphogenesis. Encycl. Plant Physiol. New Series (Shropshire, W. jr. and Mohr, H., eds.), Vol 16, pp. 704-722, Springer-Verlag, Berlin, Heidelberg, New York, Tokyo

7 Furuya, M., Ito, N., Tomizawa, K. and Schäfer, E. (1990) Planta in press

8 Holmes, M.G. and Schäfer, E. (1981) Planta 153, 267-272

9 Drumm-Herrel, H. and Mohr, H. (1978) Photochem. Photobiol. 27, 241-248
10 Oelmüller, R. and Mohr, H. (1985) Proc. Natl. Acad. Sci USA 82, 6124-6128
11 Wehmeyer, B., Cashmore, A.R. and Schäfer, E. (1990) Plant Physiol., in press
12 Kretsch, T., Ocker, B., Ehmann, B., Speth, V. and Schäfer, E. (1990) unpublished
13 Ehmann, B., Ocker, B. and Schäfer, E. (1990) submitted to Proc. Natl. Acad. Sci. USA
14 Wellmann, E. (1983) in Photomorphogenesis. Encycl. Plant Physiol. New Series (Shropshire, W. jr. and Mohr, H., eds.), Vol 16, pp. 745-756, Springer-Verlag, Berlin, Heidelberg, New York, Tokyo
15 Lam, E., Benedyk, M. and Chua, N.-H. (1989) Molec. and Cell Biol. 11, 4819-4823
16 Kasperbauer, M.J. and Reinert, R.A. (1967) Physiol. Plant 20, 977-981
17 Grill, R., Frauenhorn, I. and Schraudolf, H. (1979) Z. Pflanzenphysiol. 94, 195-200
18 Wetherell, D.F. (1969) Plant Physiol. 44, 1734-1737
19 Gottmann, K. and Schäfer, R. (1982) Plant Cell Reports 1, 221-224
20 Tanada, T. (1977) Planta 134, 57-59
21 Pratt, L.H. and Marmé, D. (1976) Plant Physiol. 58, 686-692
22 Bruns, B., Hahlbrock, K. and Schäfer, E. (1986) Planta 169, 393-398
23 Wellmann, E. (1971) Planta 101, 283-286
24 Wellmann, E. (1974) Ber. Deutsch. Bot. Ges. 87, 267-273
25 Duell-Pfaff, N. and Wellmann, E. (1982) Planta 156, 213-217
26 Ohl, S., Hahlbrock, K. and Schäfer, E. (1989) Planta 177, 228-236
27 Hartmann, K.M. (1967) Z. Naturforsch. 22b, 1172-1175
28 Briggs, W.R. and Baskin, T.I. (1988) Botanica Acta 101, 133-139
29 Schulze-Lefert, P., Becker-Andre, M., Schulz, W., Hahlbrock, K., and Dangl, J.l. (1989) The Plant Cell 1, 707-714

BLUE-LIGHT EFFECTS ON GENE EXPRESSION IN PLANT-CELL CULTURES

G. RICHTER, I. DANNHAUER, R. KALDENHOFF, N. OTTERSBACH

Institut für Botanik, Universität Hannover, Herrenhäuserstrasse 2
D-3000 Hannover 21, F.R.G.

1. INTRODUCTION

Chloroplast differentiation is a complex process which in general is dependent on the environmental factor light. To perceive it receptor systems have evolved including the well-studied red/far red light-absorbing phytochrome and the poorly characterized blue light receptors ("cryptochromes"). In addition to light the development of chloroplasts seems also to be regulated by intrinsic developmental signals, e.g. in leaf development (1, 2). In suspension cultured cells of *Nicotiana tabacum* var. *Samsun* and *Chenopodium rubrum* the differentiation of functionally active chloroplasts from leucoplast-like precursors is induced and maintained solely by blue light (3, 4). These systems offer a good chance to study the molecular mechanisms that underlie the biogenesis of these cell organelles. The present treatise will deal with the blue light control of expression of those nuclear and plastid genes which code for plastid constituents. Moreover, first evidence will be presented for the existence of genes which are rapidly and transiently expressed during the initial phase of blue light-induced chloroplast differentiation, the products of which could play a role in signal perception, transduction and signal response.

2. PROCEDURE

2.1. Materials and methods

2.1.1 *Cell culture:* Chloroplast-free callus cells of Chenopodium rubrum L. (supplied by Dr. W. Hüsemann, Münster) were grown as reported previously (5). Prior to illumination samples were taken 6 d after the transfer, i.e. in the logarithmic growth phase, suspended in fresh medium and kept in darkness for 48 h.

2.1.2 *Chlorophyll measurement* was carried out as described (6).

2.1.3 *Preparation and analysis of RNA:* Extraction, quantitation and electrophoretic separation as well as dot hybridization and Northern analysis were done as published recently (7).

2.1.4 *In-vitro translation and gel electrophoresis* were carried out as reported (7).

2.1.5 *Preparation of TAC:* The transcriptionally active complex (TAC) was purified from isolated leucoplasts, greening plastids and mature chloroplasts following the protocol of Reiss and Link (8) with minor modifications. The activity of RNA polymerase was determined as reported (9).

2.1.6 Cloning of complementary DNA (cDNA): Polyadenylated RNA (poly(A)-RNA) was isolated as described above, and the cDNA synthesized with materials from a cDNA-Synthesis Kit (Amersham) according to the suppliers' protocol. The construction of cDNA libraries from selected rapidly transcribed poly(A)-RNA species, the isolation of relevant clones and the subcloning of cDNA insertions into the plasmid vector Bluescript KS were done as published (7).

2.1.7 Sense and antisense RNA synthesis: These sequences were synthesized in vitro by T3 and T7 RNA polymerase in the presence of ^{32}P-UTP and the linearized vector Bluescript harbouring the specific cDNA insertion (10). The length of antisense RNA was determined by electrophoresis. Hbridization with antisense RNA as probe was performed according to Little and Jackson (11).

2.1.8 Sequencing of DNA: Both strands of cDNA insertions previously subcloned in the vector Bluescript (see above) were analyzed by the dideoxy-chain-determination technique with T7 polymerase (12) by using materials from a commercial T7 Sequencing Kit (Pharmacia) and following the suppliers' protocol. Sequence-data analyses and derivation of protein sequences were performed using the PC Gene Programm (Genofit). Searches for protein sequences similar to the derived amino acid sequences of the various cDNA clones were conducted with the Fast Scan Program using a Dayhoff MDM-78 matrix.

2.1.9 Hybrid-release translation: Alkali-denatured cDNA insertion (15 µg) was fixed onto nylon filters and used to select a single species of RNA from 20-40 µg poly(A)-RNA by hybridization (10). The bound RNA sequences were recovered by boiling the filters with 220 µl sterile bidestilled water for 70-80 s, and subsequently translated in the wheat-germ cell-free System (Amersham).

3. RESULTS AND DISCUSSION

3.1 Effect of light quality an transcript level

During the blue light-dependent greening process of cultured tobacco cells membrane and stroma proteins are synthesized de-novo which is reflected in the activity of the corresponding mRNAs (4). Moreover, the level of mRNAs coding for plastid proteins increases rapidly with the onset of blue light irradiation (13).

We have extended these studies to cell lines of *Chenopodium* which form chloroplasts if cultivated either in a complex medium with sucrose or in a pure mineral solution with carbon dioxide (14). With the dot hybridization technique employing cloned DNA sequences of plastid and nuclear genes (rbcS, rbcL, CAB, psbA, atpB, atbE) we have assessed mRNA species of relevant plastid proteins. Fig. 1 shows the effect of illumination time on the abundance of mRNAs

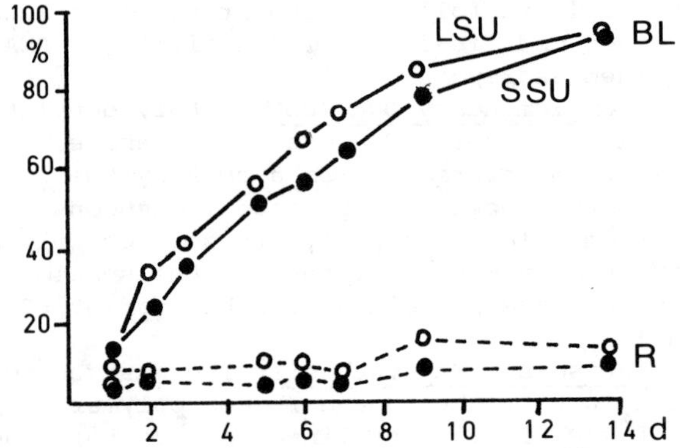

FIGURE 1. Changes in the amounts of hybridizable LSU-mRNA and SSU-mRNA in dark-grown cells of *Chenopodium rubrum* exposed to either continuous blue light (BL) or red light (R) of equal energy fluence rate (9 $W \cdot m^{-2}$). Total RNA was extracted at the times indicated and samples supplied to nitrocellulose membrane in dilution series; hybridization to the ^{32}P-labelled sequences of corresponding genes rbcL and rbcS. The data were obtained by counting the radioactivity bound to the dots. The corresponding value for the RNA preparation of fully greened cells (day 20) was set at 100%, and all other values referred to it.

encoding the large subunit (LSU) and the small subunit (SSU) of ribulose-bisphosphate carboxylase/oxygenase. A drastic increase within 14 d is observed. This time curve compares fairly well with that obtained with tobacco cells exposed to blue light (13). Treatment with red light over the same period of time gave rise only to a slight increase in the level of LSU- and SSU-mRNA. For the steady-state level of the mRNA coding for the 35 kDa precursor (PP-35) of the 32 kDa membrane protein of photosystem II a similar accumulation kinetics was obtained: A steady rise over 14 d (5). On the other hand, red light failed to bring about such an increase of PP-35 mRNA. Evidently, this light quality cannot take over the role of blue light in the induction and maintenance of chloroplast development.

3.2 Transcriptional control by blue light of plastid genes

The observed preferential expression of plastid and nuclear genes in blue light indicates either a promoted change in the transcription initiation and/or an alteration in the stability of transcripts. In order to probe into the first of the two mechanisms we have employed two cell-free transcription systems: 1. An active complex ("transcriptionally active chromosome"; TAC) from plastids of dark-grown respective light-exposed cells consisting of DNA, RNA polymerase and DNA-associated proteins (9); its capacity for catalyzing the synthesis of RNAs depended upon the presence of all four nucleotides; 2. Purified nuclei isolated from cells either cultured in darkness or in blue light respective red light (S. Bockholt, unpublished).

The activity of TAC prepared from plastids of dark-grown and blue light-treated cells (0 - 48 h) started to rise from a very low level which amounted after 48 h to about 40% of the value obtained with TAC from mature chloroplasts. By molecular hybridization employing cloned sequences of plastid genes as immobilized probes the radiolabelled transcripts among the RNAs synthesized by TAC preparations from plastids of different developmental stages could be quantified. For the ribosomal species of 16 S and 23 S as well as for the LSU and PP-35 transcripts a substantial accumulation with irradiation times was observed (Fig. 2). With isolated nuclei a similar stimulation of transcription by blue light was observed. The quantities of SSU and CAB transcripts increased with the application of blue light to the cells from which the nuclei came. This promotion of transcription, still manifest in the isolated organelles, compares fairly well with that monitored with TAC from blue light-treated cells.

These results support the notion that blue light regulates the de-novo synthesis of plastid and nuclear RNAs by promoting changes in the transcription of corresponding genes. Thus the observed accumulation of these RNA species in-vivo can be explained by an enhancement in the transcription rate of their genes.

3.3 Light-induction of genes preceding chloroplast formation

In recent studies an attempt has been made to identify genes that are rapidly and temporarily activated upon blue light irradiation of cultured cells, and their products which could be involved either in light perception, signal transduction and response or in coordination of the expression of other light-inducible genes in connection with chloroplast development (7).

The coordinate expression of nuclear and plastid genes coding for chloroplast proteins starts 12 - 36 h after administering blue light. This "lag-phase" was used to search for rapidly activated genes via their mRNAs. From the combined preparations of poly(A)-RNA isolated from dark-grown cells exposed to blue light for 6, 12 and 24 h most of the species representative for dark-grown and fully greened cells were eliminated (7). With the remaining species a cDNA library was established. By differential screening a number of positive clones were identified; they hybridized to RNAs which were rapidly formed in dark-grown cells after exposure to blue light. A first group with a high level of expression of the corresponding genes ($>$100 copies per cell) represented by the clones CRE2, CRAJ1 and CRBJ1 could be distinguished from a second group with a lower expression level: CRHC1, CRHG1, CRH301 and CRHA1. Assessment of the transcript level by dot hybridization and Northern analysis employing either the specific insertion or antisense RNA as probe revealed a rapid and transient accumulation of specific transcripts from 12 up to 4 d of blue light irradiation for most of these clones (Fig. 3). In case of CRE2-mRNA the translational activity was likewise controlled by blue quanta: The accumulation kinetics of the specific translation product compares quite well with that of the CRE2-mRNA (7).

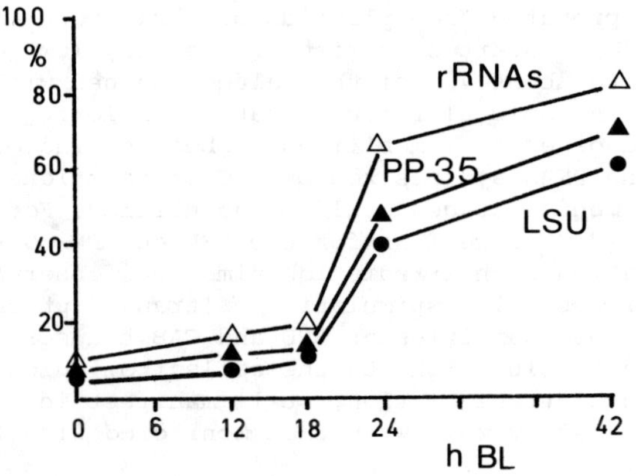

FIGURE 2. In-vitro transcription of plastid genes in TAC from plastids of dark-grown cells of Chenopodium rubrum which had been irradiated for the times indicated. The quantities of ^{32}P-labelled transcripts were determined by dot hybridization technique employing cloned sequences of plastid genes encoding LSU (rbcL), PP-35 (psbA) and the ribosomal RNA species of 16 S and 23 S. The value for the TAC preparation of fully greened cells (day 14) was set at 100%, and all other values referred to it. h BL = hours in blue light.

The protein specified by clone CRE2 has a molecular mass of 34 256 (7) and exhibits a substantial homology to the group of RLAO proteins (regulatory "acidic ribosomal proteins") in yeast and human cells (15, 16) including identical sequence elements like phosphorylation sites, leucine "zipper" motif and a predominantly hydrophilic alpha helix at the carboxy terminus.

The insertion of clone CRHC1 suggested a glycine-rich protein (GRP) of 144 amino acids and a molecular mass of 14 137 (17). The glycine residues are arranged in repeats of higher order: A stretch of 12 amino acids consisting of two slightly different hexapeptides, -GlyTyrAsnAsnGlyGly- and -GlyTyrHisAsnGlyGly-, is repeated six times, broken by one glycine residues only. Residues 9-36 probably form a transmembrane helix; the carboxy terminus of mostly hydrophilic residues seems to comprise an alpha helix, too. These structural features together with the mode of expression are unique in comparison to a group of GRPs discovered recently which are secreted from the cell and accumulate in the cell wall (18, 19).

The cDNA insertion of clone CRHG1 could encode a polypeptide of about 8 kDa (7). When the nucleotides in the coding region were compared with mRNA sequences of known proteins about 80% were identical to the mRNA of the ß-tubulin Mß5 (mouse, MMTUB5; 20). Apparently, the gene corresponding to clone CRHG1 encodes a tubulin-like protein of yet unknown function in plant cells.

We are far from understanding the significance of these early light-induced genes and the biological role of their products. They could be involved either in a sequence of events preceding the developmental process, i.e. chloroplast differentiation, or in a

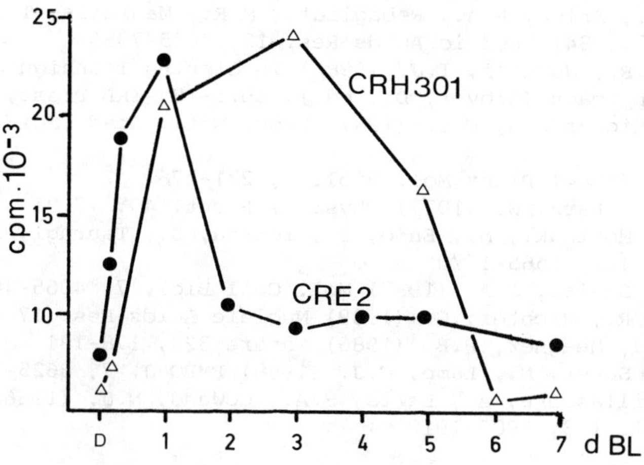

FIGURE 3. Time course of the blue light-induced changes in the amounts of mRNAs homologous to clones CRE2 and CRH301.
Hybridization was to ^{32}P-labelled antisense RNA transcribed from the cDNA insertions of clones CRE2 and CRH301 (Northern analysis). For quantitation the radioactivity bound to the hybrid bands was counted. The insertions of clones corresponding to genes without light induction served as internal standards. d BL = days in blue light.

general adaptation of the cellular activities to the impact of sudden blue light irradiation. They are novel insofar as their transcripts and deduced polypeptides have not been found so far. Further insight into their functions will come from studies using immunodetection with antibodies raised against these polypeptides after their expression in bacterial cells; these studies are in progress.

4. ADDENDUM

This studies were supported by the Deutsche Forschungsgemeinschaft and the Ministerium für Forschung und Technologie, Bonn.
We thank Mrs. I. Liebscher for her competent assistance.

REFERENCES

1 Link, G. (1988) Plant Cell Environ. 11, 329-338
2 Mullet, J.E. (1988) Ann. Rev. Plant Physiol. 39, 475-502
3 Bergmann, L., Berger, Ch. (1966) Planta 69, 58-69
4 Richter, G., Hundrieser, J., Gross, M., Schultz, S., Bottländer, K., Schneider, Ch. (1984) In Blue Light Effects in Biological Systems (Senger, H., ed.), pp.387-396, Springer, Berlin Heidelberg New York
5 Richter, G., Dudel, A., Einspanier, R., Dannhauer, I., Hüsemann, W. (1987) Planta 172, 79-87
6 Gross, M., Richter, G. (1982) Plant Cell Rep. 1, 288-290
7 Kaldenhoff, R., Richter, G. (1990) Planta 181, in press
8 Reiss, Th., Link, G. (1985) Eur. J. Biochem. 148, 207-212
9 Richter, G., Ottersbach, N. (1990) Botanica Acta 103, in press

10 Melton, D.A., Krieg, P.A., Rebagliati, M.R., Maniatis, T., Zinn, K., Green, M.R. (1984) Nucleic Acids Res. 12, 7035-7056

11 Little, P.F.R., Jackson, I.J. (1986) In Cloning Techniques; A practical approach (Glover, D., ed.), pp.1-18, IRL Press, Oxford

12 Tabor, S., Richardson, C.C. (1987) Proc. Natl. Acad. Sci. USA 84, 4767-4771

13 Richter, G. (1984) Plant Mol. Biol. 3, 271-276

14 Hüsemann, W., Barz, W. (1977) Physiol. Plant. 40, 77-81

15 Mitsui, K., Motizuki, M., Endo, Y., Yokata, S., Tsurugi, K. (1987) J. Biochem. 102, 1565-1570

16 Rich, B.E., Steitz, J.A. (1987) Mol. Cell Biol. 7, 4065-4074

17 Kaldenhoff, R., Richter, G. (1989) Nucleic Acids Res. 17, 285

18 Condit, C.M., Meagher, P.B. (1986) Nature 323, 178-181

19 Keller, B., Sauer, N., Lamb, C.J. (1988) EMBO J. 7, 3625-3633

20 Wang, D., Villasante, A., Lewis, S.A., Cowand, N.J. (1986) J. Cell Biol. 103, 1903-1910

LIGHT-DEPENDENT GROWTH AND DIFFERENTIATION PROCESSES IN PHOTOAUTOTROPHIC CELL CULTURES

W. HÜSEMANN [1], S. AMINO, K. FISCHER, H. HERZBECK & R. CALLIS [2]
1) Department of Plant Biochemistry, University of Münster, Hindenburgplatz 55, D-4400 Münster; 2) Department of Physics, University of Bremen, D-2800 Bremen, FRG

1. INTRODUCTION

Photoautotrophic growth of in vitro cultured plant cells has been achieved for about 23 different plant species. The physiological and biochemical characteristics of photoautotrophically cultured plant cells have been intensively studied (1, 2, 3, 4) and gave some valuable informations for a better understanding of basic questions about photoautotrophy.

This report describes some physiological and biochemical reactions involved in photoautotrophic growth and metabolism of cell suspension cultures from *Chenopodium rubrum*.

2. PROCEDURE

2.1 Material and methods

2.1.1 *Photoautotrophic cell cultures*: Cell suspension cultures from *Chenopodium rubrum* are grown photoautotrophically in the presence of 2% CO_2 (v/v) using the two-tier culture vessel method (5).

2.1.2 *Analytical procedures*: Measurements of cell culture growth, 14-CO_2-pulse / 12-CO_2-chase photosynthesis, metabolism of exogenously supplied radioactively labelled malate, isolation and quantification of 14-C-labelled photosynthetic compounds were performed following well established procedures (6). If necessary, the pH of the incubation medium of the cells was adjusted to 4.5 and 6.0 in the presence of 50 mM MES-NaOH-buffer.

2.1.3 *Intracellular distribution of enzyme activity*: Protoplasts were prepared from rapidly dividing cells and purified by centrifugation on a 2.5% / 5% Ficoll gradient (7, 8). Protoplasts were mechanically disrupted and the chloroplasts and mitochondria were separated by differential centrifugation on Percoll gradients; chloroplasts: 40% / 80% Percoll gradient; Mitochondria: 27% self-establishing Percoll gradient (8). Enzyme activities were measured as described (6, 7).

2.1.4 *Intracellular pH-measurements by 31-P-NMR-spectroscopy*: The 31-P spectra were recorded on a Bruker AM 360 spectrometer operating at 145.78 MHz, unlocked, and without proton decoupling.

For one spectrum 1800 scans were accumulated with a recycle time of 1 s. The tube diameter was 20 mm. Between the aquisition times the cells were supplied with a gas-mixture (air + 2% CO_2). The cells were illuminated via an optic cable. Photon flux density was 140 μE m^{-2} s^{-1}. Chemical shifts were referenced to 85% phosphoric acid. Cytoplasmic and vacuolar pH's were determined by NMR according to the chemical shifts of inorganic phosphate, using a pH calibration curve. The latter was obtained by titration of inorganic phosphate in cell sap.

3. RESULTS AND DISCUSSION

3.1 Photoautotrophic cell suspension cultures from *Chenopodium rubrum* assimilate CO_2 predominantly by $C_1 \rightarrow C_5$-carboxylation using the RUBP-carboxylase. Due to high PEP-carboxylase activity about 20% of total photosynthetically assimilated carbon is incorporated into malate. The rate of malate formation is pH-dependent. If the cells are supplied with equal amounts of 14-C-labelled HCO_3^- and the pH of the incubation medium of the cells is increased from 4.5 to 6.0, the rate of 14-C-incorporation into malate and aspartate after 10 s 14-CO_2-photosynthesis is nearly doubled at the expense of phosphoglyceric acid and sugar-phosphates. Malate is synthesized by ß-carboxylation of phosphoenol pyruvate using the PEP-carboxylase as indicated by the intramolecular 14-C-label distribution in malate (4).
It has been supposed that the increase in ß-carboxylation rates after changing the extracellular pH from 4.5 to 6.0 might be due to a pH-dependent stimulation of the PEP-carboxylase activity. This implies that changes in the extracellular pH will alter the intracellular (cytoplasmic) pH.

To study this possibility, we applied the in vivo 31-P-NMR-spectroscopy for intracellular pH measurements (Ta. 1).

TABLE 1. Intracellular pH in photoautotrophic cell cultures of *Chenopodium rubrum* as measured by in vivo 31-P-NMR methods.

pH of the incubation medium of the cells	intracellular pH cytoplasm	vacuole
4.5 ± 0.2	7.4 ± 0.2	5.3 ± 0.3
6.3 ± 0.2	7.7 ± 0.2	6.6 ± 0.3

The data are the mean of 5 independent measurements

If the incubation medium of the cells was changed from 4.5 to 6.3, the cytoplasmic pH increased for about 0.3 units, the vacuolar pH for 1.3 units. These observations corroborate experimental data, that the vacuolar pH in plant cells is more sensitive to external pH variations compared to the cytoplasmic pH which is rather tightly regulated. Possibly, the vacuole is part of a homeostatic mechanism to maintain cytoplasmic pH constant (9, 10). The induced increase in cytoplasmic pH in photoautotrophically cultured cells from 7.4 → 7.7 would be sufficient to account for the observed stimulation of ß-carboxylation and malate formation mediated by the PEP-carboxylase (4).

The pH-dependency of CO_2 assimilation is also expressed during long-term (1 h) 14-CO_2-photosynthesis. If the pH of the culture medium of the cells is changed from 4.5 to 6.0, 14-C incorporation into amino acids, organic acids, as well as into protein is substantially increased at the expense of sugar and starch. At the same time, the malate content increased from 67 µg to 440 µg per mg chlorophyll.

TABLE 2. pH-dependent changes in the metabolic flow of photosynthetically assimilated carbon in photoautotrophic cell cultures from Chenopodium rubrum

Chemical fractions	% changes in ^{14}C-label distribution during $^{14}CO_2$-pulse / $^{12}CO_2$-chase photosynthesis	
	pH of the incubation medium	
	4.5	6.0
Lipids, pigments	(+) 117	(+) 196
Amino acids	(-) 34	(+) 53
Organic acids, sugar-phosphates	(-) 60	(-) 35
Sugars	(-) 51	(-) 74
Starch	(+) 50	(+) 25
Protein	(+) 179	(+) 440
Insoluble residue	(+) 35	(+) 100

The intensity of carbon flow has been calculated from the loss (-) and increase (+) of radioactivity in the chemical fractions after a 5 h 12-CO_2-chase following a preceding 1 h 14-CO_2-photosynthesis. The data are the mean of 3 independent experiments.

Metabolic carbon flow of photosynthetically assimilated carbon has been studied by 14-CO_2-pulse / 12-CO_2-chase photosynthesis (Table 2). Changing the extracellular pH from 4.5 to 6.0 resulted in a drastic increase of carbon flow into lipids, amino acids and protein, as well as into the insoluble residue, i.e. cell wall material. Sugars and organic acids / sugar-phosphates always functioned as "carbon-sources" for divers biosynthetic pathways.

In addition, about 80% of 14-C-label in malate, synthesized during 1 h 14-CO_2 photosynthesis was turned over during the following 5 h 12-CO_2-chase, underlining its function in photosynthetic carbon metabolism. To avoid any interference of carbon flow from malate into different metabolic pathways by the metabolism of other 14-C-labeled photosynthetic intermediates, the metabolism of exogenously supplied radioactively labeled malate was measured.

TABLE 3. Metabolism of exogenously supplied ^{14}C-malate in potoautotrophic cell cultures from Chenopodium rubrum

	$(U-^{14}C)$-malate	$(4-^{14}C)$-malate
^{14}C-malate uptake :		
(% of ^{14}C-malate supplied)	92.0	92.5
^{14}C-malate metabolized :		
(% of ^{14}C-malate uptake)	95.0	94.0
^{14}C-distribution pattern :		
(% of ^{14}C-malate metabolized)		
CO_2	4.1	7.1
Citrate	3.6	1.5
Succinate	2.5	-
Fumarate	1.0	1.4
Aspartate	26.2	17.1
Alanine	27.5	24.6
Glycine, Serine	4.4	3.0
Sugar	8.8	20.0
Not identified	4.0	3.3
70%-methanol insoluble residue	18.0	23.0

Under standard culture conditions cells from exponential growth phase were exposed for 1 h to 10 mM Malate containing 37.7 KBq (U-14-C)- or 4-14-C-malate. The data are the mean of 3 independent experiments.

Uptake and metabolism of malate was linear with time, reaching about 90% within 60 min (Tab. 3). Maximum proportions of 14-C-label from malate have been recovered in amino acids, sugars. A great proportion of 14-C-radioactivity in the methanol-insoluble residue can be refered to protein. The formation of 14-C-alanine from malate requires decarboxylation as well as randomization reactions by the malic enzyme and fumarase, respectively. The formation of 14-C-labeled sugars from 14-C-malate may be accomplished by re-assimilation of 14-CO_2 evolved by decarboxylation processes.

These studies on malate metabolism suggest, that malate will serve as a source of carbon for diverse biosynthetic pathways and imply the operation of glycolytic and citric acid cycle reactions in cultured photoautotrophic cells.

The intracellular distribution of enzymes involved in malate formation und metabolism (i.e. PEP-carboxylase, malate dehydrogenase, malic enzyme) has been studied after fractionation of protoplasts (Tab. 4).

TABLE 4. Intracellular distribution of enzyme activity in photoautotrophic cell suspension cultures from Chenopodium rubrum

Enzyme	Enzyme activity		
	Protoplasts[a]	Chloroplasts[b]	Mitochondria[b]
NAD - Malate Dehydrogenase	10060	3.1	51.2
NADP-Malate Dehydrogenase	106	77.3	36.3
NAD - Malic Enzyme	67.4	--	100
NADP-Malic Enzyme	27	80.5	--
RUBP-Carboxylase	84.6	100	--
Glyceraldehyde-3-phosphate Dehydrogenase	254	98	--
PEP - Carboxylase	80.6	--	--
Catalase	46570	--	35.7
Fumarase	133.5	--	100
Cytochrome C Oxidase	21.3	--	100

a) In vitro activity of enzymes directly extracted from protoplasts without fractionation and calculated as μMol substrate (mg chl.)$^{-1}$h^{-1}
b) Recovery of enzyme activity (%)
The data are calculated with reference to RUBP-carboxylase (100%) for chloroplasts and fumarase activity (100%) for mitochondria
The date are the mean of 5 independent experiments

While isolated chloroplasts were highly purified as documented by the distribution of marker enzymes, the mitochondria preparations were obviously still contaminated with peroxisomes as indicated by catalase activity.

The intracellular localization of the NADP-malic enzyme and of the NADP-malate dehydrogenase (chloroplast fraction) as well as of the NAD-malic enzyme and NAD-malate dehydrogenase (mitochondria fraction) in photoautotrophic cell cultures from Chenopodium rubrum is quite similar to mesophyll cells. While NAD - malic enzyme is probably restricted to the mitochondria, it seems that the NAD-malate dehydrogenase will be distributed on other cell compartments too. The presence of NADP-malate dehydrogenase activity in the mitochondria might be due to substrate inspecifity of the NAD - malate dehydrogenase.

The pattern of the intracellular distribution of the NAD/NADP-dependent forms of malic enzyme and malate dehydrogenase imply, that the chloroplasts and mitochondria will be involved in specific steps of malate metabolism.

4. ADDENDUM

4.1 Under standard culture conditions photoautotrophic cell cultures from Chenopodium rubrum are grown in the presence of 2% (v/v) CO_2 in gas phase, equivalent to about 800 μMol inorganic carbon per l culture medium. That is about a 70-fold increase compared to atmospheric air. The question arises whether it will be possible to grow photoautotrophic plant cell cultures under ambient CO_2 concentrations. This question can now be affirmed. Cell cultures from Dianthus caryophyllus (11), Arachis hypogea (12), Gossypium hirsutum (13) and Chenopodium rubrum (Hüsemann, W. unpublished) are grown at ambient CO_2 concentrations, though cell growth and photosynthetic capacities of these cell lines are rather low (11, 14). Yet, they will represent a suitable system to study the "limiting factors" restricting photoautotrophic growth of cultured plant cells under atmospheric air.

Our investigations were supported by the Deutsche Forschungsgemeinschaft and by the Bundesminister für Forschung und Technologie.

REFERENCES

1 Hüsemann, W. (1985) in Cell Culture and Somatic Cell Genetics of Plants (Vasil, I K., ed.), Vol. 2, pp. 213-252, Academic Press, New York

2 Widholm, J.M. (1989) in Primary and Secondary Metabolism of Plant Cell Cultures II (Kurz, W., ed), pp. 3 - 13, Springer-Verlag, Berlin

3 Takeda, S, Sato, F. and Yamada, Y. (1989) Plant Cell Physiol. 30, 885-891

4 Hüsemann, W., Fischer, K., Mittelbach, I., Hübner, S., Richter, G. and Barz, W. (1989) in Primary and Secondary Metabolism of Plant Cell Cultures II (Kurz, W., ed.), pp. 35-46, Springer Verlag, Berlin

5 Hüsemann, W. and Barz, W. (1977) Physiol. Plant. 40, 77 - 81

6 Hüsemann, W., Herzbeck, H. and Robenek, H. (1984) Physiol. Plant. 62, 349 - 355

7 Mittelbach, I. (1988) Master's Thesis, Münster University

8 Amino, S.I. and Hüsemann, W. (1989) in Secondary Plant Products from in vitro Cultures, Book of Abstracts, Dutch-German-Workshop, pp. 73, International Agricultural Center, Waageningen

9 Guern, J., Mathieu, Y., Pean, M., Pasquiner, C., Beloeil, J. and Lallemand, J. (1986) Plant Physiol. 82, 840 - 845

10 Marré, E., Beffagna, N. and Romani, G. (1988) Botan. Acta 101, 24-31

11 Rebeille, F. (1988) Plant Science 54, 11 - 21

12 Neumann, K.H., Groß, U. and Bender, L. (1989) in Primary and Secondary Metabolism of Plant Cell Cultures II (Kurz, W., ed.), pp. 281 - 291, Springer Verlag, Berlin - Heidelberg

13 Chunhe Xu, Blair, L.C., Rogers, S.M., Godvindjee, K. and Widholm, J. M. (1988) Plant Physiol. 88, 1297 - 1302

14 Roeske, Ch. A., Widholm, J.M. and Ogren, W.L. (1989) Plant Physiol. 91, 1512 - 1519

LIGHT STIMULATED CELL PROLIFERATION AND GLYOXALASE ACTIVITY IN CALLUS CULTURES OF AMARANTHUS PANICULATUS.

TARAKNATH CHAKRAVARTI & SUDHIR K. SOPORY
Plant Research Laboratory, School of Life Sciences, Jawaharlal Nehru University, New Delhi-110067, India.

1. INTRODUCTION

There are very few reports on tissue culture of amaranths (1,2,3,4). It was shown that *Amaranthus paniculatus* develops a highly proliferative callus *in vitro* on the B5 medium supplemented with 1 ppm each 2,4-Dichlorophenoxy acetic acid (2,4-D), kinetin (Kn) and gibberellic acid (GA_3) and is therefore a good system to understand the biochemical basis of cell division and proliferation. The exact mechanism of cell proliferation and the role of light in cell proliferation in tissue culture is not known. The activity of the enzyme, glyoxalase-I (S-lactoyl GSH methyl glyoxalase EC 4.4.1.5) serves as an indicator of cell growth status. The enzyme converts α-keto aldehyde into thiol ester of glutathione which is converted to α-hydroxyl carboxylic acid by glyoxalase-II (2,5,6,7). The presence of the glyoxalase enzyme in plants has been reported (6) and a correlation was found between an increase in the enzyme activity with increased callus growth (7,8,9). In this paper we have attempted to study the role of light in cell proliferation and to check if glyoxalase activity is also stimulated by light.

2. PROCEDURE

2.1 Plant Material: The seeds of *Amaranthus paniculatus* were obtained from Indian Agricultural Research Institute, Delhi. Seeds were surface sterilized with 0.1% $HgCl_2$ and germinated on B5 medium (10) containing 0.8% agar and 2% sucrose without any hormones. Hypocotyl segments from 10 days old plantlets were inoculated on B5-medium (10) supplemented with 1mg/litre each of kinetin, gibberellic acid and 2,4-D. After 15 days the callus was subcultured on the same medium or the medium under the experimental condition and maintained at 25±1°C under constant illumination (1200 $\mu W\ Cm^{-2}$) and in darkness.

2.2 Light treatments: Light treatments were given to the different calli as per the experimental conditions. Red light (0.066 $\mu W\ Cm^{-2}$) was obtained by filtering the light from four 100 W tungsten lamps through a CBS-650 filter

(Carolina Biological Supply Co. USA); Far-Red light (0.045 μW Cm^{-2}) was obtained from a 250 W tungsten reflector lamp (Westinghouse, USA), the output of which was filtered through a CBS-750 filter (Carolina Biological Supply Co., USA); blue light (0.0015 μW Cm^{-2}) was obtained from 8,40 watts fluorescent tubes (Philips, India), light filtered through a CBS-450 filter. The growth (fresh weight) of calli was measured after 8 days of culture. The calli were transferred to darkness for 4 days and on the 5th day after the required light treatments of red, red+far-red and blue light, glyoxalase was assayed, after 4 hrs, 12 hrs and 24 hrs respectively. The experiments were repeated at least twice and ± SD was calculated.

2.3 Glyoxalase assay: The enzyme glyoxalase-I (S-lactoyl GSH methyl glyoxalase EC 4.4.1.5) was extracted by homogenizing callus (1 gm) in 0.1M sodium phosphate buffer (pH 7.0) containing 20 mg $MgSO_4.7H_2O$ and 20 ml glycerol/100 ml. The extract was centrifuged at 15000 rpm for 20 min. in SS34 rotor in Sorvall RC5 centrifuge. The supernatant was fractionated with 40-80% ammonium sulfate and the precipitate was re-dissolved in the extraction buffer, and employed for glyoxalase-I assay according to Racker (5) and as modified for plant tissues by Ramaswamy et al. (7). The assay mixture contained 100 mM sodium phosphate buffer (pH 7.5), 3.5 mM methyl glyoxal, 15 mM sodium phosphate buffer (pH 7.5), 35 mM methyl glyoxal, 15 mM magnesium sulfate and 1.7 mM glutathione. The mixture was incubated for 7 min for formation of a complex between methyl glyoxal and glutathione. The reaction was started by the addition of enzyme (0.02 ml). The formation of thioester which shows maximum absorption at 240 nm was measured in Shimadzu UV-260 spectrophotometer. The enzyme unit (IU) is defined as the amount of enzyme catalyzing the formation of 1 μmol of S-lactoylglutathione from methyl glyoxal and reduced glutathione/min. at 25°C. Specific activity of the enzyme was determined as units per gm. fr. wt.

3. RESULTS AND DISCUSSION

To check the effect of light on proliferation, callus growing on B5+2,4-D+GA_3+Kn was transferred to B5 Basal medium, (minus hormones) or to the same medium but kept in either total darkness or continuous light regime or treated under different light treatments of red, red+far-red and blue light.

3.1

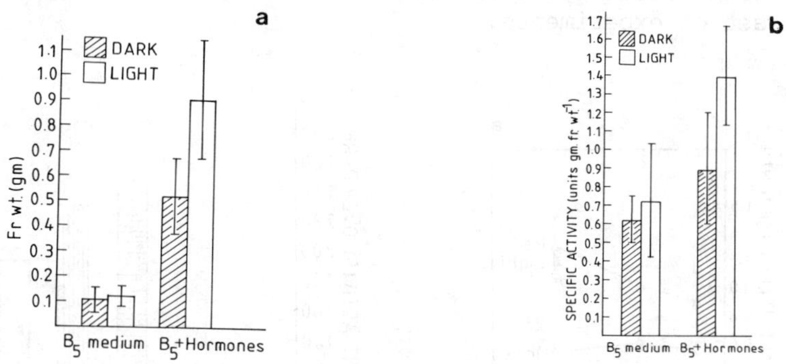

Fig.1(a) Cell proliferation in light and dark in Amaranthus paniculatus callus maintained in all three hormones or transferred from medium containing all three hormones to Basal medium. Cell proliferation expressed as gm. fr. wt.

Fig.1(b) Effect on glyoxalase-I activity in callus grown on B5 (minus hormones) and medium containing all 3 hormones under light and dark condition. After 8 days of culture the glyoxalase-I activity was measured and expressed as units gm. fr. wt.

As shown in Fig.1(a) after 8 days of culture the callus showed greater proliferation in light grown cultures. It was observed that hypocotyl segments when grown on B5 medium minus hormones showed very slow proliferation and in turn showed high proliferation when grown with all 3 hormones, indicating that hormones triggered this response. In the present study it seems that light has an effect over and above that induced by hormones. There was an 175% increase in cell proliferation in light over their corresponding dark grown cultures. If the cultures were left in darkness for long, the proliferation reduced further.

3.1.1 The cell proliferation has been related to the glyoxalase-I activity (2,6,7,8,11). Since we have shown earlier, correlation between glyoxalase activity and proliferation (2,7,8), in the present study we checked, if light also plays a role in glyoxalase activity. As evident by the data in Fig.1(b) it was observed that glyoxalase activity was higher in all 3 hormones when compared to B5 basal medium but there was a 155% increase of glyoxalase activity in light grown cultures in all 3 hormones when compared to their corresponding dark controls. The mechanism by which light enhances the enzyme activity remains to be seen, however it correlates with cell proliferation as is evident in the present set of experiments.

3.2

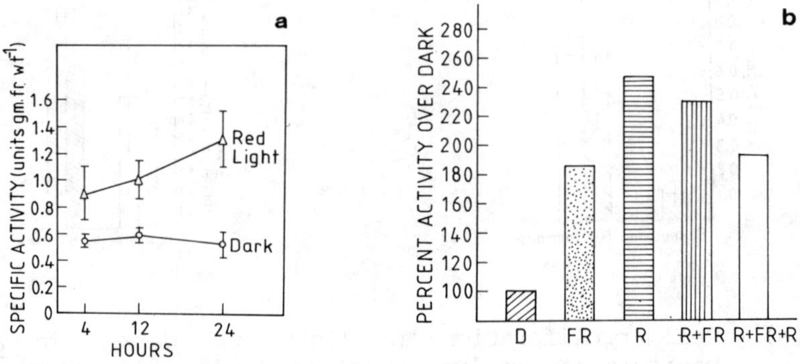

Fig.2(a) Effect on glyoxalase-I activity in **all** 3 hormones in dark and 5 min of red light treatment in <u>Amaranthus paniculatus</u> cultures. The glyoxalase activity was measured after 4, 12 and 24 hours respectively on the 5th day. The glyoxalase activity was expressed as units sp. activity/gm. fr. wt.$^{-1}$ Fig.2(b) Percentage increase in glyoxalase activity at 24 hrs, in red(R) far-red (FR), R+FR and R+FR+R light treated cultures over their corresponding dark grown cultures.

3.2.1. To check for the involvement of phytochrome in cell proliferation, the cultures were exposed to red light irradiation for 5 min and glyoxalase assay was done after 4 hr, 12 hrs and 24 hrs respectively. At 4 hrs there was a 162% increase in stimulation of glyoxalase activity compared to the dark controls and also increase in stimulation was observed in 12 and 24 hrs as shown in Fig.2(a). To check if this stimulatory effect was under phytochrome control red light and far-red light reversible experiments were done which usually demonstrate low energy response of phytochrome action. (13,14). As shown in Fig.2(b) there was no clear R/FR reversibility observed. Infact far-red alone showed 187% increase and the red light stimulation of 250% was not reversed by FR light. It seems that stimulation of glyoxalase activity could be a very low fluence response. Also it is not clear why in R+FR+R there was a further decrease in enzyme activity. This needs to be studied in detail.

3.3

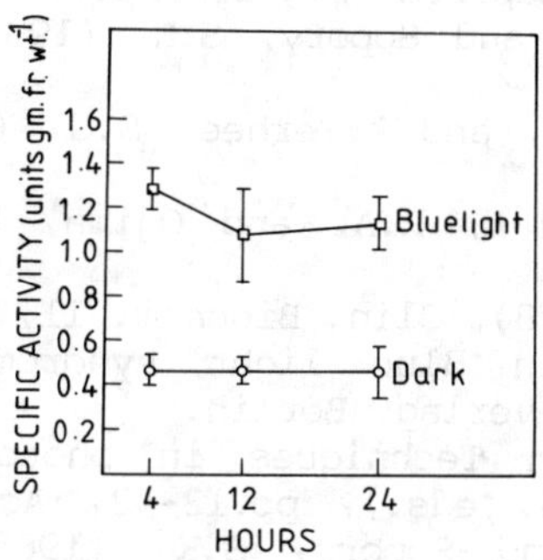

Fig.3 Effect on glyoxalase-I activity in all 3 hormones in dark and 5 min of blue light irradiation.

3.3.1 Further experiments were done to check if blue light was involved in stimulating glyoxalase activity. For this cultures were exposed to blue light treatment for 5 min and glyoxalase activity was measured after 4 hrs, 12 hrs and 24 hrs respectively. At 4 hrs there was a 278% increase in stimulation in glyoxalase activity over the dark cultures, whereas there was only 162% increase by red light. It is known that various responses are induced in plants that are operated by blue light activity via phytochrome or its own photoreceptor (12,15). It would be interesting to check in this system if the effects of blue light on the stimulation of glyoxalase is mediated via phytochrome, or a blue light receptor.

4. ACKNOWLEDGEMENT

TNC was supported by UGC fellowship. The research was partly supported by a CSIR grant to SKS. This contribution is dedicated to our lab colleague late Mr. Ashootosh Doshi.

5. REFERENCES

1 Bagga, S., Venkateshwarulu, K. and Sopory, S.K. (1987) Plant Cell Reports 6, 183-184.
2 Das, R., Bagga, S. and Sopory, S.K. (1987) Plant Science 53, 45-51.
3 Flores, H.E., Thier, A. and Galston, A.W. (1982) Am. J. Bot. 69, 1049-1054.
4 Singhal, R.S. and Kulkarni, P.R. (1988) Int. Journal of Food and Technology 23, 125-139.
5 Racker, E. (1951) J. Biol. Chem. 190, 685-696.
6 Ramaswamy, O., Pal, S. Guha-Mukherjee, S. and Sopory, S.K. (1983) Biochem. International 7, 307-318.

7 Ramaswamy, O., Pal, S. Guha-Mukherjee, S. and Sopory, S.K. (1984) Plant Cell Reports 3, 121-124.
8 Bagga, S., Das, R. and Sopory, S.K. (1987) J. Plant Physiol. 129, 149-153.
9 Sethi, U., Basu, A. and Mukherhee, S.G. (1988) Plant Science 56, 167-175.
10 Gamborg, O.L., Miller, R.A. and Ojima, K. (1968) Exp. Cell Res. 50, 151-158.
11 Nyandieka, H.S. (1978) Clin. Biochem. 11, 150-155.
12 Mohr, H. (1980) in Blue light syndrome (Senger, H. ed.), pp.97-108. Springer-verlag, Berlin.
13 Mohr, H. (1984) in Techniques in photomorphogenesis (Smith, H. and Holmes, M.G. eds.), pp.13-42, Academic Press, London.
14 Rajasekhar, V.K. and Sopory, S.K. (1985) New Phytol. 101, 251-258.
15 Sarkar, H. and Song, P.S. (1982) Photochem. Photobiol. 35, 143-247.

SPATIAL AND TEMPORAL EXPRESSION OF REGULATORY GENES IN *Zea mays*

C. TONELLI, G. CONSONNI, M. MEREGHETTI, S. DOLFINI, S. DELLAPORTA*, A. VIOTTI**, G. GAVAZZI.
Dept. of Genetics and Microbiology, Via Celoria 26, Milano 20133, ITALY
*Dept. of Biology, Yale Univ. New Haven CT USA
**Ist. Biosintesi Vegetali CNR Via Bassini 15 Milano, ITALY

1. INTRODUCTION

Flavonoids represent a class of secondary metabolites which is widespread among plants and they may exert specific functions in different processes such as pigmentation of flowers, UV protection and plant defence against pathogens (1). Pigmentation within the plant is controlled by numerous genes affecting the type, amount and distribution of pigments. The structural genes encoding the enzymes involved in the biosynthesis are expressed in a coordinate way as the result of the activity of different regulatory genes. This process is also sensitive to environmental stimuli such as light, temperature, the presence of phytopathogens, wounding and nutrient supply. Flavonoids biosynthesis thus represents an appealing model system for the analysis of gene regulation as well as for studies in developmental biology.

In maize the genetic of anthocyanin accumulation is well documented being controlled by more than 20 loci (2). The *R*, *B* and *C1* genes regulate the activity of at least three structural genes in the pathway *C2*, *A1*, *Bz1* which code for chalcone synthase, dihidroquercitin reductase and UDP-glucose: flavanol 3-O-glucosyltransferase respectively (3-4-5-6-7-8-9). Almost all maize tissues are competent to produce anthocyanin, however the timing and distribution of pigment accumulation is controlled by the regulatory genes constitution of the plant. For example *C1* and *Pl* genes, control the aleurone (*C1*) and plant tissues (*Pl*) pigmentation (10) and are molecularly homologous. Another group of related regulatory genes is represented by the *R* family consisting of the *R* complex, *Sn*, *Lc* and *B*. *R* gene sequences have been successfully employed as hybridization probes to identify *Lc*, *Sn* and *B* clones (7-8-9 and Tonelli et al. manuscript in preparation); while *Lc* and *Sn* are close to *R* (two map units *R* distal), *B* is carried on a different chromosome. Their position could be consequence of maize evolution. This paper deals with expression of *Sn*. Several *Sn* sources have been identified in bolivian material and in genetic stocks (12-13). Presence of *Sn* leads to light inducible accumulation of anthocyanins in the pericarp and cob glumes. However, if the same plant carries *Pl* (on chromosome 6) pigmentation is no longer light dependent. Studies on the activity of different enzymes of flavonoid biosynthesis in young seedlings which differ in their *Sn* constitution and which are grown either in darkness or in continuous light have shown that *Sn* increases the level of the enzymes encoded by *C2* and *Bz1* in a coordinate, light dependent manner. This suggests a transactive regulatory role of this gene on the activity of the structural genes involved in flavonoid biosynthesis (14-15). Results obtained by irradiation of *Sn* lines with light of different qualities show that blue light elicits the strongest effect in term of pigment quantity, thus suggesting a link between *Sn* and presence of a photoreceptor activated by blue light (16). Several *Sn* accessions have been studied. They all map on the long arm of the chromosome 10, about two map units *R* distal. The geographic alleles of *Sn* in our hands differ in the rate of pigment accumulation, level of pigmentation, and territory of expression. *Sn:bol3* differs from the other accessions in its potential for pigment accumulation in seedling tissues which is at least five times greater than in plants with other *Sn* alleles, and in its light-independent pigmentation of the seed integuments.

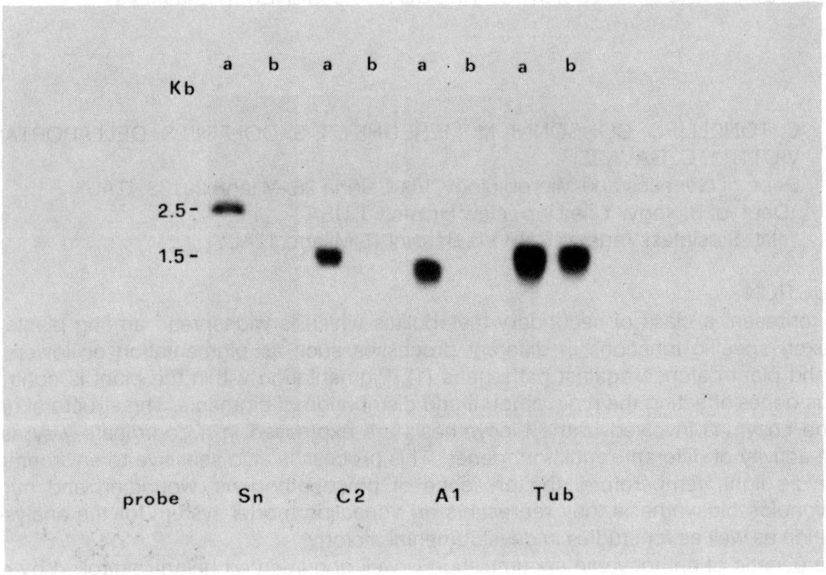

Fig. 1. Expression of *Sn* and two structural genes of anthocyanin biosynthesis by Northern blot analysis. 10 µg of total RNA extracted from scutellar nodes and mesocotyls of plants with (a) or without (b) the *Sn* gene. Seedlings were grown in darkness for five days and then exposed to continuous white light for 48 hours. The membrane was probed with 1,4 kb Pst1-EcoR1 fragment of the *Sn* cDNA (*Sn*),stripped and subsequently hybridized with a 1.5 Kb cDNA of *C2* (3) and a 700 bp *BamH*1 fragment of *A1* (4). RNA loads in each lane have been controlled (data not shown) using a maize cDNA tubulin clone (22). Film exposure for *Sn* transcript was three fold longer than for the others.

2. PROCEDURE

Materials. *Sn* is a factor lying two map units distal to *R* which confers specific tissue pigmentation following light exposure to the scutellar node, mesocotyl, leaf basis and midrib and to the seed integuments (glumes and pericarp). Three independent *Sn* accessions (*bol1, bol2, bol3*) have been identified in separate Bolivian populations collected by Dr. G. Avila (Centro Fitotecnico de Pairumani, Cochabamba) and a fourth one (*Sn:co-op*) in a stock kindly provided by the Co-op center. These *Sn* alleles were transferred to a common genetic background by repeated backcrosses into a colored version of the W22 inbred line. All four accessions are borne on an *r* marked chromosome that confers weak red pigmentation to the anthers (12-13).

Methods. RNA isolation and Northern blot analysis: seeds were allowed to germinate in darkness for 5 days at 25 ± 2°C until a mesocotyl about 8 cm long developed. Seedlings were then kept in darkness or exposed to white light for 12, 24 and 48 hrs at 21°C. As a light source Power Stars-HQ1-T400 W/DV OSRAM lamps were used with fluence rate of 68 W.m^{-2}. RNA was extracted from different plant tissues (mesocotyl, scutellar node, and leaf) by grinding with pestle and mortar in liquid N_2 and isolated by a previously described method (17). For Northern blot analysis 10 μg total mRNA was loaded on formaldehyde gels (18), and blotted after electrophoresis on Biodyne nylon membrane (Pall). Hybridizations was performed as previously described (7) for Southern blot. Filters were washed in 0.2 x SCP, 0.1% SDS at 65°C. RNA molecular weight markers were from Bethesda Research Laboratory. In situ hybridization: homozygous *rSn:bol3* developing kernels (16 days after pollination), mesocotyls, and scutellar nodes of 5 days old dark grown seedlings were exposed to light for 48 hours, fixed for 24 hours in 3:1 ethanol:acetic acid, and embedded in Paraplast Plus. Eight-micrometer sections were obtained from these tissues. ^{35}S-labeled riboprobes were synthesized using linearizing cDNA subclones, ^{35}S UTP (1300 Ci/mmole, Amersham) and T3 and T7 polymerases (IBI). Transcription, hybridization reactions, washes and autoradiography were carried as previously described (19-20). Autoradiographs were counterstained for 30 sec. with 0.025% Azur B pH 4.0.

3. RESULTS AND DISCUSSION

The cloning of the *Sn* gene has been achieved using a previously described (7) *R* genomic probe. We have demonstrated through recombinant analysis of the progeny of genetic crosses involving different *R* alleles and *Sn:bol3* a complete linkage between the fragments identified and the *Sn* gene. Genomic clones of the *Sn:bol3* covering the 5' and 3' region of the gene have been identified and characterized (manuscript in preparation).

To study the expression of this gene we compared two different *Sn* accessions: *Sn:Coop* which is strictly light dependent, and *Sn:bol3* which is partially constitutive in its expression since it conditions a weak pigmentation in dark growing seedlings. Total RNA was extracted from scutellar nodes, mesocotyls, and leaves of seedlings grown either in dark or in light, and analyzed by Northern blot . The results show that a 2.5 Kb transcript is found only in genotypes that carry *Sn* but not in *sn*. The expression of *Sn* is light regulated. In fact no transcript is found in scutellar nodes of *Sn:Coop* grown in darkness, while for the partially constitutive allele *Sn:bol3* the *Sn* transcript is also found in the scutellar nodes of dark growing seedlings.

To clone the cDNA for the 2.5 Kb *Sn* transcript, a cDNA library was established in lambda gt10 from light irradiated scutellar node mRNA of *Sn:bol3*. This library was screened with an *Sn* genomic clone, and a full size cDNA clone of 2.5 Kb has been obtained. The putative protein, deduced from the cDNA sequence has features similar to those of a transcriptional activator since it contains a large acidic domain that may be involved in transcriptional activation and a basic domain believed to be involved in DNA binding. Part of this basic domain was found to have homology to the myc-homology region found in the L-myc oncogene product and in other regulatory proteins of mammals and Drosophila (21).

Genetic and biochemical evidence indicate that *Sn*, like other *R* alleles, plays a regulatory role in anthocyanin biosynthesis, suggesting that the presence of an active *Sn* is required for the expression of the structural *A1*, *C2* and *Bz1* genes (5-8).

To confirm the regulatory role suggested by the DNA sequence, we analyzed the influence of the *Sn* gene on the *A1* and *C2* transcript levels. Total RNAs extracted from scutellar nodes of genotypes containing or lacking the *Sn* gene and exposed to light for 48 hrs were analyzed by Northern blots, hybridized with probes specific for *A1* and *C2* genes (3-4) and with the 3' end (1.4 Kb PstI-EcoRI fragment) of *Sn* cDNA. A probe for the maize tubulin gene (22) was used to control the level of mRNA in the lanes. *A1* and *C2* transcripts are detected only in the scutellar nodes where *Sn* mRNA is present (Figure 1). The accumulation of *Sn*, *A1* and *C2*, mRNA proceeds in a coordinate manner. A kinetic study showed that the maximum level of *A1* and *C2* mRNA accumulation was reached 24 hours later with respect to the maximum of *Sn*.

The spatial expression of *Sn* has been analyzed by means of *in situ* hybridization, that represents a valuable tool for the identification of specific gene expression in individual cells. Three different tissues, known to be under the regulation of *Sn* (the pericarp of the seed, the mesocotyl and the

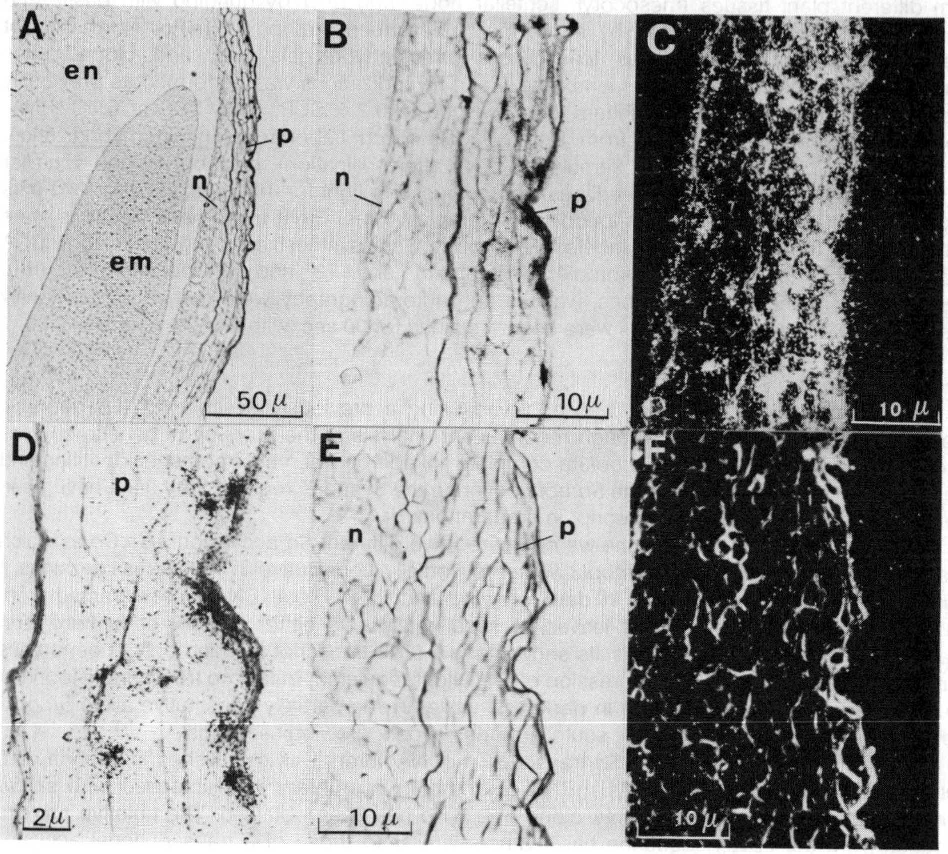

Fig. 2. Light micrographs of 8 µm sections of homozygous rSn:bol 3 developing kernels (16 days after pollination) hybridized with the antisense strand (A,B,C,D) and the sense strand (E,F) of ^{35}S-labeled RNA probes transcribed from Sn cDNA. Sections are stained with 0.025% Azur B. A strong hybridization signal, revealing the presence of Sn mRNA, is present in pericarp cells and, at a lesser degree, in nucellar tissue. C and F are dark-field views of B and E. *en* endosperm, *em* embryo, *p* pericarp, *n* nucella

scutellar node) were tested for the presence of *Sn* specific mRNA. Serial sections of the developing seed were prepared from homozygous *r Sn: bol 3* kernels at 16 days after pollination. Mesocotyls and scutellar nodes sections were obtained from 5-day-old seedlings of the same genotype grown in the dark and exposed to light for 48 hours. The *Sn* mRNA was localized by hybridizing sections with ^{35}S-labeled antisense- and sense-strand RNA probes synthesized from *Sn* cDNA. The antisense probe hybridizing with complementary mRNA, reveals the location of the transcript. On the other hand sense-strand probes containing sequences identical to mRNA show only background levels due to non-specific binding of probes to tissue sections.

Evidence of the presence of *Sn* transcripts was obtained in all tissues examined. In the developing seed (Fig. 2A), a strong hybridization signal was seen over the pericarp. *Sn* mRNA accumulated primarily in the outer layer of cells (Fig. 2B, C and D), whereas a less intense signal was detected over the nucellar tissue. Hybridization to other tissues under *Sn* regulation confirmed these results. In mesocotyls mRNA was associated largely with the external layer of cells, and the same pattern occurred in scutellar nodes (data not shown). In all the tissues, the signal was more frequently seen near cell walls and over or around nuclei because cytoplasm is reduced to a thin peripheral layer connected to the nuclear region by transvacuolar strands in highly vacuolated cells. When the same antisense-RNA probe was hybridized to sections of different tissues not under *Sn* control (like leaves), there was no visible hybridization. The specificity of binding was also confirmed by the lack of signal when the same antisense-RNA probe was hybridized to mesocotyls or scutellar nodes of genotypes lacking the *Sn* gene. Conversely, no hybridization was observed in control sections of all tissues probed with sense-RNA transcript (Fig. 2E and F). This localization of *Sn* mRNA in all tissues known to be under *Sn* control is a clear demonstration of the cell- and tissue- specific expression of this gene.

Our expression studies on *Sn* prove that *Sn:bol3*, contrary to other *Sn* accessions, is active also in darkness. This result is consistent with the observation that seedling tissues and cob glumes of *Sn:bol3* develop pigment even in darkness while all other accessions are strictly light dependent in their expression. Light is an essential signal for *Sn* activity of all known alleles except for *Sn:bol3*. In addition, the observation that the expression of the structural genes *C2*, *A1* and *Bz1* is mediated by the presence of an active *Sn* suggests the involvement of a trans acting *Sn* gene product in the coordinate activation of the genes of the flavonoid pathway.

Recently the cDNA sequence of *Lc*, a member of the *R* gene family, has been obtained (8). *Sn* cDNA sequence analysis shows a great sequence similarity with *Lc* although a large degree of polymorphism is present in the 5' transcribed but untranslated region . Taken together, the results suggest that the regulatory *R* and related *Sn*, *Lc* and *B* genes encode functionally related proteins that act as transcriptional activators of the genes encoding the flavonoid biosynthetic enzymes. The distinct patterns of pigmentation controlled by the *R*, *Sn*, *Lc*, *B* genes may reflect differences more in their promoter regions rather than in their products.

ACKNOWLEDGEMENTS

We thank Dr. U. Wienand for providing *A1* and *C2* clones and to Dr. J. Langdale for helpful suggestions for the in situ hybridization experiments. We are grateful to Prof. J. Stadler for critical reading of the manuscript and to G. Todesco for technical assistance. This work was supported by grants from Ministero dell'Agricoltura e Foreste "Progetto Sviluppo di Tecnologie Avanzate applicate alle piante" to G.G.

REFERENCES

1. Chappell, J. and Hahlbrock, K. (1984) Nature 311, 76-79.
2. Coe, E.H., Hoisington, Jr. and Neuffer, M.G. (1988) In: G.F.Sprague and J. Dudley (Ed.) "Corn and Corn Improvement" American Society of Agronomy, Inc., Madison WI U.S.A. pp. 81-258.
3. Wienand, U., Weydemann, U., Niesbach-Kloesgen, U., Peterson, P.A. and Saedler, M. (1986) Mol. Gen. Genet. 203, 202-207.

4. Schwarz-Sommer, Z., Shepherd, N., Tacke, E., Gierl, A., Rohde, W., Leclercq, L., Mattes, M., Berndtgen, R., Peterson, P.A. and Saedler, H. (1987) Embo J. 6, 287-294.
5. Dooner, H.K. and Nelson, O.E. Jr. (1977) Biochem. Genet. 15, 509-519
6. Cone, K.C., Burr, F.A. and Burr, B. (1986) Proc. Natl. Acad. Sci. USA 83, 9631-9635.
7. Dellaporta, S.L., Greenblatt, I., Kermicle, J.L., Hicks, J.B. and Wessler, S.R. (1988) In: J.P. Gustafason and R. Appels (Eds.) "Chromosome Structure and Function" Pl. Publ. Corp. pp. 263-281.
8. Ludwig, S.R., Habera, L.F., Dellaporta, S.L. and Wessler, S.R. (1989) Proc. Natl. Acad. Sci. USA 86, 7092-7096.
9. Chandler, V.L., Radicella, P.J., Chen, J. and Turks, D. (1989) Plant Cell 1, 1175-1183.
10. Cone, K.C. and Burr, B. (1989) in Genetics of flavonoids (Styles D. et al., Eds.) pp 143-145, Unicopli, Milan.
11. Styles, E.D., Ceska, O. and Seah, K.F. (1973) Can J. Genet. Cytol. 15, 59-72.
12. Gavazzi, G., Viani, I., Jacomini, E. and Helm, B. (1986) In: "Gene structure and function in higher plants" Proc. Intern. Symp. Hyderabad, India pp. 91-103.
13. Gavazzi, G., Mereghetti, M., Consonni, G. and Tonelli, C. (1990) Genetics, in press.
14. Gavazzi, G., Racchi, M.L., Mikerezi, I. and Gerats, A.G.M. (1985b) In: M. Freeling (Ed.) "Plant Genetic" A.L. Liss Inc. N.Y. UCLA Symp. Molec. Cell. Biol. New Series 35, 523-535.
15. Consonni, G., Racchi, M.L., Shammah, S. and Gavazzi, G. (1987) Maize Genet. Coop. Newsletter 60, 40.
16. Gavazzi, G., Mikerezi, I., Papinutti, P. and Tonelli, C. (1985a) Maydica 30, 309-319.
17. Van Tunen, A.J., Joes, R.E., Spelt, C.E., Van der Krol, A.R., Stuitje, A.R. and Mol, N.M. (1988) EMBO J. 7, 1257-1263.
18. Maniatis, T., Fritsch, E.F. and Sambrook, J. (1982) "Molecular Cloning: A Laboratory Manual" Cold Spring Harbor Laboratory Press, Cold Spring Harbor, N.Y.
19. Ingham, P., Howard, K.R. and Ish-Horowicz, D. (1985) Nature 318, 439-445.
20. Langdale, J.A., Rothermel, B.A. and Nelson, T. (1988) Genes and development 2, 106-115.
21. Murre, C., McCaw, P.S. and Baltimore, D. (1989) Cell 56, 777-783
22. Mereghetti, M., Consonni, G. and Tonelli, C. (1990) Maize Genet. Coop. News Letter 64, 87-88.

DEVELOPMENTAL MUTANTS IN CARROT

Mario TERZI and Fiorella LO SCHIAVO, Department of Genetics, General and Molecular Biology, University of Naples, Via Mezzocannone 8, I- 80134 Napoli, Italy.

1. INTRODUCTION

Development of the flowering plants can be divided in two parts: the first corresponds to morphogenesis and formation of the apical meristems; from this stage we pass -after desiccation and the preparation to dormancy- to germination and growth, that corresponds to development of the apical meristems, i.e. the continuous production of new plant organs, shoots, leaves, roots.

In order to study early development, the microsurgical experiments, used with such a success in animal embryology, could not be performed, due to the difficulty posed by the seed protective coats.The genetic approach was attempted as early as 1923 by Mangelsdorf (1), who described several maize strains with a recessive lethal causing seed abortion. The early work was followed in more recent times by Sheridan and co-workers (2-6) who isolated 855 mutants, of which 10% or more were assigned to specific chromosome arms using the B-A translocation technique (7).

Embryo-lethal mutants were also isolated in Arabidopsis by Muller (8) and, more recently, by Meinke (9).

Mutations affecting seed development are known since pre-Mendelian times. They affect production and proper distribution of pigments in the seed and/or production of storage material in the cotyledons and endosperm tissue (see Marx, 10). Non-lethal mutations affecting seed development may also affect accumulation of carbohydrates. Mutants defective in the production of specific seed proteins where isolated whose mutation lied in the seed protein gene itself (11) or in an unlinked regulatory locus (12).

Of immediate physiological interest are mutants that are defective in the preparation of dormancy and therefore appear as viviparous. Those in maize are particularly important for the relationship they uncovered between embryonic maturation germination, abscisic acid content and carotenoid pigments (13).

Shoot and root meristems are specified during embryogenesis. However mature plant leaf, stem and root differentiate continuosly from the apical meristems. Major morphogenetic events occur post-embryonically, after seed germination. Even floral meristems, with the reproductive organ systems, derive from a reprogramming of vegetative meristems (for general reference, see 14)

A few mutants have been isolated that affect the meristem or its function. In tomato, the reduced form of lanceolate lacks a shoot apical meristem (15) lateral suppressor cannot initiate axillary buds (16) and diageotropic does not initiate lateral roots (17).

Similar mutations have been isolated in Arabidopsis. Vegetative in pea fails to initiate flowering (18). In anantha (tomato) meristems become determined for floral function but fail to initiate floral organs (19). One flower , in Petunia (20) or floral parts, in tobacco, may emerge out of another complete flower (21).

Homeotic mutants affecting various parts of the flower (sepals replaced by leaves, petals replaced by stamens, or petals replaced by sepals and stamens replaced by carpels) or petals lacking and sepals converted to carpels were isolated in Arabidopsis (22,23). Homeotic mutants as well as mutants affecting the morphology and pattern formation of the leaf were isolated in pea; they were recently discussed in a review by Marx (24). In Antirrhinum the homeotic mutation deficiens is probably correlated with a transcription factor (Z.Schwarz-Sommer, to be published).

2. SOMATIC EMBRYOGENESIS

An alternative approach to the study of development can be made by using somatic embryogenesis. If an embryogenic culture is diluted and the medium is deprived of hormones, plant cell cultures (starting from single cells or, more frequently from clumps, which however, have been shown to derive from single cells) give rise to embryos that, morphologically, follow the same pathway of the zygotic embryo, i.e. they pass through the stages of globular, heart-shaped, torpedo-shaped and plantula (for a general reference, see 24). The fact that both processes proceed through the same sequence of developmental stages indicates that similar patterns of gene expression and similar regulatory mechanisms operate in both cases. Moreover, the fact that embryo development can occur in culture without addition of hormones is an indication that the embryo is autonomous and self-sufficient in terms of developmental programme. The adiacent tissues within the seed are only needed for nutritional reasons. All this is supposed to be true as a general rule in flowering plants. It should be stressed however, that what we are going to say in the next few pages comes entirely from carrot studies.

This fact is rather strange, because it means that callus cells, at variance with adult, differentiated cells, have acquired the ground state (25) i.e. the developmental state typical of the unfertilized egg, that permits faithful execution of morphogenesis during embryogeny. This embryonic state, or ground state, which in animals is maintained in germ line cells, in plants has to be regained. In fact, floral meristems develop from vegetative meristems which consist of somatic cells no longer embryogenic. Instead, during floral development, meiotic products that are embryogenic arise.

Although the developmental origin of the unfertilized egg requires further clarification, it is clear that the embryogenic state can be acquired by cells which have departed from the embryonic state. In fact, in carrot, the embryonic state can be regained soon after explantation of several plant body parts (e.g. phloem cells, hypocotyl tissue etc.) in a proper medium. What happens is largely unknown, but we know that treatment with auxin is necessary and several days are needed for the embryogenic potential to be acquired. In an embryogenic hypocotyl-derived suspension culture, the appearance of newly-formed pro-embryogenic masses (PEM, 26) from which somatic embryos develop under appropriate conditions has been followed. After approx. 20 days, a cell population emerges that contains, among others, clusters consisting of between 10 and 20 small, densely cytoplasmic cells, tightly adhering to each other (PEM). The fraction of the cell population containing PEM slowly increases over a period of six weeks (27) reaching a frequency which is, in different lines, between 1 and 10%.

These PEM can be concentrated in a Percoll gradient after sieving and/or picked manually and it was possible to obtain a subpopulation of single cells whose embryogenic efficiency was 90%. These single cells underwent unequal division, forming a cytoplasm-rich cell and a highly-vacuolated one. Further cell divisions occurred preferentially in the cytoplasm-rich cells resulting in the formation of a polarized cell cluster showing, e.g. polarized DNA synthesis (28).

Therefore from the initial explant to an established embryogenic culture we may have three distinct stages:
First stage: erasure of the differentiative tissue-specific programme of those cells which respond to auxin and begin to proliferate. The molecular events behind this stage are not defined. Recently it has been shown that treatment with exogenous auxin promotes DNA hypermethylation and the hypothesis was made that auxin is necessary whenever re-programming is to be made, because hypermethylation erases the previous differentiative programme (29)

Cells autotrophic for auxin (habituated cells) whether of genetic or epigenetic origin, do not hypermethylate under auxin and cannot embryogenize. An intermediate case has been reported of a mutant which can regenerate at low efficiency; it shows the presence of tumours associated with the regenerated plantlets. Curiously enough it does not show hypermethylation in the presence of auxin, nor hypomethylation in the presence of azacytidine (29).

To this first stage, a second one follows that corresponds to acquisition and maintenance of totipotency. This stage is characterized by the appearance of PEM in the cell population. The embryogenic capacity of some cells might depend on a specific molecular event -that we ignore- that may occur at the end of the first stage, or it may be the automatic consequence of the erasure of the tissue-specific programme, whereby a fraction of the cells reach the ground state, i. e. totipotency.

The third stage corresponds to the expression of the embryonic programme: When the auxin is removed and perhaps the cell density lowered, embryos are formed with the typical progression of globular, heart and torpedo. Whereas early forms require auxin, embryogenesis cannot proceed beyond globular stage if auxin is present.

Numerous workers have approached somatic embryogenesis from a biochemical point of view analyzing the third stage and recording differences between embryo and callus cells or using drugs inhibiting certain steps in order to create phenocopies.Although the earliest stages remain very interesting, owing to the experimental difficulties, the data which have accumulated concern practically only the third stage.

The rate of protein synthesis was shown to increase as did the rate of poly(A)+ RNA synthesis (see 30). DNA synthesis also increased during embryogenesis. Polyamine metabolism changed as polyamine levels increrased several times over the non-embryogenic controls (31). By analyzing the quality of the macromolecules synthesized during embryogenesis, differences were found in the 2-dimensional pattern of polyacrylamide gel electrophoresis by Sung and Okimoto (32) and Choi and Sung (33). Thomas and Wilde (34) analyzed a cDNA library and found a few clones specific for the embryos. Monoclonal antibodies were obtained against antigens unique to somatic embryos by Smith et al.(35).

All these techniques were shown to work in the system but a paradox remains, that the amount of embryo-specific functions is very small (1-2% in the case of proteins, O.5-1% in the case of cDNA) so that the problem remains on how such a complex developmental programme can be carried out by such a limited genomic information. One possible answer is that the callus cells from embryogenic lines are already early embryos; in other words, the cells that regained totipotency have expressed at least partially their developmental programme so that when we compare, say, globular embryos with cells, we are not comparing differentiated vs. undifferentiated cells, but rather two successive stages of differentiation. In fact, exponentially growing cells look more similar (according to the parameters investigated above) to early embryos than to old senescent non-embryogenic cells.

More results on this third stage come from another type of approach, i.e. to look at special classes of products (or functions) that were deemed more likely to be differentially expressed during development.

In fact, in the class of heat-shock proteins, several were found to be specifically associated with different stages of development (36). In other cases, sensitivity to drugs (5-bromo-deoxyuridine, 37, ethanol, 38, cycloheximide, 39, tunicamycin, 40) was also found to be associated with developmental stage and, in a couple of cases (azacytidine-ECP and a-amanitin) resistant mutants were isolated and characterized (29,41). Interesting enough, some of the resistances and the embryo-specific proteins were coordinately expressed, but different cell lines were expressing more or fewer functions at the callus stage, as if callus were a blocked embryonal stage and the block could occur earlier or later in the different lines.Instrumental for this type of analysis has been the purification of different embryo stages which can be achieved with different techniques (discussed in 42).

3. TEMPERATURE-SENSITIVE MUTANTS

Another,more productive approach that permits to overcome some of the limits of biochemistry consists in disrupting development from within, by making conditional mutants that, in non-permissive conditions, were arrested at some more or less specific stage. A search for such mutants, of the temperature-sensitive type, was done on haploid (43) or diploid (44-46) cells, with or without mutagenic treatment. The permissive temperature was always 24-25°C and the non-permissive one 31-33°C. Scores of mutants were isolated in each case and even after disposing of the unstable or of the leaky mutants, many more were isolated than characterized in detail.The procedures, slightly different in the details, have in common a negative selection, based on disposing of the embryos grown at the non-permissive temperature.(The successive positive selection may be done on colonies, embryos and as such may select for different types of mutants) The negative selection based on the incorporation at high temperature of drugs such as bromo-deoxy-uridine, used with success in animal cells, had only very limited application (47) in the isolation of plant developmental mutants, as that drug stops somatic embryogenesis (37). In all cases variants were found blocked at different stages as judged morphologically (43) or by studying the temperature-sensitive period (45,46).

The mutants isolated by Breton and Sung (43) belonged to six different classes: absolute or ts-emb-, incapable of forming embryos, at any temperature, or at the non-permissive one, respectively; absolute or ts-growth-, both callus and embryo development poor at any temperature, or only at the non-permissive one, respectively; habituated lines, capable of growth in embryogenic medium; 2,4-D resistant lines, capable of forming initial embryonal stages in the presence of hormones.

In general the most frequent temperature-sensitive periods in all three laboratorioes was found to be between globular and torpedo stage, with a few variants being affected at very early stages (pro-embryogenic masses).It should be mentioned however, that pre-treatment with high temperature during growth had, for some mutants, an effect on subsequent embryogenesis allowed to occur at permissive temperature. This was thought to indicate that functions necessary for embryogenesis have in some cases to be performed during callus proliferation (48).

A dominant mutation (ts59) was also isolated (49) which points to another advantage with the use of somatic vs. zygotic embryogenesis.

There are no indications that the number of mutants isolated in the different laboratories has in any way saturated the map of developmental mutants. It is possible to get some indications on the total number of loci involved in functions necessary for somatic embryogenesis (that can give rise to mutations of the ts type). This, based on the total frequency of mutation for ts-emb- mutants, compared to the frequency in known loci. This was done for different loci, with or without mutagenesis and the number of loci indicated by this rough approach turned out to be in the range 100-600. Another indication came out however, on the excess frequency of mutation in diploid, whose frequency of recessive mutants is far from being the square of the frequency found in haploid lines. This is common knowledge for both plant and animal somatic cell geneticists and the most commonly accepted interpretation lies in a sort of functional haploidization of a large number of loci (50,51). Some recent results suggest that the haploidization is a consequence of a somatic meiotic process, which is usually followed by restitution nuclei (52).

Two mutants have been characterized in molecular terms. One, ts59 is dominant in somatic cell hybrids, shows two temperature-sensitive periods, at pre-globular and heart stage and reveals a defect in the phosphorylation of a heat-shock protein. This is interesting because it assigns a role in plant cells to highly-specific kinases (that, besides self-phosphorylation, may act on one or two more peptides) in activating a function. In addition, as the peptides to be phosphorylated are of the heat-shock class, it confirms the earlier suggestion (53) that heat-shock proteins are involved in development. This mutant therefore may provide a handle in tackling some of these problems.

The other mutant, ts11c, (LoSchiavo et al.54, and in preparation) is recessive in somatic hybrids and at high temperature cannot acquire polarity, thus progressing from globular to heart stage. It is peculiar in that it can be complemented with medium conditioned by wild type embryos.The factor present in wild type conditioned medium turns out to be a glycoprotein secreted in the medium. The medium contains a relatively high number (perhaps greater than 5O) of proteins ranging in Mr from 13 to 7O kDa or more and they have to be properly glycosylated to perform their function, which is important and varied (55) . Important because if they are not properly glycosylated (as in the case of tunicamycin-treated embryos, or, if the glycans cannot be modified from the high-mannose to complex type,as in the case of ts11c) embryogenesis cannot proceed; varied because the glycoprotein necessary for the onset of embryogenesis is different from the one(s) necessary for the transition from globular to heart, and there may be others as yet unspecified. It is well possible that also in plant embryo development, as in other animal systems, secreted proteins are a way of communicating between the cells sharing a common endeavour, and of controlling and coordinating their expression. The fact that the ts mutants characterized so far are deficient in a mechanism rather than in the stability of one particular peptide and are therefore pleiotropic, may be more than a coincidence; more observations are needed to build on a firmer ground.

4. ACKNOWLEDGEMENT

Work supported by the programme "Sviluppo di Tecnologie avanzate" of the Italian Ministry of Agriculture (M.A.F.) and MURST.

REFERENCES

1 Mangelsdorf, P.C. (1923) J.Hered. 14, 119-125
2 Sheridan, W.F. (1984) Maize Genet.Coop.Newslett. 58, 95-96
3 Sheridan, W.F., Clark, J.K.(1984) Maize Genet.Coop.Newslett.58,92-95
4 Sheridan, W.F., Neuffer, M.G.(1980) Genetics 95,945-960
5 Sheridan, W.F., Neuffer, M.G.(1982) J.Hered.73,318-329
6 Neuffer, M.G., Sheridan, W.F. (1980) Genetics 95,929-944
7 Robertson, D.S. (1952) Proc.Nat.Acad.Sci. US 38,580-583
8 Muller, A.J. (1963) Biol.Zentralbl.82,133-163
9 Meinke, D.W. (1985) Theor.Appl.Genet.69,543-552
10 Marx, G.A. (1983) Ann.Rev.Plant Physiol.34,389-417

11 Goldberg, R.B., Hoschek, G.,Vodkin, L.O. (1983) Cell 33,465-475
12 Larkins, B.A., Pedersen, K., Marks, M.D., Wilson,D.R.(1984) Trends Biochem.Sci. 9, 306-308
13 Robertson, D.S. (19555) Genetics 4O,745-760
14 Sussex, I.M.(1989) Cell 56,225-229
15 Caruso, J.L.(1968) Am.J.Bot.55,1169-1176
16 Malayer, J.C., Guard, A.T. (1964) Am.J.Bot.51,140-143
17 Zobel, R.(1974) Can.J.Bot.52,735-741
18 Reid, J.B., Murfet, I.C.(1984) Ann.Bot.53,369-382
19 Paddock, E.F., Alexander, L.J.(1952) Ohio J.Sci.52,327-334
20 Gerats, A.G.M., Kaye, C., Collins, C., Malmberg, R.L. (1988) Plant Physiol.86,390-393
21 Malmberg, R.L., McIndoo, J., Hiatt, A.C., Lowe, B.A.(1985) Cold Spring Harbor Symp. 40, 475-482
22 Mayerowitz, E., Pruitt, R.E. (1985) Science 229,1214-1218
23 Komaki, M.K.,Okada, K.,Nishino, E.,Shimura, Y.(1988) Development 104,195-203
24 Marx, G.A.(1987) Plant Mol.Biol.Rep.5,311-335
25 Sung, Z.R.(1984) in Somatic Embryogenesis (Terzi, M., Pitto, L., Sung, Z.R.eds) IPRA, Rome pp.117-120
26 Halperin, W.(1966) Am.J.Bot.53,443-453
27 deVries, S.C., Booj, H., Meyerink, P., Huisman, G., Wilde, H.D., Thomas, T.L., van Kammen, A. (1988) Planta 176,196-204
28 Nomura, K., Komamine, A., (1986) New Phytol.104, 25-32
29 Lo Schiavo, F.,Pitto,L.,Giuliano, G., Torti, G.,Nuti-Ronchi, V.,Marazziti, D., Vergara, R.,Orselli, S.,Terzi, M.(1989) Theor.Appl.Genet.77,325-331
30 Nomura, K.,Komamine, A.(1986) Oxford Surv.Mol.Cell Biol. 3,456-466
31 Fienberg, A.A., Choi, J.H., Lubich, W.P.,Sung, Z.R.(1984) Planta 162,532-539
32 Sung, Z.R., Okimoto, R.(1981) Proc.Nat.Acad.Sci.US 78,3683-3687
33 Choi, J.H., Sung, Z.R.(1984) Plant Mol.Biol.Rep.2,19-25
34 Thomas, T.L., Wilde, H.D.(1985) in Somatic Embryogenesis (Terzi, M., Pitto, L., Sung, Z.R., eds.) IPRA, Rome, pp77-85
35 Smith, J.A., Choi, J.H.,Krauss, M.,Karu, A.E., Sung, Z.R.(1985) in Somatic Embryogenesis (Terzi,M., Pitto,L.,Sung,Z.R.eds.) IPRA, Rome, pp.86-94
36 Pitto, L., Lo Schiavo, F., Giuliano, G., Terzi, M.(1983) Plant Mol.Biol.2, 231-237
37 Dudits, D., Lazar, G., Bajszar, G.(1979) Cell Diff.8,135-144
38 Perata, P., Alpi, A., Lo Schiavo, F. (1986) J.Plant Physiol.126,181-186
39 Sung, Z.R., Lazar, G., Dudits, D.(1981) Plant Physiol.68,261-264
40 Lo Schiavo, F., Quesada-Allué, L.A., Sung, Z.R.(1986) Plant Sci.44,65-71
41 Pitto, L., Lo Schiavo, F., Terzi, M.(1985) Proc.Nat.Acad.Sci.US 82,2799-2803
42 Lo Schiavo, F. (1984) Plant Mol.Biol.Rep.2,15-18
43 Breton, A.M. Sung, Z.R.(1982) Dev.Biol.90,58-66
44 Giuliano, G., Lo Schiavo, F., Terzi, M.(1981) Atti Ass.Genet.Ital.28,203-205
45 Terzi, M., Giuliano, G., Lo Schiavo, F., Nuti-Ronchi, V.(1982) in Embryonic Development, part B (Burger, M.M. ed.) Alan R. Liss, New York, pp.521-534
46 Schnall, J.A., Cooke, T.J., Cress, D.E.(1988) Dev.Genet.9,49-67
47 Malmberg, R.L. (1979) Genetics 92,215-220
48 Giuliano, G., Lo Schiavo, F., Terzi, M.(1984) Theor.Appl.Genet.67,179-183
49 Lo Schiavo, F., Giuliano, G., Sung, Z.R.(1988) Plant Sci.54,157-164
50 Terzi, M.(1974) Genetics and the animal cell. Wiley-Interscience, Chichester & New York

51 Terzi, M., Sung, Z.R.(1986) CRC Crit.Rev.Biotechnol.3,303-330
52 Nuti-Ronchi, V. (1990) in Plant tissue culture: Applications and Limitations (Bhojwani, S.S., Novàk, F.G., eds.) Elsevier, Amsterdam
53 Bensaude, O., Babinet, C., Morange, M., Jacob, F.(1983) Nature, 3O5,331-333
54 Lo Schiavo, F., Giuliano, G., Terzi, M.(1985) in Somatic Embryogenesis (Terzi, M., Pitto, L., Sung, Z.R., eds.) IPRA, Rome, pp.32-34
55 deVries, S.C., Booj, H., Janssens, R., Vogels, R., Saris, L., Lo Schiavo, F., Terzi, M. van Kammen, A.(1988) Genes & Dev.2,462-476.

TISSUE AND CELL-SPECIFIC EXPRESSION OF GENES ENCODING CARROT EXTRACELLULAR PROTEINS

H. BOOIJ, P. STERK, G.A. SCHELLEKENS, A. VAN KAMMEN AND S.C. DE VRIES
Department of Molecular Biology, Agricultural University Wageningen, Dreijenlaan 3, 6703 HA Wageningen, The Netherlands

INTRODUCTION

Carrot somatic embryogenesis is an attractive model system to study early plant development because of the relative ease in which sufficiently large quantities of plant material of different developmental stages can be obtained. Embryogenic high-density cell suspensions cultures, referred to as 'S'-cultures, are maintained in B5 medium (1) containing the synthetic auxin 2,4-dichlorophenoxy acetic acid (2,4-D). Somatic embryos develop from small clusters consisting of tightly adhering cytoplasmatic cells, so-called proembryogenic masses (PEMs) in embryo ('E') cultures, a diluted phytohormone-free suspension culture enriched for PEMs. However, development is blocked at globular stage in cultures in which 2,4-D is present (proliferating ('P') cultures, indicating that dilution of the suspension culture triggers actual somatic embryogenesis, and that omission of 2,4-D is required for normal embryo development.

We have shown that initiation of somatic embryogenesis is concomitant with a marked change in the pattern of secreted proteins (2). Secondly, we have observed aberrant extracellular protein patterns in suspension cultures that have lost embryogenic potential and in non-embryogenic mutant cell cultures. The fact that somatic embryogenesis could be restored in some of these mutant cell lines by the addition of wild-type extracellular proteins indicated that one or several extracellular proteins are involved in somatic embryogenesis.

We have focussed on the cloning and characterization of genes encoding extracellular proteins. Four cDNA clones, designated PS26, PS28, PS30 and PS48, were obtained by expression screening of a carrot embryo cDNA library with antisera raised against extracellular proteins. Northern hybridization analysis showed that the genes corresponding to and further designated as the cDNA clones PS26, PS28, and PS30 were expressed in both wild-type and non-embryogenic cell lines, regardless of the differentiation state. In contrast, PS48 was only expressed in embryogenic cell lines. In plant tissues expression of PS26 was enhanced in the stem, PS28 and PS30 expression was enhanced in the root apical meristematic region and in developing flowers. Expression of PS48 was only detected in the shoot apical meristematic region and in flowers. In situ hybridization experiments revealed expression of PS48 very early in developing somatic embryos as well as striking similarities in expression between somatic and zygotic embryos.

EXPERIMENTS AND RESULTS

Expression screening. In order to clone cDNAs corresponding to genes encoding secreted proteins, antisera were raised against two SDS-polyacrylamide gel purified glycoproteins, a 27 kDA glycoprotein abundant in high-density suspension cultures containing 2,4-D and a 29 kDa glycoprotein abundant in embryo cultures. A λgt11 cDNA library from mRNA isolated from all stages of somatic embryos was screened. With the anti-27 kDa serum a positive clone referred to as PS28 and with the anti-29 kDa serum three independent positive clones designated PS26, PS30 and PS48 were obtained.

Northern hybridization. The expression patterns of the genes corresponding to the cDNAs was examined by Northern hybridization of the [^{32}P]-dATP labelled cDNAs to total RNA isolated from embryogenic and non-embryogenic carrot cell cultures and from several plant tissues as shown for PS48 in Figure 1.

We investigated expression in 'S', 'P' and 'E' cultures of *Daucus carota*: cv. 'Flakkese' SG766 Trophy, designated as cultivar '10' (embryogenic) and FG10 (2,4-D adapted; non-embryogenic), cultures of cv. 'San Valery,' designated A^+ (embryogenic), and mutant cell lines derived from A^+: $\alpha5^{II}$ (non-embryogenic) (3), $\alpha5^{IIIR}$ (an embryogenic revertant of $\alpha5^{II}$), c15 (non-embryogenic) (4), ts11c (temperature sensitive: permissive at 25°C, non-permissive at 32°C) (5) and ts11R (a revertant of ts11c; embryogenic). It turned out that all genes except PS48 were expressed in all cultures investigated. Expression of PS26 was higher in low-density cultures whereas PS28 and PS30 were not markedly regulated. In purified somatic embryos two of the four genes, PS30 and PS48, were expressed. PS48 was expressed in embryogenic cultures only. In A^+ and derived embryogenic cell lines PS48 expression was observed in 2,4-D containing 'S' and 'P' cultures, indicating that PS48 is not embryo-specific. Comparison of expression levels in suspension cultures of three '10'-lines, V10, W10 and X10, which in this order decrease in age and increase in embryogenic potential, revealed that PS48 is expressed more highly in young embryogenic cultures as measured by the presence of the relative amount of PEMs. This is further supported by the observation that in A^+ and derived embryogenic 'S' cultures, which are cultured in the presence of cytokinin, when compared to '10' cultures many more PEMs are formed resulting in the observed relatively high expression level of PS48, except in the weakly embryogenic mutant ts11. Through transformation mediated by the *Agrobacterium tumefaciens* strain LBA958 a cytokinin overproducing cell line was obtained in which cells grew as large and tight cell clusters, but which was not embryogenic. This indicates that cluster formation in itself does not give rise to PS48 expression. The finding that PS48 is expressed in purified somatic embryos and the observation that in embryo cultures of embryogenic '10' lines development of somatic embryos is much faster than in A^+ explains the higher expression level in embryo cultures of the embryogenic '10' line compared to those of A^+, $\alpha5^{IIIR}$ and ts11c.

To investigate the expression of the four genes in carrot plants we isolated total RNA from germinating seeds, from seedlings the root tip (2-3 mm) containing the root apical meristem, the root without tip, the stem without apical part, the shoot apex containing the shoot meristem and the primary leaves. From mature plants we took the stem, leaves, the developing (unfertilized) inflorescence and fertilized mature flowers. PS26 appeared to be expressed in the stem of both seedlings and mature plants. PS28 and PS30 were expressed in all plant parts examined but were enhanced in the root meristematic region and in developing flowers. Expression of PS48 was found in the shoot apical meristematic region and in the developing and mature inflorescence. It is remarkable that expression of three out of the four genes is at least enhanced in meristematic tissues, but never in all meristematic parts. The finding that cell division activity in comparable in both the root and the shoot apical meristematic regions, shown by expression of the histone 4 gene which is directly related to cell division activity (Figure 1), renders it unlikely that these genes are directly involved in cell division.

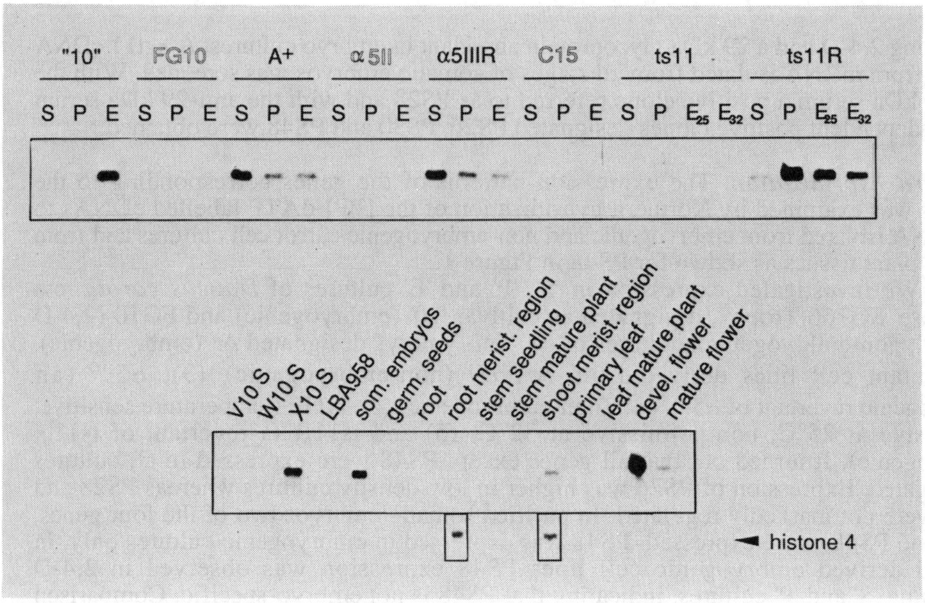

Figure 1. Expression pattern of PS48 in carrot cultures and plants.

In situ hybridization of PS48. In order to study the expression of PS48 in more detail we employed *in situ* hybridization in sectioned culture and plant material (6). Labelling was restricted to all epidermal cells in globular, heart and torpedo stage somatic embryos, and shifted in late torpedo stage embryos towards the future shoot. In plants PS48 was expressed in the L1 layer of the shoot apical meristem, except perhaps in the dividing cells in the central part, and in all surface cells of young leaf primordia. In an early stage inflorescence labelling was observed in the outer cell layer of all individual flower buds. After flower maturation PS48 was shown to be expressed in the integument cells, in areas where the two carpels have fused and in the outer cells of the spicules. Moreover, expression of PS48 was found in zygotic embryos present in developing seeds. In a heart stage embryo a labelling pattern similar to that of the somatic equivalent was observed, be it that expression was highest in the surface cells of the cotyledons and and in suspensor cells. In a globular stage embryo expression in surface cells was equally distributed. In a very early stage globular embryo estimated to contain 64 cells, strong labelling of the suspensor was found and almost no labelling of other cells, indicating that PS48 is expressed first in the suspensor of a developing zygotic embryo, possibly already after a few cell division cycles after fertilization of the egg cell. Since PS48 is such an early embryo marker during zygotic embryogenesis, it was interesting to know how early PS48 was expressed in the *in vitro* system. Therefore, we made sections through an embryogenic high-density suspension culture and investigated expression of PS48. It was found that PS48 was only expressed in proembryogenic masses. These results show that there is not only a striking similarity in expression of PS48 in both zygotic and somatic embryos, but that PS48 is the earliest embryo marker described sofar during zygotic as well as somatic embryogenesis.

DISCUSSION

Employing expression screening of a carrot somatic embryo cDNA library with antisera raised against two secreted proteins we obtained four cDNA clones. We have shown that expression of all corresponding genes is regulated in plants. Remarkably, three of these genes are expressed at relatively high levels in carrot flowers and in apical meristematic regions of either the root or the shoot, thus having a remarkable cell or tissue specificity, while in *in vitro* cultures no regulation is observed for PS28 and PS30. In contrast, expression of PS26, which is higher in low cell density cultures and of PS48, which is correlated with embryogenic potential of the culture, did not show regulation in *in vitro* cultures. Apparently, none of the genes is strictly embryo-specific, in fact only PS30 and PS48 are expressed in embryos, but not uniquely so. Upon initiation of suspension cultures, by inoculating cut hypocotyls from seedlings in 2,4-D containing B5 medium (7), PS48 is the only gene of which expression has to be reinitiated again, since the other genes are already expressed in the explant. This initiation of expression in starting cultures combined with the fact that PS48 is a marker for embryogenic potential of suspension cultures might point to a function needed very early in the development of the plant embryo.

In situ hybridization of PS48 with culture and plant material indeed confirmed that PS48 is expressed very early in development during both zygotic and somatic embryogenesis. Although striking parallels in expression pattern were observed, it must be noted that especially in the very early stages, differences in development do exist. The most important one is that somatic embryos frequently lack a suspensor. PEMs cannot be considered as a suspensor, but should be seen as embryos aborted due to culturing conditions that do not favour correct development. Only when these conditions are shifted towards those more optimal for somatic embryo development the analogy between somatic and zygotic development becomes quite striking. A second difference is the fact that in the zygotic heart stage embryo expression of PS48 is slightly more localized compared to the pattern in the somatic counterpart, i.e. more highly in surface cells of the cotyledons. In somatic embryos this is also observed, but at later stages. This leads to the conclusion that in artificial environments the spatio-temporal expression pattern is slightly altered compared to the natural situation.

In normal plant development PS48 is expressed for the first time in suspensor cells of the early globular embryo, soon followed by expression in surface cells of the embryo proper. The observed labelling patterns later in development show that PS48 is expressed in cells of the L1 layer of the shoot apical meristem and in tissues derived from those: leaf primordia and flower tissues. Thus, PS48 is expressed throughout the complete sporophytic phase of the life-cycle of the plant.

REFERENCES

1 Gamborg, O.L., Miller, R.A. and Ojima, K. (1968) Exp. Cell. Res. 50, 151-158
2 De Vries, S.C., Booij, H., Janssens, R., Vogels, R., Saris, L., LoSchiavo, F., Terzi, M. and Van Kammen, A. (1988) Genes & Development 2, 462-476
3 Vergara, M.R., Biasini, G., LoSchiavo, F. and Terzi, M. (1982) Z. Pflanzenphysiologie 107, 313-319
4 LoSchiavo, F., Giovinazzo, G. and Terzi, M. (1983) Mol. Gen. Genet. 192, 326-329
5 LoSchiavo, F., Giovinazzo, G. and Terzi, M. (1985) in Somatic Embryogenesis (Terzi, M., Pitto, L. and Sung, Z.R., ed.), pp. 32-34, IPRA, Rome
6 Cox, K.H., DeLeon, D.V., Angerer, L.M. and Angerer, R.C. (1984) Dev. Biol. 101, 485-502
7 De Vries, S.C., Booij, H., Meyerink, P., Huisman, G., Wilde, H.D., Thomas, T.L. and Van Kammen, A. (1988) Planta 176, 196-204

Microspore-derived embryos: A tool for studies in regulation of gene expression in zygotic embryos

L.A. Holbrook[1], W.R. Scowcroft[1], D.C. Taylor[2], M.K. Pomeroy[3], R.W. Wilen[4], M.M Moloney[4]

[1]Oilseeds Biotechnology Research Centre, 102-3553 31 Street N.W., Calgary, AB T2L 2K7 Canada; [2]Plant Biotechnology Research Institute, NRCC, 110 Gymnasium Rd, Saskatoon, SK S7N 0W9 Canada; [3]Plant Research Centre, Agriculture Canada, Ottawa, ON K1A 0C6 Canada; [4]Department of Biological Sciences, University of Calgary, Calgary, AB T2N 1N4 Canada.

INTRODUCTION: Although zygotic embryos of oilseeds have been extensively studied, many questions remain regarding regulation of genes encoding major storage products. For biochemical, molecular and cell biological studies, particularly at early (preglobular to heart) stages of embryogenesis, there may be many advantages to working with somatic embryo systems. A non-zygotic embryo system widely utilized for research in Brassica napus is that derived from male gametophytic microspores. The tissue culture of these microspores is well established(1) and numerous groups are utilizing the dihaploid plants derived from these cultures in breeding programs(2).

We have initiated studies into biochemical and gene regulation aspects of microspore embryogenesis(3). A comparison was made between microspore and zygotic embryos with respect to synthesis of major seed storage products and responses of the embryos to growth regulator signals. Therefore, we measured developmental changes in triacylglycerol (TAG) quantity and quality, starch accumulation, appearance of the storage proteins napin and cruciferin and of the oil body proteins, oleosins. The response of the cultured embryos to exogenously applied abscisic acid (ABA) was also examined. ABA treatment resulted in strikingly increased transcription of the storage protein genes, and also stimulated translation of a set of proteins associated with the oil body fraction. The question of the relationship between effects of ABA and osmoticum on embryogenesis was investigated using the unique advantages offered by this culturable embryo system. High osmoticum treatment employing 12.5% sorbitol affected storage protein gene transcription and oil body protein synthesis in a manner similar to ABA. However it occurred in a temporal sequence consistent with the idea that osmotic stress affects ABA accumulation resulting in the observed effects on transcription and translation.

STORAGE LIPIDS: As a first comparison of zygotic and microspore embryos, the developmental changes in yield and fatty acid composition of the TAG fraction were determined(3). Two cultivars of B. napus were evaluated for fatty acid profiles of developing microspore embryos compared with the dry seed composition of each. A high erucic acid (22:1) cultivar Reston, which characteristically has 11% (C_{20}) and 40% (C_{22}) fatty acids in TAG, was compared with Topas which has predominantly C_{18} fatty acids. An abbreviated summary of the data is shown in Table 1.

Table 1. Fatty acid composition of triacylglycerols in cotyledonary microspore embryos and in mature seed. ±SE in paratheses.

Cultivar	Source	TAG Fatty Acid Composition (mol%)						
		16:0	18:0	18:1	18:2	18:3	20:1	22:1
Reston	Embryos	4.3	1.6	30.7	9.6	8.8	10.6	31.2
		(0.2)	(0.2)	(1.7)	(0.5)	(0.8)	(2.0)	(1.8)
	Seed	3.7	1.5	20.8	13.1	6.7	11.2	39.9
		(0.3)	(0.2)	(0.1)	(0.4)	(0.4)	(0.5)	(0.4)
Topas	Embryos	5.4	2.2	62.6	18.1	7.0	a	a
	Seed	4.4	2.0	63.9	18.9	8.1	a	a
		(0.3)	(0.1)	(0.7)	(0.7)	(0.2)		

a - less than 2%

The cultivar-specific characteristics of oil quality are clearly maintained in the microspore system relative to zygotic embryos of the seed. The TAG yields are also strikingly similar at maturity when compared on a dry weight basis(3). Developmental changes in fatty acid composition (not shown) were also similar to those reported for seed development [eg. initially, proportions of linolenic acid (18:3) are very high; these decrease over time with increases in the proportion of oleic acid (18:1)]

LIPID ENZYMOLOGY: A convenient marker for embryo storage lipids in B. napus is erucic acid as it is not known to be a component of membrane lipids(4). Therefore, initial experiments were done with the cv. Reston using erucoyl-CoA as substrate. Enzyme activities measured included erucoyl-CoA synthetase, thioesterase and acyltransferase all of which demonstrated greater specific activities relative to zygotic embryo samples at the torpedo stage.

Currently, the specificity of the acyltransferases involved in TAG biosynthesis are being studied, and a highly active diacylplycerol acyltransferase is being purified and characterized.

STORAGE PROTEINS, ABA AND OSMOTICUM: The expression of storage protein genes is a characteristic of embryo maturation in many higher plant species. Napin and cruciferin, the two major storage proteins of oilseed rape, are both expressed at high levels during late embryogenesis(5). These genes are regulated during embryo development in a stage specific pattern and show regulation by ABA(6).

The expression of napin and cruciferin were therefore examined in the microspore embryo system. Little or no storage protein mRNA was detected in these embryos when cultured on the basal medium(3). The addition of 10 μM ABA resulted in a clearly detectable increase in the transcript levels of both polypeptides within two hours of application. The level of napin and cruciferin mRNA observed in these ABA treated embryos was comparable to that seen in zygotic embryos(3), and the induction by ABA was similar to the effect seen with cultured zygotic embryos(3,6). The developmental specificity of storage protein gene expression seen in zygotic embryos was maintained in the microspore embryos when cultured in the presence of ABA.

These genes were also induced by osmotic stress in both the microspore and cultured zygotic embryos(7). Culture of the microspore embryos on either 12.5% sorbitol or mannitol resulted in the accumulation of storage protein mRNA to levels similar to ABA treated embryos. This induction is probably mediated by an accumulation of ABA in the embryos in response to water stress. This possibility is supported by two findings: First, there was a lag in the induction of mRNA encoding these proteins in the osmotic treatment compared to the induction by ABA. Also, there was a 3- to 4-fold increase in endogenous ABA in these embryos within four hours of culturing in sorbitol, and this level was maintained for at least 24 hours after treatment. This increase in ABA occurred prior to the detection of storage protein mRNA in the embryos.

A third treatment of these gametophytic embryos has also been shown to induce storage protein gene expression. Jasmonic acid has been implicated in a number of stress related plant responses(8). Application of jasmonic acid (10 μM) or methyl jasmonate (30 μM) to the culture medium resulted in the induction of both napin and cruciferin mRNA in the microspore embryos. This class of growth regulators has several other ABA-like effects in these embryos as well. Both of these growth regulators inhibit precocious germination which is similar to the effect seen when cultured zygotic embryos are plated in the presence of ABA(6). Furthermore, treatment of these embryos with jasmonic acid, methyl jasmonate, or ABA seems to promote normal plantlet formation and reduced secondary embryogenesis when plated on germination medium. Given the similarities in the regulation of embryonic development demonstrated by these two classes of growth regulators, there may be a common mechanism in their mode of action. This possibility is currently being investigated.

OIL BODY PROTEINS (OLEOSINS): Oilseed storage lipids are found in subcellular organelles called oil bodies. These spherical bodies appear to have a "half-unit" membrane with a small number of proteins referred to as oleosins embedded into it(9). Oil body proteins may serve structural roles with postulated functions such as lipase binding during germination(10). In rapeseed, they appear relatively late in embryogenesis with a predominant 19kDa component that can comprise up to as much as 10% of the total seed protein(3,11).

In our experiments with microspore embryos, the synthesis of a set of oil body-associated proteins detected by SDS gel electrophoresis and autoradiography was clearly stimulated by ABA. Two major polypeptides of approximately 18 kDa and 19 kDa were more highly labelled than controls, as were three polypeptides in the 22-24 kDa range, and, to a lesser extent, one band at 30 kDa. Quite striking was a decrease in labelling of a protein at 32 kDa after ABA treatment(3). If ^{32}P orthophosphate was use for labelling, the only protein band from SDS-boiled oil bodies that appeared to be phosphorylated was this 32 kDa component. Similar effects were measured in cultured zygotic embryos if dissected from seed before 20 days post-anthesis, again showing common responses between the two types of embryos.

As with storage protein transcription, not only ABA but also both sorbitol and jasmonic acid affected an increased synthesis of oil body-associated proteins in microspore embryos. The only difference was that the decrease of the 32 kDa polypeptide seen after ABA treatment appeared to be increased with sorbitol. Therefore, for isolating preparative amounts of these oil body proteins, 2 to 4 day sorbitol-treated torpedo stage microspore embryos were produced. Individual polypeptides are being isolated for sequencing and antibody production.

SUMMARY: Over the range of parameters studied including developmental morphogenesis, TAG quality and quantity, and the response of specific genes to growth regulators and osmotic stress, there are striking similarities between *B. napus* zygotic and microspore embryo systems. The single cell origin and ability to culture microspore embryos, ease of culture manipulation to study gene induction events, in-vivo labelling capacity, and relatively high specific activity of enzymes that have been measured, argue strongly for the utility of this microspore system for basic studies in plant embryogenesis.

REFERENCES

1. Fan, Z., Armstrong, K.C., and Keller, W.A. (1988) Protoplasma 147, 191-199.

2. Siebel, J. and Pauls, K.P. (1989) Theor. Appl. Genet. 78, 473-479.

3. Taylor, D.C., Weber, N., Underhill, E.W., Pomeroy, M.K., Keller, W.A., Scowcroft, W.R., Wilen, R.W., Moloney, M.M. and Holbrook, L.A. (1990) Planta. 181, in press.

4. Roughan, P.G. and Slack, C.R. (1982) Annu. Rev. Plant Physiol. 33, 97-132.

5. Crouch, M.L., Tenbarge, K.M., Simon, A.E., Finkelstein, R. and Solberg, L. (1985) in Molecular Form and Function of the Plant Genome (van Vioten-Doting, L., Groot, G.S.P. and Hall, T.C., eds.) pp 555-567, Plenum, New York.

6. Finkelstein, R.R., Tenbarge, K.M., Shumay, J.E. and Crouch, M.L. (1985) Plant Physiol. 78, 630-636.

7. Finkelstein, R.R. and Crouch, M.L. (1986) Plant Physiol. 81, 907-912.

8. Parthier, B. (1990) J. Plant Growth Reg. 9, 57-63.

9. Qu, R. and Huang, A.H.C. (1990) J. Biol. Chem. 265, 2238-2243.

10. Vance, V.B. and Huang, A.H.C. (1987) J. Biol. chem. 262, 11275-11279.

11. Murphy, D.J., Cummins, I. and Kang, A.S. (1989) Biochem. J. 258, 285-293.

Molecular changes associated with the commitment phase of microspore embryogenesis.

Paul M. Pechan and Jeff Schell
Max-Planck-Institut für Züchtungsforschung
5000 Köln 30
West Germany

Many plant species, including economically important crops, can not be regenerated from single cells *in vitro*. Consequently one of the key areas to be investigated is the mechanism controlling induction of cell proliferation *in vitro*. These studies require that the inductive phase of cell proliferation in the plant cell system of choice is well defined and allows isolation of distinct cell populations for subsequent comparisons. *Brassica napus* cv. Topaz microspores are well suited for such studies. Microspore embryogenesis is induced from freshly isolated microspores, embryos are formed directly from microspores, without a callus phase, within 14 days of culture initiation, hormonal treatments are not required for the embryogenic process to be induced or to proceed, induction of microspore embryogenesis is dependant on the developmental stage of the microspores- thus allowing the study of cell cycle regulation of the process, and upto 70% of microspores in a given culture can undergo embryogenesis (1,2).

Microspore embryogenesis can be defined as the redirection of the pollen formation pathway to that of the embryo formation pathway. The induction process is facilitated by a temperature stress treatment: microspores can undergo embryogenesis at 32°C but usually not at 25°C. Gamma irradiation and ethanol can also facilitate this process (3). The embryogenic process is developmentally regulated: only late uninucleate to early binucleate microspores (originating from buds of approximately 4 mm in length) can undergo embryogenesis (2). The induction process consists of a number of phases: inhibition of an existing developmental pathway, acquisition of commitment to cell proliferation and commencement of cell proliferation. The objective of our investigation is to understand how this embryogenic process is initiated; namely what are the cellular and molecular events leading upto and including the first nuclear embryogenic division.

Methods

A number of comparisons were made between potentially embryogenic microspores (those at the late uninucleate to early binucleate stage of development) cultured at 25°C and 32°C, as well as between microspore populations of different developmental stages (originating from buds 3-5 mm in length) and microspore populations not capable to undergo embryogenic development. Acquisition of commitment to cell proliferation of potentially embryogenic microspores kept at 32°C for various length of time was also investigated. Isolation and culture of *B.napus* microspores is described elsewhere (1,2).

Results

Microspores become committed to the embryogenic pathway within 8 hours of culture initiation, a time period which preceeds the first embryogenic division. Based on comparisons of 1) *in vitro* translated products of mRNAs and 2) two dimensional protein gels the following observations have been made:

1. A number of high molecular weight mRNA products are absent in microspore populations not capable of undergoing embryogenic development when cultured at 32°C. However, these products are present in microspore populations developmentally too old to undergo the embryogenic process as well as in potentially embryogenic microspores cultured at 25°C. The high molecular weight mRNA products are nevertheless associated with culture initiation and heat stress because they are absent in freshly isolated microspores.
2. The amount of some proteins is reduced but of others increased in potentially embryogenic microspores cultured at 32°C. Proteins of apparent molecular weights 66,20,19 and 18 kDa have been identified as being unique to potentially embryogenic microspore cultures kept at 32°C.

Discussion

Stress alone is not sufficient to trigger *B.napus* cv.Topaz microspore embryogenesis: microspores must also be at an appropriate phase of the cell cycle (stage of development) for the process to take place. The mechanism controlling microspore embryogenesis is activated in potentially embryogenic microspores within 8 hours of culture initiation. mRNAs absent in microspores not capable of embryogenesis might be useful as markers to distinguish tissue culture responsive and tissue culture non-responsive plant varieties. However, more extensive molecular comparisons between plant varieties and species need to be carried out to confirm the general validity of this observation. The reduced amounts of some proteinsin potentially embryogenic microspores cultured at

32°C could be associated with the interruption of an existing (pollen) developmental process, what is sometimes termed the `dedifferentiation´phase. The preferential synthesis of certain proteins in potentially embryogenic microspores subsequent to a 32°C heat stress and their apparent molecular weights suggest some are possibly heat shock proteins. Whether these proteins serve to protect the microspores against the heat stress or are somehow involved in the induction of microspore embryogenesis remains to be determined. It is possible appearance of heat shock proteins in cells under stress might be one of the factors contributing to induction of cell proliferation since preferential synthesis of heat shock proteins could interrupt the synthesis of other proteins and since some heat shock proteins interact with proteins involved in cell proliferation . The current data serve as a framework for detailed molecular studies now in progress. Ultimately, these studies should provide us with a better understanding of *in vitro* and *in vivo* plant cell proliferation and with ideas how to manipulate these processes.

References

1. Keller, W. Fan, Z. Pechan, P.M. Long, N. and Grainger, J. (1988) 7th International Rapeseed Congress, Poland. Vol.1 152-157.
2. Pechan, P.M. and Keller, W.A. (1988) Physiol. Plant. 74, 377-384.
3. Pechan, P.M. and Keller, W.A. (1989) In Vitro 25,1073-74.

EXPRESSION OF CHIMERIC GENES IN DEVELOPING EMBRYOS DERIVED FROM ISOLATED MICROSPORES FROM TRANSGENIC BRASSICA NAPUS PLANTS

S.E. Radke, J.C. Turner, D. Facciotti, and V.C. Knauf
Calgene, Inc., 1920 Fifth Street, Davis, California 95616, U.S.A.

Plant embryogenesis is a process in which many physiological and morphological changes occur (for a review, see 1). We are focussing on the process of embryogeny in rapeseed in order to better understand the accumulation of storage reserve materials in the cotyledons of Brassica napus zygotic embryos.

The ability to regenerate somatic embryos from sporophytic or gametophytic cells of species such as B. napus is well established. The process of microspore-derived embryogenesis typically resembles zygotic embryogenesis in that there are globular, heart, and cotyledonary stages of development. Storage proteins, napin and cruciferin, are found in zygotic rapeseed embryos and have been detected in non-zygotic embryos from microspores and hypocotyls (2). Moreover, erucic acid containing triglycerides, typically found only in seeds of Brassica, have been observed in cotyledonary structures of microspore-derived B. campestris embryos (3).

Clearly, the embryoid structures derived from microspores involve genetic pathways otherwise unique to zygotic embryogenesis that occur during seed development. We have been using recombinant DNA methods combined with gene transfer systems in rapeseed to establish markers by which to follow normal zygotic embryogenesis. From such a knowledge base, we hope to use genetic engineering to alter embryo metabolism to either change the amounts or the types of compounds stored in the mature seed.

We have cultured microspores from transgenic rapeseed plants with marker genes in order to generate dihaploid plants homozygous for the transferred genes. The presence of marker genes for zygotic embryogenesis in developing microspore-derived embryos has allowed us to make further comparisons between zygotic and microspore-derived embryogenesis. Some unexpected observations from these experiments now lead us to speculate further about the temporal control of expression for certain genes. In addition, we have examined survival under antibiotic selection as a means to score the number of chromosomal insertions for T-DNAs.

Our first efforts to examine gene expression in both zygotic and microspore-derived embryos have focussed on two types of genes known to be involved in genetic pathways unique to embryogenesis with both directly tied to seed storage components.

Napins represent one of the two major classes of storage proteins found exclusively in Brassica seed. Acyl carrier protein (ACP) genes encode a cofactor essential for storage lipid biosynthesis. Although ACP is found in all living plant cells, we have isolated and characterized a B. campestris ACP gene (designated Bcg4-4) which has an expression pattern associated with biosynthesis of storage lipids in embryos.

Napins and ACPs found in developing embryos are both encoded by multi-gene families. Based on the use of single gene specific probes, we have established that our napin gene isolate '1-2' from B. campestris corresponds to a strong, if not the strongest, expressor in the napin gene family. In the case of ACP, gene expression is observed at a much lower level than for napin, but again the ACP gene we have chosen to characterize appears to be the most highly expressed of the ACP genes based on gene-specific probing of embryo mRNA. However, the multiplicity of ACP genes and napin genes coupled with the difficulty of comparing measurements of a napin mRNA to an ACP mRNA complicates analysis. However, it is clear from these northern blot studies that the onset and point of maximal synthesis of ACP mRNA comes earlier than napin. To the limit of our detection abilities, we can generally detect ACP mRNAs about 1-2 days earlier in embryogeny under our plant cultivation conditions. ACP quickly reaches it maximal mRNA steady state level at about 21 days after flowering which agrees well with the time course of fatty acid synthesis which requires acyl carrier protein. Total napin mRNA levels reach a peak much later, 28 or more days after flowering. One limiting factor is the degree of difficulty involved in collecting embryo material prior to the cotyledonary stage due to its small size and labor necessary for collection free from possible endosperm materials. Thus the actual 'turning on' of napin and embryo class ACP genes could be earlier; that is, assaying whole developing seeds may appear negative in some studies because most of the tissue is non-embryo.

Since we further wanted to examine the elements in napin and ACP genes that control their expression during embryogeny, we excised the gene promoter elements in such a way that we could recombine them in vitro with any gene of choice. For these analyses, we have used the beta-glucuronidase gene marker (GUS) for which there are sensitive and easy assays available to score gene expression of GUS (4).

Gene constructs were transferred to binary-type gene vector systems we previously described for Brassica napus (5). The two particular constructs used in these studies were '1813' and '1827'. In the case of 1813, there are two chimeric genes transferred: one has 1.7 kilobase (kb) napin promoter region and a 1.3 kb 3' transcription terminator region from the 1-2 napin gene flanking the GUS gene. The other is the plant selectable marker gene NPT II, a kanamycin phosphorylase gene flanked by the 35S promoter of Cauliflower mosaic virus and the 3' transcription terminator from the Transcript 7 gene from the octopine pTiA6 T-DNA. 1827 has the same selectable marker construct but the GUS gene is flanked by a 1.5 kb ACP promoter region of the Bcg4-4 ACP gene and 1.3 kb of DNA containing the transciption terminator sequences from the 1-2 napin gene.

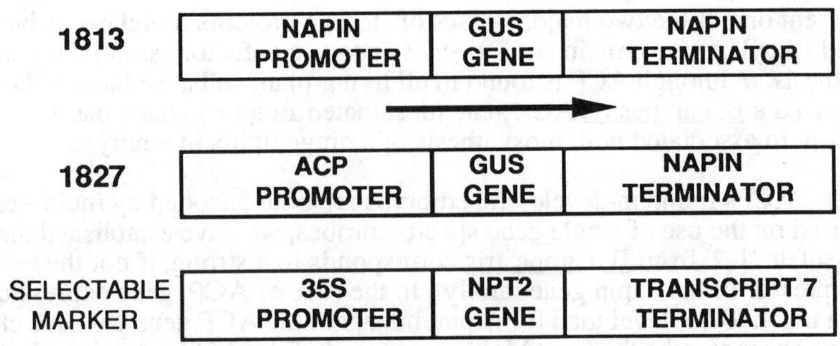

Transgenic B. napus plants were produced from B. napus varieties Global (spring-type) and Cascade (winter-type) hypocotyl explants, which were cocultivated with Agrobacterium tumefaciens strain EHA101 containing binary palsmids 1813 or 1827. Our transformation protocol which we previously described (5) was followed with these modifications. Feeder plates were made with MS medium containing 2,4-D (1 mg/l). After two days of cocultivation with Agrobacterium, hypocotyl explants were placed on B5 medium with 2,4-D (1 mg/l), kanamycin (25 mg/l) and carbenicillin (500 mg/l) for 3 to 7 days then placed on B5BZ medium containing kanamycin (25 mg/l) and carbenicillin (500 mg/l). Kanamycin (50 mg/l) was included in the shoot maturation and rooting media. Depending on the variety, transformation frequencies varied from <1 to 10 percent of the explants giving rise to at least one transformed plant. Kanamycin-selected plants were tested for NPT II activity.

Microspores were isolated and cultured using a procedure similar to the whole-bud method of Swanson et al. (6), with the following changes. Clusters of buds (0-4 mm) were crushed with a mortar and pestle. The filtrate was centrifuged for 5 min. at 200 x g, with 10 ml per wash. The buds and filtrate were also cooled during isolation. The final pellet was diluted to 200,000 microspores per ml of culture medium, then plated 7 ml / 100 x 25 mm plate, or 2.5 ml / 60 x 15 mm plate. After one day at 30°C (dark), the cultures were diluted with fresh culture medium to 100,000 microspores per ml and re-plated into twice as many plates. No further dilution of the liquid cultures was done.

Since transgenic 1813 and 1827 donor plants are hemizygous for the T-DNAs, their gametes will segregate for the inserted genes. We tested segregation for the selectable marker gene, NPT II, by plating torpedo and cotyledonary embryos on our B5BZ medium containing kanamycin (25 mg/l). Embryos were maintained at 23°C day and 19°C night temperature with a 16 hour photoperiod at 100 $uEm^{-2}s^{-1}$. After one to two weeks, segregation for kanamycin-resistance or sensitivity was scored as resistant (green) versus sensitive (bleached). Ratios are shown in Table 1. Thirteen transgenic plants exhibited 1:1 segregation ratios indicating one T-DNA locus, five plants segregated 3:1 indicating two loci, one plant segregated 7:1 indicating 3 loci, and one fit a 31:1 and 63:1 ratio indicating 5 or possibly 6 loci. Some of the plants were also tested for segregation of the GUS gene by the histochemical assay (4) in which microspore-derived developing embryos at the heart, torpedo, and early cotyledonary stages turned blue in the presence of X-gluc (1 mg/ml in 0.2 M $NaPO_4$ buffer for 16-20 hours). Segregation ratios for the GUS gene are indicated in Table 1, scored as blue

Table 1. Segregation ratios for the NPT II and GUS genes in microspore-derived embryos.

Plant #	NPT II $K^r : K^s$ [a]	Ratio[b]	GUS $G^+ : G^-$ [c]	Ratio[b]	# Loci [d]
1813-CS-1	133:32	3:1	-	-	2
1813-CS-2	144:156	1:1	19:18	1:1	1
1813-GL-1	343:9	31:1	22:0	-	5 or
"	"	63:1	-	-	6
1813-GL-2	51:49	1:1	34:37	1:1	1
1813-GL-3	97:103	1:1	30:41	1:1	1
1813-GL-4	86:114	1:1	18:21	1:1	1
1813-GL-5	74:26	3:1	36:33	1:1	2
1813-GL-6	73:27	3:1	21:4	3:1	2
1813-GL-7	103:97	1:1	112:38	3:1	2
1827-CS-1	19:17	1:1	-	-	1
1827-CS-2	85:87	1:1	-	-	1
1827-CS-3	17:19	1:1	-	-	1
1827-CS-4	91:109	1:1	55:54	1:1	1
1827-CS-5	27:9	3:1	92:51	2:1	2
1827-CS-6	85:65	1:1	-	-	1
1827-CS-8	37:35	1:1	-	-	1
1827-GL-1	18:12	1:1	-	-	1
1827-GL-2	73:27	3:1	-	-	2
1827-GL-5	104:96	1:1	-	-	1
1827-GL-6	88:12	7:1	-	-	3

a K^r= the number of kanamycin resistant embryos. K^s= the number of kanamycin sensitive embryos.

b All ratios were not significantly different than the indicated ratios when tested for goodness of fit by the X^2 test using the Yates correction.

c G^+= the number of blue embryos stained with X-gluc. G^-= the number of embryos that did not stain blue.

d The number of T-DNA insertions at different chromosomal locations.

stained embryos versus embryos which remained white or yellow in the presence of X-gluc. Of the donor plants which were tested, one plant 1827-CS-5 exhibited a ratio that did not fit 1:1 or 3:1. The segregation pattern for the kanamycin-resistance gene indicated a 3:1 ratio. Two other plants showed different segregation ratios of the NPT II and GUS genes. One, 1813-GL-5 probably contains a gus gene in one of the T-DNAs which is not expressed and the other, 1813-GL-7 probably contains one T-DNA in which the NPT II gene is silent.

Donor plants 1813-GL-3, 1813-CS-2, 1827-GL-1, and 1827-CS-4 were choosen on the basis of a 1:1 segregation pattern for GUS and NPT II in order to study timing and levels of expression of the GUS gene in developing microspore-derived embryos.

The GUS enzyme, detected by staining with X-gluc and also measured fluorometrically, was present in isolated 1827 microspores sampled one day after culture but not in freshly isolated microspores. Two days after culture, microspores isolated from 1813 plants showed GUS expression. However, the expected 1:1 ratio of microspores that could express the GUS gene was not observed when stained with X-gluc. Less than half of the cells are viable in the microspore cultures, as visualized by fluorescence microscopy after staining with fluoroscein diacetate (7). One would expect that half of the viable cells would express the GUS gene but much less has been observed. It is possible that only cells that are embryogenic express the gene or that napin and ACP promoters are induced in cultured microspores. Stages of isolated microspores which are capable of undergoing embryogenesis have been identified (8). Not all of the viable cells or cells that stain blue in the presence of X-gluc appear to be the embryogenic type, some are developmentally younger or older than the embryogenic types. In globular, heart, torpedo, and cotyledonary stage embryos, we observed continued expression of the GUS gene by the ACP and napin promoters as visualized by blue stained embryos.

T-DNA insertions can be studied using microspore-derived embryos and dihaploids. Screening for transferred genes and number of T-DNA loci can be done early in the production of dihaploids by scoring microspore-derived embryos. The segregation ratios found in developing embryos by staining with X-gluc is often a predictor of the segregation ratios found by the selectable marker gene, NPT II. There are transgenic plants in which the NPT II gene segregates as two loci (3:1) and the GUS gene segregates as one locus (1:1) and vice versa. This indicates that there are T-DNA insertions in which only part of the T-DNA is expressed possibly due to rearrangements in one but not both T-DNAs.

Crouch (2) reported that microspore-derived embryos follow normal embryo development in that embryo-specific storage proteins were present but only somewhat resembled the normal process in that protein levels were much lower in the microspore-derived tissues and appeared much earlier in development. Using the chimeric GUS gene approach, we can now show that napin (and ACP) gene expression is likely to occur even at a pre-globular stage. Crouch, however, was measuring cruciferin levels and we believe napin expression precedes cruciferin gene expression. We would like to examine GUS expression at corresponding early stages in zygotic embryogenesis using cytochemical and histological assays with thin-sectioned whole-seed material. It may be that the napin genes and even the cruciferin genes are on at early stages of embryogenesis but the assays previously used were not sensitive enough or protein was rapidly turned over. Moreover, there remain the interesting questions relating to the extent to which genes in rapeseed are expressed in an endosperm material and if this expression includes genes generally characterized as 'embryo-specific'.

The possibility exists that the reintroduction of the napin and ACP promoters into new chromosomal contexts leads to GUS expression patterns that no longer accurately represent that of the endogenous napin and ACP genes. We don't think that this is the case since all of our transgenic plant data show faithful expression of embryo-specific promoters. The process of embryogenesis in microspore-derived embryos versus zygotic embryos may differ due to environmental influences, ploidy-levels, and the ontogeny of the respective tissues.

REFERENCES

1. Goldberg, R.B., Barker, S.J. and Perez-Grau, L. (1989) Cell 56, 149-160
2. Crouch, M.L. (1982) Planta 156, 520-524
3. de la Roche, A.I. and Keller W.A. (1977) Z. Pflanzenzucht. 78, 319-326
4. Jefferson, R.A. (1987) Plant Mol. Biol. Rep. 5, 387-405
5. Radke, S.E., Andrews, B.M., Moloney, M.M., Crouch, M.L., Kridl, J.C. and Knauf, V.C. (1988) 75, 685-694
6. Swanson, E.B., Coumans, M.P., Wu, S.C., Barsby, T.L., Beversdorf, W.D. (1987) Plant Cell Rep. 6, 94-97
7. Kasten, F.H. (1981) in Staining Procedures (Clark, G., ed.) p. 94
8. Pechan P.M. and Keller, W.A. (1988) Physiol. Plant. 74, 377-384

DETECTION OF ANDROGENESIS-RELATED PROTEINS IN MAIZE

P.VERGNE*, F. RICCARDI*, M.BECKERT**, & C.DUMAS*
*. Reconnaissance Cellulaire et Amélioration des Plantes, Université Lyon 1, INRA 23879, Bât. 741, 69622 Villeurbanne Cedex, France.
**. INRA, Station d'Amélioration des Plantes, Domaine de Crouelle, 63039 Clermont-Ferrand Cedex, France.

1. INTRODUCTION

Dramatic improvements of the androgenetic response in maize (*Zea mays* L.) have been achieved recently through selection of some highly androgenic genotypes (1,2). However, the utilization of in vitro androgenesis in maize breeding still requires increased responsiveness to anther culture in germplasm with agronomic interest. Clearly, a better knowledge of the cellular and molecular mechanisms of the androgenetic induction are necessary to approach the problem of the recalcitrant genotypes. Using highly responsive lines obtained recently (2) as the basis for an experimental system, we have initiated electrophoretic studies to identify early molecular markers (proteins) of androgenesis.

It has been demonstrated previously that a cold pretreatment of maize tassels significantly enhances anther androgenetic response (3). Actually, earlier preliminary studies carried out with plants originating from the synthetic population NS (4) have shown that genotype-specific proteins could be detected in the anthers of some plants after the cold pretreatment (5). Some evidence also indicates that the anther responsiveness in cereals is partially sporophytically determined (6,7,8), i.e. that the anther wall plays an important role in the induction of microspore embryogenesis. In addition, recent studies with the carrot somatic embryogenesis system demonstrated the functional importance of extra-cellular proteins in the trigerring of the embryogenesis process (9). These data led us to investigate the variations in the whole anther protein pattern during the cold pretreatment of the tassels.

2. MATERIALS AND METHODS

2.1 Materials

The genotypes used and their anther culture responsiveness are listed in table 1.

2.2 Methods

2.2.1 In vitro androgenesis

For these experiments, tassels and anthers were handled for androgenesis as described (4). Briefly, tassels with anthers containing mid-uninucleate microspores were removed from plants prior to emergence from the whorl.

They were then wrapped in aluminium foil and stored at 7°C for 7 days. Tassels were surface-sterilized and the anthers were plated in Petri dishes containing the induction medium. After 1 week in the dark at 14°C, dishes were transferred to 28°C under dim light. The number of embryogenic anthers was recorded 40 days after tassel removal from the plant.

TABLE 1. Description of genotypes used and their response frequencies

Genotype	Type	Response Frequency		References
		Anthers plated[(a)]	Percent response[(a)]	
DH5	Inbred (DH* line)	275	45,4	(2)
DH7	Inbred (DH* line)	375	58,1	(2)
DH5 x DH7	Hybrid	1125	61,2	(2)
H99[(b)]	Inbred	800	0,0	
A188	Inbred	400	0,0	(3,10)
B73[(c)]	Inbred	225	0,0	
F1656[(d)]	Inbred	400	0,0	
Ohio 43	Inbred	200	0,0	

* Doubled haploid

(a) The number of plated anthers and the response frequencies indicated refer to the sole experiments described in this paper.

(b) In a previous study, this line consistently appeared as parent of responsive crosses (11).

(c) In a previous study, this line appeared as parent of non responsive crosses (11).

(d) Some experiments suggest that this line carries some inhibitory genes for androgenetic response (2).

2.2.2 Anther sampling and protein extraction

At different stages of the cold pretreatment samples of 30 to 60 anthers were dissected from the tassels and stored in liquid nitrogen until protein extraction.

For analysis, anthers were ground to powder in liquid nitrogen, and native soluble proteins were extracted in a Tris-HCl buffer (25 mM (pH 7,8),

5 mM DTT, 2% (w/v) insoluble PVP).

2.2.3 Isoelectric focusing(IEF)

Native proteins were separated by polyacrylamide slab gel IEF in the pH range 4,0-6,5 as described (12). 35 µg protein were loaded per lane. The separated proteins were then visualized by silver staining (13)

2.2.4 Selective two-dimensional electrophoresis

IEF gels were loaded with 150 µg protein per lane and run normally. The bands of interest were transiently visualized with $ZnCl_2$ (14) or they were located on fixed gels according to stained tracks of the same gels. They were then cut off from the gels and subjected to SDS-PAGE on 12% acrylamide slab gels according to (15). The separated proteins were detected by silver staining (13).

2.2.5 In vivo polypeptide labeling

At different stages of the cold pretreatment, anther proteins were labeled in vivo by [^{35}S]methionine incorporation. They were then processed by selective two-dimensional electrophoresis and fluorographed.

3. RESULTS AND DISCUSSION

3.1 Detection of cold induced proteins

For the highly responsive hybrid DH5 x DH7, figure 1 shows that the cold pretreatment causes the anthers to accumulate a basic protein of medium abundance (band A). This protein is not detected during the normal in vivo development at the mid- and the late bicellular pollen stages when the tassels are not sampled and are not pretreated (lanes MB and LB).

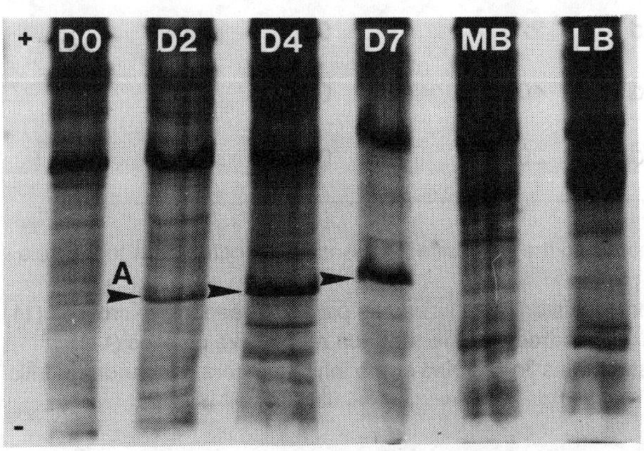

FIGURE 1. Accumulation of a basic protein in pretreated DH5 x DH7 anthers
At different stages of the cold pretreatment, anther proteins were analyzed by IEF in the pH range 4,0-6,5. Only the basic part of the gel is presented. Lanes: D0, at tassel removal; D2, after 2 days of cold treatment; D4, after 4 days of cold treatment; D7, after 7 days of cold treatment; MB, mid-bicellular stage developed in vivo; LB, late bicellular stage developed in vivo. Band A (see text) is indicated by the arrowheads.

Band A is hardly detectable at the time of tassel sampling (lane D0) and is first significantly visualized after two days of cold treatment (lane D2). The accumulation reaches a maximum after four days of treatment (lane D4) and the intensity of the band remains stable until the end of the exposition to the cold (lane D7).
These investigations were extended by comparing IEF patterns at stages day 0 (D0) and day 7 (D7) of the cold pretreatment in other competent and non-competent genotypes.
The results of the experiments are exemplified in figure 2. The line DH5, used as female in the cross DH5 x DH7, exhibits a similar pattern of accumulation of an IEF protein band (band A). On the opposite, the male parent DH7 does not display such an accumulation in the basic region of the gels. In the same way, there is some evidence that DH5 and DH7 carry separate genetic factors showing good complementation (2).
Interestingly, the anthers of the line H99, which is known for its capacity for anther culture (11,16), accumulate two proteins which focus in the basic part of the gels: one corresponding to band A, and a more intense one with a more acidic pI (band B). On the other hand, none of the non-androgenic lines tested (A188, B73, F1656 and Ohio 43) accumulated protein similar to bands A or B (as shown for A188 in fig. 2).

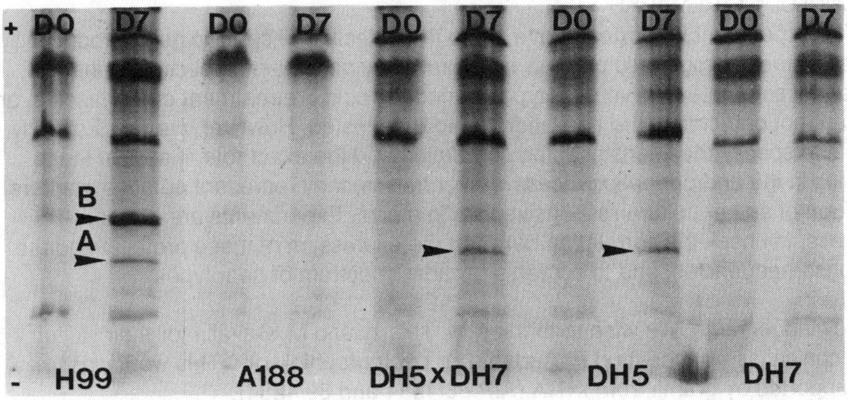

FIGURE 2. Comparison of the anther protein patterns at day 0 (D0) and day 7 (D7) of the cold pretreatment in several responsive and non responsive genotypes The proteins were separated by IEF in the pH range 4,0-6,5. Only the basic part of the gel is presented. The positions of bands A and B (see text) are shown by the arrowheads.

Thus, the cold pretreatment induces in some androgenic genotypes with different genetic backgrounds the accumulation of specific proteins in the anthers . Proteins with the same pattern of accumulation and similar pIs are not detected in the anthers of non-competent genotypes.

3.2 Characterization of the specific proteins

In order to better characterize the proteins which accumulate in the anthers of DH5 x DH7 and H99 during the cold pretreatment, we have used selective two-dimensional electrophoresis.

This analysis revealed that the major polypeptide components of bands A and B display a similar apparent molecular mass of 32 kDa. This similarity suggest that the H99 and the DH5 x DH7 polypeptides share analogous structural features.

Indeed, these results do not allow to conclude about the origin of the cold-induced accumulation of the 32 kDa polypeptides in the anthers of the responsive genotypes, i.e. wether this accumulation results from the degradation of pre-existing proteins or from protein synthesis. Thus, we finally addressed this question.

The labeling experiments demonstrated that the synthesis of the 32 kDa polypeptide actually occurs during the cold pretreatment of the tassels. Its accumulation therefore corresponds at least partly to the translation of RNA messages. However, we are presently unable to ascertain wether the acccumulation results from enhanced transcription of a specific sequence, or from protein turn-over modifications. The answer to this issue will likely come out of studies investigating differentially expressed mRNAs in cold treated and non treated anthers.

4. CONCLUDING REMARKS

The experiments reported above led to the detection of specific gene products related to the androgenetic process in maize. They accumulate specifically in the anthers of some responsive genotypes during the cold pretreatment of the tassels, and they are not detected in the non-androgenic lines tested. However, we are presently unable to specify the intensity of the correlation and the exact role, if any, of these proteins in the androgenetic process. Nevertheless, they represent putative markers of a form of anther culture responsiveness in maize. Experiments are currently in progress, to check the correlation between the expression of these proteins and the response frequencies, and to screen a broader spectrum of genotypes.

Acknowledgements. We wish to thank A.M. Thierry and M. Cavalla for their skilful technical assistance, and P. Audenis for photographic work. This work was supported by grants from INRA (AIP 88/4684 and 89/4684).

REFERENCES

1 Petolino, J.F., Jones, A.M. and Thomson, S.A. (1988) Theor. Appl. Genet. 76, 157-159
2 Barloy, D., Denis, L. and Beckert, M. (1989) Maydica 34, 303-308
3 Genovesi, A.D. and Collins, G.B. (1982) Crop Sci. 22, 1137-1144
4 Dieu, P. and Beckert, M. (1986) Maydica 31, 245-259
5 Delvallée, I. (1987) PhD Thesis, Université Lyon 1
6 Henry, Y. and De Buyser, J. (1981) Theor. Appl. Genet. 60, 77-79
7 Kolher, F. and Wenzel, G. (1985) J. Plant Physiol. 121, 181-191
8 Datta, S.K. and Wenzel, G. (1987) Plant Sci. 48, 49-54

9 De Vries, S.C., Booij, H., Janssens, R., Vogels, R., Saris, L., LoSchiavo, F., Terzi, M. and van Kammen, A. (1988) Genes Develop. 2, 462-476
10 Brettell, R.I.S., Thomas, E. and Wernicke, W. (1981) Maydica 26, 101-111
11 Petolino, J.F. and Jones, A.M. (1986) Crop Sci. 26, 1072-1074
12 Delvallée, I. and Dumas, C. (1988) J Plant Physiol. 132, 210-217
13 Morrissey, J.H. (1981) Anal. Biochem. 117, 307-310
14 Dzandu, J.K., Johnson, J.F. and Wise, G.E. (1988) Anal. Biochem. 174, 157-167
15 Laemmli, U.K. (1970) Nature 227, 680-685
16 Petolino, J.F., and Thomson, S.A. (1987) Theor. Appl. Genet. 74, 284-286

SIGNIFICANT HIGHER PROPORTIONS OF 1BL-1RS WHEAT-RYE TRANSLOCATION LINES AMONG DOUBLED HAPLOID PROGENIES DERIVED FROM F1 HYBRIDS BETWEEN TRANSLOCATED AND NON-TRANSLOCATED WHEAT CULTIVARS.

P.DEVAUX[1], P. LONNET[1] and R. JEAN[2]
1. Florimond Desprez Co., POB 41, 59242 Cappelle en Pévèle, France, 2. University of Lille 1, SN2, 59655 Villeneuve d'Ascq, France.

1. INTRODUCTION

Many plant breeding programmes have been using doubled haploid plants (DHs) for increasing their efficiency and/or for development and release of new cultivars. In common wheat, Triticum aestivum L., anther culture is the most efficient technique for obtaining haploids and DHs, and many efforts have been devoted to its improvement (for recent review, see 1). Genetic analyses have shown that male gametophyte-derived plant production is under the control of at least three independently inherited characters i.e. embryo induction rate, regeneration ability and the ratio of green to albino plants (2).

In the cytotype of wheat which carries the 1BL-1RS wheat-rye translocated chromosome, higher regeneration rate has been observed leading to the deduction that gene(s) involved in regeneration ability are located on the 1RS chromosome arm (2).

From a breeding point of view, the 1RS chromosome arm is known to carry many important genes for disease resistance (3) and has been therefore incorporated into several wheat cultivars (4, 5). Unfortunately, the 1BL-1RS transfer causes a decrease in bread-making quality (6).

Our wheat breeding programme involves some crosses between 1BL-1RS translocated cultivars and those having a normal 1B chromosome. Such F1 hybrids i.e. 1BL-1RS/ 1BL-1BS structural heterozygotes have been used for DH production and for assessing in vitro selection within male gametophyte-derived embryos in their ability to regenerate plants.

2. MATERIALS AND METHODS

The plant material for DH production consisted of seven F1 hybrids between cultivars having a normal 1B (1BL-1BS) chromosome namely Camp Remy (CR), Dartagnan (Da), Donau (Do), Fresco (F), Pernel (P), Soissons (S) and one cultivar, Apollo (A), or advanced line, FD1692 (FD), both with a homozygous 1BL-1RS translocated chromosome. All of them are winter wheats.

One hundred and forty seven fertile DHs have been developped by classical anther culture techniques (7) and subsequently analyzed for presence or absence of the 1BL-1RS translocated chromosome. Among the main systems available for characterization of 1BL-1RS translocation lines and listed in Table 1, sodium dodecyl sulfate-polyacrylamide-gel electrophoresis (SDS-PAGE) of the storage proteins has been used in the present experiment and performed on parents of the 7 F1 hybrids as well as on the 147 derived DHs. The technique was the same as described by Branlard and Le Blanc (18). Lines with the 1BL-1RS translocated chromosome can therefore be identified by the absence of 1BS-encoded ω-gliadins on such gels (Figure 1).

Figure 1. Analysis by SDS-PAGE of the storage proteins of wheat lines. Translocated 1BL-1RS lines have been identified by the absence of 1BS-encoded ω-gliadins as indicated by the arrows.

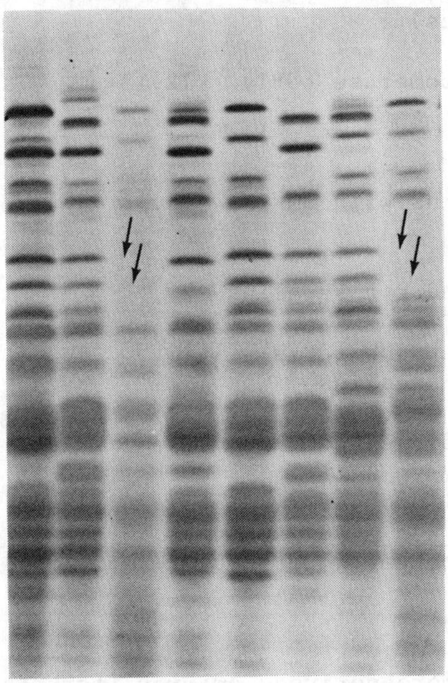

3. RESULTS AND DISCUSSION.

The frequencies of 1BL-1RS wheat-rye translocation DH lines and those having a normal 1BL-1BS chromosome are indicated in Table 2. 103 (70.1%) of the 147 DHs analyzed have been found to possess a 1BL-1RS translocated chromosome while 44 (29.9%) have a 1BL-1BS chromosome. A χ^2 test applied to the data indicates that the overall ratio translocated:non-translocated DHs is significantly different from the 1:1 expected one at the 0.001 level. Percentages of 1BL-1RS cytotypes per F1 hybrid, from which DHs have been derived, vary from 59.3% to 100%.

Our results agree closely with those reported recently (10, 14) in which 67.3% and 75% of microspore-derived plants respectively had a 1BL-1RS translocated chromosome and confirm that there is a

Table 1. Main systems for characterization of 1BL-1RS wheat-rye translocation lines.

Markers	Techniques	References
Cytological	Mitosis:	
	Nucleolus organiser competition (Feulgen, Ag-NOR banding)	8.2.4.9.10.11
	C-banding	3.11
	N-banding	9
	Meiosis in PMCs (for heterozygous 1BL-1BS/1BL-1RS lines)	3
Isozyme	Glucose phosphate isomerase (GPI)	12.13
	Peroxidase	14
Storage protein	SDS-PAGE of ω-gliadins with HMW glutenin subunits	15.16
	APAGE of gliadins and rye secalins	2
	Two-dimensional electrophoresis of gliadins and rye secalins	16.5
	HPLC of gliadins and rye secalins	(*)
RFLP	Use of pTA71, PSR161 probes	17. (**)

(*) J.A. Bietz, R.A. Graybosch, B. Lookhart, personal communication.
(**) M.D. Gale, personal communication.

preferential developpement of microspore-derived embryos for those with the 1RS chromosome arm. This difference in frequency is maintained up to the obtaining of fertile DHs, since grains collected on them have been analyzed in our experiment by SDS-PAGE. Thus, the ability for chromosome doubling, either spontaneously or after colchicine treatment, seems not to be linked to the presence or absence of the 1BL-1RS chromosome.

In a more recent study, Due _et al._ (13) have found distorted segregation of 1BL-1RS (74%) and 1BL-1BS (26%) among the green regenerants of microspore-derived embryos while a 1:1 ratio among the albino regenerants . They have made the hypothesis that the 1BS chromosome arm is acting as a semi-lethal in structures determined to become green.

From a breeding point of view, our results can be exploited as follows. From heterozygous 1BL-1BS/1BL-1RS hybrids, two kinds of DHs

Table 2. Numbers of 1BL-1RS wheat-rye translocation and normal 1BL-1BS lines among DH progenies derived from 7 F1 hybrids heterozygous 1BL-1RS/1BL-1BS.

F1 hybrid	No. of DHs with:		% 1BL-1RS	Total
	1BL-1RS	1BL-1BS		
A x CR	16	9	64.0	25
A x Da	15	1	93.8	16
A x Do	8	3	72.7	11
F x A	22	15	59.5	37
A x P	14	0	100.0	14
A x S	16	11	59.3	27
S x FD	12	5	70.6	17
Total	103	44	70.1	147

$\chi^2 = 23.68$ with 1 d.f. ($\chi^2_{.001} \geqq 10.83$)

are produced by anther culture : one carrying 1BL-1RS, the second with a normal 1B chromosome. The former can be used in feed-wheat breeding, since the 1RS chromosome arm induces a detrimental effect on bread-making quality. The latter, although in fewer proportions, can be evaluated further for quality which has become an important trait for modern cultivars.

REFERENCES

1 Picard, E. (1989) Genome 31, 1005-1013
2 Henry, Y. and De Buyser, J. (1985) Plant Cell Rep. 4, 307-310
3 Friebe, B., Zeller, F.J. and Kunzmann, R. (1987) Theor. Appl. Genet. 74, 423-425
4 Lange, W. (1988) Euphytica S, 7-25
5 Rogers, W.J., Payne, P.I. and Harinder, K. (1989) Plant Breeding 103, 89-100
6 Zeller, F.J., Günzel, G., Fischbeck, G., Gerstenkorn P. and Weipert, D. (1982) Getr. Mehl. Brot. 36, 141-143
7 Picard, E. (1984) Thèse de Doctorat d'Etat, Université Paris XI, Orsay, France
8 Miller, T.E. (1984) Can. J. Genet. Cytol. 26, 578-589
9 Rayburn, A.L. and Carver, B.F. (1988) Euphytica 38, 237-240
10 Agache, S., Bachelier, B., De Buyser, J., Henry, Y. and Snape, J.W. (1989) Theor. Appl. Genet. 77, 7-11
11 Friebe, B., Heun, M. and Bushuk, W. (1989) Theor. Appl. Genet. 78, 425-432
12 Ainsworth, C.C. and Gale, M.D. (1987) in Enzymes And Their Role In

Cereal Technology (Kruger, J.E., Lineback, D.R. and Stauffer, C.E., ed.), pp.53-82, American Association of Cereal Chemists, St Paul

13 Due, I.K., Pedersen, S., Olesen, A. and Andersen, S.B. (1990) Plant Cell Rep. (submitted)

14 Müller, G., Borschel, H., Vahl, U., Wiberg, A., Härtel, H. and Damisch W. (1989) Plant Breeding 102, 196-207

15 Payne, P.I., Holt, L.M., Hutchinson, J. and Bennett M.D. (1984) Theor. Appl. Genet. 68, 327-334

16 Payne, P.I., Nightingale, M.A., Krattiger, A.F. and Holt, L.M. (1987) J. Sci. Food Agric. 40, 51-65

17 Sharp, P.J., Chao, S., Desai, S. and Gale, M.D.(1989) Theor. Appl. Genet. 78, 342-348

18 Branlard, G. and Le Blanc, A. (1985) Agronomie 5, 467-477

DIFFERENCES IN RESPONSES TO 2,4-DICHLOROPHENOXY ACETIC ACID (2,4-D) TREATMENT BETWEEN EMBRYOGENIC AND NON-EMBRYOGENIC LINES OF ALFALFA

BÖGRE L., I. STEFANOV, M. ÁBRAHÁM, I. SOMOGYI and D. DUDITS

Institute of Genetics and Plant Physiology
Biological Research Center H.A.S.
6701 Szeged P.O.Box 521 HUNGARY

1. INTRODUCTION

Induction of the totipotent stage in cultured somatic cells with competence for initiation of somatic embryogenesis is under the control of several external and internal factors. Numerous tissue culture experiments have revealed the pivotal role of exogenously applied auxins, mainly 2,4-D in the triggering of the embryogenic pathway of somatic cells both in dicots (1, 2,) and monocots (3).

Explants from Medicago species respond to auxins with simultaneous callus and embryo formation (4). Walker et al. (5) developed a two step alfalfa tissue culture system which clearly demonstrated the importance of the auxin concentration in determining callus growth and embryo formation from somatic tissue. In alfalfa the embryogenic capability is not limited to callus cells, but embryos can be directly formed from differentiated cells as epidermis cells or mesophyll leaf protoplasts (6, 7, 8,). In protoplast cultures the induction of the first cell division was found to be critical for embryogenesis (23).

The hormone induced embryogenic response is highly species and genotype dependent. Among alfalfa clones there is a great variability in the potential for somatic embryognenesis (9, 10, 11) and this feature is genetically determined (12).

There is limited information about the key factors in auxin induced somatic embryogenesis both at molecular and cellular levels. In the present work based on two alfalfa lines we search for possible correlation between auxin response, mainly 2,4-D sensitivity, and the capability of alfalfa lines to form somatic embryos. In addition to tissue culture experiments we used two 2,4-D responsive cDNA clones, isolated from soybean, as molecular markers to monitor the effect of auxin treatment at mRNA level. The pJCW1 DNA is homologous to a gene which is activated during auxin treatment of excised soybean hypocotyl sections (13, 14). The clone, pTUO4 originated from the cDNA library made from cells grown in the presence of 2,4-D . As the DNA sequence is concerned, the protein represented by this cDNA is a cell wall protein (15). Based on the Northern blot analysis with these probes and the tissue culture behaviour of explants, we show that the somatic cells from the embryogenic alfalfa line are more sensitive toward 2,4-D treatment.

2.MATERIALS AND METHODS

Tissue culture techniques.

Seeds of Medicago varia cv. Rambler were surface sterilized by treatment with 70% (v/v) ethanol for 30 sec. and 0.1% HgCl (w/v) for 3 min. After several washes in sterile distilled water the seeds were germinated on UM (16) hormone free medium. In tests for embryogenic capability root explants were put on UM medium with 0.2 mg/l benzyladenine and 0.2;, 1;, 2;, mg/l 2,4-D. After two weeks the callus was transferred onto UM hormone free medium. The number of somatic embryos per explant was counted 2 weeks later. The selected embryogenic (A 2) and non-embryogenic (R 15) clones were propagated by nodal cutting on UM hormone free medium.

The rooting of 2 cm long shoot explants were scored after 4 weeks on UM medium containing 0-2 mg/l 2,4-D.

Leaf protoplasts were isolated and cultured as described by DEÁK et al.(17).

Measurement of ^{3}H-thymidine incorporation into leaf protoplasts.

Freshly isolated leaf protoplasts at 10^5 cells/ml cell density were distributed into 3 cm diameter plastic Petri dishes in 1 ml K75 culture medium (8) containing 1 mg/l 2,4-D and 0.5 mg/l zeatin riboside. To each Petri dish 0.37 MBq ^{3}H-labeled thymidine (Radiochemical Center,Amersham, U.K., 777 GBq/mmol) were added for 12 hours. The cells were collected by centrifugation (3 min. 100 g) in Eppendorf tubes and lysed in 1 ml cold 7% (w/v) TCA. The lysates were centrifuged and the pellets were washed by 7% TCA (w/v) subsequently with 80% (v/v) ethanol and 0.2 M perchloric acid. The pellets were extrated in hot 0.5 M perchloric acid for 30 min. After centrifugation the radioactivity was measured in 0.6 ml supernatant dissolved in 5 ml of toluole based scintillation cocktail.

2,4-D treatment of roots, RNA isolation and Northern analysis.

The roots were cut into 0.5 cm pieces and incubated in liquid UM medium containing various concentrations of 2,4-D. The samples were collected and total RNA was purified from 2 g of tissue as described by Cathala (18). Ten µg of total RNA was separated on 1,4% (w/v) formaldehyde agarose gel and blotted onto a nitrocellulose filter (19). Prehybridization and hybridization was carried out in a solution containing 50% (v/v) formamide, 5x Denharts, 5xSSC, 50 mM HEPES pH 7.5, 1 mg/l sheared chicken DNA, 0.1% (w/v) SDS for 24 hours at 42°C. $2x10^7$ cpm/ml universal primer labeled DNA probe was added to the hybridization solution. After hybridization the filters were washed 3 times at 42 °C for 15 min. in 2xSSC; 0.1% (w/v) SDS followed by one wash in 2xSSC; 0.1% (w/v) SDS at 50 °C.

cDNA clones of auxin-regulated genes isolated from soybean hypocotyl (14) and cell suspension culture (15) were kindly provided by J.L.Key (Department of Botany University of Georgia, Athens, Georgia, USA).

For an internal control in quantifying the amount of loaded RNAs we used an alfalfa cDNA clone (p35) isolated by J. Györgyey (Institute of Plant Physiology, Biological Research Center, Hungarian Academy of Sciences).

3. RESULTS

3.1 Selection of alfalfa lines with different potential to form somatic embryos

Twenty alfalfa seedlings from cultivar Rambler were tested for their capability to form somatic embryos. Extreme differences in callus growth and production of somatic embryos were found in alfalfa plants. A plant with high potential for somatic embryogenesis (A2) and a plant lacking this character (R15) was chosen for further analysis.

The two selected clones differed in 2,4-D concentra-tion, required for maximum callus growth from root explants. 2,4-D at 1 mg/l concentration was found to be optimal for callus growth in R15 tissues, but considerable callus production was observed even at 8 mg/l 2,4-D. The A2 plants showed a reduced callus formation with an optimum at 0.2 mg/l 2,4-D. Somatic embryos were produced only from A2 tissues at 0.5-4 mg/l 2,4-D concentrations. This amount of auxin already inhibited the callus growth. Frequently the somatic embryos appeared directly from the root cells. The R15 alfalfa clone did not form somatic embryos under the tested conditions.

The embryogenic and non-embryogenic behaviour was stably maintained after nodal propagation. Selfing and crossing between different lines suggests that this trait is genetically determined (data not shown).

3.2 Activation of DNA synthesis, cell divisions and embryo formation in mesophyll protoplast cultures

The relatively uniform cell populations derived from mesophyll protoplasts offer a synchronous experimental material for more detailed comparison between embryogenic (A2) and non-embryogenic (R15) clones in hormone responses. In the presence of 2,4-D, reactivation of cell cycle and induction of cell divisions are basic prerequisites for unorganized callus growth as well as for embryo differentiation.As shown by the ^{3}H-thymidine incorporation curves and cell division frequency data(Fig.1)the cell cycle was activated earlier in A2 than in R15 cells.

In addition, there were significant differences between the two lines in cell size, morphology and structure of cell divisions (Fig.2). The cells from A2 embryogenic genotype exhibited highly dense cytoplasm with reduced cell expansion and frequent asymmetric first cell division in the presence of 1 and 10 mg/l 2,4-D. Under the same conditions the R15 cells divided with considerable cell expansion and maintenance of chloroplasts. At low concentration (0.1 mg/l) of 2,4-D the cells isolated from both lines showed similar morphological features.

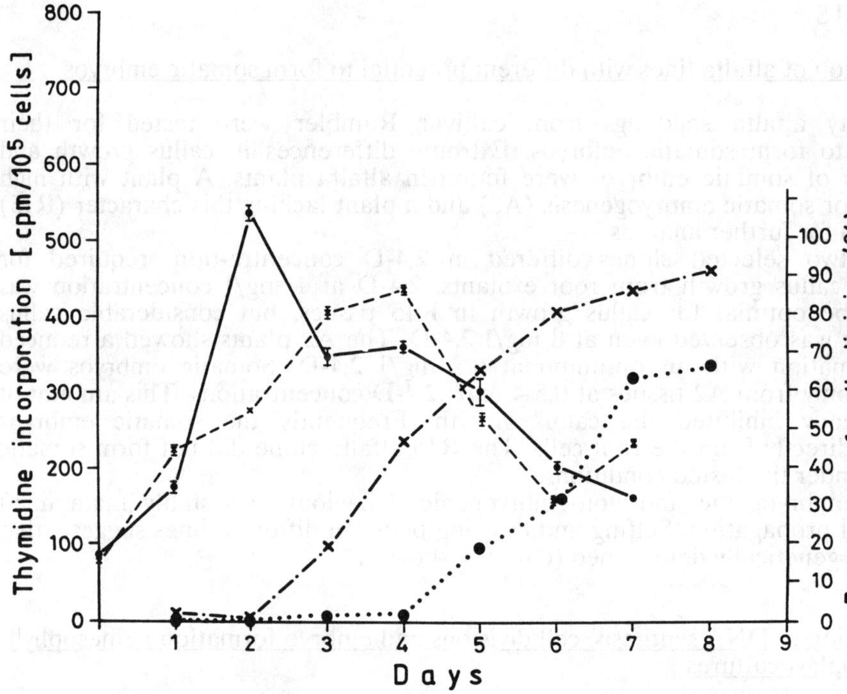

Fig. 1. <u>^{3}H-thymidine incorporation and frequency of cell division in cells from A2 embryogenic and R15 non-embryogenic genotypes</u>

——— ^{3}H thymidine incorporation, A2
x-----x ^{3}H thymidine incorporation, R15
x—·—·x cell division A2
•·······• cell division R15

Fig.2. <u>Morphological characteristics of cells during the first divisions</u>

A2 embryogenic genotype: A: freshly isolated protoplasts; B: 0.1 mg/l 2,4-D; C: 1 mg/l 2,4-D; D:10 mg/l 2,4-D

R15 non-embryogenic genotype: E: freshly isolated protoplasts; F: 0.1 mg/l 2,4-D; G: 1 mg/l 2,4-D; H: 10 mg/l 2,4-D

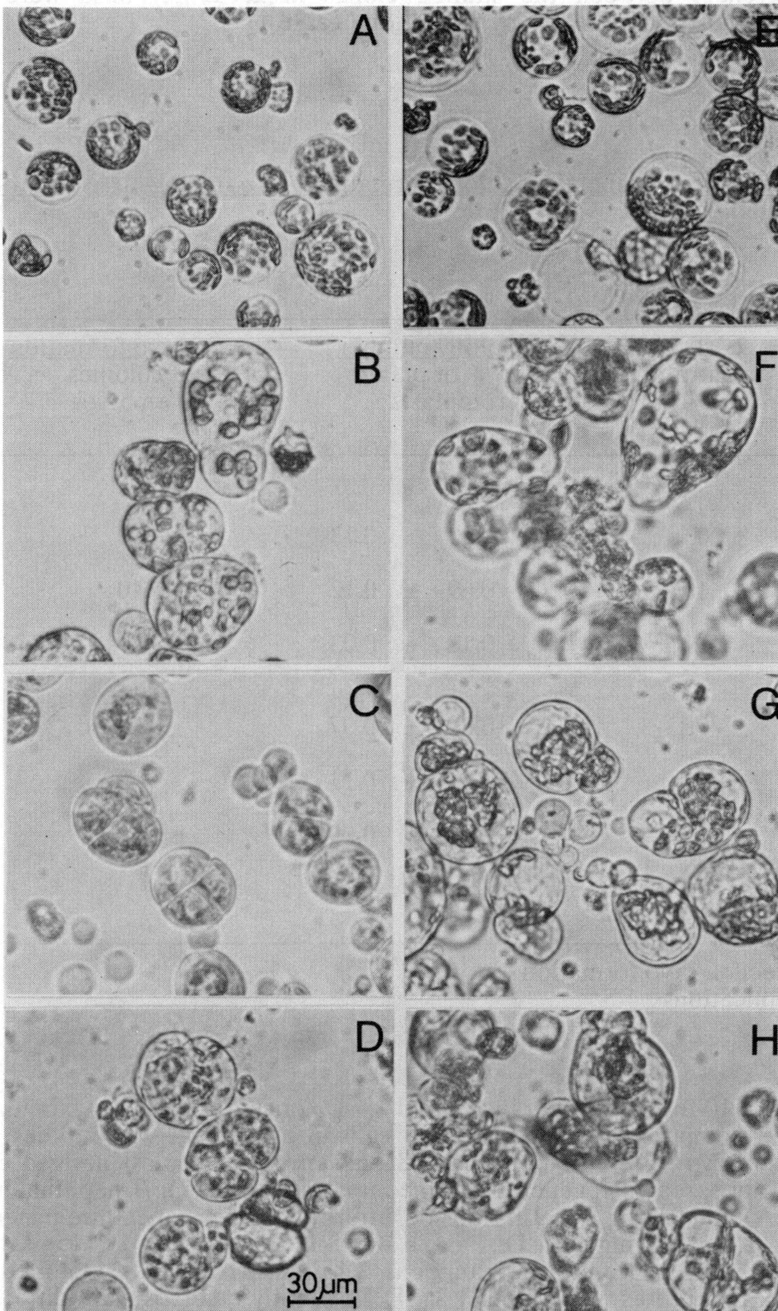
A
E
B
F
C
G
D
H
30µm

Differences between the studied clones in 2,4-D response were clearly revealed by data about the colony formation (Table 1).

Table 1.

2,4-D concentration dependent colony formation and embryogenesis in protoplast derived alfalfa cells

LINES	2,4-D concentration (mg/l)	Colony number in % of plated protoplasts	Embryo formation % of colonies with embryos
A2	0.1	1,9 ± 0.13	0
	1	0.69 ± 0.12	40
	10	0.18 ± 0.03	86*
R15	0.1	0.29 ± 0.07	0
	1	2.1 ± 0.11	0
	10	0.66 ± 0.09	0

* also direct embryo formation
three replications

There was a 10-fold difference in optimal 2,4-D concentration for maximum colony number in comparison of A2 and R15 cells.In agreement with studies on root explants embryo differentiation occured only in A2 protoplast derived colonies. Embryos appeared in regeneration medium with 0.186 mg/l naphthaleneacetic acid (NAA) and 0.5 mg/l zeatin riboside, after three weeks of culture period in the presence of 1 and 10 mg/l 2,4-D.

This is the concentration range of 2,4-D which already inhibited colony formation. In cultures with 10 mg/l 2,4-D frequently embryos were differentiated without the intervening callus formation. The R15 colonies from protoplasts failed to differentiate into embryos in all the tested culture conditions.

3.3 Differences in hormone response in Northern analysis with cDNA probes

As a molecular marker we selected two soybean cDNA probes, pJCW1 and pTUO4. Both of them hybridized to several Hind III restriction fragments of the alfalfa DNA in Southern blot analysis (data not shown). The use of these probes in Northern hybridization allowed us to monitor 2,4-D response, measured as amount of hybridizing mRNA in A2 and R15 tissues treated with auxin.

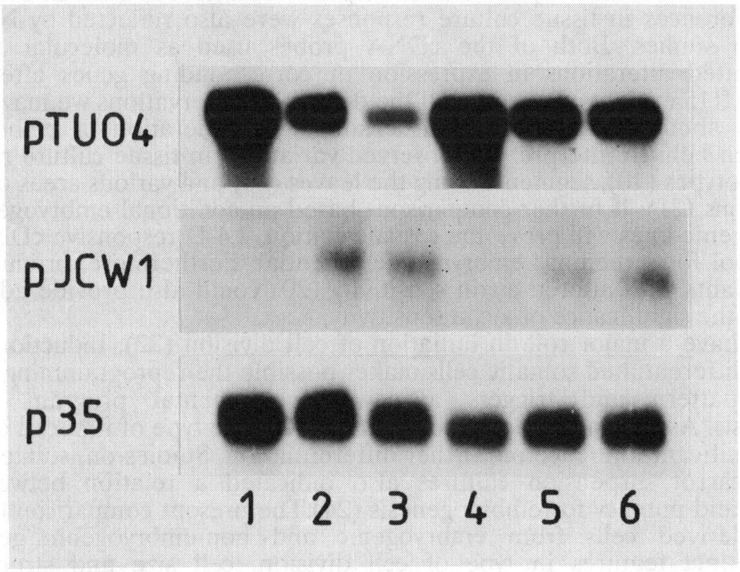

Fig.3. Northern hybridization with pTUO4, pJCW1 and p35 DNA probes to total RNA from A2 (1, 2, 3 lanes) and R15 (4, 5, 6 lanes) alfalfa roots treated with 0 (1, 4 lanes) 0.2 (2, 5 lanes) and 2 (3, 6 lanes) mg/l 2,4-D for 24 hours.

In A2 root tissues the amount of gene product homologous to pTUO4 decreased in a 2,4-D concentration dependent manner. In the case of R15 explants similar inhibition of expression could not be detected even at 2 mg/l 2,4-D.The expression of the gene homologous to pJCW1 was already stimulated by the treatment of A2 tissues with 0.2 mg/l 2,4-D. While 2 mg/l 2,4-D was necessary to induce similar level of expression in the R15 explants. As a control, a gene with constitutive expression, alfalfa cDNA probe p35, was hybridized to the same RNA samples.

4. DISCUSSION

In agreement with several previous publications which described extreme variability among alfalfa clones in tissue culture response (9, 10, 11, 12), we could identify embryogenic (A2) and non-embryogenic (R15) lines by screening a relatively small plant population from 'Rambler' alfalfa cultivar. However, the genetic nature of differences in tissue culture behaviour has not yet been established, the present comparison indicated that A2 somatic cells are more sensitive to 2,4-D treatment than tissues from R15 non-embryogenic line. This conclusion was supported by analysis of both callus growth and colony formation from protoplasts as well as by tests on rooting of shoot explants. Root formation of A2 plants was severely inhibited by 0.01 mg/l 2,4-D while in the case of R15 plants 0.1 mg/l auxin was necessary for similar degree of inhibition (unpublished).The detected differences in tissue culture responses were also reflected by Northern hybridization studies. Both of the cDNA probes used as molecular markers detected limited alterations in expression of corresponding genes after 2,4-D treatment of R15 explants. Based on all the described observations we may suggest a hypothesis about the role of auxin sensitivity in generation of embryogenic stage.This can help to interpret the observed variations in tissue culture response between genotypes (10), segments along the leaves (20) and various areas of carrot petiole sections (21). If further comparisons based on additional embryogenic and non-embryogenic lines will prove the casual relation, 2,4-D responsive cDNAs can serve as a tool for screening embryogenic potential. Furthermore, production of transgenic plants with altered auxin sensitivity (20) could also provide additional evidence for the significance of auxin sensitivity.

Auxins have a major role in initiation of cell division (22). Induction of cell divisions in differentiated somatic cells makes possible the reprogramming of gene expression pattern and triggers a new developmental program through embryogenesis. As shown by Dijak and Simmons (23) the type of first cell divisions can have significant influence on further differentiation. Studies on isolated single cells from carrot suspension cultures also indicated a relation between cell morphology and potency for embryogenesis (24).The present comparison between protoplast derived cells from embryogenic and non-embryogenic genotypes revealed distinct features in type of cell division, cell size and structure of cytoplasms. In addition we could show an interaction between genotype and 2,4-D concentration. The small cytoplasmically dense cells and later embryos appeared at 2,4-D concentration which inhibited growth of callus in the sensitive A2 cultures.

Interestingly here we found more rapid entry of the hormone treated A2 leaf protoplasts into the phase of DNA synthesis and the cell division than cells from R15 non-embryogenic line. This finding is in agreement with previous data about carrot cells, in which the most rapid rate of cell division occured at the onset of somatic embryogenesis (25).

Despite of the fact that the present studies based on only two selected alfalfa lines our work focuses the attention to the problem of auxin sensitivity as one of the possible key factors in generation of embryogenic stage in somatic plant cells.

Acknowledgements

We thank Prof. J.L.Key (Department of Botany University of Georgia, Athens, Georgia 30602) for providing the cDNA clones.

REFERENCES

1 Ammirato,P.V. (1983) in Handbook of Plant Celly Culture (Evans,D.A., Sharp,W.R., Ammirato,P.V., Yamada,Y. eds.) Vol 1., pp 82-123, Macmillan, New York

2 Sung,Z.R., Frienberg,A., Charneau,R., Borkird,C., Turner,I., Smith,I., (1984) Plant Molecular Biology Reporter **2**, 3-14

3 Vasil,I.K.(1987) J.Plant Physiol. **128**, 193-218

4 Saunders,J.W. and Bingham,E.T. (1972) Crop Sci. **12** 804-808

5 Walker,K.A., Wendeln,M.L., Jaworski,E.G. (1979) Plant Sci. Lett. **16**, 23-30

6 Dijak,M. and Brown, D.C.W.(1987) Plant Cell Tissue and Organ Culture **9**, 121-130

7 Dos Santos, A.V.P., Cutter,E.G. and Davey,M.R. (1983) Protoplasma **117**, 107-115

8 Kao,K.N. and Michayluk,M.R. (1980) Z. Pflanzenphysiol. **96**, 135-141

9 Bingham,E.T., Hurley,L.V., Keats,D.M., Sanders,J.W. (1975) Crop Sci. **15**, 719-721

10 Brown,D.C.W. and Atanassov,A. (1985) Plant Cell Tissue and Organ Culture **4**, 111-122

11 Kao,K.N. and Michayluk,M.R. (1981) In Vitro **17**, 645-648

12 Reish,B. and Bingham,E.T. (1980) Plant Sci. Lett. **20**, 71-77

13 Ainley, W.M., Walker,J.C., Nagao,R.T. and Key,J.L. (1988) J. of Biol. Chem. **263**, 1058-1066

14 Walker,J.C. and Key,J.L. (1982) Proc. Natl. Acad. Sci. USA **79**, 7185-7189

15 Hong,J.C., Nagao,R.T. and Key,J.L. (1987) J. Biol. Chem. **262**, 83667-83376

16 Uchimiya,T. and Murashige,M. (1974) Plant Phys. **54**, 936-994

17 Deák,M., Donn,G., Fehér,A. and Dudits,D. (1988) Plant Cell Reports **7**, 158-161

18 Cathala,G., Savauret,J.F., Mendez,B., West,B.L., Karin,M., Martial,J.A., Baxter,J.D. (1983) DNA **2**, 329-335

19 Manniatis,T., Fritsch,F.F., Sambrook,J. (1982) Molecular cloning : a laboratory manual. Cold Spring Harbour, New York

20 Conger,B.V., Hanning,G.E., Gray,D.J., McDaniel,J.K. (1983) Science **221**, 850-851

21 Schafer,F., Groskurt,E., Neumann,K.H. (1985) In: Somatic embryogenesis of carrot (Terzi,M., Pitto,L. Sung,Z.R.) pp. 159-171 IPRA, Rome

22 Cooke,R. and Meyer,Y. (1981) Planta **152**, 1-7

23 Dijak,M. and Simmonds,D.H. (1988) Plant Science **58**, 183-191

24 Nomura,K. and Komamine,A. (1985) Plant Physiol. **79**, 988-991

25 Fujimara,T. and Komamine,A. (1980) New Phytol. **86**, 213-218

THE COMMITMENT TO EMBRYOGENESIS, A CYTOLOGICAL APPROACH

VITTORIA NUTI RONCHI, LUCIA GIORGETTI and MARIA GRAZIA TONELLI.
ISTITUTO DI MUTAGENESI E DIFFERENZIAMENTO, CNR, VIA SVEZIA 10, 56124 - PISA, ITALY.

1. INTRODUCTION

Carrot cell cultures offers an efficient and reproducible system which allow the regeneration, through several differentiative steps, of somatic embryos fairly comparable to the zygotic embryos. Though the cultural conditions to induce embryogenesis in presence of auxin and, in its absence, to reach the subsequent phases of growth (globular, heart, torpedo-shaped embryos) are well known, still misterious and impredictable are the key events which restore totipotency to somatic cells.

Since the very beginning of the work with plant tissue cultures it was recognized that factors which cause cultured somatic cells to grow in an organized way simulate those triggering-off development of non-zygotic embryos in apomictic plants (1). In both cases somatic cells act as asexual sporocytes, decisive signal for commitment being probably assimilable to those present during floral induction in the apex. In fact, the possibility to obtain embryogenesis from reproductive organs "in vivo" (nucella, scutellum) and "in vitro" (immature inflorescences, male gametophyte) suggests that the environment and the cellular events occurring in these organs are determinant in triggering somatic embryogenesis.

Particularly interesting is, on this point, the comparison between two systems of embryos formation "in vitro": somatic embryogenesis and androgenesis. In both cases the developmental stages are perfectly comparable to zygotic embryos: both are originated from a single cell, the only difference being that, in the case of pollen derived embryos, such a cell has undergone meiosis.

The study, in carrot cell lines, of different alternative to mitosis variously affecting chromosome stability and behaviour has given the opportunity to disclose a striking incidence of cell divisions miming meiosis or segregating haploid chromosome sets (2).

These phenomena were immediately associated with the commitment to embryogenesis on the ground that a meiotic-like reductional division could likely be an essential requirement in reprogramming genomes prior to development as in pollen embryogenesis. To demonstrate the validity of the hypothesis, it has been used a rapid and efficient culture system (3) which allows to follow the acquisition of the embryogenic potential in newly initiated hypocotyl-derived cultures.

2. PROCEDURE

2.1 Materials and methods

2.1.1 <u>Cell lines and growth conditions</u>.

The cell lines of carrot (Daucus carota L) used were:

A+ = 4 years old, highly embryogenic line, initiated from seedling hypocotyle-callus of the cv. S.Valery

T2 = 2 years old, highly embryogenic, obtained from a somatic embryo of A+ line.

The culture system recently (3) developed to follow the emergence of proembryogenic masses (PEMs) was used to obtain cultures from hypocotyl sections cultured 9 d in 2,4-D containing liquid B5 medium (4). Three starting cultures were initiated from different carrot cultivars: A90 (cv Jaguar) C90 (cv S.Valery) E90 (cv Nantja).

All cultures were grown in B5 medium supplemented with 2,4-D (2.2 µM) and 6-BAP (1.1µM). Cultures were grown as previously described (5).

2.1.2 <u>Induction of embryogenesis</u>: Cells were filtered (120µm and 50 µm pore size) washed several times with hormone-free medium and resuspended in the same medium in 5 cm plates at a density of $5x10^4$ cell ml^{-1}. The total number of embryos at all stages of development was determined after 12 d.

2.1.3 Cytological analysis: It was performed on hypocotyles, cell suspensions and embryos. Cells were fixed in alcohol-acetic acid (3:1 v/v) at different times of culture, submitted to Feulgen and squashed on slides. Where necessary, haematoxylin was used as a cytoplasmic counterstain to the Feulgen nuclear purple color.

2.1.4 DNA cytophotometry: For Feulgen/DNA absorption analysis hypocotyles, cell suspension and different embryonal stages were squashed on slides after fixation on formalin 4% and digestion with 0.5% pectinase for 15': all slides from each experiments were submitted simultaneously to one hour HCl 5N hydrolysis at room temperature. Slides were made permanent after Feulgen and three 10-min washes in SO_2 water. The Feulgen/DNA absorptions in individual nuclei were measured with the two-wavelenght method (6) using the MPV Leitz. As standard, to compare results and convert Feulgen arbitrary units into picograms of DNA, human leukocytes (2C=6 pg) smeared in a corner of some slide in each experiments were measured.

3. RESULTS AND DISCUSSION

The segregating devices of somatic-meiosis and prophase-reduction have been studied in detail in the two highly embryogenic lines A+ and T2. Somatic meiosis follows quite faithfully the phases of a normal meiotic division in carrot microsporogenesis with the presence of pairing and chiasmata-like figures. Metaphase I and anaphase I segregate 9 pairs of chromosomes, but the second equational division occurs only in a number of cases, forming the tetrad. More frequently the second division is abortive and the 9 pairs of chromosomes separated by cytokinesis into two cells. Since the bipartite chromosomes separate immediately after anaphase I into their constituent chromatids, the diploid number 18 is thus restored. The second segregation device found out in carrot embryogenic cultures is a prophase reduction, the main difference from the previously described somatic meiosis being the lack of pairing and chiasmata-like figures. Prophase-like chromosomes are separated into two haploid reduced groups during different prophase stages. Both processes lead to the same results: i.e. haploidy. Moreover, Feulgen/absorption cytophotometric measurements, beside confirming the occurrence of haploid nuclei, revealed differences, compared to carrot root-tip individual nuclear content, suggesting a diminution of DNA content per nucleus. Due to the small size of the chromosomes, no good banding results have been obtained so far in carrot, therefore the integrity of the haploid complements or of the different chromosome sets has not been ascertained. On the other hand, the frequency of the phenomenon suggests a non-random spatial disposition where the parental chromosomes of each haploid set in a nucleus occupy distinct domains in interphase, as it has been demonstrated to occurr in barley premeiotic mitosis (7). We propose that similar arrangement of chromosomes, whether present in carrot cultured cells, offers both possibilities to reduce the genome to haploidy: either by simple partition of the two haploid sets, or by means of a somatic meiotic reduction, this arrangement of chromosomes favouring also chromosome pairing and thus DNA recombination. To confirm that segregational events are the prerequisite for the acquisition of embryogenic capacity, we have used the system, recently published (3), suitable for analyzing the increase of PEMs, counting the number of regenerated embryos and calculating the possible relationship between these phenomena and the meiotic-like reductional events. Cytological analysis and Feulgen/absorption cytophotometry were performed on both callusing hypocotyl segments (at different times during the first 20 days of culture) and on the suspension-cell cultures obtained. Already at the 3rd day of culture the two events alternative to mitosis, previously described for A+ and T2 carrot cell lines (2), i.e. somatic-meiosis and prophase reduction, were found both in the hypocotyles and suspension cultures. The phenomena occurred with a fairly constant frequency (1-3% of total analyzed cells) at different times of culture both in the hypocotyles and derived cultures. With an appropriate counterstain to the Feulgen dyes, it was possible to follow the increase of the PEMs derived from these reducing events, since the committed cells appeared like single round (diameter 10-12 μm) pollen mother cells (PMCs) presenting often an eccentric nucleus undergoing meiosis. The embryogenic potential of the three obtained cultures was wery high already at the first passage to hormone-free medium for A90 whereas C90 and E90 acquired the embryogenic capacity after 1 and 2 subcultures respectively.

Cytophotometric analysis beside confirming the presence of an high percentage of haploid nuclei, revealed also a diminution of total DNA per nucleus suggesting both phenomena are a prerequisite to acquire totipotency. Whether the phenomena we have discovered are true segregational events or not has to be proved by genetic analysis of homozygous segregants from an heterizygote using appropriate marker genes. On this point it is noteworthy the large amount of published data on the variability of gene expression during early development of somatic embryos, on the high frequency with which recessive mutants are found on plant cell cultures or on the loss, through meiosis, of some epigenetic phenotypes (8-10) which find a logic explanation on the assumption that genome segregation is the key event for the commitment to embryogenesis. The DNA diminution appears also an essential step in the restoration of totipotency to a somatic cell: whether the phenomenon is related to a discharge of excess (11) "somatic" DNA or to underreplication of specific sequences has to be determined.

REFERENCES

1 Steward, F.C., Ammirato, P.V. and Mapes, M.O. (1970) Ann. Bot. 34, 761-787
2 Nuti Ronchi, V. (1990) in Plant Tissue Culture: Applications and Limitations (Bhojwani, S.S. and Novak, F.S. eds), Elsevier, Amsterdam. In press.
3 de Vries, S.C., Booij, H., Meyerink, P., Huisman, G., Wilde, HD, Thomas, T.L. and Van Kammen, A.B. (1988) Planta 176, 196-204.
4 Gamborg, O.L., Meller, R.A. and Ojima, K. (1968) Exp. Cell. Res. 50, 151-158.
5 Nuti Ronchi, V., Caligo, M.A., Nozzolini, M. and Luccarini, G. (1984) Plant Cell Rep. 3, 210-214.
6 Patau, K. (1952) Chromosoma (Berl.) 5, 341-362.
7 Bennet, M.D. (1984) in Controlling Events in Meiosis (Evans, C.W. and Dickinson, H. eds), pp.87-121, Company of Biologists, Cambridge, England.
8 Evans, D.H. (1989) Trends Genetics 5, 46-50
9 Terzi, M. and Sung, Z.R. (1986) CRC Critical Reviews in Biotechnology 3, 303-330.
10. Holliday, R. (1984) in Controlling Event in Meiosis (Evans, C.W. and Dickinson, H., eds), pp.381-394, Company of Biologists. Cambridge, England.
11. Pimpinelli, S. and Goday, C. (1989) Trends Genetics 5, 310-315.

Acknowledgements: This work was supported in part by MAF (Ministry of Agricolture and Foresty) - Project on plant biotechnology and by ENEA (Ente Nazionale Energie Alternative) - CNR Joint programme on "Production of androgenic haploid plants" (L.Giorgetti is a recipient of a Montedison Spa fellowship).

IMMUNOCYTOCHEMICAL LOCALIZATION OF CALMODULIN DURING CARROT SOMATIC EMBRYOGENESIS

A.C.J. TIMMERS & J.H.N. SCHEL, DEPARTMENT OF PLANT CYTOLOGY AND MORPHOLOGY, AGRICULTURAL UNIVERSITY, ARBORETUMLAAN 4, NL-6703 BD WAGENINGEN, THE NETHERLANDS

1. INTRODUCTION

Somatic embryogenesis has become a very useful process for propagating plants. The capacity to generate somatic embryos in tissue culture however differs remarkedly between different plant species. Somatic embryos can easily be obtained from a rapidly growing suspension culture of carrot in medium with 2,4-D by reducing the number of cells and by transferring the cells into medium without 2,4-D. They develop from small clusters of cells, designated proembryogenic masses (pems) by Halperin (1), and progress through the successive stages of globular, heart, and torpedo shaped embryos. The carrot system can be used as a model system to study fundamental problems of the regulation of *in vitro* plant growth, differentiation and development (2). Answers from these studies may lead to a more sophisticated approach to the problem of creating somatic embryos in more recalcitrant species.
Modulation of the intracellular Ca^{2+} concentration has proven to be important in the initiation and coordination of a variety of cellular mechanisms involved in developmental processes (e.g. 3,4). Calmodulin, a widely distributed Ca^{2+} binding protein among all eukaryotic organisms, has been implicated as a key regulatory protein in a number of Ca^{2+} linked processes in plants (5). From an earlier study (6) we know that activated calmodulin, visualized by fluphenazine fluorescence, is distributed irregularly in proembryogenic masses of carrot, but that an obvious polar distribution is present in later embryogenic stages. Because the distributions of activated and total calmodulin are not necessarily identical we examined the distribution of total calmodulin by immunocytochemical techniques. Although calmodulin appears to be a highly conserved protein, antibodies raised against calmodulin from animal or plant sources differ in their binding capacity with plant calmodulin (7). Therefore we used three different antisera

directed against bovine or pea calmodulin. Calmodulin retains its antigenicity after fixation with paraformaldehyde and subsequent embedment in polyethyleneglycol (8). Because cryofixation and subsequent cryosectioning has often been found to improve immunocytochemical detection of antigens (9) we added the use of cryosectioning of sucrose infused plant material as an alternative for the detection of calmodulin by immunocytochemistry.

2. PROCEDURE

2.1 Materials and methods

2.1.1. Cell cultures: Plant material and culture conditions were as described elsewhere (6).

2.1.2. Specimen preparation: Cell aggregates and somatic embryos were harvested directly after the initiation of embryogenesis up to 8 days after initiation. They were fixed in 3% paraformaldehyde in culture medium for one hour and rinsed in phosphate buffered saline (PBS: 135 mM NaCl, 2.7 mM KCl, 1.5 mM KH_2PO_4, 8 mM Na_2HPO_4, pH 7.2) They were either dehydrated with ethanol (30%, 50%, 70%, 90%, 100%, each step 10 minutes) and embedded in polyethyleneglycol (PEG; mixture of 1500 and 4000, 3:2) or infiltrated with 2.3 M sucrose in PBS (0.1 M, 4 hours; 1.0 M, overnight; 2.3 M, 4 hours), quickly frozen in liquid propane and stored under liquid nitrogen
Semithin (3-5 µm) PEG sections were made on a Leitz Wetzlar microtome with a steel knife. Ribbons were broken into small pieces, picked up with an drop of 40% PEG 6000 and tipped onto poly-L-lysine coated slides (poly-L-lysine, Sigma, M_r ³ 70,000, 0,1% w/v in distilled water). Semithin cryosections (3-5 µm) were made on a Reichert Ultracut E ultramicrotome with the FC 4D chamber at -80°C, specimen temperature at -20°C and knife temperature at -35°C. Sections were picked up with a drop of 2.3 M sucrose in PBS, thawed and tipped onto poly-L-lysine coated slides.

2.1.3. Antisera: We used a commercial polyclonal antiserum directed against bovine calmodulin (Calbiochem, cat. no. 208692) and two monoclonal antibodies (designated FF7 and BE9) directed against pea calmodulin (a generous gift of dr. P. Jablonski, Canberra City). The fluorochrome conjugated antisera were respectively rabbit-anti-sheep-FITC and rabbit-anti-mouse-IgM-FITC.

2.1.4. Immunolabelling and immunofluorescence: Slides were rinsed in PBS to remove PEG or sucrose prior to incubation with the antisera for 5 min. Aldehyde blocking with NH_4Cl (0,1 M) for 5 min. and another rinsing for 5 min. preceded the labelling with the first antibody. The polyclonal serum was diluted 50 times, the monoclonal antibodies 5 times. After rinsing in PBS the slides were incubated with the appropriate FITC-conjugated antiserum and rinsed again in PBS. All incubations were done at room temperature for 45 min. Finally, sections were embedded in saturated Mowiol 4-88 in Citifluor (Citifluor Ltd, London); controls were made by omitting the first antibody incubation. Observations were made with a Nikon epifluorescence microscope, using Nikon fluor 20x, Nikon Plan 100x oil and Leitz fluor 50x water objectives. Data were recorded on Kodak Ektachrome P800/1600.

3. RESULTS AND DISCUSSION

A first series of experiments showed that indeed cryosections were superior to sections of PEG embedded material. There was a better general preservation of cellular structure and less background labelling was observed. With both methods, however, no specific labelling was found after incubation with the monoclonal antibody BE9. The labelling with the polyclonal antiserum directed against bovine calmodulin was comparable with the showed results but a somewhat more intense overall labelling disturbed the observation of the specific labelling. In control sections pems and embryos were always negative, but occasionally a clear fluorescent signal came from the wall or the cytoplasm of the large vacuolated cells. As this was also observed with untreated sections we concluded that this signal was a consequence of autofluorescence.

Figures 1 and 2 show semithin (3-5 µm) cryosections of different developmental stages of carrot somatic embryos labelled with the monoclonal antibody FF7 directed against pea calmodulin. From the globular to the torpedo shaped stage calmodulin is most obviously present in the outer layer, the protoderm, of the embryo (Figs. 1c and d, arrows). In other parts of the embryo the amount is much lower but calmodulin is still present. In pems, calmodulin is uniformly distributed and present in an amount comparable with that in the centre of somatic embryos (Fig. 1a). In these somatic embryos calmodulin is mainly localized in organelles, (pro)plastids or amyloplasts, surrounding the nucleus (Fig. 2, arrows). So somatic embryos distinguish themselves from pems by a well-defined localization of calmodulin in plastids in the protoderm. Cells with the same well-defined localization of calmodulin appearing soon after the initiation of embryogenesis (Fig. 1b, arrow) are therefore thought to be the first cells in pems that embark on a way to embryogenesis. The localization of calmodulin in plastids is not a supreme

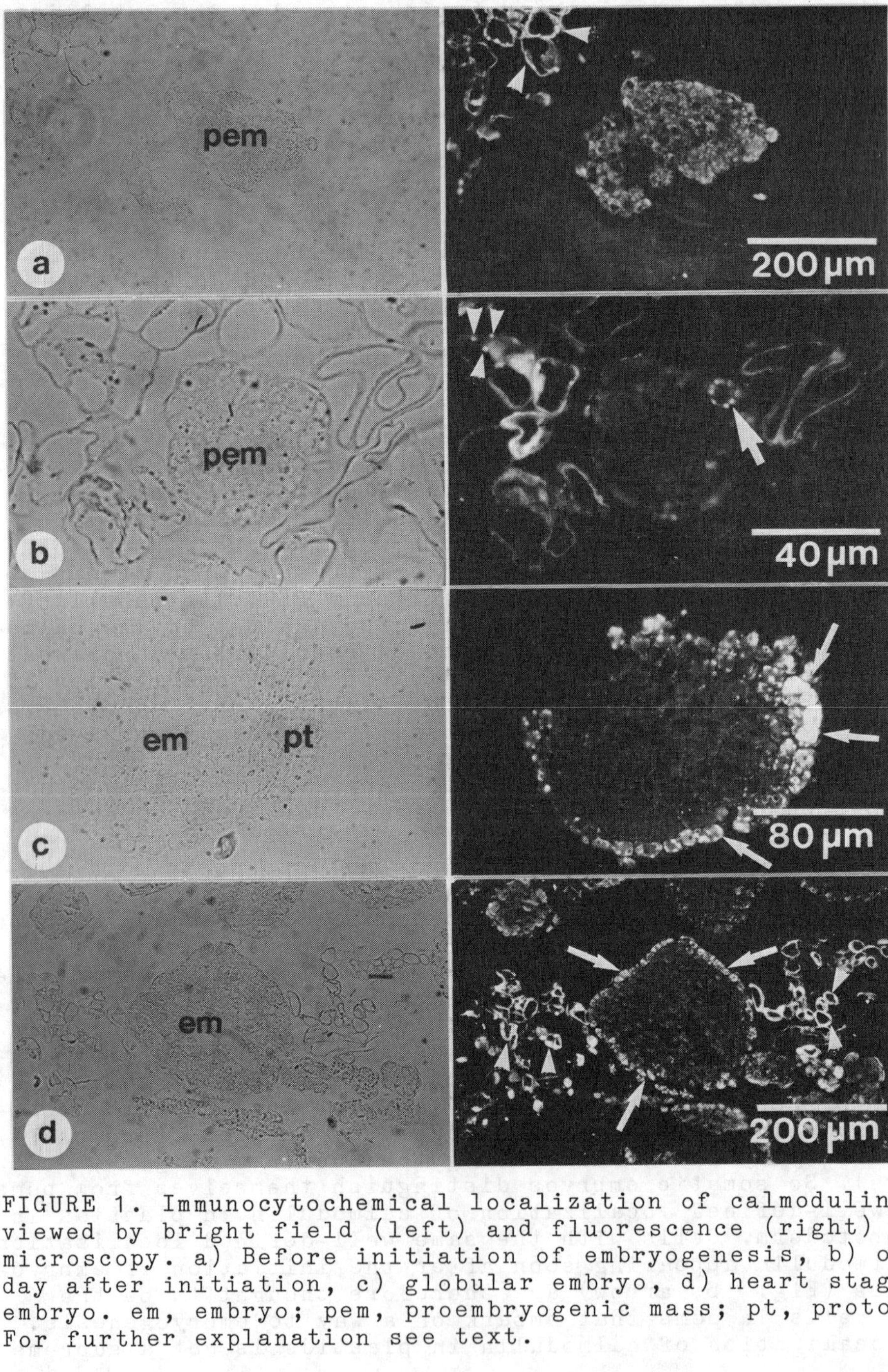

FIGURE 1. Immunocytochemical localization of calmodulin as viewed by bright field (left) and fluorescence (right) microscopy. a) Before initiation of embryogenesis, b) one day after initiation, c) globular embryo, d) heart stage embryo. em, embryo; pem, proembryogenic mass; pt, protoderm. For further explanation see text.

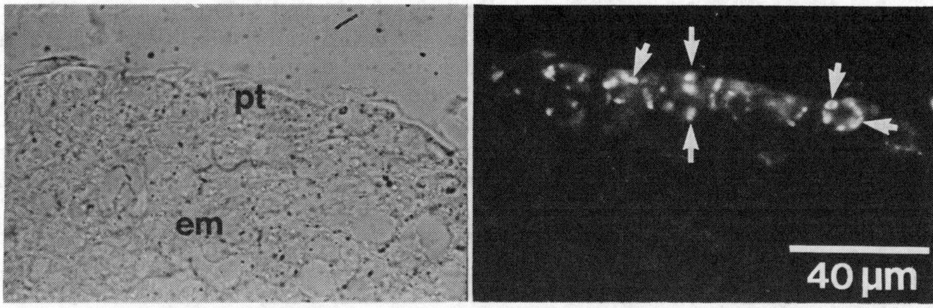

FIGURE 2. Higher magnification of a cryosection of a torpedo shaped embryo labelled with FF7. Abbreviations as in Fig. 1.

characteristic of somatic embryos. This can also be observed in the large vacuolated cells (Fig. 1b, arrowheads). These cells show in general a strong fluorescence of their cytoplasm irrespective of the presence or absence of 2,4-D in the medium (compare Fig. 1a with Fig. 1d, arrowheads).
A localization of calmodulin in plastids has been described earlier (8) in the columella of the root cap of pea seedlings. In epidermal cells of the stem and along the underside of the leaf an intense labelling of the vacuolar content was observed. We did not find any positive signal from vacuoles. In contrast with this, Lin et al. (10) did not find any signal from epidermal cells of corn root tips. In corn root sections calmodulin was found mainly in the root cap cells and in lesser amounts in metaxylem elements, stele cells surrounding metaxylem elements, apical initials and in cortical cells. Data obtained by RIA from wheat leaf cells (11) and the embryogenic axes of <u>Cicer arietinum</u> (12) show that most of the calmodulin in plant cells is present in the cytosol and that only a small fraction is associated with organelles. In our system this only seems to be valid for cells belonging to pems or cells not belonging to the embryo protoderm. In the protoderm undoubtedly most of the calmodulin is associated with plastids. We did not study the localization of calmodulin in sections of mature plants to observe if there was any difference with the localization of calmodulin in embryogenic tissue but this definitely would be very interesting.
Calmodulin can only modify specific cellular processes if it is bound to Ca^{2+}. From the globular to the torpedo shaped stage this activated calmodulin is confined to the future root side of the embryo (6). So, although calmodulin is present in all the cells belonging to the embryo protoderm it is only active in a number of cells of this cell layer. It seems plausible that only in these cells a particular Ca^{2+}/calmodulin mediated process is going on. Plastids in

pems show no lamellar organization (13,14). Such organization has only been observed in developing embryos, appearing first in the epidermal cells of globular embryos. Plastid differentiation may be an important Ca^{2+}/calmodulin mediated process during somatic embryogenesis. The embryogenic potential of cell clusters of carrot was found to be positively related to their starch content but not to their starch biosynthetic capacity (15). Therefore a first step in embryogenesis might be the breakdown of starch modulated by activated calmodulin, leading to proplastid development from which subsequently amyloplasts or chloroplasts develop.

REFERENCES

1 Halperin, W. (1966) Am. J. Bot. 53, 443-453
2 Cocking, E.C. (1987) in Plant Tissue and Cell Culture (Green, C.E., Somers, D.A., Hackett, W.P., Biesboer, D.D., eds.), Plant Biology Vol. 3, pp.3-15, Alan R Liss, New York
3 Hepler, P.K. (1988) in Proc. XIV Int. Bot. Congr. (Greuter, W., Zimmer, B. eds.), pp.225-240, Koeltz, Konigstein
4 Timmers, A.C.J. (1990) in Proc. Int. Conf. "The Impact of Biotechnology in Agriculture, 10-12 July 1989, Amiens, France, in the press
5 Hepler, P.K. and Wayne, R.O. (1985) Ann. Rev. Plant Physiol. 36, 397-439
6 Timmers, A.C.J., De Vries, S.C. and Schel, J.H.N. (1989) Protoplasma 153, 24-29
7 Watterson, D.M., Burgess, W.K., Lukas, T.J., Iverson, D., Marshak, D.R., Schleicher, M., Erickson, B.W., Fok, K.-F. and Van Eldik L.J. (1984) Adv. Cycl. Nucl. Res. 16,205-226
8 Dauwalder, M., Roux, S.J. and Hardison, L. (1986) Planta 168, 461-470
9 Tokuyasu, K.T. (1986) J. Microsc. 143,139-149
10 Lin, C.T., Sun, D., Song, G.X. and Wu, J.Y. (1986) J. Histochem. Cytochem. 34, 561-567
11 Muto, S. (1982) FEBS Lett. 147, 161-164
12 Hernandez-Nistal, J., Rodriguez, D., Nicolas, G. and Aldasoro, J.J. (1989) Physiol. Plant. 75, 255-260
13 Halperin, W. and Jensen, W.A. (1967) J. Ultrastruct. Res. 18, 428-443
14 Street, H.E. and Withers, L.A. (1974) in Tissue Culture and Plant Science (Street, H.E., ed.) pp.71-100, Academic Press, London, New York
15 Wurtele, E.S., Keller, G.L., Nikolau, B.J. and Ulrich, T.H. (1988) J. Plant Physiol. 132, 683-689

pH CONTROL OF CARROT SOMATIC EMBRYOGENESIS

DAVID L. SMITH AND ABRAHAM D. KRIKORIAN, DEPARTMENT OF BIOCHEMISTRY AND CELL BIOLOGY, STATE UNIVERSITY OF NEW YORK, STONY BROOK, NY 11794-5215 USA

INTRODUCTION

Use of synthetic auxin to initiate and manage embryogenic cultures of carrot and other species is well established (1) even though a detailed understanding of what substances like 2,4-D do in these systems is obscure. Realization of a basic understanding of what is done procedurally and why it is done, i.e. knowing what physiological or developmental controls may be in place that need to be taken into account under a given set of circumstances, has been limited both because of the empiricism involved (2) and the unavailability for in depth study of cultures that comprise a given stage or conform to a narrow range of developmental parameters. Workers generally agree that it is not easy to achieve a highly responsive, morphologically uniform embryogenic culture of any plant, including the carrot. Efforts to synchronize cultures have been even less successful (3). Clearly a better understanding of induction, requirements for maintenance in a given growth mode, and control over somatic embryogenesis is required.

The first step in a now widely adopted protocol for somatic embryogenic culture initiation in carrot has usually been to expose primary explants to an auxin such as 2,4-D, and thus foster the production of proembryonic globules (globular stage embryos). The globular stage embryos, generally identifiable by their opaque, pearly white appearance, are manually selected from among a frequently diverse complex of cell and tissue types, multiplied by any of several means *in vitro* and in this way relatively clean embryogenic stocks are obtained (4). The value of the auxins in this part of the sequence has been given special prominence since embryonic progression beyond the globular stage occurs only after *after* removal or lowering of the auxin normally used in maintaining the cultures. Continued presence of exogenous auxin is antagonistic to production of heart-stage embryos (5).

Against this brief background, it has become increasingly apparent that somatic embryogenesis can proceed, at the outset of the culture period at least, from cells that are pre-determined in the primary explant (6). Embryogenic or pre-embryonic cells may not be present in large numbers in a given explant but under appropriate conditions, a number of responsive pre-determined cells may be identified visually, multiplied and stocks built up *in vitro*. The procedure of selective subculturing of any responsive units, whether embryogenic or not, termed "repiquage" by the early tissue culture workers in France (7), has provided the basis for achieving cultures that have a fairly high response level. Even so, most "embryogenic" cultures are not uniformly embryogenic and a fairly large component of even a "clean" culture may not be responsive (cf. 2). Obviously, a culture with a high and uniform response level is to be preferred (8). To date, selective subculture combined with mechanical separation or selection by screening through sieves of known dimensions has provided the basis for obtaining cultures for study that are nominally uniform. While this approach has, of necessity been widely utilized (8), some workers have not fully recognized that molecular studies performed on cultures which are neither morphologically nor physiologically uniform have their limitations (9,10).

MATERIALS AND METHODS

We have recently described methods for obtaining embryogenic cultures of cultivated carrot that have relevance to some of the problems outlined above. The cultures in question may be reliably initiated in either of two ways. *A key feature is that a hormone-free sucrose-supplemented mineral medium is used.* The medium composition is: DS-5 salts, 1 mM NH_4Cl, 3.4 μM thiamine HCl, 57.6 mM sucrose and the pH is adjusted by addition of KOH prior to autoclaving (11). In one case, it involves selection of embryogenically competent cells from the mericarp or fruit ("seed") wall after the zygotic embryo has germinated (11). Alternatively, the mature zygotic embryo may be removed from "seeds" and wounded. In this way, competent cells are released or disclosed (12).

RESULTS AND DISCUSSION

Neither of the routes of establishing embryogenic cultures in the absence of exogenous hormone does violence to the view that pre-determined embryogenic cells may reside in the primary explant. We have not, however, encountered obviously proembryogenic cells *in situ* in the primary explants. The globular stage embryos initially generated from both sources are few in number (usually 1-10 per explant), but can be multiplied and maintained virtually completely in the preglobular proembryo stage (referred to as PGSPs for convenience) by the use of 1-5 mM NH_4^+ as the sole nitrogen source in a relatively simple medium (12). Continued development to full somatic embryo maturity is achieved when a combination of NH_4^+ and NO_3^- is provided at appropriate levels (12).

Recent experiments have extended this work and shown unambiguously that by controlling the nitrogen composition and pH of the hormone-free culture medium, one can impose major morphological consequences on this carrot system. Specifically, a pH of the hormone-free medium near 4 fosters the proliferation and multiplication of PGSPs (13). Alternatively, if the pH of the hormone-free medium is maintained at 5 to 6, and NH_4^+ is supplied at an appropriate level, all PGSPs progress to at least the globular stage and most of these, in turn, begin to develop and progress through the later stages of embryogenesis and finally give rise to fully-developed plantlets. Sustained maintenance of cultures comprised exclusively of PGSPs is dependent on medium pH throughout the culture period. Best growth of PGSP cultures occurs when the pH of the unbuffered, hormone-free medium falls from 4.5 to 4 over a 2 week period. Or, the pH may be buffered to pH 4 with MES. If the medium is buffered to sustain a pH at or above 4.5, PGSPs develop into later embryo stages. Maintenance of PGSPs occurs equally well on medium containing NH_4^+ or NH_4^+ and NO_3^-, but growth is poor with NO_3^- alone. Optimally maintained PGSP cultures show a 2 to 3 fold increase in both fresh and dry weight over a 2 week period. PGSP cultures have been continuously multiplied and maintained on hormone-free medium, without any loss of further embryogenic potential, for over 2 years.

Figures 1 and 2 summarize these events and responses diagramatically and with expanded figure legends.

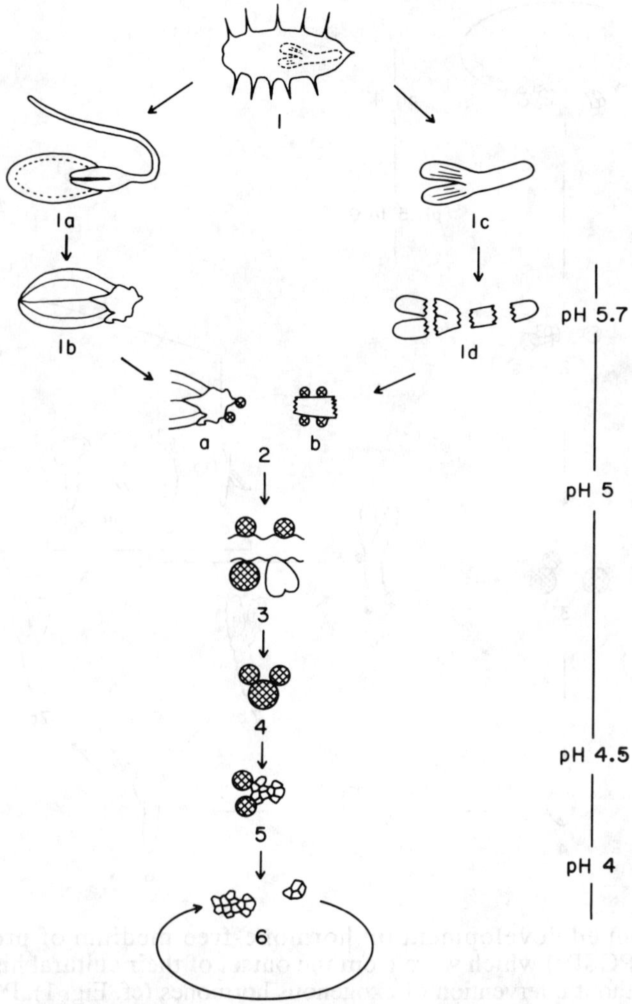

Fig. 1. Initiation and establishment of preglobular stage proembryo (PGSP) cultures on an unbuffered, hormone-free, semi-solid medium containing 1 mM NH_4^+ as the sole nitrogen source. Two tissues from mature carrot "seeds" [1] are used as the initial explants. Either excised [1c] and mechanically wounded [1d] mature zygotic embryos or the remnant fruit wall [1b] after germination of the zygotic embryo [1a] can be used to initiate globular stage somatic embryos (shown as isodiametric crosshatched globules) [2a, 2b] after 4 to 8 weeks on the same medium. The initial globular stage embryos [2a, 2b] often develop into later embryo stages [3], but as the pH of the medium drops, due to growth on medium with NH_4^+ as the sole nitrogen source, they stop maturing into later embryo stages [4] and thus yield a mixture of globular embryos and PGSPs [5]. Serial subculturing every 2 weeks onto fresh medium, titrated to pH 4.5, leads to the establishment of cultures consisting entirely of PGSPs [6].

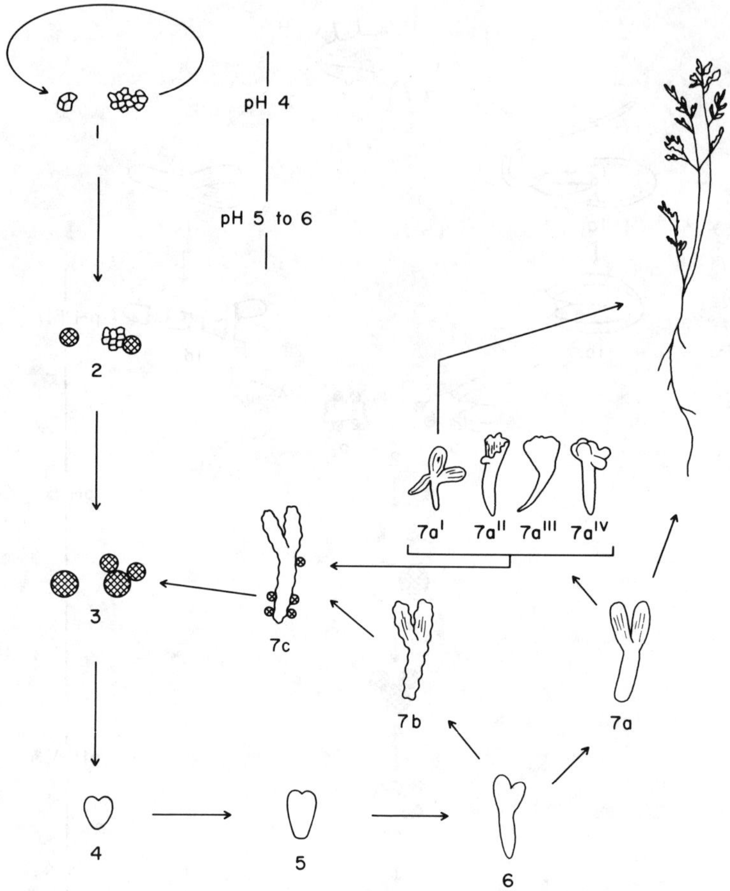

Fig. 2. Continued development on hormone-free medium of preglobular stage proembryos (PGSPs) which were from the outset of their cultural history grown and maintained without intervention of exogenous hormones (cf. Fig. 1). PGSPs maintained and multiplied at pH 4 [1] continue development into later embryo stages when grown on the same medium that has been buffered with MES at pH 5 to 6. Development to globular stage embryos is often asynchronous [2], but all PGSPs eventually develop at least to the globular stage [3]. These in turn develop to the heart [4], torpedo [5], early [6] and late [7] cotyledonary stage. Well-formed (morphologically correct) cotyledonary stage embryos [7a] give rise at a high frequency to well-formed plantlets. Many types of embryonal forms [7a^{I} - 7a^{IV}] are observed under suboptimal culture conditions or may even typify some carrot lines. These include tricotyledonous [7a^{I}], fasciated [7a^{II}], neomorphs [7a^{III}] and pluricotyledonous [7a^{IV}] forms. Of these, only the tricotyledonous embryos [7a^{I}] have been observed to give rise to well-formed plantlets. The other forms [7a^{II}, 7a^{III}, 7a^{IV}] as well as well-shaped embryos that have a poorly developed epidermis [7b] often give rise to secondary somatic embryos [7c] that can re-enter the "cycle."

The use of low pH as a tool to sustain multiplication of PGSPs provides a valuable means to obtain cultures that are both clean and highly responsive. Indeed, since the equivalent of the "repiquage" regime practiced consciously or otherwise by most workers is superseded from the very outset of the establishment of the embryogenic cultures, virtually no non-responsive cells remain in the culture. Absence of use of exogenous growth regulator in any part of the procedure to produce and establish embryogenic cultures further plays a crucial role in the protocol since only those cells are dealt with which are predetermined and habituated or at least are responsive/sensitive to the culture environment provided to them and hence are inducible to the embryogenic mode or status. Another attractive feature of this habituated system is that cells which are not easily recognizable as embryogenic, at least based on their phenotype, are obtainable in quantity. These cells, actually PGSPs, are demonstrably embryogenic, however, since a rise in pH provokes them to later stages of embryogenesis. Hitherto, totipotent cells have been assumed or presumed to be embryogenically competent or have been referred to as potentially embryogenic. The low pH-derived and -maintained PGSPs, *which have never "seen" exogenous growth regulators in their entire cultural history*, are demonstrably embryogenic.

ACKNOWLEDGEMENTS

We thank the U.S. National Aeronautics and Space Administration for supporting this work through a NASA grant "Cells, Embryos and Development in Space."

REFERENCES

1 Ammirato, P.V. (1983) in Handbook of Plant Cell Culture (Evans, D.A., Sharp, W.R., Ammirato, P.V. and Yamada, Y., eds.), vol. 1, pp. 82-123, Macmillan, New York.
2 Krikorian, A.D. (1982) Biol. Rev. 57, 151-218.
3 Ammirato, P.V. and Styer, D.J. (1985) in Biotechnology in Plant Science (Zaitlin, M., Day, P. and Hollaender, A., eds.), pp. 161-178. Academic Press, Orlando.
4 McWilliam, A.A., Smith, S.M. and Street, H.E. (1974) Ann. Bot. 38, 243-250.
5 Smith, S.M. and Street, H.E. (1974) Ann. Bot. 38, 223-241.
6 Sharp, W.R., Sondahl, M.R., Caldas, L.S. and Maraffa, S.B. (1980) Hortic. Rev. 2, 268-310.
7 Gautheret, R.J. (1959) La Culture des Tissus Végétaux. Masson, Paris.
8 Steward, F.C., Israel, H.W., Mott, R.L., Wilson, H.J. and Krikorian, A.D. (1975) Phil. Trans. Roy. Soc. B. 273, 33-53.
9 Sung, Z.R. (1985) in Plant Genetics (Freeling, M., ed.), pp. 115-128, Allen R. Liss, New York.
10 Zimmerman, J.L., El-Darwish, K., Hwang, C.H. and O'Carroll, C. (1989) in The Molecular Basis of Plant Development (Goldberg, R., ed.), pp. 143-153, Alan R. Liss, New York.
11 Smith, D.L. and Krikorian, A.D. (1988) Plant Sci. 58, 103-110.
12 Smith, D.L. and Krikorian, A.D. (1989) Amer. J. Bot. 76, 1832-1843.
13 Smith, D.L. and Krikorian, A.D. (1990) Plant Cell Reports (in press).

SOMATIC EMBRYOGENESIS IN CELERY: INDUCTION, CONTROL AND CHANGES IN POLYAMINES AND PROTEINS

A. ALTMAN, B.L. NADEL, Z. FALASH, AND N. LEVIN
Department of Horticulture and the Otto Warburg Center for Biotechnology in Agriculture, The Hebrew University of Jerusalem, P. O. Box 12, Rehovot 76100, Israel

1. INTRODUCTION

The complex, highly-conserved, developmental events which lead to formation of somatic embryos, requires ordered expression of morphogenetic stimuli. These developmental events include induction and activation of totipotent cells, resulting in formation of embryogenic cells or cell clumps, as well as subsequent bipolar differentiation and organization of roots and shoots, resulting in embryo development, maturation and plantlet formation. The two major morphogenetic stimuli are: a) *endogenous*, i.e. cellular and genotype-specific factors, and b) an array of *applied* culture conditions, i.e. medium composition and the physical environment. Whereas an extensive amount of data is available on culture conditions and on some physiological and morphological factors, little is known on the underlying biochemical and molecular mechanisms which may control somatic embryogenesis. Thus, explant source, growth regulators, other medium components and the physical-environmental conditions greatly affect the induction and development of somatic embryos in a great number of plants (1,2). The use of these 'applied' morphogenetic stimuli is largely empirical, which fostered the erroneous use of the term 'embryogenically-competent' plants in successful cases, and 'non-competent' where the embryogenic potential was not expressed. Clearly, what is lacking is an understanding of the basic control mechanisms of somatic embryogenesis: embryonic gene regulation during induction and embryo development, and reliable protein and other biochemical markers. Somatic embryo-specific proteins have been observed only recently in several plants (3,4,5). Cloning of these embryonic genes has been reported (6).

Polyamines are low-molecular weight polycations which are present in all living organisms and have a regulatory role in various developmental processes, including growth, differentiation and response to stress conditions (7). Several recent studies indicated that putrescine biosynthesis and polyamines play a role in cellular differentiation during somatic embryogenesis in carrot (8,9). In studies of celery somatic embryogenesis we found that certain inhibitors of polyamine biosynthesis inhibited various stages of somatic embryogenesis, whereas the triamine spermidine induced a major enhancement of plantlet formation (10).

Somatic embryogenesis in celery was reported and discussed previously (11). The regulation, synchronization and practical control of somatic embryogenesis in celery cell suspensions has been the subject of extensive studies in our laboratory (12,13,14,15). Some of these findings, and changes in polyamines and proteins, are reported in the following.

2. PROCEDURES

2.1 Culture conditions

Suspension cultures, see Nadel et al. (12), were transferred to liquid MS medium with 2.3 μM 2,4-D, 0.88 μM BA, 3% sucrose and 500 mg/l casein hydrolysate (CH) ('maintenance medium'). Complete plantlet regeneration was accomplished on 'regeneration medium' consisting of 0.5 strength MS, 1.15 μM kinetin, 500 mg CH, and 1.5% sucrose. Details of plantlet regeneration, mannitol and ABA treatments, mechanical sieving and plant hardening *ex-vitro* are described in Nadel et al. (12,13,14,15). For protein and polyamine studies, a comparison was made between globular-stage somatic embryos (< 800 μ) which were cultured for 2 weeks in maintenance medium, and fully-matured embryos (i.e. plantlets) which were cultured for 2 weeks in regeneration medium.

2.2 Proteins

Plant material was homogenized in a Tris grinding buffer, and the 12,000 × g supernatant was used for determination of total protein content according to Bradford. One-dimensional polyacrylamide gel electrophoresis of proteins in denaturing gels (SDS-PAGE) was performed according to Laemmli, with 10% stacking gels. Gels were stained with 1% Coommasie brilliant blue (R-250). Protein synthesis was assayed by 3 h incubation of embryo suspension with ^{14}C-leucine (2 μCi/g fresh wt). Radioactivity was measured in the TCA-insoluble fraction, using GF/C filters and extensive washing.

2.3 Polyamines

Free polyamines were extracted in 5% perchloric acid, dansylated, separated on TLC plates and measured fluorimetrically as described previously (16). Polyamine biosynthesis was assayed by incorporation of labeled precursors (16). Briefly, embryo suspensions were incubated with ^{14}C-arginine and ornithine for 3 h (2 μCi/10 ml suspension/g fresh wt), polyamines were extracted and separated as described above, and the radioactivity of the polyamine spots was counted.

3. RESULTS AND CONCLUSIONS

3.1 Induction, control and synchronization

Highly embryogenic cell lines of celery were achieved by inducing first callus, and then cell cultures in suspension, from *in vitro* cultured celery plants which were previously regenerated from somatic embryos (12,13). The proper manipulation of culture conditions, resulting in a large number of singulated, free-floating embryos, involved the use of a maintenance medium (containing, in addition to other components, 4.6 μM 2,4-D), a regeneration medium (without 2,4-D) and mannitol, in adjusted sequential order (12,15). Thus, cultures in maintenance medium alone gave rise to proembryonic clumps, and the addition of 4% mannitol permitted development of clumps of differentiated embryos as well. Subculture in regeneration medium greatly increased the development of embryo clumps and permitted the formation of singulated embryos, but adding mannitol to regeneration medium resulted in the most appropriate cultures, i.e. those composed mainly of singulated, free-floating celery embryos. It was further noted that, when added to the maintenance medium, mannitol reduced cell lysis, enhanced the formation of singulated embryos, and improved their normal differentiation (12). Regulation of embryogenesis was acheived, in addition to the culture medium by culture duration; thus, less frequent subculturing allowed embryogenic potential to be expressed earlier (13).

Embryo uniformity is most important for establishing a large-scale, efficient, somatic embryogenesis systems. Two procedures of culture synchronization were examined: the addition of ABA to the medium, and mechanical sieving (13). ABA had no effect on celery embryos maturation; however, a one-week culture of globular embryos in regeneration medium plus 0.3 μM ABA, followed by transfer to regeneration medium without ABA, increased embryo synchronization. Size segregation of embryos by sieving through metal mesh screens resulted as well in more synchronized embryo development, and embryos of 1400 μm in size provided the best growth rate combined with greater uniformity.

Further experiments, geared towards regulation of efficient large-scale somatic embryogenesis, resulted in improved procedures for long-term cold storage of embryos and for maximizing embryo conversion into plantlets. This resulted in a conversion of 89% of the globular embryos to greenhouse-established plants (14,15). This system was scaled-up, using 1 liter bubble bioreactors, to yield 75,000 plantlets per 800 ml medium (15).

3.2 Changes in proteins

Induction, regulation and development of somatic embryos is dependent on both expression of specific polypeptides and total proteins synthesis. These two aspects were investigated in globular-stage celery embryos and in fully-matured embryos, i.e. plantlets. Preliminary experiments established that total protein synthesis in plantlets was considerably higher than in globular embryos when expressed on fresh weight basis, and even more so when expressed as specific activity (Table 1). Cycloheximide inhibited leucine incorporation by more than 90%.

TABLE 1. Protein synthesis in developing somatic embryos of celery. Globular somatic embryos in maintenance medium (MC), and plantlets in regeneration medium (RC), were incubated with ^{14}C-leucine either without or with cycloheximide (CH). The TCA-insoluble macromolecular fraction was counted

	dpm · g fresh wt^{-1} CH, μg · ml^{-1}		dpm · mg protein^{-1} CH, μg · ml^{-1}		protein content, mg · g fresh wt^{-1}
	0	20	0	20	
Globular, MC	1559	185	393	47	4.0
Plantlets, RC	3607	150	2003	83	1.8

This may suggest, but does not prove, that a high degree of protein synthesis is required for plantlets growth, where cell division rate is probably higher than in globular-stage embryos. Possible changes in the expression of early and late embryonic genes were studied by comparing SDS-PAGE profiles of proteins extracted from globular embryos and plantlets (Fig. 1). The data of this experiment (and of studies in progress) indicate that at least one low molecular weight polypeptide (a) is abundant in globular embryos but is hardly detected in plantlets, whereas plantlets contain another low molecular weight polypeptide (b) which is not detected in globular embryos. A molecular weight polypeptide (about 55 kd) (c), is present in globular embryos but not in plantlets.

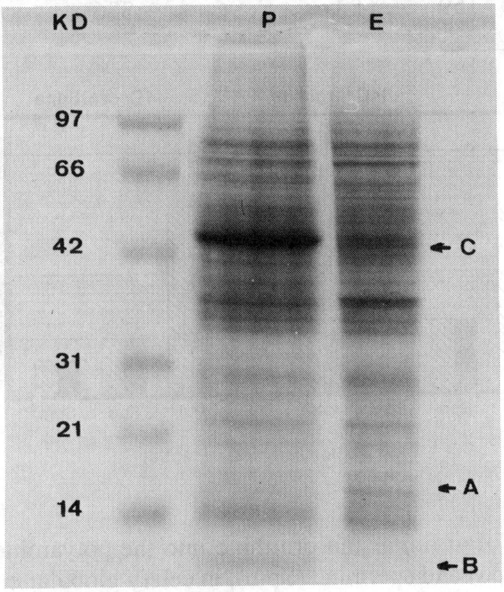

Figure 1. SDS-PAGE profiles of proteins extracted from globular embryos (E) and derived plantlets (P) of celery. E: 10 μg protein; P: 15 μg protein. Three specific polypeptides (a,b,c) are marked by arrows.

3.3 Polyamine content and biosynthesis

Previous studies indicated a role for polyamines in celery somatic embryogenesis (10). Thus, MGBG (an inhibitor of S-adenosyl methionine decarboxylase, and, thus, of spermidine and spermine synthesis) inhibited the growth of embryogenic clumps in suspension, MGBG inhibited as well shoot differentiation and plantlet growth in plated embryogenic clumps, but root formation was not inhibited (suggesting that this stage is less sensitive to inhibition of SAM-decarboxylase activity). Furthermore, addition of 0.1 mM spermidine to regeneration medium dramatically promoted and enhanced plantlet formation from embryogenic clumps. Analysis of free polyamine content in globular embryos and in plantlets revealed significant qualitative and quantitative differences (Table 2). Putrescine

TABLE 2. Free polyamine content in developing somatic embryos of celery. Globular embryos and plantlets were cultured in maintenance (MC) and regeneration (RC) media, respectively.

	nmol/g fresh wt				Ratio
	Put	Spd	Spm	Total	Put/Spd
Globular, MC	996	181	71	1248	5.5
Plantlets, RC	159	82	96	336	1.9

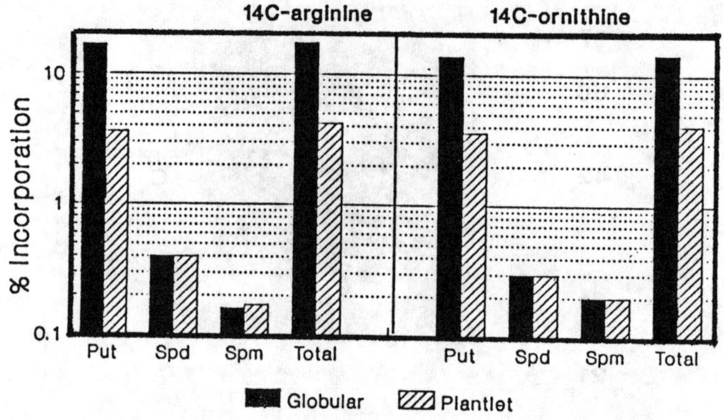

Figure 2. Incorporation of arginine and ornithine into the polyamines putrescine (PUT), spermidine (Spd) and spermine (Spm), in celery globular embryos and plantlets. Incorporation of ^{14}C-labelled precursors, as % of total uptake.

content in globular embryos was 6-fold that in plantlets, whereas the changes in spermidine content was less dramatic, resulting in high putrescine/spermidine ratio in globular embryos. This is consistent with previous studies which indicated that a high rate of cell division (which exists, probably, in actively growing plantlets) is accompanied by a low putrescine content relative to spermidine (7). The possible role of polyamine biosynthesis was further studied by comparing precursor incorporation into polyamines, in globular embryos and plantlets (Fig. 2). The incorporation of arginine and ornithine represent the two major pathways of polyamine biosynthesis in plants, i.e., via arginine decarboxylase and via ornithine decarboxylase, respectively. Data of incorporation as % of uptake (and of specific incorporation, data not shown) reveal that both enzymes operate in celery somatic embryos and plantlets. These data are also consistent with the former, showing that putrescine biosynthesis in globular embryos is considerably higher than in plantlets. The nature and role of the observed changes in polyamines content and biosynthesis, and their causal relationship to somatic embryogenesis, will be the subject of further studies.

Commercial utilization of somatic embryogenesis for clonal propagation requires a good control of 3 production stages: I. Induction of embryogenesis and plantlet establishment, II. Synchronization and improvement of the efficiency and quality of the entire embryogenesis process, III. Large-scale, industrial production of somatic embryos for agricultural use. Many of the *applied* culture conditions involved in stage I have been well documented, less so for stage II, and very little is known on stage III. The underlying *endogenous* control mechanisms, biochemical and molecular, should be the prime subject for future research, thus enabling a more intelligent, less empirical, planning of all 3 stages of somatic embryogenesis. Some culture conditions which improve stage II, in the case of celery, have been described above. The findings on changes in proteins, in addition to other studies, and on the role of polyamines, may further elucidate the endogenous regulatory mechanisms.

REFERENCES

1 Ammirato, P.V. (1983) in Handbook of Plant Tissue Culture (Evans, D.A. et al. eds.), Vol. 1, pp. 82–123, Macmillan, NY
2 Ammirato, P.V. (1989) IAPTC Newsletter, 2–16
3 Sung, Z.R. and Okimoto, R. (1983) Proc. Natl. Acad. Sci. 80, 2661–2665
4 Hahne, G., Mayer, J.E. and Lorz, H. (1988) Plant Sci. 55, 267–279
5 Stirn, S. and Jacobsen, H.J. (1987) Plant Cell Reports 6, 50–54
6 Choi, J., Liu, L.S., Borkind, C. and Sung, Z.R. (1987) Proc. Natl. Acad. Sci. 84, 1906–1910
7 Altman, A. (1989) in The Physiology of Polyamines (Bachrach, U. and Heimer, Y.M., eds.), Vol. 2, pp. 121–145, CRC, Boca Raton, Florida
8 Montague, M.J., Koppenbrink, J.W. and Jaworski, E.G. (1978) Plant Physiol. 62, 430–433
9 Feirer, R.P., Mignon, G. and Litvay, J.D. (1984) Science 223, 1433–1435
10 Altman, A., Levin, N., Cohen, P., Schneider, M. and Nadel, B. (1989) in Progress in Polyamine Research (Zappia, V. and Pegg, A.E., eds.), pp. 559–572, Plenum, NY and London
11 Orton, T.J. (1984) in Handbook of Plant Tissue Cultures (Sharp, W.R. et al., eds.), Vol. 2, pp 243–247
12. Nadel, B.L., Altman, A. and Ziv, M. (1989) Plant Cell, Tissue and Organ Culture 18, 181–189
13 Nadel, B.L., Altman, A. and Ziv, M. (1990) idib. 20, 119–124
14 Nadel, B.L., Altman, A. and Ziv, M. (1990) Scientia Hort., in press
15 Nadel, B.L., Altman, A. and Ziv, M. (1990) Acta Hortic., in press
16 Friedman, R., Altman, A. and Levin, N. (1989) Physiol. Plantarum 76, 295–302

Production and Characterization of Monoclonal Antibodies against Marker Proteins for Somatic Embryogenesis in Pea (Pisum sativum L.)

Susanne Stirn and Hans-Jörg Jacobsen
Institut für Genetik
der Universität Bonn
Kirschallee 1, 5300 Bonn 1, FRG

SUMMARY

Plant regeneration via somatic embryogenesis is now possible in pea both from immature somatic embryos, shoot apices as well as from protoplast derived calli (1,2). Pea callus cultures initiated from leaves or epicotyls cannot be routinely regenerated up to now. Therefore, we started to investigate specific biochemical aspects of induction and development of somatic embryos. We identified marker proteins only specific of embryogenic tissues by raising monoclonal antibodies against total cytosolic proteins of embryos and embryogenic calli.

Antibodies were selected which react only or preferentially with proteins from embryos or embryogenic calli. Two hybridoma cell clones have been identified producing antibodies which recognize proteins in embryogenic callus (50000 D, **IIIK1-7C5**) and/or somatic embryos (20000 D, **IIIK1-12A2**).

The monoclonal antibody **IIIK1-7C5** also recognizes a polypeptide with a molecular weight of 50000 in early globular somatic embryos of Daucus carota.

INTRODUCTION

Plant tissue culture techniques provide new possibilities for *in vitro* propagation and manipulation of plants (3, 4). The use of the *in vitro* techniques for plant breeding depends on the regeneration of plants from callus or suspension cultures.

Plant regeneration via somatic embryogenesis can be efficiently achieved only in a few plant species like carrot, tobacco or caraway (5,6), but the agronomically important pulse crops behaved recalcitrantly for a long period of time. Recently, regeneration protocols for soybean and pea were described (7,1). In these cases, immature zygotic embryos or shoot apices were the explant source for the successfull induction and regeneration of somatic embryos.

Pea callus cultures initiated from leaves or epicotyls cannot be routinely regenerated. Since variations in nutritional compounds and culture conditions have had only limited success, we started to study the biochemical events of induction and development of pea somatic embryos.

Therefore, we were looking for marker proteins only specific of tissues capable of plant regeneration, e.g. the embryogenic calli and raised monoclonal antibodies against them.
Since the yield of somatic embryos was low when we started these experiments, we decided to raise antibodies against total cytosolic proteins of embryos and embryogenic calli. Antibodies against the marker proteins were identified in two differential screenings of the hybridoma supernatants against proteins of non-embryogenic calli.

MATERIAL AND METHODS

Immature zygotic embryos as well as shoot apices of 5-day old pea seedlings were used as explants for the induction of somatic embryo formation. On MS-medium with high auxin (4 µM 2,4D), embryogenic calli developing 1-5 somatic embryos ("D4") are formed within 4-5 weeks (1). After subculturing the calli on a medium with 4.4 µM Kinetin, the calli continue to form somatic embryo ("K1") as well as meristematic centers beneath the surface of the calli (embryogenic callus "C").
As control, epicotyl segments of 5-days old pea seedlings were grown in the presence of 4.4 µM 2,4D and 11.1 µM Kinetin. Then a non-embryogenic callus without any meristematic centers is formed (N).

zygotic embryo or shoot apex
↓ 4-5 weeks, 4 µM 2,4D
callus with meristematic centers and embryos - "D4"-embryos
↓ 4-5 weeks, 4.4 µM K1
callus with meristematic centers and embryos - "BA"-embryos
↓
"C"- embryogenic callus

epicotyl segments
↓ 8-10 weeks, 4 µM 2,4D and 11.1 µM K1
callus without meristematic centers and embryos
↓
"N" non-embryogenic callus

Somatic embryos of <u>Daucus carota</u> were induced as described by (6). After 5 and 14 days in suspension culture, early globular and cotyledonary somatic embryos were harvested, respectively.
Tips ("T"), leaves ("L"), epicotyls ("E") and roots ("R") of 10-days old, pea seedlings were harvested and used for control experiments.
From these tissues soluble proteins were extracted. Embryos and tissues of young pea seedlings were ground in an agate mortar, calli were extracted in a Warring blendor (extraction buffer: 0.2 M boric acid, 2 mM borax, 0.25 M KCl, 20 mM TRIS/HCl, 4 mM diethyldithiocarbamate, 2.5% (w/v) polyclar AT, pH 6.8) (8). Both extracts were

centrifuged at 148000 x g for 2 hours at 4 °C. If nessecary, the supernatants were desalted and concentrated by ultrafiltration. The protein content was determined according to Bradford (9).

IMMUNIZATION

60-days old female balb-c mice were immunized with 100 µg protein (from embryos "D4" and "K1" and embryogenic callus "C") in 100 µl complete Freund's Adjuvant subcutaneously ("prime"). After 4 weeks immunization was repeated by intraperitoneal injection of 100 µg protein in 100 µl incomplete Freund's Adjuvant for at least three times. Before fusion of mice spleen cells with myeloma cells (SP2/O), the mice were boosted with 100 µg pure protein intraperitoneally.
The fusion was performed according to Pratt (10) with polyethyleneglycole and the selection of hybridomas was performed in HAT-medium.

ANTIBODY SCREENING

The screening protocol consisted of two subsequent ELISA-tests and one immunostain:
In the first ELISA, hybridomas producing antibodies against plant proteins were selected. Therefore, half of the wells of a microtiter plate were coated with proteins used for immunization, the other half with buffer as negative control. Subsequently, the wells were incubated with the hybridoma supernatants and rabbit-anti-mouse antibodies coupled with alkaline phosphatase (10).
In the second ELISA, antibodies were selected which reacted preferentially with proteins from embryos or embryogenic callus. Therefore, hybridoma supernatants were tested simultaneously for a reaction with proteins from embryos or embryogenic callus and proteins from non-embryogenic callus.
The washing and enhancer steps were the same as in the first ELISA (10).
The third selection step was an immunostain: SDS-PAGE was performed with proteins from embryos, embryogenic and non-embryogenic calli ("D4", "K1", "C", "N"). Separated proteins from SDS-gels were transferred to nitrocellulose by semi-dry blot. The nitrocellulose then was incubated with the hybridoma supernatants. The enhancement of the signal was performed with rabbit-anti-mouse antibodies and goat-anti-rabbit antibodies coupled with alkaline phosphatase (10). Using this scheme we assayed if the antibodies recognize proteins <u>only</u> specific of embryogenic tissue, e.g. the marker proteins for somatic embryogenesis.

RESULTS

1830 hybridoma clones resulting from seven fusions were tested. Up to now two embryo-specific clones have been identified:
The monoclonal antibody **IIIK1-12A2** recognizes a polypeptide with a molecular weight of 20000 in somatic embryos, but not in embryogenic or non-embryogenic calli (s. fig. 1b).

The monoclonal antibody **IIIK1-7C5** recognizes a polypeptide with a molecular weight of 50000 in somatic embryos <u>and</u> embryogenic calli, but not in non-embryogenic calli. Additionally, three other polypeptides were recognized in somatic (45000 D, 32000 D, 20000 D) (s. fig 1a).

Fig 1: Immunostain with the monoclonal antibodies **IIIK1-7C5** (a) **and** **IIIK1-12A2** (b)

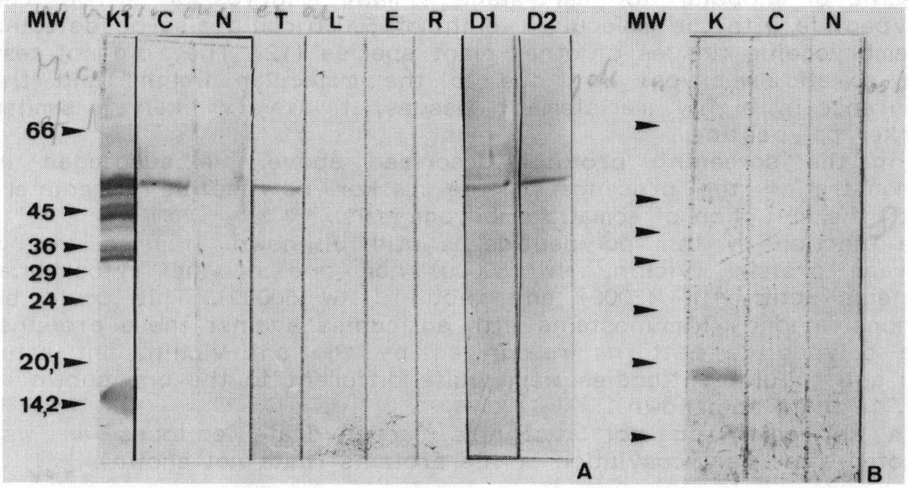

K1 = somatic embryos; C = embryogenic calli; N = non-embryogenic calli; T = tips; L = leaves; E = epicotyls; R = roots; D1 = globular somatic embryos of <u>Daucus carota</u>; D2 = cotyledonary somatic embryos of <u>Daucus carota</u>

The monoclonal antibody **IIIK1-7C5** is the most interesting one: It recognizes a polypeptide with molecular weight of 50000 in early somatic embryos of <u>Daucus carota</u> as well as in tips of 10-days old pea seedlings (s. fig. 1a)

DISCUSSION

Using the monoclonal antibody technique to identify unknown antigens, care has to be taken of the negative controls for the differential screening procedures.

Since marker proteins for early somatic embryogenesis are to be selected, the control should be a non-embryogenic callus, only differing in its regenerative capacity.

Culturing the same explants (shoot apices) on 2.4-D <u>and</u> Kinetin simultaneously as negative controls, no somatic embryos appeared after 8 weeks in culture. But histological investigations revealed meristematic clusters beneath the surface of the calli (data not shown). Therefore, we used epicotyl segments, where no <u>induction</u> of somatic embryos could be detected after 8 weeks culturing on 2.4-D and Kinetin.

We succeeded in identifying two polypeptides specific of different states of development:
The monoclonal antibody **IIIK1-12A2** recognizes a 20000 D polypeptide which appears at the globular state of embryogenesis.
IIIK1-7C5 recognizes a 50000 D polypeptide in embryogenic calli of pea, early somatic embryos of carrot and tips of young pea seedlings. Therefore, this polypeptid is specific of highly meristematic tissues.
Choi et al. (11) identified two polypeptides (20000 D and 50000 D) specific of globular to heart-stage somatic embryos of carrot. The polypeptide with the molecular weight of 50000 could also be detected in embryogenic tissues of other plant species (12). They did not test pea somatic embryos, but due to the molecular weight and the occurence in highly meristematic tissues, it is very likely a similar marker polypeptide.
Using the screening protocol described above, we succeeded in demonstrating the presence of the marker polypeptide immediately after the induction of somatic embryogenesis.
The function of this polypeptide is still unknown. It is neither a storage protein (vicilin, MW 50000) nor one of the cytoskeletal proteins actin (MW 43000) and tubulin (MW 50000). This could be demonstrated by immunostains with antibodies against these proteins: The polypeptide patterns recognized by the anti-vicilin, anti-actin and anti-tubulin antibodies were quite different to the one shown in Fig. 1a (data not shown).
Both antibodies do not recognize carbohydrate epitopes as was proofed after deglycosylation of the proteins (data not shown).

The monoclonal antibodies **IIIK1-7C5** will be used:
a) to study the time course of marker protein expression during the induction phase of somatic embryogenesis: Therefore, calli will be harvested at daily intervals after transfer to induction medium. The expression of marker proteins will be monitored using the methods of differential ELISA and immunostain.
b) to determine the cellular and subcellular localization of the marker proteins in tissue sections using immunogold-labelled antibodies.
b) to isolate the marker protein by affinity chromatography. With the help of polyclonal antibodies raised against the purified marker protein a c-DNA library will be screened (lambda gt 11) as a first step for the characterization of the marker protein genes.

The monoclonal antibody **IIIK1-7C5** is a promising tool both:
- to elucidate early events during induction of somatic embryogenesis (auxin effects, induction, competency) and
- to analyse the regenerative capacity of different genotypes and culture conditions.

This might become important in order to shorten the laborious and time-consuming method of "try and error" for the establishment of regeneration protocols used throughout tissue culture today.

Acknowledgements
This work was supported by a DECHEMA-fellowship to S.S. and a research grant of the BMFT to H.-J.J.

REFERENCES

1) Kysely, W., Myers, J.R., Lazzeri, P.A., Collins, G.B. and Jacobsen, H.-J. (1987) Plant Cell Rep. 6, 305-308
2) Lehminger-Mertens, R. and Jacobsen, H.-J. (1989) Plant Cell Rep. 8, 379-382
3) King, P. (1984) Oxfords Survey of Plant Molecular & Cell Biology 1, 7-32
4) Larkin, P.J. and Scowcroft, W.R. (1981) Theor. Appl. Genet. 60, 197-214
5) Terzi, M., Pitto, L. and Sung, Z.R. (1985) Somatic Embryogenesis in Carrots, IPRA, Roma
6) Ammirato, P.V. (1983) Handbook of Plant Cell Culture, Vol. 1, Evans, D.A., Sharp, W.R., Ammirato, P.V., Yamada, Y. (eds.), 82-123
7) Lazzeri, P.A., Hildebrand, D.F., Collins, G.B. (1986) Plant Mol. Biol. Rep. 3, 160-167
8) van der Linde, P.C.G., Bouman, H., Mennes, A.M. and Libbenga, K.R. (1984) Planta 160 102-108
9) Bradford, M. (1976) Analy. Biochem 77, 248-254
10) Pratt, L.H. (1984) Techniques in Photomorphogenesis, Academic Press, London, 201-226
11) Choi, J.H., Liu, L.-S., Borkird, C. and Sung, Z.R. (1987) Proc. Natl. Acad. Sci. USA 84, 1906-1910
12) Borkird, C., Choi, J.H., Jin, Z.-H., Franz, G., Hatzopoulos, P., Chorneau, R., Bonas, U., Pelegri, F. and Sung, Z.R. (1988) Proc. Natl. Acad. Sci. USA 85, 6399-6403

SIGNAL TRANSDUCTION AT THE MEMBRANE LEVEL OF PLANT CELLS

J. GUERN, G. EPHRITIKHINE, V. IMHOFF & J.M. PRADIER
Institut des Sciences Végétales, CNRS, Avenue de la Terrasse, 91198, Gif-sur-Yvette Cédex, France

1. INTRODUCTION

Data has accumulated during the past few years showing that at least some of the putative receptors to plant hormones, toxins and elicitors of defence reactions are located on membranes (1); functional or biochemical evidence suggest that receptors for auxins (2), fusicoccin (3) or glucans (4) could be located at the plasmalemma. Much work is needed to further characterize these receptors and their associated functions, including the mechanisms by which these signals are transduced and amplified.

Up to now, most of the efforts concerning the transduction of signals in plant membranes essentially used the concepts and techniques from the animal kingdom, particularly seeking to demonstrate that the phosphoinositol (PI) cycle and GTP-binding proteins are present and operative in plants. Aside this well conserved mechanism of signal transduction, the question of other pathways which could be specifically used by plants is discussed as well as the problem of how can be specified different signals with a paucity of transduction pathways.

2. IS THE PI CYCLE OPERATING IN PLANT CELLS ?

Analytical evidence for the presence in plant cells of the elements of the PI cycle

Several phospholipases C have been recently described in plants. This is the case for a soluble enzyme from celery stems hydrolysing phosphatidylinositol (PI) to 1,2-cyclic phosphoinositol and diacylglycerol (5) and for soluble and particulate PI hydrolytic activities from hypocotyls of soybean and bushbean, both releasing inositol phosphate (6). Celery and other higher plants also contain a particulate phosphodiesterase which hydrolyses exogenous phosphatidylinositol 4,5 diphosphate (PIP2) (7). More recently, the location at the plasmamembrane of phosphodiesterases using phosphatidylinositol and polyphosphoinositides as substrates has been demonstrated in various plants (6, 8, 9, 10). PI or PIP2-PLC activities differ in their calcium requirements and kinetic properties.

The potential PLC substrates, PIP and PIP2, are present in membranes of carrot cells, pea leaves and pulvini of *Samanea saman*

(11, 12). Very few informations concern the absolute PIP2 content of plant membranes. It is likely very low as judged by the relative proportion of labeled PIP2 estimated at less than 0.1% of the total labeled lipids in pea leaves or carrot cells (11). PI and PIP-kinase activities, which catalyze the synthesis of PIP and PIP2 have been demonstrated in plasmalemma preparations from wheat (13).

As to the products of PLC activity, one should stress that the identification and quantification of the water soluble IPn meet several difficulties in plants, specially when inositol lipids are labeled through long periods of preincubation of cells (14, 15) or organs (12) with ^{3}H-inositol, prior to the activation by a signal. These conditions (i) give low levels of the labeled substrate PIP2 and thus lead to the difficult analysis of minute amounts of IP3 and derivatives and (ii) favor the production of a variety of water soluble inositol metabolites interfering with the determination of IP3 or IP2 (15). It must be stressed also that the concerted action of phosphatases and kinases converting IP3 to a variety of water soluble inositol phosphate derivatives in animal systems (16) has still to be explored in plants. The message is that extreme caution should be exercised in the identification of inositol derivatives released under the influence of a signal. Less ambiguous situations correspond to the direct feeding of isolated membranes with labeled PIP2 where larger amounts of products of the PLC activity with a much lower diversity of metabolites could be analyzed. This permitted to establish unequivocally that the phosphodiesterase activity in plasmamembranes from *Avena* gave inositol trisphosphate (IP3) from PIP2, inositol bisphosphate (IP2) from PIP and inositol monophosphate (IP1) from PI (8). The other reaction product, diacylglycerol (DAG), potentially released from PLC-catalyzed PIP2 hydrolysis, has not been reported in *in vitro* experiments monitoring the production of IP3. Controversial reports concerning *in vivo* experiments showed either DAG fluctuations in plant cells stimulated by light (18) or unchanged level of DAG (19) in auxin-stimulated soybean hypocotyls. However these approaches should be considered as very preliminary as (i) DAG may be also obtained by PLC splitting of other phosphoinositides than PIP2 or from other phospholipid substrates (ii) DAG kinase rapidly transforms the released DAG into phosphatidic acid (PA). Much work is needed to characterize the DAG production specifically associated with PIP2 hydrolysis in plant cells.

<u>Is the PI cycle sensitive to plant signals and acting as a source of secondary messengers activating physiological processes ?</u>

Several reports describe the effect of signals on PI metabolism. A light-stimulated turnover of phospholipids, characterized by a decrease in PIP and PIP2 and a correlative increase in IP2 and IP3, have been described in *Samanea saman* pulvini (19). Osmotic shock acts as a signal modifying polyphosphoinositide metabolism in *Dunaliella salina* in such a way that the activity of PLD and PI and PIP kinases could be modified in addition to PLC (20); the type of receptor which could be involved in the perception of this type of signal is unknown. An auxin stimulation of PIP2 hydrolysis and of IP2 and IP3 release has been reported for carrot cell microsomes (21) and *Catharanthus roseus* cells (14). Finally, contradictory evidence as to the involvement of the PI cycle in the perception of elicitor signals have been published (22, 23).

The functional role of the messenger IP3 in releasing calcium from the vacuolar internal store can be considered as reasonably established. This is based on the IP3-stimulated release of calcium from intact vacuoles (24) or from vacuolar membrane vesicles (25). A remarkable effect of IP3 activating the opening of a calcium channel at the tonoplast has been recently described (26).

As discussed above, almost nothing is known as to the role of DAG acting as a secondary messenger of plant signals. Except one preliminary report describing DAG and phorbol ester (TPA) activation of a protein kinase from wheat cell homogenates (27,28), no physiological or biochemical response of plant cells to DAG or phorbol esters has been clearly demonstrated. Even the characterization of protein kinases of the C type in plants (PKC) has to be reinforced before considering them as putative effector proteins involved in plant cell signalling (28).

Perspectives

The results obtained so far, concerning the influence of signals on IP3 production, appear rather disappointing by their low intensity and transient character. This is the case for the small (x 2 at most) auxin-induced pulse-type IP3 increases which have been described, lasting for only tens of seconds to minute (19, 21). This contrasts with the higher (x 8) and more stable (up to 45 min) IP3 releases observed in some animal systems (29). Due to the cyclic organization of the phosphoinositide metabolism, without accumulation of end products, the intensity and duration of a signal-induced IP3 release should depend on the dynamics of the cycle following a PLC activation i.e. on the relative activities of PLC producing IP3, of IPn phosphatases consuming IP3 and of PI and PIP kinases replenishing the PIP2 pool. Thus, one should expect a variety of situations to occur including small and transient signal-induced IP_3 release. Judged from the hydrolysis of PIP2 externally supplied at different concentrations, PLC of plant membranes is likely strongly limited by substrate availability. Thus, to sustain a signal-induced IP3 release, the PIP2 pool must be replenished by an activation of PI and PIP kinases. Furthermore, a signal-induced IP3 release is not an absolute proof of a PLC activation as a stimulation of the PI or PIP kinases activity could give the same result. One should also consider the problem raised by the probable great importance of local conditions i.e. the fact that only the PIP2 molecules in the vicinity of PLC molecules associated with one type of receptor could be sensitive to the PLC activation. As a consequence the procedure used to label PIP2 before signal activation (long term inositol labeling of cells, short term inositol or ^{32}P labeling of isolated membranes, incubation of isolated membranes with labeled PIP2) could result in preferential labeling of PIP2 subpools with a strong influence on the relative intensity of the reaction to the applied signal.

3. GTP-BINDING PROTEINS IN PLANTS : WHAT ARE THEY FOR ?

Analytical evidence for the presence of GTP-binding proteins in plant cells

The main approach to search for GTP-binding proteins in plants is based upon the knowledge of the characteristics of the animal proteins (30) and on the use of either immunological tools built against synthetic peptides or of radiolabeled GTP to label specifically solubilized proteins from membrane fractions. All the data obtained concern membrane proteins (table 1, ref 31-40).

TABLE 1 : Overview of G-proteins in plants.
nd : not determined ; [a]Immunoblottings ; [b]$[^{32}P]$GTP binding on SDS PAGE blottings ; [c]ADP ribosylation followed by SDS PAGE separation ; [d]cDNA cloning ; [e]$[^{35}S]GTP\gamma S$ binding, K_d or I_{50} in nM ; *"K_m" in nM.

Material	Fraction	M_r(kDa)	Binding parameters	Ref
Lemna paucicostata	crude extract	nd	8[e]*	(31)
Pisum sativum	crude extract	nd	1 to 50[e]*	(32)
Daucus carota	microsomes	nd	720[e]	(33)
Cucurbita pepo	microsomes	33;50[a]	300[e]	(34)
Cucurbita pepo	microsomes	23;25;27;29[b]	nd	(35)
Funaria hygrometrica	microsomes	nd	270[e]	(36)
Pisum sativum	thylakoids	29;60[c]	nd	(37)
Zea mays	nd	21[d]	nd	(38)
Vicia faba	plasmalemma	31;37[a]	nd	(39)
Arabidopsis thaliana	""	34;38[a]	""	""
Commelina communis	""	31;36[a]	""	""
Nicotiana tabacum	plasmalemma	nd	100[e]	(40)
Arabidopsis thaliana	purified protein	nd	10[e]	(39)

It is noteworthy that the membrane peptides recognized by a G_α-common antiserum have molecular weights quite similar to those described for G_α subunits in animal membranes (41) suggesting the existence of heterotrimeric GTP-binding proteins in plants. This weak evidence is reinforced by the discovery of an ADP-ribosylation of specific peptides in thylakoids as well as the presence of a GTPase activity in these membranes. However, more direct evidence for the presence of β and γ subunits in plants is still necessary. GTP-binding sites corresponding to peptides with lower molecular weights (23 to 29 kDa) have been demonstrated in Zucchini hypocotyls (35) with characteristics which remind those of the monomeric Gn-type G-proteins (42). Such a finding is in agreement with the recent isolation in *Zea mays* of gene families encoding for proteins related to the members of the small GTP-binding protein family (43), the ras-related ypt proteins (38).

Another type of analytical evidence for the presence of GTP-binding proteins in plant cells is provided by ^{35}SGTP binding experiments performed usually on membrane fractions or on partially purified protein preparations. The results reveal (table 1) that plant and animal membranes (see for ex. 44, 45) display comparable binding characteristics.

Any evidence for coupling of GTP-binding proteins to signal receptors and to effector proteins ?

Measuring the effect of a signal on the binding of GTPγS to membrane preparations is one of the direct approach to demonstrate the coupling of GTP-binding proteins to signal receptors. This approach has been used to show that red (but also far red) light decreases (by 20-25%) the binding of [^{35}S]GTPγS to a crude extract containing membrane components of *Lemna paucicostata* (46) and that binding of GTP to thylakoids is strongly increased in the light as compared with that in the dark (47). The same approach has been used to study a potential coupling of hormone receptors to G-proteins. Preliminary results showing a reciprocal effect of auxin and GTP on their respective binding to microsomal membranes from carrot cells have been obtained (33 ; Zbell, personal communication). We also observed a limited (40% at most) and transient auxin-activation of GTP binding to plasmamembranes isolated from tobacco leaves (40).

In terms of coupling of G-proteins to putative effector proteins, the investigations have been mostly limited at a possible coupling of PIP2-PLC to G-proteins, well documented in animal systems. Somewhat contradictory results have been obtained on different plant materials. The first positive evidence for such a coupling was reported by Dillenschneider *et al.* (48) who described an increase of the formation of inositol phosphate derivatives after an addition of GTPγS to membranes from *Acer pseudoplatanus* cells. This release was di- and tris-guanine nucleotide specific and markedly reduced by the addition of GDPßS sequentially with GTP. More recently, Einspahr *et al.* (10) observed a stimulation by GTPγS of a PLC activity associated to plasmamembranes isolated from *Dunaliella salina*. Furthermore, Zbell *et al.* (33) could show a synergistic effect of auxin and GTP on the PLC activity associated to carrot cell microsomes, as well as the blockage by GDPβS of the GTP-activation. On the contrary, no effect of GTP could be detected on the PLC activity exhibited by plasmamembranes from shoots and roots of wheat seedlings (9), on the celery particulate PIP2-PLC (7) or on the *Avena sativa* plasmamembrane PIP2-PLC (8). Nothing is known as to a possible coupling of G-proteins to PLA2 and/or ionic channels.

The likely involvement of G-proteins as intermediates in more integrated responses to plant signals is suggested in several cases. The ATP-induced decrease in chlorophyll fluorescence associated with the phosphorylation of LHCII in spinach thylakoids is apparently under the control of GTP or G-cycle effectors. A pharmacological approach using different electron transfer cofactors or inhibitors showed (37) that in thylakoid membranes the photosystem II itself could function as a receptor to which the putative G-protein(s) could be coupled, the thylakoid protein kinase being possibly one effector protein. Preliminary results concerning the involvement of G-proteins in the regulation by light of the calcium-dependent swelling of etiolated wheat protoplasts have been reported (49). Finally, the fact that G-proteins could influence calcium fluxes and calcium levels through pathways different from the PI cycle is strongly suggested by the results from Allan *et al.* (50) who described a GTP-induced calcium efflux from microsomes of Zucchini hypocotyls, di- and tris-guanine nucleotide specific, drastically reduced by GDPßS and insensitive to IP3.

Perspectives

Here again the results obtained so far concerning the coupling of G-proteins to receptors of plant signals can be considered as frustrating. Of the two main strategies available to explore such a coupling (i) the direct physico chemical approach of signal effect on GTP binding and of GTP effect on hormone binding (ii) the indirect functional approach using tools perturbating the GTP cycle to show that GTP binding proteins are intermediates between signal reception and activation of effector proteins, mostly the first approach has been explored with an apparently limited success : only weak and transient signal-induced increases of GTP binding have been observed. This is likely due to a miss appreciation of the intensity of the effects which could be expected. In the simplest case, where the effect of the signal only concerns an increase in the rate of GTP binding without increase in the number of GTP binding sites or changes in their affinity for GTP, the signal-induced increase of the rate of binding is in animal systems usually weak (30-40%) and transient, its duration being limited to the time needed to saturate G binding sites with GTP (see for ex. 51).

As to the second approach, the weakness of the functional evidence for the involvement of G-related proteins in signal transduction pathways in plants is the reflect of a set of difficulties which have not been solved yet. The key problem is the lack of identified early and simple cellular responses to primary signals. The likely location of G-proteins involved in transduction pathways on the inner face of the plasmamembrane introduces a second set of difficulties linked to the necessity to have a direct access to that face of the membrane in order to use non permeant pharmacological, immunological or molecular tools. For that purpose, the protoplasts of plant cells could represent a good material, as technics like permeabilization, microinjection or electroporation may be applied. A third problem consists in our poor knowledge of the variety of G-proteins in plants (are there Gs and Gi proteins in plants ?) and in the associated lack of tools to characterize specific G-proteins (representing only a small fraction of the total GTP-binding proteins associated to a membrane) which can be coupled to one type of receptor. This problem is also reflected in the fact that, up to now, only possible functions analogous to those exhibited by heterotrimeric G-proteins in animals have been looked for in plants, nothing being known for the monomeric proteins.

4. IN SEARCH FOR OTHER TYPES OF TRANSDUCTION PATHWAYS IN PLANT MEMBRANES

People interested in demonstrating the operation of GTP-binding proteins and PI cycle in plants are often criticized for not looking at something which could be specific for plant systems. This raises two main questions (i) Do plants need specific transduction pathways in addition to those highly conserved in the biological world ? (ii) Is there any evidence for such putative specific transduction systems ?

As to the first question, considering the chain of events: signal reception - transduction - activation of an effector protein - expression of the effect, the specificity of plants can take its origin at the first step, i.e. through the type of signals and their associated receptors to react to and at the third and fourth steps through the type of proteins involved in the physiological response. As it will be discussed in chapter 5 this can be realized with only a few well conserved transduction pathways. As to the second question, even if the existence in plant cells of a signalling pathway through the phosphoinositide cycle can be considered as a reasonably well supported hypothesis, the transduction pathways used by these cells are likely more diverse, some of them being possibly specific.

Which other effector proteins than PLC could be involved ?

The role of phospholipids or more generally of phospholipid-derived metabolites is probably not limited to the classical PI cycle. Well documented evidence in animal systems show that lysophospholipids and derivatives issued from the activation of phospholipase A_2 (PLA_2) can be involved in the transduction of hormonal signals (52). Some preliminary concurring results have been presented recently for plant systems such as an auxin-increased production of lysophosphatidylcholine (suggesting an hormone-stimulated PLA_2 activity) in soybean cells or membrane fraction prepared from Zucchini hypocotyls (53) as well as the activation of the plasmalemma ATPase by lysophospholipids (54). The auxin stimulation of NADH oxidase on plasmamembranes isolated from soybean hypocotyls is well correlated to the hormone-stimulated growth of excised hypocotyls suggesting an involvement of surface oxidases in the transduction of the hormone signal (55). This has to be compared to the involvement of cell surface NADH oxidase in liver and fat cells as regards their responsiveness to insulin and glucagon (56, 57).

A third putative effector protein in plants is adenylate cyclase which produces cAMP, a potent second messenger in β-adrenergic controlled animal cells. Despite several claims, no evidence supporting any involvement of cAMP as a secondary messenger in plants has been provided up to now. The most recent critical investigation concerning the intracellular cAMP levels in plant tissues revealed that cAMP, if present, is at most at concentrations two to three orders of magnitude lower than the basal levels measured in animal tissues (58).

Calcium channels associated to plasmamembranes are considered as one of the regulatory components of the cytosolic calcium concentration in animal cells (59). Recent results in plant cells show a modulation by primary signals of the calcium fluxes across the plasmamembrane. This is the case for ABA on guard cells from *Commelina communis* (60), cytokinins and fusicoccin in *Amaranthus* protoplasts (61) or light in wheat protoplasts (49). But these evidence based on measurements of cytoplasmic Ca^{2+} levels or on the effect of drugs known to affect calcium channels do not demonstrate unequivocally that calcium channels at the plasmamembrane are involved. The presence itself of calcium channels at the plasmalemma of plant cells has first to be confirmed by more direct approaches.

Is the plasmalemma H^+-ATPase a possible effector protein ?

The possibility that target enzymes strictly specific for plants could be involved in signal transduction cannot be excluded but this remains a matter of speculation. One possible example concerns the activation of the plasmalemma ATPase by plant signals.

The plasmamembrane H^+-ATPase has a central role in the activity of a plant cell (62) and as such it can represent one of the key effector protein directly or indirectly involved in the transduction of plant signals. An interesting case is the H^+-ATPase activation by the toxin fusicoccin (FC) which seems to be rather direct as suggested by reconstitution experiments (63) where partially purified FC receptors were included into liposomes in association with purified ATPase. This experiment represents the first example where a receptor-effector protein couple is reconstituted with a well characterized short term response to the signal (proton pumping into the reconstituted vesicles). If one excludes that proteins contaminating the FC receptor or the ATPase preparations could act as transducers between the receptor and the effector protein then an original direct activation would characterize this pathway.

Changes induced by signal reception in a membrane are rarely unique but most often correspond to a set of biochemical and biophysical modifications which likely contribute to the transduction and amplification of the signal. We already discussed the fact that the ATPase can be activated by lysophospholipids and thus be a secondary effector protein of the PLA2 pathway. But the stimulation of PLA2 can in turn be at the origin of a cascade of derived messages induced by fatty acid peroxidation (64) and generation of superoxide radicals (65). On another hand, modulations of the ATPase activity can induce changes in the transmembrane electrical potential, in the cytoplasmic pH (66) and in ionic exchanges depending on electrical forces. These modifications can in turn act as secondary messages. The idea that intracellular pH could be a second messenger in plant systems has been recently presented by Felle (67) from a detailed study of auxin effects on membrane potential and on cytoplasmic pH and pCa^{2+}. Considering modifications of the membrane potential and/or of the corresponding ionic exchanges as informational events contributing to the transduction and amplification of the signal is less classical. Such modifications are amongst the earliest events induced by auxins, toxins and elicitors (68, 69). The recent report showing that vanadate, an inhibitor of the plasmalemma H^+-ATPase, mimics the effects of a fungal cell wall preparation in eliciting gene activation in peanut cell cultures (70) suggests that a change in membrane potential could be an intermediate in the conversion of the extracellular elicitor signal in intracellular messages.

5. PAUCITY OF TRANSDUCTION PATHWAYS AND SECONDARY MESSENGERS : HOW TO INSURE THE SPECIFICITY OF RESPONSE TO DIFFERENT PLANT HORMONES AND OTHER SIGNALS

The number of second messengers identified up to now in eucaryotes is really small and the number of associated pathways to transduce signals to their effector proteins is limited (71). The consequence is that different signals have to share common transduction mechanisms. This raises the question of how is insured the specificity of response to different signals .

In plants, as discussed above, the variety of transduction pathways used by hormones and other signals has still to be explored. But limited and sometimes contradictory evidence suggest that, as in animals, at least some transducing elements and secondary messengers can be shared by different signalling systems. For example, a body of evidence suggests that modulations of intracellular calcium level could be a common feature of the response to auxins (67), cytokinins (61), abscisic acid (60), elicitors of the synthesis of phytoalexins (72) and red light signals perceived by phytochrome (73). This evidence supports the idea that, in plants as in animals, cytoplasmic calcium could be a secondary messenger shared by a variety of primary signals. This raises the question of the mechanisms by which the specificity of response to these signals can be obtained.

Different degrees in the specificity of plant cell responses to signals should be distinguished:

(i) The specificity of response can be expressed in a multicellular system where cells of type A respond to one hormone or signal whereas cells of type B respond to another one. The equipment of each cell type in specific receptors and the efficiency of their coupling to the transduction pathways will determine the intensity of the response to specific signals. The same view holds for the developmental aspects of the reaction of one cell type, the reception and transduction equipment of which varies with time along cell differentiation.

(ii) The situation where one cell type could react simultaneously and specifically to different signals through the use of one transduction pathway or one secondary messenger is more difficult to analyze. The simplest case is that of a signal B interacting synergistically or antagonistically to a signal A. Such interactions can be obtained through the production of a common secondary messenger M leading to synergistic effects when the level of M stays in the suboptimal range and to antagonistic effects when M reaches supraoptimal levels.

Specifying different signals through their ability to induce the production of different amounts of secondary messengers has been demonstrated in animal systems where the selective interaction of receptors with different stimulatory G-proteins coupled to phospholipase C, can induce the production of different levels of PIP2-derived secondary messengers (74) providing the quantitative basis for specific responses. More sophisticated situations correspond to antagonisms resulting from the coupling of a common membrane effector protein to stimulatory Gs and inhibitory Gi-type proteins, Gs being associated with the A receptor and Gi with the B receptor as described in animal systems (75).

(iii) Viewing as above the signalling system of plant cells as a collection of more or less independent agonists and agonist receptors activating a few transduction pathways, is likely far too simple. For example, modulating cytoplasmic calcium levels through the release of calcium from internal stores, via the PI-PLC pathway is not equivalent in many respects to the entry of external calcium via the opening of calcium channels. In the first case, calcium increase is likely associated to DAG production and results from a receptor-mediated G-protein activation, the activated G-proteins being able to regulate several distinct effector proteins, one of which being PLC. On the other hand, rising cytoplasmic calcium through the opening of a calcium channel is not associated to DAG production but can result from more complex events such as membrane depolarization linked to pump inhibition or increases in membrane conductance by opening of K^+ or anion channels. Thus, a more complex but probably more realistic view of how is organized the transduction system is to think that the qualitative and quantitative interactions between primary signals through their associated secondary messengers constitute a network of informations equivalent to a complex signalling "language" (76) specifying the response to each signal by a special configuration of the network. The two examples detailed above show that the cytoplasmic calcium rise, common event of the two situations, is in fact associated to two different states of the information network.

(iv) Local membrane domains can contribute to the specificity of response to different signals sharing the same transduction pathways (71). It is probably not realistic to conceive that the elements constituting the signalling systems are freely flotting in membranes, sorting of these elements in specialized domains is more likely. This can be true for receptors of signals A and B preferentially located in different domains of the cell : the local release of a common secondary messenger can have different consequences if it occurs in the vicinity of the nucleus for signal A and at the opposite pole for signal B. This question is specially relevant for plant cells where the large vacuole imposes limitations on the translocation of messages in the peripheral cytoplasmic layer and cytoplasmic strands.

6. CONCLUSIONS

The study of signal transduction in plants is at its early stage and it is hoped that this review will help in developing future work in this domain.

One of the first conclusion of this review is that one can consider as reasonably convincing the evidence supporting the hypothesis that GTP-binding proteins and inositol metabolites are used by plants to transduce signals. However there is a long way to go in order to fully analyze these systems and to study their regulation. Three major steps have to be reached to make significant progress. It is needed (i) to improve the analysis of the elements of the transduction pathways (i.e. recognition of the various types of G-proteins and of the variety of inositol metabolites occuring in plants) (ii) to improve the measurements of the cytoplasmic levels of the second messengers (iii) to define better physiological responses of plant cells. Most of the transducing elements are located at the interface between the plasmalemma and the cytosol. As a consequence, one needs to access to this interface through the injection of intermediates of the pathways, analogs of these intermediates, antibodies and G-protein subunits, etc... One needs at the same time to monitor a short term response of the cells to the drug injected. In this respect of testing the functionality of transduction pathways at the cell level, plant physiologists are far behind their animal colleagues.

Aside the biochemical and pharmacological approaches, much progress has to be expected in the next few years from the analysis of genes encoding for putative elements of the transduction chains and from the controlled regulation of their expression in transgenic plants. Approaches consisting in overexpressing genes encoding for plant GTP-binding proteins or adenylate cyclase from yeast are currently undertaken.

A strong effort should be done to develop the genetical approach of transduction processes in plants. Several of the oncogenes identified in animal systems correspond to mutations affecting genes encoding for elements of transduction pathways (77). These situations demonstrate what can happen when a transduction pathway is activated independently of the signal. Nothing equivalent has been discovered in plants up to now but if the elements of the PI-cycle and the G-proteins have been, as discussed above, conserved in their function there is no reason why one should not discover the equivalent in plants of the "oncogene" mutations. This term of "oncogene" should not be taken literally. What one should find are situations where cells have acquired a phenotype of full response to a specific signal independently of the presence of this signal, whatever the response is. In this respect, the "slender" mutation of barley (78) is interesting in that sense that the phenotype given by this mutation corresponds to a typical gibberellin-stimulated growth. It is not due to an over production of endogenous gibberellins, it is not sensitive to exogenous gibberellins and is not affected by inhibitors of gibberellin biosynthesis. This mutant deserves more attention to look for any defect in the transduction pathways of the gibberellin signal. Other quite interesting situations have been described recently in *Arabidopsis* where mutants selected for their resistance to auxins have been afterwards shown to be resistant to ethylene and ABA (79). One possible interpretation of this pleiotropic resistance could be that the modified target concerns a transduction pathway common to these three signals. Resistance to supraoptimal concentrations of a common secondary messenger could fit with such a behaviour. This could provide interesting biological materials to further investigate the signal transduction pathways in plants.

REFERENCES

1 Chadwick C.M. and Garrod D.R. (1986) Hormones, receptors and cellular interactions in plants. (Cinader B., ed.), Vol. 1, pp.1-375, Cambridge Univ. Press, Cambridge, London, New York
2 Barbier-Brygoo H., Ephritikhine G., Klämbt D., Ghislain M. and Guern J.(1989) Proc. Natl. Acad. Sci. USA 86,891-895
3 Feyerabend M. and Weiler E.W. (1989) Planta 178, 282-290
4 Schmidt W.E. and Ebel J. (1987) Proc. Natl. Acad. Sci. USA 84, 4117-4121
5 Irvine R.F., Letcher A.J. and Dawson R.M.C. (1980) Biochem. J. 192, 279-283
6 Pfaffmann H., Hartmann E., Brightman A.O. and Morré D.J. (1987) Plant Physiol. 85, 1151-1155
7 McMurray W.C. and Irvine R.F. (1988) Biochem. J. 249, 877-881
8 Tate B.F., Schaller G.E., Sussman M.R. and Crain R.C. (1989) Plant Physiol. 91, 1275-1279
9 Melin P.M., Sommarin M., Sandelius A.S. and Jergil B. (1987) FEBS Lett. 223, 87-91
10 Einspahr K.J., Peeler T.C. and Thompson Jr G.A. (1989) Plant Physiol. 90, 1115-1120
11 Irvine R.F., Letcher A.J., Lander D.J., Drobak B.K., Dawson A.P. and Musgrave A. (1989) Plant Physiol. 89, 888-892
12 Coté G.G., DePass A.L., Quarmby L.M., Tate B.F., Morse M.J., Satter R.L. and Crain R.C. (1989) Plant Physiol. 90, 1422-1428
13 Sommarin M. and Sandelius A.S. (1988) Biochim. Biophys. Acta 958, 268-278
14 Ettlinger C. and Lehle L. (1988) Nature 331, 176-178
15 Rincon M., Chen Q. and Boss W.F. (1989) Plant Physiol. 89, 126-132

16 Shears S.B. (1989) Cellular Signalling, 1, 125-133
17 Morse M.J., Crain R.C., Coté G.G. and Satter R.L. (1989) Plant Physiol. 89, 724-727
18 Morré D.J., Pfaffmann H., Drobes B., Wilkinson F.E. and Hartmann E. (1989) Plant Physiol. 90, 275-279
19 Morse M.J., Crain R.C. and Satter R.L. (1987) Proc Natl. Acad. Sci. USA 84,7075-7078
20 Einspahr K.J., Peeler T.C. and Thompson Jr G.A. (1988) J. Biol. Chem. 263, 5775-5779
21 Zbell B. and Walter-Back C. (1988) J. Plant Physiol. 133, 353-360
22 Kurosaki F., Tsurusawa Y. and Nishi A. (1987) Plant Physiol. 85, 601-604
23 Strasser H., Hoffmann C., Grisebach H. and Matern U. (1986) Z. Naturforsch. 41c, 717-724
24 Ranjeva R., Carrasco A. and Boudet A.M. (1988) FEBS Lett. 230, 137-141
25 Schumaker K.S. and Sze H. (1987) J. Biol. Chem. 262, 3944-3946
26 Alexandre J., Lassalles J.P. and Kado R.T. (1990) Nature 343, 567-570
27 Olah Z. and Kiss Z. (1986) FEBS Lett. 195, 33-37
28 Blowers D.P. and Trewavas A.J. (1989) in Second Messengers in Plant Growth and Development (Boss, W.F. Morré D.J. ed.) pp.1-28, Alan R. Liss Inc , New York
29 Batty I., Nahorski S.R. (1987) Biochem. J. 247, 797-800
30 Casey P.J. and Gilman A.G. (1988) J. Biol. Chem. 263, 2577-2580
31 Hasunuma K. and Funadera K. (1987) Biochem. Biophys. Res. Comm. 143, 908-912
32 Hasunuma K., Furukawa K., Tomita K., Mukai C. and Nakamura T. (1987) Biochem. Biophys. Res. Comm. 148, 1, 133-139
33 Zbell B., Hohenadel H., Schwendemann I. and Walter-Back C. (1990) in Activation and Desensitization of Transducing Pathways. NATO ASI Series (Konijn Th. M. et al., ed.) Springer-Verlag, Berlin, Heidelberg, New York, in press
34 Jacobs M., Thelen M.P., Famdale R.W., Astle M.C. and Rubery P.H. (1988) Biochem. Biophys. Res. Comm. 155, 1478-1484
35 Drobak B.K., Allan E.F., Comerford J.G., Roberts K. and Dawson A.P. (1988) Biochem. Biophys. Res. Comm. 150, 899-903
36 Zbell B., Schwendemann I. and Bopp M. (1989) J. Plant Physiol. 134, 639-641
37 Millner P.A. and Robinson P.S. (1989) Cellular Signalling, 421-433
38 Palme K., Diefenthal T., Sander C., Vingron M. and Schell J. (1989) in The Guanine-Nucleotide Binding Proteins (Bosch L., Kraal B. and Parmeggiani A. ed.), pp. 273-284, Plenum Publishing Corporation
39 Blum W., Hinsk K.D., Schultz G. and Weiler E.W. (1988) Biochem. Biophys. Res. Comm. 156, 954-959
40 Ephritikhine G., Pradier J.M. and Guern J. (1989) NATO International Symposium "Signal Perception and Transduction in Higher Plants", Toulouse, July, 2-43
41 Freissmuth M., Casey P.J. and Gilman A.G. (1989) FASEB J. 3, 2125-2131.
42 Bhullar R.P. and Haslam R.J. (1987) Biochem. J. 245, 617-620
43 Burgoyne R.D. (1989) TIBS 14, 394-396
44 Northup J.K., Smigel M.D. and Gilman A.G. (1982) J. Biol. Chem. 257, 11416-11423

45 Kikuchi A., Yamashita T., Kawata M., Yamamoto K., Ikeda K., Tanimoto T. and Takai Y. (1988) J. Biol. Chem. 263, 2897-2904
46 Hasunuma K., Furukawa K., Funadera K., Kubota M. and Watanabe M. (1987) Photochem. and Photobiol. 46, 531-535
47 Millner P.A. (1987) FEBS Lett. 226, 155-160
48 Dillenschneider M., Hetherington A., Graziana A., Alibert G., Berta P., Haiech J. and Ranjeva R. (1986) FEBS Lett., 208, 413-417
49 Bossen M.E., Kendrick R.E. and Vredenberg W.J. (1989) NATO International Symposium "Signal Perception and Transduction in Higher Plants", Toulouse, July, 2-42
50 Allan E., Dawson A., Drobak B. and Roberts K. (1989) Cellular Signalling, 1, 23-29
51 Mousli M., Bronner C., Landry Y., Bockaert J. and Rouot B. (1990) FEBS Lett. 259, 260-262
52 Exton J.H. (1990) J. Biol. Chem. 265, 1-4
53 Sherer G.F.E. and André B. (1989) Biochem. Biophys. Res. Comm. 163, 111-117
54 Palmgren M.G. and Sommarin (1989) Plant Physiol. 90, 1009-1014
55 Morré D.J. (1989) in Second Messengers in Plant Growth and Development (Boss, W.F. Morré D.J. ed.) pp.81-114, Alan R. Liss Inc., New York
56 Goldenberg H., Crane F.L. and Morré D.J. (1979) J. Biol. Chem. 254, 2491-2498.
57 Crane F.L., Sun I., Clark M.G., Grebing M.G. and Löw H. (1985) Biochim. Biophys. Acta 811, 233-264
58 Spiteri A., Viratelle O.M., Raymond P., Rancillac M., Labouesse J. and Pradet A. (1989) Plant Physiol. 91, 624-628.
59 Berridge M.J. and Irvine R.F. (1989) Nature 341, 197-205
60 McAinsh M.R., Brownlee C. and Hetherington A.M. (1990) Nature 343, 186-188
61 Elliott D.C. and Yuguang Y. (1989) Plant Sci.65, 243-252
62 Serrano R. (1989) Annu. Rev. Plant Physiol. Plant Mol. Biol. 40, 61-94
63 Aducci P., Ballio A., Blein J.P., Fullone M.R., Rossignol M., and Scalla R. (1988) Proc. Natl. Acad. Sci. USA 85, 7849-7851
64 Leshem Y.Y. (1987) Physiol. Plantarum 69, 551-559
65 Doke N. (1985) Physiol. Plant Pathol. 27, 311-322
66 Kurkdjian A. and Guern J. (1989) Annu. Rev. Plant Physiol. Plant Mol. Biol. 40, 271-303
67 Felle H. (1989) in Second Messengers in Plant Growth and Development (Boss, W.F. Morré D.J. ed.) pp.145-166, Alan R. Liss Inc , New York
68 Ephritikhine G., Barbier-Brygoo H., Muller J.F. and Guern J. (1987) Plant Physiol. 83, 801-804
69 Pelissier B., Thibaud J.B., Grignon C. and Esquerré-Tugayé M.T. (1986) Plant Sci. USA 46, 103-109
70 Steffens M., Ettl F., Kranz D. and Kindl H. (1989) Planta 177, 160-168
71 Neer E.J. and Clapham D.E. (1988) Nature 333, 129-134
72 Stäb M.R. and Ebel J. (1987) Arch. Biochem. Biophys. 257, 416-423
73 Tagaki S.and Nagai R. (1988) Plant Physiol. 88, 228-232
74 Ashkenazi A., Peralta E.G., Winslow J.W., Ramachandran J. and Capon D.J. (1989) Cell 56, 487-493
75 Hill S.H. and Kendall D.A. (1989) Cellular Signalling 1, 135-141

76 Palme K., Diefenthal T., Hesse T., Nitschke K., Campos N., Feldwisch J., Garbers C., Hesse F., Schwonke S. and Schell J. (1989) in Signal molecules in plants and plant-microbe interactions (Lugtenberg B.J J. ed.), NATO ASI Series H , vol. 36, pp.71-74 Serie Cell Biology, Springer Verlag, Heidelberg, Berlin
77 Berridge M.J. and Irvine R.F. (1984) Nature 312, 315-321
78 Lanahan M.B. and Ho D.T-H. (1988) Planta 175, 107-114
79 Wilson A.K., Pickett F.B., Turner J.C. and Estelle M. (1990) Mol. Gen. Gen., in press

INHIBITION OF POLYAMINE BIOSYNTHESIS AFFECTS THE RHIZOGENIC RESPONSE IN TOBACCO THIN CELL LAYERS

TORRIGIANI P.^, ALTAMURA M.M.°, CAPITANI F.°, FALASCA G.°, BAGNI N.^ and SERAFINI-FRACASSINI D.^

^Dept. of Biology, University of Bologna, Bologna, Italy

°Dept. of Plant Biology, University La Sapienza, Rome, Italy

1. INTRODUCTION

Aliphatic polyamines (PA) occur throughout the plant kingdom; they are polycations capable of binding with nucleic acids, cell wall polysaccharides and proteins (1, 2). Much evidence indicates that these substances may function as growth regulators or as second messengers in plant cells (1, 3). The need for PA in sustaining cell division is well established; in general, non-growing tissues have low levels of PA, as well as of other growth substances, while actively growing tissues have higher levels (3). Although less is known about their involvement in organogenesis, elevated levels of PA and of their biosynthetic enzymes have been associated with cells undergoing changes in their state of differentiation, for example during somatic embryogenesis and during *in vitro* bud or root formation (4, 5, 6, 7, 8).

PA are derived from the activity of a series of enzymes: ornithine- (ODC) and arginine decarboxylase (ADC) for putrescine synthesis from which, in turn, spermidine and spermine are formed via S-adenosylmethionine decarboxylase (SAMDC) and aminopropyltransferases. Studies on the physiological role of PA have been performed using three approaches: 1) PA-deficient mutants 2) temporarily PA-deficient tissues 3) inhibitors of the PA biosynthetic enzymes. As far as the third approach is concerned, many studies have shown that inhibitors of one or more enzymes of PA biosynthesis cause a reduction in intracellular levels of PA, in cell division and/or organogenesis, for example during the synchronous cell cycle in *Chlorella*, during carrot somatic embryogenesis, in shoot-forming tobacco thin cell layers and during adventitious root formation in mung bean (4, 5, 8, 9, 10, 11).

In tobacco thin layer explants, the tissue culture system introduced by Tran Thanh Van et al. (12), in addition to the free PA, there are two classes of PA conjugated either to low molecular weight (TCA-soluble hydroxycinnamoylamides) or to high molecular weight (TCA-insoluble) compounds, such as proteins, whose respective physiological roles are probably different (13). In this system it has been reported that early increases in free and bound putrescine and spermidine, as well as in ADC activity (11), preceded the organization of the meristemoids which led to vegetative bud, flower bud or root formation (6, 7).

In this report the effect of α-difluoromethylornithine (DFMO), alone or in combination with α-difluoromethylarginine (DFMA), irreversible suicide inhibitors of ODC and ADC respectively, and of methylglioxal--bis-guanylhydrazone (MGBG) and cyclohexylamine (CHA), competitive inhibitors of SAMDC and spermidine synthase respectively, on free and bound PA titers and on rhizogenesis has been investigated using the tobacco thin layer system.

2. PROCEDURE

2.1 Materials and methods

2.1.1 <u>Plant material and tissue culture</u>: Plants of <u>Nicotiana tabacum</u> L. cv. Samsun were grown up to the complete green fruit stage. Thin cell layer explants (2x10 mm, 8 cell layers, about 9 mg) were excised from the stem between the 5th and the 8th node from the lowest axillary inflorescence. The explants, surface sterized with 10-fold diluted commercial bleach and then rinsed three times, were cultured, as suggested by Tran Thanh Van et al. (12) in tubes containing Murashige and Skoog (14) medium to which 0.55 mM myo-inositol, 0.3 μM thiamine HCl, 29 mM sucrose, 10 μM indole-3-butyric acid, 0.1 μM kinetin and 0.8% agar were added. The pH was adjusted to 5.6 with 1 M NaOH. In some cases the medium contained also 1 mM DFMO, 1 mM DFMO plus 1 mM DFMA, 1 mM MGBG or 2.5 mM CHA. The cultures were grown for 30 days at 27°C in continuous darkness.

2.1.2 <u>Polyamine extraction and analysis</u>: Samples (about 300 mg) were homogenized in cold 5% (w/v) trichloroacetic acid (TCA) and centrifuged at 26,000 g. The pellet was washed twice and resuspended in TCA. Replicates (0.2 ml) of this suspension and of the original supernatant were hydrolyzed in 6 M HCl for 18 h at 110°C in flame-sealed ampoules. The hydrolysates were evaporated under vacuum and resuspended in the original volume of TCA. Aliquots of the TCA supernatant (TCA-soluble free PA), of the same hydrolyzed supernatant (TCA-soluble bound PA) and of the hydrolyzed pellet suspension (TCA-insoluble bound PA) were dansylated as previously described (7) and separated by thin layer chromatography in cyclohexane:ethylacetate (3:2). Fluorescent spots, visualized under UV light, were scraped off, eluted and fluorescence measured by a Jasco FP-550 spectrofluorimeter (excitation 360 nm, emission 505 nm). The results were compared with standards dansylated in the same way.

3. RESULTS AND DISCUSSION

The time course of morphogenesis was evaluated on days 9, 15, 18, and 30 (table 1). At the end of the culture period (day 30) the percentage of control explants with roots only was about 70%. At days 18 and 30 a small percentage of control explants exhibited vegetative buds only (3%). In all cases, inhibitors delayed the appearance of roots on the explants. Almost total inhibition of the rhizogenic response was obtained by supplying DFMO plus DFMA or MGBG; DFMO alone was also strongly inhibitory, while CHA was not as effective as the other drugs. Occasional vegetative bud formation was observed in all the treatments with inhibitors except MGBG. The percentage of explants with vegetative buds only was higher under CHA treatment (18%) than DFMO plus DFMA or DFMO alone (4-5%).

TCA-soluble and TCA-insoluble bound PA were the most abundant classes and putrescine was the main polyamine (table 2). The effects of the inhibitors on PA accumulation was investigated at day 15 (table 3) when maximum PA titer occurred. Treatment with DFMO+DFMA resulted in a very strong reduction in free and bound putrescine and spermidine levels. DFMO also reduced polyamine levels especially the free and the TCA-insoluble ones. MGBG strongly reduced both putrescine and spermidine in all classes; CHA mainly reduced bound spermidine content and, to a lesser extent, the levels of putrescine.

All the inhibitors of PA biosynthesis employed in this study strongly or almost totally inhibited rhizogenesis, while leaving most of the explants viable. In particular, DFMO seemed to be slightly less inhibitory than DFMO plus DFMA to rhizogenesis and to the synthesis of the TCA-soluble polyamine conjugates. CHA was less inhibitory to putrescine accumulation and to rhizogenesis, but caused a strong reduction in bound spermidine content; this was possibly related to the appearance of caulogenesis on 18% of explants.

Inhibition of root formation and putrescine accumulation by DFMA or DFMO has been also observed in tobacco thin cell layers excised from floral branches and cultured under different conditions (15). In vegetative- or floral bud-forming thin layers, in which ADC activity rises while that of ODC declines, DFMA seemed to reduce bud initiation while DFMO bud development (11); on the contrary, CHA inhibited spermidine accumulation and floral initiation but promoted the formation of vegetative buds instead (16). A CHA-induced vegetative bud formation was observed also in the present report, concomitantly with a reduction in

Table 1. Percentage of tobacco thin cell layers with roots only, in untreated controls and in the presence of polyamine synthesis inhibitors (200 explants per treatment)

Days	Control	DFMO	DFMO+DFMA	MGBG	CHA
9	6.5	0.0	0.0	0.0	0.0
15	33.0	11.8	5.6	0.0	19.5
18	62.5	11.8	5.6	4.5	35.0
30	71.0	19.7	5.6	9.2	44.0

Table 2. TCA-soluble free (TSF), TCA-soluble- (TSB) and TCA-insoluble (TIB) bound polyamines in tobacco thin cell layers throughout the culture period. Data are expressed in nmol/g fresh weight. PU, putrescine; SD, spermidine

	TSF		TSB		TIB	
Days	PU	SD	PU	SD	PU	SD
0	14.4	10.2	125.2	37.5	100.8	29.7
9	33.1	18.2	275.4	67.6	129.9	19.5
15	242.8	37.5	440.0	256.2	407.5	126.7
18	125.7	87.1	488.0	170.1	423.9	71.9

Table 3. Percentage of TCA-soluble free (TSF), TCA-soluble- (TSB) and TCA-insoluble (TIB) bound polyamines with respect the controls in tobacco thin cell layers cultured for 15 days in the presence of PA synthesis inhibitors. PU, putrescine; SD, spermidine; tr, traces

	TSF		TSB		TIB	
Treatment	PU	SD	PU	SD	PU	SD
Control	100.0	100.0	100.0	100.0	100.0	100.0
DFMO	5.7	23.2	55.8	58.9	2.4	11.4
DFMO+DFMA	3.0	6.1	3.5	13.4	2.7	11.0
MGBG	11.2	41.3	24.8	tr	20.0	12.0
CHA	46.0	64.2	41.4	tr	39.1	tr

the rhizogenic response. In carrot embryogenic cultures DFMA was more effective than CHA in reducing PA titer and embryogenesis while DFMO was not effective (4, 8). In *Pinus radiata* shoot-forming cotyledons CHA strongly lowered spermidine and spermine levels as well as the morphogenic response (17). However the effect of CHA is generally not as dramatic as that of DFMA. MGBG has been reported to inhibit root induction in mung bean, but its effect on PA synthesis is thought to be not strictly specific (9). In our work some difference exists between the effect of DFMO alone or in combination with DFMA which indicate that both the pathways of putrescine biosynthesis are probably involved in the rhizogenic process. Different responses have been obtained using DFMO in different plant systems probably depending on the main putrescine-forming enzyme activity involved in each physiological stage considered. Until now, however, no definitive correlation has been clearly demonstrated between a particular enzyme activity and a specific physiological event.

In conclusion, the reduction in PA titer caused by inhibitors seems to be well correlated with the reduction in the organogenic response. In particular, in tobacco thin layers, when both the pathways of putrescine synthesis are blocked, or, at any rate, when endogenous putrescine is depleted, rhizogenesis is strongly inhibited. A histological analysis is under way.

4. ADDENDUM

This research was supported by funds of the Ministero della Pubblica Istruzione. Special thanks are due to the Merrel Dow Research Company, Cincinnati, for the generous gift of DFMO and DFMA.

REFERENCES

1. Bagni, N. (1989) in The Physiology of Polyamines (Bachrach, U. and Heimer, Y.M., eds), Vol. 2, pp. 107-117, CRC Press, Boca Raton, Florida
2. Evans, P.T. and Malmberg, R.L. (1989) Annu. Rev. Plant Physiol. Plant Mol. Biol. 40, 235-269
3. Bagni, N., Serafini-Fracassini, D. and Torrigiani, P. (1982) in Plant Growth Substances 1982 (Wareing, P.F., ed.), pp. 473-482, Academic Press, London
4. Fienberg, A.A., Choi, J.H., Lubich, W.P. and Sung, Z.R. (1984) Planta 162, 532-539
5. Feirer, R.B., Mignon, G. and Litvay, J.D. (1984) Science 223, 1433-1435
6. Torrigiani, P. Altamura, M.M., Pasqua, G., Monacelli, B., Serafini-Fracassini, D. and Bagni, N. (1987) Physiol. Plant. 70, 453-460
7. Torrigiani, P., Altamura, M.M., Capitani, F., Serafini-Fracassini, D. and Bagni, N. (1989) Physiol. Plant. 77, 294-301
8. Mengoli, M., Bagni, N., Luccarini, G., Nuti Ronchi V. and Serafini-

-Fracassini, D. (1989) J. Plant Physiol. 134, 389-394
9. Jarvis, B.C., Shannon, P.R.M. and Yasmin, S. (1983) Plant Cell Physiol. 24, 677-683
10. Cohen, E., (Malis)Arad, S., Heimer, Y.M. and Mizrahi, Y. (1984) Plant Physiol. 74, 385-388
11. Tiburcio, A.F., Kaur-Sawhney, R. and Galston, A.W. (1988) Plant Cell Physiol. 29, 1241-1249
12. Tran Thanh Van, K., Thi Dien, N. and Chlyah, A. (1974) Planta 119, 149-159
13. Serafini-Fracassini, D. and Mossetti, U. (1985) in Recent Progress in Polyamine Research (Selmeci, L., Brosnan, M.E. and Seiler, N., eds), pp. 551-560, Akademiai Kiado, Budapest
14. Murashige, T. and Skoog, F. (1962) Physiol. Plant. 15, 473-497
15. Tiburcio, A.F., Gendy, C.A. and Tran Thanh Van, K. (1989) Plant Cell Tissue Organ Culture 19, 43-54
16. Kaur-Sawhney, R., Tiburcio, A.F. and Galston, A.W. (1988) Planta 173, 282-284
17. Biondi, S., Bagni, N. and Sansovini, A. (1985) Physiol. Plant. 66, 41-45

PLANT GROWTH REGULATOR MANIPULATIONS IN SUGAR BEET

S.J. BROWN, K.M.A. GARTLAND, A. SLATER, J.F. HALL & M.C. ELLIOTT
Plant Molecular Biology Group, Department of Applied Biology and Biotechnology, The David Attenborough Laboratories, Leicester Polytechnic, Scraptoft, Leicester, LE7 9SU, U.K.

1. INTRODUCTION

Sugar beet (Beta vulgaris L.) is a major temperate zone crop which produces approximately 40% of the world's sucrose (1). Conventional breeding strategies although successful in the introduction of characteristics such as cold, drought and disease tolerance have been unsuccessful in increasing the fresh weight concentration of sucrose to more than 18% (2). Continued selection is unlikely to lead to further improvements in sugar yield due to the strong negative correlation between root dry mass and sugar concentration (3).

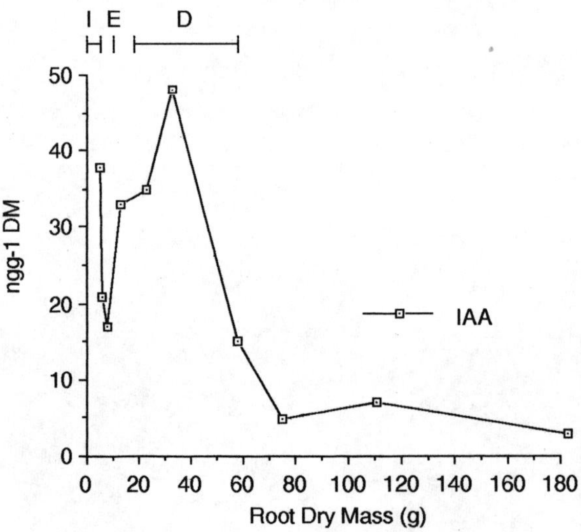

FIGURE 1. Changes in IAA levels through the growing season in the sugar beet storage organ.
I - period of cambial initiation
E - period of rapid cell enlargement
D - period of maximal cell division

The sugar beet storage root has up to twelve concentric secondary cambial rings. These are produced early in root development. Usually only the first five to seven secondary cambia make significant contributions to the tissue complement of the mature root. The, outer cambial rings make little contribution to the girth or sucrose storage capacity of the root (4). Studies of the changing plant growth regulator levels during development have implicated IAA and cytokinins in cambial initiation and activation and characteristic hormone profiles are associated with particular phases of cell division and/or enlargement (Fig. 1). Modifications of phytohormone status may enhance cell division in the outer cambial rings so that the number of storage parenchyma cells separated from the phloem by optimally short transport paths is increased. This will increase the total sucrose storage capacity of the root (3,4). One means of achieving such changes is by the introduction of selected genes for plant growth regulator biosynthesis enzymes. The Ri plasmid of _Agrobacterium rhizogenes_ harbours such genes.

Ri plasmid TR-DNA gene 1 encodes tryptophan-2-monooxygenase, which converts tryptophan to indole-3-acetamide (IAM). Gene 2 encodes IAM hydrolase, responsible for the conversion of IAM into the active auxin indole-3-acetic acid (IAA). Here we describe the effects of transferring these auxin biosynthetic genes, as part of the pRi A4b plasmid, into the sugar beet genome.

2. MATERIALS AND METHODS

Sterile seedlings of sugar beet (_Beta vulgaris_ cv. Bella) were obtained as follows.

2.1 Seed Sterilisation and Germination -

Seeds were washed in 1% v/v Teepol, rinsed thoroughly with sterile water and transferred to 50% v/v Domestos (Lever Bros.) for 90 mins. Following 2x20 mins. washes in sterile distilled water, seeds were briefly rinsed a further five times.

Seeds were germinated on Murashige and Skoog (MS) medium (5), containing 3% w/v sucrose and 0.8% w/v agar, at 25 °C. under artificial daylight conditions.

2.2 Steckling Cultures

Apical segments from germinated seedlings were excised and transferred to MS-medium supplemented with 0.5 mg/dm^3 6-benzyladenine (BAP), 3% w/v sucrose and 0.8% w/v agar (MSBA medium). Stecklings were transferred to fresh medium every four weeks.

2.3 _Agrobacterium_ Infections

Petiole sections (2.5 cm long) were punctured with a 21G needle dipped in an overnight culture of _A. tumefaciens_ C58C1 harbouring pRi A4b, and cultured on MSBA medium for four weeks. Individual hairy root tips were excised and cultured separately. Bacteria were removed by several passages on half-strength Gamborg's B5 medium (6), containing 4% w/v sucrose, 0.6% w/v agar and 0.5 g/dm^3 carbenicillin.

2.4 Hairy Root Cultures

Putative transformed roots were maintained in the dark on half-strength B5 medium without antibiotics, and sub-cultured every four weeks, by the removal of four growing tips to a new culture vessel. After at least two antibiotic free sub-cultures, hairy roots were transferred to liquid half-strength B5 medium to promote rapid growth, at 25 ^{o}C, with 60 r.p.m. shaking.

2.5 Callus Culture

Callus was initiated from hairy roots using MS-medium supplemented with 0.5 mg/dm^3 2,4-D, and 0.5 mg/dm^3 kinetin, and grown in the dark at 25 ^{o}C. Once established, callus was transferred to similar medium additionally containing 0.5 mg/dm^3 BAP, at 25°C. in the light.

2.6 Agropine Testing

Putative transformed roots and seedling roots were tested for the presence of agropine and mannopine as described by Morgan *et al.* (7).

2.7 DNA Preparation

DNA from putative transformed material, and untransformed plants was extracted using the method of Shure (8). pLJ85 cosmid DNA for use as a probe in DNA-DNA hybridisations was prepared according to Perbal (9).

2.8 DNA-DNA Hybridisations

Nucleic acid hybridisations were performed using the non-radioactive digoxigenin labelling system from Boehringer Mannheim.

2.9 IAA and IAM Analyses

Hairy root and seedling root material was frozen in liquid nitrogen and ground in methanol. Known amounts of ^{3}H-IAA and ^{14}C-IAM were added as internal standards and samples shaken at 4°C. overnight. Extracts were filtered under vacuum and the volume reduced to 1 cm^3 at 35°C under reduced pressure. K_2HPO_4 was added to pH 8.0, and samples partitioned thrice against ethyl acetate to extract IAM. Aqueous samples were rapidly adjusted to pH 3.0 with 6 mol dm^{-3} HCl and partitioned three times against diethyl ether to extract IAA. 1 cm^3 K_2HPO_4 was added to each sample and the volume reduced to the aqueous phase at 35°C under reduced pressure. Samples were passed through a polyvinyl-polypyrrolidone column and partitioned against ethyl acetate (IAM) or diethyl ether (IAA). Extracts were stored at -70°C. to freeze out water which was removed following passage through Whatman 1PS filter paper, and reduced to 1 cm^3 volume at 35°C under reduced pressure. IAA samples were purified by reverse phase ion pair HPLC as described by Blakesley *et al.* (10). Quantification was by HPLC using spectrophotofluorimetric detection, using an excitation wavelength of 445 nm and 480 nm emission wavelength of 2-methyl indolo-α-pyrone (2-MiP) derivatised IAA (11). The 2-MiP derivative was separated using the mobile phase of a methanol and water (pH 3.5) gradient, using a semi-preparative 5 microm Hypersil O.D.S. column (250x10 mm internal diameter). Non-derivatised IAM was analysed using a similar procedure, with spectrophotofluorimetric detection and excitation and emission wavelengths of 280 nm and 350 nm respectively.

3. RESULTS AND DISCUSSION

Cultivar Bella petiole sections inoculated with A. tumefaciens C58Cl carrying pRi A4b produced fluffy white roots, with many root hairs, from the site of inoculation. On clonally subculturing these roots on hormone free medium, rapid proliferation was observed. Wounded uninfected petioles failed to produce such roots. Seedling roots were unable to grow rapidly in the absence of hormones for extended periods. The hormone independent growth of pRi A4b incited hairy roots occurred when both liquid and solidified media were used.

Paper electrophoresis revealed the presence of both agropine and mannopine in each of eight putative hairy root clones, each derived from individual tips from separate petioles. Neither agropine nor mannopine has been detected in seedling root, or leaf material, in the course of these experiments (Fig.2).

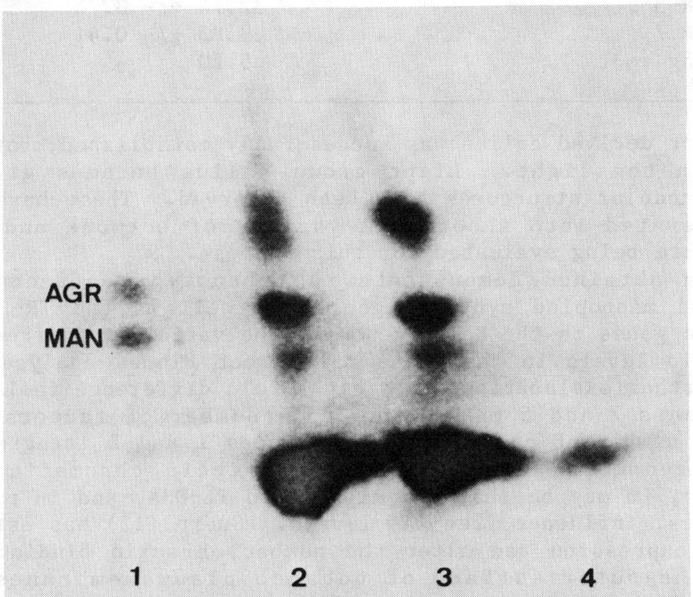

FIGURE 2. Agropine Test of Sugar Beet cv. Bella hairy roots, and seedling root.
Track 1: Agropine (Agr), Mannopine (Man) Standards
Track 2: Hairy root clone 3
Track 3: Hairy root clone 7
Track 4: Seedling root

DNA-DNA hybridisations confirmed the presence of at least part of the pRi A4b TR-DNA in each of these clones. The extent of TL-DNA integration into the sugar beet genome in these hairy roots is being investigated further.

IAA and IAM gave peaks with characteristic retention times of 7 min 35 sec and 14 min 39 sec respectively. Data accumulated from triplicate IAA analyses are shown in Table 1.

Precise quantification of IAM concentrations is currently rendered difficult by comigrating peaks. Alternative HPLC programmes are being evaluated to increase the reliability of this procedure.

TABLE 1. Indole-3-acetic acid content of sugar beet tissues

Clone	ng IAA/g fresh weight
HR 3	6.15 +/- 1.73
HR 5	32.74 +/- 3.75
HR 7	3.85 +/- 0.41
Seedling root	5.20

Hairy root derived callus was successfully established, both in the dark and in the light. Light grown callus becomes green, and occasional nodular structures have been observed. These have not yet been regenerated into shoots. A variety of hormone and thermal treatments are being evaluated for this purpose.

The data obtained demonstrate the phenotypic effects of the agropine and mannopine synthase genes, as well as the TR-DNA auxin biosynthetic genes on the Bella genome. The variation in free indole-3-acetic acid levels in the three hairy root clones analysed so far requires further explanation. The eight-fold difference in IAA levels bewteen clones 7 and 5 may be due to a number of factors. These include the number of copies of TR-DNA genes 1 and 2 integrated into the Bella genome, their position, and their chromatin status. Additionally, it may be that the Ri plasmid TL-DNA, and in particular the rol genes, influence free IAA levels. Guern (12) has established that rol B expression can alter the number of auxin binding protein sites on the outer surface of tobacco plasmamembranes. This sensitivity-increasing effect, together with expression of rol A and rol C , may exert some influence upon the TR-DNA auxin biosynthetic genes, or upon the sugar beet's own IAA biosynthetic circuits. These possibilities are currently being investigated, beginning with an evaluation of the TL-DNA content of these hairy root clones. Increased understanding of the regulation of IAA biosynthesis and metabolism in genetically manipulated sugar beet cells and plants and of organ specific promoter sequences may facilitate precise control of cambial ring activation, and increased sucrose storage in this important crop plant.

REFERENCES

1. Fick, G.W. Loomis, R.S. and Williams, W.A. (1975) Crop Physiology (Evans, L.T., ed.), pp.260-295, Cambridge University Press.
2. Milford, G.F.J. Biscoe, P.V. Jaggard, K.W. Scott, R.K. and Draycott, A.P. (1980) Opportunities for Increasing Crop Yields

(Hurd, R.G. Biscoe, P.V. and Dennis, C., eds.), pp.77-83, Pitman, London.
3. Hosford, D.J. Lenton, J.R. Milford, G.F.J. Pocock, T.O. and Elliott, M.C. (1984) Plant Growth Regulation 2, 371-380.
4. Elliott, M.C. Hosford, D.J. Lenton, J.R. Milford, G.F.J. Pocock, T.O. Smith, J.I. Lawrence, D.K. and Firby, D.J. (1983) Growth Regulators In Root Development (Jackson, M.B. and Stead, A.D. eds.), pp.25-33,
5. Murashige, T and Skooge, F. (1962) Physiol.Plant. 15, 473-497.
6. Gamborg, O.L. Miller, R.A. and Ojima, K. (1968) Exp. Cell Res. 50, 151-158.
7. Morgan, A.J. Cox, P.N. Turner, D.A. Peel, E. Davey, M.R. Gartland, K.M.A. Mulligan, B.J. (1987) Plant Science 49, 37-49
8. Shure, M. Wessler, S. and Fedoroff, N. (1983) Cell. 35, 225-233
9. Perbal, B. (1988) in A Practical Guide to Molecular Cloning Wiley and Sons Inc U.S.A.
10. Blakesley, D. Allsopp, A.J.A. Hall, J.F. Weston, G.D. and Elliott, M.C. (1984) J.Chromatography. 294, 480-484.
11. Blakesley, D. Hall, J.F. Weston, G.D. and Elliott, M.C. (1983) J.Chromatography. 258, 155-164
12. Guern J. S.E.B. Symposium: Hormone Perception and Signal Transduction, Nottingham, 1989 (in press)

REGULATION OF CYTOKININ CATABOLISM IN TOBACCO CALLUS CULTURES

V.MOTYKA, M.KAMÍNEK

Institute of Experimental Botany, Czechoslovak Academy of Sciences, Ke dvoru 16, 166 30 Prague 6, Czechoslovakia

1. INTRODUCTION

The control of growth and differentiation in plant tissue cultures involves complex interactions between auxins and cytokinins. The endogenous levels of these phytohormones are regulated by means of various mechanisms controlling their biosynthesis and degradation.

The cytokinin catabolism in plant tissues is governed by specific enzyme, cytokinin oxidase. This enzyme catalyzes the oxidative degradation of cytokinins bearing unsaturated N^6-isoprenoid side chains to form adenine or its riboside and the corresponding side chain aldehyde /1,2/. The cytokinin oxidase activity was detected in a number of plant tissues /3-8/ and a homogenously purified enzyme has been isolated recently from *Zea mays* kernels /9/.

In *Phaseolus* callus tissue the cytokinin oxidase activity is enhanced *in vivo* by exogenous application of purinyl and urea-type cytokinins /6,8/. This effect of externally applied cytokinins on the enzyme activity is reported also for tobacco cell suspension and soybean callus tissue /10, 11/. The increase in cytokinin oxidase activity in plant tissues occurs rapidly after application of cytokinin and is sensitive to antibiotic inhibitors of transcription and translation /6/. Cytokinin degradation in tobacco pith explants was found to be also increased by externally applied auxins /12/.

The increase of cytokinin oxidase activity in plant tissues following the exogenous application of growth regulators is closely related to the control of plant growth and morphogenesis and may be one of reasons of failure to stimulate cell division and/or organ differentiation in some plant tissue cultures. Recent report about the existence of two molecular forms of cytokinin oxidase with different localization in plant cells /8/ makes this problem more complex. With respect to the complexity of the problem possibilities of regulation of cytokinin oxidase activity in tobacco callus cultures by exogenous cytokinins and auxins were examined in the present study.

2. PROCEDURE

2.1. Materials and methods

2.1.1. Tissue culture: Callus tissues of tobacco (*Nicotiana tabacum* L.cv.Wisconsin 38) 26 days old were selected for use. Tissues were cultivated in dark at 26°C on medium of Murashige and Skoog /13/ containing auxin (5.4 μM NAA) and cytokinin (1.3 μM BAP).

2.1.2. Extraction and assay of cytokinin oxidase: The enzyme preparation was extracted and partially purified by a modified method of Chatfield and Armstrong /6/. The activity of cytokinin oxidase was examined using an assay based on the conversion of /2,8-^{3}H/-N^{6}-(Δ^{2}-isopentenyl)adenine to adenine.

2.1.3. Regulation of cytokinin oxidase activity in vivo: Cytokinins and auxins were dissolved in water and 0.05 M $NaHCO_3$, respectively, and 100 μM solutions were applied directly to the surface of callus tissues (0.2 ml/g tissue). The tissues were incubated at 26°C in dark for 6 h (except for time course studies). Following incubation the cytokinin oxidase was extracted, its enzyme activity was determined and compared with the activity of control (application of water, resp. 0.05 M $NaHCO_3$, in time 0 h).
Simultaneously with the examination of cytokinin oxidase activity the total protein was determined by the protein-dye binding assay of Bradford /14/.

3. RESULTS AND DISCUSSION

The cytokinin oxidase activity in tobacco callus tissue was found to be regulated *in vivo* by exogenous application of cytokinins. The surface treatment of the tissue with a cytokinin solution resulted in a significant and relatively rapid increase in cytokinin oxidase activity. The results of treatment with *trans*-zeatin are shown in Fig.1. The cytokinin oxidase activity was enhanced as early as 2 h after zeatin application (up to 160%) and reached the maximum 6 h after the treatment (201%).

With respect to these results the incubation period of 6 h was used in investigation of regulatory effects

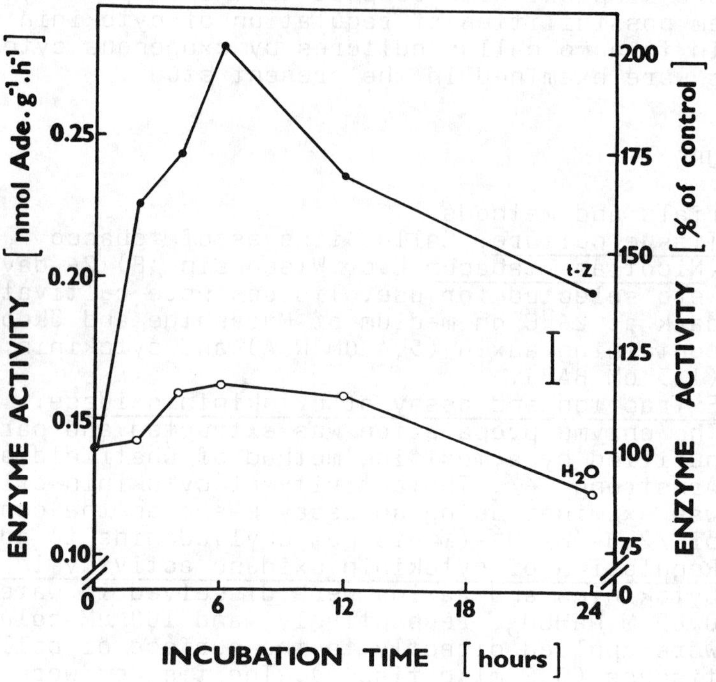

FIGURE 1. Effect of exogenously applied cytokinin trans-zeatin (t-Z) to callus tissue of tobacco (Nicotiana tabacum L.cv.Wisconsin 38) on cytokinin oxidase activity.

of other cytokinins and auxins on enzyme activity in vivo. In agreement with the regulation of the enzyme activity in Phaseolus callus tissues /6,8/ all tested cytokinin--active substances including purinyl nonsubstrate and urea-type cytokinins (benzyladenine=BAP, /m-hydroxybenzyl/-adenosine=/mOH/[9R]BAP, Thidiazuron) were effective in enhancement of cytokinin oxidase activity in tobacco callus tissue (table 1.). The effectiveness of nonsubstrate exogenous cytokinins suggests their significance in the regulation of endogenous cytokinins representing active pool of natural cytokinins in plant cells. Although the knowledge on regulatory mechanisms of cytokinin degradation is still limited, the stimulatory effect of externally applied cytokinins on cytokinin oxidase activity is also reported for other plant tissues /10,11/.

With respect to the reported dependence of cytokinin oxidase enhancement on protein synthesis /6,15/ total

TABLE 1. Effect of exogenously applied cytokinins to callus tissue of tobacco (Nicotiana tabacum L.cv.Wisconsin 38) on cytokinin oxidase activity and protein content.

COMPOUND	CYTOKININ OXIDASE ACTIVITY AFTER 6 HOURS OF INCUBATION (% of control in time 0)	PROTEIN CONTENT AFTER 6 HOURS OF INCUBATION (% of control in time 0)
H_2O	116	104
trans-zeatin	201	-
BAP	161	114
/mOH/[9R]BAP	157	110
Thidiazuron	156	-

The cytokinin oxidase activity of control (H_2O) in time 0 h was 0.14 nmol/g.h (100 %).
The protein content was determined as described by BRADFORD /14/ on the basis of absorbance (595 nm) of the protein-Coomassie blue complex, A_{595} of control (H_2O) in time 0 h was 1.17 (100 %).

TABLE 2. Effect of exogenously applied auxins to callus tissue of tobacco (Nicotiana tabacum L.cv.Wisconsin 38) on cytokinin oxidase activity and protein content.

COMPOUND	CYTOKININ OXIDASE ACTIVITY AFTER 6 HOURS OF INCUBATION (% of control in time 0)	PROTEIN CONTENT AFTER 6 HOURS OF INCUBATION (% of control in time 0)
$NaHCO_3$ buffer	97	101
IAA	95	100
IBA	92	-
NAA	90	95

The cytokinin oxidase activity of control ($NaHCO_3$ buffer) in time 0 h was 0.15 nmol/g.h (100 %).
The protein content was determined as in table 1., A_{595} of control ($NaHCO_3$ buffer) in time 0 h was 1.14 (100 %).

protein content in crude preparations was determined after the application of exogenous cytokinins (benzyladenine, /m-hydroxybenzyl/adenosine) to the surface of tobacco tissue. Our results suggest that the cytokinin—induced enhancement of total protein participates only partly on the increase of the enzyme activity (table 1.).

The surface treatment of tobacco callus tissue by exogenous auxins had no effect on cytokinin oxidase activity and proteosynthesis (table 2.). Palni et al./12/ studied an influence of externally applied auxins on zeatin riboside degradation in tobacco pith explants. They found an inverse relation between auxin concentration in the incubation medium and cytokinin stability suggesting the mediating role of cytokinin oxidase in this process. The difference between Palni´s and our results is probably due to different arrangements of experiments and use of different substrates and plant tissues.

The increase of cytokinin oxidase activity after the external application of cytokinins may be one of reasons of failure to stimulate cell division and/or organ differentiation in some plant tissue cultures by exogenous cytokinins. Mechanisms participating in hormonal regulation of these processes are likely to be subject to multiple controls. The treatment of plant tissues by synthetic cytokinins enhances both degradation and accumulation /16-18/ of natural endogenous cytokinins. All regulatory mechanisms involved are closely related to the control of plant growth and morphogenesis.

REFERENCES

1. Brownlee,B.G., Hall,R.H., Whitty,C.D. (1975) Can.J. Biochem. 53, 37-41
2. McGaw,B.A., Horgan,R. (1983) Phytochemistry 22, 1103-1105
3. Pačes,V., Werstiuk,E., Hall,R.H. (1971) Plant Physiol.48, 775-778
4. Whitty,C.D., Hall,R.H. (1974) Can.J.Biochem. 52, 789-799
5. McGaw,B.A., Horgan,R. (1983) Planta 159, 30-37
6. Chatfield,J.M., Armstrong,D.J. (1986) Plant Physiol. 80, 493-499
7. Laloue,M., Fox,J.E. (1989) Plant Physiol. 90, 899-906
8. Kamínek,M., Armstrong,D.J. (1990) Plant Physiol. in press
9. Burch,L.R., Horgan,R. (1989) Phytochemistry 28, 1313-1319.
10. Terrine,C., Laloue,M. (1980) Plant Physiol.65, 1090-1095
11. Palmer,M.V., Palni,L.M.S. (1987) J.Plant Physiol. 126, 365-371
12. Palni,L.M.S., Burch,L.R., Horgan,R. (1988) Planta 174, 231-234
13. Murashige,T., Skoog,F. (1962) Physiol.Plant. 15, 473-497
14. Bradford,M.M. (1976) Anal.Biochem. 72, 248-254
15. Szweykowska,A., Gwozdz,E., Spychala,M. (1981) in Metabolism and Molecular Activities of Cytokinins (Guern,J., Péaud-Lenoël,C., eds.), pp.212-217, Springer-Verlag

16.Thomas,J.C., Katterman,F.R. (1986) Plant Physiol. 81, 681-683
17.Hansen,C.E., Meins,F.Jr., Aebi,R. (1987) Planta 172, 520-525
18.Vaňková,R., Kamínek,M., Eder,J., Vaněk,T. (1987) J.Plant Growth Regul. 6, 147-157

DIFFERENTIAL SENSITIVITY TO AUXIN AT THE PLASMA MEMBRANE LEVEL

Michel ROSSIGNOL, Véronique SANTONI, Wojciech SZPONARSKI and Gérard VANSUYT, INRA, Physiopathologie Végétale, BV 1540, 21034 Dijon cedex, France

1. INTRODUCTION

Among growth regulators, auxins constitute the group for which the most convincing evidences were raised for an action at the membrane level. Particularly, several data argue for an action on the plasma membrane H^+-ATPase (1-3) and it is now largely assumed that auxin stimulates the ATPase. In addition, this effect would involve auxin binding proteins (1-3). However one intriguing point concerns the concentrations used to demonstrate the stimulation. For instance the ATPase activity is stimulated by 1 µM IAA in membrane vesicles from zucchini hypocotyls (4) but by 100 µM IAA in membrane vesicles from pumpkin hypocotyls (5). With tobacco protoplasts, maximal hyperpolarization is observed in the presence of 5 µM NAA, and depolarization with 100 µM NAA (6). On the other hand, in studies with artificial phospholipid membranes, millimolar auxin concentrations were found to alter permeability properties (7-9). The question to know to what auxin concentrations the plasma membrane is sensitive appears so quite puzzling.

In the present work, auxin effects on the proton translocation by the plasma membrane H^+-ATPase were systematically investigated according to the hormone concentration. Dose-response curves were raised for purified plasma membrane vesicles from tobacco leaves as well as for reconstituted systems.

2. MATERIAL and METHODS

Plasma membrane vesicles were purified from tobacco leaves by two phase partitioning according to described procedures (10). ATP hydrolysis by these vesicles was nearly fully sensitive to vanadate and dependent on the presence of magnesium. Fractions were almost completly devoided of phosphatase activity and of phosphohydrolase activities originating from endomembranes as assessed by the effects of molybdate (phosphatase), nitrate (tonoplast), azide or oligomycin (mitochondria) or IDP (Golgi apparatus).

Plasma membrane H^+-ATPase was solubilized from corn roots as previously described (11) and purified by centrifugation over glycerol gradient. Crude auxin binding protein preparations were obtained according to ref 12 and chromatographied onto DEAE-cellulose. Reconstitutions were performed by an octyglucoside dilution procedure as already published (13).

Proton translocation was measured by the initial fluorescence quenching rate of the the dye 9-amino-6-choro-3-methoxy-acridine (11-13). Experiments were performed in 10 mM MES/tris buffer at pH 6.5, containing 0.1 M potassium salt, and $MgSO_4$, and started by the addition of ATP. For relaxation experiments, the medium contained in addition 25 mM glucose and hexokinase was added at the steady state (42 $U.ml^{-1}$); the proton passive efflux was monitored until complete dissipation of the gradient. Auxins were added from a 100 times concentrated solution in ethanol or acetonitrile and controls included 1% ethanol or acetonitrile. When used, valinomycine was added at a concentration of 5.10^{-7} M. All fluorescence measurements were performed at 25°C on a SLM 8000 spectrofluorometer.

3. RESULTS

3.1. *Dose-response curves are bell-shaped*

Auxin effects on the proton translocation were first investigated for tobacco leave plasma membrane vesicles. IAA was found to stimulate the proton translocation for

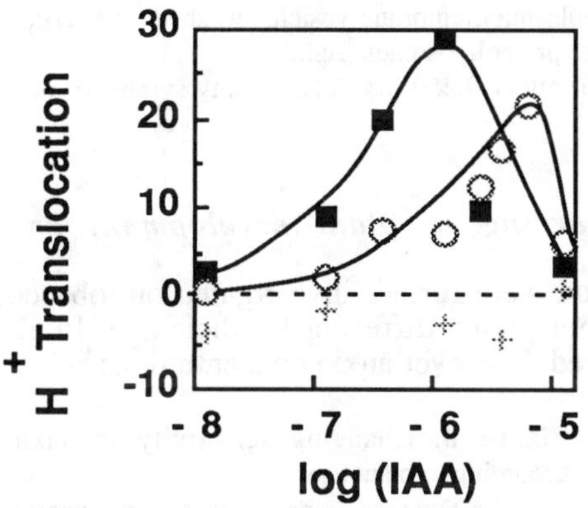

Figure 1. Auxin-induced stimulation of the proton translocation activity.

Stimulation of the H^+ translocation activity was measured by the percentage over control in the absence of auxin for tobacco leave plasma membrane vesicles (•) or purified plasma membrane ATPase co-reconstituted with a crude auxin binding protein extract (o). + control experiment with acetic acid.

concentration ranging between 10^{-7} and 10^{-5} M. However biphasic responses were systematically observed and stimulation disappeared for the highest concentrations (Figure 1, squares). No effect could be observed with a weak acid of similar pKa but devoided of biological activity (Figure 1, crosses). When corn root plasma membrane H^+-ATPase was co-reconstituted with a crude membrane protein extract containing auxin binding protein, a biphasic respons was also obtained (Figure 1, circles).

3.2. *Stimulation involves proteic intermediaries*

When plama membrane vesicles were washed with Triton X-100, a treatment which is known to remove auxin binding proteins from membranes (14), no stimulation could be observed (Figure 2, left). In fact a slight inhibition was measured, suggesting that the detergent removed or inactivated some intermediary. On the other hand, when corn root plasma membrane H^+-ATPase was reconstituted alone into proteoliposomes, no effect at all was observed (Figure 2, right). This indicates that the auxin-induced stimulation requires some intermediary which was present in the crude auxin binding protein extract.

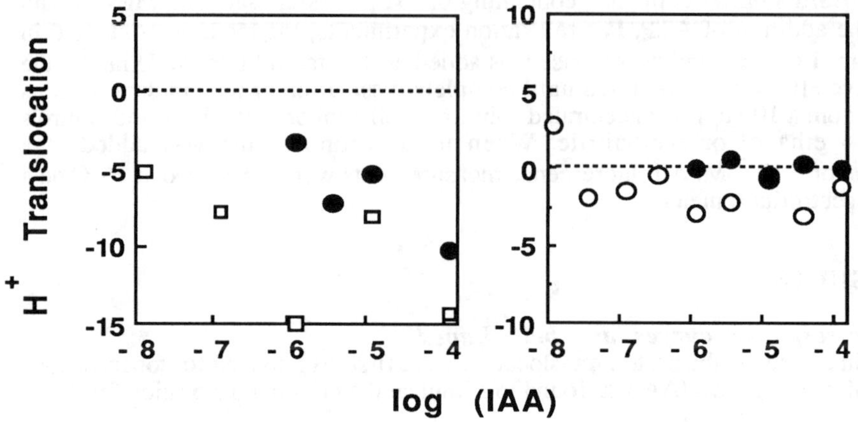

Figure 2. Auxin effects on the proton translocation by plasma membrane vesicles washed with Triton X-100 (left) or on purified ATPase reconstituted alone into proteoliposomes (right).

Experiments were performed in the presence of either 0.250 Mg-ATP (empty symbols) or 3 mM Mg-ATP (full symbols).

3.3. *Sensitivity to auxin changes according to plant development and culture conditions*

Auxin effects on the proton translocation were further investigated on tobacco leave plasma membrane vesicles prepared from plants differing by their age. In all experiments biphasic responses were still observed. However auxin concentrations stimulating the ATPase were found to depend on the plant age (Figure 3).Thus, the sensitivity to auxin, monitored by the auxin concentration inducing the maximal stimulation, appeared to depend on the development state of the plant. In addition, submillimolar concentrations inhibited the proton translocation.

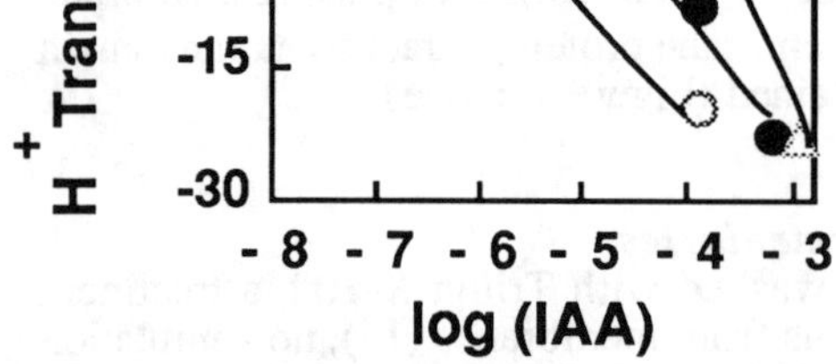

Figure 3. Changing sensitivity to auxin according to plant age.

Dose-response curves for proton translocation as a function of auxin concentration in the case of plasma membrane vesicles prepared from plants at different ages during the induction to flowering (61 days-old points), induced to flowering (68 day-old, circles) or flowering (91 day-old, triangles).

Systematic investigation was then performed according to the plant age, in the standard culture conditions (culture in greenhouse during 45 days and then in environment chamber : 20°C, 16000 lux, 16 hr light, 75% relative humidity). Figure 4 (full symbols) shows that an high sensitivity to auxin was acquired during the induction to flowering period (between days 58 and 65 in these culture conditions). By contrast, vegetative or flowering plants (respectively younger than 58 days or older than 90 days in these conditions) exhibited a 1000-fold lower sensitivity.

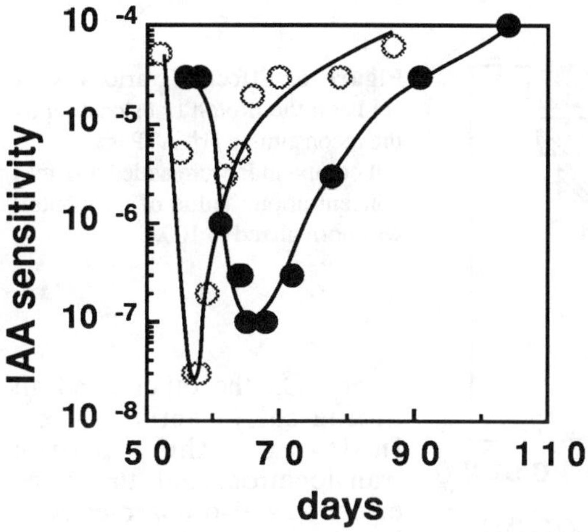

Figure 4. Changing sensitivity to auxin according to plant age and culture conditions.

Plot of the auxin concentrations inducing the maximal stimulation as a function of plant age for two sets of culture conditions : greenhouse for 45 days and then environment chamber (full symbols), or environment chamber during the whole culture (empty symbols).

When culture is completely performed in environment chamber, the induction to flowering is hastened. In these conditions the acquisition of a high sensitivity to auxin was observed to occure between days 52 and 58 (Figure 4, empty symbols). As in standard conditions, this high sensitivity disappeared when plants became flowering.

3.4. *Submillimolar auxin concentrations exert a non-specific inhibitory effect*

The inbitory effect observed at submillimolar auxin concentrations was studied in reconstituted systems containing only the H^+-ATPase so that no stimulation could occur (see Figure 2b). None of the three auxins IAA, 1-NAA or 2,4-D affected the ATP hydrolysis activity of the enzyme (Figure 5, open symbols). This corroborated the above conclusion that auxins do not affect directly the ATPase. However submillimolar auxins

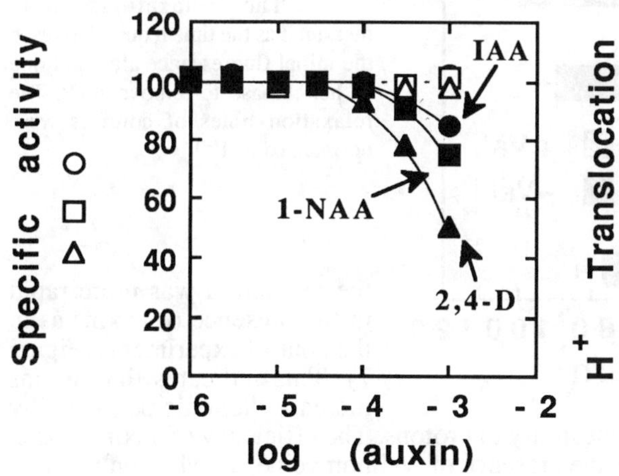

Figure 5. Auxin effects on the H^+-ATPase reconstituted alone into proteoliposomes.

Purified corn root plasma membrane ATPase was reconstituted into pure phospholipid vesicles. Auxin effects on the ATP hydrolysis activity (open symbols) and on the proton translocation activity (full symbols). Values of control were normalized to 100.

were found to inhibit at the same time the proton translocation activity (Figure 5, full symbols). Effects of other weak acids on the proton translocation activity are summarized in Figure 6. When used at 1 mM, nor acetic acid nor benzoic acid had any inhibitory

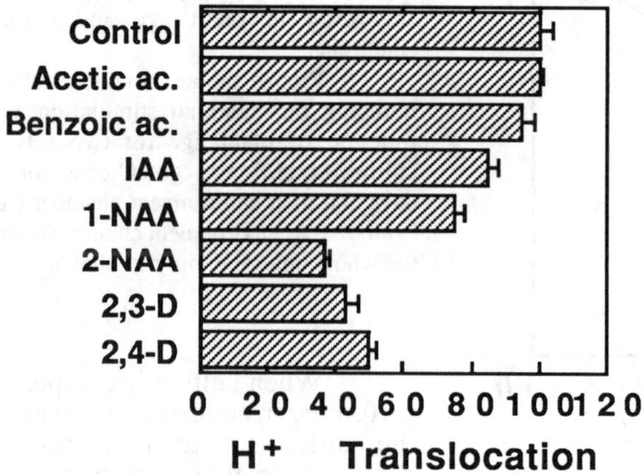

Figure 6. Effect of various weak acids on the proton translocation by the reconstituted H+-ATPase. All compounds were added at 1 mM concentration. Value of the control was normalized to 100.

effect. On the other hand all biologically active auxin inhibited the proton translocation, but the same effect was also observed with inactive analogs.

3.5. *Submillimolar auxins increase the membrane permeability to protons*

As the ATP hydrolysis activity of the enzyme remained still unsensitive to auxins, even at 1 mM, we tested the hypothesis of an auxin induced change of the permeability to protons. For this purpose we measured the time required for complete relaxation of the proton gradient. Measurements were performed both in the presence or in the absence of valinomycine in order to distinguish between auxin effects on the permeability to protons or on other ions which contribute to collapse the electrochemical gradient. In all situations

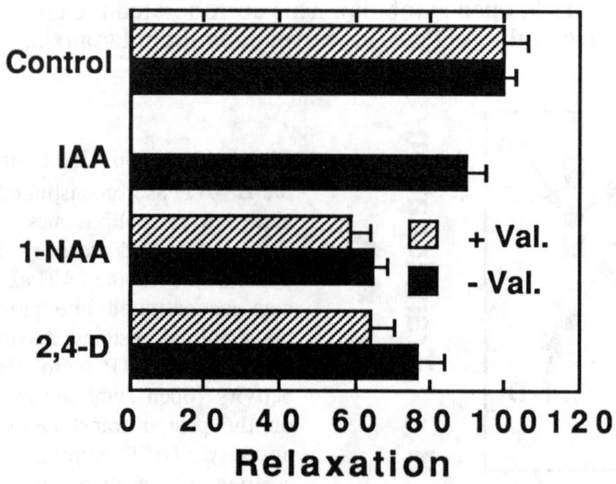

Figure 7. Auxin effects on the permeability to protons.

The relaxations were measured as the time required to reach the initial fluorescence after addition of hexokinase to remove ATP. The relaxation times of controls were normalized to 100.

the relaxation was more rapid in the presence of auxins as in the control experiments (Figure 7). This indicates that auxins increase the ionic permeability of membranes and especially the permeability to protons. The efficiency of auxins was in the order 1-NAA > 2,4-D > IAA. In the presence of valinomycine, *ie* when only the permeability to protons is limiting, no significative difference could be detected between 1-NAA and 2,4-D. This shows that both auxins induce identical permeability increase to protons. By the same way this suggests that the sequence above is mainly due to differential auxin-induced permeability increase to other ions.

4. DISCUSSION and CONCLUSIONS

Auxins appear do not exert any direct effect on the plant plasma membrane H^+-ATPase (Figure 2 and Figure 5 open symbols). However, according to the hormone concentration, the ATPase may be either stimulated or inhibited. This indicates that auxins act by several ways. The stimulatory effect involves specific proteins and is observed for low auxin concentrations. On the other hand the inhibitory effect is observed only for submillimolar auxin concentrations and concerns the permeability barrier. Therefore this effect involves likely perturbations of the lipid moiety of membranes. In addition it should be pointed out that, in many cases, the stimulation disappears for auxins concentrations which are not effective on the permeability barrier (Figure 3). This suggests the presence of an other kind of regulatory effect whose nature remains obscure.

An other major conclusion is that the auxin-induced stimulation of the H^+-ATPase depends on the physiological state of plants. As this state is clearly influenced by culture conditions, the auxin effect also depends on culture conditions (Figure 4). This stimulatory effect appears so to be a way to monitor the sensitivity to hormone. Accordingly the sensitivity seems to be high only at the end of the induction to flowering period. This shows that the sensitivity to hormones results not only from genetical properties but is also regulated in a complex way. In this view it should be emphasized that approaches at the membrane level offer new opportunities for biochemical characterization of this regulation of the sensitivity to hormones.

REFERENCES

1 Libbenga, K.R., Maan, A.C., Van der Linde, P.G.C. and Mennes, A.M. (1986) in Hormone, receptors and cellular interactions in plants (Chadwick, C.M. and Garrod, D.R., eds), pp.1-68, Cambridge University Press, Cambridge

2 Libbenga, K.R. and Mennes, A.M. (1988) in Plant hormone and their role in plant growth and development (Davies, P.J., ed), pp.194-221, Kluwer Academic Publishers, Dordrecht

3 Morré, D.J. (1989) in Second messengers in plant growth and development (Boss, W.F. and Morré D.J., eds), pp81-114, Alan R. Liss, New York

4 Scherer, G.F.E. (1984) Planta 161, 394-397

5 Scherer, G.F.E. (1981) Planta 151, 434-438

6 Ephritikhine, G., Barbier-Brigoo, H., Muller, J.F. and Guern, J. (1987) Plantn Physiol. 83, 801-804

7 Jones, G.P. and Paleg, L.G. (1984) Biochemistry 23, 1521-1524

8 Hester, P. and Stillwell, W. (1984) Biochim. Biophys. Acta 770, 105-107

9 Stillwell, W., Brengle, B., Belcher, D. and Wassall, S.R. (1987) Phytochemistry 26, 3145-3150

10 Santoni, V., Vansuyt, G. and Rossignol, M. (1990) Plant Sci. 68, 33-38

11 St Marty-Fleurence, F., Bourdil, I., Rossignol, M. and Blein, J.P. (1988) Plant Sci. 54, 177-184

12 Klämbt, D. and Löbler M. (1987) in Plant hormone receptors (Klämbt, D., ed), pp.261-263, Springer-Verlag, Berlin

13 Rossignol, M., Kervella, V. and Blein, J.P. (1989) Plant Physiol. Biochem. 27, 437-441

14 Nakamura, C. and Ono, H. (1988) Plant Physiol. 88, 685-689

RÔLES OF THE PRIMARY CELL WALL IN MORPHOGENESIS

Stephen C. FRY

Department of Botany, University of Edinburgh, The King's Buildings, Mayfield Road, Edinburgh, EH9 3JH, U.K.

1 INTRODUCTION

Primary cell walls play important and diverse rôles, both mechanical and regulatory, in morphogenetic processes [1]. An understanding of these rôles may improve our ability to induce the formation of organised tissue, either by organogenesis or by embryogenesis [2], in plant tissue cultures. The purpose of the present paper is therefore to review these rôles of the cell wall and to identify areas of our understanding that stand out as requiring further clarification. The review focuses on the <u>early</u> stages of morphogenesis, during which the basic characteristics which constitute organisation are imposed on previously disorganised tissue. Thus, we shall be concerned with the features of, for instance, a globular embryo and how they differ from those of a rapidly growing, undifferentiated cell culture [2]. Of course, the production of a globular embryo, although necessary, is not sufficient for embryogenesis.

The factors that influence the progression of morphogenesis in plant cell and callus cultures are extremely diverse. It is unfortunately the case that at present each new species (and often cultivar) that is brought into culture must have its cultural conditions optimised empirically. Major variables include the composition of the culture medium [e.g. the type, concentrations and ratios of phytohormones present, the nature of the carbon source (e.g. sucrose *vs.* glucose *vs.* maltose), the form in which nitrogen is supplied (e.g. NO_3^- *vs.* NH_4^+) and the pH], the genetic make-up of the plant under investigation, the length of time for which it has been maintained in culture (and thus possibly undergone heritable changes, e.g. a loss of totipotency), the cell density and prevalence of cell/cell contact, and the light/dark régime employed.

Given this multiplicity of external and genetic factors, a scientist who focuses on one particular part of the cell — the cell wall — will not expect to solve *all* the problems of inducing morphogenesis in recalcitrant cultures! However, there now exists sufficient evidence to allow the unequivocal conclusion to be drawn that the cell wall is one of the most important foci in the regulation of morphogenesis, and one that has until recently been seriously overlooked. The cell wall enthusiast therefore has something new to contribute to the study of morphogenesis.

2 RÔLES OF THE CELL WALL IN THE MECHANICS OF MORPHOGENESIS

2.1 Background

Mechanically, the cell wall governs the cell's shape, size and adhesiveness [1,3,4]. These characters are the very features which serve to distinguish the cells of young organised structures (organs and embryoids) from those of the calli (or suspension cultures) from which they are produced. Young organised structures resemble disorganised calli in exhibiting rapid cell division; however, they differ from calli principally in having smaller and more isodiametric cells and tighter cell-to-cell contact. It is also worth noting that attempts to isolate protoplasts from recalcitrant tissue cultures depend very much on the breakdown of mechanical features of the cell wall. Similarly, attempts to regenerate walled cells from protoplasts depend on the re-establishment of the wall's mechanical features [5]. For all these reasons, a real understanding of the cell wall and its assembly is valuable to applied work in plant tissue culture.

2.2 Imposition of a small cell size during morphogenesis

The average cell size of a population of plant cells is governed by the balance of cell division and cell expansion: it is instructive to ask which of these two processes is primarily responsible for the small cell size that characterises organised tissue. Since organised tissues do not consistently exceed calli or suspension cultures in their rates of cell division (indeed, undifferentiated cell cultures exhibit some of the most rapid cell division seen in the plant kingdom), the small cell size characteristic of organised tissues seems likely to be due primarily to a failure of daughter cells to expand after cytokinesis rather than to any enhancement of the cells' ability to divide rapidly.

If this analysis is correct, then any attempt to enhance morphogenesis in an active callus or cell culture seeks (explicitly or implicitly) not only to maximise cell division, but also simultaneously to minimise the expansion of the dividing cells.

Having identified the curbing of cell expansion as significant, we can ask what factor(s) limit the expansion of cells in an organised, rapidly-dividing tissue such as a successful globular embryo. Since cell expansion is driven by the osmotic uptake of water (which generates the turgor pressure) and resisted by the physical properties of the cell wall, a small cell size could *a priori* be imposed either by the maintenance of a low osmotic pressure in the cell sap or by an inability of the cell wall to deform under turgor pressure. Although there appear to have been no direct studies of osmotic or turgor pressure or of the relevant physical properties of the cell wall (extensibility and yield threshold) in relation to morphogenesis in cultured cells, it is a well established generalisation in plant physiology that cell expansion is usually limited by the (in-)extensibility of the cell wall [6], which sub-cellular compartment is therefore highlighted. This applies particularly to the large body of work in which the effects of phytohormones on expansion growth has been investigated [6,6a].

The chemical basis for observed changes in the physical properties of the cell wall is poorly understood. Since loosening and tightening of the cell wall have not so far been shown to be consistently associated with any particular change in wall polymer composition, attention has been focused instead on the making and breaking of the interpolymeric cross-links that bind together the individual polymers of the wall — polymers which, in isolation, are largely water-soluble. This work has been reviewed [3,7]; here I draw attention to three extracellular enzymes that might fruitfully be examined for causal relationships to the changes in wall plasticity thought to be necessary for embryogenesis.

2.2.1 <u>Cellulase</u>: Despite its name, cellulase is usually better able to hydrolyse hemicelluloses than to hydrolyse cellulose [8]. In particular, xyloglucan (which is the major hemicellulose of the primary cell walls of Dicots and which has a β-(1→4)-ᴅ-glucan backbone identical to that of cellulose) appears to be the main substrate of cellulase *in vivo* [7-9]. If, as has been proposed, xylolgucan molecules serve a "tethering" rôle, holding together adjacent microfibrils [7], then the enzyme that cleaves xyloglucan could have a major impact on the physical properties of the cell wall, loosening it [9a]. Besides cellulase, an enzyme has been discovered in germinated *Tropaeolum* seeds that specifically hydrolyses the backbone of xyloglucan [10]; it will be of interest to determine how widespread enzymes of this type are. There is evidence that cellulase does not simply hydrolyse the xyloglucan chains [this would unduly weaken the wall], but that it can catalyse an endotransglycosylation reaction [which would temporarily allow wall creep (expansion) with a minimum loss of strength] [11,12]. In endotransglycosylation, the substrate polysaccharide chain is cut and one portion is transiently attached to the enzyme; this new "polysaccharidyl" fragment is then transferred from the enzyme not on to a water molecule (as in the case of hydrolysis) but on to an alcoholic (—OH) group, e.g. of another polysaccharide chain [13]. The net result is no change in mean molecular weight, as can be seen in the scheme,

ooooooooooooooooooooooooooooo□ + Є + ●●●●●●●●●●●●●●●●●●■ ↔

[ooooooooooooooooooo~Є + oooooooooo□ + ●●●●●●●●●●●●●●●●●●■ ↔]

ooooooooooooooooooo●●●●●●●●●●●●●●●●●●■ + oooooooooo□ + Є

where:
circles represent sugar residues,
squares represent reducing termini and
Є = enzyme.

Endotransglycosylation is therefore very difficult to detect *in vivo*. So far we have two pieces of evidence for its occurrence:

(a) $[^3H]$Oligosaccharides of xyloglucan ($[^3H]$XGOs), of molecular weight ~1200-1500, when infiltrated into the walls of living cells, become incorporated into extracellular xyloglucan-like polysaccharides without being taken up into the cytoplasm [11]. The reducing terminus of the $[^3H]$XGO remains a reducing terminus after incorporation [12]and the bond between the $[^3H]$XGO moiety and the polymer is alkali-stable; these facts strongly suggest that a glycosidic bond has formed between an endogenous polysaccharide and the exogenous oligosaccharide. Although oligosaccharides are not structural components of the cell wall, they will nevertheless permeate it and may thus participate in reactions normally catalysed by wall enzymes; thus, we speculate that xyloglucans undergo similar endotransglycosylation reactions with other polymeric xyloglucan molecules during wall expansion *in vivo* [12].

(b) When high-M_r non-radioactive xyloglucan [M_r = molecular weight $\approx 2 \times 10^5$] is mixed with moderate-M_r radioactive xyloglucan [$M_r \approx 5 \times 10^4$] *in vitro* in the presence of a crude preparation of *Phaseolus* cellulase, the initial products include high-M_r radioactive xyloglucan, suggesting that endotransglycosylation occurred [14]; final products were all low-M_r, showing that, ultimately, hydrolysis predominated over transglycosylation under the *in-vitro* conditions employed.

It would seem worthwhile exploring the occurrence and significance of this type of reaction *in vivo* in the walls of cells undergoing (or failing to undergo) embryogenesis.

2.2.2 Extracellular Peroxidase: A reaction that may tighten the cell wall, restricting cell expansion, is the peroxidase-catalysed oxidative coupling of phenolic side-chains on the structural polymers of the primary cell wall [3,15,15a]. These phenolics include the tyrosine residues of extensin and possible other wall glycoproteins, the hydroxycinnamic acid residues (mainly ferulic and *p*-coumaric acids) of certain xylans and pectins, the *p*-hydroxybenzoic acid residues (of certain xylans?) and possibly the phenylpropanoid moieties of cutin [1,16,17]. That oxidative phenolic coupling reactions occur in the cell wall *in vivo* is evidenced by the finding in wall hydrolysates of phenolic dimers e.g. isodityrosine [18], diferulic acid [19]and various less well characterised substances exhibiting many of the properties of phenolic oxidation products [20]. The reactions basically resemble those of lignin synthesis, but they occur in non-lignified primary cell walls. In addition, phenolic dimerisation reactions may occur without the help of peroxidase [21]. If the phenolic oxidation involves an *inter*molecular coupling (and there is strong circumstantial evidence for this), then the process will tie together neighbouring polymer (polysaccharide and/or glycoprotein) molecules of the cell wall and this could clearly reduce its extensibility [3].

These observations suggest that it would be relevant to our understanding of embryogenesis to study the production and secretion of peroxidase, the incorporation of phenolic side-chains into wall polysaccharides, and the generation of extracellular H_2O_2 — the oxidant required for the coupling reaction.

2.2.3 Pectin methylesterase: This extracellular enzyme catalyses the hydrolysis of the methyl ester groups from pectic polysaccharide chains. It tends to do so in a concerted manner, attacking a series of contiguous D-galacturonic acid (GalA) residues, and probably stopping when it encounters an intervening L-rhamnose (Rha) residue. The result of the action of this enzyme on an initially highly esterified pectin molecule is therefore to yield a polysaccharide chain in which there are "runs" of acidic residues interspersed with "runs" of neutral residues:

```
 CH3   CH3   CH3   CH3   CH3   CH3        CH3   CH3   CH3   CH3   CH3   CH3   CH3
  |     |     |     |     |     |          |     |     |     |     |     |     |
...GalA-GalA-GalA-GalA-GalA-GalA-Rha-GalA-GalA-GalA-GalA-GalA-GalA-GalA...

+ 6 H2O →
  H     H     H     H     H     H         CH3   CH3   CH3   CH3   CH3   CH3   CH3
  |     |     |     |     |     |          |     |     |     |     |     |     |
...GalA-GalA-GalA-GalA-GalA-GalA-Rha-GalA-GalA-GalA-GalA-GalA-GalA-GalA...

+ 6 CH3OH
```

Pectin methylesterase has a high degree of specificity with respect to both the acyl moiety (which must be a sizeable oligomer of D-galacturonic acid; the trisaccharide is not adequate) and the alcohol moiety (methyl ≫ ethyl). However, its specificity for H_2O does not appear to have been tested thoroughly, and the possibility deserves examination that the enzyme can catalyse a transacylation reaction [22] analogous to the endotransglycosylation proposed for cellulase (see above). The hydrolytic action of pectin methylesterase would generate domains that were suitable for Ca^{2+}-bridging, and this would be expected to limit cell expansion; if transacylation also occurs, the result could be to form covalent cross-links (esters or secondary amides) which would strengthen the trend. Work is in progress to test the occurrence of such bonding.

2.3 Imposition of tissue coherency during morphogenesis

Similarly, the tight cell-to-cell contact characteristic of organised tissues is imposed by the polymeric "glues" of the cell wall and associated middle lamella. What chemical features characterise the tightly adhesive walls of organised tissues? An understanding of these features is not only relevant to attempts to increase cell-to-cell adhesion (e.g. in the induction of morphogenesis in callus and cell cultures), but also to attempts to decrease it (e.g. in the maintenance of well-disaggregated cell suspension cultures suitable for biochemical and mutant selection work).

The middle lamella is established at cytokinesis, and its deposition can be observed to result from the coalescence of vesicles, presumably laden with matrix polymers, at the phragmoplast [1,4]. Very soon in the process of cytokinesis, however, cellulose deposition resumes on the opposing plasma membranes of the two daughter cells. It is uncertain whether there is usually a clear layer of cellulose-free material in the middle lamella, or whether the deposition of cellulose usually begins so early that the microfibrils produced by the two daughter cells become very closely associated with each other and thus become bonded together by the matrix polymers such that ~~that~~ they effectively form a single wall [23]. The truth probably lies between these two extremes.

2.3.1 <u>Rôle of pectins</u>: Treatment of tissues with crude commercial preparations of pectinase often causes maceration, yielding isolated cells. Occasionally a similar result can be obtained with chelating agents. For this reason, the nature of the "glue" in the middle lamella area is widely believed to be mainly pectic — a mixture of α-D-galacturonic acid-rich polysaccharides capable of self-association via at least two different types of non-covalent junction-zone: one Ca^{2+}-dependent and the other (responsible for the gelling of jams) Ca^{2+}-independent [24]. Because of the difficulty of extracting pectic polysaccharides from the cell wall by non-degradative techniques (at room temperature and moderate pH), the possibility of various types covalent bonding has also been mooted [3,22], including (a) the oxidative coupling of the phenolic groups that occur on some pectins as side-chains, and (b) the participation of the —COOH group of D-galacturonic acid in ester or secondary amide linkages. The production of any of these bonds could be a factor governing the cohesion of tissue necessary for morphogenesis and therefore deserves thorough exploration.

2.3.2 <u>Rôle of extensin</u>: Extensin, a glycoprotein, is another major component of the primary cell wall matrix in cultured Dicot cells. However, it seems unlikely to play a significant rôle in maintaining cell/cell adhesion because extensin molecules do not span the middle lamella [25]. The extensin secreted by any given cell cannot reach the primary wall of its sister cell, and therefore seems unlikely to be responsible for anchoring a cell to its sister cell.

2.3.3 <u>Rôle of hemicelluloses</u>: The timing of the production of hemicelluloses in the middle lamella and following on from cytokinesis has not been rigorously studied [26]. However, the fact that hemicelluloses can hydrogen-bond strongly to cellulose may make them significant in bonding the microfibrils of one cell to those of its sister [7]. The kinetics, quantity and quality of hemicellulose deposition during the cell cycle therefore deserves re-investigation [26] as a possible control point in embryogenesis and organogenesis.

2.3.4 <u>The possible rôle of a "cuticle"</u>: It has been noted that embryoids, even in the very early globular stages, possess an external structure resembling a cuticle [27]. This zone is apparently diagnostic of embryogenesis, and characterised by the presence of phenolic-reacting material (as is cutin [17]). The possibility that this pseudo-cuticle is a necessary part of embryogenesis, restricting cell expansion and cell separation in the embryoid, deserves careful attention.

3 RÔLES OF CELL WALL FRAGMENTS IN THE REGULATION OF MORPHOGENESIS

3.1 Background

At a very different level, morphogenesis may also be regulated by oligosaccharide fragments of the cell wall, some of which accumulate spontaneously in spent culture media. The general significance of extracellular glycoconjugates in numerous control processes is widely recognised [1,28-35]. Evidence has, for example, been presented for the control of cell elongation by molecular fragments derived from xyloglucan. Recent research on the mode of action of growth-promoting xyloglucan fragments suggests that they activate cellulase, a wall-bound enzyme whose possible significance has been mentioned above.

An oligosaccharin is a particular oligosaccharide that exerts biological effect(s) on plant cells other than through its rôle as a carbon and/or energy source. Strictly, this definition might include sucrose, whose effect on morphogenesis often differs considerably from that of glucose (despite the fact that sucrose and glucose are presumably both available to the cells as carbon sources), and possibly maltose, which permits much more regeneration in certain tissues than does glucose or other commonly used carbon sources. However, oligosaccharins as ordinarily considered are the soluble (relatively low-M_r) products of the partial degradation of cell wall polysaccharides [34]. All known oligosaccharin activities have been discovered in factitious oligosaccharides produced by the *in-vitro* action of enzymes, acids or alkalis on isolated cell walls or polysaccharides. However, at least one oligosaccharin orginally discovered in this way (XG9 — see below) has subsequently been found to occur in spent culture media [36]. It will be important in the next few years to see what other oligosaccharides occur in spent media and whether any of them are (new?) oligosaccharins.

3.2 Xyloglucan oligosaccharides (XGOs) as growth regulators

3.2.1 Inhibition of elongation growth: The prolonged action of cellulase on xyloglucan generates a limited number of specific oligosaccharides (XGOs) mainly composed of the sugars D-glucose (Glc), D-xylose (Xyl), D-galactose (Gal) and L-fucose (Fuc) [35]. One of these, XG9 {Glc_4·Xyl_3·Gal·Fuc}, can antagonise the growth-promoting action of 2,4-dichlorophenoxyacetic acid (2,4-D) in the standard pea stem segment straight-growth bioassay [38-41]. It exerts its action optimally at 1-10 nM, and loses activity (i.e. restores 2,4-D-promoted growth!) at 0.1-1.0 μM. The growth-inhibiting activity is dependent on the fucose residue, as shown by the inactivity of XG8 {Glc_4·Xyl_3·Gal} and XG7 {Glc_4·Xyl_3} [40]. It is independent of the terminal xylose residues as shown by the activity of 10 nM XG5 {Glc_2·Xyl·Gal·Fuc} and 2'-fucosyl-lactose {Glc·Gal·Fuc; obtained from human milk} [39]. XG9 also antagonises the growth-promoting action of pH-4.5 buffer [42]. Other than this, the means by which XGOs antagonise the action of 2,4-D are unknown.

3.2.2 Promotion of elongation growth: Certain XGOs will also *promote* elongation in segments from etiolated pea stems [14] and, to this extent, they mimic auxin. This effect differs in at least two major respects from the growth inhibition mentioned above: (a) optimal growth-promoting action is seen at ~1 μM XGO; and (b) it is most effective with XG9n {$Glc_4 \cdot Xyl_3 \cdot Gal_2$} and XG8, although it is also seen with XG7 and XG9. It thus seems likely that the growth-promotion and the growth inhibition involve very different mechanisms.

The growth promotion appears to be related to an ability of XGOs to "activate" plant cellulases in their action on xyloglucan. This effect is seen in *in-vitro* viscometric assays: it is only seen with xyloglucan as substrate (not carboxymethylcellulose), and only with plant cellulases (not fungal) [14,43]. The effect therefore seems to have evolved to rather precise specifications. Most significantly, XG9n and XG8 are more effective than XG7 and XG9 in this cellulase "activation" assay.

The "activation" of cellulase by XGOs is not a simple promotion of the hydrolysis of xyloglucan chains, as the same effect is not observed when the cellulase's action is monitored by the production of new reducing termini [14]. This suggests that the effect is a promotion of endotransglycosylation — see above. It remains to be seen whether the added XGOs themselves participate in transglycosylation during these assays. It is interesting that studies of both the mechanical processes of wall loosening and the oligosaccharin control thereof lead to the same end-point: endotransglycosylation. It seems possible that what started out as a mechanical process has been refined (evolutionarily) into a control process.

3.3 Other oligosaccharins relevant to morphogenesis

Oligogalacturonides have also been reported to influence auxin-induced growth, although at higher (~mM) concentrations [44]. Oligogalacturonides are also well known to elicit the synthesis of phytoalexins in many plant cells [32,45]; it remains to be seen whether this evocation of secondary metabolism [46] has consequences for embryogenesis.

Other, less well defined oligosaccharins have been reported to influence morphogenesis in thin cell layers, swaying the balance between tendencies to generate shoots, roots, flowers and calli [47]. It is urgent to extend this work and to define the structures of the oligosaccharides (or other substances) involved in this system.

4 CONCLUSIONS

Cell walls provide the nuts and bolts by which tissues, organs and embryoids are built up. Understanding the control of this building process requires a more thorough knowledge than we currently possess of the chemical architecture of the cell wall. The time is ripe for a concerted and sustained effort (a) to define the interpolymeric cross-links that hold together the structural molecules of the cell wall, (b) to explore the processes which make and break these cross-links, and (c) to investigate how these processes can be regulated to manipulate the behaviour of cells in culture.

Interestingly, it is also becoming clear that molecular fragments of the cell wall, sometimes at vanishingly low concentrations, can themselves *regulate* the wall building processes. Few of these oligosaccharins have been studied in any detail, and basic questions remain wide open such as: (a) how many different processes can be regulated by how many different oligosaccharins? (b) what controls the production of oligosaccharins *in vivo*? (c) how do oligosaccharins act *in vivo*?

The possibility, mentioned earlier, that endotransglycosylation is central to both the wall building processes and to a sophisticated regulatory mechanism is particularly intriguing.

ACKNOWLEDGEMENTS

The author is grateful to Mrs J.G. Miller for excellent technical assistance, and thanks the Agricultural and Food Research Council for financial support of his research.

REFERENCES

1 Fry, S.C. (1988) The Growing Plant Cell Wall: Chemical and Metabolic Analysis, 333 pp., Longman, London {ISBN 0-582-01897-8}
2 Street, H.E. (ed.) (1977) Plant Tissue and Cell Culture, 2nd edn., 614 pp., Blackwell, Oxford {ISBN 0-632-09010-3}
3 Fry, S.C. (1986) Ann. Rev. Plant Physiol. 37, 165-186
4 Frey-Wyssling, A. (1976) The Plant Cell Wall, Gebr. Borntraeger, Berlin
5 Hanke, D.E. and Northcote, D.H. (1974) J. Cell Sci. 14, 29-50
6 Cleland, R.E. (1981) in Encyclopedia of Plant Physiology, Vol 13B, (Tanner, W. and Loewus, F.A., eds.), pp. 255-276, Springer, Berlin
6a Taiz, L. (1984) Ann. Rev. Plant Physiol. 35, 585-657
7 Fry, S.C. (1989) Physiol. Plant. 75, 532-536
8 Hayashi, T., Wong, Y. and Maclachlan, G. (1983) Plant Physiol. 75, 605-610
9 Byrne, H., Christou, N.V., Verma, D.P.S. and Maclachlan, G.A. (1975) J. biol. Chem. 250, 1012-1018
9a Davies, E. and Özbay, O. (1980) Z. Pflanzenphysiol. 99, 461-469
10 Edwards, M., Dea, I.C.M., Bulpin, P.V. and Reid, J.S.G. (1986) J. biol. Chem. 261, 9489-9494
11 Baydoun, E.A-H. and Fry, S.C. (1989) J. Plant Physiol. 134, 453-459
12 Smith, R.C. and Fry, S.C. (1989) in Fifth Cell Wall Meeting Book of Abstracts and Programme (Fry, S.C., Brett, C.T. and Reid, J.S.G., eds.),

Abstract No. 139, Scottish Cell Wall Group, Edinburgh
13 Nilsson, K.G.I. (1988) Tibtech 6, 256-264
14 McDougall, G.J. and Fry, S.C. (1990) Plant Physiol., in press.
15 Fry, S.C. (1979) Planta 146, 343-351
15a Negrel, J. and Lherminier, J. (1987) Planta 172, 494-501
16 Fry, S.C. (1986) in Molecular and Physiological Aspects of Plant Peroxidases (Greppin, H., Penel, C. and Gaspar, T., eds.), pp. 169-182, University of Geneva Press
17 Riederer, M. and Schönherr, J. (1988) Planta 174, 127-138
18 Fry, S.C. (1982) Biochem. J. 204, 449-455
19 Markwalder, H-U. and Neukom, H. (1976) Phytochemistry 15, 836-837
20 Fry, S.C. (1984) Phytochemistry 23, 59-64
21 Hartley, R.D., Whatley, F.R. and Harris, P.J. (1988) Phytochemistry 27, 349-351
22 Brown, J.A. and Fry, S.C. (1989) in Fifth Cell Wall Meeting Book of Abstracts and Programme (Fry, S.C., Brett, C.T. and Reid, J.S.G., eds.), Abstract No. 156, Scottish Cell Wall Group, Edinburgh
23 Jeffree, C.E., Dale, J.E. and Fry, S.C. (1986) Protoplasma 132, 90-98
24 Jarvis, M.C. (1984) Plant Cell Env 7, 153-164
25 Stafstrom, J.P. and Staehelin, L.A. (1988) Planta 174, 321-332
26 Amino, S., Takeuchi, Y. and Komamine, A. (1984) Physiol. Plant. 60, 326-332
27 Olesen, P., Buck, E. and Keimer, B. (1988) in Sexual Reproduction in Higher Plants (Cresti, M. *et al.*, eds.), pp. 107-112, Springer, Berlin
28 De Vries, S.C., Booij, H., Janssens, R., Vogels, R., Saris, L., LoSchiavo, F., Terzi, M. and van Kammen, A. (1988) Genes & Development 2, 462-476
29 Knox, J.P., Day, S. and Roberts, K. (1989) Development 106, 47-56
30 Hahne, G. and Lörz, H. (1988) J. Plant Physiol. 132, 345-350
31 Waffenschmidt, S., Spessert, R. and Jaenicke, L. (1988) Planta 175, 513-519
32 Bolwell, G.P., Robbins, M.P. and Dixon, R.A. (1985) Eur. J. Biochem. 148, 571-578
33 Brockaert, W.F. and Peumans, W.J. (1988) Physiol. Plant. 74, 740-744
34 Darvill, A.G. and Albersheim, P. (1985) Scientific American
35 Fry, S.C. (1989) J. exp. Bot. 40, 1-11
36 Fry, S.C. (1986) Planta 169, 443-453
37 York, W.S., Darvill, A.G. and Albersheim, P. (1984) Plant Physiol 75, 295-297
38 McDougall, G.J. and Fry, S.C. (1988) Planta 175, 412-416
39 McDougall, G.J. and Fry, S.C. (1989) J. exp. Bot. 40, 233-238
40 McDougall, G.J. and Fry, S.C. (1989) Plant Physiol. 89, 883-887
41 Emmerling, M. and Seitz, H-U. (1989) in Fifth Cell Wall Meeting Book of Abstracts and Programme (Fry, S.C., Brett, C.T. and Reid, J.S.G., eds.), Abstract No. 47, Scottish Cell Wall Group, Edinburgh
42 Lorences, E.P., McDougall, G.J. and Fry, S.C. (1990) Physiol. Plant., in press
43 Farkas, V. and Maclachlan, G. (1988) Carbohydr. Res. 184, 213-219
44 Branca, C., De Lorenzo, G. and Cervone, F. (1988) Physiol. Plant. 72, 499-504
45 Nothnagel, E.A., McNeil, M., Albersheim, P. and Dell, A. (1983) Plant Physiol. 71, 916-926
46 Funk, C., Gügler, K. and Brodelius, P. (1987) Phytochemistry 26, 401-405
47 Tran Thanh Van, K., Toubart, P., Cousson, A., Darvill, A.G., Gollin, D.J., Chelf, P. and Albersheim, P. (1985) Nature 314, 615-617

SIGNALS INFLUENCING CELL ELONGATION, CELL ENLARGMENT, CELL DIVISION AND MORPHOGENESIS.

K. TRAN THANH VAN AND S. MUTAFTSCHIEV (1),
Laboratoire de Physiologie végétale, C.N.R.S. 91.190 Gif sur Yvette FRANCE
Centre de Biochimie et de Biologie moléculaire végétale, C.N.R.S. 31, Chemin Joseph Aiguier, 13009 Marseille FRANCE (1)

1. INTRODUCTION

Cell elongation, enlargment, division and morphogenesis were shown to be affected by signals of diverse nature such as plant growth regulators, pH, light, temperature...Recent data have shown that cell wall oligosaccharides influence i) cell enlargment, division and organogenesis in tobacco thin cell layer (1) and ii) elongation on wheat coleoptile segments (2).

It was hypothesized that factors such as pH, auxin, cytokinin (1,3) and ionic extra/intracellular environment (3), can among other effects, activate specific hydrolytic cell wall enzymes which could in turn, release biologically active oligosaccharides. These, by mechanisms which are still unknown, affect cell growth and developmental processes.

In order to study these mechanisms, the identification and purification of the biologically active oligosaccharides are a prerequisite step. To allow this step to be acheived, biological systems with reproducible responses are the only recourse. Several systems, pea internodes (4), Lemna plants (5) were used. We have developed the two following systems: the tobacco thin cell layers (TCL) which have been proved to be a suitable model system (though not yet ideal) for the study of the effect of oligosaccharides on cell enlargment, division and organogenesis (1) and ii) wheat coleoptile segments on which elongation can be induced by either IAA alone or cell wall oligosaccharides alone (2). Another difficult problem to be also solved first was the range of concentrations to be used since the oligosaccharides currently available were not purified and the composition of the mixture can differ from one preparation to the other . There was no solution to this vicious circle. In fact, our first trials for which we were advised to use concentrations hundred time higher then the active ones, were negative. This is to mention the number of difficulties which must be overcome when studing the biological effects of compounds not yet well identifed but highly interesting.

In this paper, we report i) the influence of oligosaccharides on cell division and organogenesis in tobacco TCL in relation to *pH variations during the time course of culture,* ii) the effect of *the degree of polymerisation* of oligosaccharides on the elongation of wheat coleoptile segments and iii) the comparison of the *IAA-induced* and of the *oligosaccharides-induced elongation* in relation to a possible involvment of *cell wall hydrolytic enzymes* acting in cell wall loosening process such as β 1-3 endoglucanase. Its implication was investigated using specific antibodies.

2. MATERIALS AND METHODS

Sycamore cell wall fragments released by endopolygalacturonase (EPG fragments), or by basic hydrolysis (B fragments) were provided by P. Albersheim's group, CCRC, Athens. Cell wall fractions from *Rubus fruticosus* cell suspensions were provided by J. P. Joseleau's group, CERMAV, Grenoble.

The IAA or oligosaccharide induced elongation/ enlargment was performed on wheat coleoptile segments and tobacco TCL systems which were previously described (1,2).

The effect of enzyme/antibody interaction on IAA and oligosaccharide-mediated growth was studied. Different series of 10 segments (8 mm length) of coleoptile and 8 TCL explants (5mm length), were preincubated during 60 minutes in Na citrate buffer (1mM), pH 5.8. This step was followed by an incubation during 90 minutes in rabbit antiserum (provided by B. Fritig's group, IBMP, Strasbourg) against tobacco β 1-3 glucanase (Agl) and nonimmun serum (Nis). The dilutions used were: 1/1000 and 1/5000 . Coleoptile segments and TCL explants were then allowed to grow on buffer solutions containing IAA (10^{-5} M), B fragments (10^{-5} mg /ml and 10^{-6} mg/ml). The elongation or enlargement was measured after 20h (for coleoptiles) and 48h (for tobacco TCL) of incubation at 25 °C in the dark. Results expressed were the average from 4 separates experiments for wheat coleoptile and from 3 separate experiments for tobacco TCL.

3. RESULTS AND DISCUSSION

3.1. Influence of culture medium-pH on cell division and morphogenetic effects of oligosaccharides in tobacco TCL.

Results shown in Table 1 indicated i) the variation of pH at day 0, 7 and 10 in medium buffered or not by 50 mM of MES (Morpholino-ethane sulphuric acid) and ii) the consecutive changes induced by EPG fragments supplied (from 3.10^{-5} mg/ml to 10^{-7} mg/ml) to root inducing medium containing IBA (7μM) and kinetin (0.2 μM). When the medium was not buffered, tobacco TCL formed 100 % of root with an average of 15 roots per TCL explant. On a buffered medium, no root was formed, flowers (30 % of TCL explant) mixed to vegetative buds (70 %) were obtained.

Addition of 10^{-7} or 10^{-6} mg/ml of EPG fragments did not induce any change in the organogenesis on unbuffered medium. At 10^{-5} mg/ml and 2.10^{-5} mg/ml, the percentage of TCL explant forming roots decreased while the percentage TCL explant forming callus increased. In contrast, addition of EPG fragment at 10^{-7} mg/ml increased the percentage TCL explant forming flowers. At 10^{-6} mg/ml, 80 % of TCL explant formed flowers, 10 % floral shoot and 10 % callus. The percentage of TCL explant forming callus increased up to 70 % and floral shoot replaced flowers.

The following remarks could be drawn from these results. Organogenic differentiation`depends on culture medium pH variations and possibly on intracellular pH. The effect of oligosaccharide was pH and dose dependant.

Previously obtained results (not shown) have shown that the effect of oligosaccharides depends on the relative concentration of IBA and kinetin and on the time at which oligosaccharides were added to the culture medium. This suggests that there may exist a sensitive phase towards oligosaccharides and that this sensitive phase could be related to the extra and intracellular pH.

TABLE I. Influence of EPG fragments and pH variations* on the in vitro control of morphogenesis** in tobacco thin cell layer explants.

Initial medium pH	MES (mM)	pH at day			EPG fragments (mg ml^{-1})					
		0	7	10	0	10^{-7}	10^{-6}	10^{-5}	2.10^{-5}	3.10^{-5}
5.95	0	5.95	4.9	5.30	100R_{15}	100R_{15}	100R_{15}	60±10 R_8 40±10 C	40±5 R_8 60±5 C	nd
5.95	50	5.95	5.92	5.90	30±5 F_2 70±5 B_3	60±5 F_2 40±5 B_5	80± 5 F_6 10±5 F_s 10±5 C	40±5 F_s 20±5 B_2 60±5 C	50± 5 F_s 60±5 C	45±5 F_s 70± 5 C

* Medium containing IBA (7μM) and kinetin (0.2μM).

** in percentage of explants forming roots (R), Flowers (F), Floral shoots (F_s), vegetative buds (B), callus (C).

MES : Morpholino ethane sulphuric acid.

Subscripts give the average number of organs formed per explant.

nd.: not determined.

[after Mutaftschiev, Cousson and Tran Thanh Van]

3.2. Influence of the oligosaccharides-degree of polymerization on wheat coleoptile elongation.

Cell-wall fragments purified from Rubus fructicosus cell suspensions were tested for their activity on cell elongation according to their degree of polymerization (from Dp 3 to Dp 10) at concentrations ranging from 10^{-1} to 10^{-9} mg/ml. Figure 1 shows results obtained with Dp 7, 8 and 9. Polymers of Dp 7 and 9 were found to be active at 10^{-9} mg/ml and at 10^{-8} mg/ml respectively while the other polymers were not active (results not shown except for Dp 8).

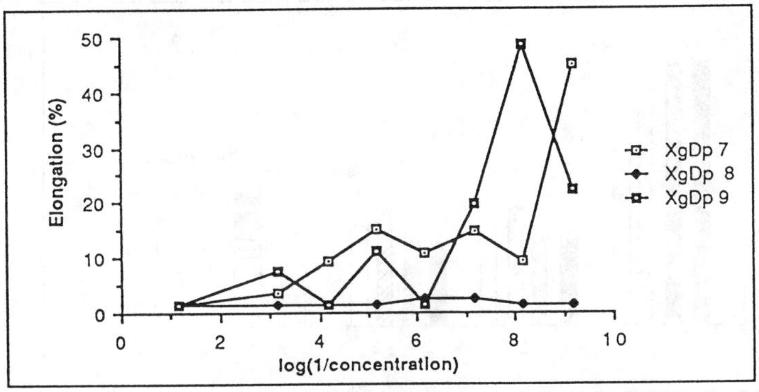

Figure 1. Wheat coleoptile elongation induced by decreasing concentrations (10^{-1}to10^{-9} mg/ml) of oligosaccharide polymers (different degree of polymerization). Results are the average means of 5 separate experiments.

3.3. Influence of antibodies

The results obtained showed that elongation or enlargement induced by IAA or by oligosaccharides were differently affected by antiglucanase antibodies. These antibodies, when supplied alone, had no impact on cell elongation of wheat coleoptile and tobacco TCL. IAA-induced elongation of wheat coleoptile was inhibited in presence of 1/1000 and 1/5000 of antiglucanase antibodies (Fig. 2), while oligosaccharide- induced elongation was stimulated. The inhibition of IAA-mediated elongation of wheat coleoptile segments was accompanied by characteristic curvatures suggesting that the elongation was inhibited at the sites of binding of antibodies to the β 1-3 endoglucanase (Fig. 3. a and b). As this enzyme was purified from tobacco plantcs, we have extended this study to tobacco TCL.

On tobacco TCL, similar responses were observed. In the presence of antiglucanase antibodies (1/5000), enlargement as well as a slight elongation were initiate. The enlargment induced by EPG fragments was amplified

1.7 times after a treatment by the glucanase antibodies (1/5000). The IAA alone did not affect TCL enlargment. However, an important enlargment was inhanced (1.7 times) after a treatment by the same antibodies. One striking observation was that the TCL showed an important curvature with a convexe epidermal side due to a dissymetric growth pattern: the elongation of the cortical side of the TCL explant being inhibited. This dissymetric pattern of elongation could be due to the antibody binding to the cortical cells and not to the epidermis side because of the presence of the cuticule.

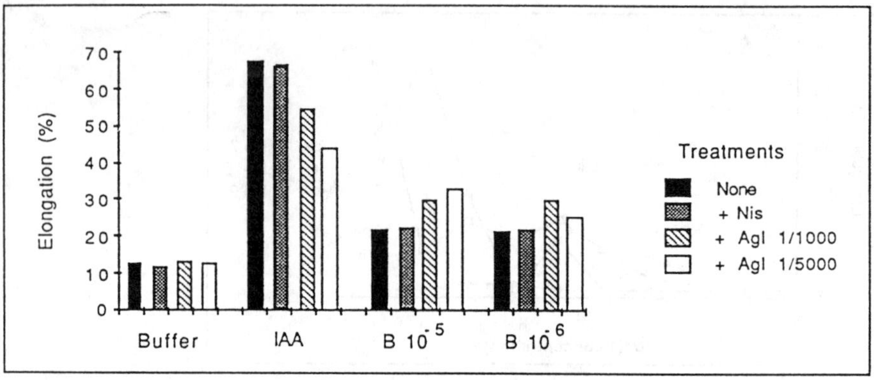

FIGURE 2. Effect of β 1-3 endoglucanase antibodies on auxin-induced and cell wall oligosaccharide-induced elongation of wheat coleoptile segments. Series of segments were incubated in Na citrate buffer (1 mM, ph 5,8) , nonimmun serum (Nis) and two dilutions (1/1000 and 1/5000) of β 1-3 endoglucanase rabbit antiserum (Agl). Samples were subsequently treated with IAA 10^{-5}M, oligosaccharide fragments B 10^{-5} mg/ml and 10^{-6} mg/ml .

It seems that in both systems, IAA- induced elongation may, at least in part, process through glucanase which may act by itself, causing cell wall loosening. Glucanase activity was previously shown to be affected by auxin and cytokinin (6). On the other hand, oligosaccharides also initiate elongation, but there was no curvature in any of the two systems studied. Furthemore, the oligosaccharide-induced elongation was not inhibited by antiglucanase antibodies. Such enzyme/ antibody interaction could result in a decrease in the *in situ* production of active cell wall fragments. Thus,the increase in the elongation rate observed in wheat coleoptiles pretreated with glucanase antiserum and with oligosaccharides, could be due to an optimal concentration of exogenous oligosaccharides supplied. Endogenous formation of oligosaccharides was demonstrated in spinach cell suspensions (7).

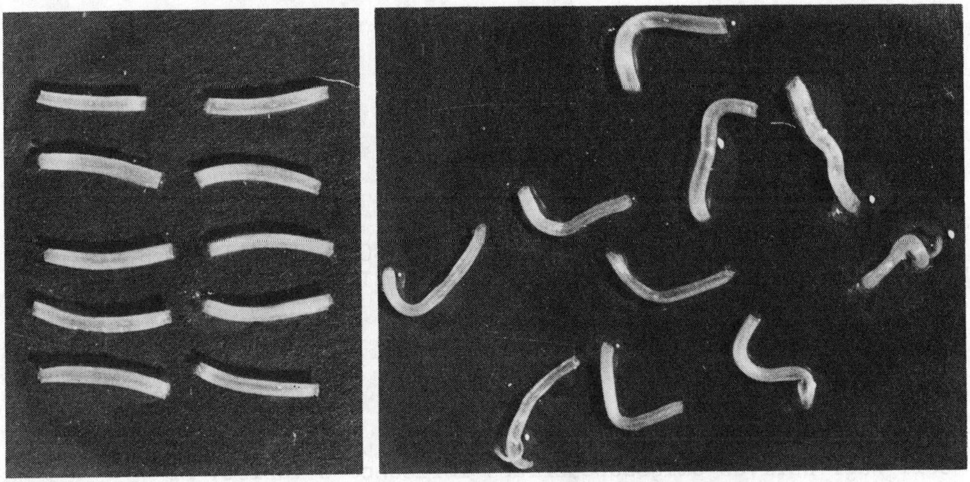

FIGURE 3. a and b. Wheat coleoptile segments treated with:
a) IAA 10^{-5} M
b) rabbit β-endoglucanase antiserum and IAA 10^{-5} M

4. CONCLUSION

The effects of oligosaccharides are influenced by the variation of the culture medium pH which can affect the intracellular pH, the ionic environment and the activity of hydrolytic cell wall enzymes.

Part of the mechanisms of action of IAA and of oligosaccharides on cell elongation seems to proceed through different pathways: IAA can cause elongation via, in particular, the glucanase activity, whereas effect of exogenous oligosaccharides on elongation or enlargment seems not to be affected by the enzyme/antibody interaction.

Among the different xyloglucan oligomer tested, Dp 7 and 9 were found to be active on wheat coleoptile elongation. The negative results obtained with Dp 8 call for further investigations since, as reported here, the effects of oligosaccharides could be masked by external/ internal pH variations, ionic microenvironment...The specific size requirement for the expression of biological functions suggests a specific signalling mechanism.

REFERENCES

1 Tran Thanh Van, K., Toubart, P., Cousson, A., Darwill, A. G., Gollin, D. J., Chelf, P. and Albersheim, P. (1985) Nature 314, 615- 617

2 Mutaftschiev, S., Cousson, A. and Tran Thanh Van, K. (1987) British Plant Growth Regulator Group, Monograph 16

3 Tran Thanh Van, K., Richard, L. and Gendy, A. in Nato Biolotechnology Series (Durzan, Rodriguez eds.) in press

4 York,W.S.,Darwill,A.G. and Albersheim,P.(1984)Plant Physiol. 75,295-297

5 Gollin,D.J., Darwill,A.G. and Albersheim, P. (1984) Biol. Cell 51, 275-280

6 Mohnen,D.,Shinshi,H., Felix,G.and Meins,F.J (1985) EMBO J. 4,1663-1635

7 Fry, S. C. (1986) Planta 169, 443-453

EFFECT OF A XYLOGLUCAN-DERIVED NONASACCHARIDE ON CARROT PROTOPLASTS

M. EMMERLING, H. WARNECK, H.U. SEITZ, BOTANISCHES INSTITUT DER UNIVERSITÄT, AUF DER MORGENSTELLE 1, 7400 TÜBINGEN, FRG

1. INTRODUCTION

Cell wall-derived oligosaccharides influence a variety of morphological and developmental processes in plants.

A nonasaccharide XG9 ($Glc_4\ Xyl_3$ Gal Fuc) can be obtained from xyloglucan by controlled action of cellulases *in vitro*. This nonasaccharide acts as an anti-auxin in 2,4-D promoted growth in etiolated pea stems with maximum inhibition at nanomolar concentrations (1;2).

We were interested in the influence of this XG9 on the regeneration of plant protoplasts. The main aim of this work was to confirm the anti-auxin activity of XG9 in a totally different biological system that requires the presence of 2,4-D.

2. MATERIAL AND METHODS

2.1. Cell cultures: *Daucus carota* L. suspension-cultured cells were maintained as previously described (3).

2.2. XG9-preparation, pea stem bioassay, protoplast preparation and culture as well as carbohydrate analysis were performed as described (4). Glycan synthases were assayed according to (5).

3. RESULTS AND DISCUSSION

XG9 was prepared from cell walls of 7 days old suspension-cultured carrot cells. The treatment of the 4M KOH-soluble portion of the cell wall preparation with KJ/J_2 resulted in the precipitation of 10-20% carbohydrate, most likely consisting of an arabinogalactan. The purified hemicellulose mainly composed of xylose, glucose and fucose was submitted to partial hydrolysis by the action of Cellulase Onozuka R 10. The 50% ethanol-soluble portion of the cellulase-digested hemicellulose was separated on Sephadex G 25 SF. The elution profile showed peaks eluting at K_{AV}-values approximately corresponding to starch-derived maltononaose, maltoheptaose, maltotriose, and maltose. The peaks corresponding to maltononaose and maltoheptaose were pooled and further characterized. To determine the purity of the nonasaccharide peak, a thin layer chromatography was performed. Two spots could be detected, one of which had a R_f equal to maltononaose and one similar to maltoheptaose. This second substance was only present in traces (up to 10%).

The nonasaccharide was then submitted to monosaccharide analysis. Glucose, xylose, galactose and fucose could be identified as the main sugars with only little contamination by arabinose and mannose. Monosaccharide analysis of the heptasaccharide fraction resulted in glucose and xylose (4.3:3 molecular ratio) and minor amounts of arabinose, fucose (ca. 1 mole-%), mannose and galactose (ca. 4 mole-%). All analytical data including the results of methylation analysis indicated that the nonasaccharide peak eluted from the Sephadex columns consisted almost exclusively of XG9 corresponding to the proposed structure (6;7).

To verify the identity of the isolated nonasaccharide as the biologically active XG9, we tested its effect in the pea stem bioassay in a concentration range from 10^{-11}M to 10^{-7}M (see Fig. 1). The experiments were performed with the pea cultivar "Feltham's First".

In all assays containing XG9 significant inhibition of 2,4-D-promoted growth was observed, exhibiting a sharp concentration optimum of the nonasaccharide varying between 10^{-10}M to 10^{-8}M depending on the preparation. In control assays lacking in exogenous 2,4-D the increase of pea stem growth was 1.2 to 2.0 mm, whereas the growth response in assays containing 1µM 2,4-D was 5.0 to 6.5 mm.

Our results proved the strong anti-auxin activity of the obtained XG9-preparation and confirmed former findings (2;1). These observations further support the hypothesis that xyloglucan and xyloglucan-fragments play an important role in growth regulation.

To confirm the anti-auxin activity of XG9 in a totally different biological system that requires 2,4-D, we chose the regeneration of carrot protoplasts. The success of protoplast regeneration partially depends on the presence of exogenous phytohormones. In our experiments 2,4-D was the only phytohormone in µM concentrations.

The results of the assays containing 2,4-D plus varying concentrations of XG9 from 10^{-7}M to 10^{-11}M were compared to controls regenerated with or without 2,4-D. Protoplasts cultivated for four days in medium containing 4.52µM 2,4-D showed a viability of 84.5% ($\pm$5%) as estimated using Phenosafranin, whereas the absence of 2,4-D lead to a decrease in viability to 62.1% ($\pm$2.7%). When protoplasts were cultivated in medium containing 4.52µM 2,4-D and XG9 concentrations ranging from 10^{-7}M to 10^{-11}M the viability was suppressed depending on the applied XG9 concentrations (see Fig. 2). Nanomolar concentrations lead to maximum suppression of viability of 75% compared to the control. At higher and lower concentrations of XG9 the effect was much less marked. These results indicate that the nonasaccharide-preparation was not cytotoxic. XG9 seems to influence the viability of protoplasts in a different manner than the absence of 2,4-D as can be concluded from the action of glucose-6-phosphate dehydrogenase, an enzyme of the primary metabolism. The activity of this enzyme was markedly reduced to 1/3 by the absence of 2,4-D while it retained its activity completly when the cells were cultivated in the presence of 2,4-D and varying concentrations of XG9.

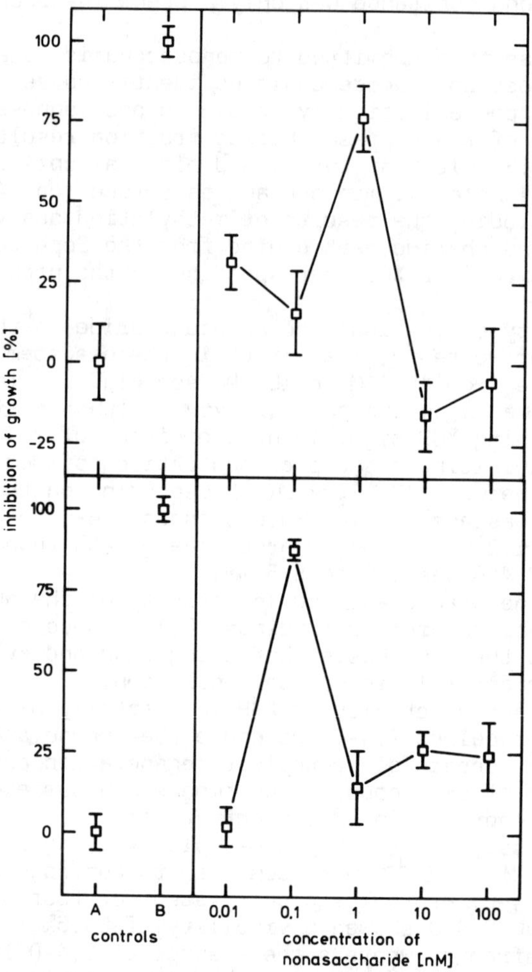

Figure 1. Biological activity of nonasaccharide preparations in the pea stem bioassay. Nonasaccharides were obtained in two independent experiments (above/below) and applied to pea stems in different concentrations. Control A was grown with 1µM 2,4-D, control B without 2,4-D.

The success of the regeneration of protoplasts not only depends on the viability of the cells, but also on their ability to synthesize a new wall. When the monosaccharide composition of the cell wall preparations was analyzed the sugar compositions displayed the typical patterns of regenerating protoplasts (four days). The only significant event was the 50% reduction of the mannose-content of the cell wall when 2,4-D was omitted. A significant effect of XG9 on the cell wall composition could not be pointed out.

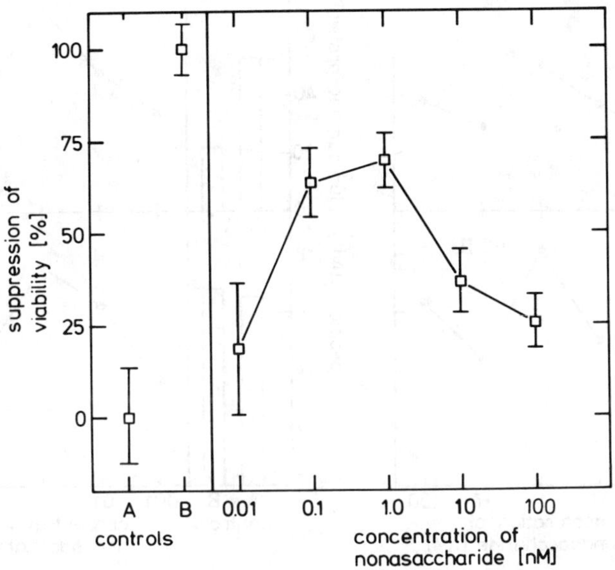

Figure 2. Influence of nonasaccharide on viability of regenerating carrot protoplasts. The viability of cells grown with 4,52μM 2,4-D was 84.5% $\pm$ 5% (control A), cells grown in the absence of 2,4-D showed a viability of 62.1% $\pm$ 2.7% (control B).

Figure 3 shows the effects of 2,4-D-supplementation and of XG9 in varying concentrations on the activities of glycan synthases likely to be involved in cell wall synthesis. Except for the GSI-G$^+$, an enzyme that <u>in vitro</u> synthesizes a glucomannan (5) and was shown to do so <u>in vivo</u> (8) the activities of the enzymes were increased to different extents by the omission of 2,4-D. The addition of XG9 resulted in increasing activities of all four enzymes examined, with maximum activation at nanomolar XG9 concentrations. Higher and lower concentrations were less effective. Lower concentrations (10^{-11}M) seemed to inhibit enzyme activities compared to the control experiments containing only 2,4-D.

All the presented results show that a xyloglucan-derived nonasaccharide preparation is able to influence several processes in regenerating carrot protoplasts. The effect of XG9 on protoplast viability and the activity of glycan synthases might be called auxin-antagonistic, but the anti-auxin properties of XG9 could not definitely be proved.

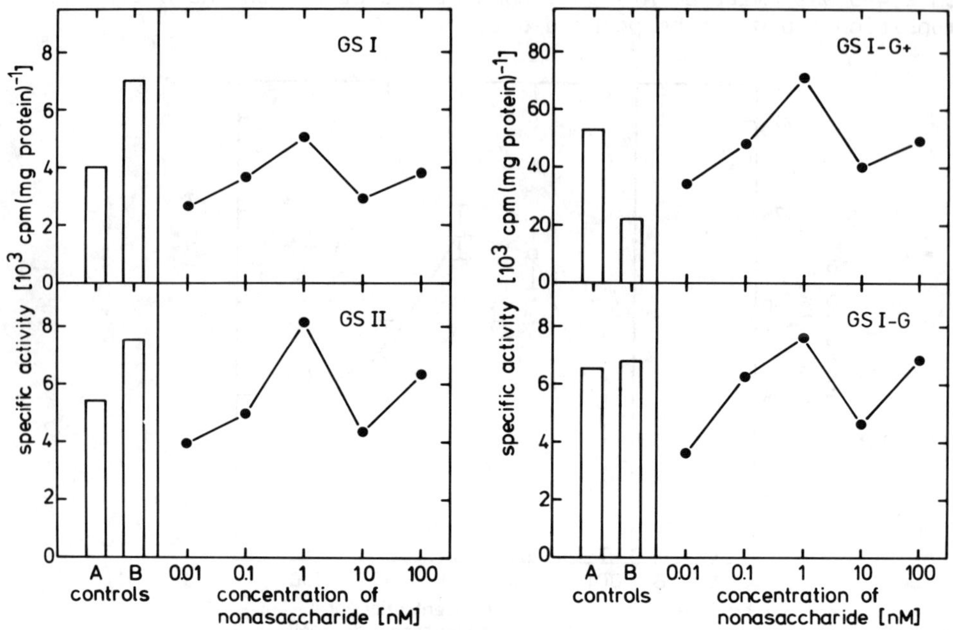

Figure 3. Influence of nonasaccharide on the activities of glycan synthases of regenerating protoplasts. Control A, cultivation in the presence of 2,4-D; control B in the absence of 2,4-D. The other assays were performed with 4,52µM 2,4-D plus varying concentrations of XG9. GS I = glucan synthase using µM UDP-Glucose; GS II = glucan synthase using mM UDP-Glucose; GS I-G = glucan µM GDP-Glucose; GS I-G^{+} = glycan synthase using µM GDP-Glucose and uM GDP-Mannose.

Since all phenomena exhibit an optimum concentration of XG9 of 10^{-9}M with higher concentrations being less effective a simple cytotoxic effect of XG9 can be excluded. The optimum concentration of 10^{-9}M is found for all XG9-effects including the anti-auxin activity in the pea stem bioassays. Further investigations to unravel the action of this oligosaccharide more precisely are in progress.

REFERENCES

1 McDougall, G.J. and Fry, S.C. (1988) Planta 175, 412-416
2 York, W.S., Darvill, A.G., Albersheim, P. (1984) Plant. Physiol. 75, 295-297
3 Seitz, H.U., Seitz, U., Alfermann, W. (1985) in Pflanzliche Gewebekultur. Ein Praktikum. Gustav Fischer, Stuttgart New York
4 Emmerling, M., Seitz, H.U., Planta, in press
5 Ingold, E., Quentmeier, H., Seitz, H.U. (1987) J. Plant Physiol. 129, 327-339
6 Kato, Y., Matsuda, K. (1980) Agric. Biol. Chem. 44, 1759-1766
7 Valent, B.S., Darvill, A.G. McNeil, M., Robertsen, B.K., Albersheim P. (1980) Carbohydr. Res. 79, 165-192
8 Emmerling, M. (1990) Ph.D. Thesis, University of Tübingen

CHARACTERIZATION OF GP57, AN AUXIN-REGULATED AND DERMAL-TISSUE-LOCALIZED EXTRACELLULAR GLYCOPROTEIN OF CARROT

SHINOBU SATOH, UNIVERSITY OF TSUKUBA, INSTITUTE OF BIOLOGICAL SCIENCES, TSUKUBA, IBARAKI 305, JAPAN

1. INTRODUCTION

Small carrot cell clusters suspended in auxin-containing medium can form somatic embryos at a high rate, when they are transferred to auxin-free medium. As a result of this finding, cultured carrot cells have been used as suitable material to investigate the mechanism of differentiation and biochemical events occurring during embryogenesis. In animals, some quantitative and qualitative difference in extracellular glycoproteins have been reported between certain organs and *in vitro* cultured cells. Some of these glycoproteins have been shown to play important roles in cell-to-cell interactions as growth factors or as matrix components, and consequently in the differentiation of cells.

I have reported variations in extracellular glycoproteins released by carrot cells which were cultured under different conditions (1). The appearance or disappearance of two glycoproteins of Mr 65,000 (GP65) and Mr 57,000 (GP57) was closely related to the formation of somatic embryos. GP65 was released from embryo-forming cells cultured without 2,4-dichlorophenoxyacetic acid (2,4-D). GP57 was released from the embryogenic cells, if they were cultured in a medium with 2,4-D, in which they cannot form somatic embryos. Auxin-autotrophic non-embryogenic cells, which formed callus with or without auxin, released GP57, but not GP65, even in the absence of auxin. Since the release of GP57 occurs only when auxin has endogenously or exogenously been supplied to the cultured cells, this system is suitable for studies of auxin action on either synthesis or release of glycoproteins, and for studying the physiological functions of glycoproteins specifically involved in morphogenesis.

I report here the biochemical and physiological properties of GP57, and discuss the implication of the results in the morphogenesis or the other physiological phenomena.

2. MATERIALS AND METHODS

2.1 Purification and chemical analysis: GP57 was purified by

DEAE-Sepharose, hydroxylapatite and Mono P column chromatographies from the 70%-ethanol precipitate of culture medium of non-embryogenic carrot cells (Daucus carota L. cv. US-Harumakigosun) grown with 2,4-D (1 mg/l)(2). Deglycosylation of the purified GP57 was performed using trifluoromethanesulfonic acid (TFMS). Determination of pI was done by isoelectric focusing on sucrose gradient. Amino-acid and sugar analysis were done as described before (2).

2.2 Immunological analysis: Monoclonal antibody (mAb) directed against the polypeptide of GP57 was prepared using purified GP57 as antigen (2). Immunoblotting and immunocytochemistry were done using the mAb (2).

2.3 Screening of functions: 1)(Growth-modulating activity) Protoplasts of non-embryogenic callus were feeded with [methyl-^{3}H]thymidine (2.5 µCi/ml) in the presence or absence of purified GP57 (0.1 mg/ml), anti-GP57 serum (5%) or anti-GP57 mAb (1 mg/ml), and then the incorporation into acid-insoluble DNA fraction was measured. 2)(Adventitious root formation) Hypocotyl cuttings with cotyledons from 10-day-old carrot seedlings were incubated with or without anti-GP57 mAb (0.1 mg/ml) or anti-GP57 ascitic fluid (10%) for 1-2 weeks, and then the adventitious root formation was observed. 3)(Lectin activity) Human (A,B,AB,O), rabbit or sheep blood was mixed with purified GP57 (1-100 µg/ml) and incubated at 25 °C, and then the aggregation was observed. 4)(Anti-microorganism activity) The culture medium of bacterium (Klebsiella aerogenes) or fungus (Cladosporium fulvum) was added with purified GP57, and the growth was observed after incubation. 5)(Enzyme activity) Purified GP57 or crude culture medium proteins were run on isoelectric focusing polyacrylamide gel, and then the active stainings for peroxidase and esterase were done using amino-ethyl carbazol and α+β-NAA as substrates, respectively.

2.4 cDNA cloning and northern blotting: cDNA was synthesized from the poly[A]RNA of 10-day-old seedling's roots using cDNA synthesis system plus (Amersham), and the cDNA library was made with λgt11 using cDNA cloning system (Amersham). The first screening was immunologically done using anti-GP57 serum. DNA was prepared from the positive clones and subjected to southern blotting using the oligonucleotides deduced from the two parts of amino-acid sequences of GP57 as probes. The cDNA of final positive clone was cloned into pUC118. The DNA was sequenced by the dideoxy chain termination method. The total RNA was extracted from cultured cells or plant parts by SDS-phenol and subjected to northern blotting using the cloned cDNA fragment as a probe.

3. RESULTS AND DISCUSSION

3.1 Chemical properties: GP57 was purified by the sequential column chromatographies with a yield of about 0.5 mg/l culture medium of non-embryogenic callus cultured with 2,4-D. Its apparent Mr was estimated to be 57,000 by SDS-PAGE and 50,000 by gel filtration. GP57 contained 14% (w/w) carbohydrate; the Mr of the peptide portion was estimated to be 55,000 after deglycosylation by TFMS. GP57 is composed of two polypeptides with the same Mr and with very similar amino-acid composition but different pI values, 8.8 and 9.5. Both are rich in aspartic acid, serine and threonine, and may possess N-linked oligosaccharide chains, including fucose and xylose. The extinction coefficient was 12.4 at 280 nm and there was no absorbance in the visible region.

3.2 Localization: Because GP57 had xylose-containing glycans which was highly immunogenic, mAb against polypeptide of GP57 was prepared and used for all experiments. Immunoblotting of medium proteins (Fig. 1) shows that GP57 was released to the media from non-embryogenic callus cultured both with or without 2,4-D. Additionally, GP57 was released from calli-forming embryogenic cells incubated with 2,4-D. No GP57 was released from embryogenic cells incubated without 2,4-D which formed somatic embryos. Only a small amount of GP57 was detected in both embryogenic and non-embryogenic callus itself cultured with 2,4-D (data not shown). This means that almost all of GP57 was released into the medium, not retained in the cell wall of cultured cells. Because non-embryogenic callus is thought to be auxin-autotrophic, the synthesis of GP57 may be regulated by auxin endogenously or exogenously supplied in the culture system. GP57 was also detected in the Ti-plasmid induced crown galls and Ri-plasmid induced hairy roots (data not shown).

Subsequently, the presence of GP57 in intact plant was tested by immunoblotting and immunocytochemistry. At the start of life cycle, GP57 was localized in the interspace between embryo and endosperm of dry seed, especially around radicle (Fig. 2 A). After germination, the amount of GP57 in the seedling decreased gradually until 4 days after sowing, but GP57 appeared again in the epidermis and endodermis of basal part of young root (B), not in the apical part. In three-month-old plants, GP57 was found in the periderm of the taproot (C) and in the epidermis of the petiole of mature leaves, but not in the phloem or xylem.

3.3 Functions: From the localization pattern of GP57, three putative functions, growth factor, anti-microorganism substance, cell-wall-forming enzyme, were proposed.

1) Because the presence of GP57 was correlated to unorganized callus-state of cells, effect of GP57 on the

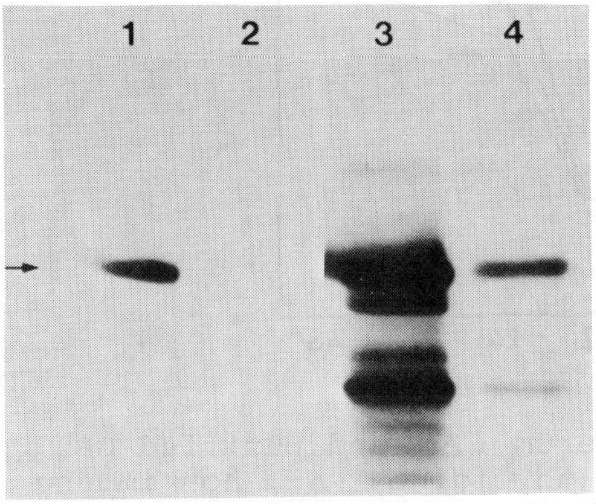

Fig. 1. Detection of GP57 (arrow) in the medium proteins of embryogenic callus with (1) or without (2) 2,4-D, and of non-embryogenic callus with (3) or without (4) 2,4-D.

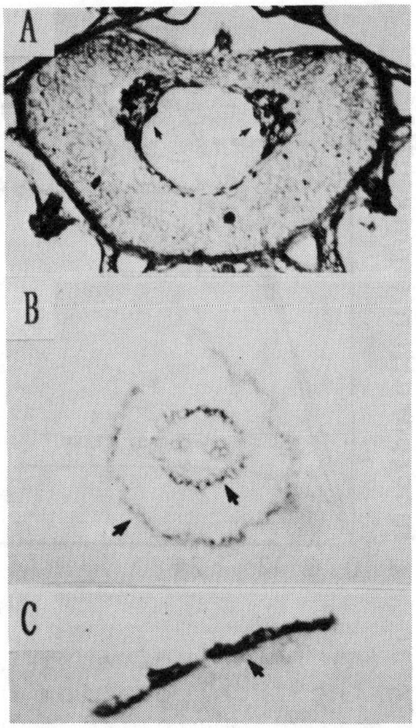

Fig. 2. Immunocytochemical detection of GP57 (arrows) in dry seed (A) and root of seedling (B) or of three-month-old plant(C).

callus growth was tested. As shown in Fig. 3 A, anti-GP57 serum strongly inhibited the incorporation of [^{3}H]thymidine in the protoplast of callus. However, purified GP57 or mAb against polypeptide showed no effect on it. Because the antiserum was raised against whole GP57 containing highly immunogenic glycans, there may be two possibilities that a) anti-glycan antibody bound to the glycans of cell membrane proteins and inhibited the cell functions, b) GP57 was indispensable for the growth of callus.

2) The mAb did not affect the adventitious root formation from the hypocotyl cutting and growth of root.

3) Because GP57 was localized in dermal tissues and around embryo in seed, GP57 may play role in defense mechanisms. However, no lectin, anti-bacterium or anti-fungus activity was found in purified GP57.

4) Because the dermal tissues are known to be sometimes suberized, some suberin-related enzymes were tested. However, no peroxidase and esterase activity was detected in GP57.

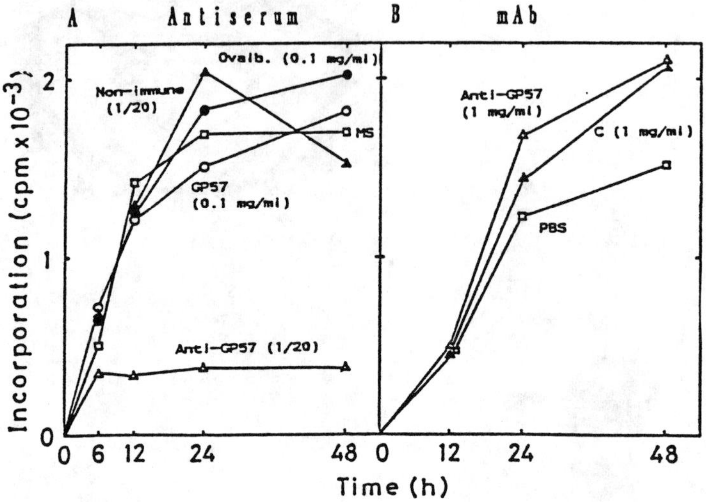

Fig. 3. (A) Effect of anti-GP57 serum (△) and purified GP57 (○) on the incorporation of [^{3}H]thymidine. ▲, non-immune serum; ●, ovalbumin; □, none. (B) △, anti-GP57 mAb; ▲, control mAb; □, phosphate buffered saline.

3.4 Molecular biological approach: The function of GP57 is still unknown after the screening described above. More careful and wider screening should be needed. On the other hand, knowledge about the structure of the protein should be important for the elucidation of biochemical and

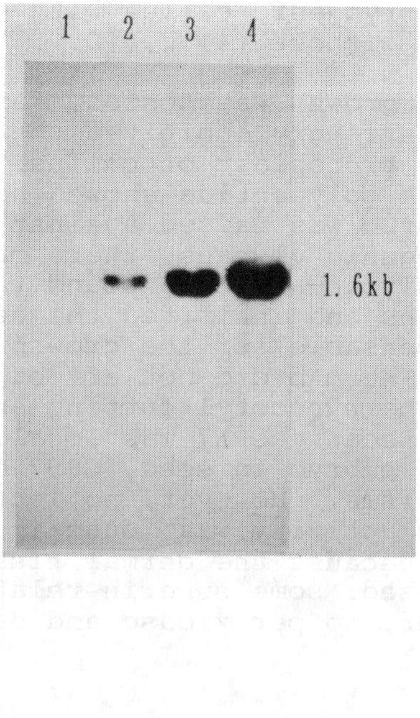

Fig. 4. Northern blotting of total RNA from somatic embryo (1), embryogenic callus (2), non-embryogenic callus (3) or young root (4) with GP57 cDNA.

physiological function of GP57. Therefore, the cDNA library of young carrot root was screened, and 1.5 kbp cDNA clone which reacted to anti-GP57 serum and two kinds of oligonucleotides was obtained. Fig. 4 shows the northern blotting using the cloned cDNA as a probe. A 1.6 kb band was detected strongly in root and non-embryogenic callus RNAs, weakly in embryogenic callus RNA, and very weakly in somatic embryo RNA. The distribution of RNA was the same as that of protein. The sequence analysis of 1 kb coding region of the cDNA showed no homology in the known proteins. Further, the full length clone was obtained from the cDNA library of non-embryogenic callus and the sequencing is under going. If no sequence homology will be found between GP57 and known proteins, the analysis of transgenic carrot plants, whose GP57 is overexpressed or supressed, should be required for the understanding of the function of GP57.

The cDNA cloning work was done in parts with Dr. A. Sturm and Prof. M.J. Chrispeels in University of California, San Diego.

4. ACKNOWLEDGEMENT

This work was supported in parts by Grant-in-Aid for Special Research on Priority Areas (01660002, Cellular and Molecular Basis for Reproductive Processes in Plants) from the Ministry of Education, Science and Culture, Japan.

5. REFERENCES

1 Satoh, S., Kamada, H., Harada, H. and Fujii, T. (1986) Plant Physiol. 81, 931-933
2 Satoh, S. and Fujii, T. (1988) Planta 175, 364-373

MOLECULAR ASPECTS OF THE CELL CYCLE IN *CATHARANTHUS ROSEUS* SYNCHRONOUS CELL DIVISION CULTURES

HIROAKI KODAMA, MASAKI ITO AND ATSUSHI KOMAMINE
BIOLOGICAL INSTITUTE, FACULTY OF SCIENCE, TOHOKU UNIVERSITY, SENDAI, 980, JAPAN

1. INTRODUCTION

One of the most fundamental problems in cellular biology is the elucidation of molecular mechanisms that control the cell cycle. The cell cycle consists of the G_1, S, G_2 and mitotic phases, and is controlled by the ordered expression of genes.

In higher plant cells, only a few studies dealing with gene expression during the cell cycle have been reported. The cells of the higher plant, *Catharanthus roseus* L. (G) Don. (periwinkle), grown in suspension, has been used in several investigations of cell proliferation. Cell proliferation of *C. roseus* is strictly regulated by phosphate in the medium, and cell division can be synchronized by the double phosphate starvation method of Amino et al. (1). In this system, cells were arrested in the early G_1 phase after phosphate starvation, and then re-addition of phosphate induced synchronous division of cells. In such a system, 70-80 % of cells usually divided within 4 hours. The DNA content of synchronized cells of *C. roseus* was also measured by flow cytometrical analysis (2). The gene expression during the cell cycle has been analyzed using the synchronous cultures of *C. roseus*. The variations in the population of mRNAs were investigated by translation in vitro followed by two-dimensional gel electrophoresis, and it was revealed that the levels of four species of mRNA were changed transiently during the cell cycle (3). In addition, two mRNAs were identified that were detected commonly and preferentially in actively cycling cells (4). These results indicate that some of genes are expressed specifically during the certain phases of the cell cycle in *C. roseus* cells.

In this paper, we report the isolation, sequences and expression patterns of cDNAs corresponded to the mRNAs that vary in level during the cell cycle of *C. roseus* cells. Furthermore, we also show the isolation of a cDNA of proliferating-cell nuclear antigen (PCNA) and its expression during the cell cycle. PCNA was initially recognized as a nuclear antigen which reacts with autoimmune sera from a certain group of patients with systemic lupus erythematosus. This antigen is an acidic, nuclear, nonhistone protein and is present preferentially in the nuclei of proliferating mammalian cells, especially cells in the S phase. In fact, PCNA has been recognized as an auxiliary protein of DNA polymerase-δ (5,6). Recently, Suzuka et al. (7) demonstrated that a gene for PCNA is also present in the genomes of higher plants. A PCNA-related clone (pCJ-1) was isolated from rice genomic DNA and the partial nucleotide sequence was determined. Using rice genomic clone, pCJ-1, we have isolated the cDNA encoding plant PCNA, and investigated its expression during the cell cycle in synchronous cultures of _C. roseus_.

2. MATERIALS AND METHODS

2.1 Cells and synchronized growth

For synchronization by the double phosphate starvation method, suspension cultures of strain B of _Catharanthus roseus_ L. (G) Don. were used and a cDNA library was constructed from poly(A)$^+$RNA isolated from the cells of strain B in the S phase. In addition, another strain (TN21) was used. This second strain was initiated from the culture of anthers of _Catharanthus roseus_ L. cv. Little-Pinky in our laboratory in 1987. The growth of cells of strain TN21 can be regulated by the level of auxin in the medium. These strains (strain B and strain TN21) have been maintained at 27°C in the dark in the Murashige-Skoog medium with 3% (w/v) sucrose. The concentrations of 2,4-dichlorophenoxyacetic acid (2,4-D) were 2.2 µM and 4.4 µM in cultures of strain B and strain TN21, respectively. Cells were subcultured at intervals of 7 days.

Cells of strain B were synchronized in the early G_1 phase by the double phosphate starvation method. Cells were cultured in phosphate-free medium at an initial cell density of about 3×10^5 cells/ml for 2 days. Phosphate was then added to the medium at a final concentration of 0.14 mM. After 18 hours the cells were washed with phosphate-free medium and transferred to phosphate-free medium at a cell density of about 2×10^5 cells/ml. Phosphate was added again after 2 days in culture, and the synchronous cell division was observed.

When DNA synthesis and subsequent mitosis were to be induced synchronously by the addition of 2,4-D, the cells of strain TN21 were utilized. Ten-day-old cultures of strain TN21, at the stationary phase, were transferred to 2,4-D-free medium. After cells were cultured for 2 days in this medium, they accumulated exclusively in the G_1 phase, and the addition of 2,4-D at a final concentration of 4.4

μM induced synchronized synthesis of DNA and cell division.

2.2 Purification of RNA and construction of cDNA library

Total RNA was prepared by using 2M LiCl, and poly(A)$^+$ RNA was prepared from total RNA by chromatography on oligothymidylic acid-cellulose. A λgt11 cDNA library was constructed from poly(A)$^+$RNA isolated from the cells in the S phase.

2.3 Differential screening

The λgt11-cDNA library was plated at a low plaque density and screened by differential hybridization. Duplicate nylon-membrane replicas were lifted from each of plates. About 100,000 plaques on nylon membranes were screened by hybridization with ^{32}P-labeled single-stranded cDNA synthesized from the poly(A)$^+$RNA isolated either from cells in the S phase or from cells during cytokinesis. Another 100,000 plaques were screened with ^{32}P-labeled probes prepared from the poly(A)$^+$RNA isolated from the cells in the S phase and from phosphate-starved cells.

3. RESULTS

3.1 Differential screening of the cDNA library prepared from poly(A)$^+$RNA isolated from the cells in the S phase

C. roseus cells were synchronized by the double phosphate starvation method. Cells in the S phase were collected 15 hours after the addition of phosphate, and poly(A)$^+$RNA was isolated. A λgt11 cDNA library was screened for genes that are preferentially expressed at the S phase by differential plaque hybridization, as described in Materials and Methods. Of approximately 200,000 plaques screened, 3 independent cDNA clones were selected. These 3 representative cDNA inserts were subcloned into pBluescript KS M13$^+$ and designated as pCYC02, 07 and 17, respectively. The genes, which are correspond to cDNA clones, pCYC02, 07 and 17 are designated as *cyc02*, *07* and *17*, respectively.

3.2 Isolation of a cDNA clone for periwinkle PCNA

The cDNA library was screened with the labeled 5-kb *HindIII* fragment of pCJ-1, which contains PCNA-related gene sequences. Out of 120,000 plaques screened, only one plaque was found to hybridize strongly. The *EcoRI* insert from a selected recombinant phage DNA was subcloned into the unique site of *EcoRI* of the plasmid vector, pBluescript KS M13$^+$.

3.3 Sequence of cDNAs for *cyc02*, *07*, *17* and periwinkle PCNA

Full length cDNA for <u>cyc02</u>, <u>07</u>, <u>17</u> and periwinkle PCNA were isolated and their nucleotide sequences were determined.

The coding region of <u>cyc02</u> cDNA contains 303 nucleotides, equivalent to 101 codons, and encodes a protein with a molecular weight of 11067 daltons. The predicted amino acid sequence of <u>cyc02</u> suggests that the entire <u>cyc02</u> protein is basic. No sequence similarities were found between the nucleotide sequences of <u>cyc02</u> and other sequences in the EMBL Data Library.

The <u>cyc07</u> cDNA contains an open reading flame of 735 nucleotides, encoding a highly basic protein of 245 amino acids. The nucleotide sequence of <u>cyc07</u> shows homology (about 65%) with genomic sequence of an open reading flame which is found downstream of SIR3 gene of budding yeast, <u>Saccharomyces cerevisiae</u>. The gene encoded by this open reading flame has not been identified in yeast. The predicted amino acid sequence of yeast genomic homologous region is compared with that of <u>cyc07</u>. Surprisingly, about 67% homology was observed between the predicted amino acid sequence of <u>cyc07</u> and that of yeast.

The <u>cyc17</u> cDNA contains an open reading flame of 651 nucleotides, and encodes 217 amino acids. High content of proline residue (about 42%) and basic amino acids (about 16%) were observed. The 17 repeated amino acid sequences, Ser-Pro-Pro-Pro-Pro, were observed. It is still unclear whether these proline residues are hydroxylated or not. However, the structure enriched with proline residue shows a homology to the proteins so called "extensin".

The cDNA encoding periwinkle PCNA shows an open reading flame of 804 nucleotides, encoding a protein 268 amino acids. When conservative substitutions of amino acids were included, a high degree of similarity (about 85% homology) was observed between the predicted amino acid sequence of periwinkle PCNA and that of human PCNA.

3.4 Expression of isolated cDNAs during the cell cycle in a culture of cells synchronized by the double phosphate starvation method

To determine the time course of appearance and decay of mRNAs for isolated genes during the cell cycle, total RNA was isolated at various phases of the cell cycle in <u>C. roseus</u> cells in a synchronized culture induced by the double phosphate starvation method.

The <u>cyc02</u> mRNA was present in cells arrested at the G_1 phase by phosphate starvation. After the phosphate addition, the level of <u>cyc02</u> mRNA declined, and increased again at the G_1/S boundary of the cell cycle.

The level of <u>cyc07</u> mRNA increased dramatically, but transiently, at the G_1/S boundary. The levels of both <u>cyc02</u> and <u>cyc07</u> mRNAs declined at the late S phase or G_2

phase.

The pattern of the time course of expression of cyc17 was quite similar to that of cyc02. The cyc17 mRNA was present in the phosphate-starved cells and in the S phase cells.

The periwinkle PCNA mRNA was undetectable in phosphate-starved cells and cells in the early G_1 phase. An increase in expression of periwinkle PCNA mRNA was detected at the G_1/S boundary and reached a maximum at the S phase.

3.5 Expression of isolated cDNAs during the cell cycle in a culture of cells synchronized by starvation and subsequent addition of auxin

The expression of isolated genes was investigated in another synchronized cell-division system. Cells of strain TN21 of *C. roseus* were arrested at the G_1 phase by starvation for 2,4-D, and subsequent addition of 2,4-D induced the synchronous progression of the cell cycle.

The cyc02 mRNA was detected at high level in the 2,4-D-starved cells. When 2,4-D was added, the level of cyc02 mRNA declined in the G_1 phase and again increased slightly at the S phase.

The cyc07 mRNA was present preferentially in the cells in the S phase. In cells in the G_2 phase, the level of cyc07 mRNA decreased.

The cyc17 mRNA increased in level preferentially and transiently at the G_1/S boundary.

The expression of periwinkle PCNA mRNA was induced at the G_1/S boundary, reached a maximum level at the S phase, and declined at the G_2 phase.

4. DISCUSSION

Four cell-cycle-dependent cDNA clones were isolated in this investigation using suspension cultures of *C. roseus* cells. The synchrony was induced by the two different methods, namely the double phosphate starvation method and starvation of 2,4-D. In these two different synchronous cultures, 4 genes, cyc02, 07, 17 and a gene for periwinkle PCNA, were expressed transiently during the cell cycle. The cyc02 mRNA disappeared when cells began to progress the cell cycle synchronously, and increased at the early S phase. The cyc07 was expressed only at the S phase, and encodes a highly basic protein. The homologous sequence to cyc07 is found in the genome of yeast. If the homologous region of yeast is also expressed, the product of cyc07 may play an important and essential role in the maintenance of progression of the cell cycle. The isolation of the homologous region of yeast is now undertaken. The cyc17 encodes an extensin-like protein. The function of extensin is still unclear, but since the expression of the cyc17 is cell-cycle-dependent, it is suggested that cyc17 may have a close relation to the cell growth. It was demonstrated in this report that plant PCNA was very similar in primary structure to a mammalian PCNA and was expressed specifically in the S phase. PCNA has been recognized as an auxiliary protein of DNA polymerase-δ in mammalian cells. Therefore, it is probable that plant PCNA is required during the S phase for the regulation of the activity of PCNA-dependent DNA polymerase-δ.

5. ACKNOWLEDGMENTS

We wish to thank Drs. K. Nakamura, and T. Hattori, Faculty of Agriculture, Nagoya University, for their useful advice and cooperation. We thank Dr. Suzuka, Department of Molecular Biology, National Institute of Agrobiological Resources, for the generous gift of rice genomic clone, pCJ-1.

REFERENCES

1 Amino, S., Fujimura, T. and Komamine A. (1983) Physiol. Plant. 59, 393-396
2 Ando, S., Shimizu, T., Kodama, H., Amino, S. and Komamine, A. (1987) Agric. Biol. Chem. 51, 1443-1445
3 Kodama, H., Kawakami, N., Watanabe, A. and Komamine, A. (1989) Plant Physiol. 89, 910-917
4 Kodama, H., Ando, S. and Komamine, A. (1990) Physiol. Plant. in press
5 Bravo, R., Frank, R., Blundell, P.A. and MacDonald-Bravo, H. (1987) Nature 326, 515-517
6 Prelich, G., Tan, C., Kostura, M., Mathews, M.B., So, A.G., Downey, K.M. and Stillman, B. (1987) Nature 326, 517-520
7 Suzuka, I., Daidoji, H., Matuoka, M., Kadowaki, K., Takasaki, Y., Nakane, P.K. and Moriuchi, T. (1989) Proc. Natl. Acad. Sci. USA, 86, 3189-3193

MICROTUBULES AND THE FIRST CELL CYCLE IN CULTURED MESOPHYLL PROTOPLASTS OF NICOTIANA

W. Wernicke, M. Rös and G. Jung
Institut für Allgemeine Botanik, Johannes Gutenberg-Universität, P.O.Box 3980, D-6500 Mainz, FRG

1. INTRODUCTION

When a protoplast is excised and cultured in vitro, an organized pattern of cell divisions must be reestablished to permit organogenesis, including plant regeneration. However, in a number of species protoplasts have proved to be very recalcitrant with regard to cell division and/or plant regeneration. In an attempt to find means to overcome the regeneration problem in recalcitrant species we are investigating the processes leading to cell division and morphogenesis in mesophyll protoplasts of the model system _Nicotiana_. Of particular interest are the initiation of the cell cycle and the coordination of progress through the cycles as a prerequisite for subsequent morphogenesis. Mesophyll protoplasts are highly vacuolated. In vacuolated cells a plate of cytoplasm (phragmosome) must be formed prior to mitosis in order to permit positioning of the nucleus and bisecting of the cell through the vacuole during cytokinesis (1). Thus, the phragmosome determines the orientation of the plane of cell division. Another structure which is thought to be involved in determining the plane of cell division, is a cortical band of microtubules commonly termed the preprophase band of microtubules (PPB). The term refers to the time when the bands first appear and not to the life span, which includes prophase (2). There seems to be a close temporal and spatial correlation between the establishment of phragmosomes and PPBs (2,3,4). Cells growing in an organized fashion usually form PPBs, although PPBs are not necessarily restricted to programmed cell divisions which lead to spatially defined morphogenesis (5). Interestingly, it was recently reported that bands of microtubules reminiscent of PPBs occur in high frequencies during prophase of the first division cycle of _Medicago_ and _Nicotiana_ mesophyll protoplasts (6).

The first part of the project was concerned with the development of fast and convenient means to study progress of protoplast populations through the first mitotic cycle. One of the purposes was to identify the stage of G2 to prophase transition. This could be the crucial period during which the plane of cell division and thus, perhaps, the onset of organized growth is determined. In the second part we investigated the behaviour of the microtubular cytoskeleton in a highly regenerative *Nicotiana* variety (7) by immunofluorescence microscopy. Special attention was paid to the potential occurrence of PPBs during the first cycle in relation to cytokinesis.

2. MATERIALS AND METHODS

2.1 Protoplast culture

Plants of *Nicotiana tabacum* var. *Atropurpurea* were grown under axenic conditions from surface sterilized seeds. Protoplasts were isolated from expanding leaves of ca 6 weeks-old plants by overnight incubation of leaf strips in an enzyme solution containing 0.2 % Cellulase "Onozuka" R-10, 0.2 % Macerozyme R-10 and 0.4 M sucrose. The protoplasts were purified and cultured in darkness at a density of 3-5 x 10^4 protoplasts ml^{-1} in liquid MS-medium supplemented with 1 µM NAA and 1 µM BAP. Each leaf was treated separately as an individual population of cells. In five separate preliminary control experiments a plating efficiency of 80 ± 10 % was obtained.

2.2. DNA staining, cytofluorimetry and detection of BrdU labelled DNA

Protoplasts and/or protoplast-derived cells were fixed in 1.5 % paraformaldehyde in phosphate buffer for 30 min. Triton X-100 extraction (0.5-1 % w/v in TE buffer) of the cells after fixation was essential before staining DNA for cytofluorimetry and, advantageous before detection of a potential BrdU label. DNA was stained with DAPI (0.1 µg ml^{-1} in TE buffer) for 15 min. The cell walls could be counterstained with Calcofluor white M2R (not recommended when DNA is stained for subsequent cytofluorimetry). The specimen were mounted in TRIS medium (8) to reduce fading during cytofluorimetry (Zeiss MPM 200 microscope-photometer). However, fading only became negligible when the light intensity of the excitation light was distinctly reduced with grey filters. BrdU-labelled DNA was detected employing an antibody solution (Amersham, RPN202) consisting of a monoclonal BrdU-specific antibody and a nuclease for DNA denaturation. The bound monoclonal antibody was localized either with a FITC-labelled anti-mouse antibody or, in order to enhance the sensitivity of detection, with a biotin-labelled antibody followed by FITC-streptavidin.

2.3 Staining of the microtubular cytoskeleton

Cells were preincubated in 5-10 % DMSO (not possible with freshly isolated protoplasts) for 30 min to stabilize microtubules, fixed in paraformaldehyde (1.5 %, 5 mM EGTA, 20 mM phosphate buffer), rinsed in MSB (50 mM phosphate buffer, 5 mM EGTA, 1 mM $MgSO_4$, 0.2 M mannitol) and immobilized on poly-l-lysine coated slides. Standard procedures were applied for localizing microtubules. A biotin-labelled anti-mouse antibody/FITC-streptavidin system was used to enhance sensitivity of detection of the bound tubulin-specific antibody. Cells were counterstained for DNA with DAPI and, in some instances, also with Calcofluor white for cell wall.

3. RESULTS AND DISCUSSION

Calcofluor staining indicated substantial cell wall regeneration as early as 2 hours after plating. However, the speed in cell wall regeneration seemed to differ between protoplasts. The differences in stainability became less distinct after 6 hours. Cell wall regeneration appeared to be complete after 24 hour, at least according to the staining. About the same time the first cell divisions were observed. Figure 1.a. shows the time course of progress through cell division. Yet, for the investigation concerning the cytoskeleton, it was important to determine progress through the individual phases of the mitotic cycle.

The relative DNA content of nuclei was measured by cytofluorimetry with the aim of determining the changes in the ratio of G1 to G2 cells. The relative fluorescence of 80-90 nuclei was plotted in a histogram. When the two peaks relating to G1 and G2 were observed, a line was drawn through the minimum between the peaks. The portion with the lower intensity was defined as G1/S-phase, the remaining as S/G2-phase. No attempt was made in these experiments to distinguish between G- and S-phase nuclei. The percentage of S-phase nuclei was determined by labelling DNA with the thymidine-analogue BrdU, which was subsequently immunolocalized. This method proved to be a very simple and fast means of detecting DNA replication.

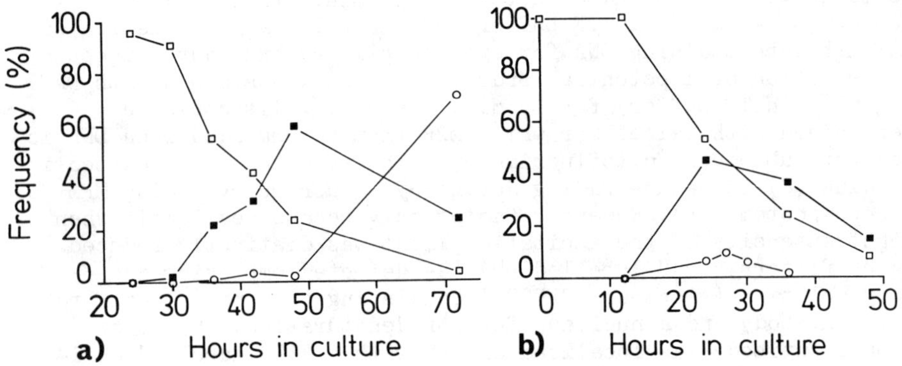

FIGURE 1. Progress of Nicotiana mesophyll protoplast populations through cell cycle. a) Plating efficiencies determined after staining of nuclei with DAPI and of cell wall with Calcofluor (24 and 48 h: mean of 2 separate populations; remaining: single measurements). Single, uninucleate cells -□-, two-celled colonies (separating walls apparently complete) -■-, colonies with more than 2 nuclei (the exact number of cells could not always be identified) -○-. b) Changes in phases of mitotic cycle determined by cytofluorimetry: G1/S-phase -□-, S/G2-phase -■-, and by BrdU labelling (1 hour pulse, different population): S-phase -○-.

In a preliminary experiment continuous labelling (BrdU added during plating) revealed onset of S-phase between 12 and 24 hours after plating. However, BrdU proved to be inhibitory, at least during further development, even at the lowest concentration (10 μM) required for immunodetection. Pulse labelling (100 μM BrdU for 1 h before sampling) showed that a maximum (ca 9 %) of S-nuclei occurred between

20 and 30 hours. The relatively low percentage of S-phase nuclei in the populations confirmed that the cytofluorimetric measurements gave reasonable estimates of the numbers of cells progressing from G1 to G2. The results are summarized in Fig. 1.b.

The highest number of G2 cells and rapid transition through the first cycle was observed between 24 and 42 hours after plating. The changes in the pattern of the microtubular cytoskeleton were investigated during this culture period because it covered the range where PPBs should be found in the highest numbers, if any. Some of the characteristic microtubule patterns observed are shown in Fig. 2. The results are summarized in Table 1. It must be stressed that the classification is arbitrary to a certain extent, because the changes in the patterns were gradual.

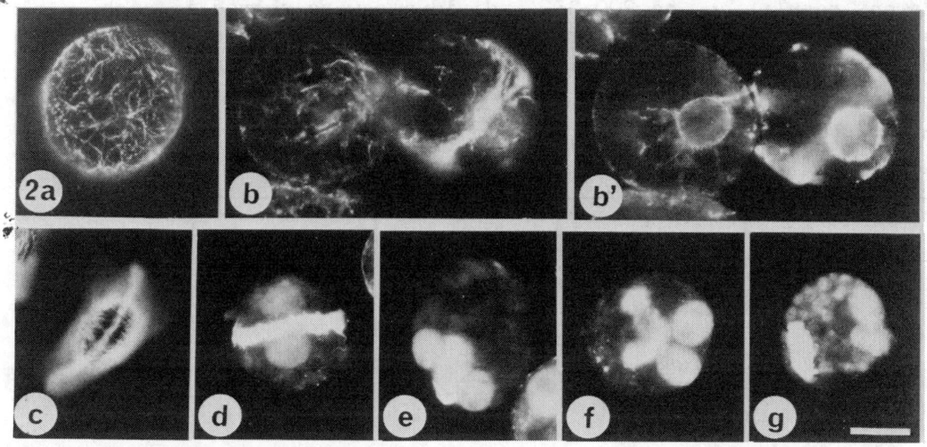

FIGURE 2. Tubulin visualized by immunofluorescence; cells derived from protoplasts, except (c). a) Cortical microtubules. b) Periphery of two cells, one with patchy, diffuse fluorescence, the other with band-like alignment (this was one of the sharpest bands found). b') Same as (b) but plane of nuclei in focus, note perinuclear fluorescence. c) Narrow PPB in root meristem cell. d-g) Phragmoplast microtubuli in various positions relative to periphery and nuclei: normal position (d,e), apparently abnormal position (f,g). Double exposure to show position of DAPI stained nuclei. Bar: 20 μM

A random network of cortical microtubules predominated during the first 30 hours (Fig 2.a.). Nevertheless, in advanced cultures a perinuclear fluorescence appeared with strands radiating towards the cortex, where a network was still observed (Fig. 2.b.,c.). In prophase the radiating microtubule strands and the microtubules in the cortex became less distinct until most fluorescence was associated with the nucleus, possibly in the future spindle regions. Band like alignments in the cortex reminiscent of PPBs were found predominantly in prophase cells. Basically. this behaviour was very similar to what has alread been described by Meijer and Simmonds (6). However, bands were found in much lower frequencies. When root meristems representing organized growth during normal ontogenesis of Nicotiana were tested, 20-40 % of PPBs were found in preprophase cells and a band could be observed in more than 95 % of all prophase cells. Incidentally, the bands in prophase cells of the root meristem were considerably less diffuse than the bands found in protoplast cultures (compare e.g. Figs. 2.b. and 2.c.). Arrangements of metaphase and anaphase microtubules in the protoplast-derived cells were similar to those in root meristems. Yet, deviations again were observed in telophase and cytokinesis.

TABLE 1. Patterns observed after tubulin labelling in correlation with stages of the nuclear cycle. Samples were harvested 24, 30, 36 and 42 hours after plating; data were pooled as shown. Each sample represents a different population, ca 200-350 individuals were scored in each sample. PNR = perinuclear radiating, PNF = perinuclear fluorescence (6). Data in brackets: phragmoplast microtubuli in apparently abnormal position.

Stage		Frequency (%)	
Nuclei	Microtubules	24/30 h	36/42 h
Uninucleate	Cortical	73.1	49.1
	PNR/PNF	13.5	10.4
	Band	0.2	0
Prophase	Cortical	0.2	1.9
	PNR/PNF	0.2	3.5
	Band	0.2	1.2
Meta-/Anaphase	Spindle	3.1	3.3
	Band	0	0
Telophase	Phragmoplast	0.2	0.2
Binucleate	Phragmoplast	2.1 (0.3)	5.6 (0.9)
≥ Binucleate	Miscellaneous	7.2	24.8

In meristem cells cytokinesis appeared to be completed in late telophase. In the cultures, however, phragmoplast microtubuli were still observed when nuclei with distinct nucleoli were already reestablished. In addition, phragmoplast microtubuli were found in apparently abnormal regions ("displaced" phragmoplasts (9)), particularly when the nuclei were obviously in an eccentric position (Figs. 2.d.- 2.g.). One explanation could be that the still growing edges of the cell plates, with the phragmoplast microtubuli, had not yet reached the far side of the cell (see also (3)). Sometimes it seemed that the phragmoplast had to squirm to pave a way through the cytoplasmic bridges of a presumptive phragmosome (Fig. 2.f.). The figures show relatively small cells, which were probably rich in cytoplasm. The problem could even be greater in larger, more vacuolated cells. "Accidents" seemed to occur, because incomplete separating cell walls without phragmoplast microtubuli were observed. The phragmoplast abnormalities occurred much more frequently than in the cultures described previously (9). At first sight this seemed to be intriguing. It might be argued that our protoplast populations were in a better state, because the plating efficiency was higher and because the cell cycle was entered more rapidly. However, it could just be the high speed of reentering the nuclear cycle which caused the abnormalities. Phragmosome formation and nuclear cycle are not necessarily fully coordinated and can be uncoupled (3). The same seems to apply to PPB formation and nuclear cycle (8). The diffuse texture of the putative PPBs found, and the comparatively short appearance, if not even lack of PPBs in many cells, could be an indicator of inadequate coordination between phragmosome formation and the nuclear cycle. A cell must face considerable strategic problems separating the two daughter nuclei smoothly, if the establishment of a perfect phragmosome lags behind the nuclear cycle. Lack of coordination seems to be a major problem e.g. during phragmosome formation and cell division in mesophyll protoplasts of the cereal *Avena* (G.Jung and W.Wernicke, unpublished). Incidentally, incomplete cytokinesis resulting from imperfect phragmosomes could be a common phenomenon in tissue culture (10). The inadequate coordination may have important consequences for the regeneration potential. It is well known that many of the protoplasts, which divide initially, fail to develop further in a regular fashion and e.g. abort after a few divisions or show somaclonal variation if divisions continue. Careful monitoring of the cycle and adapting the culture conditions so that all processes are fully coordinated could eliminate many problems inherent in tissue culture.

REFERENCES

1 Sinnott, E.W. and Bloch, R. (1940) Proc. Natl. Acad. Sci., USA 26, 223-227
2 Gunning, B.E.S. and Wick, S.M. (1985) J. Cell Sci. Suppl 2, 157-179
3 Venverloo, C.J. and Libbenga, K.R. (1987) Z. Pflanzenphysiol. 131, 267-284
4 Lloyd, C.W. and Traas, J.A. (1988) Development 102, 211-221
5 Gorst, J., Wernicke, W. and Gunning, B.E.S. (1986) Protoplasma 134, 130-140
6 Meijer, E.G.M. and Simmonds, D.H. (1988) Physiol. Plant. 74, 225-232
7 Wernicke, W. and Thomas, E. (1980) Plant Sci. Lett. 17, 401-407
8 Mineyuki, Y., Wick, S.M., and Gunning, B.E.S. (1988) Planta 174, 518-526
9 Meijer, E.G.M., Keller, W.A. and Simmonds, D.H. (1988) Physiol. Plant. 74, 233-239
10 Mahlberg, P.G., Turner, F.R., Walkinshaw, C., Venketeswaran, S. and Mehrotra, B. (1975) Bot. Gaz. 136, 189-195

Section C
Secondary Metabolites

POSSIBILITIES FOR IMPROVING YIELDS OF SECONDARY METABOLITES IN PLANT CELL CULTURES

Y. YAMADA & T. HASHIMOTO
Research Center for Cell and Tissue Culture, Faculty of Agriculture, Kyoto University, Kyoto 606, Japan

1. INTRODUCTION

Since the possibility of producing useful secondary products in plant cell culture was first recognized in 1970s, a number of researchers in both academic and industrial circles have been struggling to apply the great biosynthetic potential of plant cells to commercial production. However, despite more than 15 years of effort, the compounds that are now, or are about to be, produced commercially are very limited in number. What have we learned in the vast majority of unsuccessful cases? Can we make any rational and predictable strategies to improve yields in cell culture based on past experience?

From our collective experiences of the last two decades, we have learned that selection of stable, high-producing cell lines and optimization of culture conditions are both important. We know that elicitation of plant cells can induce or hasten the timing of biosynthesis of some compounds. Organized cultures, especially root cultures, have been found to be useful in some cases when undifferentiated cell cultures fail to synthesize compounds of interest. Most of these approaches, however, are empirical and underlying cellular mechanisms are usually left unexplored. Even though we have accumulated experience in cell manipulation and bioprocessing, our understanding of the biosynthesis of secondary products and its *in vivo* regulation is poor.

A problem of the field of secondary metabolite production may be that there are so many different subjects for "biotechnological research". The rates of synthesis of secondary compounds are readily influenced by a variety of factors (but rarely enhanced to a point that guarantee commercialization). Additionally, there are so many interesting compounds to study and there is little knowledge of their biosynthesis or its regulation. It is a pity that the study of secondary product biosynthesis tends to spread horizontally to almost all groups of compounds and it may be inappropriate to propose a few "model" compounds whose biosynthesis could be tackled intensively by many scientists from different backgrounds. However, to cope with rapid advances in molecular biology, it is becoming ever more important that our research should be concentrated on several groups of potentially useful compounds for which highly productive tissue culture systems have already been established.

2. TARGET COMPOUNDS

It may not matter what compounds we study for basic research but, if we expect to find future applications and to receive financial support from industries, the compounds should have potential commercial value. There are two often overlooked factors which we should keep in mind when we consider the potential commercial value of a compound. First, good marketing strategies can create new markets. Even when we have at hand sufficient production of compounds that are not used commercially, we may be able to create novel and more extensive uses for these compounds. A good example is shikonin production in cell culture, which was the first commercial success in secondary metabolite production. Shikonin and its derivatives extracted from the root of *Lithospermum erythrorhizon* have traditionally been used in Japan for treatment of burns and hemorrhoids, as well as for bright red dyes. Because the traditionally small market for shikonin was easily saturated by the high productivity of the culture system developed by Mitsui Petrochemical Industries (1), Mitsui and its marketing partner Kanebo decided to market the product in lipsticks dubbed "Bio-Lips" in 1984. Fancy advertisements on TV and in other mass media greatly helped to boost the sales of the first "bio"-cosmetics in Japan. The second factor we should keep in mind when evaluating commercial potential is the impact of the regulatory laws of a country. Many oriental and some western medicines are used as raw plant tissues or crude herbal extracts that contain a mixture of pharmaceutically active compounds. The wholesale prices of these unprocessed materials are sometimes higher than the sum of the pure constituents. Marketing medicinal plant cells or tissues cultured *in vitro* and their crude cellular extracts as food or medicine, however, is strictly regulated in many countries because of the possible influence of ingredients of the culture media (e.g. phytohormones) on human health. For this reason, the cost of purifying desired compounds from a crude extract on a commercial scale must be included in our calculations, and efficient downstream processing of natural products from plant cell culture must be considered as an essential part of commercialization.

3. TYPE OF CULTURES

3.1. Cell Culture

Despite the frequently reported instability of production of some compounds in cell cultures, certain classes of secondary metabolites are produced stably in undifferentiated plant cells. Notable examples in relation to commercial production are shikonin from *Lithospermum erythrorhizon*, berberine from *Coptis japonica* and sanguinarine from *Papaver* species. The fact that these compounds are produced in undifferentiated cells does not necessarily mean that their biosynthesis in the plant is not organ- or tissue-specific. Several tissue-specific genes such as SAM synthetase (2) are known to be expressed at high levels in callus, although they are only expressed in restricted tissues of the plant. It should be interesting and feasible to study the tissue- and cell-specific expression of the above compounds in the whole plant, especially because cDNAs of, or antibodies against, biosynthetic enzymes are not so difficult to obtain from highly productive cell culture systems.

3.2. Organ Culture

Recent re-evaluation of root culture as a production system for secondary metabolites has led an increasing number of scientists to use an organized culture system. Many compounds that are scarcely synthesized in undifferentiated cells are produced at higher levels in cultured roots (3). These roots grow vigorously in the presence of an appropriate concentration of auxin or when transformed with *Agrobacterium rhizogenes*. Large scale fermentors have been developed for the culture of roots and shoots (3). Jar fermentors are designed so that roots are anchored to a solid structure within the fermenter, and are aerated by bubbling humid air through a sparger placed at the bottom of the tank. Few studies on scaled-up root culture have been published, but root cultures (actually cell aggregates with profuse regenerating roots) of *Panax ginseng* have been grown successfully in a 20,000-liter tank by Nitto Electric Co., Japan, for saponin production (4). The potential of organ cultures to be useful for production of secondary compounds has been increased now that *A. rhizogenes* can be used to introduce genes of interest to hairy root cultures. Biosynthesis genes of secondary product formation can be engineered to be expressed in the root and flanked by T-DNA border sequences in an expression plasmid. Inserted genes in the plasmid are co-transferred at a high frequency to hairy roots along with the wild-type Ri T-DNA genes. This allows many new approaches to be used as more biosynthesis genes are isolated and become available. Some possible strategies are discussed in the next section in the broader context of whole plants, and tissue and organ culture.

Shoot cultures may also prove to be useful sources for some compounds. Certain secondary products are produced preferentially in shoots or leaves, and these compounds are logical candidates for production in shoot cultures. In the case of *Digitalis*, shoot cultures have indeed been found to contain larger amounts of cardiac glycosides than do root cultures or cell-suspension cultures (5, 6). However, except in the cases of a few species (3), rapid and large scale propagation of cultured shoots in liquid medium remains difficult.

4. POSSIBLE STRATEGIES FOR YIELD IMPROVEMENT

Over the next several years, we should see the application of molecular biology for improving yields of secondary products both in whole plants and in plant tissue and cell cultures. We here list several such strategies that may soon become possible and give some examples from the biosynthesis of tropane alkaloids to illustrate how the strategies could be applied. The biosynthetic pathway and its enzymes are shown in Fig. 1.

4.1. Overproduction of Simple Precursors of the Secondary Compounds

The biosynthesis of amino acids in plants is primarily controlled by feedback inhibition of one or a few regulated enzymes in each biosynthetic pathway by the end product amino acids from that pathway. The regulated enzymes usually catalyze the first reaction in the pathway common to a group of amino acids or the reaction after a branch point to the biosynthesis of a particular amino acid. Molecular cloning of such regulatory enzymes has been reported from several higher plants, including our isolation of two cDNAs of dihydrodipicolinate synthase, a key enzyme in lysine biosynthesis (Fig. 2), from wheat (7). Site directed mutagenesis or random mutagenesis, followed by functional screening, may enable us to engineer mutant enzymes that are no longer feedback inhibited by the end product amino acids. Transgenic plant cells expressing such mutant enzymes would overproduce the amino acids whose synthesis the wild-type enzyme controls. Because the biosynthesis rate of secondary products usually is not controlled by the supply of amino acid precursors, overproduction of precursor amino acids in normally secondary product low-producing cells or tissues would not cause product formation. This point was shown in the case of indole alkaloid production in tryptophan-overproducing _Catharanthus roseus_ (8, 9). It is hoped, however, that in high producing cells increased supply of primary precursors may improve the end product formation still further.

4.2. Increasing the Concentration of Rate-limiting Enzymes

Most enzymes in a given pathway of secondary metabolism are coordinately regulated and there are no clear rate-limiting enzymes as found in primary metabolism. However, the catalytic activities of individual enzymes in a pathway often vary considerably, which may result in modest accumulation of some intermediates unless metabolic channeling or compartmentation occurs. To cite an example from tropane alkaloid biosynthesis, several tropane alkaloid-producing plants (e.g. many _Datura_, _Scopolia_ and _Atropa_ species) accumulate hyoscyamine as a major alkaloid with scopolamine present in smaller quantities. In these plants, the enzymes catalyzing oxidative conversion of hyoscyamine to scopolamine (two or, possibly, three enzymes may be involved) are apparently limiting. Hyoscyamine 6β-hydroxylase, the enzyme that catalyzes the first reaction in this transformation, is a promising candidate that, when expressed at high levels in hyoscyamine-accumulating plants, would result in increased oxidation of hyoscyamine, and higher levels of scopolamine. A cDNA of the hydroxylase has been cloned from _Hyoscyamus niger_. A plant expression vector was constructed so that the cDNA is expressed under the control of the CaMV 35S promoter. We are currently introducing this gene to hyoscyamine-accumulating plants by Ti- and Ri-based binary systems (10). One variable in this approach is the choice of the promoter. In addition to being relatively strong, the promoter to be used must function in the cells in which the substrate of the introduced enzyme (in this case, hyoscyamine) is synthesized or accumulates in an accessible metabolic pool. Our immunohistochemical studies showed the hydroxylase to be localized in the pericycle of the root (10), indicating that hyoscyamine should be present there. With the 35S promoter, the introduced gene will be expressed in almost all of the cells of transgenic plants. We do not yet know the consequences of this superfluous expression: the expressed enzyme may just exist with no catalysis in most of the cells, or may have access to the hyoscyamine during translocation of the alkaloid from the root to the aerial parts. If precise expression of the enzyme in the cells in which it naturally occurs is desired, a pericycle-specific promoter should be used, and the best candidate is the promoter of the hydroxylase gene itself.

4.3. Creating a New Branch from a Preexisting Pathway

A major limitation for modifying an existing biosynthetic pathway by introducing a foreign enzyme is the substrate specificity of the introduced enzyme, as it must act on an intermediate in the pathway of interest. We can look for modification enzymes in the degradation system in secondary product-utilizing microorganisms or the xenobiotic detoxifying (microsomal) system in animals. However, a more promising approach might be to search in other plant species for enzymes that have different substrate specificities. A well known example is the generation of a new petunia flower color by transformation of a petunia mutant with a maize dihydroflavonol 4-reductase gene (11). The experiment was based on the different substrate specificities of the reductases from petunia and maize on dihydrokaempferol substrate.

In tropane alkaloid biosynthesis, such strategies could also be applied. The diamine oxidases of tropane and tobacco alkaloid-producing plants have stronger affinity for N-methylated- than non-N-methylated-diamines as substrates (12, 13). By coupling with strong activities of putrescine N-methyltransferase, the oxidases regulate the biosynthesis of N-methylated alkaloids. In contrast, diamine oxidases from other, more common plants (e.g. pea) oxidize symmetrical diamines more rapidly than N-methylated diamines (14). Thus, introducing a pea diamine oxidase, for example, to the alkaloid producing plants may alter the flow of metabolites to increased formation of non-N-methylated alkaloids. Another example of enzymes that catalyze similar reactions but differ markedly in their substrate specificities is found in the oxidases that stereospecifically dehydrogenate ring C of [S]-tetrahydroprotoberberines in _Berberis_ and _Coptis_ (15, 16). As more enzymes of secondary metabolism are characterized from different species, our ability to create new branches from preexisting pathways will increase rapidly.

4.4. Reducing the Rate of an Existing Side Reaction

The basis of antisense mRNA technology is that an RNA strand complementary to the target mRNA is expressed as the product of a transgene, and a hybrid duplex forms between the two RNAs, inhibiting translation of the target mRNA. The technology has been used successfully to suppress flower color formation in petunia (17, 18), and should be useful for inhibiting undesirable side reactions that may limit the formation of desired compounds. Drawing again on tropane alkaloid biosynthesis for an example, in several Hyoscyamus root cultures the activities of pseudotropine-forming tropinone reductase are higher than the activities of tropine-forming tropinone reductase (19). It is therefore postulated that the two reductases are competing for the same substrate tropinone and that the stronger pseudotropine-forming activity may be competing for tropinone which could otherwise be utilized as a precursor of hyoscyamine and scopolamine. If the side reaction leading to pseudotropine can be suppressed by expressing the antisense RNA of the undesired reductase, it may result in increased alkaloid formation.

4.5. Manipulation of Regulatory Genes

Co-ordinate tissue- and cell-specific expression of structural genes for each group of secondary metabolites is believed to be controlled by one or more regulatory genes. The products of the regulatory genes may be DNA-binding proteins that function as transcriptional activators or repressors. Several such genes that control flavonoid biosynthesis have been cloned (mainly from maize) (20-23). Although regulatory genes for other groups of secondary products have not been identified, we can expect rapid progress in this field. When regulatory genes controlling the genes for enzymes catalyzing the biosynthesis of useful secondary products are isolated, an obvious experiment will be to express such genes constitutively in cell culture or in every cell of a whole plant. The expression of these regulatory genes in undifferentiated cells may allow us to decouple biochemical differentiation from morphological differentiation, a decoupling which, at present, has only been achieved in a few cases. Whether we can solve the long-standing problem of differentiation in cell culture will depend on future advances in the studies on the molecular mechanisms of tissue-specific expression.

4.6. Selection of Regulatory Mutants

Regulatory gene products are believed to act as transcriptional regulators by binding to promoter regions of the structural genes they control. When calli are induced from a transgenic plant that expresses a selectable marker gene (e.g. antibiotic-resistant genes) under the control of a tissue-specific promoter of a secondary metabolism gene, and then mutagenized, it may be possible to select calli with increased expression of the introduced (resistant) marker. In such resistant mutant cells, the expression of the tissue-specific promoter might be changed to a tissue-independent manner. All the secondary metabolism genes under the control of similarly regulated promoters might also show altered, tissue-independent expression. Although the strategy is straightforward, we do not know whether such de-regulated mutants can be obtained by simple mutagenesis. In maize flavonoid biosynthesis, several regulatory genes exhibit extensive allelic diversity with respect to the tissue specificity (23). Molecular analyses of such alleles will provide a more sound basis for estimating our chances for producing mutant regulatory genes with altered tissue-specificity.

5. CONCLUDING REMARKS

The ability of plant cells to synthesize a vast array of secondary products in a temporal and cell-specific manner far exceeds the biosynthetic diversity of the other kingdoms. Using some of the strategies listed above and others of which we have not yet thought, we can begin to unravel the cellular mechanisms that allow such controlled expression of biosynthesis genes. A good combination of basic research and technological applications will continue to support advances in studies of secondary product biosynthesis.

ACKNOWLEDGEMENT

We thank J. H. Fitchen for his useful suggestions and linguistic help for preparation of this manuscript.

REFERENCES

1. Fujita, Y. (1988) in Applications of Plant Cell and Tissue Culture (Yamada, Y., chairman), Ciba Foundation Symposium 137, pp.228-238, John Wiley & Sons, Chichester
2. Peleman, J., Boerjan, W., Engler, G., Seurinck, J., Botterman, J., Alliotte, T., Van Montagu, M. and Inzé (1989) The Plant Cell **1**, 81-93
3. Hashimoto, T. and Yamada, Y. (1990) in Methods in Plant Biochemistry (Hostettmann, K., ed.), Vol. 9, Chapter 2, Academic Press, London (in press)
4. Ushiyama, K. (1989) in Abstracts for the 1989 International Chemical Congress of Pacific Basin Societies, Honolulu, Hawaii, No.256 in Bioscience and Technology section
5. Lui, J.H.C. and Staba, E.J. (1979) Phytochemistry **18**, 1913-1916
6. Hagimori, M., Matsumoto, T. and Obi, Y. (1982) Plant Physiol. **69**, 653-656
7. Kaneko, T., Hashimoto, T., Kumpaisal, R. and Yamada, Y. (submitted)
8. Scott, A.I., Mizukami, H. and Lee, S.-L. (1979) Phytochemistry **18**, 795-798
9. Schallenberg, J. and Berlin, J. (1979) Z. Naturforsch. **34C**, 541-545
10. Hashimoto, T., Matsuda, J., Okabe, S., Amano, Y., Yun, D.J., Hayashi, A. and Yamada, Y. (1990) in this volume of IAPTC Abstracts
11. Meyer, P., Heidmann, I., Forkmann, G. and Saedler, H. (1987) Nature **330**, 677-678
12. Hashimoto, T., Mitani, A. and Yamada, Y. (1990) Plant Physiol. **93**, (in press)
13. Mizusaki, S., Tanabe, Y., Noguchi, M., and Tamaki, E. (1972) Phytochemistry **11**, 2757-2762

14. Frydman, R.B., Ruiz, O.R., Kreisel and M., Bachrach, U. (1987) FEBS Lett. **219**, 380-386
15. Galneder, E., Rueffer, M., Wanner, G., Tabata, M. and Zenk, M.H. (1988) Plant Cell Rep. **7**, 1-4
16. Yamada, Y. and Okada, N. (1985) Phytochemistry **24**, 63-65
17. van der Krol, A.R., Lenting, P.E., Veenstra, J., van der Meer, I.M., Gerats, A.G.M., Mol, J.N.M. and Stuitje, A.R. (1988) Nature **333**, 866-869
18. van der Krol, A.R., Mur, L.A., de Lange, P., Gerats, A.G.M., Mol, J.N.M. and Stuije, A.R. (1990) Mol. Gen. Genet. **220**, 204-212
19. Dräger, B., Hashimoto, T. and Yamada, Y. (1988) Agric. Biol. Chem. **52**, 2663-2667
20. Paz-Ares, J., Ghosal, D., Wienand, U., Peterson, P. and Saedler, H. (1987) EMBO J. **6**, 3553-3558
21. Dellaporta, S., Greenblatt, I., Kermicle, J., Hicks, J.B. and Wessler, S. (1988) in Chromosome Structure and Function: Impact of New Concepts (Gustafson, J.P. and Appels, R., eds.), pp.263-282, Plenum Press, New York
22. Ludwig, S.R., Habera, L.F., Dellaporta, S.L. and Wessler, S.R. (1989) Proc. Natl. Acad. Sci. USA **86**, 7092-7096
23. Chandler, V.L., Radicella, J.P., Robbins, T.P., Chen, J. and Turks, D. (1989) The Plant Cell **1**, 1175-1183

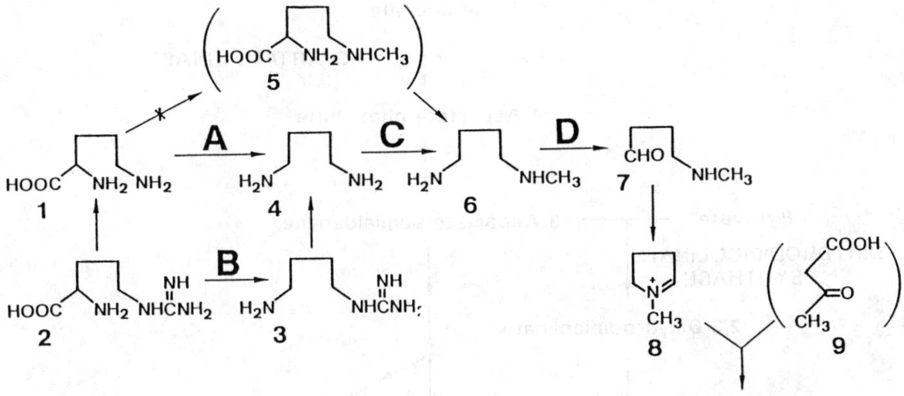

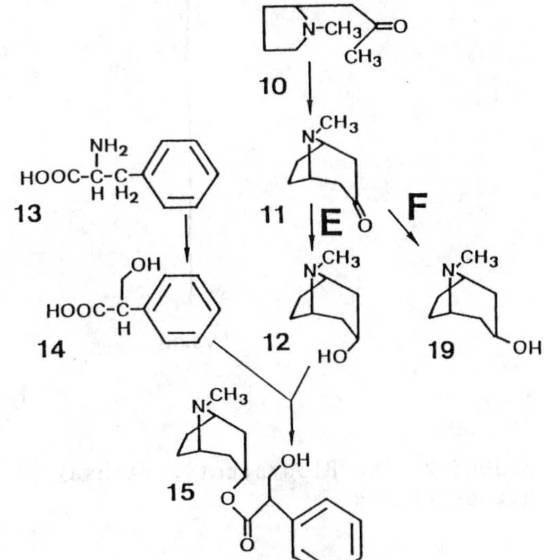

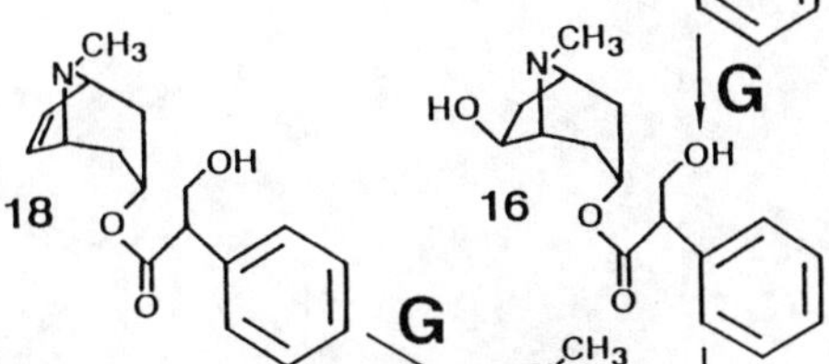

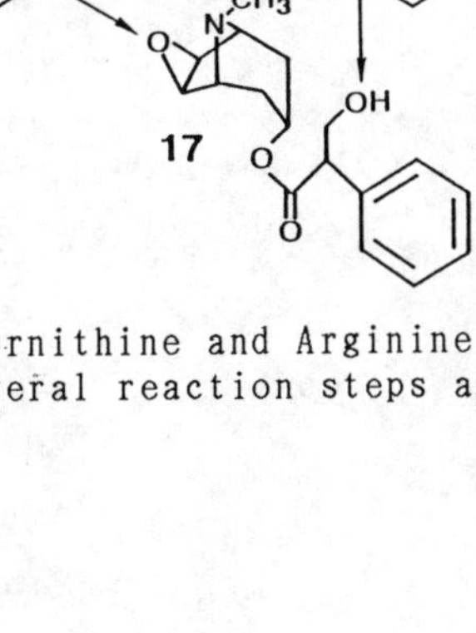

1. L-ornithine
2. L-arginine
3. agmatine
4. putrescine
5. δ-N-methyl-L-ornithine
6. N-methylputrescine
7. 4-methylaminobutanal
8. N-methyl-1-pyrrolinium cation
9. acetoacetic acid
10. hygrine
11. tropinone
12. tropine
13. L-phenylalanine
14. [S]-tropic acid
15. [S]-hyoscyamine
16. 6β-hydroxyhyoscyamine
17. scopolamine
18. 6,7-dehydrohyoscyamine
19. pseudotropine

A. Ornithine decarboxylase
B. Agrinine decarboxylase
C. Putrescine N-methyltransferase
D. Diamine oxidase
E. Tropinone reductase (tropine-forming)
F. Tropinone reductase (psudotropine-forming)
G. Hyoscyamine 6β-hydroxylase

FIGURE 1. The Metabolic Pathways from Ornithine and Arginine to Scopolamine. Enzymes that catalyze several reaction steps are represented alphabetically (A-G).

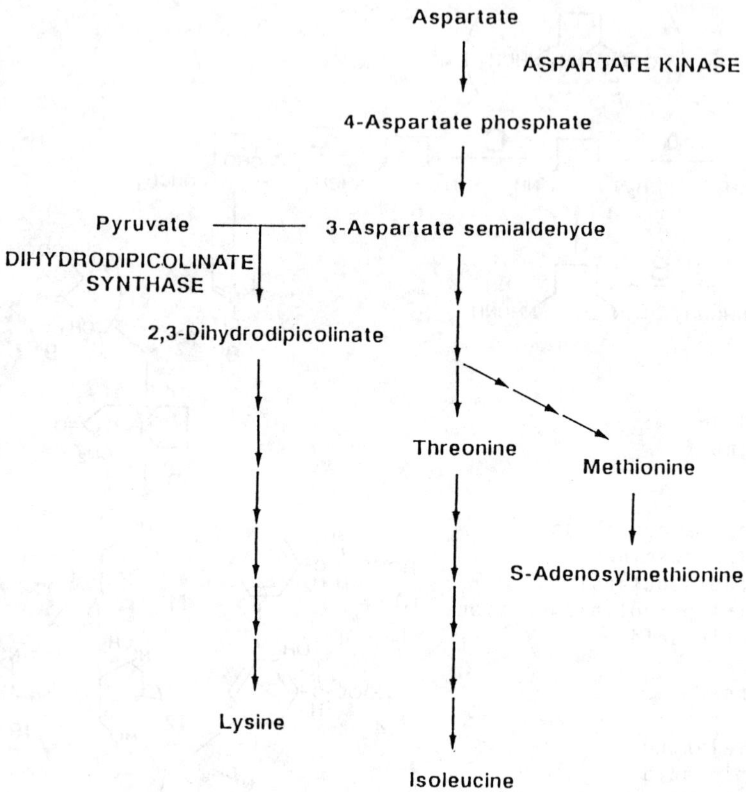

FIGURE 2. The Biosynthetic Pathway of Aspartate Family Amino Acids in Higher Plants.

THE TOXIC COMPARTMENT OF PLANT CELLS

PHILIPPE MATILE, Dept. of Plant Biology, University of Zürich, Zollikerstrasse 107, CH-8008 Zürich, Switzerland

1. INTRODUCTION

Spinach has a mild taste and is not astringent. Its eatableness is due to low contents of those secondary products of plant metabolism which are abundant in unpalatable tissues of wild plant species. Normally homogenates turn brown and enzymes are prone to be inactivated as a consequence of binding to tannins and other phenolics. Spinach leaves rank among the most favoured sources of enzymes because such undesirable effects of tissue disintegration are scarce. In the case of a majority of other plant tissues the safe isolation of proteins requires the use of sophisticated procedures employing media which contain protective ingredients for the binding of phenolics (1).

Phenolics merely provide a well known example for illustrating a general phenomenon: plant cells contain compounds which are potentially harmful as they interfere, in one way or another, with vitally important functions of metabolism. Obviously, the living cell is well protected against endogenous autotoxic compounds. Among the mechanisms responsible for detoxification the sequestration of secondary compounds in a "Toxic compartment of plant cells" (2) is the most conspicuous one. This compartment is largely (not exclusively) identical with the vacuoles.

2. POTENTIALLY AUTOTOXIC COMPOUNDS

Selected examples of potential poisoning of plant cells by endogenous compounds are compiled in Table 1. The mechanisms of poisoning appear to be as manifold as are the chemical structures of compounds which are potentially harmful. There is virtually no part of cellular metabolism which is not, in one species or another, potentially endangered by endogenous substances. Moreover, these toxic compounds quite often occur in large quantities.

Malonic acid is a classical inhibitor of succinate dehydrogenase. In soybean this respiratory poison is the dominant organic acid, twenty times more abundant than succinic acid (3) yet "shortness of breath" does not occur in soybean tissues. Rotenon is a well known insecticide from the root of *Derris elliptica*; it inhibits mitochondrial electron transport specifically as does the phytoalexin glyceollin which is produced in infected soybean and is known to block electron transport in soybean mitochondria (4). Likewise it has been demonstrated that another isoflavonoid, pisatin, a phytoalexin of pea, causes the lysis of pea protoplasts when added to the suspending medium (5). Loss of membrane integrity is also caused by a number of other potentially autotoxic compounds such as the natural detergents, saponins. The example of avenacosides, sterolglycosides of oat, demonstrates that harmful metabolites may be produced upon the disrupture of cells and tissues:

in this case the glycosides are hydrolyzed by a ß-glucosidase to yield the toxic aglycones (6). Obviously, *in vivo* poisoning is prevented as the enzyme has no access to the substrate. In a similar way toxic isothiocyanates are produced from glucosinolates when tissues of Cruciferae are crushed. Widespread in the plant kingdom are plants producing cyanogenic glucosides; again the toxic principle, HCN, is generated enzymatically upon the breakdown of cellular and subcellular separation of potentially toxic compounds and catabolizing enzymes.

TABLE 1. Examples of potentially harmful compounds and causes of toxicity

Compound	Cause of toxicity
Organic acids: malic, citric acid	acidity
malonic acid	inhibitor of succinic dehydrogenase
Amino acids: canavanine	structural analogue of arginine
Amines: serotonin	structural analogue of tryptophane
Benzopyrane lactones: coumarin	uncoupler
Furanocoumarins: psoralen	photodynamic compound
Flavonoids: kaempferol	inhibits oxidative phosphorylation
Isoflavonoids: rotenone, glyceollin	inhibit mitochondrial electron transport
Polymeric flavonoids: condensed tannins	interaction with proteins and glycoproteins
Benzoic acid derivatives: gallotannins	
Glucosinolates: sinigrin ⟶ allyl NCS	cytotoxic, covalent binding to DNA
Cyanogenic glycosides: dhurrin ⟶ HCN	inhibitor of cytochrome oxidase
Sterols: saponins	detergents, membrane deterioration
Alkaloids: colchicine	blocks mitosis
berberine	interaction with DNA
sanguinarine	deterioration of membranes
Porphyrins: chlorophyll	photodynamic compound
Proteins: unspecific phosphatases	interference with intermediary sugar metabolism
proteases	cleavage of enzyme proteins
ricin, reserve protein of castor bean	inhibitor of protein synthesis

3. DETOXIFICATION: THE ROLE OF VACUOLES

Vacuoles represent the most conspicuous storage compartment of secondary compounds. The cell sap which may occupy more than 90% of the total cell volume is astonishingly absorbable. In young shoots of *Serghum bicolor*, the cyanogenic glycoside, dhurrin, makes up as much as 30% of the dry mass and the compound is exclusively located in the vacuoles of epidermal cells (17,32). Water solubility is a prerequisite of vacuolar solutes. Apolar secondary compounds are usually conjugated with polar molecules, glycosilation being the most frequent mode of conferring adequate polarity.

In the case of coloured phenolics such as anthocyanins, the vacuolar location can be observed directly in the light microscope, e.g. in epidermal cells of petals. Localization at the level of electron microscopy requires adequate fixation in the presence of caffeine which forms precipitated complexes with phenolics; the subsequent staining with Fe^{2+} reveals a strict confinement of tannins in non-plasmatic subcellular compartments including the central vacuoles (7). The same principle of precipitation has been employed for demonstrating the fusion of vacuoles upon the artificial combination of protoplasts (8) of *Happlopappus* (vacuolar phenolics) and *Cataranthus* (vacuolar alkaloids). In cultured cells of *Ocimum basilicum* the newly produced Lamiacean tannins are continuously sequestered in the vacuoles as shown in a morphometric study (9); only in the stationary phase of the culture, compartmentation breaks down in dead cells and phenolics diffuse into the cytosol. Considering the toxic nature of phenolics it may be stated, with slight exaggeration, that the distance between life and death of individual cells is 7.5 nm, the thickness of the tonoplast which separates poisonous substances from the cytosolic proteins.

TABLE 2. Localization of secondary metabolites in isolated vacuoles

Compound	Plant species	Reference
o-Coumaric acid glucoside	*Melilotus alba*	10
Anthocyanin	*Hippeastrum, Tulipa*	11
Anthocyanin, flavonol glucosides, hydroxycinnamic acid esters	*Vitis vinifera*	12
Shikimic acid	*Fagopyrum esculentum*	13
Avenacosides	*Avena sativa*	14
Cardenolide glycoside	*Convallaria majalis*	15
Cardiac glycosides	*Digitalis lanata*	16
Dhurrin	*Sorghum bicolor*	17
Sinigrin, gluconasturtin	*Armoracia lapathifolia*	18
Glucobrassicin	*Armoracia rusticana*	19
	Brassica napus	44[1]
Gentiopicroside	*Gentiana lutea*	20
Nicotine	*Nicotiana tabacum*	21
Sanguinarine, coptisine	*Chelidonium majus*	22
Thebaine, sanguinarine	*Papaver bracteatum*	39
Serpentine	*Catharanthus roseus*	23
S-Reticuline	*Fumaria capreolata*	24
Senecionine-N-Oxide	*Senecio vulgaris*	25
Betanin	*Beta vulgaris*	26
Capsaicine	*Capsicum annuum*	27

[1] Cytochemical localization in protein storage vacuoles of seeds

In the past decade methods for the isolation of vacuoles have been employed for quantitative assessments of subcellular distributions of secondary compounds. It appears from Table 2 that a variety of glycosides and alkaloids has been localized in vacuoles. Although the calculations of vacuolar/extravacuolar distributions (either based on number of isolated vacuoles or on vacuolar marker enzymes) are associated with considerable errors, the data suggest that the bulk, if not the totality of the compounds in question are sequestered in the cell sap. Compartmentation in vacuoles is also responsible for the protection of the cytoplasm from dangerous enzymes such as phosphatases and other hydrolases. Ricin is a toxic glycoprotein of castor bean endosperm which inhibits protein synthesis in 80 S ribosomes; its compartmentation in the storage vacuoles (protein bodies) protects cytoplasmic protein synthesis during germination (28, 29). A newly discovered class of toxic proteins, the pathogenesis-related thionins, occurs, in barley leaves, in too closely related groups which are located in the vacuoles and in the cell walls, respectively (30). A similar dual location of proteins in the cell wall and in the cell sap is also typical of a number of hydrolytic enzymes. Indeed, the space outside the plasmalemma can be conceived as an inside-out vacuole. This conception may even be consistent with the apoplastic localization of secondary metabolites such as flavonoids. In *Chrysoplenium americanum* polymethylated flavonol glycosides have been located predominantly in the cell walls and the vacuoles were found to contain only a minor fraction of these compounds (31). Admittedly, work with protoplasts is inevitably associated with the loss of extracellular fluids. Yet, there are good reasons to believe that in most cases the secondary compounds are not excreted but rather accumulatd intracellularly. This view is supported by results concerning the subcellular distribution of o-coumaric acid glucoside and corresponding ß-glucosidase in leaves of *Melilotus albus*. The glycoside has been localized in the vacuoles of epidermal cells; it is separated by two membranes, tonoplast and plasmalemma, from the ß-glucosidase located in the outermost wall layer of mesophyll cells (10). Obviously, the glycoside is stable only if accumulated intracellularly and toxic coumarin is produced only upon crushing of leaf tissue and breakdown of compartmentation. Analogous examples of differential localization of potentially toxic glycosides and corresponding catabolic enzymes are also illustrated in Fig. 1. In *Sorghum bicolor*, dhurrin is accumulated in large quantities in the vacuoles of epidermal cells (17,32). Two enzymes responsible for cyanogenesis, ß-glucosidase and hydroxynitrile lyase have been localized in the chloroplasts and in the cytosol, respectively, of mesophyll cells (33). The storage tissue of horseradish taproots contain large amounts of sinigrin, gluconasturtin, and a number of minor glucosinolates which are compartmentalized in the vacuoles (18). Thioglucosidase appears to be a cytosolic enzyme (34) and, hence, the conversion of glucosinolates to isothiocyanates requires the demolition of subcellular compartmentation, e.g. upon chewing horseradish tissue.

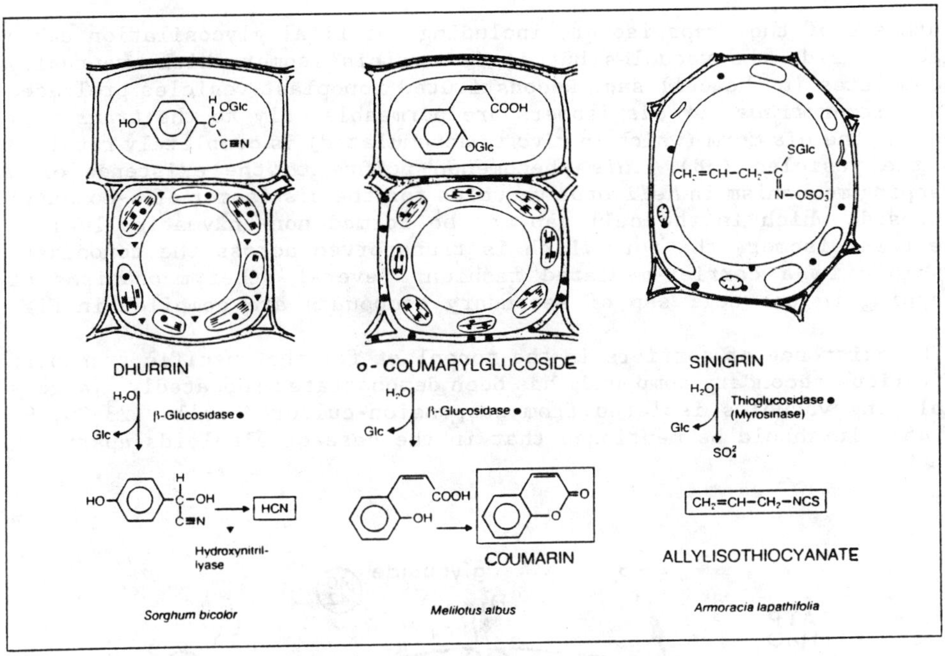

FIGURE 1. Examples of differential localization of potentially toxic glycosides in vacuoles and corresponding catabolic enzymes in other compartments. For references see text.

4. ACCUMULATION OF SECONDARY COMPOUNDS IN VACUOLES

The accumulation and final deposition in vacuoles of secondary compounds pose the problem of transport across the tonoplast and retention in the cell sap. Early hypotheses that final steps of biosynthesis such as glycosilation are located at the tonoplast and thus may be associated with group translocation across the membrane have not been confirmed. In many cases it has been shown that the entire biosynthesis takes place in extravacuolar compartments. However, there are exceptions as e.g. in *Capsicum* fruits where both, the pungent principle, capsaicinoid, and capsaicinoid synthase have been localized in isolated vacuoles (35). Another case of enzymatic synthesis within the cell sap, the transacylation of hydroxycinnamic glucose-esters to malic acid-esters has been discovered in protoplasts of *Raphanus sativus* leaves (36). In general, however, the final products of biosynthesis are transported into the vacuoles and, in one way or another, are locked in the cell sap.

One of the most intensively studied system is sweet clover and its potentially toxic compound, o-coumaryl glucoside (37). The entire synthesis of the *trans* isomer, including the final glycosilation takes place outside the vacuoles but it is the *cis* isomer which eventually accumulates in the cell sap. Reconstituted tonoplast vesicles preloaded with either *trans*- or *cis* isomers are permeable only to the *trans* form whereas the *cis* form (which *in vivo* is accumulated) is completely retained in the vesicles (38). This phenomenon points to the existence of a trapping mechanism in *Melilotus* vacuoles for the *cis* isomer of o-coumaryl glucoside which in the cell gap may be formed non-enzymatically from the *trans* isomer, the form which is transported across the tonoplast, possibly in a carrier-mediated fashion. Several other mechanisms of trapping in the cell sap of secondary compounds are compiled in Fig. 2.

The existence of carriers in the tonoplast for the specific transport of various secondary compounds has been demonstrated repeatedly, largely employing vacuoles isolated from suspension-cultured cells (23-25,39-43,45). It should be mentioned that in the case of alkaloids apart

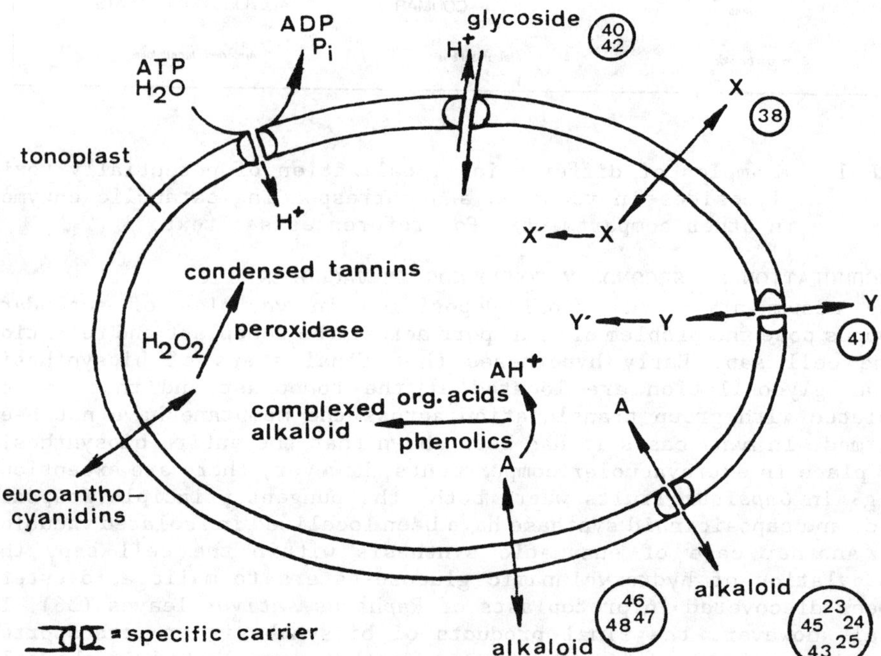

FIGURE 2. Mechanisms of transport across the tonoplast and retention in the cell sap of secondary metabolites. Encircled numbers indicate references. The formation of condensed tannins in the cell sap is a speculative analogy to peroxidase-catalysed lignin formation.

from carrier-mediated transport across the tonoplast, uncatalized diffusion has also been observed (22,46,47). Like neutral red, the classical vital stain of vacuoles, alkaloids may pass the tonoplast as lipophilic free bases; in the acidic environment of the vacuolar fluid they are eventually protonated and trapped as cations for which the membrane is impermeable. Moreover, binding in the cell sap to phenolics or organic acids may occur (47, 48). Mechanism of accumulation in the vacuoles as dependent on acidic cell saps and on pH gradient between cell sap and cytosol, respectively, depend indirectly on the tonoplastic proton-pumping ATPase which is involved in the regulation of vacuolar pH (49). In case of secondary active transport of compounds into vacuoles, ATP and tonoplastic ATPase, respectively, are involved directly in the energetization of the process.

The internal deposition of secondary compounds in the vacuoles is a serious handicap in the employment of cultured cells for the industrial production of valuable substances. Excretion into the medium would be of great biotechnological significance. The elucidation of carrier-mediated transport processes across the tonoplast of glycosides (16, 40-42) or alkaloids (23-25, 43) may open possibilities of changing internal into external excretion. Accumulation in vacuoles appears to be associated with the expression of genes encoding for specific permeases residing in the tonoplast. It can be expected that the targeting of proteins to distinct membranes will soon be elucidated at the molecular level. Knowledge about the requirement of trafficing proteins specifically to the plasmalemma will eventually allow the engineering of cell lines possessing plasmalemma-located carriers. In such cell lines the desired compounds would be excreted into the medium and harvesting would be possible in a non-destructive fashion.

5. FINAL VERSUS TEMPORARY DEPOSITION IN VACUOLES

The high retention of secondary compounds in isolated vacuoles which has been observed in many instances suggests that deposition may be final and last for the entire lifespan of the cell. This may be the rule but there are exceptions as e.g. the cyanogenic glucoside, linamarin, which, during gemination of *Phaseolus lunatus* seeds gradually disappears from the cotyledons and appears to be translocated to growing tissues of the seedling (50). Removal from the vacuoles and transport (in the phloem?) must proceed in such a way that cyanogenesis is avoided. The catabolic key enzyme, ß-glucosidase, has been localized in the apoplast (51) and, hence, apoplastic phloem loading would be associated with the initiation of cyanogenesis. In a similar system, *Hevea brasiliensis*, cyanogenesis appears to be avoided by glycosilation of linamarin to the diglucoside linustatin which is resistant to linamarase (52).

6. CHLOROPHYLL: DETOXIFICATION OF A PHOTOSENSITIZER

Free porphyrins are highly toxic when illuminated in the presence of oxygen. In chloroplasts the situation is particularly intriguing as light is both the indispensable source of energy and oxygen is a product of the light reaction. In the pigment-protein complexes, chlorophyll is associated with carotenoids which are potent quenchers of activated oxygen. The effect of herbicides aimed at the inhibition of carotenoid synthesis, demonstrates the suicide power of photodynamically unprotected chlorophyll.

In senescent leaves the protection of chlorophyll is abolished upon the dismantling of thylakoids and the production of free chlorophyllide. Yet, viability of the senescent mesophyll is very important because the orderly breakdown of cellular constituents and the withdrawal of organic compounds from the leaves can take place only in compartmentalized cells. Hence, the chloroplasts not only represent a potentially toxic compartment, they must also be equipped with a potent machinery for the detoxification of photodynamic free porphyrins.

The recent discovery of chlorophyll catabolites in senescent leaves of barley and other species (53) has allowed the search for reactions taking place in chloroplasts and yielding compounds known to be derived from green pigments. One of these catabolites, a lipofuscin-like blue-fluorescent compound (of yet unknown chemical structure) could be identified as being produced in senescent chloroplasts. The reaction turned out to depend strictly on ATP as well as on oxygen (54). ATP appears to have a dual function; one of them concerns the export of the chlorophyll catabolite from the chloroplasts. The analysis of subcellular distribution of the catabolites has shown that these compounds are finally accumulated in the vacuoles (53). Hence, vacuoles not only represent a toxic compartment, they also store the products of detoxification in the chloroplasts.

7. CONCLUDING REMARKS

Secondary products of plant metabolism represent a centre piece of ecological biochemistry (55). They constitute a web of chemical signals which determine or influence interactions between plants and other organisms within the biocoenosis. Coevolution has generated, on the one hand, species-specific chemical constitutions of plants and, on the other hand, corresponding behaviours of herbivores and pollinators or strategies of pathogenic microbes. Coevolution includes also various detoxification mechanisms, last but not least within the plant cells which produce potentially toxic compounds. Subcellular accumulation in non-plasmatic compartments, particularly in vacuoles appears to represent the most striking protection of the cytoplasm from poisoning by endogenous metabolites.

REFERENCES

1 Loomis, W.D. (1966) Phytochemistry 5, 423-438
2 Matile, Ph. (1984) Naturwissenschaften 71, 18-24
3 Stumpf, O.K. and Burris, R.H. (1981) Plant Physiol. 68, 989-991
4 Boydston, R., Paxton, J.D. and Koeppe, D.E. (1983) Plant Physiol. 72, 151-155
5 Shiraishi, T., Oku, H., Isono, M. and Ouchi, S. (1975) Plant Cell Physiol. 16, 939-942
6 Laudenbach, U. and Kesselmeier, J. (1982) Z.Naturforsch. 37, 1095-1099
7 Mueller, W.C. and Greenwood, A.D. (1978) J. Exp. Bot. 29, 757-764
8 Constabel, F., Koblitz, H., Kirkpatrick, J.W. and Rambold, S. (1980) Can. J. Bot. 1032-1034

9 Lang, E., Amelunxen, H., Friedrich, H. and Hörster, H. (1979) Europ. J. Cell Biol. 19, 67-72
10 Oba, K., Conn, E.LE., Canut, H. and Boudet, A.M. (1981) Plant Physiol. 68, 1359-1363
11 Wagner, G.J. (1979) Plant Physiol. 64, 88-93
12 Moskowitz, A.H. and Hrazdina, G. (1981) Plant Physiol. 68, 686-692
13 Hollaender-Czytko, H. and Amrhein, N. (1983) Plant Sci. Lett. 29, 89-96
14 Kesselmeier, J. and Urban, B. (1983) Protoplasma 114, 133-140
15 Loeffelhardt, W., Kopp, B. and Kubelka, W. (1979) Phytochemistry 18, 1289-1291
16 Kreis, W. and Reinhard, E. (1987) J. Plant Physiol. 128, 311-326
17 Saunders, J.A. and Conn E.LE. (1978) Plant Physiol. 61, 154-157
18 Grob, K. and Matile, Ph. (1979) Plant Sci. Lett. 14, 327-335
19 Helminger, J., Rausch, T. and Hilgenberg, W. (1983) Plant Physiol. 58, 302-310
20 Keller, F. (1986) J. Plant Physiol. 122, 473-476
21 Saunders, J.A. (1979) Plant Physiol. 64, 74-78
22 Matile, Ph., Jans, B. and Rickenbacher, R. (1970) Biochem. Physiol. Pflanzen 161, 447-458
23 Deus-Neumann, B. and Zenk, M.H. (1984) Planta 162, 250-260
24 Deus-Neumann, B. and Zenk, M.H. (1986) Planta 167, 44-53
25 Ehmke, A., v. Borstel, K. and Hartmann, T. (1988) Planta 176, 83-90
26 Leigh, R.A., Rees, T., Fuller, W.A. and Banfield, J. (1979) Bichem. J. 178, 539-547
27 Fujiwake, H., Suzuki, T. and Iwai, K. (1980) Plant Cell Physiol. 21, 1023-1030
28 Tully, R.E. and Beevers, H. (1976) Plant Physiol. 58, 710-716
29 Youle, R.J. and Huang, A.H.C. (1976) Plant Physiol. 58, 703-709
30 Reimann-Philipp, U., Schrader, G., Martinoia, E., Barkholt, V. and Apel, K. (1989) J. Biol. Chem. 264, 8978-8984
31 Ibrahim, R.K., de Luca, V. Khouri, H., Latchinian, L., Brisson, L. and Charest, P.M. (1987) Phytochemistry 26, 1237-1246
32 Kojima, M., Poulton, J.E., Thayer, S.S. and Conn, E.E. (1979) Plant Physiol. 63, 1022-1028
33 Thayer, S.S. and Conn, E.E. (1981) Plant Physiol. 67, 617-622
34 Lüthy, B. and Matile, Ph. (1984) Biochem. Physiol. Pflanzen 179-5-12
35 Fujiwake, H., Suzuki, T. and Iwai, K. (1982) Agr. Biol. Chem. Tokyo 46, 2685-2690
36 Strack, D., and Sharma, V. (1985) Physiol. Plantarum 65, 45-50
37 Boudet, A.M., Alibert, G. and Marigo, G. (1984) Ann. Proc. Phytochem. Soc. Europe 24, 29-47
38 Rataboul, P., Alibert, G., Boller, Th. and Boudet, A.M. (1985) Biochim. Biophys. Acta 816, 25-36
39 Kutchan, T.M., Rush, M. and Coscia, C.J. (1986) Plant Physiol. 81, 161-166
40 Hopp, W. and Seitz, H.U. (1987) Planta 170, 74-85
41 Matern, U., Reichenbach, C. and Heller, W. (1986) Planta 167, 183-189
42 Werner, Ch. and Matile, Ph. (1985) J. Plant Physiol. 118, 237-249
43 Mende, P. and Wink, M. (1987) J. Plant Physiol. 129, 229-242
44 Yiu, S.H., Collins, F.W., Fulcher, R.G. and Altosaar, I. (1984)

Can. J. Plant Sci. 64, 869-878
45 Homeyer, B.C. and Roberts, M.F. (1984) Z. Naturforsch. 39 c, 876-881
46 Renaudin, J.-P. (1989) Plant Physiol. Biochem. 27, 613-621
47 Matile, Ph. (1974) Nova Acta Leopoldina suppl. Nr. 7, 139-156
48 Neumann, D., Krauss, G.Hieke, M. and Gröger, D. (1983) Planta medica 48, 20-23
49 Kurkdjian A. and Guern J. (1989) Annu. Rev. Plant Physiol. Plant Mol. Biol. 40, 271-303
50 Clegg, D.O. and Conn, E.E. (1979) Nature 278, 343-344
51 Frehner, M. and Conn, E.E. (1987) Plant Physiol. 84, 1296-1300
52 Selmar, D., Lieberei, R., Biehl, B. and Voigt, J. (1987) Plant Physiol. 83, 557-563
53 Matile, Ph., Düggelin, Th., Schellenberg, M., Rentsch, D., Bortlik, K., Peisker, Ch. and Thomas, H. (1989) Plant Physiol. Biochem. 27, 595-604
54 Schellenberg, M., Matile, Ph., and Thomas, H. (1990) J. Plant Physiol. (in press)
55 Harborne, J.B. (1987) Introduction to ecological biochemistry 3rd ed. Academic Press, New York

TRANSPORT AND ACCUMULATION OF SECONDARY METABOLITES

PETER E. BRODELIUS*, INSTITUTE OF BIOTECHNOLOGY, ETH-HÖNGGERBERG, CH-8093 ZÜRICH, SWITZERLAND

1. INTRODUCTION

In intact plants secondary metabolites are either accumulated within the biosynthetic cells or they are excreted. Some excreted products are incorporated into cell walls of the biosynthetic cells but more often they are transported to and accumulated within specialized storage cells. The development of biosynthetic and storage cells are linked to differentiation of the plant. Storage cells may be located in the same organ as the cells synthesizing the secondary metabolite or they may be found in another organ requiring transport over a relatively long distance. Dawson's classical experiments in the early 40s elegantly demonstrated the latter situation [1,2]. Using reciprocal grafts between tobacco and tomato Dawson could show that nicotine is synthesized in the roots of tobacco and then transported to the leaves where the alkaloid is accumulated. Secondary products are under certain conditions accumulated within the intercellular space. This is particularly true for phytoalexins produced upon infection by a pathogen.

A similar pattern of secondary metabolite synthesis and transport can be found in biosynthetically active cell cultures as schematically shown in Figure 1. The secondary metabolites are frequently stored within the vacuoles of the producing cells. However, there are also a number of examples of excretion of the secondary metabolite into the growth medium. Our understanding of the processes directing a secondary metabolite into the vacuolar space or into the extracellular medium is still relatively limited.

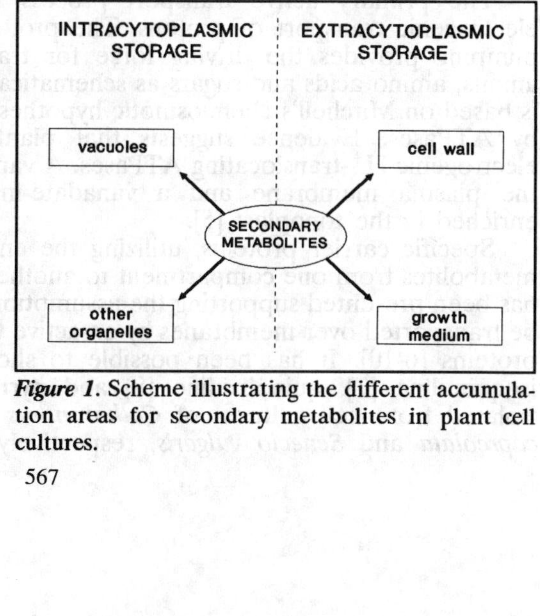

***Figure 1*. Scheme illustrating the different accumulation areas for secondary metabolites in plant cell cultures.**

*Present address: Department of Plant Biochemistry, University of Lund, P.O.Box 7007, S-22007 Lund, Sweden.

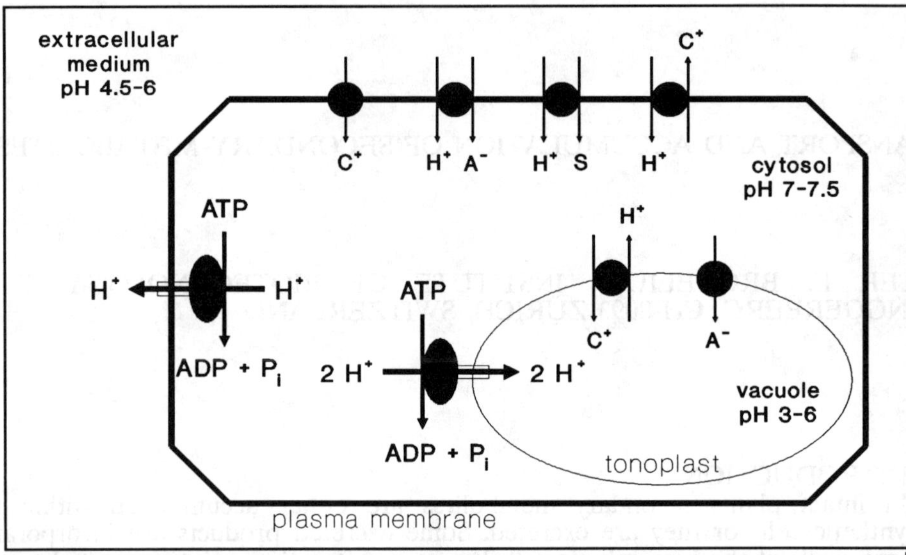

Figure 2. **Schematic diagram of working model showing the function and location of H^+-translocating ATPases in a plant cell. An electrochemical gradient of protons may be coupled to transport of other solutes (C^+, A^- and S refer to cation, anion and organic solutes, respectively. (Adopted from [3])**

2. MEMBRANE TRANSPORT

Two membranes, i.e. the plasma membrane separating the cell interior from the external medium and the tonoplast surrounding the vacuole, are of particular importance for the transport and accumulation of secondary metabolites. Both active energy-requiring and passive transport over these membranes have been suggested.

2.1 Active transport

The primary active transport process in plant cells is thought to be an electrogenic transport of protons. The proton motive force generated by proton pumping provides the driving force for transport of solutes, including cations, anions, amino acids and sugars as schematically illustrated in Figure 2. This model is based on Mitchell's chemiosmotic hypothesis [4]. Proton gradients are generated by ATPases. Evidence suggests that plant cells have at least two types of electrogenic H^+-translocating ATPases. A vanadate-sensitive ATPase is enriched in the plasma membrane and a vanadate-insensitive, anion-sensitive ATPase is enriched in the tonoplast [5].

Specific carrier proteins, utilizing the energy of the pH gradient, translocate metabolites from one compartment to another. During recent years some evidence has been presented supporting the assumption that secondary metabolites may also be transported over membranes by an active transport system involving such carrier proteins [6-10]. It has been possible to show the specific uptake of indole [6], isoquinoline [7], quinolizidine [8] and pyrrolizidine alkaloids [9] into vacuoles isolated from cell cultures of *Catharanthus roseus, Lupinus polyphyllus, Fumaria capreolata* and *Senecio vulgaris*, respectively. The following conclusions strongly

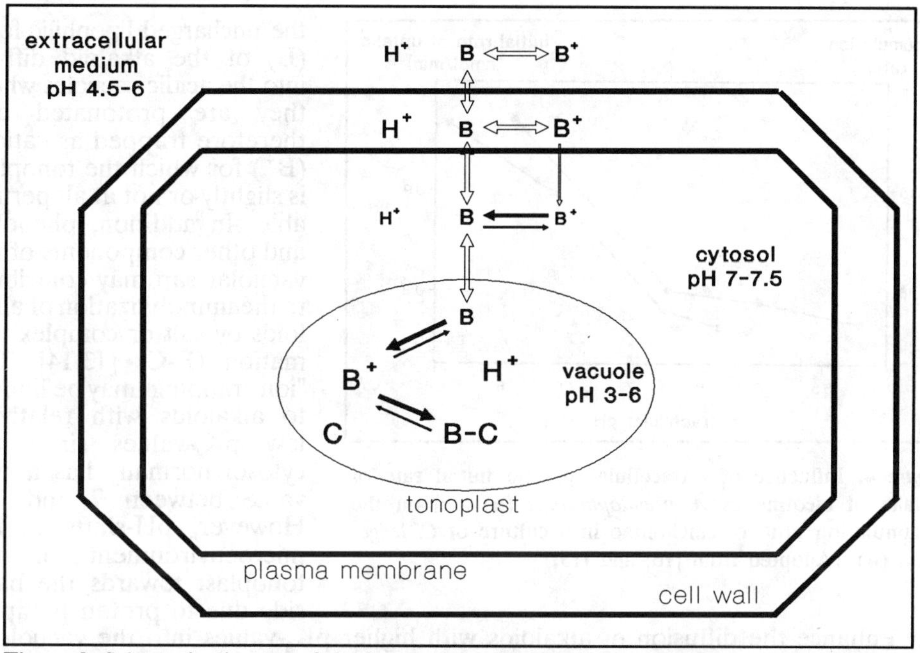

Figure 3. Schematic diagram of the ion-trap mechanism for accumulation of an alkaloid in the vacuole. B and B^+ are uncharged and charged forms of the alkaloid, respectively. H^+ in the proton and C represents complexing sites. (Adopted from [12])

supported carrier-mediated active transport:

- the uptake of externally added alkaloids followed saturation kinetics
- ATPase inhibitors blocked the uptake of alkaloids
- stimulation of alkaloid uptake by ATP
- uptake of species-specific alkaloids

Active transport of alkaloids into the extracellular medium has also been observed [10]. A variant cell line of *Thalictrum minus* produce large amounts of isoquinoline alkaloids which are spontaneously released into the growth medium. However, ATPase inhibitors blocked alkaloid efflux indicating an energy-requiring transport system.

It may be concluded that at least in some instances transport of secondary metabolites over membranes is an active, energy-requiring process which is specific for the substrate. The integration of specific carrier proteins into the tonoplast or into the plasma membrane can thus direct secondary metabolites either to the vacuole or to the extracellular medium. Such a carrier protein has not yet been isolated.

2.2 Passive transport

It has been proposed that neutral alkaloid molecules can freely diffuse across membranes, but protonated alkaloid cations cannot [11,12]. Alkaloids can thus accumulate in acidic compartments (e.g. vacuoles) within the cells by an "ion-trapping" mechanism as schematically shown on Figure 3. According to this model

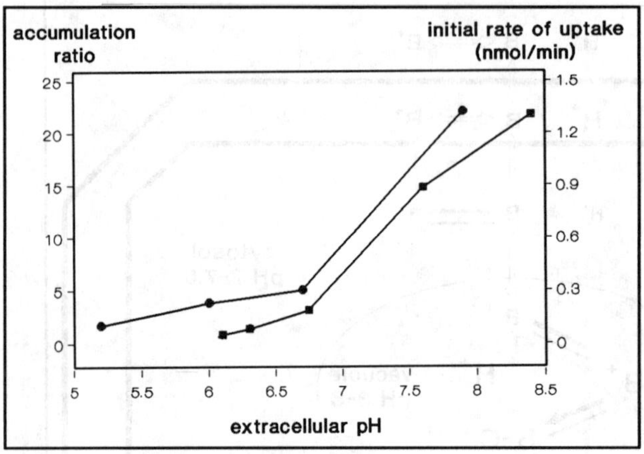

Figure 4. Influence of extracellular pH on initial rate of uptake of nicotine by *A. pseudoplatanus* (●) and on the accumulation ratio of cinchonine in a culture of *C. ledgeriana* (▪). (Adopted from [16] and [15])

the uncharged lipophilic form (B) of the alkaloid diffuse into the acidic vacuoles where they are protonated and therefore trapped as cations (B^+) for which the tonoplast is slightly or not at all permeable. In addition, phenolics and other components of the vacuolar sap may contribute to the immobilization of alkaloids by salt or complex formation (B-C) [13,14]. The "ion-trapping" may be limited to alkaloids with relatively low pK_a-values since the cytosol normally has a pH-value between 7 and 7.5. However, pH-shifts in the microenvironment of the tonoplast towards the basic side due to proton pumping may enhance the diffusion of alkaloids with higher pK_a-values into the vacuoles.

The ion-trap mechanism has been proposed to be operative in the uptake from the medium of externally added indole [11,12] and quinoline [15] alkaloids and nicotine [16] by cultivated cells of *C. roseus*, *Cinchona ledgeriana* and *Acer pseudoplantanus*, respectively. The initial uptake rate and the accumulation ratio (i.e. the ratio between the intracellular (C_i) and the extracellular (C_e) concentration of the fed substance) increased with increasing extracellular pH as shown in Figure 4. A higher than from "ion-trapping" alone predicted intracellular concentration of secondary metabolite was observed in some of the studies. The accumulation ratio may be shifted either by an active transport within the cells or by the deposition of the product as an immobilized complex (c.f. Fig. 3).

In conclusion, using intact cells it is relatively uncomplicated to study the passive transport over the plasma membrane while it may be difficult to study such a transport over the tonoplast. Also here isolated vacuoles would give a better experimental system. However, a combination of active and passive transport of secondary metabolites within cultivated plant cells appears to be likely.

3. RELEASE OF INTRACELLULARLY STORED SECONDARY METABOLITES BY PERMEABILIZATION

The biotechnological utilization of plant cell cultures (especially immobilized cells) is hampered by the fact that secondary metabolites are stored intracellularly. Various attempts have been made to make cell membranes (both the plasma membrane and the tonoplast) more permeable for the release of such intracellularly stored metabolites into the medium. However, until now no general procedure has been elaborated.

3.1 Chemical permeabilization

Various surface active chemicals (e.g. organic solvents and detergents) have

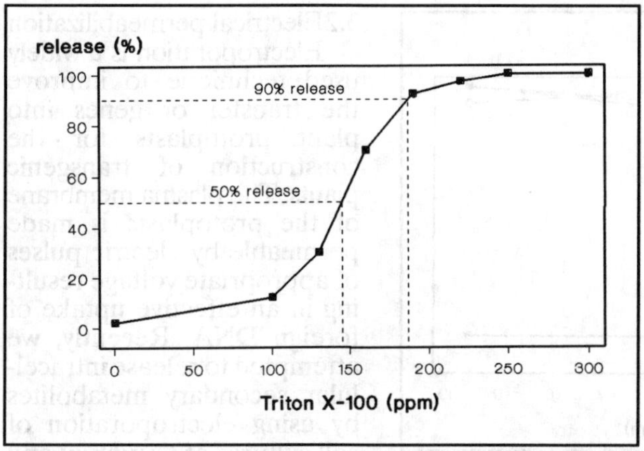

Figure 5. **Relative release of vacuolar P_i from cultivated cells of *C. roseus* as function of Triton X-100. (Adopted from [19])**

been employed to make plant cell membarnes permeable to various solutes [17]. Dimethylsulfoxide (DMSO) has been used to release indole alkaloids from immobilized cells of *C. roseus* [18]. After essentially quantitative release of the product the cells appeared to be viable. A repeated permeabilization could be carried out and a cyclic process involving product formation, permeabilization and cell recovery was suggested. This procedure would allow the reutilization of the expensive biomass. However, more extensive studies have shown that cell cultures of *C. roseus* are exceptional model systems [19]. The amount of permeabilizing agent required for quantitative release of intracellular products may be determined for any cell culture as shown in Figure 5. The concentration of some permeabilizing agents required to release 50 and 90% of the product from three different cell cultures are listed in Table 1. The viability of cells was very low after such permeabilization. This is most likely not due to the toxicity of the permeabilization agent but to the destruction of cell compartmentation and the release of intracellular toxic substances such as proteases and phenolics.

In summarizing, permeabilization by chemical substances may result in a quantitative release of intracellularly stored secondary metabolites but a very low, if any, cell viability is generally observed. This observation has also been made by other invetigators attempting chemical permeabilization of plant cells. Chemical permeabilization appears therefore not to be applicable to plant cells in the same manner as to microbial cells [20].

Table 1. **Concentration of various permeabilizing agents required for the release of 50 or 90% of intracellularly stored products from cultivated plant cells. The release of P_i, betacyanins and berberine from cells of *Catharanthus roseus*, *Chenopodium rubrum* and *Thalictrum rugosum*, respectively, was estimated. (From [19])**

Permeabilizing agent	*C. roseus* 50%	*C. roseus* 90%	*C. rubrum* 50%	*C. rubrum* 90%	*T. rugosum* 50%	*T. rugosum* 90%
DMSO (%)	3	7	10	35	13	30
Phenethylalcohol (%)	1.07	1.16	0.86	0.98	0.60	0.80
Chloroform (% sat.)	53	66	54	64	50	67
Triton X-100 (ppm)	140	185	185	230	140	210
Hexadecyltrimethylammonium bromide (ppm)	44	72	22	84	24	60

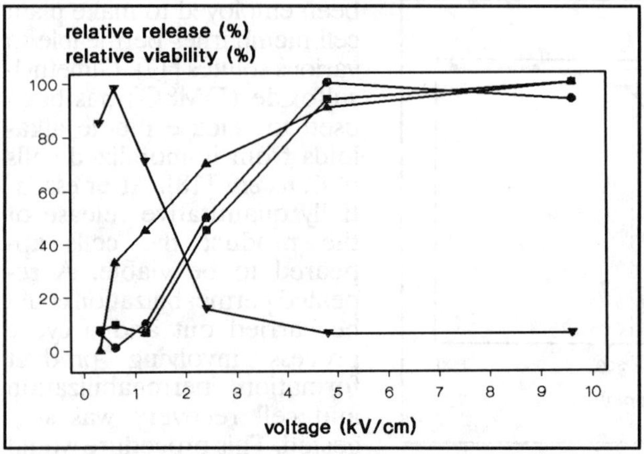

Figure 6. Relative release of berberine from freely suspended (■) agarose-entrapped (●) and alginate-entrapped (▲) cells of *T. rugosum* and relative cell viability (▼) of freely suspended cells as function of voltage applied during electroporation. (Adopted from [21])

3.2Electrical permeabilization

Electroporation is a widely used technique to improve the transfer of genes into plant protoplasts for the construction of transgenic plants. The plasma membrane of the protoplasts is made permeable by electric pulses of appropriate voltage resulting in an effective uptake of foreign DNA. Recently, we attempted to release intracellular secondary metabolites by using electroporation of cell cultures of *C. rubrum* and *T. rugosum* [21]. These two cell lines showed different response to the electroporation. The former cell culture required around 10 kV/cm for quantitative release of betacyanins from the vacuoles while the latter cell culture only required 5 kV/cm for release of berberine. Figure 6 shows the release of berberine from *T. rugosum* and cell viability (colony formation) as function of voltage used for the electroporation. From this graph it is evident that electroporation resulting in product release also leads to decreased cell viability.

The release of berberine from alginate- and agarose-immobilized cells of *T. rugosum* was also investigated. Cells entrapped in agarose responded to electroporation in the same way as freely suspensded cells, while alginate-entrapped cells released significant amount of berberine at lower voltages as shown in Figure 6. The differences observed were ascribed to the different ion character of the two polymers.

3.3 Ultrasonical permeabilization

In one study ultrasonication (1.02 MHz) has been used to release a secondary product (betanin) from suspension cultured cells (*Beta vulgaris*) [22]. The quantity and rate of release of the betacyanin increased in proportion to the length of the treatment. The investigators concluded that ultrasound stimulate the release of secondary metabolites from vacuoles by a non-thermal, caviation-mediated phenomenon.

3.4 Iontophoretic release

An iontophoretic release of intracellular alkaloids from cells of *C. roseus* has been reported [23]. In a specially designed device the cells were subjected to a low constant current for 18 h. Ajmalicine and yohimbine were during this treatment continuously released into the medium at a constant rate. An increase in current from 1 to 2 mA essentially doubled the amount of alkaloids released. No statements regarding the viability of the treated cells are made.

3.5 Conclusions regarding permeabilization

The results so far obtained utilizing various permeabilization techniques are relatively discouraging. It appears very difficult, if not imposible, to release vacuolar substances into the medium by permeabilization of the plasma membrane and the tonoplast without killing the cells. Other approaches to release intracellular secondary metabolites must be evaluated.

4. PRODUCT ACCUMULATION AFTER ELICITATION OF CELLS

Induction of secondary metabolism of cultivated plant cells by addition of biotic elicitors to the cultivation medium results in an increased productivity. Elicitor-induced secondary products are frequently at least partly released into the medium [24-27]. Consequently, elicitation is a very attractive approach for biotechnological production of secondary metabolites.

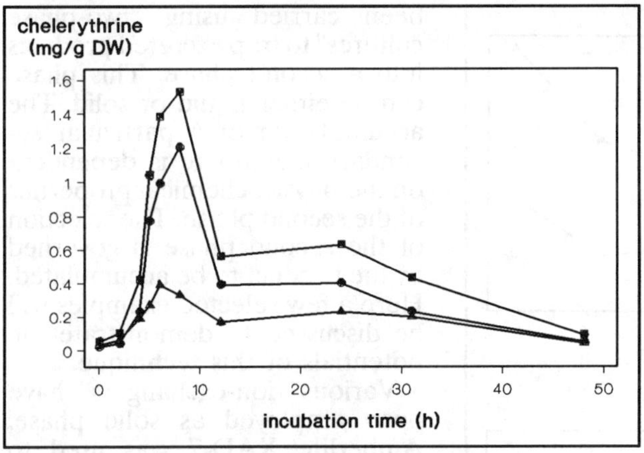

Figure 7. Time course of chelerythrine formation in elicitor-treated cells of *Eschscholtzia californica.* (▪) *Total extractable chelerythrine;* (•) *amount of chelerythrine found in medium;* (▴) *amount of chelerythrine found in cells.* (Adopted from [24])

In our own laboratory we have investigated the effects of elicitation on benzophenanthridine alkaloid formation and accumulation in cell cultures of *Eschscholtzia californica* [24]. A number of different benzophenanthridine alkaloids (e.g. sanguinarine, chelerythrine and macarpine) are produced by elicited cells. Figure 7 demonstrates the distribution of chelerythrine in cells and medium after elicitation. Between 70 and 80% of the alkaloid can be isolated from the medium at high total alkaloid levels. The same pattern was observed for sanguinarine. The disappearance of the alkaloids may be due to degradation or due to metabolization to other alkaloids (e.g. chelilutine and chelirubine). A continuous harvesting of excreted alkaloids as outlined below can be employed to prevent degradation of valuable secondary metabolites.

The reason for transport of secondary metabolites over the plasma membrane in elicited cells has not been established. Release of a significant portion of the extracellular product by cell lysis has been ruled out. The concomitant induction of carrier proteins with other elicitor-induced proteins may deserve attention.

Ultrastructural changes have been observed in association with the accumulation and secretion of sanguinarine in elictor-treated cells of *Papaver bracteatum* [28]. In elicited cells the alkaloid appears to be stored in electron-dense deposits (osmiophilic aggregates) which occur on the tonoplast and as freely floating bodies in vacuoles. However, most of the alkaloid (73% of total) was recovered from the medium in a 100 g sedimenting, cell-free, particulate fraction. This pellet consisted of electron-dense material resembling those found associated to vacuoles of intact plants. This is an interesting observation and extended studies may give information about the mechanism by which secondary metabolites are translocated in elicited plant cells.

5. EXTERNAL ACCUMULATION OF SECONDARY METABOLITES

Plant cells excrete secondary metabolites into the medium under certain conditions as outlined above. However, enzymatic or nonenzymatic degradation of extracellular products may occur. A continuous withdrawal of released product(s) from the growth medium prevents such a degradation. Futhermore, by removing the metabolite(s) any feed-back product inhibition regulating the biosynthetic pathway may be eliminated. This can result in an increased productivity. In addition, potentially toxic substances are removed from the environment of the cells in such a procedure.

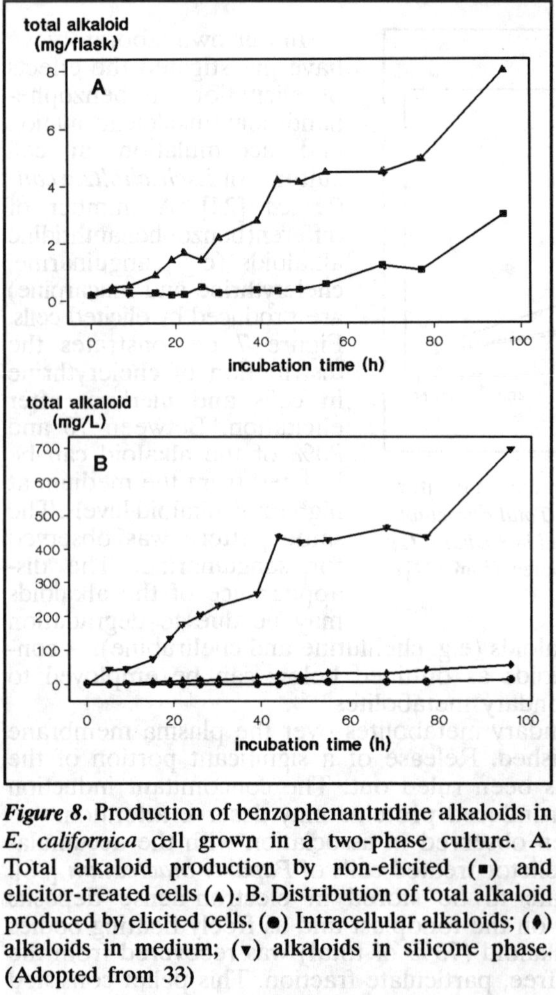

Figure 8. Production of benzophenantridine alkaloids in *E. californica* cell grown in a two-phase culture. A. Total alkaloid production by non-elicited (■) and elicitor-treated cells (▲). B. Distribution of total alkaloid produced by elicited cells. (●) Intracellular alkaloids; (♦) alkaloids in medium; (▼) alkaloids in silicone phase. (Adopted from 33)

A number of investigations have been carried using "two-phase cultures" to trap excreted products into a second phase. This phase can be either liquid or solid. The accumulation of a particular secondary metabolite is dependent on the physicochemical properties of the second phase. The selection of the second phase is governed by the product to be accumulated. Here a few selected examples will be discussed to demonstrate the potentials of this technique.

Various ion-exchangers have been employed as solid phase. Amberlite XAD-7 was used to adsorbe indole alkaloids (ajmalicine and serpentine) [29] and anthraquinones [30] from the medium of *C. roseus* and *C. ledgeriana*, respectively. In the latter study it was shown that for maximum yields both the amount of adsorbent used and the time after subcultivation at which it is added to the system is critical. The production of anthraquinones by *Cinchona* culture could be increased from around 1 mg/l/day to around 20 mg/l/day.

Activated charcoal [31] and reversed-phase silica [32] are other examples of solid phases employed to produce increased amounts secondary metabolites. In the latter case the production of valepotriates by cell cultures of *Valeri*ana wallichii was increased up to 4-fold and around 95% of total product was recovered from the silica phase.

Various liquid second phases have also been tested for the extraction of secondary metabolites from the medium. A promising approach has been taken by Pedersen's group. A dimethyl siloxane polymer, a silicone based antifoam fluid, has been employed to accumulate benzophenanthridine alkaloids (sanguinarine, chelerythrine, chelirubine and macarpine) produced by elicited cells of *E. californica* [33]. Cell growth was essentially unaffected by the polymer at concentrations below 23% (v/v). Using this amount of polymer the production of benzophenanthridine alkaloids could be increased 3-fold without elicitation. Figure 8A shows the total production of alkaloids by non-elicited and elicited cells and Figure 8B shows the distribution of the alkaloids produced by elicited cells. Elicitation increases the productivity further and a total productivity of 700 mg/l is achieved. Without elicitation and extraction into the second phase the corresponding value is around 40 mg/l. Figure 8B clearly demonstrates that almost quantitative accumulation in the polymeric phase may be achieved by selection of an appropriate second phase.

The combination of elicitation and two-phase culture is an important contribution for the development of biotechnological production of secondary metabolites. This approach may eliminate some of the bottlenecks for industrial applications of plant cell cultures.

REFERENCES

1 Dawson, R.F. (1941) Science 94, 396-397
2 Dawson, R.F. (1942) Amer. J. Bot. 29, 813-815
3 Sze, H. (1984) Physiol. Plant. 61, 683-691
4 Mitchell, P. (1976) Biochem. Soc. Trans. 4, 399-430
5 Rea, P.A. and Sanders, D. (1987) Physiol. Plant. 71, 131-141
6 Deus-Neumann, B. and Zenk, M.H. (1984) Planta 162, 250-260
7 Deus-Neumann, B. and Zenk, M.H. (1986) Planta 167, 44-53
8 Mende, P. and Wink, M. (1987) J. Plant Physiol. 129, 229-242
9 Ehmke, A., von Borstel, K. and Hartmann, T. (1988) Planta 176, 83-90
10 Yamamoto, H., Suzuki, M., Suga, Y., Fukui, H. and Tabata, M. (1987) Plant Cell Rep. 6, 356-359
11 Neumann, D., Krauss, G., Hieke, M. and Gröder, D. (1983) Planta Med. 48, 20-23
12 Renaudin, J.-P. and Guern, J. (1982) Physiol. Veg. 20, 533-547
13 Jans, B.P. (1973) Ber. Schweiz. Bot. Ges. 83, 306-344
14 Matile, P. (1984) Naturwissenschaften 71, 18-24
15 Parr, A.J., Robins, R.J. and Rhodes, M.J.C. (1986) Physiol. Veg. 24, 419-429
16 Kurkdjian, A. (1982) Physiol. Veg. 20, 73-83
17 Felix, H., Brodelius, P. and Mosbach, K. (1981) Anal. Biochem. 116, 462-470
18 Brodelius, P. and Nilsson, K. (1983) Eur. J. Appl. Microbiol. Biotechnol. 17, 275-280
19 Brodelius, P. (1988) Appl. Microbiol. Biotechnol. 27, 561-566
20 Felix, H. (1982) Anal. Biochem. 120, 211-234
21 Brodelius, P.E., Funk, C. and Shillito, R.D. (1988) Plant Cell Rep. 7, 186-188
22 Kilby, N.J. and Hunter, C.S. (1986) Poster no. 375 presented at Vi International Congress of Plant Tissue and Cell Culture, August 3-8, 1986, Minneapolis, Minn, USA

23 Pu, H.T., Yang, R.Y.K. and Saus, F.L. (1989) Biotechnol. Lett. 11, 83-86
24 Collinge, M.A. and Brodelius, P.E. (1989) Phytochemistry 28, 1101-1104
25 Tyler, R.T., Eilert, U., Rijnders, C.O.M., Roewer, I.A. and Kurz, W.G.W. (1988) Plant Cell Rep. 7, 410-413
26 Barz, W., Daniel, S., Hinderer, W., Jaques, U., Kessmann, H., Köster, J. and Tiemann, K. (1988) in Plant Cell Biotechnology (Pais, M.M.S., Mavituna, F. and Novais, J.M., eds.) pp. 211-230, Springer-Verlag, Heidelberg
27 Eilert, U. (1989) in Primary and Secondary Metabolism of Plant Cell Cultures II (Kurz, W.G.W., ed.) pp. 219-228, Springer-Verlag, Heidelberg
28 Cline, S.D. and Coscia, C.J. (1989) Planta 178, 303-314
29 Payne, G.F., Payne, N.N., Shuler, K.L. and Asada, M. (1988) Biotechnol. Lett. 10, 187-192
30 Robins, R.J. and Rhodes, M.J.C. (1986) Appl. Microbiol. Biotechnol. 24, 35-41
31 Knoop, B. and Beiderbeck, R. (1983) Z. Naturforsch. 38c, 484-486
32 Becker, H. and Herold, S. (1983) Planta Med. 49, 191-192
33 Byun, S.Y., Pedersen, H. and Chin, C.-K. (1990) Personal communication

ACCUMULATION OF AJMALICINE AND SERPENTINE IN VACUOLES ISOLATED FROM *CATHARANTHUS ROSEUS* CELL SUSPENSION CULTURES.

T.J.M.Blom*, M.Sierra**, F.van Iren*, R.Verpoorte**, and K.R.Libbenga*

Biotechnology Delft Leiden, Project-Group Plant Cell Biotechnology,
*Department of Plant Molecular Biology, Leiden University,
Nonnensteeg 3, 2311 VJ Leiden.
**Division of Pharmacognosy, Centre for Bio-Pharmaceutical Sciences,
Leiden University, P.O.Box 9502, 2300 RA Leiden. The Netherlands.

1. INTRODUCTION

An important function of the central vacuole of plant cells is the storage and sequestration of secondary metabolites including alkaloids. The way alkaloids are accumulated is not yet clear. Several authors have found evidence for metabolic active carriers. (1,2,3,4,5). Others report diffusion of the (non-protonated) neutral form of the alkaloids and ion trap of the (protonated) charged form inside the acidic vacuole (6,7,8) Recently the oxidation of senecionine to senecionine-N-oxide as a storage form of alkaloids has been discussed (9). In this paper we report studies on the accumulation of two alkaloids, ajmalicine and serpentine, in isolated vacuoles from *Catharanthus roseus* cell suspensions. These two alkaloids, although closely related, differ considerable in physical properties which might be reflected in transport behaviour.

2. PROCEDURE

2.1 *Catharanthus roseus* (L.) G. Don cell suspension MP183D (kept in dark) and MP183L (kept in continuous light) were subcultured every 7 days on LS-medium (10) containing 30 g/l sucrose, 2 mg/l NAA, and 0.2 mg/l kinetin. at 25°C on a gyrotory shaker at 120 rpm.

2.2 After lyophilization the alkaloids were extracted from the cell mass with chloroform using yohimbine as an internal standard. They were analyzed on HPLC (column: Waters-μ-Bondapak Phenyl; mobile phase: phosphate buffer (pH3.6)/aceto-nitrile/methoxy-ethanol=80/15/5 % v/v; flow: 2 ml/min; detection: UV-absorbance at 254 nm).

2.3 Protoplasts were obtained by digestion of the cell walls with a buffer containing 0.5 M mannitol (pH 5.5), 1% w/v cellulase R10, and 0.33% w/v macerozyme R10. Vacuoles were obtained from protoplasts with an osmoshock procedure by a five fold dilution in a lysis buffer (10 mM Hepes pH 7.4, 0.2 M mannitol, 0.03% PEG-4000, 5 mM EDTA and 8% ficoll-400) and they were purified by centrifugation on a step gradient of 2 ml of a buffer with 4% ficoll-400. and 1 ml of a buffer but containing no ficoll. The vacuoles were harvested in the uppermost layer. Purity of the vacuoles was checked with marker enzymes.

2.4 Accumulation of labelled compounds was assayed using vacuoles suspended in buffer containing either; [^{14}C]ajmalicine (kindly provided by Prof. J. Guern, CNRS, Gif sur Yvette, France), [^{14}C]benzoic (probe for the pH gradient) acid, or [^{3}H]TPP^{+} (probe for the potential difference) and [^{3}H]water (used to estimate the volume of vacuoles). The extra-vacuolar volume was determined by dual-labelling with [^{14}C]dextrane and [^{3}H]water. The vacuolar content was recovered with the oil filtration technique (11). and the radioactivity of the fractions was measured by dual label scintillation counting.

2.5 Peroxidases were isolated from crude cell extracts in a phosphate buffer (pH 6.1) containing PVPP. The enzymes were precipitated with 40-90 % ammonium sulphate and the samples were desalted on a PD-10 column. Separation of enzymes on FPLC; column Pharmacia-Mono-Q, mobile phase; 20 mM tris buffer (pH 8.0), and a 0 - 1 mM KCl gradient. The incubations of ajmalicine with peroxidases extracted from cell suspensions or from isolated vacuoles were performed as described by Fillipelli (12).

3. RESULTS AND DISCUSSION.

The two *Catharanthus roseus* cell suspension cultures used differed in alkaloid production. In MP183L ajmalicine and a considerable amount of serpentine was produced. The amount of ajmalicine per gram freshweight of cells increased from 5 days after the start of the subcultures, reached a maximum after 9 days and was again on its initial level at day 12. The serpentine content initially was low but increased from day 5 to day 12 and stayed at a constant high level. Such a time course in alkaloid production can be expected when ajmalicine is converted into serpentine. Cell-line MP183D produced mainly ajmalicine and only a small amount of serpentine. The ajmalicine level increased and reached a maximum in the same period as was found in MP183L but it stayed high while almost no serpentine was produced. As ajmalicine is the precursor for serpentine, the different behaviour in alkaloid production of the two cell lines can be explained from a difference in activity of the enzymes involved in the conversion.

Ajmalicine added to vacuoles isolated from *C.roseus* cell-line MP183D was rapidly taken up and accumulated against a concentration gradient. In equilibrium the accumulation-ratio (the quotient of vacuolar and medium concentrations) was about 1.3 (table 1). Addition of nigericin (an ionophore that disrupts the pH gradient but not the electrical potential difference) reduced the accumulation-ratio. Addition of valinomycin (an ionophore with an effect inverse to that of nigericin) had no effect. These results indicate that the ajmalicine accumulation was regulated by the trans-tonoplast pH gradient and not by the electrical potential difference. Cross checks with the accumulation of the pH probe benzoic acid and with the membrane potential probe TPP+, supported this conclusion. Further support came from calculations of the internal pH with the use of the ajmalicine accumulation (13) and with benzoic acid accumulation. Both methods resulted in about the same values (results not shown)

Table 1. The influence of ionophores and MgATP on the accumulation of ajmalicine, benzoic acid, or TPP+ in isolated vacuoles from *C.roseus* cell-line MP183D

Ionophores	without MgATP			5 mM MgATP
	Ajmalicine accumulation ratio	Benzoic Acid accu.ratio	TPP+/area membrane *)	Ajmalicine accumulation ratio
None	1.3 ±0.2	0.5 ±0.0	7.3 ±1.8	7.3 ±0.8
2μM Nigericin	0.9 ±0.2	1.2 ±0.1	6.2 ±0.7	1.8 ±0.1
2μM Valinomycin	1.4 ±0.2	0.5 ±0.1	2.7 ±0.5	5.0 ±1.2
2+2μM Val.+ Nig.	0.8 ±0.1	1.1 ±0.1	2.5 ±0.3	1.4 ±0.3

*)arbitrary units

Accumulation of ajmalicine in vacuoles isolated from the same cell-line increased significantly after addition of MgATP (table 1 and 2). This increase was also found after addition of both MgATP and the plasmalemma ATPase inhibitor sodium vanadate, but it was not found when vanadate was replaced by the general ATPase inhibitor DCCD or by the tonoplast ATPase inhibitor potassium nitrate. These results demonstrate that the increase of ajmalicine accumulation was due to the activity of tonoplast ATPases. Like in the situation without ATP, the influence of the pH gradient and the membrane potential difference on ajmalicine accumulation was tested by addition of ionophores. As is shown in table 1 a low ajmalicine accumulation was found after addition of MgATP and nigericin. Since nigericin affects the pH gradient, the increased ajmalicine accumulation after ATP addition must have been caused by an increased pH gradient and not by stimulation of an ajmalicine transporting carrier. Valinomycin in contrast with the situation without ATP had some effect. The reduced ajmalicine accumulation after addition of valinomycin probably also was caused by a decrease in pH gradient. This can be explained by considering that the proton motive force PMF is a function of both the membrane potential difference and the pH gradient (14). It also must be taken into account that the membrane potential difference is negative (TPP+ accumulates). After addition of the ionophore, the PMF finally will be the same as before addition, under the assumption that valinomycin only affects the membrane potential difference and not the maximum contribution of the ATPases or the proton-efflux to the PMF. In the assumed situation the loss of membrane potential difference will be compensated by a lower pH gradient. If valinomycin inhibits ATPase activity or causes leakage of protons it will lower the PMF. As a consequence the pH gradient will even be more reduced than in the assumed situation.

From both *C.roseus* cell-lines alkaloid extracts of cells, protoplasts, and vacuoles were made. The extract of MP183D cells contained ajmalicine, but ajmalicine was not found in the extracts of isolated protoplasts or vacuoles. Apparently, it was lost during the isolation procedure. The extracts of cells from line MP183L contained serpentine, which was also found in considerable amounts in the extracts of vacuoles and protoplasts (results not shown). Obviously serpentine remained inside the vacuoles. Such a retention can be explained by the properties of the two alkaloid. Ajmalicine (in its non-protonated form) is a neutral molecule which may diffuse, while the charged serpentine almost cannot diffuse across membranes.

Table 2. Effect of 5 mM MgATP on the accumulation of ajmalicine in isolated vacuoles of *C.roseus* cell-line MP183D and the influence on accumulation of ATPase inhibitors

	Ajmalicine uptake % ±sd
MgCl2	100 (def.)
MgATP	239 ±39
MgATP + 90 μM Na3VO4	237 ±32
MgATP + 1 μM DCCD	118 ±5
MgATP + 50 mM KNO3	106 ±4

After it was found that ajmalicine can be oxidized to serpentine by horseradish peroxidases (Blom et al. unpublished), the relation between serpentine content of the cells and peroxidase activity was investigated. During fourteen days after the start of subcultures of *C.roseus* cell-line MP183L. The specific activity of basic peroxidases rose in parallel with the serpentine content. In isolated vacuoles from this cell-line we found only basic peroxidases. With the peroxidases extracted from these vacuoles 34% of the added ajmalicine was converted into the only detectable product serpentine. The peroxidase activity per million vacuoles from cell-line MP183L was 91 mUnits while the activity per million vacuoles of cell-line MP183D was only 4 mUnits. By comparing the difference in serpentine production between the two cell-lines with the difference in basic peroxidase activity and by considering the capacity of basic peroxidase to convert ajmalicine into serpentine we came to the following conclusions; Ajmalicine is converted into serpentine inside the vacuoles by basic peroxidases. The difference in serpentine content of the cells of the two cell-lines is caused by a difference in peroxidase activity of their vacuoles.

Table 3. Vacuolar concentration of serpentine (μM) in mixtures containing: isolated vacuoles from *C.roseus*, 100 μM ajmalicine, 5 mM MgATP, and 100 μM hydrogen peroxide.

	before incubation	after 16 h incubation
disrupted vacuoles	1.47	1.52
intact vacuoles	1.50	6.17

To check the validity of the conclusion about the vacuole being the organelle in which ajmalicine is transformed into serpentine an *in situ* experiment was carried out. Ajmalicine was added to a reaction mixture containing hydrogen peroxide (essential because without this electron acceptor no reaction was noticed) MgATP and either disrupted or intact vacuoles from *C.roseus* cell-line MP183L. As is shown in table 3, serpentine was only formed after incubation with intact vacuoles and not in a mixture with disrupted vacuoles.

4. CONCLUDING REMARKS.

The obtained results fit the following hypothesis describing the transport, accumulation, and bio-conversion of ajmalicine. Ajmalicine forms a pH dependent equilibrium between a neutral (non-protonated) and a charged (protonated) form. Neutral ajmalicine diffuses across the tonoplast towards concentration equilibrium. Since there is a pH gradient across the tonoplast with the lowest pH inside the vacuole the concentration of protonated ajmalicine inside will be higher than outside. By this ion trap ajmalicine accumulates inside the vacuoles. Therefore the accumulation can be increased by stimulation of protonpumps with MgATP. The bio-conversion of ajmalicine into its charged anhydronium compound serpentine by peroxidases inside the vacuoles may be used as an alkaloid trap.

5. REFERENCES

1 Deus-Neumann,B., Zenk,M.H. (1984) Planta: 162, 250-260.
2 Deus-Neumann,B., Zenk,M.H. (1986) Planta: 167, 44-53.
3 Mende,P., Wink,M. (1987) J. Plant Physiol.: 129, 229-242.
4 Yamamoto,H., Suzuki,M., Suga,Y., Fukui,H., Tabata,M. (1987) Plant Cell Rep: 6, 356-359.
5 Yamamoto,H., Suzuki,M., Kitamura,T., Fukui,H., Tabata,M. (1989) Plant Cell Rep.: 8, 361-364.
6 McCaskill,D.G., Martin,L.D., Scott,I.A. (1988) Plant Physiol.: 87, 402-408.
7 Renaudin,J.P. (1989) Plant Physiol. Biochem.: 27.
8 Renaudin,J.P., Guern,J. (1982) Physiol. Vég.: 20, 533-547.
9 Ehmke,A., von Borstel,K., Hartmann,T. (1988) Planta: 176, 83-90.
10 Linsmaier,E.M., Skoog,F. (1965) Physiol. Plant.: 18, 100-127.
11 Martinoia,E., Flugge,F.U., Kaiser,G., Heber,U., Heldt,W.H. (1985) Biochim. Biophys. Acta: 806, 311-319.
12 Fillipelli,F.J., (1986) in Ph.D.Thesis, (Rosazza,J.P.,sup.vis), University of Iowa
13 Waddell,W.J., Butler,T.C. (1959) J. Clin. Invest.: 38 p. 720.
14 Sze,H. (1985) Annu. Rev. Plant. Physiol.: 36, 175-208.

RECOVERY OF SECONDARY METABOLITES WITH MINIMAL LOSS OF CELL VIABILITY

P.J. WEATHERS, A. DIIORIO, R. CHEETHAM & M. O'LEARY, Department of Biology and Biotechnology, Worcester Polytechnic Institute, Worcester, MA 01609 USA

1. INTRODUCTION

The high cost of producing natural products from plant tissues in culture are due in part to slow growth and production rates associated with plant tissue, and the extraction of products. Hairy roots produced by transformation with *Agrobacterium rhizogenes*, offer rapid, stable growth and product synthesis (1,2).

Mild heating (30-40°C) can permeabilize cell membranes allowing efflux of intravacuolar components (3), not readily released into the surrounding culture medium (4). However, viability of plant tissues after mild heat treatments, for example, hairy roots of beet and safflower, have not been studied.

We report here results on the feasibility of using mild heat and calcium to extract secondary compounds from plants while maintaining viability.

2. PROCEDURE

2.1 Materials and Methods

2.1.1 <u>Cultures:</u> Hairy root cultures of safflower (*Carthamus tinctorius*), and beet (*Beta vulgaris* Detroit dark red), were maintained on solid media (MS salts (5) with 3% sucrose and 0.2% Gelrite). After transformation of *Beta vulgaris* with *Agrobacterium rhizogenes* strain 15834, excised beet hairy roots were cultured on solid media with carbenicillin to obtain clean explants (6). Inocula for all experiments were from 2 week old cultures grown on solid media.

2.1.2 <u>Heat Treatments:</u> Hairy roots in flasks were incubated for 1 hr at temperatures ranging from 25-55°C. Tissues were removed and assayed daily for respiratory CO_2 via gas chromatography. Biomass increases were also assayed.

Beet root disks were incubated in MS media for up to 1 hr at temperatures ranging from 25-55°C. After heat treatment and assay of media for pigments (7), disks were assayed for respiratory CO_2 as above. All cultures were verified axenic to preclude measurement of CO_2 from bacterial instead of plant respiration.

Beet root disks were also heat treated for 8, 16, and 60 min. at 40°C in test tubes. Each 1/4 time period, the media was removed, assayed for betacyanins and replaced with pre-heated fresh media. Controls used an equal volume without rinsing over the same time period.

3. RESULTS AND DISCUSSION

3.1 In all experiments, increasing the temperature and duration of heat, increases the release of secondary product. Efflux of UV absorbent material in safflower (polyacetylenes) followed the same general pattern of heat release as in beets, yet was not quantitated due to the large volumes required for such quantization. Viability was maintained using mild treatment temperatures (35°C) with addition of calcium post-heat treatment. Heating in excess of 40°C, while causing more product to be released, reduced viability. Calcium aids in stabilizing the membrane (3) and, thus, improves viability (Figs. 1a and 1b).

Heat treatment of beet root disks causes release of betanin, a red pigment stored in vacuoles (Fig.1a). Ten per cent of the total pigment was released after a 35°C treatment and up to 60% was released after a 55°C treatment for 45 min. The amount of pigment released at 55°C, although high, was not 100% due to the high extracellular concentration of betanin during and after this heat treatment. Betanin release can be increased by lowering the concentration gradient restricting pigment diffusion into the surrounding culture media. In Figure 1a, a ratio of 100 mg dry wt beet root disks/ml media yielded a 60% release of total pigment after 45 min of heat at 55°C while a ratio of 10 mg dry wt hairy root tissue/ml media yielded 90% release of total pigment (Fig.1b).

In Table 1, as near continuous removal of released pigment is approached (four 2 minute rinses over an 8 minute time period), pigment release relative to the control (no rinses in a 20 ml volume) increases.

Furthermore, the shorter the duration of the heat treatment, the more viability is retained. Beet root disks treated at 40°C for 45 min showed a 38% reduction in viability vs. those treated for only 15 min, as determined by CO_2 production 1 day after treatment. These results show that a semi-continuous rinsing process is optimal not only for pigment release, but also for retaining viability.

Carbon dioxide production, which is used as a viability assay, drops off drastically with treatments of 45°C and higher. Addition of $CaCl_2$ post-heat treatment improved CO_2 production compared to disks placed in B5 media without $CaCl_2$. When additional $CaCl_2$ was placed in the media prior to heat treatment, no pigment was released after treatment at 25, 35 and 40°C, and only minimal amounts (about 10% of total pigment) were released after treatment at 45°C (Fig.1a). For hairy roots, betanin release in beets is 10%, 15% and 90% after 1 hr of treatment at 35, 40 and 55°C, respectively (Fig.1b).

Viability of safflower and beet hairy roots, measured as carbon dioxide production 1 day after heat treatment, shows increased CO_2 production after 35°C treatments in comparison to the control (25°C). Treatment temperatures in excess of 40°C appear to be damaging to the tissue, however, the use of $CaCl_2$ did improve carbon dioxide production after heat treatment at all temperatures except 55°C.

After heat treatments, safflower hairy roots were cultured for 1 week in MS media with an additional 20mM $CaCl_2$ and the subsequent increase in biomass was measured. A series of control flasks were cultured in MS media with and without additional $CaCl_2$ at 25°C. Hairy roots treated at 45 and 55°C were discolored and the final dry weight was less than the starting inoculum, as were cultures treated with dichloromethane and 9.5% ethanol for 1 hr at 25°C. The tissue treated at 40°C (1 hr) did not grow yet appeared healthy. Beet hairy root tissue treated at 40°C showed no loss of viability (measured as cell mass increases) when the standard B5 vitamin mixture was included in the post-heat treatment growth media.

The use of $CaCl_2$ post-heat treatment caused a reduction in growth at 25°C, and at all temperatures compared to the control without $CaCl_2$. However, cultures at 35°C + $CaCl_2$ showed improved growth (1.2

generations/week) over the cultures at 25°C + $CaCl_2$ (0.8 gens./week), as well as the 35°C treated control without additional $CaCl_2$ (1.0 gens./week). Tissue culture media was designed to be optimal for growth and the addition of excess $CaCl_2$ may have altered the growth potential of the media. Further experiments using shorter exposures to excess $CaCl_2$ may control this phenomenon.

Use of organic solvents on plant cells for the purpose of permeablization have resulted in irreparable damage to the tissue (8,9). Incubation at 35°C for 1 hr and subsequent culturing for 1 week showed improved growth of the tissue over the untreated control while releasing about 10% of the total stored product. Heat treatment at 40°C for 1 hr does not affect viability yet causes up to 15% of total product to be released. Product release can be increased by rapidly or continuously rinsing away the released product. These results suggest that a cyclic process, with periods of product release followed by a short recovery period, could yield more product from a continuous culture extraction process, eliminating the costly separation steps of harvesting and primary extractions from homogenized root tissue.

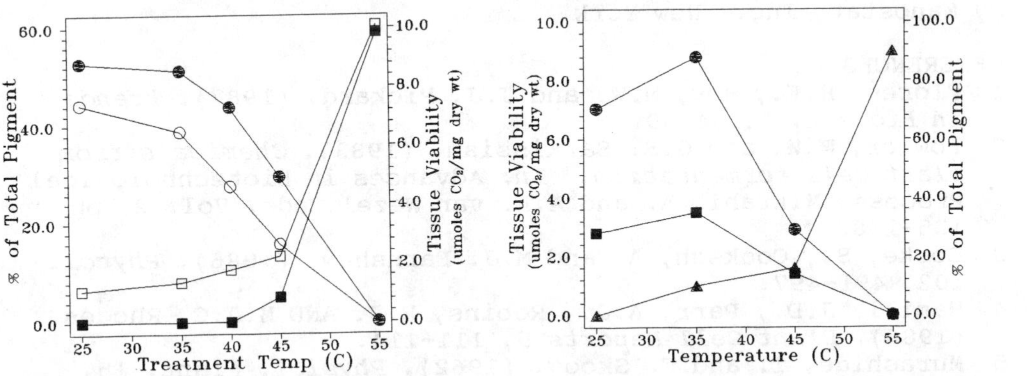

Figure 1a) Viability of and pigment efflux from heat treated beet root disks. The duration of all heat treatments was 45 min. □, % of total pigment B5 media; ■ % of total pigment B5 media + 20mM $CaCl_2$; ○, CO_2 production in B5 media; ●, CO_2 production in B5 media + 20mM $CaCl_2$. b) Viability of beet and safflower hairy roots after 1 hr heat treatment in B5 media, post-heat treatment with B5 + 20mM $CaCl_2$. CO_2 production was measured after 1 day in stoppered flasks. ■, safflower tissue viability; ●, beet tissue viability; ▲, betanin release in beets.

Table 1. Betanin efflux from heat treated beet root disk with periodic rinsing.

Extraction Time (min)	Number of 5 ml Rinses	Volume (ml) Initial	Volume (ml) Final	Bet. Released µg/g dry wt
8[a]	4	5	20	76.8
8	0	20	20	54.8
16[a]	4	5	20	121
16	0	20	20	104
60[a]	4	5	20	134
60	0	5	20	132

[a] Four media exchanges, equally spaced in time were performed. The MS media containing released betanins was removed and replaced with an equal volume of pre-heated MS media.

Approximately 10 mg dry wt tissue/ml was used. All tissues were exposed to a total volume of 20 ml.

4. ACKNOWLEGEMENTS

We would like to thank Drs. Hector E. Flores (Penn. State Univ.) and J.D. Hammill (AFRC Inst. of Food Research, Norwich) for supplying strains of Agrobacterium and hairy root cultures of safflower and beet. This work was funded in part by Manostat, Inc., New York.

REFERENCES

1 Flores, H.E., Hoy, M.W. and J.J. Pickard. (1987). *Trends in Biotech.* 5, 64-69.
2 Fowler, M.W. and G.S. Sarkissian. (1983). *Chemicals from plant cell fermentation. In:* Advances in Biotechnological Process, Mizrahi, A. and A.L. van Wezel, eds. Vol. 2, pp. 135-158.
3 Cooke, S., Cookson, A. and M.J. Earnshaw. (1986). *Phytol.* 102, 491-497.
4 Hamill, J.D., Parr, A.J., Robins, R.J. AND M.J.C. Rhodes. (1986). *Plant Cell Reports* 5, 111-114.
5 Murashige, T. and F. Skoog. (1962). *Physiol. Plant.* 15, 473-482.
6 Flores, H.E., Hoy, M.W. and J.J. Pickard. (1987). *Trends in Biotech*. 5, 64-69.
7 Saguy, I., Kopelman, I.J. and S. Mizrahi. (1978). *J. Food Sci.* 43, 124-127.
8 Brodelius, P. (1988). *Appl. Microbiol. Biotechnol.* 27, 561-566.
9 Parr, A.J., Robins, R.J. and M.J.C. Rhodes. (1986). Secondary Metabolism in Plant Cell Cultures, P. Morris Ed., Cambridge University Press.

CELLULAR ASPECTS OF THIOPHENE ACCUMULATION IN RELATION TO ROOT DIFFERENTIATION

R.R.J. ARROO, H. KENKEL, H. MEIJERS, A.F. CROES, G.J. WULLEMS,
NOVAPLANT CELL BIOTECHNOLOGY GROUP, DEPT. OF EXP. BOTANY, UNIVERSITY, TOERNOOIVELD, 6525 ED NIJMEGEN, THE NETHERLANDS

1. INTRODUCTION

A great variety of compounds that are not primarily involved in growth and functioning of the plant cell, are synthesized via the complex pathways of secondary metabolism. There is considerable evidence for a link between the accumulation of these substances and the developmental state of the cell. Synthesis of secondary metabolites is in many cases confined to differentiated cells and tissues (1). Roots are a good example of this synthetic capacity of differentiated tissues as these organs are highly structured and, in many species, are a rich source of secondary compounds (2). There are more reasons why roots are an excellent model system to study the relation between morphogenesis and secondary metabolism. Excised roots can be grown indefinitely as primary roots in tissue culture (3). Moreover, their growth rate and morphology can be manipulated by external factors (3) and by transformation with *Agrobacterium* (4).

Accumulation of a secondary product in a complex organ such as a root, results from the balance between synthesis and breakdown, and may also be influenced by transportation. For a more detailed study of the relation between morphological and chemical differentiation it is, therefore, essential to know to what extent these processes contribute to product accumulation.

We are using root cultures of *Tagetes* (marigold) to study the link between the accumulation of secondary metabolites and root morphology (5,6). Root cultures of these plants accumulate thiophenes (7), heterocyclic sulfurous compounds with a strong nematicidal activity (8). Thiophenes are easily quantified even in minute amounts (5,6). Synthesis and breakdown can be monitored by labeling with [^{35}S]sulfate (9). In normal roots and in roots transformed with *Agrobacterium rhizogenes* the accumulation of thiophenes and the morphology of the roots are modified by indoleacetic acid (IAA). Thiophene content is reduced by high IAA concentrations that stimulate branching (6) but the effect of low hormone concentrations is not known yet.

In the present study the effects of IAA on root morphology and thiophene content were investigated over a wide range of IAA concentrations. Moreover, biosynthesis and breakdown of thiophenes were quantified in exponentially growing root tissues.

2. MATERIALS AND METHODS

2.1. Root cultures

Untransformed root cultures of *T. patula* L. were maintained on Gamborg's B5 medium (10) supplemented with 3% (w/v) sucrose and 10^{-8} M IAA. The roots were incubated on a rotary shaker at 100 rpm at 25°C in the dark. Roots transformed by *Agrobacterium rhizogenes* LBA 9402 (pRi 1855) were grown under the same conditions without IAA. The kinetics of growth and thiophene accumulation in liquid culture was followed in medium inoculated with root tips of approximately 1 cm length. For the determination of the effect of IAA on tip elongation, branching, and thiophene accumulation, the medium was solidified with 0.2 % (w/v) Gelrite and IAA was added at a range of concentrations.

2.2. Thiophene analysis

Thiophenes were extracted from the roots in 50% (v/v) ethanol and purified by partitioning between the ethanol and a 1:1 mixture of hexane and *t*-butylmethylether (6). Thiophenes were recovered from the hexane-ether, subjected to high-performance liquid chromatography on a C_{18} column and quantified in the eluate on basis of UV absorption (6).

2.3. Tracer studies

Root tips from 14-days old roots in the exponential growth phase were labeled with [^{35}S]sulfate, 37 TBq. mol^{-1} (Amersham) at a radioactive concentration of 0.74 $MBq.ml^{-1}$. To minimize the effect of the internal sulfate pool on label dilution, the sulfate concentration in the medium was raised to 30 mM when thiophene synthesis was measured. Thiophene breakdown was monitored by labeling roots with carrier-free [^{35}S]sulfate in sulfate-free medium for 2 h. The roots were then transfered to medium containing 30 mM unlabeled sulfate, and the radioactivity in thiophenes was followed during 7 h.

3. RESULTS AND DISCUSSION

The amount of thiophene accumulated in roots is strongly coupled to the growth phase of the culture (Fig. 1). During mid-exponential phase, the accumulation rate exeeded the increase of biomass and, consequently, the thiophene content (in µmol. g FW^{-1}) also increased. The amount of thiophene per unit biomass stabilized at a lower level during the transition to stationary phase. The total amount of thiophene (in µmol.$flask^{-1}$) continued to increase until the beginning of the stationary phase when growth ceased. From that point on the amount in the root culture steadily declined.

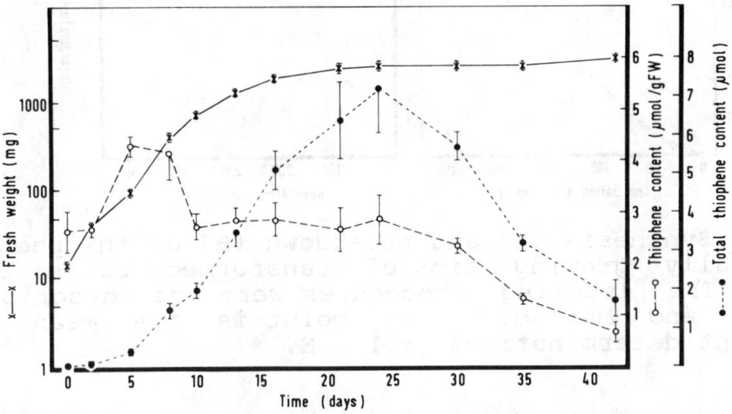

Fig. 1. Fresh weight, thiophene content and total thiophene in transformed roots of *T. patula* during culture. Each point is the mean of 4 independent determinations ± 1 S.E.

To gain insight in the relation between synthesis, degradation and accumulation, we fed radioactive sulfate to exponentially growing root tips. Thiophenes were rapidly and effectively labeled. After extraction more than 90% of the radioactivity in the hexane-ether fraction was incorporated in thiophenes. The rate of sulfur incorporation increased after a lag until it reached a stable value after approximately 2 h (Fig. 2A). The slow start was probably due to limited uptake and the time needed to saturate the immediate precursor pool. The rate of thiophene synthesis was calculated from the linear part of the curve and amounted to 12 nmol.h^{-1}.g FW^{-1}. Degradation of thiophenes was quantified in a pulse-chase experiment. Root tips were labeled with carrier-free [^{35}S]sulfate and then transfered to standard culture medium with an elevated sulfate concentration. No decrease in the radioactivity in thiophenes was detected within 7 h, which shows that no thiophene degradation took place during late exponential growth. This indicates that thiophenes are stable up to the stationary phase and that in the earlier growth stages synthesis is the only factor determining the rate of accumulation.

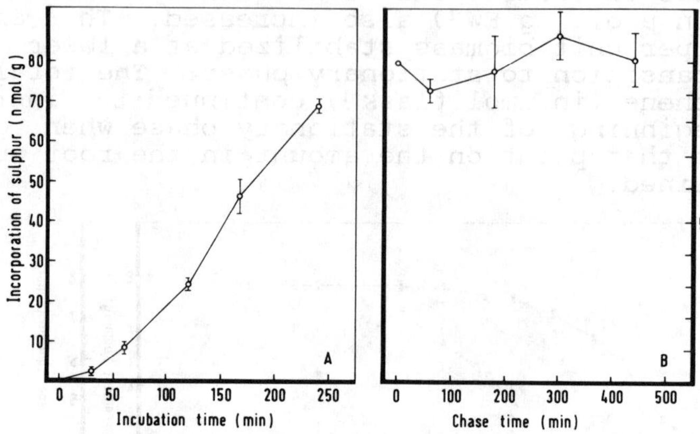

Fig. 2. Synthesis (A) and breakdown (B) of thiophenes in exponentially growing tips of transformed roots of *T. patula*. The labeling procedures were as described in Materials and Methods. Each point is the mean of 4 independent determinations ± 1 S.E.

As an approach to manipulate thiophene metabolism in relation to root morphology, we studied thiophene accumulation, root tip extension, and branching at a wide range of IAA concentrations. The auxin concentration optimal for branching in untransformed roots appeared to be thousandfold higher than the optimal concentration for thiophene accumulation (Fig. 3). These two phenomena were inversely related in the range of IAA concentrations between 10^{-7} M and 10^{-4} M. In contrast, a direct relation between tip growth and thiophene content is suggested by the concurrent inhibition of both processes by the same IAA concentrations.

An open question concerning the relation between morphology and secondary product accumulation is at which developmental state of the tissue thiophene accumulation occurs. The linear differentiation gradient along the root axis, ranging from the meristematic tip through the elongation zone to the fully differentiated parts, provides a unique opportunity to tackle this problem. Localization of thiophene accumulation at the cellular level with thiophene-directed antibodies will be a promising future approach.

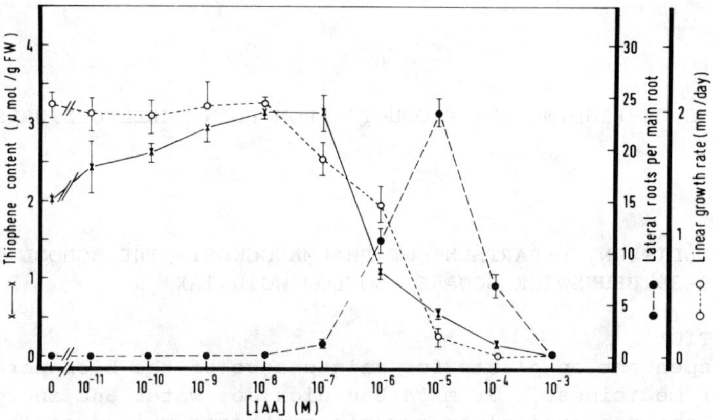

Fig. 3. Effect of IAA concentration on thiophene content, tip elongation, and branching of untransformed *T. patula* roots. Each point is the mean of at least 4 independent determinations ± 1 S.E.

REFERENCES

1 Yeoman, M.M. (1987) Ann. Bot. 60, Suppl. 4, 157-174

2 Flores, H.E. (1987) in Applications of Biotechnology in Agricultural Chemistry (Mumma, R.O., Lebaron, H., Honeycutt, R.C., Duesing, J.H., eds.), pp. 67-86, American Chemical Society, Washington, D.C.

3 Butcher, D.N. and Street, H.E. (1964) Bot. Rev. 30, 513-586

4 Rhodes, M.J.C., Robins, R.J., Hamill, J.D., Parr, A.J., and Walton, N.J. (1987) IAPTC Newsletter 53, 2-15

5 Croes, A.F., Aarts, A.M., Bosveld, M., Breteler, H., and Wullems, G.J. (1989) Physiol. Plant. 76, 205-210

6 Croes, A.F., Van den Berg, A.J.R., Bosveld, M., Breteler, H., and Wullems G.J. (1989) Planta 179, 43-51

7 Bohlman, F., Burkhardt, T., and Zdero C. (1973) Naturally Occurring Acetylenes, Academic Press, London

8 Uhlenbroek, J.H. and Bijloo, J.D. (1960) Rec. Trav. Chim. Pays Bas 79, 1181-1196

9 Schulte, K.E., Henke, G., Rücker, G., and Foerster, S. Tetrahedron 24, 1899-1903

10 Gamborg, O.L. (1970) Plant Physiol. 45, 372-375

POSSIBILITIES OF FINDING NEW PRODUCTS FROM PLANT CELL CULTURES

J. DAVID PHILLIPSON, DEPARTMENT OF PHARMACOGNOSY, THE SCHOOL OF PHARMACY, 29-39 BRUNSWICK SQUARE, LONDON WC1N 1AX

1. INTRODUCTION

Man is dependent on plants for maintenance of the biosphere, for food and for medicines. From carbon dioxide, water and inorganic ions, plant enzymes manipulate organic syntheses which result in a complex array of natural product molecules, many of which have provided challenges for the most able of synthetic chemists. The rapid development of plant biotechnology, coupled with the growing fear that planet earth is losing plant species at an alarming rate, has helped to focus attention on alternatives to whole plants for the production of natural products which are either of direct or of potential benefit to man. Plant cell culture techniques have been utilised to produce a range of secondary metabolites (1-3 and references therein). There have been notable successes, particularly the commercial production of the naphthoquinone pigment shikonin from cultures of *Lithospermum ruderale*, biomass production of ginseng and tobacco as well as the large-scale growth of plant cultures to obtain digoxin, geraniol and rosmarinic acid. To counter-balance these successes there has been the frustration invoked by the failure of plant cell cultures to produce consistent high yields of the clinically useful vinblastine and vincristine from *Catharanthus roseus* or morphine and codeine from *Papaver somniferum*. The economic significance and usefulness of plants and their constituents is particularly relevant to medicines and has recently been reviewed (4). The purpose of this chapter is to highlight the successful production of diverse natural products which are of pharmacological interest, to give examples of novel chemical structures produced by plant cell cultures and not as yet reported from the parent plant and to indicate some of the areas where scientific investigation should be encouraged.

2. SECONDARY PRODUCTS FROM PLANT CELL CULTURES

Many of the secondary plant products which are either used in medicine or are of pharmacological interest, are not readily synthesised and hence continue to be obtained from whole plants. Cell cultures afford possible alternative sources of such compounds and some examples from four major groups of secondary metabolites, namely, alkaloids, terpenoids, lignans and quinones, will be briefly considered.

2.1 Alkaloids

This diverse group of natural products has provided many clinically useful compounds with differing pharmacological activities. More than 200 alkaloids have been produced by plant cell cultures and for more than 30 of these the yields produced have been equal to or higher than the parent plant (1).

2.1.1 Indole and related alkaloids: More than 100 tryptophan-derived alkaloids have been isolated from cell cultures of 14 species of only 10 genera (1). There remain other genera as yet not investigated for their ability to produce indole alkaloids from cell cultures. Although over 90% of the cell culture-derived alkaloids are produced from tryptamine and secologanin, a number of simple indoles of the β-carboline and canthin-6-one types have been isolated. In our laboratories, plant cell cultures of *Ailanthus altissima* (5) and of *Brucea javanica* (6,7) have produced canthin-6-one alkaloids in yields which are far higher than those of the parent plant. *A. altissima* cultures produce canthin-6-one and 1-methoxycanthin-6-one (see Fig. 1) as major alkaloids although the yields and relative proportions obtained depend on the source of seed from which the cultures were produced. Similar findings have been noted for the canthin-6-one alkaloids of *Brucea javanica* cell cultures. Canthin-6-one and 11-hydroxycanthin-6-one (see Fig. 1) are the major alkaloids in cultures derived from Chinese seeds whereas canthin-6-one is absent from cultures derived from Thai seeds and the major alkaloids are 5-methoxycanthin-6-one and 11-hydroxycanthin-6-one (see Fig. 1).

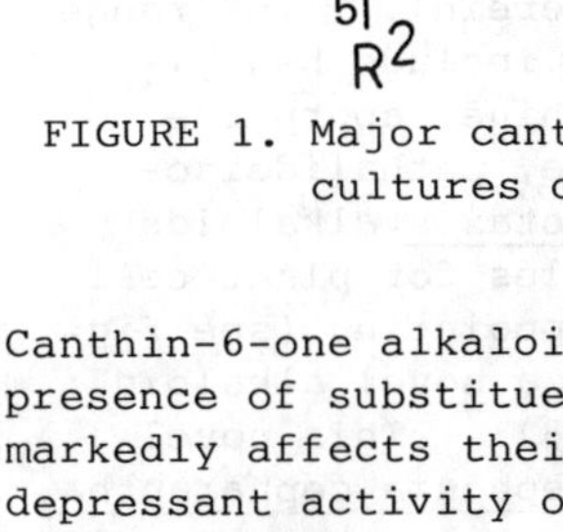

	R^1	R^2	R^3
canthin-6-one	H	H	H
1-methoxycanthin-6-one	H	H	OMe
5-methoxycanthin-6-one	H	OMe	H
11-hydroxycanthin-6-one	OH	H	H

FIGURE 1. Major canthin-6-one alkaloids produced by cell cultures of *Ailanthus altissima* and *Brucea javanica*

Canthin-6-one alkaloids are not readily synthesised and the presence of substituents around the tetracyclic aromatic structure markedly affects their biological properties (5,8-10). The CNS depressant activity of *Ailanthus altissima* has been investigated using bio-assay guided fractionation of extracts from root bark and from cell cultures (11). 4,5-Dihydrocanthin-6-one was identified as the major active principle from cell cultures producing 60% reduction in motility and 80% reduction in sleeping time at a dose of 3 mg kg^{-1} in mice.

Over 300 secologanin-tryptamine derived indole alkaloids have been isolated from whole plants of Apocynaceae, Rubiaceae and Loganiaceae. Plant cell cultures have produced a range of structural types including corynanthean, aspidospermatan, ibogan, strychnan, plumeran, eburnan, vallesiachotaman, apparicine, indoloquinolizidinone and quinoline-quinuclidine (1). Some 19 novel indole alkaloids have been produced from cell cultures in recent years (12) and the potential for new alkaloids is considerable.

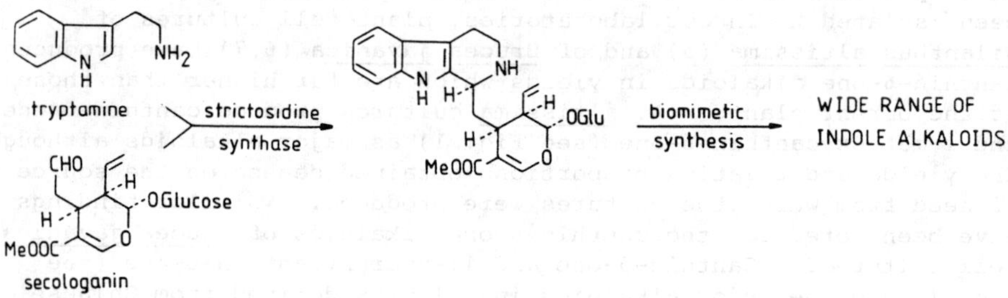

FIGURE 2. Potential of strictosidine synthase expressed in Escherchia coli to produce indole alkaloids

There are many biologically active indole alkaloids which could be produced potentially from cell cultures. One such alkaloid of current interest is yeuchukene, from Murraya paniculata, which inhibits implantation of the fertilised ovum in experimental animals (see 3 and references therein).

2.1.2 Isoquinoline alkaloids: More than 60 isoquinoline alkaloids have been isolated from plant cell cultures obtained from 35 species of 20 genera (see 1 and references therein). The range of structural types produced by cell cultures include benzylisoquinoline, bisbenzylisoquinoline, proaporphine, aporphine, protoberberine, protopine, benzophenanthridine, phthalideisoquinoline, morphinan, aristolactam and Cephalotaxus alkaloids (1). Morphine and codeine are target molecules for plant cell culture production but results have been disappointing (see Fig. 3). Norcepharadione serves as an example of a novel alkaloid which has been produced from cell cultures (18). This novel oxoaporphine was obtained from cultures of Stephania cepharantha.

Extensive biosynthetic studies have elucidated the major pathways to a number of key isoquinoline alkaloids. Dopamine and 4-hydroxyphenylacetaldehyde are the precursors of the 1-benzyltetrahydroisoquinoline alkaloid norcoclaurine and further selective hydroxylations and methylations lead to (S)-reticuline, a key intermediate in the formation of berberine. All of the biosynthetic steps leading to berberine have been elucidated at the enzyme level. (S)-reticuline is converted to the (R)-isomer which is the precursor of the morphinan alkaloids in Papaver somniferum (see Fig. 3; 1,2 and references therein).

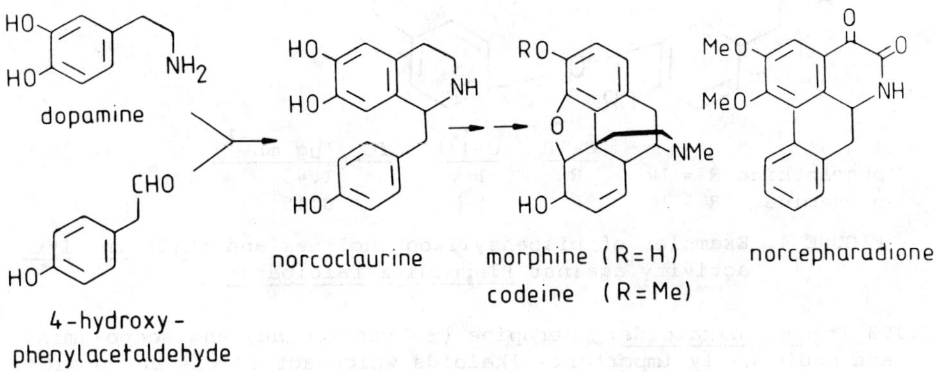

FIGURE 3. Precursors of the clinically useful morphine and codeine; the novel norcepharadione derived from plant cell cultures

Bisbenzylisoquinoline alkaloids are formed via oxidative coupling of monomeric benzyltetrahydroisoquinolines involving some 26 different types of linkage (19). Biological activities which have been reported for the dimers include anticancer, anti-parasitic and muscle relaxant. It is not easy to synthesise many of these types of dimer which exist in plants and hence for biological evaluation resource has to be made to isolation from plant sources. During an investigation of 12 medicinal plants from 9 families which are used in Sierra Leone for the treatment of malaria, we selected Triclisia patens (Menispermaceae) on the basis of its high activity against Plasmodium falciparum in vitro (20). The active principles are bisbenzylisoquinoline alkaloids and it has subsequently been established that the stereochemistry at C-1 and C-1' as well as the type of dimeric linkage and aromatic substituents are important requirements for antiplasmodial activity (see Fig. 4). High-yielding cell lines of a series of bisbenzyliqouinoline-producing cultures are needed in order to obtain sufficient quantities of alkaloids for more detailed investigation of biological activities. The enzymes responsible for individual dimerisation linkages require to be characterised and hopefully their c DNA will be cloned and their activity expressed in either yeasts or bacteria so that different monomers can be linked in specific ways. Harringtonine and homoharringtonine are of current interest in cancer chemotherapy and their supply remains limited. Their production by plant cell cultures of Cephalotaxus harringtonia has been reviewed (21).

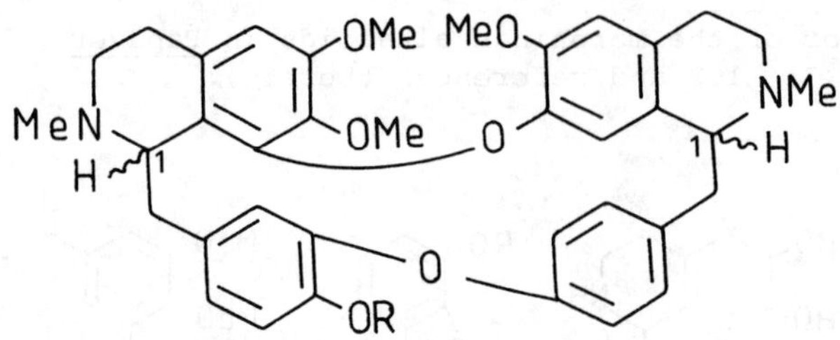

		C-1H	C-1'H	IC_{50} ($\mu g\ mL^{-1}$)
phaeanthine	R = Me	R	R	1.43
pycnamine	R = H	R	R	0.15

FIGURE 4. Examples of bisbenzylisoquinolines and their *in vitro* activity against *Plasmodium falciparum*.

2.1.3 *Tropane alkaloids*: Atropine (±-hyoscyamine) and scopolamine are medicinally important alkaloids which act on the autonomic nervous system. They have been target molecules for plant cell culture production and their occurrence in cell cultures has been reviewed (1). Tropane alkaloids are biosynthesised in the roots of several species of Solanaceae and this family in particular is amenable to the formation of transformed "hairy-roots" via *Agrobacterium rhizogenes* infection (22 and references therein). In our laboratories, a high alkaloid-yielding hybrid of *Datura candida* has been transformed to produce hairy roots containing 0.68% d.w. of total alkaloid (22). These root cultures produce scopolamine as 17% of their total alkaloid content. GC-MS has been used to identify 19 alkaloids present in these roots and of particular interest are the new alkaloids 3-hydroxy-6-propionyloxytropane and 3-hydroxy-6-butyryloxytropane (23) (see Fig. 5).

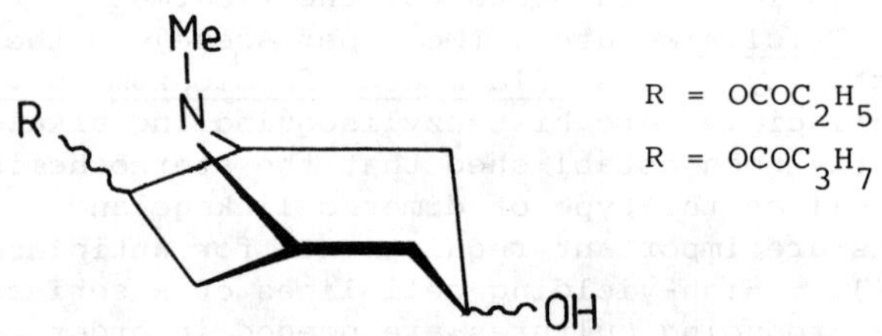

FIGURE 5. Novel tropane alkaloids isolated from hairy root cultures of a *Datura candida* hybrid

The ready production of transformed roots of solanaceous species and the capacity of alkaloid producing plants to product tropane alkaloids in such cultures indicates that every possibility exists for identifying new tropane alkaloids via this technique. One clone selected from hairy root cultures of *Duboisia leichardtii* has been selected on the basis of growth rate, lateral branching, opine production and high scopolamine content. Under prescribed conditions this clone yields 78 mg l^{-1} of scopolamine in a four week period (24).

2.1.4 Miscellaneous alkaloids: There is considerable scope for research into the production of alkaloids of novel structure or of interesting biological action from other groups of alkaloids. One such example is the indolizidine alkaloid castanospermine which is obtained from *Castanospermam australe* and which has activity against the AIDS-associated HIV-1 virus. Other interesting groups of alkaloid include quinolizidines, the anthranilate derived furanoquinolines, acridones and quinolines as well as steroidal alkaloids and purines.

2.2 Terpenoids

More than 26 different pharmacological properties have been attributed to individual mono-, sesqui-, di- and tri-terpenes. A brief consideration of the potential of plant cell cultures to product biologically active terpenoids has been published (25). In this chapter a few pertinent examples will be mentioned.

2.2.1 Sesquiterpenes: One of the most intriguing sesquiterpenes of current clinical interest is artemisinin (qinghaosu) the antimalarial drug obtained from *Artemisia annua*. A review of the plant, extraction, chemical, biological, toxicological and clinical properties of artemisinin has been published (26). Artemisinin, artemether and sodium artesunate are used clinically in China for the treatment of chloroquine-resistant malaria and it is anticipated that other countries will also utilise these novel antimalarial drugs. The total synthesis of artemisinin has been achieved but may not be economic for large scale production. Hence artemisinin will continue to be obtained from the plant or possibly from cell cultures. Low yields of artemisinin have been reported from roots and rooted plantlets derived from callus cultures of *A. annua* (27). In our laboratories we have not been able to detect artemisinin in cell cultures. Callus and suspension cultures extracted at monthly intervals were extracted and evaluated for their *in vitro* activity against *Plasmodium falciparum* (28). Low levels of activity were obtained but this activity is not attributed to artemisinin. There are many examples of sesquiterpenes which are of interest because of differing biological activities and these are candidates for plant cell culture production.

2.2.2. Diterpenes: Forskolin (coleonol) from *Coleus forskohlii* activates adenylate cyclase and is of interest as a cardiotonic and anti-hypertensive. The phorbol esters of *Euphorbia* species substitute for diacylglycerol in the activation of protein C kinases and are used to investigate the mechanisms involved in inflammation and cancer. Gingkolides from *Gingko biloba* inhibit the binding of platelet aggregation factor (PAF) to platelets and to lymphocytes. They are of interest as potential drugs for the treatment of inflammation, allergies and asthma. Diterpenes of current interest as anticancer drugs include tripdiolide from *Tripterygium wilfordii* and taxol from *Taxus baccata*.

2.2.3 Triterpenes and quassinoids: Triterpenes for steroid production, cardenolides from Digitalis species and ginsenosides from Panax species are of considerable economic importance and hence are prime targets for plant tissue culture production. The production of high-yielding strains and of novel structure formation are feasible by cell culture techniques. The quassinoids are a group of bitter terpenoids which are derived from triterpenes in some species ofthe Simaroubaceae. As a group of closely related natural products they show a diverse range of biological activities which include anti-cancer, anti-inflammatory, antiprotozoal, antiviral and insecticide. These interesting molecules usually occur in mixtures of closely related compounds found in species which are mainly tropical. Considerable scientific effort has gone into the synthesis of such compounds but the routes are long and the yields are low. Novel quassinoids as well as larger quantities of specific quassinoids for further studies are urgently required. In our laboratories, we have established that some quassinoids (e.g. brusatol, holacanthone) have potent activity against Plasmodium falciparum (29). The biosynthetically closely related limonoids obtained from species of Meliaceae and Rutaceae also have diverse biological activities and are targets for plant cell culture techniques.

2.3 Lignans

Podophyllotoxin from podophyllum resin is highly active against cancer cell lines but is markedly toxic to normal cells. Chemical modification of the natural lignan has led to the clinically useful anticancer drugs teniposide and etoposide. The synthesis of podophyllum-type lignans is difficult, particularly in view of their chiral centres and new routes to lignans are required. More than 30 lignans are known to exert powerful action against tumour cell lines and other lignans are potent inhibitors of PAF, thus making candidates for the development of antiallergic and anti-inflammatory drugs. Lignans are clearly targets for plant cell culture production.

2.4 Quinones

Shikonin is the only natural product from a higher plant which is currently produced commercially by plant cell culture techniques. The closely related naphthoquinone lapachol has antimicrobial and anticancer activities whilst β-lapachone and allyl-β-lapachone are potent inhibitors in vitro of Trypanosome cruzi. Plumbagin is active in vitro as an antifungal and antiprotozoal agent. Quinones are of interest for the development of antiprotozoal drugs and the Wellcome compound BW58C (2-(4-t-butylcyclohexyl)-3-hydroxy-1,4-naphthoquinone is more active than chloroquine against P. falciparum in vitro and is active in vivo against P. berghei (30 and references therein). A number of anthraquinones have been produced by cell cultures of Cinchona ledgeriana and yet recent work in our laboratories has resulted in the isolation of four novel anthraquinones from cultures of this species (31).

3. CONCLUSIONS

Plant cell cultures have produced a wide range of secondary products and it has been reported that 85 novel compounds have been isolated from some 30 different plant cultures (12). The novel compounds included some 23 alkaloids, 19 terpenoids, 30 quinones and 11 phenyl propanoids. Seventy of these 85 compounds were isolated in the five years immediately prior to 1989. It is estimated that 65-75% of plant species are located in tropical rainforests and their widespread destruction has led to the gloomy prediction that some 50,000 species may be extinct within the next 13 years. Plant cell cultures may be used for the storage of these gene pools and the regeneration of endangered species as well as alternatives to the whole plant for molecule hunters. The majority of higher plants have not been investigated systematically for their chemical constituents or for pharmacological activity and for those plants which are not readily available either because of geography or because they are endangered, plant cell cultures could well be investigated.

The difficulties in producing economic syntheses for natural products often rich in chiral centres, means that plants will continue to be utilised as sources of chemicals. For pharmacological evaluation and new drug development, high-yielding cell cultures need to be developed for biological testing of active compounds. Our knowledge of existing biologically active molecules enables us to provide lists of target molecules for production by plant cell cultures. The application of genetic engineering techniques with higher plant enzymes being produced in yeasts and bacteria opens the way for obtaining key intermediates for the synthesis of natural product molecules. The potential of cell cultures to provide new molecules is obvious and it is up to scientists to persuade funding bodies that such research should continue for the benefit of mankind.

REFERENCES

1 Hay, C.A., Anderson, L.A., Roberts, M.F. and Phillipson, J.D. (1988) in Biotechnology in Agriculture (Mizrahi, A., ed.) pp97-140, Liss, New York

2 Anderson, L.A., Phillipson, J.D. and Roberts, M.F. (1985) in Advances in Biochemical Engineering/Biotechnology (Fiechter, A., ed.) pp1-36, Springer, Berlin

3 Phillipson, J.D. (1990) in Secondary Compounds from Plant Tissue Culture (Charlwood, B.V. and Rhodes, M.J.C., eds.) Oxford University Press

4 Principe, P.P. (1989) in Economic and Medicinal Plant Research, Vol. 3 (Wagner, H., Hikino, H. and Farnsworth, N.R., eds.) pp1-19, Academic London

5 Anderson, L.A., Harris, A. and Phillipson, J.D. (1983) J. Nat. Prod. 46, 374-378

6 Liu, K.C.-S., Yang, S.-L., Roberts, M.F. and Phillipson, J.D. (1990) Phytochemistry 29, 141-143

7 Liu, K.C.-S., Yang, S.-L., Roberts, M.F. and Phillipson, J.D. Plant Cell Reports, in press

8 Arisawa, M., Kinghorn, A.D., Cordell, G.A. and Farnsworth, N.R.

(1983) J. Nat. Prod. 46, 222-225
9 Fukamiya, N., Okano, M., Aratani, T., Negoro, K., Lin, Y. M. and Lee, K.-H. (1987) Planta Med. 53, 140-143
10 Towers, G.H.N. and Abramowski, Z. (1983) J. Nat. Prod. 46, 576-581
11 Crespi Perellino, N., Guicciardi, A., Minghetti, A. and Speroni, E. (1988) Pharmacol. Res. Commun. 20, 45-48
12 Ruyter, C.M. and Stöckigt, J. (1989) GIT Fachz. Lab. 33, 283-293
13 Kutchan, T.M. (1989) Febs Lett. 257, 127-130
14 Brown, R.T., Leonard, J. and Sleigh, S.K. (1977) J. Chem. Soc. Chem. Comm. 636-638
15 Brown, R.T. and Leonard, J. (1979) J. Chem. Soc. Chem. Comm. 877-879
16 Brown, R.T. and Pratt, S.B. (1980) J. Chem. Soc. Chem. Comm. 165-167
17 Pfitzner, U. and Zenk, M.H. (1989) Planta Med. 55, 525-530
18 Akasu, M., Itokawa, H. and Fujita, M. (1975) Phytochemistry 14, 1673-1674
19 Schiff, P.L. (1985) In The Chemistry and Biology of Isoquinoline Alkaloids (Phillipson, J.D., Roberts, M.F. and Zenk, M.H., eds.) pp126-139, Springer, Berlin
20 Partridge, S.J., Russell, P.F., Kirby, G.C., Bray, D.H., Warhurst, D.C., Phillipson, J.D., O'Neill, M.J. and Schiff, P.L. (1988) J. Pharm. Pharmacol. 40, 53P
21 Misawa, M., and Hakanishi, T.M. (1984) in Biotechnology in Agriculture and Forestry 4 (Bajaj, Y.P.S., ed.) pp191-208
22 Christen, P., Roberts, M.F., Phillipson, J.D. and Evans, W.C. (1989) Plant Cell Reports, 8, 75-77
23 Christen, P., Roberts, M.F., Phillipson, J.D. and Evans, W.C. (1990) Plant Cell Reports, in press
24 Mano, Y., Ohkawa, H. and Yamada, Y. (1989) Plant Science 59, 191-201
25 Sticher, O. (1977) in New Natural Products and Plant Drugs with Pharmacological, Biological or Therapeutical Activity (Wagner, H. and Wolff, P., eds.) pp137, Springer, Berlin
26 Trigg, P.I. (1989) in Economic and Medicinal Plant Research, Vol. 3 (Wagner, H., Hikino, H. and Farnsworth, N.R., eds.) pp20-56, Academic, London
27 Nair, M., Acton, N. and Klayman, D.L. (1986) J. Nat. Prod. 49, 504-507
28 Tawfiq, N.K., Anderson, L.A., Roberts, M.F., Phillipson, J.D., Bray, D.H. and Warhurst, D.C. (1989) Plant Cell Reports 8, 425-428
29 Phillipson, J.D. and O'Neill, M.J. (1987) in Biologically Active Natural Products (Hostettmann, K. and Lea, P.J., eds.) pp49-64
30 Wright, C.W. and Phillipson, J.D. (1990) Phytotherapy Research, in press
31 Tawfiq, N.K. (1990) Tissue Culture of Antimalarial Plants, M. Phil. Thesis, University of London

LARGE-SCALE PRODUCTION OF NEW BIOLOGICALLY ACTIVE COMPOUNDS IN PLANT-CELL CULTURES

K. WESTPHAL, DIVERSA GESELLSCHAFT FÜR BIO- UND VERFAHRENSTECHNIK MBH, ALTER POSTWEG 1, 2070 AHRENSBURG, WEST GERMANY

The large-scale fermentation of plant-cell cultures has been realized for the first time in West Germany. Founded at the beginning of 1988 the Hamburg company, DIVERSA Gesellschaft für Bio- und Verfahrenstechnik mbH, a 100% subsidiary of BATIG Gesellschaft für Beteiligungen mbH, took over the biotechnological activities of BATCF with a view to developing and operating procedures for the biotechnological use of plant-cell cultures on a large scale. As this technology involves technological and economic risks, this project has been financially supported by the Federal Ministry of Research and Technology, under No. TBE/1-034003-038629 0, since 1st May 1985.

In 1986 work started on setting up a five-stage fermentation plant in the grounds of the BAT factory in Ahrensburg near Hamburg. (1)

The plan for the fermentation cascade on a production scale of up to 75,000 litres was based on several years' experience gained from a special pilot plant of up to 300 litres which had been optimized for plant cells. The core of the plant consists of five stirring tanks with volumes of 75 litres, 750 litres, 7,500 litres, 15,000 litres and 75,000 litres. The two small units were designed as a compact plant whilst the other three are equipped as large-scale fermentation units.

All the fermenters have vessels with an antifoaming agent and also, with the exception of the 75 litre tank, for a nutrient solution concentrate. An acid-alkaline station is also set up centrally as is the tank for preparing nutrient solutions via constant sterilisation.

All parts of the plant are separately equipped with their own control panels.

The control and regulation concept is based on a three-level structure: the first level is a maintenance and repair level for emergency action. This is followed by a decentralized control and regulation level with retrievable function flows and with interlocks. A Simatic S5 control was used as hardware building block. There is a process guiding system with its own multifunctional units available for the central control panels of all the control parameters and for the documentation of every kind of user intervention and process flow. The system is safeguarded by a non-destructible power supply.

In connection with weekly protection of data the process guiding system provides the standard necessary for GMP production.

The medium is sterilised in situ in the 75 litre and 750 litre fermenters. The larger fermenters - 7,500, 15,000 and 75,000 litres - are first sterilised empty and then filled under sterile conditions using constant-flow sterilisation equipment.

The smooth use of continuous sterilisation is helped by the fact that clear nutrient solutions are used which have a basic composition as laid down by Linsmaier and Skoog (2) and are optimized in line with the specific conditions of the culture being used. The materials used are subjected to a quality test and are weighed via a computer-controlled weighing system in such a way that the desired nutrient solution compositions have to be observed accurately. Demineralized water is used to prepare the nutrient solutions.

By including the nutrient solution preparation in the control flow the fermenters can to a large extent be fed automatically.

There are various possibilities for the logistics of transfer to the next cascade stage thanks to flexibly designed tubing.

First of all, the usual forward transfer via all the cascade stages is possible.

With its flexible transfer possibilities, however, the existing configuration also permits reverse transfers which means that each of the three large-scale fermenters can be used as an inoculation stage for the other two.

In this way the greatest possible production safety is achieved and this means that where disruptions occur and a tank is no longer sterile, the otherwise lengthy breeding up to the 75,000 litre reactor can be decisively shortened.

The equipment also offers the possibility of semicontinuous operation and can thus, when necessary, operate with differing volumes.

This method of operation reduces the growth phase by harvesting a large part of the fermenter contents and adding new nutrient solution to the remaining, actively growing cells.

When constructing the fermenter, particular care was taken in selecting those components which, as we know from experience with the pilot plant, encourage the breeding of plant-cell cultures. This applies, inter alia, to the stirring system for which the following criteria were taken into account:

- to create process conditions in the fermenter which are as identical as possible on every spot at all times,

- to treat the plant cells with care by applying little shearing stress, and

- by ensuring good dispersion of the gas bubbles.

As in the test results in the pilot fermenter, these criteria were also achieved in the large fermenters by using a multi-stage stirring organ of large diameter. The stirring systems are so designed that the formation of foam is reduced. (3)

In order to ensure adequate ventilation of the cell cultures rheological models were modified and from them a ventilation system developed.

As this is a pilot plant in spite of the large volumes involved, we intend - in addition to the usual measuring equipment (temperature, pH, pressure, oxygen partial pressure) - to introduce options for installing further sensors.
The growth of the cells can thus be followed up both by on-line measurement and in addition by regular off-line examination of samples.

In the latter, fresh weight, dry weight, refractive index, osmosis, redox potential and, in individual cases, the metabolite yield are determined.

The date for a transfer must be determined very accurately from the growth curve so as to ensure that fermentation runs successfully.

The precondition for successful realization of large-scale fermentation was the know-how which had previously been gained on the laboratory and pilot scale:

For this, highly productive calli were obtained from selected cell material and were transferred to suspension cultures. With these cultures a comprehensive series of tests was carried out. Initially, mediums were varied under the aspect of growth acceleration in view of the scheduled large-scale fermentation. (4)

When measuring the conductivity of the nutrient solution, we repeatedly observed that, up to the end of the growth phase, only approx. 50% - 60% of the inorganic medium components was absorbed by the cells.

In numerous tests the influence of these variants on the course of the culture, the cell yield and on productivity was determined. In the course of these tests it transpired that a general reduction in macro-nutrient salts (20%) proved to be disadvantageous whilst an increase in phosphates led to a slight increase in cell yield and to relatively strong growth acceleration.

In these tests, with the exception of $MgSO_4$, reduction of individual macro-nutrient salts proved to have no effect.

In further tests, on the other hand, simplification of the medium composition with a view to cutting costs was the main target.

To sum up, one can say that the growth medium developed for Echinacea purpurea has proved its worth in the long term. In the last few years the time needed for the culture with 10% inoculation has decreased from 21.5 to 7.3 days.

The findings thus made in the laboratory were then translated on to the pilot scale. Here too, the time needed for fermentation on the 50 litre and 300 litre scale was reduced from 19 days in 1986 to 9 days in 1989.

The measurement data obtained during fermenter runs in the pilot cascade (55 - 300 litres) were used for model calculations which are intended to lead to better understanding of the fermentation process and to make piloting of the process easier. This formal description of the fermentation process using mathematical equations is still being continued today and initial findings have already been applied. (5)

On the basis of experience gained in the way described, the first fermentations of cell cultures were carried out at the beginning of 1987.

From the very outset it was clear that the cultivation of plant cells up to a scale of 60 cbm is fundamentally possible. It was, however, clear that other improvements in the process would have to be made in view of the different fermentation times and the yields in the fermenters.

In the meantime the fermentation of plant cells in batches with forward and reverse transfer possibilities has almost become routine. The introduction of O_2 control, in particular, has led to optimization of the process in the last few months.

In the normal fermentation run the amount of air fed in has so far been stipulated as a fixed target figure for the flow regulator. As the cell cultures need oxygen according to their growth phase, very different use is made of the constant amount of air which is fed in.

In Phase 1 there was surplus oxygen during the growth of the culture.

In Phase 2 good use was made of the oxygen supplied, dependent on growth.

In Phase 3 the supply of oxygen was inadequate and this, together with improverished nutrients in the nutrient solution, meant that growth ceased.

This point in time was then selected as the time to harvest.

In the course of testing the fermenter cascade various filling levels and, during semi-continuous operation which is described separately, differing volumes were used. This called for a constant change in the ventilation rate. With the aid of a computer the relevant air throughput per filling level (which was determined by high-pressure measurement of the differences) was laid down as target figure for the regulator. For the fermentation process this does not yet involve any changes apart from the fact that operational safety was greatly increased.

In a further step efforts were made to adjust the input of air to the conditions of the culture. The above-mentioned works were used as theoretical basis for this.

To realize the process an oxygen probe was added to the system.

Thus there was a decisive change in the 3 defined phases:

In Phase 1 there is only a little surplus air because a minimun air volume is laid down.

In Phase 2 there are optimum oxygen conditions for the plant cells.

In Phase 3 a decrease in available oxygen can again be observed because, owing to the mechanical conditions (diameter of the conductor and valve), no greater oxygen input is possible.

Analysis of the laboratory figures showed us that cell yield rose by approx. 10% to approx. 200 g/l moist weight due to these changes in the process. In addition, follow-up tests are to examine whether use of the oxygen control will further shorten the growth period.

In addition to experience with flexible handling of the transfer possibilities, as described above, great success was achieved with semi-continuous operation.

The following results can be derived from the tests:

1. Semi-continuous operation of the equipment with plant cells is possible with Echinacea and it can be assumed that semi-continuous operation with plant cells is possible in general.

2. With semi-continuous operation the productivity of the fermenters can be increased considerably because the otherwise customary lag phase can be cut substantially thanks to the greater density of inoculations and the physiological condition of the cultures.

   We were able to show that even with long-term cultivation, no degeneration effects occurred in the fermenters. On the contrary, the cell suspensions became better, i.e. the cells were mainly single cells and/or cell filaments and no longer microcalli.

   Use of this operation also means that there are no cleaning and maintenance intermissions which are necessary in batch operations.

   The yield per time unit is clearly higher. By approx. 1/3rd for the 75 cbm fermenter and by almost double for the 15 cbm fermenter, in relation to the total growth duration.

   Limiting factors for semi-continuous fermenter operation are not physiological factors but merely incalculable technical defects. Semi-continuous fermentations of up to 80 days were carried out.

From the tissue cultures of Echinacea purpurea immunologically active polysaccharides were isolated and their structure largely defined. They are two neutral polysaccharides and one acid polysaccharide. The two neutral ones were found to be fucogalactoxyloglucanes with molecular weights of 10,000 D and 25,000 D, whilst the acid polysaccharide with a molecular weight of 75,000 D was identified as being an arabinogalactane with a very complex structure. All three polysaccarides are metabolites of the primary cell wall. (6)

For further tests large quantities of these compounds had to be produced in a reproducible manner.

To solve this problem a process was developed which, as a first step, separates solid from liquid with a filter press and then provides for mild concentration of the active substances, the result being that precipitation can then be carried out with more favourable ratios of yield/volume. (7)

After separating off the cell mass the resultant product solution is purified and then, using the membrane filter, is freed of coarse colloids and mucins of high molecular weight (micorfiltration). The solution is subsequently concentrated in the ultrafilter. The abscission cut is chosen in such a way that the target products are enriched in their retention form. Concentration factors of 50 to 200 can easily be achieved.

The concentrate which has thus been thickened can now easily be subjected to further treatment in the usual way on laboratory or pilot scale. Even in large-volume fermentation the scaling-up of these follow-up steps presents no problem because, due to the wide possibilities of concentration, the dimensions of the resultant volumes and quantities for processing remain very clear.

The yields produced depend on the quality of the relevant initial fermentation stock and can amount to roughly 10 to 100 ppm for the individual fractions, related to the original suspension volume.

After the first successful cycles with Echinacea purpurea, the fermenter cascade was run with Rauwolfia serpentina up to the 75 cbm stage. Here too, we tested semi-continuous operation. No notable formation of active substances in the strong growing cultures was observed.

In addition, two attempts were made to breed cultures of Panax ginseng in the fermenter cascade. These were relatively young, not optimized cultures which had been freshly obtained from calli. Contrary to Echinacea and Rauwolfia no comprehensive optimization of the cell cultures on a laboratory scale was made. This is probably why there were considerable difficulties with the culture in the fermenter. The first fermentation was stopped after 4 weeks when no major growth was to be seen. In a second test the Ginseng culture also grew extremely slowly. For this reason medium even had to be added, because growth was monitored continuously via off-line samples. The yield from this culture proved to be relatively low. The presence of ginsenosides was proved. Further tests in this direction have been postponed for the time being.

To sum up it can be said that the large-scale fermentation of plant-cell cultures has been achieved for the first time in West Germany.

The problems connected with piloting the process were successfully solved and we succeeded for the first time in propagating plant-cell cultures under sterile conditions in fermenters of up to 75,000 litres at a justifiable cost and then in isolating the desired target substances.

The object of further work will be to push forward the processes of biotechnological procedures for the production of further valuable plant raw materials.

REFERENCES

(1) Rittershaus, E., et al. (1989): Bio Engineering 1/89, 5. Jahrg., pp. 8 - 10

(2) Linsmaier, E.M.; Skoog, F.: Physiol. Plant Vol. 18, pp. 100 - 127

(3) Rittershaus, E., et al. (1989): Bio Engineering 2/89, 5. Jahrg. pp. 28 - 34

(4) Willberg, B. pers. com. 1990

(5) Rittershaus, E., et al. (1989): Bio Engineering 5/89, 5. Jahrg. pp. 54 - 70

(6) Wagner, H., et al. (1989): Z. Phytoth. 10, Vol. 2 pp. 35 - 38

(7) Rittershaus, E., et al. (1989): Bio Engineering 3+4/89, 5. Jahrg. pp. 51 - 65

TERPENE PRODUCTION BY LIVERWORTS OF IN VITRO CULTURES

HANS BECKER
Pharmakognosie und Analytische Phytochemie der
Universität des Saarlandes, D 6600 Saarbrücken, FRG

1. INTRODUCTION

Generally, in vitro cultures of plants, plant organs, and plant cells are industrially acceptable if the price of the desired product is very high or if the conventional production of the plant causes problems. Liverworts meet at least the second criterion. They contain a large variety of secondary products, especially mono-, sesqui- and diterpenes, lunularic acid derived phenoles and flavonoids (e.g. 1, 2, 3 for review). Some of these products exert interesting biological activities. In many cases, the amount of isolated products does not allow for biological screening. In vitro culture of liverworts is the method of choice in obtaining sufficient amounts of secondary products and in analysing species unavailable in nature in quantities necessary for chemical analysis. In contrast to higher plants, the study of in vitro cultures of liverworts for the production of secondary compounds is rather scarce. Only in 1981, Takeda and Katoh (4) reported on the production of volatile compounds by suspension cultured cells of Calypogeia azurea. We first started our work on the production of secondary metabolites on in vitro culture of liverworts in 1985. In the following, some of the results with these cultures will be reported.

2. MATERIALS AND METHODS

2.1 Fossombronia pusilla

Plants containing spore capsules were collected in the Saarland. The intacts spore capsules were surface sterilized with 0.1 % benzalkonium chloride for 10 minutes, followed by 5 minutes with 1 % sodium hypochlorite. The capsules were washed with sterile water and transferred to glass tubes (16 x 3 cm), containing 10 ml of Gamborg B5 agar medium. Developing plantlets were transferred to Erlenmeyer flasks, containing 50 ml of B5 agar medium, and then subcultured every four weeks. Temperature was set at 23° C with continuous light about 4500 lux.

The cultures were harvested, dried at room temperature and extracted exhaustively with dichloromethane. Individual compounds were separated by a combination of column chromatography and HPLC, using silicagel as stationary phase and petrolether/ethylacetate as eluents. Structure elucidation was mainly obtained by 1H- and 13C-NMR spectroscopy.

2.2 Ricciocarpos natans
Aseptic cultures were provided by the Institute of Botany, Czecho-Slovak Academy of Sciences, Culture Collection of Autotrophic Organisms, Treboon, CSFR. The cultures had been isolated in 1947 and kept on a salt medium since then. They were maintained in our laboratory on Gamborg B5 liquid medium (5) with 10 g/l sucrose under continuous illumination of about 2000 lux at 22± 1° C. Experimentally, growth was also tested for the following media with the respective macroelements: Lorenzen et al. (6), Benecke (7), Knop b(7), Knudsen (8), MSK 2 (9), and White (10).

Fresh plant material was extracted with dichloromethane. Terpenes were isolated by a combination of column chromatography with silicagel and sephadex LH 20, and finally with HPLC. Detailed spectroscopic studies yielded the structure of seven terpenes.

2.3 Reboulia hemisphaerica
Axenic cultures of Reboulia hemisphaerica were provided by the Culture Collection of Czecho-Slovak academy of Sciences (see 2.2). They had been isolated in 1941 and kept on a salt medium since then. In our laboratory they were maintained on Gamborg B5 medium or MS-medium, both had trace elements of MSK-2-medium (9) and additions of 2 % sucrose.

Callus was induced on MS-medium either with 3 % sucrose, 2 mg/l α-Naphthalene acetic acid (NAA) and 1 mg/l kinetin (KIN), or on the same basal medium with 4 % glucose.
Fresh plant material was extracted with diethylether and subsequently fractionated by a combination of chromatographic methods. Detailed spectroscopic studies yieldes the structure of five terpenes.

3 RESULTS

3.1 Fossombronia pusilla
Of the large number of individual spores from 40 spore capsules, only two within two separate tubes germinated. They grew to globules of about 5 mm diameter within 6 weeks. After subculturing they reached a diameter of about 8 mm after 10 weeks. Individual globules were divided now and subcultured every 4 weeks. The two lines derived from the two spores were kept separately, they were similar in growth and production characteristics. The cultures were kept on Gamborg B5 medium (5) with 2 % sucrose added, 125 mg/l citric acid monohydrate and 12.5 µg/l vitamin B12. Within 4 weeks, agar cultures as well as liquid cultures showed a fivefold increase under these conditions.

The cultures on agar were highly differentiated into plantlets of a similar morphology as those of field collected plants. They contained about 25 oil bodies per cell. Submerse cultures tended to grow in globular aggregates with thick cell walls. The formation of leaf-like and stem-like structure could be suppressed on Knop medium with 4 % glucose. However, rhizoids still developed and growth was very poor with a doubling time of 25 days.

Redifferentiation could be induced by transferring the cultures to the standard B5 medium with 2 % sucrose. The cultures contained the diterpenedialdehydes perrottetianal A and perrottetianal B, first described by Asakawa et al. (11) for the liverwort Porella perrottetiana, the sesquiterpene lactone α-(-)-santonin, hitherto only known from higher plants, and a newly found perrottianal-derivative: 8-hydroxy-9-hydroperrottetianal. The differentiated agar and liquid cultures as well as collected material showed similar amounts of the two main products perrottetianal A and perrottetianal B, whereas the highly undifferentiated cultures contained only one tenth of these amounts (see table 1).

	collected plants	differentiated agar cultures	differentiated liquid cultures	undifferentiated agar cultures
P A	74	80	87	8
P B	90	87	91	11

Table 1: Content (mg/g dry weight) of the two diterpenedialdehydes perrottetianal A (P A) and perrottetianal B (P B) in cultivated and collected plant material of Fossombronia pusilla.

A petrol ether extract of the differentiated cultures (100 µg/ml) as well as the isolated compounds (100 µg/ml) showed good antibacterial activity against E. coli, Klebsiella pneumoniae, Staphylococcus aureus, Bacillus subtilis.

3.2 Ricciocarpos natans

Of the 7 tested media, cultures grew best on Gamborg B 5 medium (5). Starting from an inoculum of 1.5 g, the cultures reached 18.5 g fresh weight within a period of 35 days (see table 2), and thus was threefold, compared to cultures on Lorenzen medium. The differences in dry weight

	Benecke	Gamborg	Knop	Knudsen	Lorenzen	MSK	White
fresh weight / g	6.9	18.5	8.1	11.0	5.5	8.0	7.0
dry weight / mg	640	860	720	750	550	720	660

Table 2: Fresh weight and dry weight of axenic cultures of R. natans after a culture period of 35 days. Inoculum was 1.5 g (fresh weight) and 70 mg (dry weight).

were less pronounced but still obvious (12). Differences were also observed in the appearances of the cultures reaching the stationary phase. Plants on media of Benecke, Knop b, Lorenzen, and White turned from dark green to light green and subsequently turned brown. Plants on B5, MSK 2, and Knudsen Medium remained dark green for as long as 35 days.

Ono et al. (13) reported on the induction of callus in R. natans by the adding 4 % glucose. We could not induce callus with 4 % glucose or sucrose. The plantlets, however, showed significant morphological changes. The size of thalli was reduced, and the surface was curved to the ventral size; the number of rhizoids were reduced.
Chemical analysis (Wurzel and Becker, Phytochemistry in press) revealed the presence of three sesquiterpenes with novel structural elements, hitherto unfound in liverworts, which we named ricciocarpin A, ricciocarpin B, and ricciofuranol. We further determined (-)-limonene, two cuparene derivatives, cuprenolide and cuprenolidol, and phytosterol. Terpenoid content showed a net increase when the cultures reached the stationary phase. Cultures on media with nitrogen or phosphorous deficiencies had a lower terpenoid content.

3.3 Reboulia hemisphaerica
Cultures on B5 agar medium showed different morphological behaviour in dependency of kind and concentration of the sugar. The broadest thallus (20 mm) was obtained with 2 % sucrose. With 0.5 % galactose the thallus reached only 5 mm in width. The amount of oil cells was highest (20 to 30/mm²) with 1 % sucrose and low (5 to 10/mm²) without any additional sugar. However, in cultures kept in dark and 0.5 % sucrose added, oil cells were rarely seen (0 to 3/mm²). Fresh weight increase was better with a 18:6 light regime (sixfold within 40 days), than with continuous light (threefold within 40 days). Callus could be induced either with a combination of phytohormones (2mg/l NAA + 1mg/l KIN) or with a high sugar (4 % glucose) concentration. Suspension cultures were derived from the callus cultures on agar. They could be maintained under light as well as in the dark.

The main sesquiterpenoid of the cultures, gymnomitr-8(12)-en-9α-ol (G-ol), has recently been found in collected plant material of R. hemisphaerica (14). It was quantified for different cultures. In agar cultures with different sugars and sugar concentrations, the maximum content of G-ol (0.75 g/100 g dry weight) was with 1 % sucrose, the lowest content (0.62 g/100 g) with 0.5 % galactose.
In differentiated suspension cultures, the amount of G-ol was nearly the same (0.7 g/100 g) as in the agar culture. In undifferentiated suspension cultures (with 4 % glucose), G-ol reached only about one tenth (0.06/100 g) of the differentiated cultures. Suspension cultures kept in the dark did not produce any sesquiterpene.

4. ABSTRACT

It is possible to grow plant material of various liverworts using in vitro cultures. The cultures produce at least some terpenes known in field collected plants in similar amounts. Undifferentiated cultures produce only small amounts of terpenes. Dark grown cultures of Reboulia hemisphaerica did not form any sesquiterpenes in detectable amounts.

5. ACKNOWLEDGEMENT

The above mentioned experiments were carried out in cooperation with Ms Martina Sauerwein (F. pusilla), Mr Gerald Wurzel (R. natans), and Mr Rui Morais (R. hemisphaerica). To all of them I express my sincerest thanks for their commitments. Detailed results, including growth parameters, structure elucidation, and biological activities are in press.

The author thanks the Institute of Botany, Czecho-Slovak Academy of Sciences for providing cultures of Ricciocarpos natans and Reboulia hemisphaerica.

REFERENCES

1 Huneck, S. (1983) in Manual of Bryology (Schuster, R.M., ed.) Vol. 1, pp. 1-116, Hattori Bot. Lab. Nichinan, Miyazaki

2 Asakawa, Y. (1982) in Fortschr. Chem. Org. Naturstoffe (Herz, W., Grisebach, H., and Kirby, G.W., eds.) pp. 1-285, Springer, Wien, New York

3 Zinsmeister, H.D., and Mues, R. (1987) GIT Fachz.Lab. 499-512

4 Takeda, R., and Katoh, K. (1981) Planta 151, 525-530

5 Gamborg, O.L., Miller, R.A., and Ojima, K. (1968) Exp. Cell Res. 50, 151-158

6 Lorenzen, H., Kaiser, U., and Foerster, M. (1981) Ber. Dtsch. Bot. Ges. 94, 719-725

7 Bopp, M. (1983) in Manual of Bryology (Schuster, R.M., ed.) Vol. 1, pp. 276-324, Hattori Bot. Lab. Nichinan, Miyazaki

8 Basile, D.V., and Basile, M.R. (1988) in Methods in Bryology (Glime, J.M., ed.) pp. 1-16, Hattori Bot. Lab. Nichinan, Miyazaki

9 Katoh, K., Ishikawa, M., Miyake, K., Ohta, Y., Hirose, Y., and Iwamura, T. (1980) Physiol.Plant 49, 241-247

10 Sing, M., and Krikorian, A.D. (1981) Ann. Bot. 47, 133-139

11 Asakawa, Y., Toyota, M., and Takemoto, T. (1979) Phytochemistry 18, 1681-1685

12 Wurzel, G., and Becker, H. (1990) Z. Naturforsch. 45c, 13-18

13 Ono, K., Murasaki, Y., and Takamiya, M. (1988) J. Hattori Bot. Lab. 65, 391-401

14 Morais, R.M.S.C., Harrison, L.J., and Becker, H. (1988) J. Chem. Research (S) 380-381

IMMOBILIZED *GINKGO BILOBA* CELL CULTURES

D. J. CARRIER[1], P. COULOMBE[2], M. MANCINI, R. NEUFELD[1], M. WEBER[1] and J. ARCHAMBAULT. Biotechnology Research Institute, National Research Council of Canada, 6100 Royalmount Ave., Montreal, Canada H4P 2R2. [1]McGill University, Chemical Engineering Department, 3480 University St., Montreal, Canada H3A 2A7. [2]Alternatives, 148 Colonnade Road Unit 7, Nepean, Canada K2E 7R7

1. INTRODUCTION

Ginkgolides are unique C_{20} lactone cage molecules (Fig.1) extracted from the leaves of the slow growing Ginkgo biloba tree [1]. These phytochemicals have pharmacological activity against Platelet Activating Factor (PAF) [2]. Of these three compounds, Ginkgolide B (GB) is the most efficient PAF antagonist. The synthesis of these complex products requires multi-step procedures which are difficult to scale up economically [3].

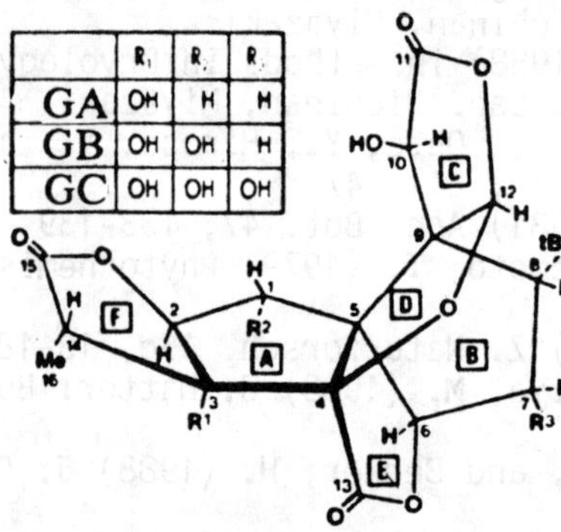

Figure 1. Ginkgolides A, B and C

The culture of G. biloba cells represents an alternative for this production which has not yet been assessed [4]. This communication addresses the production of ginkgolides by G. biloba cells cultured in bioreactors.

2. PROCEDURE

2.1 Materials: G. biloba trees were field grown (Botanical Gardens, Montreal, Canada). Leaves were picked in late August 1988 and stored at -80° for subsequent extraction. Seeds for initiation of callus cultures were kindly provided by Prof. W. Yates (University of Indiana, Indianapolis, Indiana, USA).

Ginkgolide A (GA) and ginkgolide B (GB) were kindly provided by Dr. P. Braquet (Institut Henri Beaufour, Les Ulis, France).

2.2 Culture Methods: G. biloba suspension cultures were obtained as reported previously [5] and maintained in MS medium supplemented with 1 $mg.L^{-1}$ naphthaleneacetic acid (NAA), 0.1 $mg.L^{-1}$ kinetin (K) and 30 $g.L^{-1}$ sucrose. The pH of this medium was adjusted to 6.0 prior to sterilization (121°, 1 bar, 25 minutes). Delong flasks (500 mL) containing 100 ml were inoculated with 30 mL 10 day old suspension. The suspensions were agitated at 120 RPM and maintained at 25° under constant illumination. G. biloba cells were also grown in 2 L and 6 L immobilization bioreactors [7] in the same medium. Cells were readily immobilized on material AO7 (Texel, Ste-Marie de Beauce, Canada). The immobilized cultures were aerated at a rate of 0.05 (2 L) and 0.1 (6 L) VVM with air supplemented with 2% (V/V) CO_2 (dissolved oxygen concentration > 60%) and maintained at 26° under constant illumination. These cultures were inoculated (20%(V/V)) with 10 day old shake flasks grown suspension cultures.

2.3 Analytical: Biomass and extracellular carbohydrate, ammonium, phosphate and nitrate concentrations were analysed as described elsewhere [5]. The biomass of shake flask cultures from Experiment 1 was lyophilized and extracted with ethanol in a Soxhlet apparatus for 24 hr. The biomass from Experiment 2 was extracted with 1.5 L ethanol for 24 hr. In both cases, the solvent was evaporated under vacuum and the extract was purified as described elsewhere [8]. An equal amount of residue and of a silylating agent (Trisil BSA DMF, Pierce Chemicals USA) were vortexed and heated for 1 hr at 73°. The resulting silylated samples were analysed by Gas Chromatography (GC) using an SE-30 W.C.O.T. capillary column (30mx0.25mm I.D.) maintained at 295°. The injector and FID detector were kept at 350°. The biomass from the 6 L immobilization and the second series of shake flask cultures (Experiment 3a and b) as well as G. biloba leaves were lyophilized and extracted in 500 mL acetone for 16 hr. For each sample, the solvent was evaporated and the residue was silylated. Gas Chromatography-Mass Spectrometry (GC-MS) analysis was performed using a capillary column (DB1, 30mx0.25mm I.D.) and an increasing temperature profile (200° to 285° at a rate of 10° per minute). This system was coupled to a VG ZAB-HF mass spectrometer. An on-column injector was used. Results were acquired in the selected ion recording (SIR) mode and processed by a VG11/250 data system.

3. RESULTS AND DISCUSSION

G. biloba cells were grown in shake flasks (Experiment 1) in MS medium supplemented with 1 $mg.L^{-1}$ NAA and 0.1 $mg.L^{-1}$ K (N_1K_1MS) medium. As shown in Fig. 2, biomass dry weight concentration increased to 12 $g.L^{-1}$. After 16-18 days, all extracellular ammonium, phosphate and carbohydrate had been taken up by the cells while 15 mM nitrate remained unconsumed. After 18 days, cells were harvested and lyophilized and extracted with ethanol for 24 hr. The extract was silylated and analysed with GC. GA and GB compounds were not extract was silylated and analysed with GC. GA and GB compounds were not detected. G. biloba leaves were reported to yield 0.005% dry weight of GB [9]. We found a 0.08% yield of GB in leaves. Since lower product yields are generally expected from first plant cell culture studies, the amount of ginkgolide produced in shake flask cultures may not have been sufficient to allow detection using GC analysis.

In order to overcome this problem, larger quantities of cells were grown in surface immobilization 2 L bioreactors using the same medium formulation (Experiment 2). G. biloba cells attached readily to the matrix within 24 hr. The culture was stopped at extracellular carbohydrate depletion (day 26). At that time, the immobilization matrix was covered by a thick (ca. 3 cm) emerald green biofilm. Extracellular nutrients consumption rates were 40% slower (Fig.3) than observed for shake flask cultures (Fig.2). At harvest, the biomass loaded matrix was immersed in ethanol for 24 hr. The ethanolic phase was evaporated, purified [8] and silylated for GC analysis. Small peaks were detected at the same retention times as GB. G. biloba cells were also grown in a 6 L immobilization bioreactor using the same medium formulation (Experiment 3a). At the end of the 31 day cultivation period, the biomass was removed from the matrix, lyophilized and extracted with acetone for 16 hr. The solvent

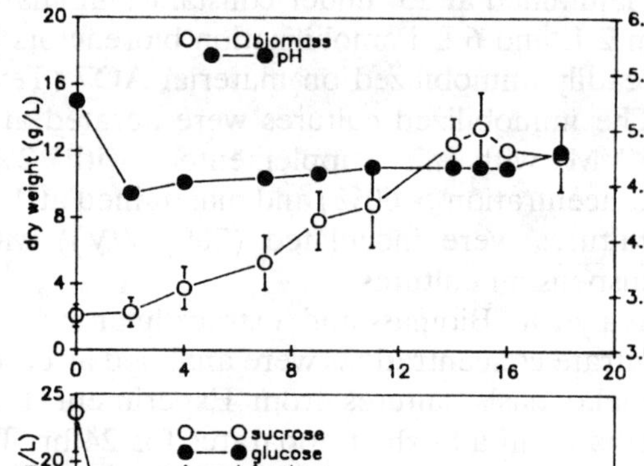

Figure 2. Growth of G. biloba in shake flask. Every point is a mean of 12 samples. Error bars represent 90% confidence intervals.

phase was evaporated and the crude extract silylated. At the same time, G. biloba cells were grown in shake flasks using the same medium (Experiment 3b). After 27 days, the biomass was harvested and treated as above.

High resolution GC-MS analysis indicated yields of 0.0000007% and 0.0000013% GA from the immobilized and suspension cultured biomass respectively. In both cases, GB was detected in trace amounts. GB may be more difficult to detect because steric hinderance limiting the efficiency of silylation.

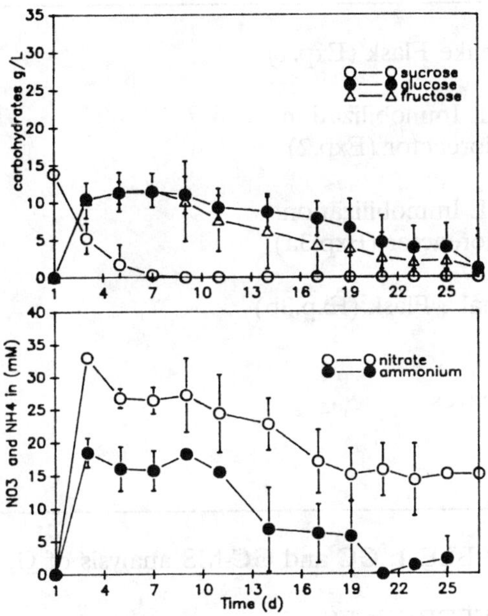

Figure 3. Growth of G. biloba in immobilization bioreactor. Every point is a mean of 2 samples. Error bars represent 90% confidence intervals.

4. CONCLUSION

Results from this first study show that G. biloba cells grown in vitro do produce ginkgolides although at much lower yields than observed in the plant (Table 1). Obviously, much optimization work remains to be performed before the potential of this production system can be assessed. Productive cell lines need to be selected using sensitive semi-automated immuno-analytical screening techniques. Product induction schemes (elicitation, ginkgolide production medium (GPM), transformed cultures etc.) have to be devised and evaluated. Already, preliminary studies using a first GPM have resulted in a 10 fold increase in GA yield. Enzymology and production kinetics have to be defined. The cultures discussed in this work may not have been harvested at peak production. Lastly, process design and integration may prove more efficient than harvesting and extracting leaves for GA and GB large scale procduction.

MODE OF CULTIVATION	METHOD OF ANALYSIS	
	GC	GC-MS
Shake Flask (Exp.1)	Not-detected	
2 L Immobilization Bioreactor (Exp.2)	Trace amounts	
6 L Immobilization Bioreactor (Exp.3a)		0.0000007% GA Trace of GB
Shake Flask (Exp.3b)		0.0000013% GA Trace of GB
Leaves		0.25% GA 0.083% of GB

TABLE 1: GC and GC-MS analysis of G. biloba biomass

REFERENCES

1 Drieu, K. (1986) Press. Med. 15, 1455-1457

2 Braquet, P., Spinnewyn, B., Braquet, M., Bourgain, R., Taylor, J., Etienne, A. and Drieu, K. (1985) Blood. Vess. 16, 558-572

3 Corey, E., Kang, M., Desai, M., Ghosh, A. and Houpis, I. (1988) J. Am. Chem. S. 110, 649-650

4 Nakanishi, K. and Habaguchi, K. (1971) J. Am. Chem. S. 93, 3546-3547

5 Carrier, D., Cosentino, G., Neufeld, R., Rho, D., Weber, M. and Archambault, J. (1990) Plant. Cell. Rep. (in press)

6 Murashige, T. and Skoog, F. (1962) Physiol. Plant. 15, 473-493

7 Archambault, J., Volesky, B. and Kurz, W. (1990) Biotech. Bioeng. 35, 702-711

8 Lobstein-Guth, A., Briançon, F. and Anton, R. (1983) J. Chromat. 267, 431-438

9 Okabe, K., Yamada, K., Yamamura, S. and Takada, S. (1967) J. Chem. Soc. (C), 2201-2206

PRODUCTION OF TERPENES BY TRANSFORMED DIFFERENTIATED SHOOT CULTURES OF MENTHA PIPERITA CITRATA AND M.PIPERITA VULGARIS

A. Spencer, J. D. Hamill, J. Reynolds, M. J. C. Rhodes

Plant Biotechnology Group, AFRC Institute of Food Research, Colney Lane, Norwich, NR4 7UA, England

Introduction

Transformed root cultures formed following infection with Agrobacterium rhizogenes have proved to be a stable, productive system for the study of root-derived secondary products[1]. We are investigating the use of Agrobacterium-derived shooty teratoma cultures for the production of essential oils in plant tissues. In this study we used two species of mint, Mentha piperita var citrata (Ehrh.) Brig. (M.citrata) (Bergamot mint) and Mentha piperita vulgaris (M.piperita) (Black peppermint) which have very different spectra of terpene compounds in their essential oils[2]. Shooty teratomas are produced in tobacco explants using A.tumefaciens strains carrying mutations in either of the auxin genes, tms1, or tms2 [3,4] and this has been exploited to produce shoot cultures active in biotransformation reactions[5]. The use of such strains with mint does not result in the formation of shooty teratomas, but it was found that wild type nopaline strains of A.tumefaciens were effective in producing shooty teratomas in both tobacco[6] and mint[7]. The present paper describes the properties of shoot cultures derived following transformation of the two mint species by the A.tumefaciens nopaline strain T37.

Methods

The M.citrata was a commercial variety obtained locally while the M.piperita was kindly supplied by Dr. K. Svoboda, West of Scotland Agricultural College. The bacterial strain used A.tumefacien T37 was supplied by Dr. G. Ooms of Rothamsted Experimental Station, Harpenden. The methods for the establishment, maintenance of cultures and measurement of growth kinetics are as described[7]. The extraction of terpenes from M.citrata and M.piperita and their quantitation by GLC are as previously reported[7]. The chromatograms shown in Fig F were of essential oils isolated by super critical CO_2 extraction of leaf and shooty teratoma tissue, at 200 bars.

Results

The shooty teratomas formed following infection of stem tissues of both M.citrata and M.piperita were established in cultures, and were conclusively demonstrated to be free of contaminating bacteria. The two shoot cultures are rather different in appearance [Fig A and B]. The M.citrata culture has a central core of callus from which shoots continuously emerge while the M.piperita culture is entirely organised and shoots radiate from stolon like structures. In both cultures the stem/leaf ratio is quite high. Both stem and leaf tissue were shown by SEM to have oil-glands but the density of glands was higher on the leaf tissue. Fig C shows that both cultures, but not their parent plant tissues produce nopaline thus confirming their transformed status. Furthermore Southern blot analysis (Fig D) showed the presence of a fragment in the digest of the shooty teratoma tissue which was strongly homologous to a ^{32}P-labelled probe for the ipt gene of the T-DNA of A.tumefaciens. This fragment was about 2 Kb in size indicating that the region of the T-DNA was transferred intact into the DNA of the cultured tissue. Fig E shows the growth curves of both M.citrata and M.piperita cultures in liquid media.

Fig F shows GL Chromatograms of the spectrum of terpenes in the essential oils of M.citrata (FI) and M.piperita (FII) and of the shooty teratomas derived from them (figures FIII and FIV respectively). In M.citrata the spectra of terpenes are similar in both plants and in the shooty cultures derived from them. In this species there are only two major components, linalool and linalyl acetate which comprise over 85% of the oil. The ratio of linalool/linalyl acetate is somewhat lower in the shoot cultures (0.60) compared to the leaf of the plant (0.89). The absolute levels of oil produced per unit weight are also lower in the shoot cultures[6]. This probably reflects that the low leaf/stem ratio in this culture, as leaves had a higher content of terpenes than the rest of the culture.

The oil of M.piperita is much more complex than that of M.citrata and neither linalool nor linalyl acetate are present in significant amounts. In contrast the major components are menthol and menthone, and there are several other components present at significant levels. The ratio of menthone/menthol is higher in the shooty teratomas than in the parent plant (Table 1). Although the absolute amounts of menthone and menthol produced per gram of culture is only just over 10% of that present in the leaf. It should be noted that the leaf produces a much higher oil yield than the average of the whole plant. The culture also is mainly composed of low-oil bearing stem tissue which will also account for the lower total yield. An important objective in the improvement of such cultures would be to increase the leaf/stem ratio, which would involve the use of both genetic and physiological approaches.

In conclusion we have demonstrated that teratoma shoot cultures of Mentha species can be developed and grown in vitro and that these cultures synthesise terpenes as expected. The spectrum of terpenes produced in culture in general reflects that of the parent plant but in both species

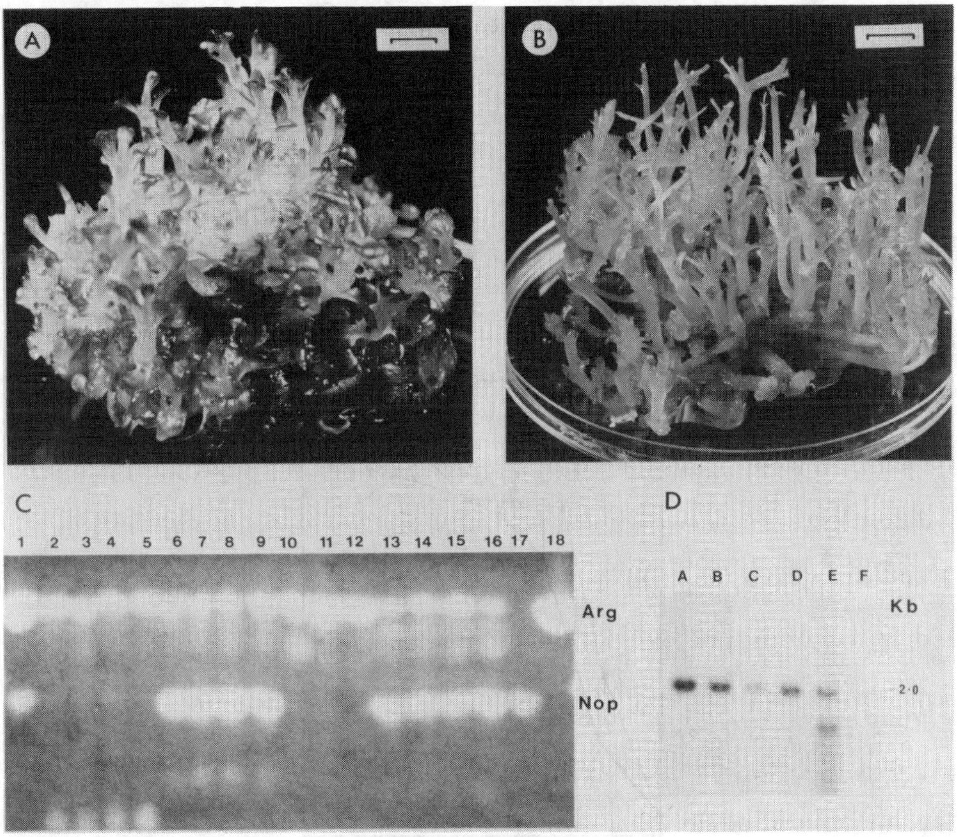

Fig A. Axenic shooty teratoma of M.citrata transformed by A.tumefaciens strain T.37. Bar represents 7mm.
Fig B. Axenic shooty teratoma of M.piperita transformed by A.tumefaciens strains T.37. Bar represents 8mm.
Fig C. An electrophoretogram showing the presence of nopaline in extracts from both shooty teratomas transformed by T.37; M.citrata tracks 6-9, M.piperita tracks 13-16, and the absence of nopaline in extracts from both parent plants; M.citrata tracks 2-5, M.piperita tracks 10-12. Track 1 nopaline and arginine, track 17 nopaline, track 18 arginine.

Fig D. Southern blot of DNA extracted from M.citrata shoot cultures transformed by T.37(A-E) and the parent plant material (F) probed for the presence of the ipt gene.

Table 1 Comparison of Terpene accumulation in leaf tissue and in transformed shoot cultures of two Mentha species.
The tabulated results are the means of replicates from the same sample.

	µg terpenes/g.f.wt.		
M.piperita ("Black Mint")			
Tissue	Menthone	Menthol	Ratio
Leaf	1536±510	1628±510	0.94
Transformed Shoot Culture	240±25	90±65	2.67

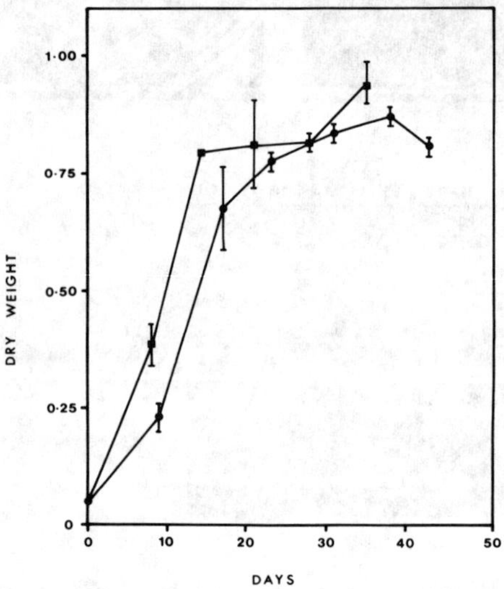

Fig E. Growth curves (Dry Matter accumulation) of M.citrata and M.piperita shooty teratomas, grown in 50 ml liquid media at 25°C, in a 16 hr day with a light intensity of 840 lux. M.citrata ● M.piperita ■ .

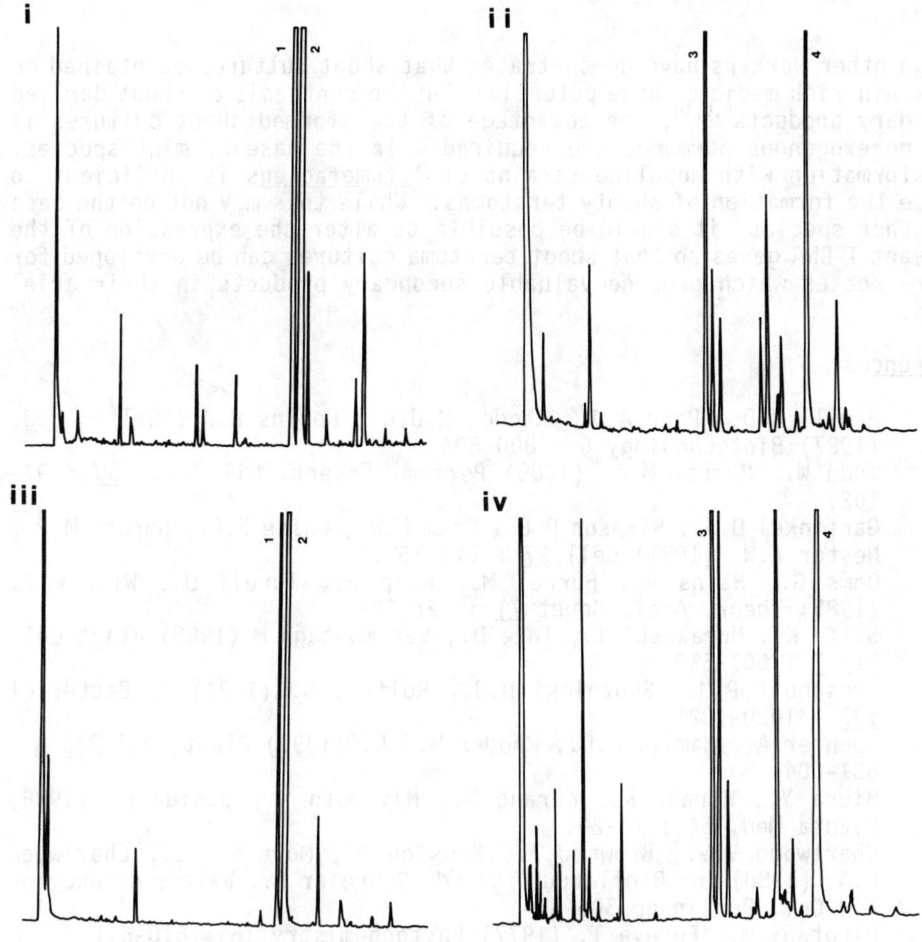

Fig F. G.C. Chromatograms of essential oils extracted from (i) M.citrata untransformed plant (ii) M.piperita untransformed plant (iii) M.citrata shooty teratoma (iv) M.piperita shooty teratoma. (A) Linalool (B) Linalyl acetate (C) Menthone (D) Menthol.

the overall yield of oil in the shooty cultures is lower than in leaf tissue.

Though other workers have demonstrated that shoot cultures maintained on cytokinin-rich media do have potential for the synthesis of shoot derived secondary products[8,9,10], the advantage of transformed shoot cultures is that no exogenous hormones are required. In the case of mint species, transformation with nopaline strains of A.tumefaciens is sufficient to induce the formation of shooty teratomas. While this may not be the case for other species, it should be possible to alter the expression of the relevant T-DNA genes so that shoot teratoma cultures can be developed for other species which produce valuable secondary products in their ariel parts.

References

1) Hamill J.D., Parr A.J., Rhodes M.J.C., Robins R.J., Walton N.J. (1987) Biotechnology 5 : 800-804.
2) Todd W., Murray M.J. (1968) Perfum. Essent. Oil Rec., 59 : 97-102.
3) Garfinkel D.J., Simpson R.B., Ream L.W., White F.F., Gordon M.P., Nester E.W. (1981) Cell 27 : 143-153.
4) Ooms G., Bains A., Burrel M., Karp A., Turell D., Wilcox E. (1985) Theor. Appl. Genet 71 : 325-329.
5) Saito K., Murakoshi I., Inze D., Van Montagu M (1989) Plant Cell Rep 7 : 607-610.
6) Gresshoff P.M., Skotnicki M.J., Rolfe S.G. (1984) J. Bacteriol 137 : 1020-1021.
7) Spencer A., Hamill J.D., Rhodes M.J.C. (1990) Plant Cell Rep 8 : 601-604.
8) Miura Y., Hirata K., Kurano N., Miyamoto K., Uchida K. (1988) Planta Med. 54 : 18-20.
9) Charlwood V.V., Brown J.T., Mouston C., Morris G.S., Charlwood K.A. (1988) In: Bioflavour '87 Ed. Schreier P., Walter de Gruyter and Co., Berlin pp 305-314.
10) Hirotani M., Furuya T. (1977) Phytochemistry 16 : 610-611.

ISOLATION AND DISTRIBUTION OF 7β-HYDROXYHYOSCYAMINE IN SOLANACEOUS PLANTS

K. SHIMOMURA and K. ISHIMARU

Tsukuba Medicinal Plant Research Station, National Institute of Hygienic Sciences, Tsukuba , Ibaraki, 305 Japan

1. INTRODUCTION

In the course of our systematic chemical and metabolic studies on tropane alkaloids and tissue culture products of solanaceous plants [1], we have isolated a new tropane alkaloid 7β-hydroxyhyoscyamine (**1**) [2] together with 6β-hydroxyhyoscyamine (**2**) [3], scopolamine (**3**), hyoscyamine (**4**), norhyoscyamine (**5**) and tropine (**6**) from the leaves of *Duboisia* hybrid M-II-8-6 and the hairy root of *Hyoscyamus albus*. Simultaneous determination of these alkaloids (**1**-**4**) by HPLC was established and the levels of **1**-**4** in the plants and in the cultures of several solanaceous plants (*Scopolia lurida, S. tangutica, Hyoscyamus niger, H. albus, Atropa belladonna, Datura innoxia, Duboisia* hybrid M-II-8-6, *D.* M-II-8-14) were determined.

2. RESULTS AND DISCUSSION

2.1. Isolation of the alkaloids

(From the leaves of *Duboisia* hybrid M-II-8-6) The dried leaves of *Duboisia* hybrid M-II-8-6 (crossbred between *D. myoporoides* and *D. leichhardtii*) were extracted with MeOH-28 % NH_4OH (99:1) and the extract, after concentration, was partitioned with $CHCl_3$ and 0.25 M H_2SO_4. The aqueous solution, after adjusted to pH 10, was extracted with $CHCl_3$. The $CHCl_3$ layer, after evaporated to dryness, was applied to a combination of Fuji-gel ODS G3 and Bondapak C_{18} porasil B column chromatographies and preparative TLC to afford **1**-**6**.

(From the hairy roots of *Hyoscyamus albus*) The hairy roots of *Hyoscyamus albus* were induced by the leaf disc method co-cultured with *Agrobacterium rhizogenes* strain 15834 and after removed the bacteria, the axenic hairy roots were cultured on Woody Plant (WP) medium [4] containing 3 % sucrose without phytohormones. The hairy roots were transferred to WP liquid medium containing 5 % sucrose without phytohormones in the dark at 25 ℃ in air-lift type fermenters and harvested after 4 week culture. The hairy roots, after lyophilization, were extracted with $CHCl_3$-MeOH-28 % NH_4OH (15:5:1) and the extract, followed by a similar chromatographic separation as mentioned above gave **1**-**3**.

2.2. Structure determination of the alkaloids

2-**6** were identified as 6β-hydroxyhyoscyamine, scopolamine, hyoscyamine, norhyoscyamine and tropine, respectively, by comparisons of their physical and spectral data with those of authentic samples.

1, colorless needles, mp 67 ℃, $[\alpha]_D$ + 1.2 ° (acetone), FABMS m/z 306 $[M+H]^+$, high resolution EIMS m/z 305.1627 $[C_{17}H_{23}O_4N]^+$, showed ^1H and ^{13}CNMR spectra which were similar to those of **2** [2]. Acid hydrolysis of **1** with 10 % HCl gave 3α,7β–dihydroxytropane HCl and (-)-tropic acid. Furthermore, treatment of (-)-scopolamine HBr with Raney Ni in H_2O under H_2 at atmospheric pressure [5] afforded **1** together with **2** and **4**. From these spectroscopic and chemical evidence **1** was characterized as 7β-hydroxyhyoscyamine.

2.3. Procedure for HPLC analysis

Samples were extracted with $CHCl_3$-MeOH-28 % NH_4OH (15:5:1) in a sonicator (20 min) at room temparature. The each extract, after filtration, was concentrated and partitioned with 1 N H_2SO_4 and $CHCl_3$. Aqueous layer, after adjusted to pH 10, was extracted with $CHCl_3$. The $CHCl_3$ layer, dried with Na_2SO_4 and filtered, was evaporated to dryness. Samples thus obtained were dissolved with MeOH and injected to HPLC. HPLC conditions were as followes: column, TSK-gel ODS-120A (4.6 mm ID x 250 mm); mobile phase, CH_3CN-10 mM SDS (pH 3.3 adjusted with AcOH, 4:6); column temp., 40℃; flow rate, 1.1 ml/min; detect, UV 215 nm; t_R: **1** (13.42 min), **2** (14.14 min), **3** (14.81 min), **4** (18.86 min).

2.4. Results of HPLC analysis

The levels (% dry wt) of **1**-**4** in the plants, adventitious and/or hairy roots of *Scopolia lurida, S. tangutica, Hyoscyamus niger, H. albus, Atropa belladonna, Datura innoxia, Duboisia* hybrid M-II-8-6 and *D.* M-II-8-14 were shown in Table 1.

Table 1. Tropane alkaloids (**1-4**) in plants and tissue cultures of solanaceous spp.

Material	Alkaloid content (% DW)			
	1	**2**	**3**	**4**
<Plants>				
Scopolia lurida (leaf)	trace	0.060	0.223	0.110
(stem)	trace	0.017	0.077	0.077
(root)	trace	0.011	0.025	0.087
(rootlet)	trace	0.011	0.033	0.187
(root cortex)	trace	0.010	0.056	0.118
(root stele)	trace	0.011	0.028	0.091
Scopolia tangutica (leaf)	trace	0.004	0.033	0.164
(stem)	trace	0.004	0.007	0.140
(root)	0.005	0.059	0.035	0.561
(rootlet)	0.034	0.290	0.034	1.213
(root cortex)	0.010	0.133	0.131	1.003
(root stele)	0.003	0.027	0.012	0.442
Duboisia hybrid M-II-8-6 (leaf)	0.001	0.024	0.486	1.683
Duboisia hybrid M-II-8-14 (leaf)	trace	0.009	0.159	2.517
<Adventitious roots>				
Scopolia lurida				
(MS, 0.5 mg/l IAA, 8W)	n.d.	0.005	0.073	0.026
(MS, 0.1 mg/l NAA, 8W)	trace	0.003	0.132	0.054
Scopolia tangutica				
(MS, 0.5 mg/l IAA, 8W)	n.d.	0.005	0.005	0.011
(MS, 0.1 mg/l NAA, 8W)	n.d.	0.004	0.005	0.004
Duboisia M-II-8-14				
(MS, 0.5 mg/l IAA, 4W)	trace	0.076	0.167	0.301
Datura innoxia				
(MS, 0.5 mg/l IAA, 6W)	n.d.	n.d.	0.006	0.360
Hyoscyamus niger				
A (1/2 MS, 6W)	0.004	0.183	0.271	0.461
B (1/2 MS, 6W)	0.003	0.102	0.497	0.351

Table 1. Continued.

Material	Alkaloid content (% DW)			
	1	**2**	**3**	**4**
Hyoscyamus albus				
(MS, 4W)	0.019	0.245	0.440	0.827
(1/2 MS, 4W)	0.023	0.398	0.584	1.147
(B5, 4W)	0.023	0.113	0.568	0.445
(WP, 4W)	0.028	0.306	0.640	0.392
(RC, 4W)	0.030	0.320	0.311	0.698
<Hairy roots>				
Scopolia tangutica				
A (1/2 MS, 8W)	trace	0.014	0.013	0.041
B (1/2 MS, 8W)	trace	0.014	0.018	0.052
C (MS, 4W)	trace	0.003	0.019	0.046
Atropa belladonna				
(MS, 4W)	n.d.	0.011	0.002	0.360
Datura innoxia				
(MS, 4W)	trace	0.037	0.035	0.172
Duboisia M-II-8-6				
(1/2 MS, 8W)	trace	0.123	0.247	0.214
Hyoscyamus niger				
A (1/2 MS, 5W)	0.005	0.246	0.059	0.795
B (1/2 MS, 5W)	n.d.	0.071	0.086	0.459
C (1/2 MS, 5W)	0.004	0.248	0.250	0.482
D (MS, 4W)	0.007	0.155	0.018	1.251
Hyoscyamus albus				
(WP, 5W)	0.014	0.166	0.455	0.338

n.d.: not detected.

In the plants of *Scopolia lurida*, the content of **1** was very low (trace amount), on the other hand, *S. tangutica* , in its root, rootlet and root bark, contained 0.005-0.034 % for 1. *S. tangutica* also showed the high levels of **2** and **4** which were ca. 10 times higher than those in *S. lurida*. *Duboisia* hybrid M-II-8-6 and M-II-8-14 leaves contained a large amount of **4** (1.683 % and 2.517 %, respectively) showing that the enzyme activities catalysing the biosynthesis of **4** was remarkably high in this genus. Taking account of the low content of **2** (intermediate in the convertion of **4** into **3**) compared with high content of **3**, in this hybrid plants, **4** seemed to be easily converted to **3**. In the leaf of *D*.hybrid M-II-8-14, **1** was not detected.

The adventitious roots of *Hyoscyamus albus* showed the highest level of the production of **1** and **2** among the species we examined. Furthermore, when cultured in 1/2 Murashige-Skoog (MS) medium [6], the amount of **4** in this root remarkably increased. The adventitious roots of *Hyoscyamus niger* exhibited higher production of **3** (0.271 or 0.497 %) than that of the hairy root (0.018-0.250 %). *Datura innoxia, Scopolia tangutica* and *S. lurida* (cultured in MS medium containing IAA) adventitious roots did not produce any detectable level of **1**.

The hairy roots of *Scopolia tangutica* showed relatively large amount of **1**-**4** in comparison with those of its normal root. Among the hairy roots we investigated (*Scopolia tangutica, Atropa belladonna, Datura innoxia, Duboisia* hybrid M-II-8-6, *Hyoscyamus niger* and *H. albus*) the hairy root of *H. albus* produced the highest levels of **1** and **3** and the hairy root of *H. niger* , especially cultured in MS medium, accumulated the highest content of **4**.

Time course of the contents of **1**-**4** in the adventitious and hairy roots of *Hyoscyamus albus* were also determined and shown in Figure 1. The contents of **1** and **2** were almost plateau in both the adventitious and hairy roots after 3rd week of culture (**1**: 0.01-0.02 %, **2**: 0.14-0.25 %). In the adventitious root, the production of **3** and **4** began to increase when the roots reached the logarithmic phase of growth (2W-4W). After reached the highest level (0.878 %, 4W) the production of **4** gradually decreased, while the production of **3** continued to increase even after the growth had reached the maximum level. In the case of the hairy root culture, the content of **4** was ca. 0.38-0.50 % throughout the culture period, whereas the production of **3** began to increase after 2nd week culture and continued to augment until the end of the culture. This evidence suggested that in this plant the transformation with *Agrobacterium rhizogenes* caused the enhancement of the enzyme activity catalysing the conversion of **4** into **3**.

Thus the contents of tropane alkaloids (**1**-**4**) in solanaceous plants and cultures *in vitro* were specifically different. The results mentioned above may be useful for the chemotaxonomic and biosynthetic studies of these solanaceous plants.

Figure 1. Contents (% DW) of **1-4** in the adventitious and hairy roots of *Hyoscyamus albus*.

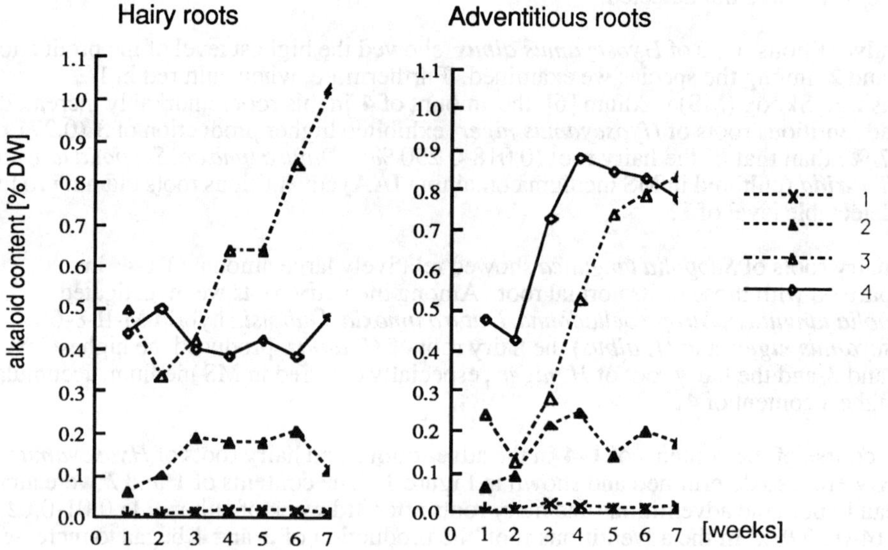

REFERENCES

1. Kamada, H., Okamura, N., Satake, M., Harada, H. and Shimomura, K. (1986) Plant Cell Rep., **5**, 239-242.
2. Ishimaru, K. and Shimomura, K. (1989) Phytochem., **28**, 3507-3509.
3. Hashimoto, T. and Yamada, Y. (1986) Plant Physiol., **81**, 619-625.
4. Lloyd, G. B. and McCown, B. H. (1980) in Proc. of Intnal. Plant Propag. Soc., **30**, 421-427.
5. Fodor, G. and Kovacs, O. (1953) J. Chem. Soc., 2341-2344.
6. Murashige, T. and Skoog, F. (1962) Physiol. Plant. **15**, 473-497.

BIOCONVERSION OF XENOBIOTICS (PESTICIDES) IN PLANT CELL CULTURES

W. BARZ, M. JORDAN & G. METSCHULAT
Department of Plant Biochemistry, Westfälische Wilhelms-Universität, Hindenburgplatz 55, D-4400 Münster, FRG

1. INTRODUCTION

Higher plants possess a pronounced ability for the metabolism and degradation of xenobiotics. Numerous pesticides (i.e. fungicides, insecticides, herbicides, growth regulators) have well been investigated in plants and plant cell cultures. Plant metabolism of xenobiotic compounds may lead to 1) degradation, 2) conjugation of lipophilic molecules with hydrophilic moieties to yield polar products suitable for vacuolar storage, or 3) oxidative polymerization into insoluable cellular structures. Plant cell suspension cultures have repeatedly been demonstrated as very suitable model systems for pesticide metabolism (1-3). Such cells rapidly take up foreign compounds, possess high rates of metabolite formation and grow free from microorganisms. Cell cultures are especially suitable for investigations on polar hydrophilic metabolites. The spectrum of such metabolites in plants and plant cell cultures is in principle identical though quantitative differences may occur.

2. RESULTS

2.1. Phase I and Phase II reactions

Xenobiotic metabolism in plant cells proceeds through two different partially linked stages.

The Phase I reactions introduce functional groups (-OH,-NH,-SH) in lipophilic substrates by reductive, oxidative or hydrolytic enzymes. Hydrolytic reactions (i.e.esterases, amidases) are quite common and the enzymes are often found as multiple isoforms. The greatest interest in Phase I should be placed on the various oxidation reactions (epoxidation, O- or N-dealkylation, aryl or alkyl hydroxilation, N- or S-oxidation) because they appear to be catalyzed by cytochrome P450 monooxygenases. These microsomal enzymes, extremely well investigated in mammalian systems, occur constitutively but they also seem to be inducible in plants by wounding, microbial infection, various forms of stress and possibly by xenobiotic substrates (4). Position-specific aryl hydroxilation reactions (6-hydroxy- versus 8-hydroxy compound) of the herbicide bentazon in various cell cultures reflected differences found between resistant and susceptible

plant species (5).
The wide range of Phase II reactions comprises numerous transferase reactions in which predominantly glucosyl or several other glycosyl moieties are being attached to the functional groups formed in Phase I. Various amino acids, malonic acid or glutathione residues are also transfered in this phase so that in some cases (i.e. 2,4D or other growth regulators) complex mixtures of metabolites are formed (1,6). Plant cell cultures are especially suitable to elucidate Phase II reactions and to investigate the transferases involved.

2.2 Glutathione conjugates
Numerous pesticides are converted to glutathione (GSH) conjugates in higher plants and in plant cell cultures. The structures in Fig. 1 demonstrate some of the most important reactions observed with plant GSH S-transferases, i.e. removal of halogen substituents from heterocyclic, aromatic and aliphatic substrates, cleavage of diphenyl ethers and fission of sulphoxides. The glutathione conjugates are subsequently metabolized by removal of the glutamate and glycine moieties, N-malonylation, or rearrangement to N-conjugates. In general, a complex mixture of compounds is formed which also depends on the plant species involved (6). Glutathione conjugation is particularly important in the metabolism of herbicides because it frequently results in the selective detoxification of the xenobiotic and thus in species tolerance against the herbicide.

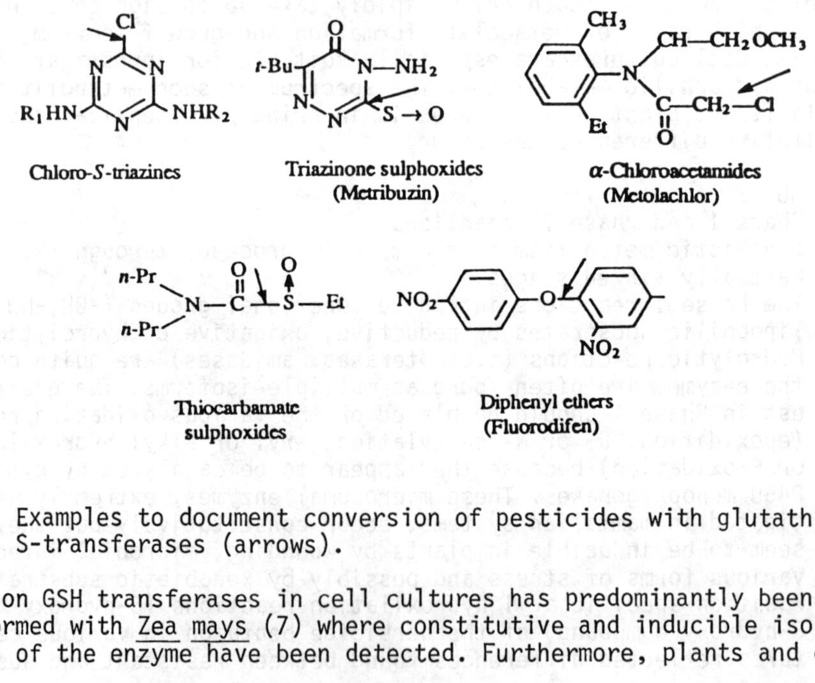

FIGURE 1. Examples to document conversion of pesticides with glutathione S-transferases (arrows).

Work on GSH transferases in cell cultures has predominantly been performed with Zea mays (7) where constitutive and inducible isoforms of the enzyme have been detected. Furthermore, plants and cell cultures seem to differ in the expression of the various isoforms and their subunits (8). Thus, in case of a particular cell line the rapid metabolism of a chlorine containing herbicide is explained by the high concentration of the enzyme.

2.3 Malonyl conjugates

Among the polar metabolites of xenobiotics isolated from plant cell cultures malonyl conjugates have been found (9,10) as 6-O-malonylglucosides and as malonylamides of aromatic, heterocyclic or aliphatic secondary amines (Fig.2), of D-amino acids and as N-malonylcysteine derivatives. Malonylation is considered to be a final step in xenobiotic metabolism with the conjugate either accumulating in vacuoles or excreted from the cells (11). It is unknown at present

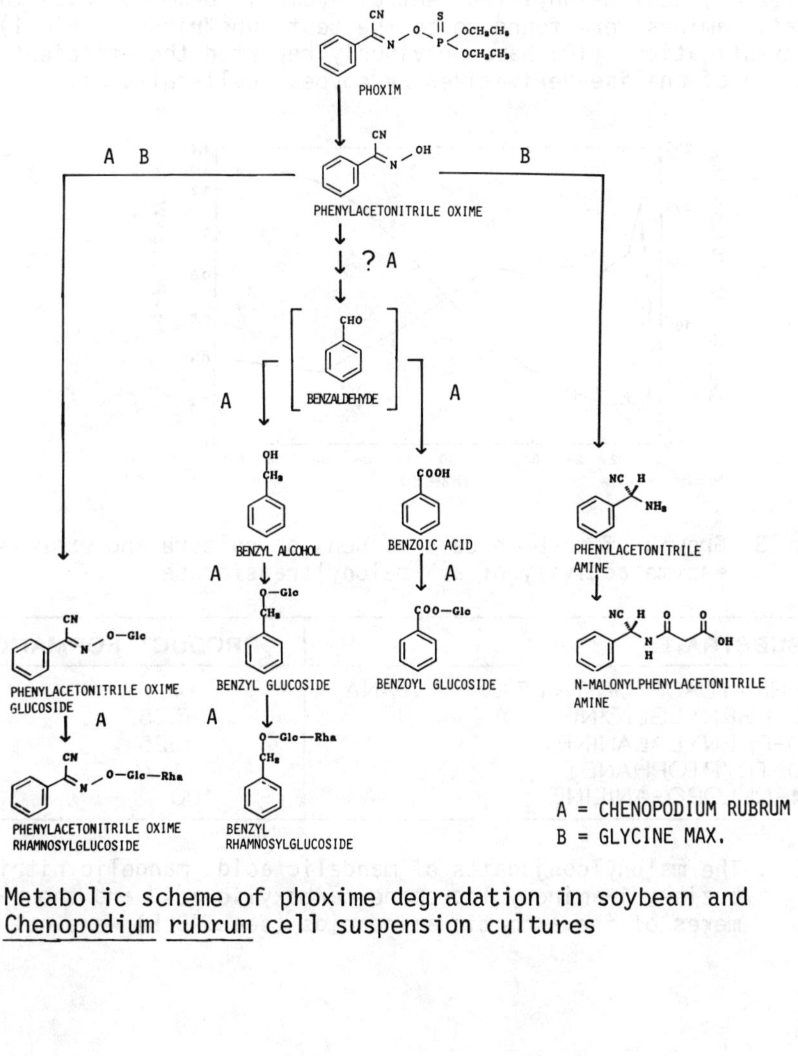

FIGURE 2. Metabolic scheme of phoxime degradation in soybean and Chenopodium rubrum cell suspension cultures

whether these excretion phenomena in cell cultures are related to translocation processes of the intact plant.

Investigations on the metabolism of the insecticide phoxime in soybean cell suspension cultures (11) have shown that the N-malonylphenylacetonitrile amine (Fig.2) is the major metabolite which is quantitatively excreted into the nutrient medium; the phenylacetonitrile oxime glucoside (Fig.2) is only transiently found in the cells. Analysis of the stereochemistry of the N-malonylated amine (11) led to the determination of the R-configuration so that this N-conjugation reaction is highly reminiscent of the plant-specific detoxification mechanism for D-amino acids (9). The N-malonyltransferase for this amine has meanwhile been purified to apparent homogeneity. Protein was extracted from heterotrophic soybean cell cultures 6 hours after inoculation into fresh medium. Fig.3 demonstrates the obvious induction of enzyme activity upon transfer of cells. The purified enzyme (M_r 52.000 dalton, Km phenylacetonitrile amine 2 mM, Km malonyl CoA 0.32 mM) well malonylated neutral aromatic D-amino acids though aromatic amines were found to be the best substrates (Table 1). Other investigations (10) have previously reported the efficient N-malonylation of aniline derivatives by soybean cell cultures.

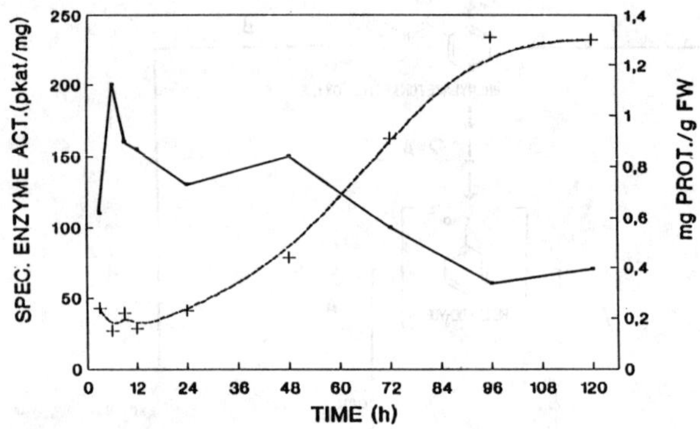

FIGURE 3. Growth of soybean cell suspension culture and expression of enzyme activity of a N-malonyltransferase

SUBSTRATE	PRODUCT FORMATION (%)
PHENYLACETONITRILEAMINE (PANA)	14
D-PHENYLGLYCINE.	6,25
D-PHENYLALANINE	2,25
D-TRYPTOPHANE	1
4-CHLORO-ANILINE	100

TABLE 1. The malonylconjugates of mandelic acid, mandelic nitrile, metribuzine, 1-aminocyclopropane carboxylic acid and the L-enantiomeres of the aromatic aminoacids (see TABLE) were not detected

2.4. Heterotrophic versus photoautotrophic cultures
Investigations on xenobiotic metabolism in plant cell cultures have so far only been performed with heterotrophic cultures though photoautotrophic cell suspension cultures of higher plants are now available for xenobiotic metabolism to be investigated. Such cultures possess specific properties which result from the presence of functioning chloroplasts and from their high similarity to mesophyll cells (12,13). Recent reports (14,15) have, for instance, shown that such photoautotrophic cell cultures are very suitable systems to study the effects of herbicides or other phytotoxic compounds on photosynthesis and connected areas of cellular metabolism. Biochemical properties thought to be specific for the photoautotrophic form of a cell culture can easily be identified by reversion of such a culture to the photomixotrophic or the heterotrophic type of culture and vice versa (13).
Heterotrophic, photomixotrophic and photoautotrophic _Catharanthus roseus_ cell suspension cultures showed no quantitative or qualitative differences in the metabolism of p-nitrophenol and its conversion to the glycosyl conjugates shown in Fig. 4. However, an O-malonyltransferse specific for the malonylation of p-nitrophenyl-O-β-D-glucoside in position 6 (F. Sato and W. Barz, unpublished) was preferentially expressed in photoautotrophic cells and only barely detectable in heterotrophic cells.

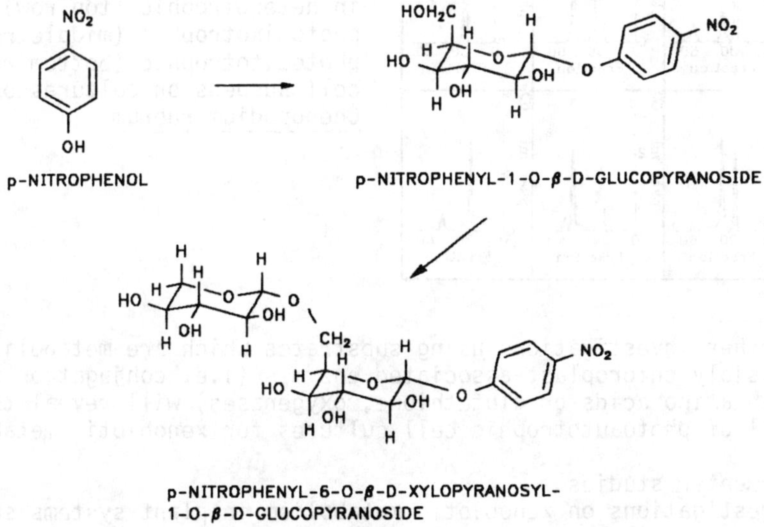

FIGURE 4. Conjugation sequence of p-nitrophenol in _Catharanthus roseus_ cell suspension cultures

Detailed investigations on the degradation of the insecticide phoxime in _Chenopodium rubrum_ cell cultures (Fig.2) have demonstrated significant differences between heterotrophic and photoautotrophic cells. The former mainly produce the phenylacetonitrile oxime conjugates whereas the photosynthetically active cells preferentially produce benzyl alcohol and its conjugates. Thus, the two forms of cell cultures differ in the type of aglycone but not in the conjugating sugar moieties. Further differences between these three Chenopodium rubrum cell suspension cultures has been observed in the expression of UDP-dependent glucosyltransferases found to be involved in the glucosydation of phenylacetonitrile oxime, benzyl alcohol and benzoic acid (Fig.5). The data demonstrate the multiplicity in these cells of glucosyltransferases accepting the aforementioned substrates and also show the differences in the enzyme pattern between green and non-green cells.

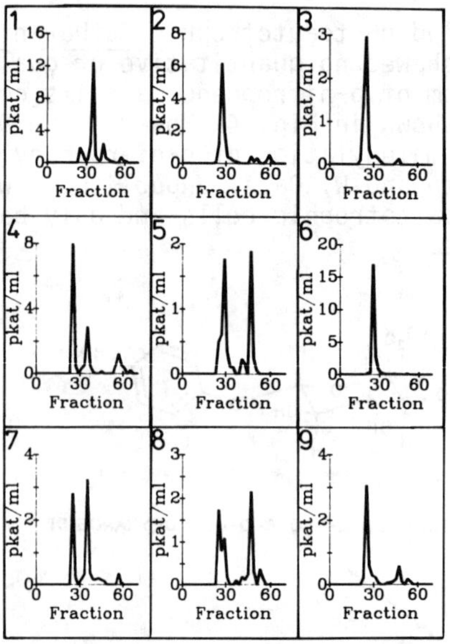

FIGURE 5

FPLC diagrams to document multiplicity of UDP-glucosyltransferases for phenylacetonitrile oxime (graphs 1,4,7), benzyl alcohol (graphs 2,5,8), and benzoic acid (graphs 3,6,9) in heterotrophic (top row), photomixotrophic (middle row), and photoautotrophic (bottom row) cell suspension cultures of Chenopodium rubrum

Further investigations using substrates which are metabolized by possibly chloroplast-associated enzymes (i.e. conjugation reactions with amino acids or glutathione, oxygenases) will reveal the potential of photoautotrophic cell cultures for xenobiotic metabolism.

2.5. Fermenter studies

Investigations on xenobiotic metabolism in plant systems should include determination of the toxicological and pharmacological properties of the intermediates or the final metabolites. In case of complicated structures which may be difficult to obtain by organic synthesis isolation from fermenter-grown cell cultures represents a suitable alternative. Investigations in our laboratory aim at establishing a relevant fermenter process for xenobiotics. The airlift fermenter (15 l medium) previously described for biotransformation experiments (16) has been used.

During investigations on the insecticide phoxime (Fig.2) in fermenter-grown soybean cell cultures (11) the critical steps in the process were the time point and the rate of substrate application. The insecticide, dissolved in dimethylsulphoxide, had to be applied during the second half of the linear growth phase or in the early stationary phase because otherwise excessive browning of cells and cessation of growth were observed. Furthermore, the total amount of substrate (158 mMol) to be added to the fermenter was batchwise applied over a period of appr. 3 days. Under such circumstances the product, N-malonyl phenylacetonitrile amine (Fig.2), was recovered from the nutrient medium in a 12% yield. Further quantitative improvement of this transformation reaction will require the addition of substrate in a continuous mode parallel to growth so that any disturbance of the physiology of the cells is avoided.

As a second example of xenobiotic conversion the conjugation reaction of p-nitrophenol to a xylosylglucoside via the glucoside (Fig.4) has been studied in heterotrophic cultures of Catharanthus roseus. These cultures metabolize the phenolic substrate exceptionally well and the glucoside intermediate is completely converted to the disaccharide upon longer incubation. However, the p-nitrophenol has considerable cytotoxic properties which lead to complete inhibition of substrate uptake at concentrations higher than 0.8 mMol (17). The present status of the fermenter process developed for p-nitrophenol conversion is illustrated in Fig. 6 and 7.

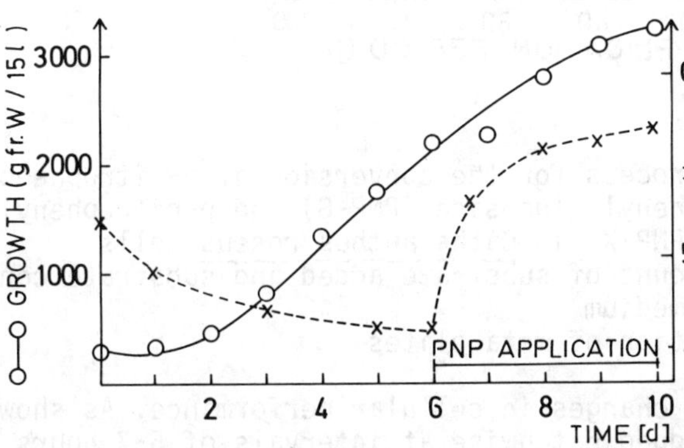

FIGURE 6. Growth of Catharanthus roseus cell culture in a 15 l airlift-fermenter and change of pH value during application of p-nitrophenol (PNP)

Heterotrophic C. roseus cells in the airlift fermenter readily reach appr. 215 g fr.w. per litre. Substrate application was started at day 6 of the growth cycle and performed for up to 92 hours. During this period the pH value of the nutrient medium rapidly rose to 5.7

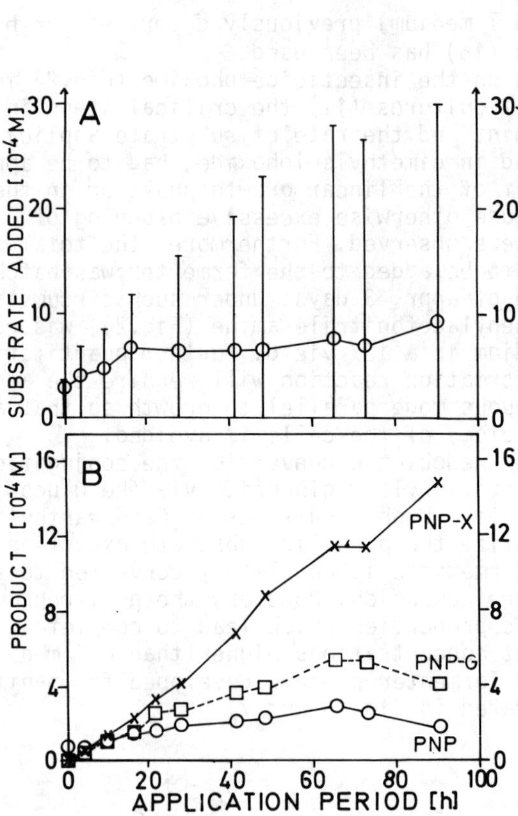

FIGURE 7. Fermenter process for the conversion of p-nitrophenol (PNP) to p-nitrophenyl glucoside (PNP-G) and p-nitrophenyl xylosylglucoside (PNP-X) in Catharanthus roseus cells
A. Total amount of substrate added and substrate concentration in medium
B. Accumulation of metabolites

with no detectable changes in cellular performance. As shown in Fig. 7A substrate was added batchwise at intervals of 5-7 hours so that its concentration in the medium was kept low. The data in Fig. 7B document that the intracellular level of PNP never rose above a critical limit and that the terminal xylosylglucoside accumulated at a linear rate. After 92 hours 50% conversion of the phenol to the disaccharide was observed.

The present experience with degradation and conjugation of xenobiotics in fermenter-grown cell cultures indicates that the application of substrate represents the critical step and that it must be performed during longer periods parallel to growth and in accordance with the rates of uptake and conversion. Similar observations have been made with biotransformation reactions of various natural products (18).

3. DICUSSION

Comprehensive investigations in recent years have unveiled the full potential of plant cell cultures for xenobiotic metabolism. Use of heterotrophic and photoautotrophic cells offer the possibility to investigate two basically different types of cell cultures. This will undoubtedly be of importance for future work. Finally, plant cell cultures can easily be standardized as routine test systems for the ecotoxicological evaluation of xenobiotics (2,19) and for determination of the biodegradability of synthetic compounds.

Acknowledgement
Financial support by Minister für Wissenschaft und Forschung, Düsseldorf, and Bundesminister für Forschung und Technologie, Bonn.

REFERENCES

1 Mumma, R.O. and Davidonis, G.H. (1983) in Progress in Pesticide Biochemistry and Toxicology (Hutson, D.H., Roberts, T.R., eds.) Vol. 3, pp. 255-278, Wiley & Sons, Chichester
2 Harms, H. and Langebartels, C. (1986) Plant Sci. 45, 157-165
3 Grossmann, K. (1988) in Advances in Cell Culture (Maramorosch, K., ed.) Vol. 6, pp. 89-136, Academic Press, Orlando San Francisco
4 Higashi, K. (1988) Mutation Research 197, 273-288.
5 Sterling, T.M. and Balke, N.E. (1989) Pesticide Biochem. Physiol. 34, 39-48
6 Owen, W.J. (1989) in Herbicides and Plant Metabolism (Dodge, A.D., ed.) pp. 171-198, Cambridge University Press, Cambridge
7 Edwards, R. and Owen, W.J. (1986) Planta 169, 208-215
8 Edwards, R. and Owen, W.J. (1988) Planta 175, 99-106
9 Barz, W., Köster, J., Weltring, K.-M. and Strack, D. (1985) Ann. Rev. Phytochem. Soc. Europe 25, 307-347
10 Winkler, R. and Sandermann, H. (1989) Pesticide Biochem. Physiol. 33, 239-248
11 Höhl, U. (1988) Doctoral Thesis, Münster University
12 Hüsemann, W., Fischer, K., Mittelbach, I., Hübner, S., Richter, G. and Barz, W. (1989) in Primary and Secondary Metabolism of Plant Cell Cultures II (Kurz, W.G.W., ed.) pp. 35-46, Springer-Verlag, Berlin Heidelberg
13 Barz, W. (1988) BioEngineering 4, 145-149
14 Sato, F., Takeda, S. and Yamada, Y. (1988) Plant Cell Rep. 6, 401-404
15 Thiemann, J., Nieswandt, A. and Barz, W. (1989) Plant Cell Rep. 8, 399-402
16 Höhl, U., Upmeier, B. and Barz, W. (1988) Appl. Microbiol. Biotechnol. 28, 319-323
17 Hein, D. (1989) Master Thesis, Münster University
18 Reinhard, E. and Alfermann, W. (1980) Adv. Biochem. Engineering 16, 49-84
19 Swisher, B.A. (1987) Amer. Chem. Soc. Ser. 334, 18-40

BIOCONVERSION OF PRECURSORS OCCURRING IN PLANTS AND OF RELATED SYNTHETIC COMPOUNDS

Niesko Pras
Laboratory for Pharmacognosy, University Centre for Pharmacy, Groningen University, A. Deusinglaan 2, 9713 AW Groningen, The Netherlands.

I. INTRODUCTION

The nearly unlimited enzymatic potential of plant cell cultures can basically be employed for bioconversion purposes. In this paper, bioconversion is defined as the modification of added precursors into more valuable products, such as pharmaceuticals and fine chemicals. These bioconversion products belong to groups of secondary metabolites or related compounds. In general, freely suspended as well as entrapped plant cells or isolated enzymes can be used as biocatalytic systems. From an economical view, the application of plant enzymes is only worthwhile when the modification of the precursor can not or at least less effectively, be achieved by chemical synthesis or by the use of micro-organisms. There are two main reasons to choose plant enzymes for bioconversion purposes; firstly, these enzymes can catalyze reactions stereospecifically and secondly, they can perform modifications, that are not easily to carry out by chemical synthesis or by micro-organisms.

The bioconversion of naturally occurring or related synthetic precursors by several biocatalytic systems is presented for the production of two pharmaceutically important groups of compounds: catechols and lignans, i.e. podophyllotoxins. Since mass transfer limitations may be present in entrapped cell systems, kinetic aspects are discussed. Finally, we focus on a new method for the bioconversion of water-unsoluble precursors.

II. BIOCONVERSION BY ENTRAPPED PLANT CELLS

The great interest for the entrapment of plant cells has been stimulated by the potential advantages of entrapped systems over freely suspended cells for the production of valuable compounds. Entrapment may have profitable physiological effects, may enable prolonged use of biomass, protects cells against shear forces in reaction vessels and guarantees the separation between biomass and medium. It has been shown that a large number of higher plant cells can be entrapped in a variety of polymeric matrices [1]. Successfully applied bioconversions cited in the literature are reductions by *Papaver* and *Mentha* cells and hydroxylations by *Daucus*, *Digitalis* and *Mucuna* cells.

For entrapped cells of *Mucuna pruriens* L. (Fabaceae), the bioconversion spectrum, defined as the range of compounds that can be converted, as well as related kinetic aspects have been studied. When cells of *M. pruriens* are entrapped in calcium alginate, they are able to *ortho*-hydroxylate the precursor L-tyrosine (a monophenol) into the anti-Parkinson drug L-DOPA (a catechol), which is released directly into the medium. The reaction is catalyzed by a phenoloxidase, E.C. 1.14.18.1 (**Fig. 1**):

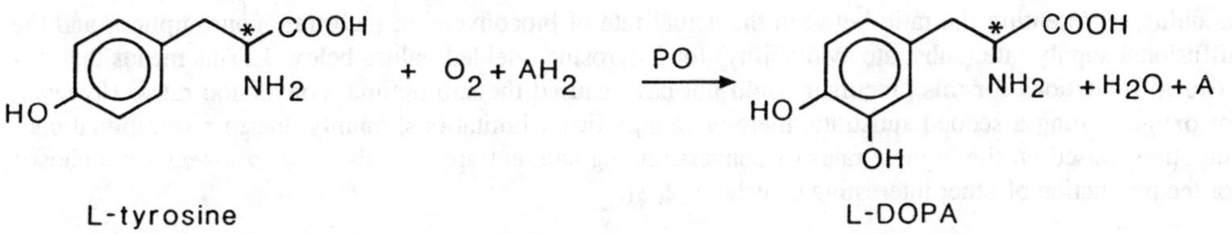

Fig.1: Reaction scheme for the conversion of L-tyrosine into L-DOPA by *Mucuna*-phenoloxidase (PO): O_2 is second substrate, sodium ascorbate (AH_2) serves as cofactor and as reductant [2].

The bioconversion of L-tyrosine into L-DOPA has been used for kinetic studies with cells entrapped in calcium alginate, calcium pectinate, agarose or gelatine. In all cases L-DOPA was released directly into the medium (**Fig. 2**). Based on Michaelis-Menten kinetics it was found that the maximum rate of conversion

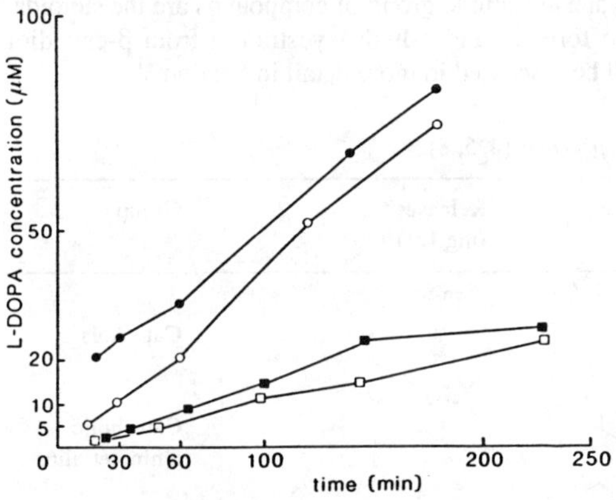

Fig. 2: Course of appearance of L-DOPA in the medium for bioconversion by alginate- (●), pectinate- (○), agarose- (■), and gelatine- (□) entrapped cells of *M. pruriens* at 1.5 mM L-tyrosine concentration.

for each entrapped cell system was lower than the rate calculated for a cell homogenate. Determination of the effective diffusion and partition coefficients using entrapped, glutaric aldehyde inactivated cells, showed that L-tyrosine could diffuse freely into each matrix (**Table 1**). Calculation of an observable

Table 1. Characteristic parameters of entrapped cells of *M. pruriens*[a] [2].

Matrix	O_c (mg h^{-1} g^{-1})	D_e ($.10^{-10} m^2 s^{-1}$)	K_p	φ_{tyr}	φ_{oxy}
Alginate	0.65	3.1	0.87	0.38	12.9
Pectinate	0.45	4.2	1.02	0.23	9.9
Agarose	0.42	4.4	0.94	0.08	11.8
Gelatine	0.31	2.9	0.68	0.08	9.1

[a] Oxygen consumption O_c, effective diffusion coefficient D_e of L-tyrosine, distribution coefficient K_p, and observable moduli φ for L-tyrosine and oxygen. Oxygen consumption expressed as mg O_2 consumed h^{-1} g^{-1} cells, fresh weight. Distribution coefficient corrected for excluded bead volume. Observable moduli calculated at L-tyrosine concentration 2x 10^{-3} mol L^{-1}, and at oxygen concentration 0.24x 10^{-3} mol L^{-1}.

modulus, representing the ratio between the actual rate of bioconversion (substrate consumption) and the diffusional supply rate (substrate availability) for L-tyrosine, yielded values below 1. This means that diffusional limitations for this precursor could not have caused the sub-optimal conversion rates. However, for oxygen, being a second substrate, there were significant limitations, mainly due to respirational consumption. Based on the highest rates of conversion, alginate-entrapped cells of *M. pruriens* were chosen for the production of other interesting catechols [2, 3].

A wide range of precursors, naturally occurring as well as synthetic, has been tested. The precursors and products are listed in **Table 2**. The products belong to several chemical groups of compounds with an increasing structural complexity. From 14 simple monocyclic monophenols, the corresponding catechols were formed; naturally occurring compounds such as L-DOPA and dopamine, but also synthetic organic acids and alcohols [4]. The 2-aminotetralins and hexahydronaphthoxazines are new dopaminergics. The chemical synthesis of the mono-hydroxylated derivatives is relatively easy, whereas the related catechols are more difficult to prepare. Regioselective *ortho*-hydroxylation using our *Mucuna* cells may result in new pharmaceuticals. From 12 bi- and tri-cyclic phenols, corresponding catechols were formed; mostly two isomers (**Table 2**) [5]. So far, the performance of bioconversions was facilitated, because the precursors involved are all water-soluble. An example of a water-unsoluble group of compounds are the steroids. Alginate-entrapped cells of *M. pruriens* were able to form 2- and 4-hydroxyestradiol from β-estradiol complexed with cyclodextrins. This bioconversion will be discussed in more detail in Section V.

Table 2. Bioconversions by alginate-entrapped cells of *M. pruriens* [4, 5, 8].

Precursor	Product		Release[a] (mg L^{-1})	Group
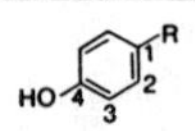	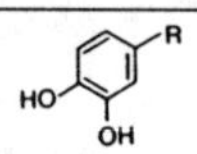		20-160	Catechols
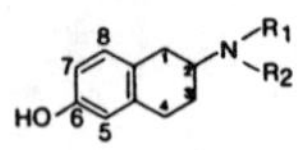	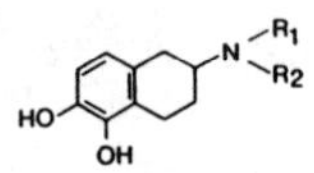	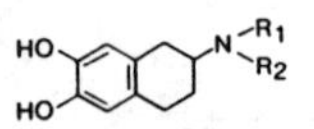	4-13	Catecholic aminotetralins
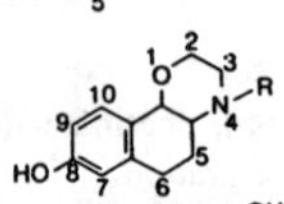	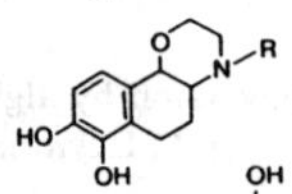	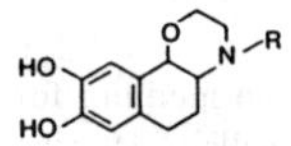	2-5	Catecholic hexahydro-naphthoxazins
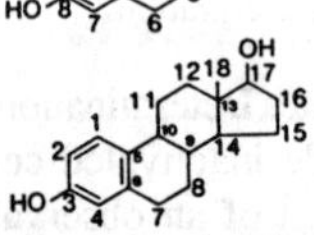	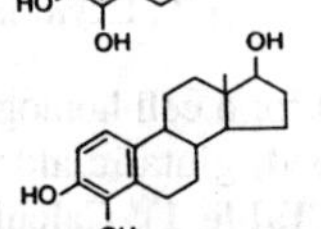	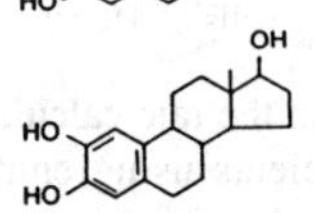	0.1-0.6[b]	Catecholic estrogens

[a] = at 2 mM substrate concentration [b] = complexed with β-cyclodextrin

It can be concluded that alginate-entrapped cells of *M. pruriens* have a broad bioconversion spectrum. All precursors tested were also converted by *Mucuna*-phenoloxidase preparations, showing the very low substrate specificity of this enzyme and explaining the broad spectrum of the cell system.

Also for the production of podophyllotoxins by entrapped cells, calcium alginate has been chosen as the matrix. The impact of these lignans on pharmacy is discussed in Section III. Several precursors of the lignan biosynthesis were fed to alginate-entrapped cells of *Podophyllum hexandrum* Royle (Berberidaceae) and *Linum flavum* L. (Linaceae) and have to undergo multi-step bioconversions for lignan formation. For both cell lines, no bioconversion was measured, while the endogenous synthesis of podophyllotoxin and 5-methoxypodophyllotoxin, respectively ceased. Apparently, both cell culture species experience the entrapped state as rather unpleasant. This suggestion is supported by a number of results presented in the next Section, where it is shown that freely suspended cells of both species can indeed convert some precursors into podophyllotoxins.

III. BIOCONVERSION BY FREELY SUSPENDED PLANT CELLS

The application of freely suspended cells has its own benefits. Firstly, precursors can be supplied directly to the cultures; no entrapment procedure is needed. Secondly, mass transfer limitations are less likely to occur, and, finally, cell viability and physiology are not influenced by a matrix. Many attempts have been made to produce valuable compounds by adding precursors to several culture species. Often the precursor undergoes more than one bioconversion and complex mixtures of (unknown) products are formed. A number of important one-step bioconversions by freely suspended cells have been reported in literature; hydroxylations by *Glycyrrhiza* and *Daucus* cultures, glucosylation by *Datura* cultures and even stereospecific reduction by *Mentha* cultures. For the formation of podophyllotoxins by *Podophyllum* and *Linum* cultures, multi-step bioconversions should be performed.

Podophyllotoxin is used as the starting compound for the chemical modification into etoposide (VP-16-213) and teniposide (VM-26), both being applied successfully as anti-tumour agents (**Fig.3**). Due to the limited availability of *P. hexandrum* plants, there is a shortage of *Podophyllum* resin (podophyllin). Together with the complex isolation, podophyllotoxin is an expensive compound. Attempts for larger-scale preparation by chemical synthesis failed so far. Therefore, the biotechnological production may form an attractive route to obtain this lignan.

Recently, we initiated cell suspension cultures of *P. hexandrum*, that are able to accumulate podophyllotoxin up to 0.1% on a dry weight basis [6]. In the present study we investigated the bioconversion capacity of these cultures. For a proper choice of precursors, knowledge of the routing is required. In **Fig. 4** a simplified scheme of the podophyllotoxin biosynthesis is depicted: from phenylalanine (I), coniferyl alcohol (II) is synthesized by the action of phenylpropanoid (PPR) enzymes. Coniferyl alcohol can be converted into coniferin (III) and vice-versa. A peroxidase converts (II) into coniferyl alcohol radicals (IV), which dimerize and are synthesized further to lignans such as podophyllotoxin (V) or pinoresinol (VI).

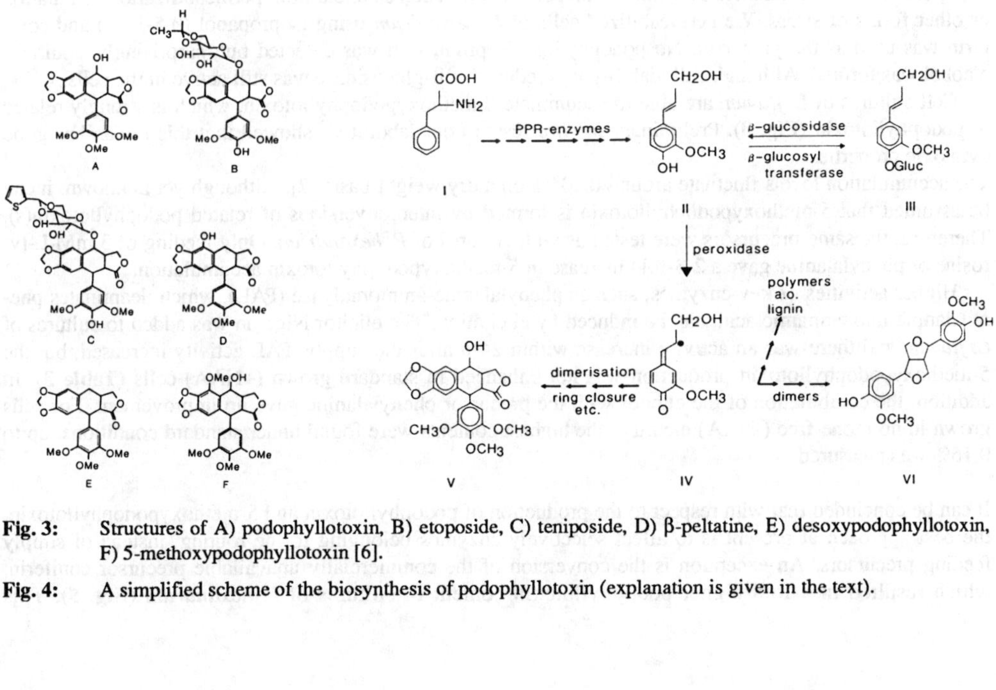

Fig. 3: Structures of A) podophyllotoxin, B) etoposide, C) teniposide, D) β-peltatine, E) desoxypodophyllotoxin, F) 5-methoxypodophyllotoxin [6].

Fig. 4: A simplified scheme of the biosynthesis of podophyllotoxin (explanation is given in the text).

The precursors tested were the following intermediates: L-phenylalanine, L-tyrosine, *trans*-cinnamic acid, *trans*-coumaric acid, *trans*-ferulic acid, *trans*-caffeic acid, coniferin and one related synthetic compound, 3,4-methylenedioxycinnamic acid. Only coniferin was found to exhibit a positive and reproducible effect on the podophyllotoxin accumulation (**Fig. 5**).

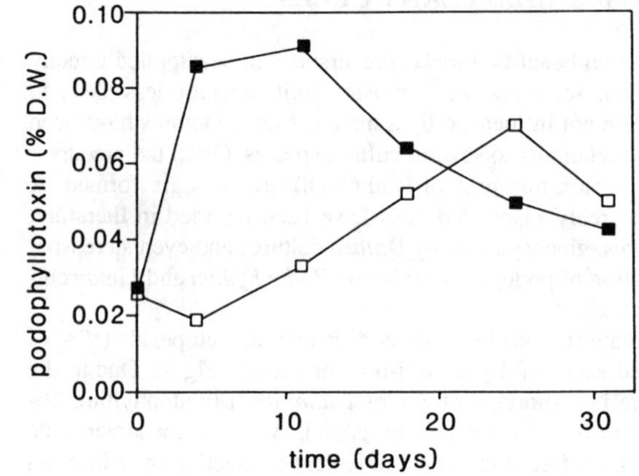

Fig. 5: Time course of the podophyllotoxin content on a dry weight basis in a suspension culture of *P. hexandrum*; standard-grown (□) and after feeding of 2.1 mM coniferin (■).

The optimal precursor concentrations were in the range of 0.8-2.1 mM. Another important observation made for these cultures was the presence of extracellular β-glucosidase and peroxidase activities under standard culture conditions. When cell-free medium was incubated with coniferin, pinoresinol was formed. This multi-step bioconversion fits in the biosynthetic scheme (**Fig 4**).

There are several manners to influence bioconversions, such as elicitation, permeabilization, irradiation or other forms of stress. We permeabilized cells of *P. hexandrum* using isopropanol (0.5-10%) and coniferin was used as the precursor. No podophyllotoxin production was detected but, surprisingly, coniferyl alcohol was formed. Although cell viability was reduced, a β-glucosidase was still active in the cells.

Cell cultures of *L. flavum* are able to accumulate 5-methoxypodophyllotoxin, which is strongly related to podophyllotoxin (**Fig. 3**). Preliminary *in vitro* tests in our laboratory showed that this lignan has good cytotoxic properties.

The accumulation levels fluctuate around 0.004% on a dry weight basis [7]. Although yet unknown, it can be assumed that 5-methoxypodophyllotoxin is formed by interconversions of related podophyllotoxin(s). Therefore, the same precursors were tested as with cultures of *P. hexandrum.* Only feeding of 3 mM L-tyrosine or phenylalanine gave a 2-3-fold increase in 5-methoxypodophyllotoxin accumulation.

Higher activities of key-enzymes, such as phenylalanine-ammonialyase (PAL), which deaminates phenylalanine into cinnamic acid, can be induced by elicitation. The elicitor Nigeran was added to cultures of *L. flavum* and there was an activity increase within 24 h after the supply. PAL-activity increased, but the 5-methoxypodophyllotoxin production was not enhanced in standard-grown (+NAA) cells (**Table 3**). In addition, the combination of the elicitor with the precursor phenylalanine gave no improvement. For cells grown in hormone-free (-NAA) medium, the highest contents were found under standard conditions; up to 0.16% was measured.

It can be concluded that with respect to the production of podophyllotoxin and 5-methoxypodophyllotoxin, the best approach at present is to affect selectively enzymes belonging to the routing, instead of simply feeding precursors. An exception is the conversion of the commercially unavailable precursor coniferin, which resulted in 4-fold higher podophyllotoxin content in cultures of *P. hexandrum* (**Fig. 5**). This

Table 3. Highest levels of PAL activity and maximal 5-MPT content observed in 8 to 11-day-old cell suspensions of *L. flavum* grown under different experimental conditions [7].

Conditions	PAL (μkat kg^{-1} protein)		5-MPT (% DW)	
	+NAA	-NAA	+NAA	-NAA
Standard	91.5	17.2	0.004	0.161
+ Elicitor	169.4	97.1	0.002	0.059
+ 3mM Phenylalanine	12.7	31.5	0.013	0.161
The combination	41.1	40.0	0.003	0.073

availability problem has already been solved, since we recently measured coniferin contents up to 12% on a dry weight basis in *L. flavum* cultures and a rapid, rather efficient isolation procedure was developed (W. van Uden, unpublished).

IV. BIOCONVERSION BY ENZYME PREPARATIONS

When enzyme preparations are chosen for bioconversion purposes, some criteria have to be met. It is necessary that sufficient amounts of enzyme can be isolated easily from cell cultures. The enzyme involved must be characterized; on the basis of substrate specificity and the co-factors (enzymes) required, controlled bioconversions can be carried out. Furthermore, prolonged stability of activity during the bioconversion process and reusability are preferable. The plant enzymes that have been described in literature up to now, usually catalyze one-step reactions. Examples are the stereospecific condensation by strictosidine synthase, transesterification by rosmarinic acid synthase, and hydroxylations by digitoxin 12β-hydroxylase, 6β-hyoscyamine hydroxylase and *Mucuna*-phenoloxidase. In the following the continuous production of a pharmaceutical, the dopaminergic 7,8-dihydroxy N-di-n-propyl 2-aminotetralin (7,8-$(OH)^2$ DPAT) by regioselective *ortho*-hydroxylation of 7-hydroxy N-di-n-propyl 2-aminotetralin (7-OH DPAT) using a *Mucuna*-phenoloxidase preparation will be discussed.

Although alginate-entrapped cells of *M. pruriens* can produce considerable quantities of several catechols (**Table 2**), kinetic studies revealed that this system functioned sub-optimally [2, 3]. For this reason, the use of phenoloxidase preparations instead of living cells appeared to be a better approach.

Table 4. Formation of 6,7- and 7,8-$(OH)_2$ DPAT by bioconversion.

Substrate	Product	v_o (nmol h^{-1} mg^{-1} protein)		B_e (mg L^{-1})	
		EC[a]	PPP[b]	EC	PPP
7-OH DPAT	6,7-$(OH)_2$ DPAT	4.8	7.1	2.4	2.3
	7,8-$(OH)_2$ DPAT	4.4	111.3	2.2	41.1
8-OH DPAT	7,8-$(OH)_2$ DPAT	- - -	0.64	- - -	0.26

[a] EC = entrapped cells [b] PPP = partially purified phenoloxidase

Since suspension cultures of *M. pruriens* possess high phenoloxidase activities, they form an adequate source for the isolation of this enzyme. From kg-quantities of cells, a 200-fold purified phenoloxidase preparation could be isolated using a simple $(NH_4)_2SO_4$-procedure with an activity yield of 25%. It is well-known that phenoloxidases are relatively stable enzymes and therefore prolonged periods of bioconversion are to be expected. In order to prevent activity losses after isolation, we used the phenoloxidase preparation in dialysis tubing instead of the adsorbed or entrapped enzyme. The continuous flow system consisted of an airlift fermentor coupled with an aluminium oxide column for selective product isolation. In this fermentor, sufficient oxygen supply and mixing conditions are guaranteed. Based on initial rates of bioconversion (v_o) and bioconversion efficiencies (B_e) 7-OH DPAT was chosen as the precursor for the production of 7,8-$(OH)_2$ DPAT (**Table 4**). In a standard assay, using the partially purified phenoloxidase, 41.1 mg L^{-1} product was formed after 24 h (=B_e). Then, the bioconversion stopped, probably due to product inhibition because the product formed was not removed, and no oxygen limitations occurred. From **Fig. 6** it can be seen that with the system product formation continued for at least 50 h. After uncoupling the column, ca. 120 mg of the desired product, 7,8-$(OH)_2$ DPAT, and less than 4 mg of its isomer, 6,7-$(OH)_2$ DPAT, eluted respectively, using 0.5 M acetic acid. The phenoloxidase preparation could be reused displaying 85% of its original activity.

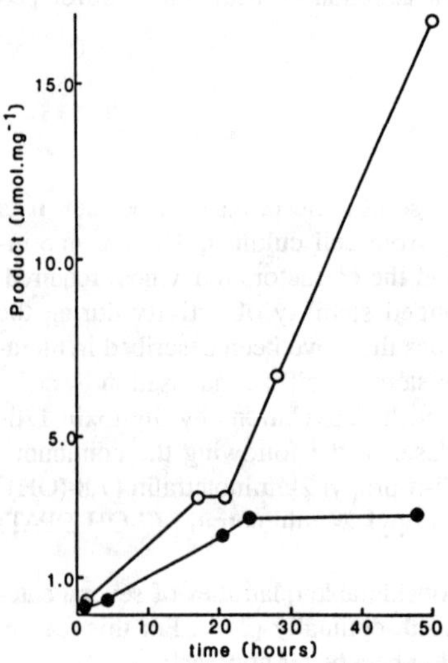

Fig. 6: Time course of the formation of 7,8-$(OH)_2$ DPAT by means of the continuous flow system (○) and the static system (●) using purified phenoloxidase. Product formation was expressed as µmol per mg protein.

On basis of these results, a definite choice for enzyme preparations as a biocatalytic system in order to produce catechols seems justified. However, it should be kept in mind that the applicability depends on the balance between the activity losses introduced by the enzyme isolation procedure and the superiority of the bioconversion efficiency of the resulting preparation compared to the (entrapped) cell system.

V. BIOCONVERSION OF WATER-UNSOLUBLE PRECURSORS

Several useful precursors are not or very poorly soluble in aqueous media and reaction rates are therefore very low. So far, poorly water-soluble precursors have been applied in two-phase systems. Many plant cell species hardly converted the precursors in the presence of organic phases, often as a result of a dramatic decrease of cell viability. Moreover, enzyme activities are negatively influenced in these systems. A new approach to solve this problem is to combine the advantages of apolar systems (higher solubility) and of aqueous systems (maintainance of cell viability) by carrying out bioconversions in the presence of clathrating agents, such as cyclodextrins. Cyclodextrins are cyclic oligosaccharides that are able to form inclusion complexes with a variety of apolar ligands (**Fig. 7**). Through complexation, the physical-chemical properties of the ligands are changed, including their solubility in aqueous media.

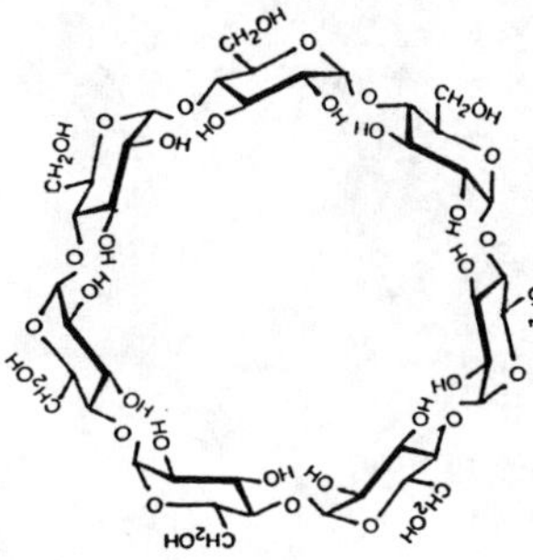

Fig. 7: Structural formula of β-cyclodextrin.

In this Section the applicability of β-cyclodextrin (β-CD) is demonstrated for the regioselective *ortho*-hydroxylation of β-estradiol (E_2) (**Table 2**) and the bioconversion of coniferyl alcohol into podophyllotoxin [8, 9]. Complexation of E_2 with β-CD resulted in enhanced solubility, the limit was 660 μM (1700 mg L^{-1} !). The solubility of uncomplexed E_2 was only 12 μM: no bioconversion could be measured using the systems listed in **Table 5**. When E_2 was fed as a complex, the efficiency of the bioconversion, after 72 h, increased in the following order: freely suspended cells (0%), alginate-entrapped cells (1%, 4-hydroxyestradiol= 4 OH-E_2), cell homogenate (6% 4 OH-E_2, 1% 2 hydroxyestradiol= 2 OH-E_2), phenoloxidase preparation (40% 4 OH-E_2, 3%, 2 OH-E_2) (**Table 5**). The amounts of 4 OH-E_2 formed by *Mucuna*-phenoloxidase can compete with those obtained with micro-organisms, such as *Aspergillus alliaceus* [8]. The finding that even a steroid structure can serve as a precursor for our phenoloxidase confirms its very low substrate specificity.

Table 5. The bioconversion of E_2 into 2 OH-E_2 and 4 OH-E_2 in different biocatalytic systems, expressed as nmol product formed per mg protein (mean ± s.d.; n=3) after 72 h [8].

Biocatalytic system	2 OH-E_2	4 OH-E_2
Freely suspended cells	0	0
Entrapped cells	6.5 ± 1.2	37.6 ± 6.9
Cell homogenate	12.6 ± 3.2	62.1 ± 9.4
Phenoloxidase preparation*	60.8 ± 8.7	688.3 ± 83.5

* Purification fold ca. 185.

The second application of β-CD deals with the improvement of the podophyllotoxin production by cell cultures of *P. hexandrum* [9]. In Section III, it was demonstrated that feeding of the water-soluble precursor coniferin enhanced the production of this lignan (**Fig. 5**). Coniferyl alcohol is a key intermediate in the biosynthetic pathway of podophyllotoxin (**Fig. 4**) and it is poorly water-soluble. Standard-grown cells of *P. hexandrum* endogenously accumulate podophyllotoxin in contents ranging from 0.001 to 0.002% on a dry weight basis (**Fig. 8**).

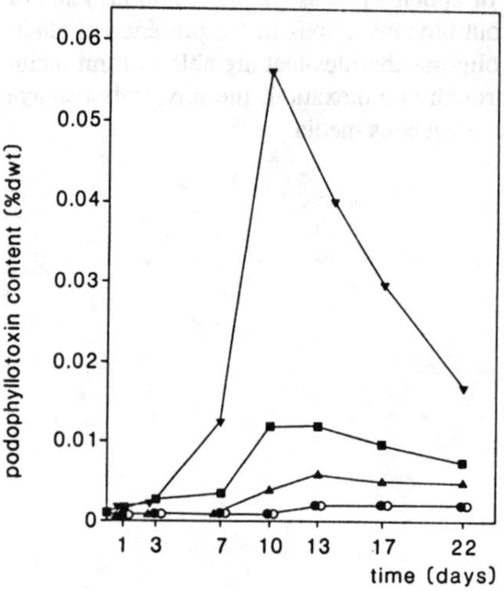

Fig. 8: Time course of the podophyllotoxin content, on a dry weight basis, in suspension-grown cultures of *P. hexandrum*; control (●), 3 mM β-CD (○), 3 mM coniferyl alcohol/β-CD complex (■), 3 mM coniferyl alcohol (▲) and 3 mM coniferin (▼) [9].

The precursor feeding experiments have been performed with the highest possible coniferyl alcohol/β-CD concentration. Feeding of 3 mM resulted in an enhanced podophyllotoxin accumulation with a maximum of 0.012% at day 10, while 3 mM coniferin gave 0.056% at that point of time. Non-complexed coniferyl alcohol suspended in the medium in a concentration of 3 mM gave 0.006% at day 13. The β-CD itself had no influence on the culture characteristics, nor was it metabolized by the cells. It can be concluded that cyclodextrins may be applied successfully in plant cell biotechnology because of their solubilizing action, leading to an extension of bioconversion possibilities [9].

VI. CONCLUSIONS AND PROSPECTS

From the results in this paper, it is clear that two pharmaceutically important groups of compounds, the catechols and podophyllotoxins, can be produced by bioconversion. Entrapped as well as freely suspended cells or enzyme preparations were used as biocatalytic systems. Although promising results have been obtained, the employment of plant enzymes for larger-scale production of valuable compounds is generally not yet achievable. The plant cell cultures initiated so far, have mainly been investigated with respect to secondary metabolites. Detailed screening on enzyme activities, even using non-producing cultures, might yield new bioconversions. Nevertheless, only fundamental research on biosynthetic routings can give us powerful tools for an optimal exploration of the enzymatic potential in the plant kingdom. In this way, the production of valuable compounds by bioconversion, next to biosynthesis and *de-novo* synthesis, may play an important role outside the non-profit laboratories in future.

REFERENCES

[1]. Hall, R.D., Holden, M.A. and Yeoman, M.M. (1988) in Medicinal and Aromatic Plants I (Bajaj, Y.P.S., ed.), pp. 136-153, Springer Verlag, Berlin Heidelberg

[2]. Pras, N., Hesselink, P.G.M., ten Tusscher, J. and Malingré, Th.M. (1989) Biotechnol. Bioeng. *34*, 214-222

[3]. Pras, N., Hesselink, P.G.M., Guikema, W.M. and Malingré, Th.M. (1988) Biotechnol. Bioeng. *33*, 1461-1468

[4]. Pras, N., Wichers, H.J., Bruins, A.P. and Malingré, Th.M. (1988) Plant Cell Tissue Organ Culture *13*, 15-26

[5]. Pras, N., Booi, G.E., Dijkstra, D., Horn, A.S. and Malingré, Th.M. (1990) Plant Cell Tissue Organ Culture, *21*, 9-15

[6]. Van Uden, W., Pras, N., Visser, J.F. and Malingré, Th.M. (1989) Plant Cell Reports *8*, 165-168

[7]. Van Uden, W., Pras, N., Vossebeld, E.M., Mol, J.N.M. and Malingré, Th.M. (1990) Plant Cell Tissue Organ Culture *20*, 81-87

[8]. Woerdenbag, H.J., Pras, N., Frijlink, H.W., Lerk, C.F. and Malingré, Th.M. (1990) Phytochemistry *29*, 1551-1554

[9]. Woerdenbag, H.J., Van Uden, W., Frijlink, H.W., Lerk, C.F., Pras, N. and Malingré, Th.M. (1990) Plant Cell Reports, in press

BIOCONVERSION OF XENOBIOTICS IN DIFFERENT PLANT SYSTEMS - CELL SUSPENSION CULTURES, ROOT CULTURES AND INTACT PLANTS

HANS HARMS and ELKE KOTTUTZ, Institute of Plant Nutrition and Soil Science, Federal Agricultural Research Centre (FAL), Bundesallee 50, D-3300 Braunschweig

1. INTRODUCTION

In vitro techniques using undifferentiated plant cell suspension cultures or differentiated root cultures have been most commonly applied for studying the fate of xenobiotics in plants. Although in vitro systems offer many advantages over whole plants there are still questions about the validity of extrapolating xenobiotic metabolism in vitro into what occurs in vivo. Therefore, in the investigations reported here, we have compared xenobiotic metabolism in cell cultures, root cultures and in intact plants.

2. MATERIALS

2.1 Cell suspension cultures: Various cell suspension cultures of different monocots such as Triticum aestivum L, Hordeum vulgare L, Pennisetum americanum L, and dicots such as Glycine max L, Daucus carota L, Solanum tuberosum L were cultured as described (1, 2) and used for metabolism tests during the last 48 hours of the late logarithmic growth phase.

2.2 Root cultures: By transformation of Lupinus hartwegii plants by Agrobacterium rhizogenes, a root culture was obtained which was cultivated in MS-medium. Because of the organized structure and the development of secondary cell walls, root cultures are a good stepping stone between cell culture and whole plants. Furthermore, root cultures are not very sensitive to microbial contamination. It is, therefore, possible to draw a parallel between aseptic conditions and those that would be found in insterile experiments. The metabolism tests were as those described for cell culture except that the incubation period with ^{14}C-labelled xenobiotics was lengthed to 3 days.

2.3 Aseptically grown plants: Wheat cv. Heines Koga and tomato cv. Money maker plants were aseptically cultured using described methods (3). After 2 to 3 weeks of growth the test compounds were added to the nutrient solution. After 5 days treatment the plants were harvested and analyzed.

2.4 Uptake of xenobiotics by plants from soil: To investigate the uptake of xenobiotics by crop plants from soils, a special technique originally designed by (4) was used. The plants were grown in a rectangular plexiglass pot containing 1 kg soil polluted with 5 or 10 ppm of the selected ^{14}C-labelled contaminants. To follow the evolution of $^{14}CO_2$ or other labelled volatiles from degradation processes in the soil and to distinguish between uptake of the original compound and a possible photosynthetic incorporation of released $^{14}CO_2$, the root and shoot spheres were separated from each other by an airtight lid. The shoots pass through appropriate openings, their stems being sealed-in with silicon rubber. A circulating air stream allows continuous sampling of the gases from the root zone for subsequent analysis. The incubation period lasted 4 weeks.

2.5 Metabolism test and analysis for bound residues: The uptake and metabolic fate of xenobiotics were studied by standardized methods (5). The bound or non-extractable residues were analysed by a sequential fractionation procedure (6).

3. RESULTS AND DISCUSSION

3.1 Metabolism of pentachlorophenol[ring-U-^{14}C] and 4-chloroaniline-[ring-U-^{14}C] in wheat cell suspension cultures and whole plants

The validity of extrapolating data obtained with cell culture techniques to those of intact plants is still a matter of debate. In order to compare these two systems, cell suspension cultures and wheat seedlings of the same cultivar were incubated with pentachlorophenol and 4-chloroaniline. The metabolic fate of these compounds in the two differently differentiated plant systems are summarized in table 1.

TABLE 1. Metabolism of pentachlorophenol and 4-chloroaniline in wheat cell suspension cultures and intact plants

Fractions	% of applied radioactivity					
	Pentachlorophenol			4-Chloroaniline		
	parent compound	metabolites	bound residues	parent compound	metabolites	bound residues
Cell cultures						
cells	0.7	48.2	41.3	3.0	6.1	72.2
medium	0.6	5.8		3.9	8.8	
Intact plants						
roots	11.3	28.6	15.2	6.9	2.9	60.9
shoots	1.1	3.5	1.0	0.2	0.6	0.7
nutr. sol.	35.6	7.2		4.2	12.4	

The compounds were taken up and metabolized by both plant systems. ^{14}C-label of both compounds was transported from the roots into the shoots of the intact wheat plants. Cell cultures adsorbed pentachlorophenol very rapidly (2) and formed high amounts of polar metabolites which were mainly associated with the cells. 41 % of the radiolabel was converted (via the conjugate fraction) into the non-extractable residue fraction. ^{14}C-label was bound mainly to lignin and to a high molecular weight hemicellulose fraction. Polar conjugates could also be extracted from roots and shoots, and PCP glycosides were also predominant as found in the cultured cells. More than 16 % of the total radioactivity from shoots and roots were found as bound residues. These were fractionated into several wall components and yielded a pattern similar to that in cell culture.

In cell cultures, 72 % of 4-chloroaniline was detected in the bound residue fraction (5). Further studies (7) showed that this high proportion of ^{14}C-label was mainly associated to the pectin and lignin fractions of the cell wall. 4-Chloroaniline was also rapidly absorbed by the roots of wheat plants and metabolized to polar conjugates. The radioactivity of the bound residue fraction reached more than 61 % which is comparable to the percentage found in cell cultures.

3.2 Metabolism of anthracene [9-^{14}C] and phenanthrene [9-^{14}C] in cell suspension cultures, root cultures, and crop plants grown under either septic or aseptic conditions

Previous experiments (8) showed that the uptake of polycyclic aromatic hydrocarbons depended not only on the size of the molecule but also on the structure of the particular chemical. The compounds, anthracene and phenanthrene, tested here, are widespread priority organic pollutants. Both are composed of three fused benzene rings but of different molecular configuration, thus providing a good comparison for these tests.

<u>Investigations with cell suspension cultures</u>

Previous investigations have also shown different metabolic behaviour of monocotyledonous and dicotyledonous plants. In order to understand these effects better, cell cultures of various crop plants were compared. The rates of uptake and turnover are summarized in table 2.

The assimilation rate of anthracene and phenanthrene was very similar for all cultures tested, varying for anthracene between 60 to 70 % and for phenanthrene between 71 % (monocots) and 37 % (dicots), respectively. The highest amount of both compounds was taken up by wheat cultures, which metabolized 50 % of anthracene and 11 % of phenanthrene. Similar results were obtained with pennisetum cultures. Tomato cultures adsorbed more than 80 % of the applied phenanthrene, but most of this radioactivity was identified as

TABLE 2. Metabolism of anthracene and phenanthrene in cell suspension cultures of various crop plants (n.d. = not detectable)

Cell cultures	Anthracene — % of applied radioactivity					
	parent compound		metabolites		bound residues	
	cells	medium	cells	medium	cells	medium
Barley	25.7	0.8	31.2	8.3	4.6	-
Wheat	33.2	0.3	51.4	6.2	10.0	-
Pennisetum	n.d.	3.1	52.3	8.4	11.3	-
Soybean	56.9	0.1	34.4	15.2	9.8	-
Carrot	2.9	n.d.	31.2	9.2	26.8	-
	Phenanthrene					
Barley	53.4	2.5	9.2	2.0	1.5	-
Wheat	60.1	4.5	11.0	1.1	7.9	-
Soybean	29.4	1.9	8.3	8.0	19.9	-
Carrot	37.3	1.4	9.3	2.1	11.1	-
Tomato	80.4	2.1	6.9	2.1	0.4	-

the parent compound. In the nutrient solutions most of the radioactivity could be attributed to the metabolites. Soybean cultures seem to metabolize both compounds very rapidly, transfering the metabolites back to the nutrient solution.

Investigations with aseptic root cultures and intact plants

While cell suspension cultures easily adsorb exogenously applied compounds by diffusion, uptake of xenobiotics into root cultures and plants represents an active process. The summarized results from these experiments are demonstrated in table 3.

TABLE 3. Metabolism of anthracene and phenanthrene in root cultures of lupin and aseptically grown tomato plants

Fractions	% of applied radioactivity					
	Anthracene			Phenanthrene		
	parent compound	metabolites	bound residues	parent compound	metabolites	bound residues
Root culture						
roots	6.3	26.4	39.7	42.1	11.5	17.5
medium	1.6	6.8	-	4.7	0.1	-
Aseptic plants						
roots	5.4	2.9	3.6	7.4	5.0	2.9
shoots	3.2	4.3	0.9	7.2	0.3	0.6
nutr. sol.	18.0	40.3	-	36.7	12.1	-

Although the uptake of the 3-ring-system was the same for both, the distribution of the parent compound and bound residues was considerably different. Remarkably it was found that 40 % of the linear anthracene was bound in the non-extractable residue whereas 6 % remained in the soluble cell extract. In treatments with the non-linear phenanthrene only 17 % was incorporated into the residue, but as much as 42 % was left in the extracts. A substantial difference could also be seen in the amount of polar metabolites produced: 43 % more metabolites were found in anthracene test cultures than in those for phenanthrene. The uptake of the added compounds by aseptically grown tomato plants was also almost the same for both chemicals. TLC examination of the extracts showed that with phenanthrene most of the radioactivity was related to the parent compound, whereas with anthracene half of the radioactivity could be attributed to polar metabolites. Only small amounts of radioactivity could still be detected in the non-extractable residue with more being bound in the roots than in the shoots.

Comparing the three different aseptic plant test systems - cell suspension culture, root culture and intact plants - it is obvious, that all three systems metabolize anthracene to a higher extent than phenanthrene. The degree of metabolism is quantitatively different in the systems but follows the same pattern.

Experiments with insterile grown plants

In order to get an idea about the soil-plant transfer, anthracene and phenanthrene were added to the soil of insterile grown plants.

TABLE 4. Soil-plant transfer and degradation of anthracene and phenanthrene in sterile tomato plants

	% of applied radioactivity after 4 weeks of incubation			
	shoots	roots	CO_2-produced	soil*
Anthracene				
Control without plant			7.7	80.2
5 ppm	0.9	0.5	42.4	40.8
10 ppm	0.8	0.6	42.4	36.8
Phenanthrene				
Control without plant			36.8	32.0
5 ppm	1.2	0.7	39.1	35.1
10 ppm	1.7	0.6	15.5	50.6

* at the end of experiment

Microbial degradation of anthracene within the containers with plants was about 6 times that of containers without plants. This could be due to extensive root growth within soil and therefore wide distribution of plant exudates. Although varied concentrations of anthracene were applied the amount of radioactive CO_2 released was the same, almost half the total of applied concentration. In experiments with phenanthrene, the differences in microbial and plant degradation levels did not occur, but where doubled concentrations of phenanthrene were used, radioactive CO_2-production was markably reduced, possibly due to toxic levels of the chemical. The transfer factor plant/soil for anthracene 5 ppm and 10 ppm was 0.014. For phenanthrene the factors were 0.019 and 0.023, respectively.

4. CONCLUSIONS

Plant cell cultures have been used for an ecotoxicological evaluation of chemicals and compared with root cultures and with intact plants grown either under aseptic or insterile conditions. All plant test systems were able to metabolize the compounds by common metabolic pathways. Qualitatively the metabolites were the same in all systems. However, using cell cultures the results may be obtained much quicker with less analytical expense. They are therefore a useful system to obtain rapid evidence of the ecotoxicological behaviour of chemicals in plants.

REFERENCES

1 Harms, H. (1973) Landbauforsch. Völkenrode 23, 127-132

2 Langebartels, C. and Harms, H. (1984) Z. Pflanzenphysiol. 113, 201-211

3 Langebartels, C. and Harms, H. (1986) Angew. Botanik 60, 113-123

4 Helal, H.M. and Sauerbeck, D.R. (1984) Plant and Soil 76, 175-182

5 Harms, H. and Langebartels, C. (1986) Plant Science 45, 157-165

6 Langebartels, C. and Harms, H. (1985) Ecotox. Environ. Safety 10, 268

7 Harms, H. and Langebartels, C. (1988) in Methoden zur ökotoxikologischen Bewertung von Chemikalien (Verfondern, M. and Scheele, B. eds.) Jül.-Spez. 441, 131-146

8 Harms, H. (1981) Landbauforsch. Völkenrode 31, 1-6

METABOLISM OF POLYCHLORINATED BIPHENYLS (PCBS) BY PLANT TISSUE CULTURES

INCHOEL LEE AND JOHN S. FLETCHER, DEPARTMENT OF BOTANY AND MICROBIOLOGY, UNIV. OF OKLAHOMA, NORMAN, OK 73019-0245 U.S.A.

1. INTRODUCTION

The environmental persistence of polychlorinated biphenyls (PCBs) and their demonstrated toxicity to animals has made this class of compounds a focal point of environmental concern around the world. PCB molecules consist of a biphenyl nucleus carrying 1 to 10 chlorines (see Fig. 1); hence there are 209 possible PCB congeners that differ in the number and position of the chlorines. Up until 1976, when they were banned in the U.S., a mixture of approximately 100 PCB isomers was routinely used commercially under the common name aroclor. Their primary use was in electrical equipment. In the U.S. they are included on the EPA Priority Pollutant List, and are known to be present in 25% of the nation's superfund sites.

Several studies have shown that plant-soil systems are capable of degrading PCBs, with the resulting accumulation of breakdown products in both the plants and the soil (1-3). However, since these studies were performed under field conditions, it is not certain what portion of the degradation was due to plant metabolism and what portion to that of soil microbes which have been demonstrated in pure culture to degrade PCBs (4). We resolved this issue by using axenic plant cultures (5,6) to show that plant cells are capable of metabolizing ^{14}C labeled PCBs without any involvement of microbes. The current study was conducted to determine the influence of chlorine content and the position of substitution on PCB metabolism.

2. PROCEDURE

2.1 Materials and Methods

2.1.1 Cells: This investigation was performed with nonphotosynthetic suspension cultures of *Rosa* cv. Paul's Scarlet grown in the dark at 25°C on a rotary shaker at 150 rpm. Individual cultures provided with 80 ml of MPR medium (7) were transferred and harvested for analysis as previously described (8).

2.1.2 Metabolism Assay: Approximately 0.5g of rapidly expanding 9-day-old cells was placed in 2 ml of sterile medium held in a 12 ml culture tube. A mixture of 10 PCB congeners (10 μl acetone solution) was added aseptically to each culture tube. The final concentration of each congener was 5 μM. Two different mixtures (see Fig. 2) possessing a combination of 19 different congeners were examined. The specific congeners examined were the same as those studied by Bedard et al. (4). The mixtures provided a broad spectrum of PCBs with different chlorine contents, patterns of substitution, and non-overlapping GC retention times. Control cultures were killed by adding perchloric acid before the addition of PCBs. After a 96 h incubation period an aliquot was removed from each culture and plated on bacterial medium to check for bacterial contamination. The assay was then stopped with perchloric acid, and parent compounds were extracted by adding 8 ml of a hexane - Triton X-100 mixture (1% v/v) and 0.5g of Na_2SO_4 to each culture tube. The tubes were shaken vigorously overnight in a horizontal position. Following phase separation the hexane layer was sampled and GC analysis conducted. Preliminary studies conducted with and without dead cells showed that the recovery of each congener was at least 95% of the starting material.

2.1.3 Analysis: The extent to which each compound was metabolized was determined by comparing the amounts of each congener recovered from the living cultures with that recovered from the dead cultures (see Fig. 2). The percentages shown in Table 1 are average values taken from two experiments each with triplicate incubations for living and dead cells.

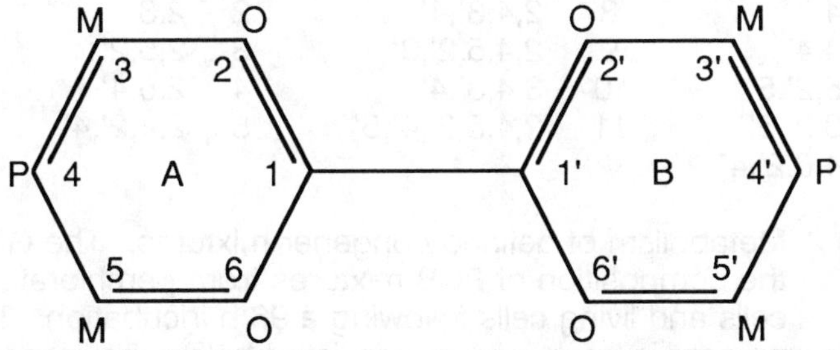

Figure 1. Biphenyl structure showing the numbering of each ring, and positions designated as ortho (O), meta (M) and para (P) where chlorine substitution can occur in PCBs.

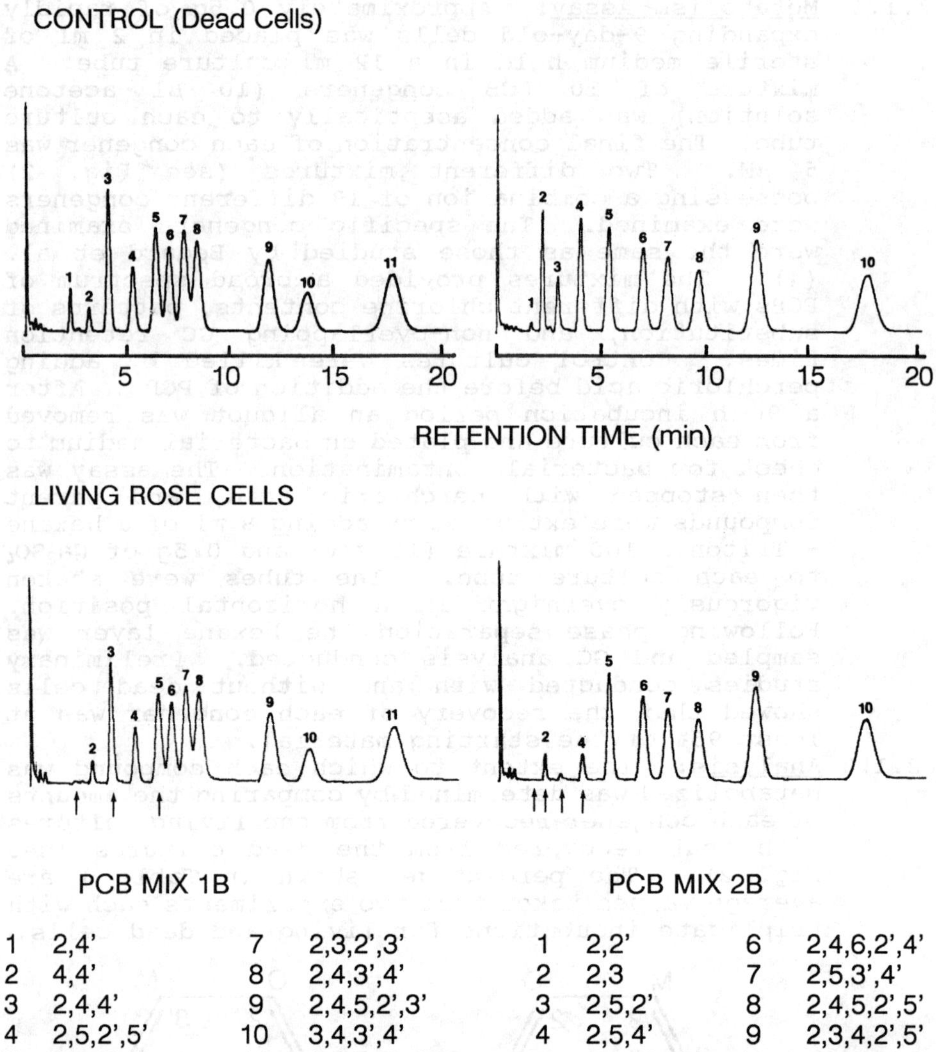

PCB MIX 1B

1	2,4'	7	2,3,2',3'
2	4,4'	8	2,4,3',4'
3	2,4,4'	9	2,4,5,2',3'
4	2,5,2',5'	10	3,4,3',4'
5	2,3,2',5'	11	2,4,5,2',4',5'
6	2,4,6,2',4'		

PCB MIX 2B

1	2,2'	6	2,4,6,2',4'
2	2,3	7	2,5,3',4'
3	2,5,2'	8	2,4,5,2',5'
4	2,5,4'	9	2,3,4,2',5'
5	2,4,2',4'	10	2,4,5,2',4',5'

Figure 2. Metabolism of defined congener mixtures. The GC profiles show the composition of PCB mixtures from perchlorate-killed control cells and living cells following a 96 h incubation. The compounds present in each mixture are listed below the respective profiles. Arrows mark congeners which were metabolized by > 20%.

3. RESULTS AND DISCUSSION

3.1 Seven of the 19 congeners which were examined were metabolized by amounts ranging from 20 to over 80% of the starting material (see Table 1). Two congeners 3,4,3'4' and 2,5,2'5' appeared to be metabolized to a small degree, 5-10%. Any metabolism less than 5% was not distinguishable with the assay used. In a similar study conducted by Bedard et al. with the same mixtures

TABLE 1. Metabolism of polychorinated biphenyls by 9-day-old rose cells during a 96 h incubation.

	PCB CONGENER	H_2O SOLUBILITY[a] (M)	log K_{ow}[b]	METABOLISM
	2,3	-	-	+++[c]
	2,4'	4.71×10^{-6}	5.32	++++
	2,5,4'	-	5.77	+++
	2,2'	5.41×10^{-6}	5.02	++++
Open	2,3,2'3'	1.16×10^{-7}	6.67	-
ortho, meta	2,5,2'	1.72×10^{-6}	5.64	++++
and	2,3,2'5'	5.80×10^{-7}	6.67	+
meta, para	2,4,5,2',3'	-	-	
sites	2,5,3',4'	1.40×10^{-7}	6.39	-
	2,3,4,2',5'	1.38×10^{-8}	6.85	-
	4,4'	2.96×10^{-7}	5.28	-
Open	2,4,4'	6.68×10^{-7}	5.74	+
o,m sites	2,4,3',4'	1.99×10^{-7}	6.67	-
	2,4,2',4'	2.33×10^{-7}	6.44	-
	3,4,3',4'	5.99×10^{-7}	6.52	-
	2,4,6,2',4'	-	-	-
Open	2,5,2',5'	8.97×10^{-8}	6.26	-
m,p sites	2,4,5,2',5'	3.65×10^{-8}	6.85	-
Blocked o,m and m,p Sites	2,4,5,2',4',5'	8.32×10^{-9}	7.44	-

[a]Values taken from reference 9.
[b]Values taken from reference 10.
[c]Symbols represent the percent of the starting material which was metabolized: +20-39, ++ 40-59, +++ 60-79, and ++++ 80-100. Starting mixtures possessed 5 μM concentrations of each congener.

of PCBs (4), they found that the number of different congeners metabolized by 25 different bacterial strains ranged from 4 to 18 with a mean of 8.

3.2 The 7 compounds which underwent a substantial amount of metabolism were the most water soluble congeners (see Table 1). The position of chlorine substitution also appeared to influence their susceptibility to metabolism since all of the congeners experiencing greater than 80% metabolism were structurally similar, they possessed one ring with Cl substitution only at the 2-postion (see Fig. 1). Most of the metabolized congeners had lower log k_{ows} than those which were not metabolized, but there was one exception, the 2,3,2'5' congener whose log K_{ow} is 6.67 was metabolized, whereas 4,4' PCB with a log K_{ow} of 5.28 was not altered. It is interesting that the simpler 4,4' congener was not metabolized, whereas the slightly more complex 2,4,4' was altered. Bedard et al. (4) observed the same phenomenon with 4 different bacterial strains, and have suggested that chlorine substitution at the ortho position facilitates enzymatic attack in some bacteria. In keeping with this hypothesis, all of the PCBs which were metabolized by the rose cells possessed at least 1 ortho substitution.

4. CONCLUSIONS

4.1 Rose cells have been demonstrated to metabolize an assortment of PCBs. Chlorine substitution at the ortho position, and increased solubility in H_2O appeared to favor the metabolism of PCBs.

REFERENCES

1. Moza, P. Weisgerber, W. Klein, W. and Korte, F. (1974) Bull. Environ. Contam. Toxicol. 12:541-546
2. Moza, P. Kilzer, L. Weisgerber, I. and Klein, W. (1976) Bull. Environ. Contam. Toxicol. 16:454-463
3. Moza, P. Scheunert, I. Klein, W. and Korte, F. (1979) J. Agric. Food Chem. 27:1120-1124
4. Bedard, D.L. Uterman, R. Bopp, L.H. Brennan, M.J. Haberl, M.L. and Johnson, C. (1986) Appl. Environ. Microbiol. 51:761-768
5. Fletcher, J.S. Groeger, A.W. and J.C. McFarlane (1987) Bull. Environ. Contam. Toxicol. 39:960-965
6. Fletcher, J. Groeger, A. McCrady, J. and McFarlane, J. (1987) Biotech. Let. 9:817-820
7. Nesius, K. and Fletcher, J. (1972) Planta 106:173-176
8. Mohanty, B. and Fletcher, J. (1978) Phy. Pl. 42:221-225
9. Leifer, A. (1983) in EPA-560/5-83-025, U.S. Environmental Protection Agency, Washington, D.C.
10. Erickson, M.D. (1986) in Analytical Chemistry of PCBs, pp. 11-12, Butterworth, Boston

METABOLISM OF METRIBUZIN IN TOMATO CELL CULTURES BY A GLUCOSYLTRANSFERASE

D. G. DAVIS, P. A. OLSON, H. R. SWANSON AND D. S. FREAR,
USDA Biosciences Research Laboratory, State University Station,
Fargo, ND, USA 58105.

1. INTRODUCTION

Plant cell cultures have been used to study the metabolism of herbicides for a number of years (1,2,3) in an effort to replace whole plants with more convenient and controlled systems. Isolation of metabolites from cultures has proven to be more convenient than with intact plants and cell cultures produce many of the same metabolites that are formed by intact plants (2,3).

The asymmetric triazinone herbicide metribuzin (4-amino-6-tert-butyl-3(methylthio)-1,2,4-triazin-5(4H)-one) controls many weeds in a variety of crop species (4). Metribuzin is an effective inhibitor of photosynthesis (4) but it also inhibits the growth of non-photosynthetic callus of tomato (<u>Lycopersicon esculentum</u> L.) (5) and tobacco (Davis, Olson & Frear, unpublished) and the viability of cell suspensions of soybean (6).

Differential absorption and metabolism of metribuzin have been proposed as mechanisms for greater tolerance of winter wheat than of downy brome grass (7), and for differences in tolerance between diploid and tetraploid soybean plants and cell cultures (8). Metribuzin is degraded to deaminated, diketo or deaminated diketo compounds in ivyleaf morningglory and soybean (8,9) and is also metabolized to glucose conjugates in tomato (10,11).

In tomato, a glucose conjugate is formed rapidly by the enzyme UDP-glucose:metribuzin N-glucosyltransferase (MGT) (10,11). A possible mechanism for tolerance of metribuzin by plants is a rapid conjugation of the herbicide with glucose and consequent inactivation and/or compartmentation before it can reach the chloroplasts. Evidence of a positive correlation between MGT activity in tomato plants and its differential herbicide tolerance has been published (10).

The metabolism of metribuzin by MGT in green plants has been reported, but its activity in callus cultures or cell suspensions has not been characterized. This study was undertaken to obtain cell suspension cultures with significant enzyme activity for studies on MGT activity and kinetics, or for possible inclusion in genetic transformation studies for the development of metribuzin resistant crop plants.

2. PROCEDURE

2.1 Materials and methods

2.1.1 Callus and Cell Suspensions: Greenhouse-grown tomato plants were used for initiation of callus. Stems or petioles of four to six week old plants were sterilized with 70% ethanol (10 sec), 1.6% Na hypochlorite (10 min) followed by three rinses in filter sterilized deionized water.

Media used were B5 (12) or MS (13) salts with 2% sucrose and 0.7% agar. Calli were grown in six cm dia. plastic Petri plates (wrapped with Parafilm(R) or in glass bottles (with bakelite screw caps) at 26C either in darkness or under a 14 h photoperiod (70 µE m^{-2} s^{-1}). Various auxin/cytokinin combinations were tested. Initially, most of the cultures were dark in appearance, and the cultures that grew the best were selected for subculturing at monthly intervals onto the same medium in which they were initiated.

For suspension cultures, seven month old calli were placed into 250 cm^3 Erlenmeyer flasks containing 50 cm^3 of the same medium (without agar) in which the callus was developed and shaken on a rotary shaker (100 cycles min^{-1}), 26C, under 70 µE m^{-2} s^{-1} fluorescent light, 14 h photoperiod. Dark-grown cultures were grown in flasks wrapped with two layers of aluminum foil and placed onto the same shakers. Subcultures were made at about weekly intervals when satisfactory growth was observed.

2.2.2 Assay for MGT activity: 1 to 3 g of callus tissue or cells were homogenized in a motor driven tissue grinder with a buffer (0.1 M Kphosphate, pH 7.8, 0.001 M EDTA and 0.2 M isoascorbate) and centrifuged 30 min at 10-15 X 10^3 g. The supernatant (2.5 cm^3) was placed on a Sephadex PD-10(R) column (9 cm^3 volume) equilibrated with 0.05 M Kphosphate, pH 7.8, containing 5 mol m^{-3} EDTA and 5 mol m^{-3} dithiothreitol. The protein was eluted with 3.5 cm^3 of equilibration buffer.

Approximately 0.6 cm^3 of the eluant (200 to 400 µg protein) were added to the reaction mixture: 0.1 cm^3 0.5 M Kphosphate pH 7.8, 0.1 cm^3 metribuzin (5 µg) 0.1 cm^3 ^{14}C-metribuzin (50,000-100,000 dpm) and 0.1 cm^3 UDPG (500 µg). Controls were run without enzyme. After 60 min at 30 C 1 cm^3 of 0.5 N Na acetate, pH 4.0, and 1 cm^3 of 0.1 N HCl were added and the mixture was extracted 2 to 3 times with ethyl ether. An aliquot (1 cm^3) of the aqueous phase was counted in a liquid scintillation counter. Enzyme activity was calculated as nanomoles of product (N-glucoside) formed (mg protein) $^{-1}$ h^{-1}. Protein content was determined with the Pierce(R) protein assay reagent (Coomassie Blue G-250) and bovine serum albumin as a standard.

3. RESULTS AND DISCUSSION

3.1 Several cultivars of tomato with varying susceptibilities to metribuzin (5,14,15) were used for the establishment of callus, but little detectable MGT activity was observed in many of the established callus tissues. Only the results of the 'Sheyenne' cultivar are reported here. 'Sheyenne' is intermediate in its susceptibility to metribuzin and has been used in previous work with MGT (10).

Cultures initiated from leaves and petioles grew very slowly on MS medium containing various combinations of 2,4-dichlorophenoxyacetic acid (2,4-D), benzylaminopurine, kinetin (kin) and naphthyleneacetic acid (NAA). Cultures grown on B5 medium with the same combinations of growth regulators grew somewhat better. Leaf-derived calli grown in darkness on B5 salts and vitamins with a 6 g m^{-3} 2,4-D, 0.2 g m^{-3} kinetin and 0.4 g m^{-3} NAA were rated with only fair growth but had the highest MGT activity (3.2 units (mg protein)$^{-1}$ h^{-1} out of seven callus cultures tested.

Callus grown in B5 medium containing 3 g m^{-3} 2,4-D had moderate MGT activity that seemed to be related to the growth of the culture, as judged by the size of the calli and the browning of the tissue. Dark brown cultures had MGT activities of 0.4 units (mg protein)$^{-1}$ h^{-1}, whereas light tan cultures that appeared to grow well averaged 3.5 units (mg protein)$^{-1}$ h^{-1}.

Cell suspensions were difficult to establish from the callus, but after numerous attempts, suspensions were obtained successfully from callus derived from leaves and petioles in both MS and B5 media. All successful cultures contained 2,4-D at either 3 or 6 g m^{-3}, with or without kinetin or NAA (Table 1). Cell suspensions derived from the leaves and grown in B5 medium with 3 g m^{-3} 2,4-D (B5-3 medium) proved to be the most successful cell suspension, and had the most consistently high MGT activity over a long time. Omission of 2,4-D resulted in lack of cell growth and no MGT activity. Replacement of 2,4-D by other auxins has not resulted in significant MGT activity whether growth occurred or not.

TABLE 1. MGT activity in cell suspension cultures of 'Sheyenne' tomato under 14 h photoperiod. 70 µE m^{-2} s^{-1}.

Origin	Medium	Growth Regulators 2,4-D	Kin	NAA	Range of MGT activities at intervals during the indicated subcultures (SC):	
		(----g m^{-3}	-----	)	(SC)	(units h^{-1} (mg protein)$^{-1}$)
Leaf	B5	3	0.0	0.0	2-21	3.2 - 6.4
Leaf	B5	3	0.2	0.0	3-34	0.7 - 3.7
Leaf	B5	6	0.2	0.4	2-34	2.5 - 5.9
Leaf	MS	3	0.0	0.0	10	2.1
Leaf	MS	3	0.2	0.0	2-17	1.3 - 2.7
Petiole	B5	3	0.2	0.0	3-17	1.9 - 2.5
Petiole	MS	3	0.2	0.0	3-34	0.9 - 3.2
Petiole	MS	6	0.2	0.4	3-34	1.8 - 5.2

Activity of MGT in green leaves increased several fold during the first few hours into the photoperiod (Frear & Swanson, unpublished). However, light was found to have no effect on the activity of MGT from cell cultures. Cell suspension cultures grown in B5-3 medium in total darkness for 15 days or for up to 141 days had MGT activities of about 5 units (mg protein)$^{-1}$ h^{-1}. Similar cultures exposed to light had activities of 0.5, 1.5 and 1.1 units (mg protein)$^{-1}$ h^{-1} at 0, 6 and 9 hours of light, respectively. Although these results led us to believe that light may be increasing MGT activity, several experiments indicated these values were within the experimental range of error, so an apparent increase due to light was not consistent.

MGT activity in cell suspension cultures grown in B5-3 medium reached maximum level 3 days after subculturing into the same medium and declined with age and soluble protein (Table 2). Growing cell cultures are actively dividing at 3 days (16,17) so MGT activity may be associated with cell division. Dry weight accumulation of tomato cell suspensions was maximum at 14 days, at which time MGT activity was about one-half that at 3 days. Cell cultures normally contain many senescent cells at that stage of development (16,17), so MGT activity was low. At 23 days MGT activity was not detected and the cultures had only 12.5% of the soluble protein content as at 3 days of culture (Table 2).

TABLE 2. MGT activity in 'Sheyenne' tomato cell suspension cultures as a function of growth.

Time	Soluble protein	MGT specific activity	
(d)	(mg (g FW)$^{-1}$)	(units h^{-1} (mg protein)$^{-1}$)	(units h^{-1} (g FW)$^{-1}$)
3	4.8	4.8	23
10	2.5	3.0	8
14	1.2	2.3	3
23	1.4	0.0	0

It is possible that the MGT enzyme in cell cultures is a totally different enzyme from that in green leaves, since their response to the light differs markedly. However, the isolated reaction product obtained with a crude tomato cell culture enzyme preparation was identified tentatively as a metribuzin N-glucoside conjugate by cochromatography (TLC and HPLC) with a reference conjugate isolated from metribuzin-treated tomato plants and from reactions with MGT extracted from tomato leaves (10). Fast atom bombardment (FAB) mass spectrometric analyses indicated that the reaction products from the different sources were identical. The MGT from cell cultures may be an isozyme of the MGT from green leaves. Other possibilities are that endogenous inhibitors exist in the cell cultures or that endogenous stimulators of activity exist in green leaves.

The results indicate that cell suspension cultures contain an enzyme that catalyzes the formation of an N-glucoside of metribuzin, but the enzyme levels are too low for the cultures to replace green plants as a source of enzyme for kinetic studies or for use in biotechnology. The lack of response to light and the apparent requirement for 2,4-D in cell culture are properties of the enzyme that appear to differ from that of green plants and need to be elucidated; especially since other auxins could not replace 2,4-D in this system.

REFERENCES

1 Ray, T.B. and Still, C.C. (1975) Pestic. Biochem. Physiol. 5, 171-177
2 Davis, D.G., Hodgson, R.H., Dusbabek, K.E. and Hoffer, B.L. (1978) Physiol. Plant. 44, 87-91
3 Dusky, J.A., Davis, D.G. and Shimabukuro, R.H. (1982) Physiol. Plant. 54, 490-494
4 Weed Science Society of America (1983) Herbicide Handbook, fifth edition, pp. 317-321, Champaign, IL, U.S.A.

5 Harrison, H.F., Bhatt, P. and Fassuliotis, G. (1983) Weed Sci. 31, 533-536
6 Oswald, T.H., Smith, A.F. and Phillips, D.V. (1978) Pestic. Biochem. Physiol. 8, 73-83
7 Devlin, D., Gealy, D.R. and Morrow, L.A. (1987) Weed Sci. 35, 741-745
8 Abusteit, E. O., Corbin, F.T., Schmitt, D.P., Burton, J.W., Worsham, A.D. and Thompson, L. (1985) Weed Sci. 33, 618-628
9 Klamath, E.E., Fedtke, C. and Kuhbauch, W.C. (1989) Weed Sci. 37. 517-520
10 Frear, D.S., Mansager, E.R., Swanson, H.R. and Tanaka, F.S. (1983) Pestic. Biochem. Physiol. 19, 270-281
11 Frear, D.S., Swanson, H.R. and Mansager, E.R. (1985) Pestic. Biochem. Physiol. 23, 56-65
12 Gamborg, O.L., Miller, R.A. and Ojima, K. (1968) Exper. Cell Res. 50, 151-158
13 Murashige, T. and Skoog, F. (1962) Physiol. Plant. 15, 473-497
14 Gawronski, S.W. (1983) Weed Sci. 31, 525-527
15 Stephenson, G.R., McLeod, J.E. and Phatak, S.C. (1976) Weed Sci. 24,161-165
16 Brezeanu, A. and Davis, D.G. (1979) Protoplasma 98, 163-176
17 Davis, D.G. and Brezeanu, A. (1979) Can. J. Bot. 57, 2006-2020

CONVERSION OF GLUCOSINOLATE SYNTHETIC PRECURSORS IN *BRASSICA* CULTURES

J.W.D. GROOTWASSINK, A.D. KOLENOVSKY, J.J. BALSEVICH, J.C.JAIN AND E.W. UNDERHILL
National Research Council, Plant Biotechnology Institute, Saskatoon, Saskatchewan, Canada, S7N 0W9.

1. INTRODUCTION

Basic glucosinolates (GS) are synthesized from amino acids in two phases: 1. carbon chain extension and 2. main group formation, comprising N-oxidation, decarboxylation, sulfur incorporation, glucosylation and sulfation (Fig. 1). Conventional breeding of low-GS oilseed varieties has relied exclusively on an undefined 'leaky' metabolic block in the chain extension, originally found in *Brassica napus* cv 'Bronowski'. We are exploring whether additional low-GS variability can be created by genetic modification of the enzyme system assembling the main group. One of the approaches being pursued is mutagenesis and isolation of biochemical mutant cell lines. Screening of tissue cultures based on endogenous GS content is not feasible because levels are extremely low or non-detectable. In contrast, the biosynthetic enzymes downstream from N-oxidation are expressed at similar levels in cultures as in whole plants [1]. Concomitantly, the lack of GS precursors could be overcome by supplementation of growth media with chemical equivalents. The enzymes for main group assembly have turned out to be non-specific for the side chain, thus allowing conversion of synthetic analogs [2]. Here we report the conversion of 2-nitrobenzaldoxime to its GS end-product and similar non-GS by-products. The stability of these artificial product pools was studied under different nutritional and physical conditions.

2. MATERIALS and METHODS

In vitro cultivation. Callus and cell cultures initiated from hypocotyl explants of *Brassica juncea* cv Cutlass were maintained on K-627-1 and B5 medium (Sigma) as described previously [3]. *B. nigra* cell cultures (kindly provided by Dr. G. Séguin-Swartz, Agriculture Canada, Saskatoon) were grown alternately in 2MS and 1B5 medium. Precursor feedings were carried out by direct addition of ethanol solutions of 2-nitrobenzaldoxime (Aldrich) to cultures.

Metabolite analysis. Cells were extracted in boiling water, filtered and the filtrate applied to DEAE-Sephadex A-25 (Pharmacia) columns. The columns were washed twice with water and loaded with purified arylsulfatase (Type H-1, Sigma). The sulfatase-loaded columns were kept overnight at room temperature and the desulfated metabolites eluted with water. All enzyme activities in the eluate were inactivated by heating in a boiling water bath. 2-Nitrophenol galactoside (ONPG) was added as internal standard. The eluate was subjected to HPLC separation and constituents detected by UV absorption at 226 nm [3]. Chemical identities were established by co-elution with authentic compounds, combined with specific enzymatic hydrolysis of samples with thioglucosidase and ß-glucosidase (Sigma) prior to HPLC. Proton NMR spectrometry was used to resolve structural details of the bonds between moieties of the conjugation products [3]).

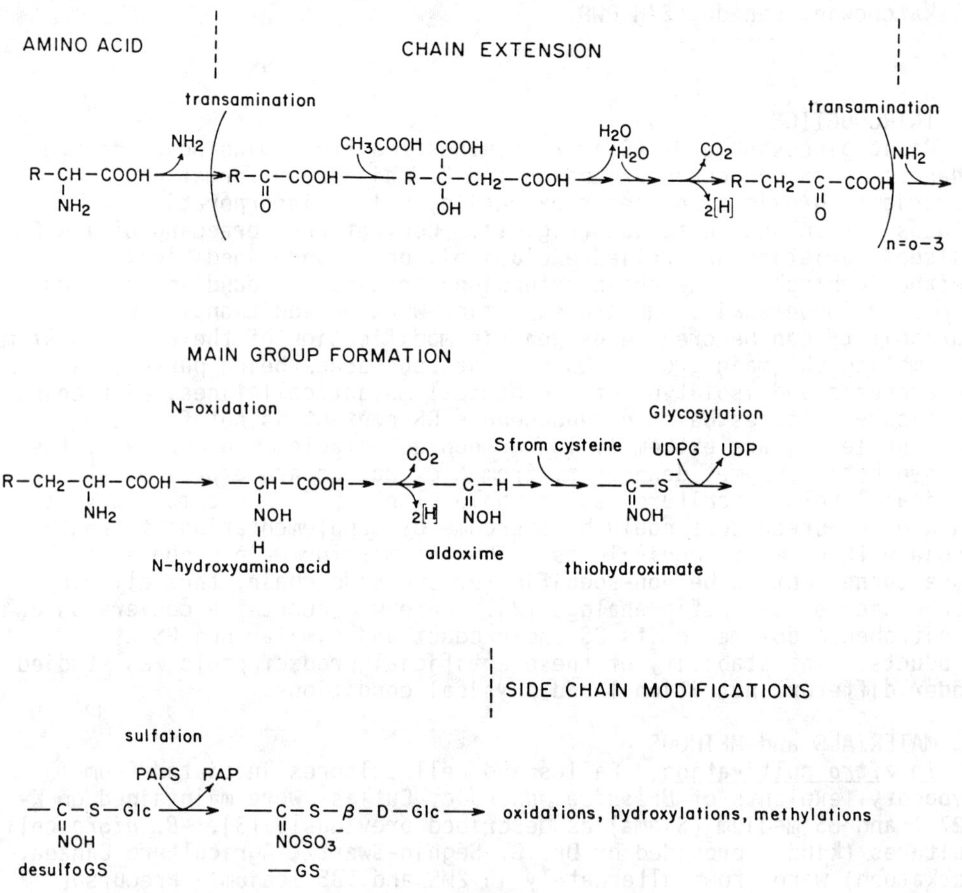

Fig. 1. Stylized glucosinolate (GS) biosynthetic pathway

3. RESULTS and DISCUSSION

Endogenous GS. Although aliphatic GS, e.g. 2-propenyl-, 3-butenyl- and 2-hydroxy-3-butenylGS derived from methionine, generally dominate in *Brassica* plants and seeds, indole GS derived from tryptophan dominate in tissue culture. In fact, aliphatics become undetectable in serial subcultures soon after initiation from explants. Medium manipulation has been unsuccessful in sustaining production. Indole GS levels also dropped during subculture. Production improvement was achieved on only one occasion, with *B. nigra*, but the response was transient (Fig. 2). Perturbations caused by changes in hormone and salt concentrations seemed responsible for temporary increases in indole GS.

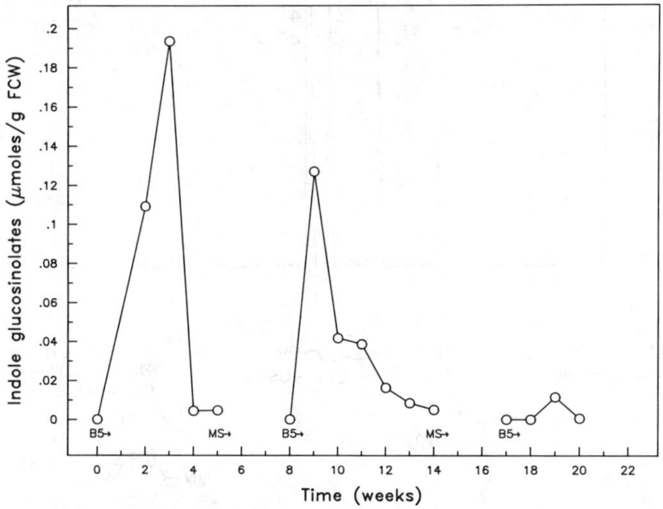

Fig. 2. Indole glucosinolate production in *B. nigra* suspension cultures grown successively in 2MS and 1B5 medium.

2-nitrobenzaldoxime feeding. Providing precursors in the form of excess methionine has not yielded any measurable levels of aliphatic GS in *Brassica* cell culture. At present therefore, marking of GS biosynthesis is limited to main group assembly by feeding intermediates past the amino acid stage [2]. Attention has focussed on the precursor 2-nitrobenzaldoxime for its ability to generate a chromogenic GS [4]. Besides conversion to 2-nitrophenylGS, other sulfated end-products are generated (Fig. 3). Isolation and structural analysis has shown that the aldoxime can be glucosylated directly followed by sulfation of the glucoside moiety. In addition, the aldoxime can be hydrolyzed and reduced to its alcohol, which is glucosylated and sulfated. The enzymes catalyzing these conjugation reactions seem to pre-exist, but are different from the activities of GS biosynthesis. It is not clear whether sulfatoglucosides are formed under normal physiological conditions or only during stress caused by feeding excess toxic aldoximes. Detoxification to sulfatoglucosides was found with a variety of aldoximes and in a range of crop species [3].

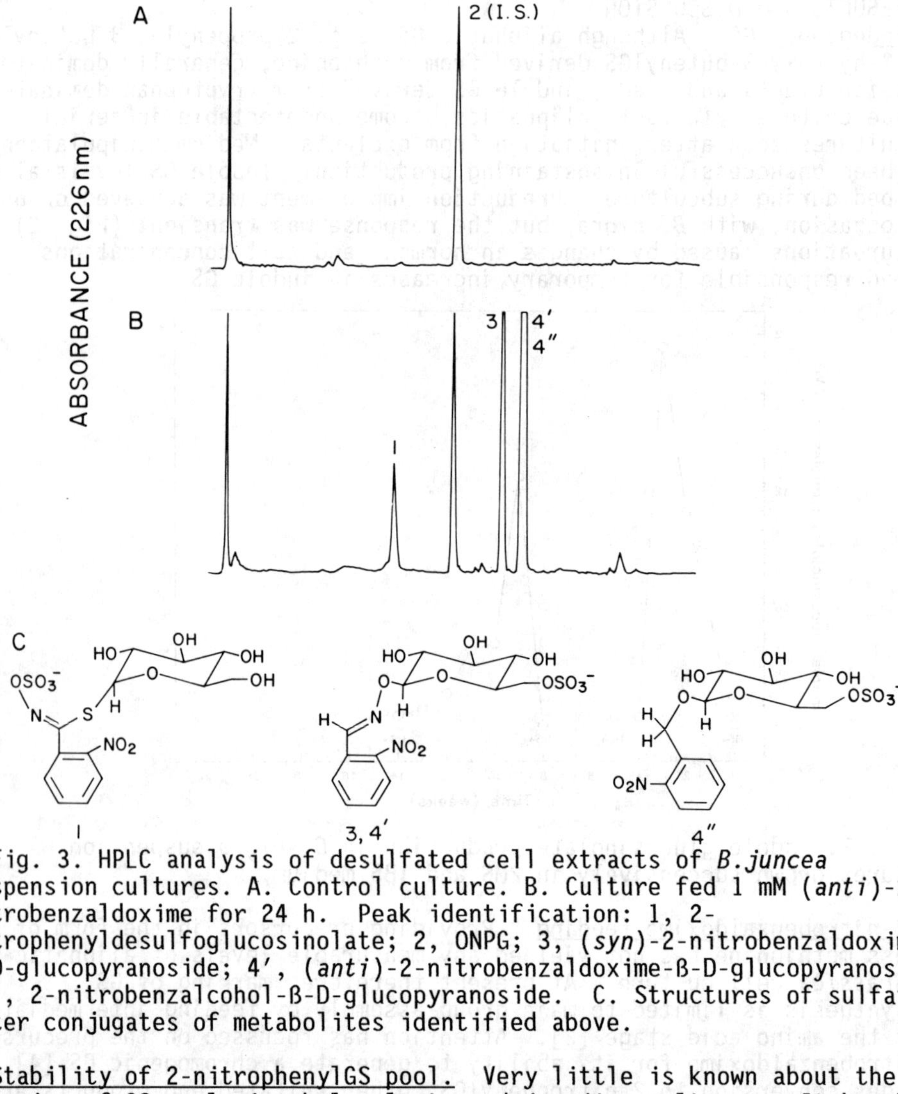

Fig. 3. HPLC analysis of desulfated cell extracts of *B.juncea* suspension cultures. A. Control culture. B. Culture fed 1 mM (*anti*)-2-nitrobenzaldoxime for 24 h. Peak identification: 1, 2-nitrophenyldesulfoglucosinolate; 2, ONPG; 3, (*syn*)-2-nitrobenzaldoxime-ß-D-glucopyranoside; 4′, (*anti*)-2-nitrobenzaldoxime-ß-D-glucopyranoside; 4′′, 2-nitrobenzalcohol-ß-D-glucopyranoside. C. Structures of sulfate ester conjugates of metabolites identified above.

Stability of 2-nitrophenylGS pool. Very little is known about the dynamics of GS pools in whole plants and *in vitro* cultures. It has been suggested that sulfate availability affects the production and turnover [5]. We have examined the metabolic stability by using cells pre-loaded with 2-phenylglucosinolate (and attendant sulfatoglucosides) as inoculum for media with a range of sulfate concentrations (Fig. 4 and 5). The results show that in the presence of excess sulfate, GS (and sulfatoglucoside) pools remained static. Thus, in the absence of *de novo* synthesis, the pools represent simple storage, perhaps for cellular isolation of toxic compounds. Under conditions of sulfate starvation (below 900 μM), cell growth occurred at the expense of sulfated metabolites.

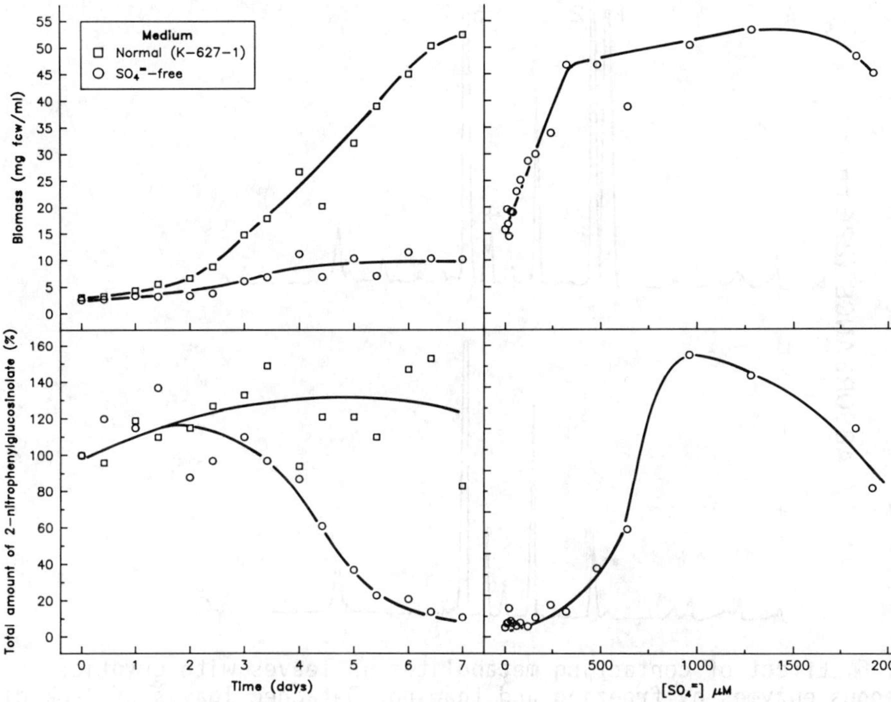

Fig. 4 Fig. 5

Fig.4. Maintenance of 2-nitrophenylGS pool during cultivation of *B. juncea* cells in normal and sulfate-free medium. Cells grown in K-627-1 medium were fed 2-nitrobenzaldoxime (0.4 mM) for 24 h, washed with and transferred to fresh K-627-1 medium containing normal levels of sulfate (1800 μM) or no sulfate.

Fig. 5. Effect of medium sulfate concentration on growth and maintenance of 2-nitrophenylGS pool in *B. juncea* cell culture. Cells were grown for 7 days after preparation as described in Fig. 4.

<u>Susceptibility of GS and sulfatoglucosides to endogenous leaf enzymes</u>. It is generally assumed that the function of GS lies in the breakdown products (cyanates and isothiocyanates) generated by the enzyme thioglucosidase (myrosinase). Mechanical disruption of plant tissue appears required for enzymatic action. Attempts to compromise cell integrity more gently with organic solvents, salts, dehydration and high temperature have failed to cause *in situ* degradation of GS. Physical cell disruption by freezing and thawing showed rapid disappearance of the artificial 2-nitrophenylGS and the endogenous 3-butenylGS in *B. campestris* leaves (Fig. 6). However, sulfatoglucosides were not affected, suggesting they accumulated as dead-end detoxification products, similar to glucosylated herbicides in resistant plants [6].

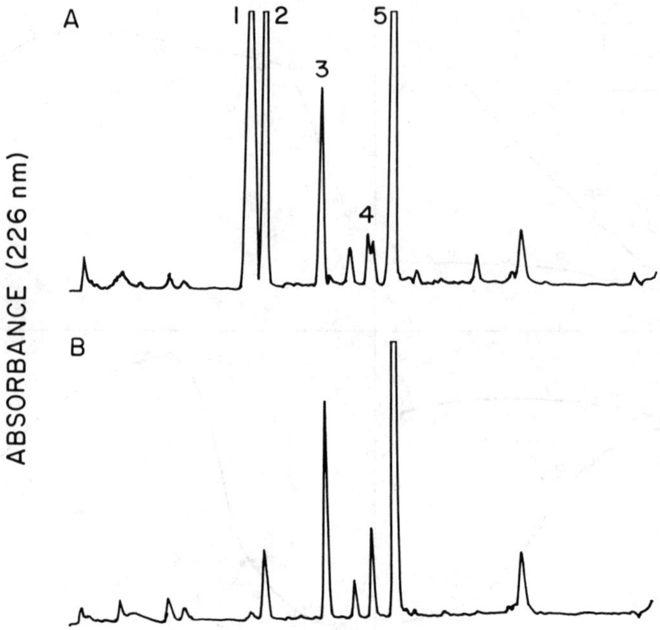

Fig. 6. Effect of contacting metabolites in leaves with cryptic endogenous enzymes by freezing and thawing. Detached leaves of 4-wk-old plantlets of *B. campestris* cv R-500 were fed 2-nitrobenzaldoxime (1 mM) for 3 days. A. Control leaves. B. Leaves subjected to two cycles of freezing/thawing followed by 15 min incubation at 30°C. HPLC peak identification: 1, 3-butenyldesulfoGS; 2, 2-nitrophenyldesulfoGS; 3, ONPG; 4, (*syn*)-2-nitrobenzaldoxime glucoside; 5, (*anti*)-2-nitrobenzaldoxime glucoside

4. REFERENCES

1. Jain, J.C., GrootWassink, J.W.D., Kolenovsky, A.D. and Underhill (1990) Phytochem. (in press)

2. GrootWassink, J.W.D., Kolenovsky, A.D., Jain, J.C., Michayluk, M.R., Chatson, K.B., Reed, D.W., Giblin, E.M. and Underhill, E.W. (1987) Crucifer Improvement Cooperative Proceedings, Crucifer Genetics Workshop IV, Madison p.30

3. GrootWassink, J.W.D., Balsevich, J.J. and Kolenovsky, A.D. (1990) Plant Sci. 66, 11-20

4. GrootWassink, J.W.D., Nelson, L.A.K., Kolenovsky, A.D. and Underhill E.W. (1988) Eucarpia Congress-Genetic Manipulation in Plant Breeding-Biotechnology for the Plant Breeder, Helsingor, Denmark.

5. Wetter, L.R., Giblin, E.M. and Underhill, E.W. (1988) Second Canadian Plant Tissue Culture and Genetic Engineering Workshop, Ottawa,, Canada

6. Frear, D.S., Mansager, E.R., Swanson, H.R. and Tanaka, F.S. (1983) Pesticide Biochem. Phys. 19, 270-281

POSSIBILITIES, PROBLEMS, AND PITFALLS OF LARGE-SCALE PLANT CELL CULTURES.

H.J.G. TEN HOOPEN, W.M. VAN GULIK, J.J. MEIJER.

Biotechnology Delft Leiden BDL
Project Group Plant Cell Biotechnology
Department of Biochemical Engineering, Delft University of Technology, Julianalaan 67, 2628 BC Delft, The Netherlands.

1. INTRODUCTION

There have been many studies published on aspects of large-scale culture of plant cells. However, large-scale culture has been defined quite differently by the various authors, ranging from a few liters to about 70 m³. In fact, it appears that some authors interpret large-scale cultures primarily as cultures directed to a commercial aim, whereas others emphasize 'large'. These interpretations do not always correlate. If somebody should force a break-through in the production of vincristine by cell cultures of *Catharanthus roseus* increasing the productivity per liter of culture to the level obtained with commercial shikonin production, a rather small 2-l bioreactor would be quite sufficient for commercial production. On the other hand, probably the largest bioreactors especially designed for plant cell cultures (1) are primarily aimed at process development.

In this paper, both aspect will be considered. In the section on 'possibilities' the commercial aspect will be discussed, and in the section on 'problems' the emphasis will be laid on 'large'. However, in both cases it is assumed that the equipment could be scaled-up; therefore stirred vessels or air-lift bioreactors are discussed but not shake flasks. In the section on 'pitfalls' all three types of culture vessels are involved.

2. POSSIBILITIES OF LARGE-SCALE CULTURES

Large-scale culture of plant cells may involve several practical problems, nevertheless it appears technically possible.

Already in 1977, Noguchi *et.al.* (2) cultivated tobacco cells in an impeller-stirred tank reactor of 20 m³. Schiel and Berlin (3) scaled-up cell cultures of *C. roseus* to volumes of 5 m³ using conventional reactors with flat-blade impellers. The German company DIVERSA constructed in 1986 a fermentation cascade consisting of five stirred tank fermenters of 0.075, 0.75, 7.5. 15, and 75 m³ (gross volume). This plant has been built to study all aspects of the economic production of secondary metabolites by large-scale cultivation of plant cells.

DIVERSA claims that it has designed a low-shear impeller system capable of providing adequate homogeneity and oxygen transfer. Several cell lines, including *Echinacea purpurea*, were cultivated on a scale of 60 m^3 without contamination problems (1).
Rather than the practicality the economic feasibility of the production of secondary metabolites on a large scale is much more a point for discussion. Various studies have been published in this field (4,5,6). They all reached the same conclusion: with the state of the art of current technology, only a few products are potential candidates for an economically feasible production of plant cell biotechnology. A second general conclusion is that the low productivity per volume is an important bottle-neck and fundamental research should be directed towards the enhancement of this productivity.Until now the number of commercial products from plant cell biotechnology is still very small.
An aspect disregarded in most studies is the prospects of the various product groups from plant cells. Most authors mention the fields of interest such as pharmaceuticals, fragrances, flavours, pigments, pesticides, pheromones and enzymes. However, the acceptance, regulatory problems and, costs for testing for products from a new technology vary strongly for the various product groups. This point could be illustrated by the fact that the first commercial product, shikonin, has been introduced primarily as a pigment and in cosmetics, and not as a pharmaceutical.

3. PROBLEMS OF LARGE SCALE CULTURES

To pin-point the technological problems involved in large-scale culturing of plant cells, the outline for designing a production process shown in figure 1, might be helpful. If it is decided on the basis of both fundamental and feasibility studies to design a large-scale process bioreactor, the necessary size must be estimated on the basis of required production volume and data on growth and production kinetics. The types of bioreactor and process must be chosen, and the behaviour of the process on a large scale predicted by scale-up studies. The design should be developed in interaction with the down-stream processing

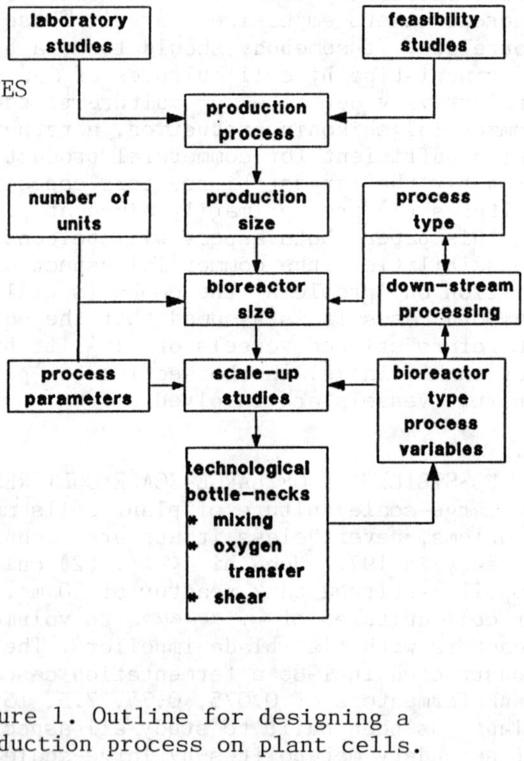

Figure 1. Outline for designing a production process on plant cells.

possibilities.
Problems in the following fields will be discussed here:

* growth and production kinetics
* process type
* bioreactor type
* scale-up
* downstream processing

3.1. Growth kinetics.

In the literature a lot of data has been reported on the growth and production kinetics of plant cell cultures. However, most of the data is not suitable for the design of a large-scale process because of the experimental set-up used to collect the data (mostly batch cultures in shake flasks). A more convenient methodology is continuous

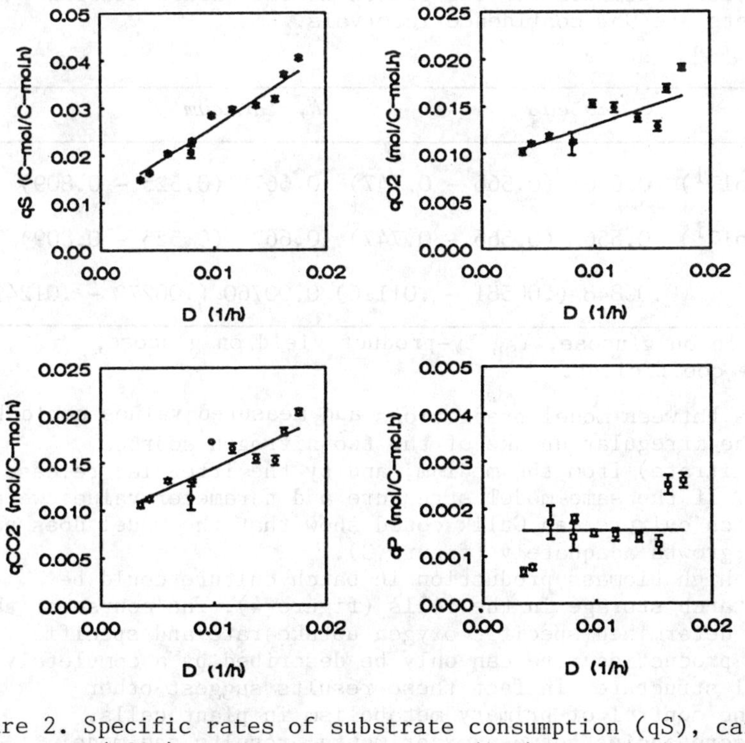

Figure 2. Specific rates of substrate consumption (qS), carbon dioxide production (qCO2), oxygen consumption (qO2), and (by)product formation (qP) during steady state growth as a function of dilution rate for *C. roseus*. Vertical bars represent the 95% confidence intervals. Solis lines represent the fit of the model to the data.

culture using the same type of bioreactor as planned for the large-scale process. In continuous culture, growth and production can be studied under steady state conditions. A few continuous culture studies on plant cells have been reported. However, in most cases the equipment did not perform properly, *e.g.* the biomass concentration in the effluent and in the bioreactor was not identical. Furthermore most studies were limited to one or a few steady states. Recently van Gulik (7) reported a comprehensive study on the growth kinetics of *C. roseus* and *Nicotiana tabacum* in glucose-limited continuous cultures. It was found in this study that growth in continuous culture could be described rather well with a currently available model for microbial growth. In figure 2 the results for *C.roseus* are shown.
Important process parameters in this model are the yield of biomass on glucose and the substrate uptake for maintenance. Cultures of *C. roseus* and *N. tabacum* showed comparable values for these parameters (Table 1).

TABLE 1. Parameter estimates for the growth model. Values between brackets are 95% confidence intervals.

Parameter	*C. roseus*	*N. tabacum*
Y_{sx} (C-mol.C-mol^{-1})	0.656 (0.566 - 0.747)	0.667 (0.525 - 0.809)
Y_{sp} (C-mol.C-mol^{-1})	0.656 (0.566 - 0.747)	0.667 (0.525 - 0.809)
m_s (h^{-1})	0.00848 (.00581 - .01116)	0.00760 (.00272 - .01249)

Y_{sx} biomass yield on glucose, Y_{sp} by-product yield on glucose, m_s maintenance coefficient.

Some deviations between model predictions and measured values could be explained by the irregular uptake of the two nitrogen sources (ammonium and nitrate) from the medium, and by the irregular release of by-products. If the same model structure and parameter values were used for a batch culture van Gulik could show that the model does not describe batch growth adequately (figure 3).
The unexpected high biomass production in batch culture could be explained by starch storage in the cells (figure 4). The course of the experimentally determined specific oxygen uptake rate and specific carbon dioxide production rate can only be described by a completely different model structure. In fact these results suggest other concepts for the control of primary metabolism in plant cells.
Lastly, the discrepancies between experimental results and model predictions suggested the occurence of at least two glucose uptake systems. These are a low-affinity system, active at high glucose concentrations in batch culture, and a high-affinity system active at low glucose concentrations in glucose-limited continuous culture. This phenomenon is also known, for instance, in bakers yeast (8).

It might be stressed that the by-product formation in table 1 concerns the growth-linked excretion of carbohydrates. Growth kinetics of secondary metabolites is not studied until now.

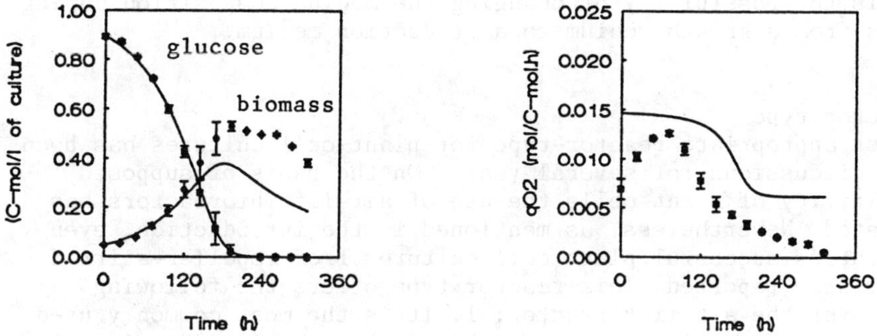

Figure 3. Growth of *C. roseus* cells in batch culture. Solid curves represent the model predictions (7).

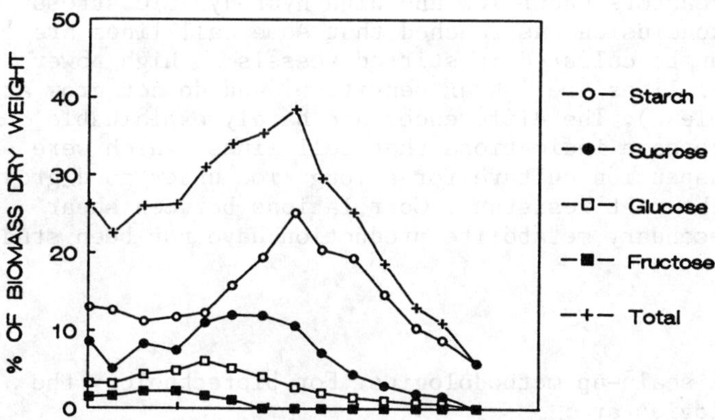

Figure 4. Changes in the amounts of intracellular carbohydrates of *C. roseus* cells during growth in batch culture (7).

3.2 Process type

Although continuous culture can theoretically provide the highest biomass productivity, this system is, in most cases, less acceptable for secondary metabolite production. In many cases product formation starts up in the stationary phase. Batch culture and, especially, fed-batch culture appears a more appropiate choice. In a fed-batch culture a higher biomass concentration can be obtained. An additional advantage is the possibility of changing the medium composition during the process from a growth medium to a production medium.

3.3 Bioreactor type

The most appropriate reactor-type for plant cell cultures has been subject of discussions for several years. On the basis of supposed shear-sensitivity of plant cells the use of air-lift bioreactors has been advocated. Nevertheless, as mentioned in the introduction, even in the seventies succesful plant cell cultures from impeller-stirred bioreactors were reported. This reactor type offers the following advantages over the air-lift reactor: 1. it is the most commonly used bioreactor in fermentation industry, which makes the introduction of a plant cell process more easily accepted and,

2. the stirred bioreactor has two control possibilities: stirrer speed and aeration rate instead of only aration rate in the case of the air-lift bioreactor.

Recently studies were published by Leckie (9) and Meijer (10) on shear sensitivity of plant cells. This type of research is complicated by the fact that the various shear forces in a bioreactor and the effects on the cells by these forces are difficult to investigate. Both studies therefore focussed on culturing several cell lines in stirred tank bioreactors under low and high hydrodynamic stress situations. The conclusion was reached that some cell lines are 'shear resistant' and can be cultured in stirred vessels at high power inputs. Other cell lines are 'shear sensitive' and do not grow at high power inputs (table 2). The differences are hardly explainable although there are some indications that cell lines, which were subcultured in suspension culture for a long time under good growth conditions, are the most resistant. Correlations between shear resistance and secondary metabolite production have not been studied yet.

3.4. Scaling up

There are several scale-up methodologies. For biotechnology the workable methodologies are:

the scale-down method in combination with some rules of thumb (6). In this methodology the ruling regime at large scale is determined either experimentally or by estimation using physical correlations and rules of thumb. Then the bottle-neck situation at large scale is created at laboratory scale. In this small-scale equipment the bottle-neck situation is studied and eventually obtained improvements are implemented at large-scale.

TABLE 2. Shear sensitivity of various cell lines (3-l vessel, 4.5 cm turbine impeller) (10)

Cell line	Impeller speed rpm	Culture type	Shear resistant
Catharanthus roseus	250	batch	+
C. roseus	1000	batch	+
C. roseus	250	continuous	+
C. roseus	1000	continuous	+
Nicotiana tabacum	250	batch	+
N. tabacum	1000	batch	+
Tabernaemontana divaricata	250	batch	+
T. divaricata	1000	batch	−
Cinchona robusta	250	batch	+
C. robusta	1000	batch	−

This methodology was applied on the simulation of a large-scale production process of secondary metabolite by plant cells. Adequate oxygen transfer to provide optimal growth was chosen as design criterion. On the basis of currently accepted process parameters from literature it could be shown that at a large scale mixing problems might be anticipated. Although there is no uniformity in the scale-up rules for shear effects, the assumption is acceptable that shear-resistant cell lines would withstand a still increased power input; therefore the mixing problems might be overcome.

3.5 Down-stream processing

This area is not the most intensively studied in plant cell biotechnology. This could be justified by the view that down-stream processing is not interesting as long as there is no economic production and by the idea that the existing procedures for extraction of products from plant material could be adopted. However it is a good strategy to aim at an interaction between development of the growth and production units and down-stream processing. Furthermore some interesting alternatives concerning release of products by plant cells and separation of the products by extraction or absorption during the production process have been reported (11,12). A special feature of these techniques could be stimulation of the production by lowering of the product concentration in the broth, although the involvement of product inhibition has not been proven.

4. PITFALLS OF LARGE-SCALE CULTURES

Severals pitfalls have already been mentioned above. In this paragraph one rather general pitfall will be stressed. Almost all research in plant cell biotechnology starts with experiments in petri dishes and shake flasks. With respect to suspension cultures a wealth of experimental results have been produced from shake flask cultures. However, if one wants to translate these results to a larger scale, starting up experiments in stirred vessels or air-lift fermenters problems arise. Biomass growth rate and product formation rates may change completely, either in a positive, or in a negative direction (13,14). This could be caused by the fact that shake flasks are aerated only through the surface and furthermore in most cases the overall gas exchange in the flask is hampered by the flask closures used by most researchers. In table 3 the composition of the gas phase in a shake flask closed with cotton wool and aluminium foil is shown during growth of a cell culture. This gas phase composition does not resemble the gas phase in a bioreactor. Furthermore one should realise that also the plant hormone ethene may have accumulated. Kumar (15) estimated up to 20 ppm ethene in shake flasks with growing plant cells. Therefore studies directed to commercial production of plant products should be performed in an early stage in bioreactors of the type expected to be used at large scale.

TABLE 3. Composition of the gas phase over 60 ml suspension of growing cells in a 300-ml flask (closure cotton wool and Al-foil)

Age of the culture	% oxygen	% carbon dioxide	dry weight, g/l
0	21	0	3
1	20	3	3
3	16	9	6
5	14	11	11.5
7	13	11	16.5

REFERENCES

1. BINE Project Info-service nr.13. (1988) Biotechnologische Nutzung von pflanzlichen Zellkulturen im technischen Maszstab. Fachinformationszentrum Energie, Physik, Mathematik. Karlsruhe, FRG.
2. Noguchi, M. *et. al.* (1977) in Plant tissue culture and its biotechnological application (W. Barz, E. Reinhard and M.H. Zenk, Eds.).pp.85-94, Springer Verlag, Berlin, FRG.
3. Schiel, O. and Berlin, J. (1987) Plant Cell, Tissue and Organ Culture 8,153-161.
4. Goldstein, W.E., Lasure, L.L. and Ingle, M.B. (1980) in Plant tissue culture as a source of biochemicals (J.E. Staba, ed.). pp.191-

5. Veltkamp, E., Breteler, H., Huizing, H.J., and Bertola, M.A. (1985) Plantebiotechnologie in Nederland. Nationale Raad voor Landbouwkundig Onderzoek, Den Haag, Nederland.
6. Ten Hoopen, H.J.G., Van Gulik, W.M., Meijer, J.J., and K.Ch.A.M. Luyben (1988) in Proceedings of the 8th International Biotechnology Symposium (Durand,G., Bobichon,L.and Florent,J. eds.). pp.179-192. Soc. Franc. Microbiol., Paris, France.
7. Van Gulik, W.M. (1990). Growth kinetics of plant cells in suspension culture. Thesis Delft, The Netherlands.
8. Postma, E. (1990). Regulation of sugar transport in yeasts; a continuous culture study. Thesis Delft, The Netherlands.
9. Leckie, F., Scragg, A.H. and Cliffe, K.C. Enzyme Microb. Technol. (submitted).
10. Meijer, J.J. (1989). Effects of hydrodynamic and chemical/osmotic stress on plant cells in a stirred bioreactor. Thesis Delft, The Netherlands. 11. Mavituna, F., Park, J.M., Wilkinson, A.K., and Williams, P.D. (1987) in Process possibilities for plant and animal cell cultures (Webb, C. and Mavituna, F. eds.). pp.93-115. Ellis Horwood Publ., UK.
12. Payne, G.F., Payne, N.N., Shuler, M.L., and Asada, M. (1988). Biotechnology Letters 10, 187-192.
13. Moo-Young, M. and Chisty, Y. (1988) in Proceedings of the 8th International Biotechnology Symposium (Durand,G., Bobichon,L.and Florent,J. eds.). pp.454-466. Soc. Franc. Microbiol., Paris, France.
14. Scragg, A.H., Morris, P., Allan, E.J., Bond, P., and Fowler, M.W. (1987). Enzyme Microb. Technol. 9, 619-624.
15. Kumar, P.P., Joy IV, R.W., and Thorpe, T.A. (1989). J.Plant Physiol. 135,592-596.

BIOSYNTHESIS OF SANGUINARINE BY ELICITATION OF SURFACE IMMOBILIZED CELLS OF PAPAVER SOMNIFERUM L.

W.G.W. KURZ, N.L. PAIVA[1], AND R.T. TYLER[2]

Plant Biotechnology Institute, National Research Council of Canada, Saskatoon, Saskatchewan, Canada S7N 0W9; ([1]Present address: The Noble Foundation, P.O. Box 2180, Ardmore, Oklahoma, USA 73402; [2]Present address: Dept. of Applied Microbiology and Food Science, University of Saskatchewan, Saskatoon, Saskatchewan, Canada S7N 0W0).

1. INTRODUCTION

Upon exposure to a variety of elicitors, suspension cultures of the Papaver somniferum L. cell line PBI #2009, initiated more than ten years ago (2), produce the benzophenanthridine alkaloids, sanguinarine and dihydrosanguinarine (1). Cells of this line, which when unelicited accumulate, at best, only trace quantities of these alkaloids, release a portion of the product into the culture medium, suggesting a possible long-term, continuous use of the cell material. Because of the potential use of sanguinarine as an antibiotic in oral hygiene products (3), an effort was made to develop efficient production methodology by eliciting and re-eliciting surface immobilized cells. The results of these investigations are presented here.

2. PROCEDURE

2.1 Materials and methods

2.1.1 Cell immobilization: A novel technique of surface immobilization was employed (4). Cells were immobilized on a fabric of loosely woven polyester fibres arranged in a spiral configuration on a stainless steel support frame, which was placed in a 2-litre glass vessel fitted with a lid having multiple inlet and outlet ports for reactor attachments. The air inlet was attached to a glass sparger extending down the center of the matrix structure with a magnetic stir bar below this structure providing medium circulation. The reactor was placed on a magnetic stir plate and the temperature (25°C) electronically controlled by either heating from the stir plate or cooling via coils wrapped around the reactor vessel. The cell suspension (2.5% v/v) to be immobilized was poured into the sterilized reactor which contained the selected culture medium (usually 2000 ml B5 with 0.1 mg/l NAA). To compensate for the low inoculum ratio and ensure the survival of the sensitive Papaver cells, 150 ml of preconditioned medium i.e. medium in which Papaver cells had been grown for 6 days, was added to the culture medium. Within 10-15 minutes, with stirring and aeration, the cells were immobilized by entrapment in pockets among the fibres or by fibres catching in crevices of cell aggregates, and within 6 days of culture the entire matrix was covered with cell material

2.1.2 Suspension cultures: For purposes of comparison, cell suspension cultures were grown in identical bioreactors; however, the stainless steel support frame and the polyester material were omitted. Instead, a glass draft tube (65mm x 80mm) was suspended by stainless steel wires in the center of the reactor and an air sparger extended to the base of the draft tube. A raised stir bar was used to eliminate grinding of cells against the reactor bottom. To prevent cell loss from the reactor during medium exchange procedures, extra coarse sintered glass filters were used to siphon spent medium from the reactor vessel.

2.1.3 Elicitor treatment: The general conditions of elicitor treatment have been described earlier (1). The eliciting agent used in the experiments described here was solubilized chitin, prepared by autoclaving a suspension of purified chitin (Sigma, C-3641) (10mg/ml H_2O) for 30 minutes followed by successive filtrations through Whatman #1 and Millipore 0.45μm filters and a further autoclaving for 25 minutes. 0.5 ml solubilized chitin/100 ml culture was used as eliciting agent in all experiments.

2.1.4 Alkaloid analysis: Sanguinarine and dihydrosanguinarine were extracted and determined according to procedures described earlier (5).

2.1.5 Papaver somniferum L. cv. Marianne cell line PBI #2009 was maintained in B5 medium (6) containing 20 g/l sucrose and supplemented as follows: 1B5C with 1 mg/l 2,4-dicholorophenoxyacetic acid (2,4-D) and 1 g/l casein hydrolyzate, or B5/0.1N with 0.1 mg/l α-naphthaleneacetic acid (NAA). Subculturing was performed at weekly intervals and the cultures maintained at 25°C in continuous dim light on rotary shakers (150 rpm). Sub-line PBI #2009-LA was derived by transferring B5C stock suspension cultures to liquid B5/0.1N medium for 12 passages. Thereafter, the culture was kept on solidified B5/0.1N medium and transferred at 2-3 week intervals by selecting firm and dense pieces of calli. After 6 months, this culture was returned to liquid medium and the culture maintained by selectively transferring large aggregates only.

2.1.6 Biomass estimation of immobilized cultures was made without removing the cells from the immobilization material. After disassembling the bioreactor, the entire frame was allowed to drain for 45 minutes, weighed and the wet fresh weight (wfw) determined by subtracting the weights of frame and polyester material. By several procedures, it was demonstrated that filtered fresh weight of cells was equal to 60% of wfw. After drying the entire assembly at 55°C for 48 hours, the uncorrected dry weight (udw) was determined. A correction factor, which was obtained by multiplying 40% of wfw by the total carbohydrate concentration in the medium at the time of harvest, was subtracted from the udw to obtain the corrected dry weight (dw).

2.1.7 Preparation of resin cartridge: Amberlite XAD-4 resin (Rohm and Haas) was precleaned by soaking in acidified MeOH (5 ml conc. HCl/l MeOH) overnight and rinsing in several volumes of double distilled water, followed by autoclaving in water for 1 hour. 180 g resin was poured as a slurry into a cartridge chamber (45 x 150 mm) and the cartridge autoclaved for 1 hr.

2.1.8 Elution of products from the resin was accomplished with several volumes of acidified MeOH. Most of the sanguinarine eluted within the first 500-1000 ml.

2.1.9 Cell viability was assessed by fluoresceine diacetate staining (7).

3. RESULTS AND DISCUSSION

The immobilization procedure, which has worked well with cell cultures from other species such as Catharanthus roseus, required the use of nurse medium to compensate for the low inoculum size in order to achieve successful growth in bioreactors. We also noticed that when very fine cell suspensions were used as inocula, the cells were killed upon elicitation. We, therefore, selected for a new sub-line (PBI #2009 LA) which grew in larger cell aggregates and which was used throughout this investigation. Care had to be taken, however, that stock cultures were transferred according to the procedures described in Methods, as this line tended to select toward faster growing and smaller cell aggregates, which if used as inoculum for bioreactors adversely affected the alkaloid yield (Table 1).

Table 1: Total sanguinarine production by Papaver somniferum cells immobilized on polyester fabric using as inoculum cells maintained as suspension culture for 2 or 3 months.

TIME IN SUSPENSION	PRODUCTION (MG/G FW)
2 mo.	3.6
3 mo.	0.1

(10 ml elicitor preparation/2000 ml culture, added at 7.7 days; medium exchanged after 2 days; cells harvested 6.5 days after exchange).

A comparison between suspended and immobilized cells both elicited 7.5 days after inoculation (Table 2), showed that sanguinarine production by immobilized cells was twice that of cell suspensions. While both cultures exhibited browning, the immobilized cells were fully retained on the matrix and the medium remained free of cell debris. In comparison, the medium of the suspension reactor contained considerable cell debris and cell mass adhered to the reactor wall. Suspension culture gave better product release; however, this may have been due primarily to cell breakage.

Table 2: Sanguinarine production by immobilized and suspended Papaver somniferum cells using "2-month-old", aggregated cells as inoculum.

	SANGUINARINE (mg)				PRODUCTION (mg/g fw)	FRESH WEIGHT (g)
	MEDIUM		CELLS	TOTAL		
	EXCHANGE	FINAL				
Immobilized	9	6	648 (97%)	690	3.6	180
Suspension	3	16	200 (89%)	230	1.6	125

(10 ml elicitor preparation/2000 ml culture, added at 7.7 days; medium exchanged after 2 days; cells harvested 6.5 days after exchange).

Time of addition of elicitor was another important factor in sanguinarine production. Table 3 shows results for bioreactors which were elicited 5.8, 7.0 and 8.5 days after inoculation. Two days after elicitation, the medium was exchanged and eight days after this exchange the reactors were harvested. The reactor elicited after 5.8 days produced the highest amount of sanguinarine and 50% of the cells were viable at the time of harvest. This production was more than three times that seen for elicitation at 7 days and eight times higher than elicitation at 8.5 days. Specific productivity was similarly affected. A fourth reactor inoculated at 7 days with especially large cell aggregates produced significantly more sanguinarine than a 7 day reactor inoculated with standard inoculum, which confirms the beneficial effect of larger cell aggregates in this type of a culture system. In all cases, however, more than 95% of the product was retained within the cell.

Table 3: Effect of time of addition of elicitor on final sanguinarine production.

TIME OF ADDITION (days after inoculation)	SANGUINARINE (mg) IN MEDIUM 2 DAYS	10 DAYS	IN CELLS 10 DAYS	TOTAL	PRODUCTION (mg/g fw)
5.8	8 (1%)	11 (1%)	822 (98%)	841	4.0
7.0	2 (1%)	2 (1%)	242 (99%)	246	1.0
8.5	1 (1%)	tr	106 (99%)	107	0.4
7.0*	7 (2%)	3 (1%)	344 (97%)	354	1.4

(10 ml elicitor preparation/2000 ml culture medium; exchanged 2 days after elicitation; cells harvested 10 days after elicitation).
*Large cell aggregates

Figures 1 and 2 show the time course of sanguinarine synthesis after elicitation. Accumulation during the first two days was minimal and increased steadily thereafter. By comparing the values for dry weight, fresh weight and specific productivity, it is apparent that the increase was due to an increase in specific production and not biomass alone. The cell viability at the end of these experiments was estimated to be 60%, which might be acceptable for batch suspension processes where all cell material is harvested. Such values are definitely too low for a semicontinuous immobilized process where continued high cell viability and product release is needed. When a reactor was elicited 3 days after inoculation, the total and specific production was only half of that in reactors elicited 7 days after inoculation; however more than 90% of the cells remained viable.

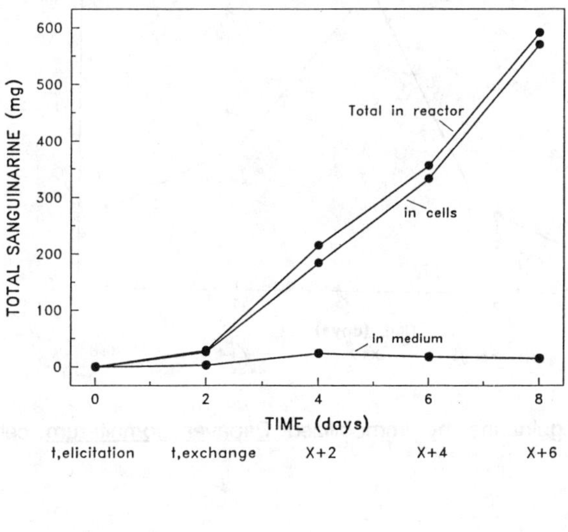

Figure 1

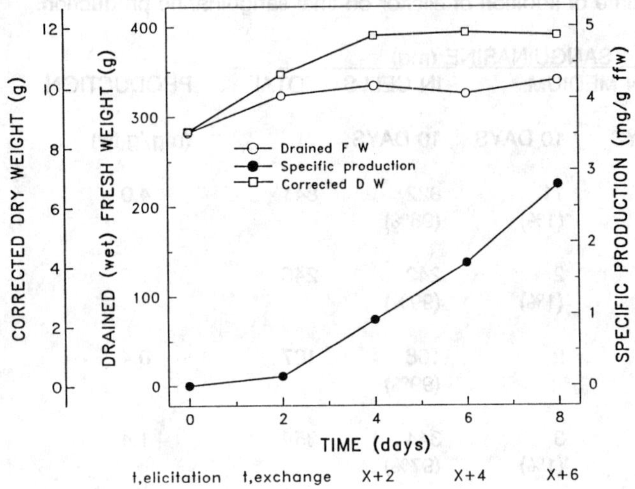

Figure 2

Figures 1 and 2: **Growth and sanguinarine production characteristics of immobilized Papaver somniferum cells in 2 litre bioreactors.**

Figure 3 shows that the peak of sanguinarine release into the medium, 34 mg/l, occurred 4 days after elicitation and declined rapidly thereafter. This might be due to re-absorption, since we have found that when sanguinarine was added to non-elicited cells in suspension culture, considerable absorption was taking place.

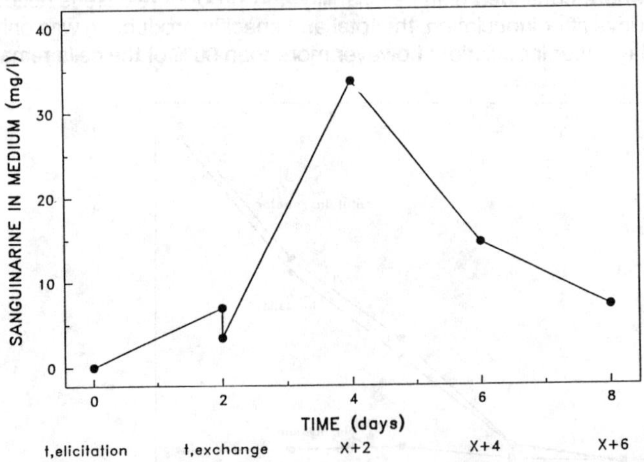

Figure 3: **Release of sanguinarine by immobilized Papaver somniferum cells in 2 litre bioreactors.**

To counteract this, multiple exchanges of medium at shorter time intervals were implemented (Table 4). In comparison with a control reactor, the percentage of alkaloid recovered from the medium (29% vs 6%) as well as the total alkaloid recovered were significantly higher. This method, however, had several process-related drawbacks including increased volume of media, a more dilute product stream, and lower cell viability, probably due to the stress induced by medium exchange.

Table 4: Effect of single versus multiple medium exchanges on sanguinarine release.

	MEDIUM	CELLS	TOTAL
MULTIPLE EXCHANGES AT 2,3,4,5 & 6 DAYS POST-ELICITATION	125 mg (29%)	305 mg (71%)	430 mg
SINGLE EXCHANGE AT 2 DAYS POST-ELICITATION	20 mg (6%)	311 mg (94%)	331 mg

Cells were harvested 8 days post-elicitation.

A second approach to increasing sanguinarine release involved the use of a hydrophobic resin to adsorb sanguinarine from the medium. Rather than draining and refilling the reactor, the medium was pumped via an external loop through the resin cartridge and recycled back to the reactor. A hydrophobic resin was chosen to adsorb not only sanguinarine but also toxic phenolics without affecting other medium components such as cell nutrients.

Table 5: Effect of product adsorption on resin and medium recycling on the release of sanguinarine.

	SANGUINARINE (mg)			
	MEDIUM	CELLS	RESIN	TOTAL
MEDIUM EXCHANGE ONLY	39 (9%)	376 (91%)	--	414
MEDIUM EXCHANGE PLUS PRODUCT ADSORPTION ON RESIN	9 (3%)	43 (18%)	191 (79%)	243

Table 5 shows that by this method, almost 80% of the total sanguinarine produced could be adsorbed onto the resin from where it could be eluted with methanol-HCl. The amount of product remaining in the cells was drastically reduced, the cells showed almost no browning, and their viability was close to 100%. Unfortunately, the total yield of sanguinarine was lower in the resin-treated reactor. Further optimization and re-elicitation should improve yields. Recent selections of new cell sub-lines have improved sanguinarine yields to 130 mg/g dw.

REFERENCES

1. Eilert, U., Kurz, W.G.W. and Constabel, F. (1985). J. Plant Physiol. 119, 65-76.
2. Tam, W.H.J., Constabel, F. and Kurz, W.G.W. (1980) Phytochemistry 19, 486-487.
3. Southard, G.L., Boulware, R.T., Walboru, D.R., Groznik, W.J., Thorne, E.E. and Yankell, S.L. (1984) J. Am. Dent. Assoc. 108, 338-341.
4. Archambault, J., Volesky, B. and Kurz, W.G.W. (1990). Biotech and Bioeng 35, 702-711.
5. Tyler, R.T., Eilert, U., Rijnders, C.O.M., Roewer, I.A. and Kurz, W.G.W. (1988) Plant Cell Reports 7, 410-413.
6. Gamborg, O.L., Miller, R.A. and Ojima, K. (1968) Exp. Cell Res. 50, 151-158.
7. Kartha, K.K. (1982) in Plant Tissue Culture Methods (Wetter, L.R. and Constabel F., eds.) pp. 25-31, National Research Council of Canada, Saskatoon, Sask. NRCC 19876.

THE EFFECT OF CONTINUOUS HIGH SHEAR STRESS ON PLANT CELL SUSPENSION CULTURES.

F. LECKIE[1], A. H. SCRAGG[1] and K. C. CLIFFE[2]

[1]Department of Molecular Biology and Biotechnology, Department of Mechanical and Process Engineering[2], University of Sheffield, Sheffield, S10 2TN, U.K.

1. INTRODUCTION

Plant cell suspension cultures have been shown to yield many different and economically important chemicals[1] and the development of a process to produce these compounds will require the mass cultivation of plant cells. Mass cultivation in a conventional stirred tank bioreactor involves mixing by use of an impeller. The impeller is the major source of shear stress in the bioreactor and is often quoted as the reason for the failure of the growth of cultures.[2] This failure has encouraged the use of pneumatically mixed bioreactors such as the airlift bioreactor.[3,4]

The effects of shear stress on plant cell suspensions has only recently been investigated[5] but only on the effects of short term exposures. The work presented here investigates the effects of long term shear stress on the growth and alkaloid accummulation (serpentine and ajmalicine) by suspension cultures of *Catharanthus roseus*.

2. MATERIALS AND METHODS

The *C.roseus* cell ID1 was used for all experiments and had been selected for high alkaloid accumulation.[6] The growth conditions, medium, alkaloid extraction and estimation have been described previously.[7] The cultures were grown in 3l and 12l stirred tank bioreactors as described.[8] In all cases the initial oxygen transfer coefficient (Kla) was set at $10.0h^{-1}$ and samples taken at intervals to determine dry weight, wet weight and alkaloid accumulation.[7]

3. RESULTS

The *C.roseus* culture was grown in a 12l stirred tank bioreactor at impeller speeds of 100, 300, 500 and 700 rpm. The growth and alkaloid accumulation data is given in Table 1, and is compared with data from shake flasks. As the impeller speed increased the growth rate was reduced but the maximum dry weights were little changed and compared well with those from shake flasks. Figure 1a compares the growth patterns of *C.roseus* grown in a shake flask and in a bioreactor at an impeller speed of 700 rpm, where it can be seen that the wet weights were most affected by the impeller speed.

Table 1 The growth and alkaloid accumulation data for cultures grown in bioreactors at increasing impeller speed

Speed rpm	Max wet weight g.l-1	Max dry weight g.l-1	Growth rate hr-1	Cellular alkaloid mg.l-1	Medium alkaloid mg.l-1	Total alkaloid mg.l-1	%	Productivity mg.l-1.day-1
100	200	12.5	0.013	8.9	13.2	22.1	60	2.2
300	239	13.6	0.012	15.0	14.2	29.2	49	2.9
500	177	8.6	0.008	10.3	1.9	12.2	16	1.2
700	142	9.5	0.010	7.0	2.0	9.0	22	0.9
Shake Flask	200	10.7	0.015	39.9	0.3	40.0	0.3	3.1

% = the percentage of the total alkaloid value contributed by the medium alkaloid value.

The effects of shear stress on alkaloid accumulation in the C.roseus cultures is given in Table 1. The total alkaloid yield was less in the bioreactor grown cultures when compared with shake flasks and the yield also decreased at the higher impeller speeds. However, alkaloid productivity at the impeller speed of 300 rpm were close to that achieved in shake flasks. There was little evidence that increased impeller speeds increased the release of alkaloids to the medium.

Table 2 The growth and alkaloid accumulation data for cultures grown in bioreactors fitted with different impeller designs at increasing impeller speed

Speed rpm	Impeller design	Max wet weight g.l-1	Max dry rate g.l-1	Growth alkaloid hr-1	Cellular alkaloid mg.l-1	Medium mg.l-1	Total mg.l-1	%	Productivity mg.l-1day-1
100	RT	200	12.5	0.013	8.9	13.2	22.1	60	2.1
100	30	245	12.1	0.018	8.6	3.0	11.6	26	0.6
100	60	2.2	11.9	0.016	23.1	1.7	24.8	7	1.5
300	RT	239	13.6	0.012	15.0	14.2	29.2	49	2.9
300	30	142	10.0	0.014	0.7	7.2	7.9	9	0.7
300	60	165	8.5	0.009	14.0	2.1	16.1	13	1.7
1000	RT	64	6.3	0.004	3.1	7.3	10.4	70	1.2
1000	30	73	6.7	0.005	2.8	2.7	5.5	49	0.5
1000	60	109	7.4	0.006	2.4	21.8	24.2	90	3.4

% = the percentage of the total alkaloid value contributed by the medium alkaloid value
RT = Rushton turbine, 30 = 30° inclined impeller, 60 = 60° inclined impeller

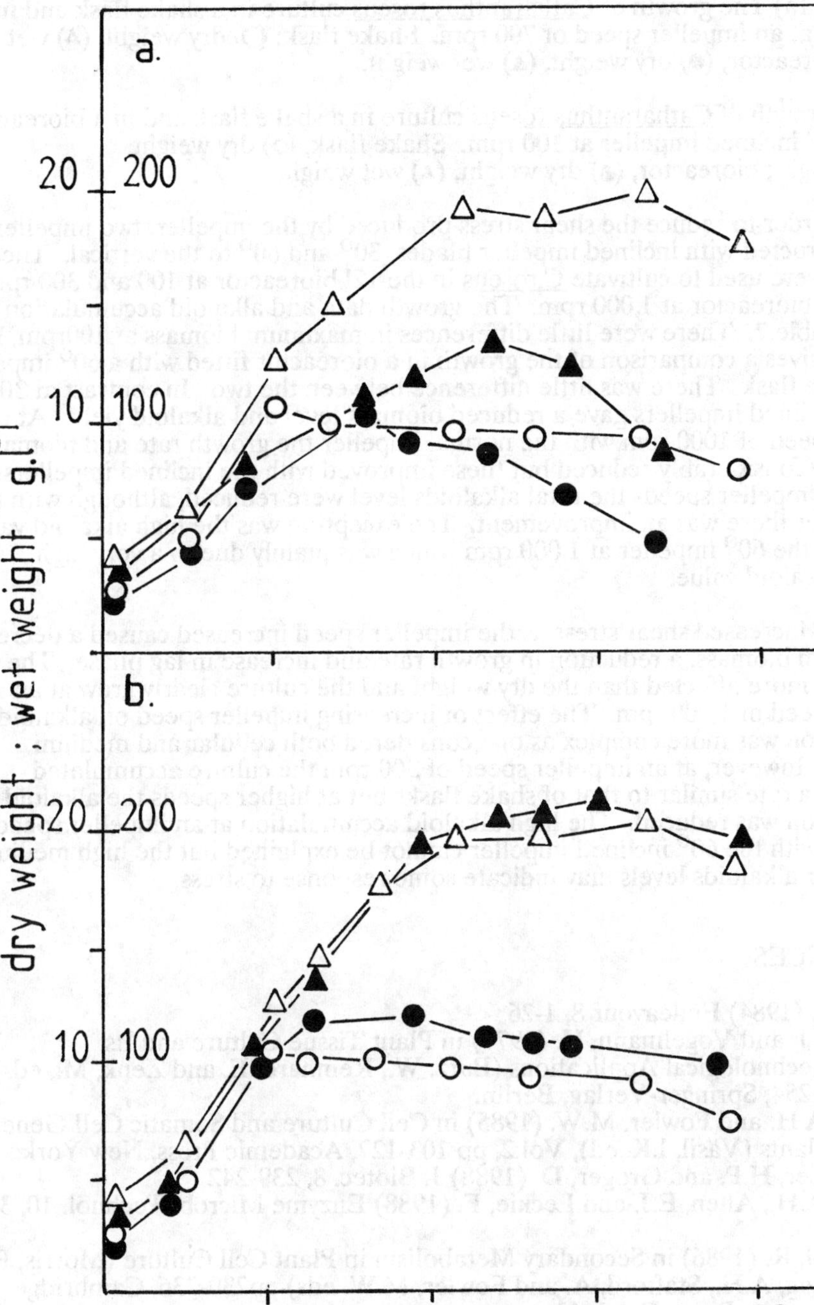
a.
20
200
10
100
dry weight · wet weight g.l-1
b.
20
200
10
100
200
400
time (hours)

FIGURE (1a) The growth of Catharanthus roseus culture in a shake flask and in a bioreactor at an impeller speed of 700 rpm. Shake flask; (○) dry weight, (△) wet weight; bioreactor, (●) dry weight, (▲) wet weight.

(1b) The growth of Catharanthus roseus culture in a shake flask and in a bioreactor with an 60° inclined impeller at 100 rpm. Shake flask, (○) dry weight, (△) wet weight; bioreactor, (●) dry weight, (▲) wet weight.

In order to reduce the shear stress produced by the impeller, two impellers were constructed with inclined impeller blades, 30° and 60° to the vertical. These impellers were used to cultivate C.roseus in the 12l bioreactor at 100 and 300 rpm and in a 3l bioreactor at 1,000 rpm. The growth data and alkaloid accumulation is given in Table 2. There were little differences in maximum biomass at 100rpm, but Figure 1b gives a comparison of the growth in a bioreactor fitted with a 60° impeller and a shake flask. There was little difference between the two. In contrast at 300 rpm the inclined impellers gave a reduced biomass level and alkaloid yield. At an impeller speed of 1000 rpm with the normal impeller the growth rate and biomass levels were considerably reduced but these improved with the inclined impellers. At the higher impeller speeds the total alkaloids level were reduced, although with the 60° impeller there was an improvement. The exception was the high alkaloid value found with the 60° impeller at 1,000 rpm which was mainly due to a very high medium alkaloid value.

The increased shear stress as the impeller speed increased caused a decrease in maximum biomass, a reduction in growth rate and increase in lag phase. The wet weight was more affected than the dry weight and the culture clearly grew at an impeller speed of 1,000 rpm. The effect of increasing impeller speed on alkaloid accumulation was more complex as one considered both cellular and medium alkaloids. However, at an impeller speed of 300 rpm the culture accumulated alkaloid at a rate similar to that of shake flasks but at higher speeds the alkaloid accumulation was reduced. The high alkaloid accumulation at an impeller speed of 1,000 rpm with the 60° inclined impeller cannot be explained but the high medium and cellular alkaloids levels may indicate some response to stress.

REFERENCES

1. Berlin, J. (1984) Endeavour 8, 1-26
2. Wagner, J. and Vogelmann, H. (1977) in Plant Tissue Culture and Its Biotechnological Applications (Barz, W., Reinhard, E. and Zenk, M., eds) pp 245-254, Springer-Verlag, Berlin.
3. Scragg, A.H. and Fowler, M.W. (1985) in Cell Culture and Somatic Cell Genetics of Plants (Vasil, I.K.,ed), Vol.2, pp 103-127, Academic Press, New York.
4. Schmauder, H.P. and Groger, D. (1988) J. Biotec. 8, 239-242
5. Scragg, A.H., Allen, E.J. and Leckie, F. (1988) Enzyme Microb. Technol. 10, 361-367
6. Cresswell, R. (1986) in Secondary Metabolism in Plant Cell Culture (Morris, P., Scragg, A.H., Stafford, A. and Fowler, M.W. eds) pp230-236, Cambridge University Press, Cambridge.
7. Scragg, A.H., Cresswell, R., Ashton, S., York, A., Bond, P. and Fowler, M.W. (1988) Enzyme Microb. Technol. 10, 532-536.

8. Boyson, F., Cliffe, K.R., Leckie, F. and Scragg, A.H. (1988) in 2nd International Conference on Bioreactor Fluid Dynamics (King, R., ed) pp 245-259, Elsevier, London.

APPLICATION OF THE LIQUID-IMPELLED LOOP REACTOR FOR THE PRODUCTION OF ANTHRAQUINONES BY PLANT CELL CULTURES

Buitelaar, R.M., Susaeta, I. and Tramper, J.
Wageningen Agricultural University
Department of Food Science, Food and Bioprocess Engineering Group
P.O. Box 8129, 6700 EV WAGENINGEN
The Netherlands

SUMMARY

The Liquid-impelled Loop Reactor (LLR), a new type of density-difference mixed bioreactor, can be an interesting tool for the application of biocatalysis in two-phase systems. The LLR combines the advantages of the use of an ALR with those of two-phase systems.

These advantages will be more evident when the biocatalysts used are not inhibited by the solvent used in this type of fermenter and when the product formed will preferentially be transfered into the organic phase.

The cells used in this study are *Rubia tinctorum* and *Morinda citrifolia*, that produce anthraquinones, rather hydrophobic compounds.

First, organic solvents having no harmful effect on growth or production as compared to a one-phase system were selected. Then growth and production kinetics of these cells in the presence of the selected organic solvents were characterized.

The results of these experiments were used to start the LLR with hexadecane as the second phase and to compare the results with the performance of an ALR.

INTRODUCTION

Cells and products

Plant cells can in potential be used for the production of many interesting compounds. Until now, however, not many processes based on plant cell cultures have been commercialized. One of the major reasons for this delay is the low productivity of plant cell cultures (1). Moreover, most of the products are stored intracellularly (1), which hampers work-up. When, on the contrary, they are excreted into the medium, the concentration often is very low, which also raises the costs of the down-stream processing. In the present study the compounds of interest are anthraquinones (see Figure 1), produced by cell cultures of *Rubia tinctorum* and *Morinda citrifolia.* Anthraquinones are natural dyes, but they also have antimicrobial properties. These suspension cultures produce a mixture of about 20 different anthraquinones. The anthraquinones produced by these cell cultures are highly substituted in the A-ring, making them moderately soluble in water (2). In general, these anthraquinones are released into the medium to a certain extent, which may cause some problems. The products may be rapidly degraded in the extracellular medium (3,4). Also, the product concentration in the fermentation broth may become inhibitory for the cells and therefore it is desirable to maintain only low levels of product in the medium. Concentration thus usually is necessary before purification can take place.

Figure 1

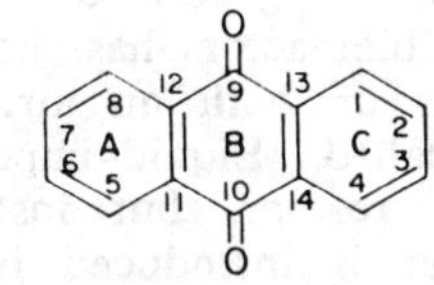

Anthraquinone

Solvents

To overcome these problems, many approaches are possible. Among them is the use of organic solvents in a two-phase system. The use of organic solvents in biocatalysis is becoming increasingly important (5). There are many possible advantages (6); some of them could also apply for secondary metabolite production with plant cells. For instance, substrate and/or product inhibition can be lowered or prevented and in situ concentration of the product can be achieved, which can lower the costs of down-stream processing. However, organic solvents can have a negative effect on the activity of the biocatalysts. As was indicated by Freeman (7) in an overview of solvent effects on multiphase biocatalysis, Brink and Tramper (8) were the first to attempt to structure the effects of organic solvents on the activity of biocatalysts. Laane et al. (9) proposed the use of the log P value of a solvent to predict the influence of the solvent on the activity retention of the biocatalyst. P is the partition coefficient of a compound over a standard octanol/water two-phase system. It can be determined experimentally, but it can also be calculated using the system of hydrophobic fragmental constants developed by Rekker and de Kort (10). From the results of many experiments with enzymes and microorganisms, Laane et al. (9) proposed some general rules: if log P of a solvent is lower than 2, the effect of the solvents on the biocatalyst will be harmful; if log P is higher than about 4, there will be no negative effect and solvents with a log P value between 2 and 4 will behave rather unpredictably. These rules can be applied for enzymes and various cell types, like yeast, bacterial and plant cells (11). In order to facilitate work-up, the partition of the produced anthraquinones into the organic phase will be an advantage. This partition will be promoted by matching the log P values of these products and the organic phase (9).

Immobilization

Immobilization of the cells can have a number of advantages. When slow growing cells are used in continuous systems, the cells can be retained in the bioreactor without limiting the possible dilution rates. Also separation of the immobilized cells from the broth is much easier. Immobilization prevents contact with the interface and thus the aggregation and clotting of the cells at the interface (12). Hence, if interfacial deactivation is the likely mechanism to occur, immobilization will provide protection against deactivation in a two-phase system. Immobilization can also have a stimulating effect on the production and/or excretion of secondary metabolites by plant cell cultures (13, 14, 15). Combining all these factors made it very interesting to test the possible effects of immobilization in calcium alginate on our system as well.

Bioreactor

When plant cells are to be used on a larger scale, the choice of an appropriate bioreactor is an important factor in the design of the process. The use of airlift (loop) fermenters has often been proposed for plant cell cultures (1), because of the low shear stress generated by these reactor types. In our department, a new type of bioreactor has been developed (patent pending), which combines the advantages of both an airlift loop reactor and the use of organic solvents. This so called Liquid-impelled Loop Reactor (LLR) is comparable to an airlift loop reactor, but instead of air, an organic solvent that is not miscible with water is introduced into the reactor (6). This solvent can either be lighter or heavier than water. The introduction of the solvent and the density difference will cause circulation and mixing in the bioreactor.

In this paper, the results are presented of studies with free and immobilized cells of *Rubia tinctorum* and *Morinda citrifolia.* First, suitable solvents were selected for this work. The influence of immobilization was then considered and finally, these results were used to start experiments with the LLR.

MATERIALS AND METHODS

Cells

The cell cultures of *Rubia tinctorum* and *Morinda citrifolia* were a kind gift from the Division Pharmacognosy of the Center for Biopharmaceutical Sciences in Leiden, The Netherlands. They were maintained by weekly subculturing in 250 ml Erlenmeyer flasks with 50 ml medium on a rotary shaker (100 rpm) in the dark at about 24°C. The medium used was Gamborg's B5 (16), supplemented with (per dm^3): 20 g sucrose, 2 mg 2,4-D, 0.5 mg NAA, 0.5 mg IAA and 0.2 mg kinetine for *Rubia tinctorum* and with (per dm^3): 20 g sucrose and 1.86 mg NAA for *Morinda citrifolia.*

Growth was determined by measuring the packed cell volume (17) in the culture flasks at regular time intervals.

Immobilization

The cells were immobilized in calcium alginate according to the procedure described elsewhere (18).

Solvent testing

The solvents tested are listed in Table I (see Results and Discussion section).

Toxicity tests were carried out by adding 1, 5 or 10% (v/v) of the solvent to the medium and following the growth and anthraquinone production with time and comparing them with controls without solvent.

Anthraquinone extraction and analysis

For the extraction of the anthraquinones, 60-80 mg of cells are extracted with 80% ethanol. The analysis is done by measuring the absorption of the extract on a spectrophotometer at 432 nm (19). Alizarine (1,2-dihydroxy-anthraquinone) is used as the reference compound to make calibration curves. To determine the amount of anthraquinones in the medium, x ml of medium was diluted with x ml 80% ethanol and the absorbance of this mixture was measured and compared with a calibration curve of alizarin in 40% ethanol.

Bioreactors

The LLR used in this study has been described elsewhere (6,18). Hexadecane was used as the second phase. Because the density of hexadecane is lower than that of medium, the hexadecane was introduced in the bottom section of the LLR with separation taking place in the top section. An airlift loop reactor was used as a control.

Table I
Solvents tested with their log P values and growth results for both cell types.

Solvent	log P	Growth 1%	Growth 5%	Growth 10%
Alkanes				
Hexane	3.5	+	-	-
Heptane	4.0	-	nd	nd
Octane	4.5	-	nd	nd
Decane	5.6	+	+	+
Dodecane	6.6	+	+	+
Hexadecane	8.8	+	+	+
Phthalates				
Diethylphthalate	3.3	-	nd	nd
Dibutylphthalate	5.4	+	-	-
Dioctylphthalate	9.6	+	+	+
Didecylphthalate	9.8	+	+	+
Alcohol				
Nonanol	3.4	-	nd	nd
Ether				
Diisopentylether	3.9	-	nd	nd
Perfluorcarbon				
FC40	about 11	+	+	+

Legend:
+ is growth
- is no growth
nd is not determined

RESULTS AND DISCUSSION

In Table I, the results of the growth experiments can be seen. As can be seen, the log P value of the solvent is a good indication for the toxicity of the solvent for the cells. All the solvents that were toxic to the cells had a log P value lower than 5 and all the solvents with a log P value higher than 5 were not toxic. Laane et al. (9) suggested that the difference between harmless and unpredictable behaviour would be at a log P value of 4, but we have observed before (18) that 5 is a better value. The solvents that had proved to be non-toxic in a concentration of 1% were tested in higher concentrations, 5 and 10%. Table I also shows the results of these experiments, and it can be concluded that most of the solvents that have no negative effect on cells in a concentration of 1% also have no negative effects when applied in higher concentrations.

Not only the cell growth is an important factor; also the production must be at an acceptable level. Moreover, for a succesful applicaton of a two-phase system, the partition coefficient of the product (C_{org}/C_{medium}) must be high. Those two factors were tested as well with the solvents that sustained growth. The production was not affected by adding a second phase that has no negative effects on the growth of the cells. The production of *Rubia tinctorum* is always about 4 μmol/g cells and the production of *Morinda citrifolia* is always at 1 μmol/g cells. When the cells are grown in medium, only 1-5% of the total anthraquinone production is excreted into the medium. This has been observed before with those two cell lines (20). When two-phase systems are used, a much higher part of the anthraquinones is excreted. The results of the experiments to determine the partition coefficient are given in Table II.

Table II
Partition coefficients for alizarine in the tested solvents and their densities.

Solvent	Partition coefficient (C_{org}/C_{medium})	Density (kg/m^3)
Hexane	2.4	0.66
Heptane	2.1	0.68
Octane	2.1	0.70
Decane	2.1	0.73
Dodecane	1.9	0.75
Hexadecane	2.0	0.77
Diethylphthalate	inf*	1.23
Dibutylphthalate	>4^	1.05
Dioctylphthalate	>4^	0.99
Didecylphthalate	>4^	0.96
Nonanol	inf*	0.85
Diisopentylether	1.3	0.77
FC40	0.7	1.87

* : all the alizarine goes into the organic phase
^ : no phase separation occurred, so no quantitative measurement was possible

For the final selection of suitable solvents, the density difference with aqueous medium was also taken into account. Because the circulation in the LLR is driven by the density difference of the two phases, this difference must be as high as possible in order to obtain the best mixing and phase separation. For this reason, the phthalates cannot be used in the LLR. FC40 has a very favourable density difference with medium, but the anthraquinones do not dissolve substantially in this organic phase. So, hexadecane has been used for further studies.
The LLR was tested with free and immobilized cells of *Rubia tinctorum* and *Morinda citrifolia.* As a control, an ALR was used. The results of these experiments show that the LLR is an interesting system for the study of the production of secondary metabolites with plant cell cultures. Many aspects of the process still have to be optimized, like the ratio medium/solvent, batch vs. continuous flow for both phases, and so on.

REFERENCES

1 Scragg, A.H. and Fowler, M.W., in: Cell culture and somatic cell genetics of plants, Constabel, F. and Vasil, I.K. (eds.), vol. 2, 103-126

2 Robins, R.J. and Rhodes, M.J.C., 1986, Appl. Microbiol. Biotechnol., 24: 35-41

3 Wink, M. and Hartmann, T., 1982, Z. Naturforsch. 37: 369-375

4 Robins, R.J., Furze, J.M. and Rhodes, M.J.C., 1985, Phytochemistry 24:709-714

5 Laane, C., Tramper, J. and Lilly, M.D. (eds.), 1987, Biocatalysis in organic media, Elsevier, Amsterdam

6 Tramper, J., Wolters, I. and Verlaan, P., 1987, in: Biocatalysis in organic media, Laane, C., Tramper, J. and Lilly, M.D. (eds.), 311-316

7 Freeman, A., 1986, Trends in Biotechnology, vol. 4, no. 3 [26], 53-54

8 Brink, L.E.S. and Tramper, J., 1985, Biotechn. Bioeng., vol. XXVII, 1258-1269

9 Laane, C., Boeren, S., Vos, K. and Veeger, C., 1987, Biotechnology and Bioengineering, vol. XXX, 81-87

10 Rekker, R.F. and de Kort, H.M., 1979, Eur. J. Med. Chem.-Chimica Therapeutica, 14, no. 6, 479-488

11 Buitelaar, R.M., Hulst, A.C. and Tramper, J., 1988, Biotechn. Techn., vol 2, no 2, 109-114

12 Brink, L.E.S. and Tramper, J., 1985, in: Johnson, A. (ed.), 'Modelling and control of biotechnological processes', Proceedings of the 1st IFAC symposium, Noordwijkerhout, The Netherlands, Pergamon Press, Oxford, 111-117

13 Asada,M. and Shuler, M.L., 1989, Appl. Microbiol. Biotechnol., 30, 475-481

14 Wichers, H.J., Malingré, T.M. and Huizing, H.J., 1983, Planta, 158, 482-486

15 Bringi, V. and Shuler, M.L., 1990,in: Bont, J.A.M. de, Visser, J., Mattiasson, B. and Tramper, J. (eds.), 'Physiology of immobilized cells', Elsevier, Amsterdam

16 Gamborg O.L., Miller, R.A. and Ojima, K., 1968, Exp. Cell Res., 50:151-158

17 Gilissen, L.J.W., Hänisch ten Cate, C.H. and Keen, B., 1983, Plant Cell Reports, 2: 232-235

18 Buitelaar, R.M., Vermuë, M.H., Schlatmann, J.E. and Tramper, J., 1990, submitted

19 Zenk, M.H., El-Shagi, H. and Schulte, U., 1975, Planta med., suppl., 79-101

20 Robins, R.J., Payne, J. and Rhodes, M.J.C., 1986, Phytochemistry, vol. 25 no 10, 2327-2334

THE CULTIVATION OF TRANSFORMED ROOTS FROM LABORATORY TO PILOT PLANT

P.D.G.WILSON[1], M.G.HILTON[2], P.T.H.MEEHAN[1], C.R.WASPE[1], and M.J.C. RHODES[2]
[1]Biological Process Engineering Group, [2]Plant Biotechnology Group, AFRC Institute of Food Research, Colney Lane, Norwich, NR4 7UA, U.K.

1. INTRODUCTION

Root cultures formed following infection of explant material with *Agrobacterium rhizogenes* have been proposed as a system for the *in vitro* production of valuable plant metabolites. A major hurdle to commercial exploitation by this method is the development of bioreactor systems for the large scale cultivation of these so-called "hairy roots".

We have investigated the growth of transformed roots in various laboratory-scale bioreactors, and have developed a pilot plant research bioreactor in order to address problems of scale-up that would be difficult to solve at the laboratory scale. This paper describes a two litre bioreactor for detailed study of growth kinetics, a 10 litre system based on a modified stirred tank and a 500 litre pilot plant.

2. DROPLET PHASE BIOREACTOR SYSTEM

A two litre hairy root bioreactor system has been developed to study the growth kinetics of transformed roots. A schematic diagram of the reactor system is shown in Figure 1. Growth medium is recycled from a two litre magnetically stirred vessel (L.H. Fermentation Ltd., U.K.) via a peristaltic pump to a liquid distributor in the top of the growth vessel. The growth vessel has a capacity of 2 litres and is constructed from standard glass pipe components (Corning Glass Ltd., U.K.) and a stainless steel head plate (L.H. Fermentation Ltd., U.K.). The liquid droplets formed by the distributor, trickle over the growing roots, which are supported by a stainless steel mesh. Upon reaching the base of the growth vessel, the liquid is returned to the medium vessel by means of another peristaltic pump. The growth vessel is maintained free of excess liquid by setting the drain pump to a higher flow rate than the inlet pump. Air is aspirated through both the growth vessel and the medium vessel. The growth vessel is

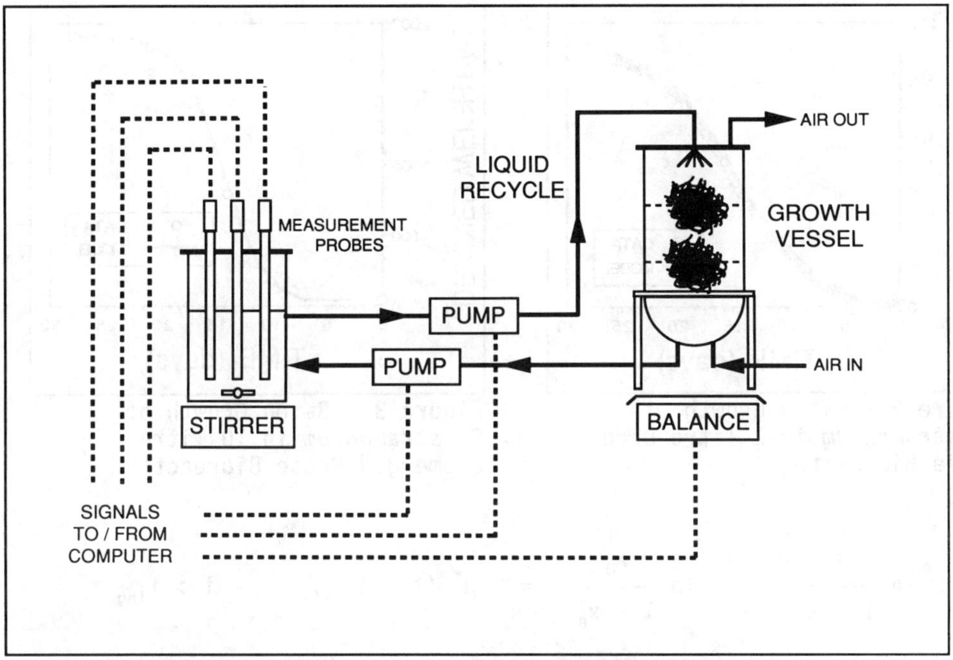

Figure 1 - Schematic Diagram of 2 litre Droplet Phase Bioreactor

permanently situated on an electronic balance. This allows the weight of roots and surface liquid to be continuously measured, knowing the weight of the empty vessel. Both peristaltic pumps and the electronic balance are connected to a process control computer system (Magus-08, Measurement Systems Ltd., U.K.). At predetermined times (typically every six hours), the feed pump is switched off, allowing the liquid on the surface of the roots to drain and be returned to the medium vessel. Measurement of the mass of the roots during this draining process has shown that a constant weight is obtained within 15 minutes. After this time, the mass of the drained roots is recorded by the computer system, and the feed pump restarted. Measurement probes from the system (pH, dissolved oxygen content of the medium, and the temperature in each vessel) are also connected to the computer system, allowing accurate monitoring of environmental parameters during the growth cycle.

Figure 2 shows the batch growth of a transformed root culture of *Datura stramonium*. The growth medium comprised 3% (w/v) sucrose and half strength Gamborg's B5 salts. Growth was carried out at 25°C. The growth of the roots can be well represented by a logistics growth model of the form :

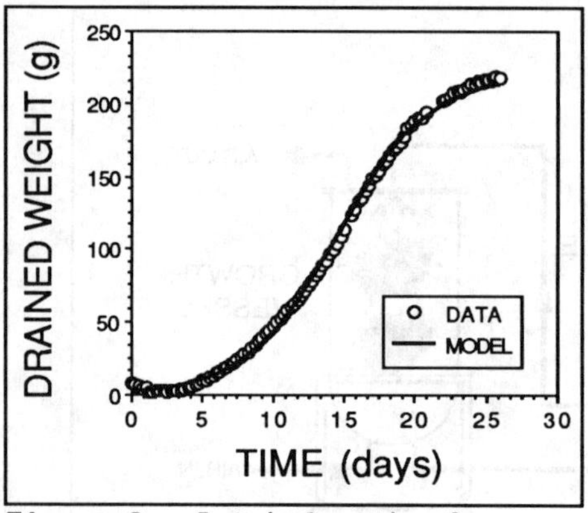

Figure 2 - Batch Growth of *D. stramonium* in 2 litre Droplet Phase Bioreactor

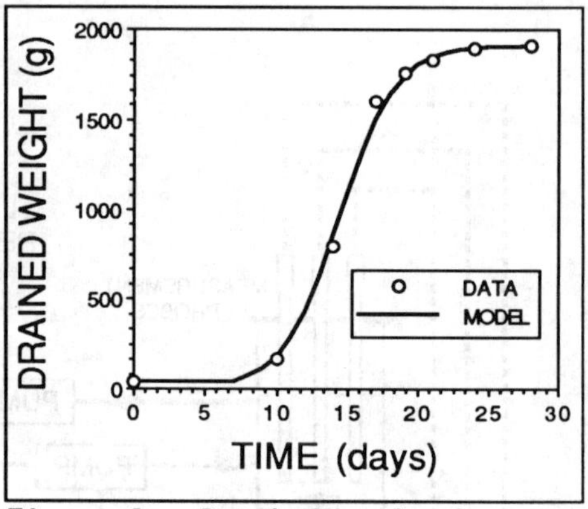

Figure 3 - Batch Growth of *D. stramonium* in 10 litre Submerged Phase Bioreactor

$$\ln \frac{x}{1 - x} - \ln \frac{x_0}{1 - x_0} = \mu^* (t - t_{lag}) \qquad t > t_{lag}$$

$$x = x_0 \qquad t \leq t_{lag}$$

where x is the cell mass expressed as a fraction of the final mass, x_0 is the initial fractional cell mass, μ^* is a growth rate parameter and t is the time from inoculation. t_{lag} represents a time lag before growth, should this be present. Using a linearising transformation as indicated by the form of this equation, growth parameters for the data in Figure 2 were determined. These values, together with biomass yield and alkaloid productivity data from the experiment are presented in Table I.

3. SUBMERGED PHASE BIOREACTOR SYSTEM

A submerged culture bioreactor has also been used for the cultivation of transformed roots, comprising a cylindrical glass vessel with a single radial flow impeller situated at the base. A stainless steel mesh prevents contact of the roots with the impeller and provides immobilisation sites. Transformed roots of *D. stramonium* were grown in the vessel under conditions identical to those described above. The growth of roots was followed by periodically draining the liquid from the vessel and weighing the bioreactor. The weight of roots could thus be determined from knowledge of the weight of the empty vessel. This method, along with a more detailed description of the vessel is described elsewhere [1]. Data from this experiment are

Table I - Results of Batch Growth Experiments in Droplet and Submerged Fermenters

PARAMETER	DROPLET	SUBMERGED	UNITS
μ^*	0.31	0.51	day^{-1}
t_{lag}	0	7	days
Biomass Yield	123	121	g WW (litre medium)$^{-1}$
	10.3	9.8	g DW (litre medium)$^{-1}$
Hyoscyamine	500	460	μg (g WW)$^{-1}$

WW = Wet Weight (Drained of free liquid, but not blotted)
DW = Dry Weight (Dried to constant weight at 110°C)

shown in Figure 3. A logistics growth model was also fitted to these data. Estimates of the model parameters and yields from the culture are presented in Table I. Compared with the droplet reactor described above, the growth rate was higher, although there was a significant lag before growth began. The explanation for the differences in these two bioreactor systems may be due to a medium conditioning phenomenon, and is currently under investigation.

4. PILOT PLANT ROOT BIOREACTOR

Based on our experience of growing transformed roots at the laboratory scale [1-4], we have identified the main problems of scale-up to be (i) the aseptic transfer of root material between inoculum and growth vessels, (ii) obtaining an even distribution of roots throughout the growth vessel, and (iii), the subsequent harvesting of the biomass. In order to tackle these problems, we have designed a bioreactor pilot plant specifically for the growth of transformed roots. The fermenter was designed to be run as a droplet or mist-phase reactor, but could also be used for submerged culture. The system comprises a 500 litre growth vessel, connected to a 50 litre medium holding vessel in a similar arrangement to that shown in Figure 1 for the two litre system. Growth medium is fed from the 50 litre vessel via a diaphragm pump to two sets of spray nozzles located immediately below the headplate of the growth vessel. Each set of spray nozzles produces a different droplet size, and may be fed independently, depending on the growth requirements of the roots. To ensure even distribution of the inoculum, the growth vessel contains an immobilisation matrix, comprised of a series of stainless steel barbs, to which roots may attach. The barbs are linked together in the form of a chain, which may easily be removed from the biomass after harvesting. The matrix is illustrated schematically in Figure 4, and described in more detail in a recent patent application [5]. The growth vessel itself (Figure 5) has an aspect ratio of 2:1, and has tapered walls (the larger diameter at the top)

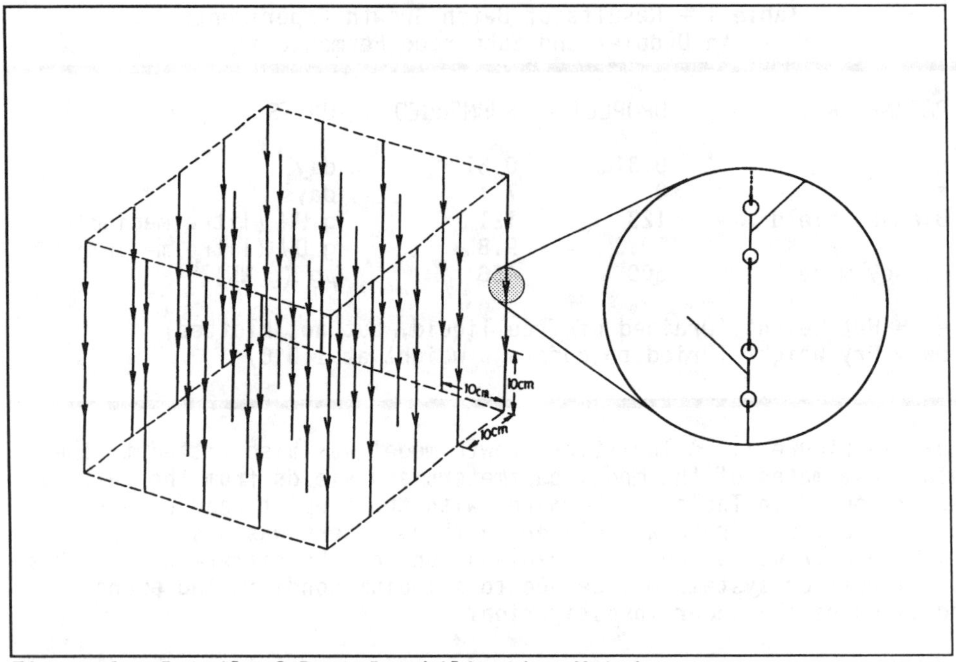

Figure 4 - Detail of Root Immobilisation Matrix

Figure 5 - 500 litre Pilot Scale Bioreactor

to facilitate the harvesting of densely packed biomass, as achieved at the laboratory scale [1]. Sightglasses are provided, covering the full height of the reactor to enable root distribution and growth to be observed directly.

A result of running the vessel as a droplet or mist-phase reactor is that it is possible to measure the extent of growth on-line by weighing the vessel contents. To this end, the entire vessel is mounted on a series of load cells, with all connections to the vessel being made by flexible hoses. This allows the contents of the vessel to be weighed to a precision of 100g. Sterile air may be supplied to the vessel through a number of ports. Initial estimates of the oxygen mass transfer rate indicate that it is so high as not to be limiting. This is perhaps not surprising, considering the large interfacial area of the droplets. Liquid is returned to the medium holding vessel by means of a peristaltic pump.

The pilot plant bioreactor is currently being evaluated, and results from system will be published as soon as they are available.

5. ACKNOWLEDGEMENTS

We are most grateful to the Institute's Engineering staff for their work on this project, to our Photography Department for producing the Figures, to Ms. A. Peerless and Ms. E. Bent for alkaloid analysis, to Mr. K. Painting for assistance with Figure 4, and to Mr. R.H. Thirkettle for his contribution to the design of the vessel.

6. REFERENCES

[1] Hilton, M.G. and Rhodes, M.J.C. (1990), 'Growth and hyoscyamine production of hairy root cultures of *Datura stramonium* in a modified stirred tank reactor', Appl. Microbiol. Biotechnol. (In Press)

[2] Rhodes, M.J.C., Hilton, M.G., Parr, A.J., Hamill, J.D., and Robins, R.J., (1986), Biotech. Lett., (8) 415-420

[3] Wilson, P.D.G., Hilton, M.G., Robins, R.J., and Rhodes, M.J.C. (1987) In: Moody, G.W. and Baker, P.B. (Eds.) Bioreactors and Biotransformations. Elsevier, London, pp 38-51

[4] Hilton, M.G., Wilson, P.D.G., Robins, R.J. and Rhodes, M.J.C.(1988) In: Robins, R.J. and Rhodes M.J.C. (Eds) Manipulating Secondary Metabolism in Culture, Cambridge University Press, Cambridge, U.K., pp 239-245

[5] Wilson, P.D.G., Hilton, M.G., Steer, D.C., Waspe, C.R., Robins, R.J., Rhodes, M.J.C. (1989), PCT Patent Application PCT/GB89/00510

PRODUCTION OF DEACETYLLANATOSIDE C BY DIGITALIS LANATA CELL CULTURES

WOLFGANG KREIS & ERNST REINHARD, PHARMAZEUTISCHES INSTITUT, AUF DER MORGENSTELLE 8, D-7400 TÜBINGEN, FRG

1. INTRODUCTION

Digitalis lanata cell suspension cultures are not able to synthesize cardiac glycosides de novo but can be used to modify exogenous ones. Among the conversion reactions observed, the regio- and stereospecific 12β-hydroxylation of digitoxin derivatives to their respective digoxin derivatives is of special interest (1). In recent years, research carried out by Reinhard and co-workers focussed on a process in which β-methyldigitoxin was used as the substrate for 12β-hydroxylation. The product, β-methyldigoxin, appeared in the spent medium where levels of 500-700 mg dm^{-3} could be reached. A semi-continuous process for the production of β-methyldigoxin was devised and run successfully in a 300 dm^3 airlift bioreactor (2).

In 1986, we reported for the first time on the efficient 12β-hydroxylation of digitoxin and the development of a two-stage method for the production of deacetyllanatoside C, a digoxin derivative (DC, see Fig.1) (3). DC produced upon incubation with digitoxin is not excreted into the culture medium but instead accumulates in the cells (up to 2.2% of the dry weight). Maximal production rates were achieved in 8% glucose solution and cultivation at high cell densities was found conducive to high DC yields. Other factors, like the moment of transfer of cells into production medium, or the moment and mode of substrate supply were also investigated and optimized (4).

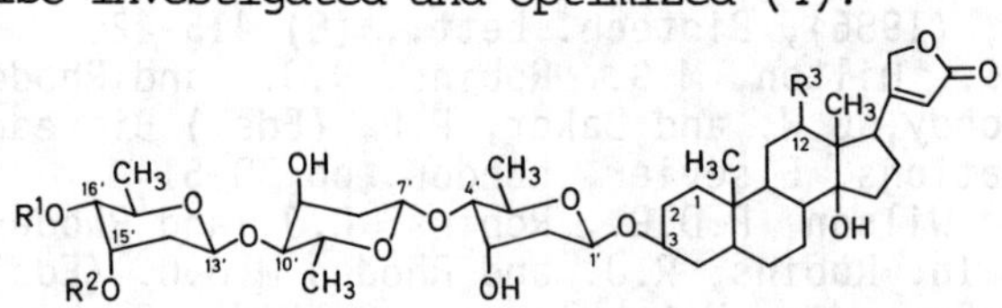

FIGURE 1. Structures of the cardenolides mentioned in the text.

$R^1 = R^2 = R^3 = H$:	Digitoxin
$R^1 = R^2 = H$, $R^3 = OH$:	Digoxin
$R^1 = CH_3$, $R^2 = R^3 = H$:	β-Methyldigitoxin
$R^1 = CH_3$, $R^2 = H$, $R^3 = OH$:	β-Methyldigoxin
R^1 = glucosyl-, $R^2 = R^3 = H$:	Purpureaglycoside A
R^1 = glucosyl-, $R^2 = H$, $R^3 = OH$:	Deacetyllanatoside C

With a view to improve the DC process further, we conducted a series of selection experiments in order to establish stable high-yield cell lines. The experiments which yielded the cell line W.1.4 are described here. This particular cell line was used in biotransformation experiments on the 20 dm^3 scale (5) and was also employed as the cell culture material in a two-stage process for the semicontinuous production of DC (6).

2. PROCEDURE

2.1 Materials and methods

2.1.1 Culture conditions: The growth media used were based on MS medium (4, 6). Cell suspensions (300 cm^3) were grown in 1 dm^3 Erlenmeyer flasks on orbital shakers in the dark and were maintained as described (4, 7). Growth and medium parameters were determined as described in previous publications (6, 8).

2.1.2 Cell-aggregate cloning: The *D. lanata* cell line K 3 OHD (4) was used as the starting material for 2 selection experiments carried out in 1986. The repeated cell-aggregate selection method described by Yamamoto et al. (9) was employed to produce variant D. lanata cell lines (6).

2.1.3 Test 1: Callus pieces, 2 g (fresh weight) each, were cultivated in 20 cm^3 modified MS medium, termed OH medium (4). Digitoxin (100 mg dm^{-3}) was added and after 4 d of incubation the cells were harvested for cardenolide determination (10). Cell lines with high DC levels were propagated as stock callus cultures on agarized (0.8%) OH medium under continuous dim light at 21°C. Suspension cultures were produced from high-yield callus, as well as from randomly-selected low-yield ones.

2.1.4 Test 2: Suspension-cultured cells, 8 g (fresh weight), were transferred to 30 cm^3 of 8% glucose solution, pH 5.5 and digitoxin (500 mg dm^{-3}) was added after 3 d of incubation. After another 4 d, the suspension was mixed with an equal amount of methanol, the mixture then sonified for 30 min at 40°C and the lysate analysed for its cardenolide pattern.

2.1.5 Bioreactors: The bioreactor experiments were carried out in 20 dm^3 airlift bioreactors. Details on the bioreactor configuration and bioreactor-related equipment have been published elsewhere (2, 6, 8, 11).

3. RESULTS AND DISCUSSION

3.1. Two-stage process for the semicontinuous production of DC by *D. lanata* cells: The bioreactor set-up used as well as details on this novel cultivation mode were published recently (4). *D. lanata* cells were grown in a 20 dm^3 airlift bioreactor (termed growth reactor; working volume 12 dm^3) for 7 d, after which 9 dm^3 of the suspension were pumped into a second bioreactor (termed production reactor; working volume 17 dm^3) containing 8 dm^3 of 17% glucose. The suspension left in the growth reactor was diluted with fresh culture medium in order to initiate the next growth cycle. After another 7 d, the cell suspension in the production reactor was harvested for product isolation, whereas the cell suspension in the growth reactor was treated as described above. Growth and product formation, as well as the changes in several medium parameters were monitored during the whole experiment (Fig.2). Table 1 summarizes the conditions chosen for semicontinuous DC production in the two-stage process devised (6).

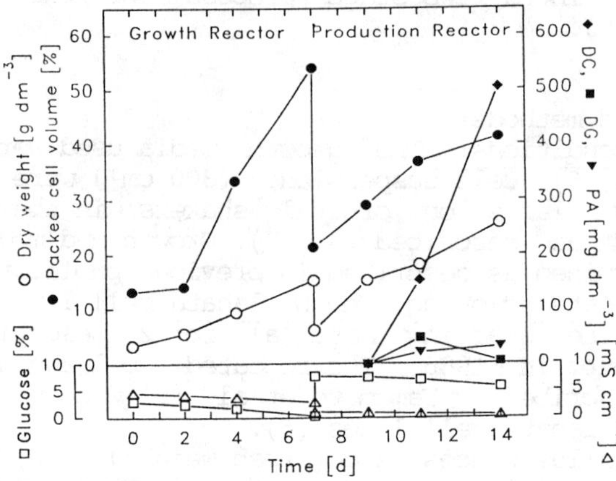

FIGURE 2. Semicontinuous production of deacetyllanatoside C in 20 dm^3 airlift bioreactors. DC, deacetyllanatoside C; DG, digoxin; PA, purpureaglycoside A.

TABLE 1. Conditions for semicontinuous production of deacetyllanatoside C by Digitalis lanata cells cultivated in 20 dm^3 airlift bioreactors.

Cell line used:		W.1.4 (cell-aggregate cloning)
Growth	Cycle:	7 d
	Medium:	OH-medium (modified MS)
	Inoculum:	3.25-3.75 g (dry weight) dm^{-3}
	Aeration:	Sterile air: vvm = 0.27-0.42
Production	Cycle:	7 d
	Medium:	8% glucose, pH 5.5
	Inoculum:	5.2-6.0 g (dry weight) dm^{-3}
	Aeration:	Sterile air: vvm = 0.39-0.49
	Pre-incubation phase:	2 d
	Substrate supply:	700-750 mg dm^{-3}

3.2. Biotransformation characteristics of D. lanata cell lines derived from line K 3 OHD by cell-aggregate cloning: Since the selection and characterization of an appropriate high-yield cell line was an important facet of the establishment of our two-stage process for DC production, the selection procedure used and the results obtained will be described in some detail.

3.2.1 Selection protocol. Two selection experiments were carried out employing the cell-aggregate cloning method (10). The D. lanata line K 3 OHD (4) was used as the starting material. Fine cell aggregates were plated on agar medium and during this first cloning step, an average plating efficiency of 68% was obtained. Repeated cloning resulted in increased plating efficiency; on and after the third cloning step, all of the respective sub-clones survived. Our cloning experiments yielded more than 2,000 well-growing calli. Considerable variation was found among the new cell lines, with respect to callus growth, colour and friability. Eighty callus lines were selected and checked for their ability to biotransform digitoxin (Test 1, see PROCEDURE). The cardenolide patterns varied considerably and only 7 of the aggregate clones tested yielded more DC than the mother cell line (Fig.3). After transfer to liquid medium 7 out of 32 clones investigated did not grow and eventually died, including the high-yield clones W.1.6 and W.2.4.8.

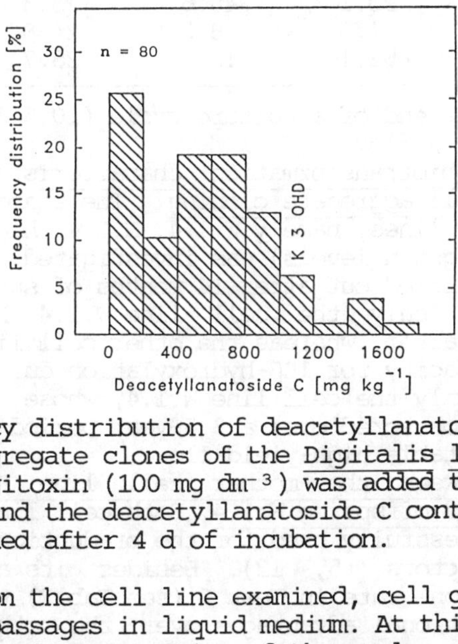

FIGURE 3. Frequency distribution of deacetyllanatoside C formation in cell-aggregate clones of the Digitalis lanata cell line K 3 OHD. Digitoxin (100 mg dm^{-3}) was added to submerged callus pieces and the deacetyllanatoside C content of the cells determined after 4 d of incubation.

Depending on the cell line examined, cell growth had stabilized after 6-10 passages in liquid medium. At this point of time, the biotransformation capacities of the newly established suspension cultures were determined (Test 2, see PROCEDURE). Cell lines which showed high DC formation in Test 1 generally formed high amounts of the target compound when tested in the two-stage production system. However, perhaps because of the relatively small sample size and the fact that different test systems were employed, no clear correlation was seen ($n = 25$, $r = 0.477$). The data scored in Table 2 illustrate that a) only few aggregate clones showed higher 12β-hydroxylation rates than the mother cell line (W.1.4, H.2.4, W.8, W.8.5) and b) clone-specific biotransformation patterns were obtained, with cell lines producing mainly DC (e.g. H.2.4), digoxin (e.g. W.8.5) or purpureaglycoside A (e.g. W.2.4.1).

TABLE 2. Biotransformation of digitoxin (500 mg dm^{-3}) by aggregate clones of Digitalis lanata K 3 OHD (Test 2, see PROCEDURE).

Cell line	Relative cardenolide content (%)				Packed cell volume[1] (%)
	Digitoxin	Purpurea-glycoside A	Digoxin	Deacetyl-lanatoside C	
K 3 OHD	5.0	15.0	25.4	55.6	37
W.1.1	0.9	18.0	38.4	42.7	25
W.1.4	0.0	6.6	32.8	60.6	28
H.2.4	1.0	8.5	9.2	81.3	27
H.3.3	3.6	7.9	10.5	78.0	27
W.8.5	2.8	12.2	55.7	29.3	28
W.10.5	8.0	20.1	48.6	23.3	26
W.2.4.7	6.2	61.0	8.0	24.8	36
W.2.4.1	5.8	52.1	13.4	28.7	--

[1] determined at the end of a culture cycle (10.5 d).

3.2.2 Stability of biotransformation characteristics in cell lines obtained by cell-aggregate cloning. The stability of the most promising cell lines, namely W.1.4, H.2.4 (high DC levels) and W.8.5 (high digoxin levels) was investigated. Biotransformation experiments carried out after 14 months of sub-cultivation (5) indicated that only the cell line W.1.4 had retained its high-yield character, whereas the other cell lines had partially lost their capacity for 12β-hydroxylation during prolonged sub-cultivation. Only the cell line W.1.4, whose biotransformation capacity was monitored over a 3.5-year period (6), can hence be regarded as a stable high-yield clone.

3.2.3 Comparison between the mother cell line K 3 OHD and its aggregate-cloned offspring W.1.4. Both cell lines, W.1.4 and K 3 OHD, were successfully used for the production of DC in 20 dm^3 airlift bioreactors (5, 12). Besides its about 15% higher 12β-hydroxylation rate in Test 2 (see Tab.2) the selected line W.1.4 has additional advantages over the mother cell line K 3 OHD: The highest digitoxin level that K 3 OHD was able to biotransform completely was 750 mg dm^{-3}, 618 mg dm^{-3} being hydroxylated to DC and digoxin after 11 d of incubation (4). Not only was the cell line W.1.4 able to convert 900 mg dm^{-3} digitoxin, 725 mg dm^{-3} being biotransformed to 12β-hydroxylated compounds after 5 d of incubation, but also the initial hydroxylation rate was found twice as high as that achieved with K 3 OHD. Moreover, the pre-incubation phase, i.e. the period of cultivation in production medium before addition of substrate, could be shortened to 2 d when using line W.1.4 (6), instead of 3 d which were necessary for line K 3 OHD (4).

Repeated cell-aggregate cloning has proven to be a simple and efficient method for the selection of high-yield cell lines (e.g. 9, 13, 14). It is, however, much easier to check cell clones for high product levels (colour, fluorescence, immunoassay, TLC, GC, HPLC) than to screen them for improved biotransformation potential, since this is sort of a "hidden" character, which can not be evaluated directly. The test system used in the selection protocol described here was based on digitoxin biotransformation experiments conducted in an early stage of callus growth. From the data obtained, the new cell lines' biotransformation capacities could be evaluated. Although the selection protocol employed yielded the high-producing cell line W.1.4. it had some short-comings. Relatively large sample sizes were necessary and the biotransformation tests took a long time. Therefore, we have meanwhile developed other evaluation methods, including enzyme assays and micro-biotransformation assays requiring by far less material and time (W. Kreis, Th. Baumann, E. Schiller, unpublished).

The cell line W.1.4 had maintained its high-yield character over long periods of time. This was an important prerequisite for its successful use in long-term biotransformation experiments, some of which lasted more than 100 d (6). Using cell line W.1.4, 12β-hydroxylation rates as high as 2.1 mg dm^{-3} h^{-1} were achieved in long-term bioreactor experiments (6) with maximal conversion rates of 12 mg dm^{-3} h^{-1}. These are the highest 12β-hydroxylation rates achieved so far with free or immobilized D. lanata cells.

4. ADDENDUM

4.1 Acknowledgement: The financial support from the Fonds der chemischen Industrie is gratefully acknowledged. We thank Mr. D. Schwantag for his expert technical assistance.

REFERENCES

1 Reinhard, E., Boy, H.M., Kaiser, F. (1974) Planta Med. Suppl. 163-168

2 Reinhard, E., Kreis, W., Barthlen, U., Helmbold, U. (1989) Biotech. Bioeng. 34, 502-508

3 Kreis, W., Reinhard, E. (1986) Planta Med. 51, 418-419

4 Kreis, W., Reinhard, E. (1988) Planta Med. 53, 95-100

5 Kreis, W., Zhu, W., Reinhard, E. (1989) Biotech. Lett. 11, 325-330

6 Kreis, W., Reinhard, E. (1990) J. Biotech. (in press)

7 Kreis, W., Reinhard, E. (1985) J. Plant Physiol. 121, 385-390

8 Alfermann, A.W., Bergmann, W., Figur, C., Schwantag, D., Schuller, I., Reinhard, E. (1983) in Plant Biotechnology (Mantell, S.H, Smith, H., eds) pp. 67-74, Cambridge Univ. Press, Cambridge

9 Yamamoto, Y., Mizuguchi, R., Yamada, Y. (1982) Theor. Appl. Genet. 61, 113-116

10 Kreis, W., May, U., Reinhard, E. (1986) Plant Cell Rep. 5 442-445

11 Wahl, J. (1978) Doctoral thesis, University of Tübingen

12 Kreis, W. (1987) Doctoral thesis, University of Tübingen

13 Watanabe, K., Yano, S., Yamada, Y. (1982) Phytochem. 21, 513-516

14 Sato, F., Yamada, Y. (1983) Phytochem. 23, 281-285

USE OF GENETIC ENGINEERING TO IMPROVE YIELDS IN CELL CULTURES, e.g. (ANTI)SENSE DNA TECHNOLOGY

J.N.M. Mol[1], P. de Lange[1], A. Oostdam[2] and LH.W. van der Plas[2], DEPTS. OF GENETICS[1] AND PLANT PHYSIOLOGY[2], FREE UNIVERSITY, De Boelelaan 1087, 1081 HV AMSTERDAM, TELEFAX 31-20-429202.

1. INTRODUCTION

Many valuable plant products, such as secondary metabolites, are synthesized via multiple enzymic steps that usually are part of complex, branched biosynthesis pathways [1]. To enhance the yield of a desired plant product, a specific channeling of precursors into specific products would be desirable. Such channeling might be obtained by blocking unwanted sidebranches. This could be accomplished by classical mutagenesis, however this is a rather troublesome procedure. A more specific procedure makes use of antisense genes targeted against the gene(s) encoding the first enzymatic step of the side-branch pathway. On the other hand, product formation may be enhanced by over-expressing those genes whose products are rate-limiting in the synthesis of the desired metabolites. We studied the feasability of such product channeling in plants [2-5]. As a first model system we chose the flower pigmentation pathway of *Petunia hybrida*. Flower pigmentation provides a nice model system since changes in phenotype are readily scored by eye. The pigments in petunia flowers are derived from the phenylpropanoid pathway and are comprised of anthocyanins and flavonols.

In this paper we will give a brief overview of what is known about (anti)sense regulation in plants and describe two experimental modelsystems that should allow one to assess the feasability of the (anti)sense approach in plant tissue culture systems.

2. PROCEDURES

All materials and methods used have been described elsewhere [2 - 6].

3. RESULTS AND DISCUSSION

3.1 *Bacterial antisense genes in plants*

The first report on artificial antisense regulation of gene expression came from Ecker & Davis [7]. They described effective transient inhibition of chloramphenicol transacetyl transferase (CAT) activity in carrot cells by co-introduction of sense and antisense cat genes in protoplasts. By double transformation of tobacco plants with nos and antisense nos genes, NOS activity can be modulated [8]. Since these initial data, other bacterial genes such as bar [9] and gus [10] have been succesfully down-regulated by antisense techniques.

3.2 *Plant (anti)sense genes in plants*

The first authentic plant gene succesfully down-regulated by antisense technology was the gene for chalcone synthase (chs) encoding the key enzyme of flavonoid biosynthesis [2]. As expected, evenly-reduced pigmentation up to fully white was obtained in independent transformants. Unexpectedly, pigmentation patterns were observed in rings and sectors that showed variation with light and hormone regime [3]. The different phenotypes observed in independent transformants are attributed to position effects of the antisense gene on gene expression. The effects of different promoters, antisense gene fragments and chromosomal location have been described [4].

Attempts to enhance coloration by introducing sense chs or dfr genes in Petunia had an inverse effect [5, 11]. Antisense-like effects were observed ranging in phenotype from patterns to fully white. Transcript analysis showed that transcript levels of both the endogenous and the transgene were specifically reduced leaving the levels for other flavonoid-specific genes unaltered. We call this phenomenon co-suppression, since an incoming gene shuts off itself and a resident copy of that same gene. The implications of this phenomenon, that to our knowledge has not been reported before in plants, for channeling studies is obvious. Its mechanism is under study.

Since the first report on antisense repression of gene expression more succesfull attempts have been reported using the polygalacturonase (pg) gene that is involved in fruit softening and the gene encoding the small subunit of ribulose 1,5 bisphosphate carboxylase (ss-Rubisco). Limited success has been reported in the inhibition of plant RNA virus replication (cf. 12 for an overview). At the last UCLA Meeting on "Molecular Strategies for Crop Improvement" four more examples have been given: (i) the manipulation of starch in potatoes by gbss antisense RNA, (ii) the inhibition of lignin synthesis in tobacco expressing cad antisense RNA, (iii) resistance to gemini virus infection, (iv) restoration of engineered male sterility by antisense rol RNA.

3.3 (Anti)sense genes in plant cell culture

3.3.1. ß-glucuronidase (GUS) modelsystem

From the previous paragraphs one might expect that branched pathways may be shut off in plant tissue culture systems. As an initial attempt to test this idea we constructed different sense and antisense gus fusions (CaMV 35S, chs, nos promoters; different 3' tails) and tested those in different combinations in a petunia leaf-based protoplast system. In a first set of experiments combinations of sense and antisense gus constructs gave severe reduction of GUS activity in transgenic expression assays relative to controls. We are in the process of refining those experiments and compare the transient data to *in planta* results. We feel that these experiments could shed more light on the mechanism of (anti)sense gene expression and provide the necessary information to effectively shut off authentic plant genes in tissue cultures. The phenylpropanoid system provides a good test system.

3.3.2. The phenylpropanoid system

The phenylpropanoid pathway is one of the best studied secondary metabolite routes. The precursor for all phenylpropanoids is phenylalanine that is synthesized via the shikimate pathway. Phenylalanine is converted into 4-coumaroyl CoA by the concerted action of phenylalanine ammonia-lyase (PAL), cinnamate-4-hydroxylase (C4H) and 4-coumarate CoA ligase (4CL). The lignin pathway branches off from coumaric acid and/or coumaroyl CoA whereas the flavonoids branch off from 4-coumaroyl CoA. The exact branchpoint for the lignan pathway is not completely elucidated.

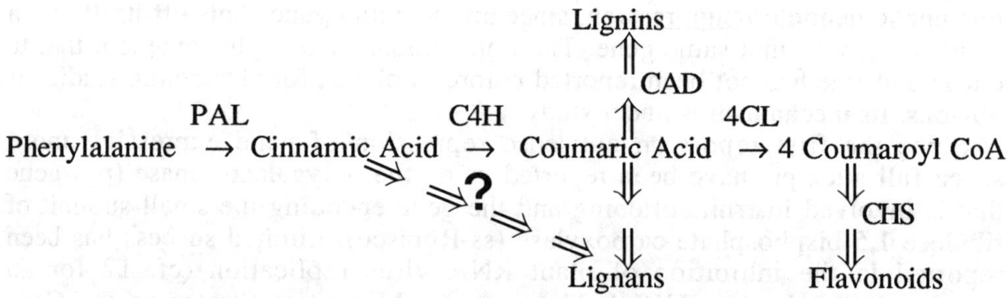

Podophyllotoxin belongs to the lignan cytostatics. It is extracted from the rhizomes of *Podophyllum hexandrum* and *Podophyllum peltatum* and used for the production of etiposide and teniposide which are potent chemotherapeutic agents. Since podophyllotoxin is rather expensive and difficult to isolate in high quantities from wild-grown plants its production in plant tissue cultures may be a good alternative. Since tissue culture of *Podophyllum* is extremely difficult, cultures from Linum flavum have been established that produce the 5-methoxy derivative of podophyllotoxin (5-MPT) which has also cytostatic activity [6].

5-MPT is mainly produced in roots of *L. flavum*. From literature it is known that the production of secondary metabolites in callus or cell suspension generally is lower than in whole plants. However, high production of secondary metabolites has been reported in hairy root cultures [13]. We have established hairy root cultures of *L. flavum* by infection of leaf explants with *A. rhizogenes*. The 5-MPT-content of these root cultures is 5-10 fold higher than that of an untransformed cell suspension culture [6]. We conclude that hairy roots are the best experimental system for *in vitro* production of 5-MPT.

The availability of the gene encoding cinnamyl alcohol dehydrogenase (cad) offers a unique opportunity to decrease the lignin content of *L. flavum* hairy roots and examine the effect on 5-MPT production [14]. Such experiments will be undertaken.

3.3.3. Future prospects

Antisense may turn out to be a powerful tool to manipulate not only the production of secondary metabolites such as phenylpropanoids, carotenoids, terpenes, alkaloids etc. but also the production of other substances such as fatty acids, starch a.o. Hopefully, the use of ribozymes [15] will improve the efficiency even more.

REFERENCES

1. Harborne, J.B. The flavonoids. Harborne JB (ed.) Chapman and Hall, London.
2. Krol van der, A.R., Lenting, P.E., Veenstra, J., Meer van der, I.M., Koes, R.E., Gerats, A.G.M., Mol, J.N.M. en Stuitje, A.R. (1988) Nature 333, 866-869.
3. Krol, van der, A.R., de Lange , P., Gerats, A.G.M., Mol, J.N.M. and Stuitje, A.R. (1990) Mol. Gen. Genet. 220, 204-212.
4. Krol, van der A.R., Mur, L.A., de Lange, P., Mol, J.N.M. and Stuitje, A.R. (1990) Plant Mol. Biol. 14, 457-466.
5. Krol, van der A.R., Mur, L.A., Beld, M., Mol, J.N.M. and Stuitje, A.R. (1990) Plant Cell 2, 291-299.
6. Oostdam, A., van der Plas, L.H.W. and Mol, J.N.M. (1990) Proceedings of the 3rd Netherlands Biotech. Congress (NBC-3), Amsterdam pp. 255-259.
7. Ecker, J.R. and Davis, R.W. (1986) Proc. Natl. Acad. Sci USA 83, 4300-4304.
8. Rothstein, S.J., DiMaio, J., Strand, J.M. and Rice, D. (1987) Proc. Natl. Acad. Sci. USA 84, 8439-8443.
9. Cornelissen, M. and Vanderwiele, M. (1989) Nucl. Acids Res. 17, 833-843.
10. Robert, L.S., Donaldson, P.A., Ladaigne, C., Altosaar, I., Arnison, P.G. and Fabijanski, S.F. (1989) Plant Mol. Biol. 13, 399-409.

11. Napoli, C., Lemieux, C. and Jorgensen, R. (1990) Plant Cell 2, 279-289.
12. Krol van der, A.R., Mol, J.N.M. and Stuitje, A.R, (1988) Gene 72, 45-50.
13. Hamill, J.D., Parr, A.J., Rhodes, M.C.J., Robbins, R.J. and Walton, N.J. (1987) Biotechnology 5, 800-804.
14. Walter, M.H., Grima-Pettenati, J., Grand, C., Bordet, A.M. and Lamb, C. (1988) Proc. Natl. Acad Sci. USA 85, 5546-5550.
15. Haseloff, J. and Gerlach, W.L. (1988) Nature 334, 585-591.

STRATEGIES TO IMPROVE YIELDS OF SECONDARY METABOLITES TO INDUSTRIALLY INTERESTING LEVELS

FRANK DiCOSMO, CENTRE FOR PLANT BIOTECHNOLOGY, DEPARTMENT OF BOTANY; INSTITUTE OF BIOMEDICAL ENGINEERING; AND DEPARTMENT OF MECHANICAL ENGINEERING, UNIVERSITY OF TORONTO, TORONTO, ONTARIO, M5S 3B2, CANADA

ABSTRACT

Plants yield biochemicals used as medicines, and compounds important to the food, flavourings, cosmetic and fine chemicals industries. Because of decreasing plant resources it is crucial to discover alternative sources of these valuable compounds; plant cell cultures promise complementary and competitive production systems. However, industrial production of secondary metabolites is now only commercially viable when the market value exceeds U.S. one-thousand to several thousand dollars per kilogram of product. It is therefore essential to develop production strategies using cell culture systems that yield economically viable levels of biochemicals.

High-yielding or industrially attractive systems can be devised by the cultivation of organ cultures. Alternatively, suspension cultured cells selected for high productivity, coupled with media optimization and high-density culture offer other possibilities for commercialization. Some inorganic and organic agents that stimulate product synthesis and/or accumulation give encouraging results, but production is variable and transient. Immobilized cells have potential for long-term synthesis of biochemicals and industrially attractive yields; but these benefits are only achieved when the product is excreted into the liquid phase. Cell immobilization may also allow for semi-continuous or continuous biotransformation processes. Indeed, the most reliable methods for ensuring industrial and competitive production of high-value-added products is by chemical and/or enzyme-mediated syntheses. For example, the fascile synthesis of α-3', 4'-anhydrovinblastine can be achieved in industrially attractive yields.

1. INTRODUCTION

The important commodity biochemicals produced by plants that are used in health care, food production and industry include, vinblastine and vincristine, codeine, morphine, digoxin and quinine that are valued drugs; quinine, carvone, capsaicin and others find uses as flavourings and essences. Other fine chemicals such as shikonin, geraniol and rose oil are used in the cosmetic industry. Certainly, decreased plant resources due to natural depredation, ruthless exploitation and increased demand raise concerns regarding the continued availability of commercially important biochemicals from intact plants. Plant cell cultures represent an alternative and complementary source of these compounds, commonly referred to as secondary metabolites. However, cultured plant cells usually fail to produce, or yield low levels of biochemicals characteristic of the intact plant that are of great interest to industry (1-4). It is crucial for industry that alternative sources and production methods of these materials are found and competitively established.

The chemical synthesis of industrially important biochemicals is perhaps the most obvious approach. Because of their structural complexity, however, the total synthesis of many secondary compounds is not presently commercially feasible. The use of genetic engineering to produce various secondary metabolites is impeded by our lack of knowledge concerning the often complex and multi-branched, enzyme pathways leading to the majority of the products. The biosynthesis and accumulation of increased levels of the desired compounds or even novel products can sometimes be achieved by other techniques that include: 1) screening and selection of cell lines with high yields, high density cell culture and subsequent optimization of the growth conditions; 2) treatment of cells with specific organic or inorganic agents; 3) cell immobilization; and 4) enzyme-mediated biosynthesis and chemical synthesis. Recent advances suggest that these strategies will contribute significantly to yield commercially viable industrial production systems.

What do we mean when we speak of improving yields of secondary metabolites to industrially interesting levels? How much product is required to determine the scaling up of a particular process? What level of scale up or production hardware will be required to achieve the desired amount? The simple answer to these questions is that process efficiency and product demand will determine what level of plant-derived biochemical will be considered industrially viable. For example, cost analysis for the production of ajmalicine using a *Catharanthus roseus* cell culture system indicates that the alkaloid would cost about U.S. $3,000/kg (5). However, when intact roots are used as the source of ajmalicine the cost is about $600/kg. Codeine and morphine are valued at about $1,300/kg; vinblastine and vincristine have been estimated to cost about $5,000 per gram (wholesale) and up to $20,000/g retail. The global retail market in dollars ($ U.S.) has been estimated to be 5 million for ajmalicine, 90-100 million for codeine, and 50-75 million for vincristine/vinblastine. Therefore, not all production systems require the same scale. Advances in optimization of production strategies for many plant derived biochemicals indicate that the largest production volume may be 1,000 litres. The production of secondary metabolites in bioreactors has recently been reviewed (2); plant cells can be cultured in volumes of up to 75,000 litres. It is therefore important for companies to plan wisely the production technology that yields not only commercially viable levels of product, but also the capacity to make the quantity of product needed.

This brief review is not intended to be exhaustive; it provides and discusses some recent examples of techniques used to achieve industrially interesting levels of secondary metabolite synthesis. Problems and opportunities for biosynthesis and industrial production of plant biochemicals have been authoritatively reviewed elsewhere (2,3,5-10).

2. STRATEGIES FOR IMPROVING PRODUCTION OF SECONDARY METABOLITES

2.1 Shoot cultures

Industrial scale propagation of specific plant tissues known to produce and accumulate substantial yields of secondary metabolites may lead to commercial production methods. Plant cell culture systems, such as shoot and root cultures offer industrially attractive methods for the production of a wide variety of plant-derived biochemicals. These compounds are not usually synthesized in undifferentiated cultured cells. Perhaps not surprisingly, when undifferentiated cells are induced to differentiate into either shoot or root organs product accumulation also ensues. For example, shoot cultures of *Catharanthus roseus* yielded up to 2.6 μg α-3', 4'-anhydrovinblastine per gram fresh weight of shoots. The leaves from intact plants produced 9.1 μg of the dimeric alkaloid anhydrovinblastine per gram fresh leaves (11). Shoot cultures of *Artemisia annua* produce the antimalarial constituent artemisinin; typically 1-5 mg/100 g of dried leafy shoots are produced. On the other hand, intact *A. annua* plants yield 50-500 mg/100 g of leaf tissue (E.J. Staba, personal communication). The production of insectidial chemicals, such as pyrethrin from shoot cultures of *Chrysanthemum cinerariaefolium* has also produced encouraging results. Zieg *et al.* (12) established shoot cultures of *C. cinerariaefolium* some of which produced up to 341.8 mg of pyrethrin per 100 g of dried shoot tissue. Similarly, the production of quinine and quinidine are stimulated when *Cinchona ledgeriana* is cultivated as shoot organ cultures (13). Thirty-two week-old cultures produced about 350 mg of alkaloids per 100 g of tissue. The total amount and concentrations of specific alkaloids in shoot cultures were similar to those found in 1 year old intact plants.

2.2 Root organ culture

Root cultures are able to produce a wide variety of commodity biochemicals, such as biocidal thiophene derivatives from *Tagetes patula* cv. French dwarf, isoquinoline alkaloids from *Cinchona ledgeriana* (15), indole alkaloids from *C. roseus* (16), tropane alkaloids from *Hyoscyamus niger* (17) and *Datura candida* (18), and enzymes (e.g. horseradish peroxidase) from *Armoracia rusticana* (19). Flores *et al.* (14) reviewed the production of secondary metabolites using root cultures and pointed out the enormous utility of this system.

Root cultures of *Cinchona ledgeriana* will synthesize and accumulate quinine and quinidine; feeding tryptophan (a precursor of quinine and quinidine) to *C. ledgeriana* root cultures resulted in a five-fold increase in the yields of these quinoline alkaloids (15).

Hashimoto and Yamada (16) regenerated and selected root cultures of *H. niger* that produced high levels of the tropane alkaloids hyoscyamine and scopolamine; the cultures synthesized significantly more scopolamine than leaves of the intact plant. Root cultures of *Datura candida* yield up to 0.68% (dry weight) of tropane alkaloids (hyoscyamine + scopolamine); this is approximately three times the amount found in intact plants (18). Although shoot and root cultures provide a vast repository of useful biochemicals, often yielding amounts of these compounds greater than levels found in intact plants they are usually slow growing and require extensive subculturing to obtain the biomass required for industrially attractive production processes. The use of "hairy-root" cultures (transformed with *Agrobacterium rhizogenes*), however, offers a method for achieving increased growth rates of root cultures (14,19) concomitant with biochemical production comparable to intact plants. For example, Taya *et al.* (19) developed a two stage process for the production of peroxidase from hairy-root cultures of *Armoracia rusticana* (horseradish). Hairy-root cultures normally have peroxidase (POD) activity comparable to intact horseradish roots. However, using a two stage process in which the first stage was devoted to generation of biomass, and the second stage included high concentrations of Polypepton which resulted in the enhancement of POD activity to 5.4 times found in cultures in Polypepton-free media.

2.3 Cell suspension cultures

Plant cell suspension cultures are widely recognized as an industrially attractive and valuable sources of secondary metabolites because: 1) of the relatively rapid propagation of biomass, 2) of the relatively homogeneous cell types in the culture system, and 3) of the ability to manipulate the culture environment relatively easily, such as adding adjuvants to the growth media. However, few commercial successes have been reported. This is simply because cultured plant cells usually fail to produce the products sought after by industry.

Several strategies have been developed to increase the yields of secondary metabolites in suspension cultures. These techniques include: 1) screening cell lines for high yields and optimization of culture conditions, and 2) the use of specific biotic or abiotic agents to stimulate product accumulation.

2.3.1 <u>Screening of cell lines and optimization of production</u>:

Screening of cultured cell lines, such as *Catharanthus roseus* (20), *Lithospermum erythrorhizon* (21), and *Coptis japonica* (22) for high yields of ajmalicine, shikonin and berberine, respectively has resulted in the establishment of cells with increased productivities. Optimization of the growth media and culture conditions for *C. roseus* (20), *L. erythrorhizon* (21), and *Coptis japonica* (22-24) allowed the development of a two-stage culture process for increased production of indole alkaloids, commercial production of shikonin (25) and high density cultures of *Coptis japonica* for berberine production, respectively. Use of these strategies has allowed for the industrial production of shikonin to levels approaching 4g of shikonin per litre of culture media (25). Similarly, the cultivation of *Coptis japonica* cells in high density cultures resulted in yields of 3.5 g berberine per litre of growth medium (24).

2.3.2 Use of biotic agents: Various agents that are usually components of fungal cell walls, have been used to stimulate secondary metabolite production in cultured cells. These materials have been termed "elicitors". DiCosmo and Misawa (26) provided a review of methods useful for eliciting the biosynthesis and accumulation of secondary metabolites in cell cultures. Recently, Funk *et al.* (27) used a carbohydrate elicitor to enhance berberine production in *Thalictrum rugosum* cells. Similarly, sanguinarine was stimulated by the addition of fungal elicitor to cultures of *Papaver somniferum* (28). Other systems have also been described (8-10,26). Although such systems have obvious practical potential for the production of useful biochemicals, important and fundamental problems remain. For example, the elicitor-mediated production is transient, and deteriorates relatively rapidly. Also the level and rates of productivity are usually insufficient for industrial application.

2.4 Immobilized cells

The term "cell immobilization" describes the retention of cells either within a matrix, or on an inert support material, such as calcium alginate and polyurethane foam, respectively. The benefits to be derived from cell (biocatalyst) immobilization include: 1) development of high cell densities with high biocatalytic activity, 2) the possibility for reuse of the biocatalyst, 3) the possibility for development of continuous production methods, 4) maintenance of the cells in their most productive phase for extended periods, and 5) removal of the product from the liquid phase easily (when the product is excreted, e.g. peroxidase). The techniques, details and advantages of plant cell immobilization have been authoritatively discussed in numerous reviews (29-32). However, the practical application of plant cell immobilization for industrial scale production of biochemicals remains to be demonstrated. Indeed, the author is not aware of the industrial production of any plant cell secondary metabolite using immobilized cells; Fontanel and Tabata (2) made the same observation.

Advances in immobilization technologies may allow the development of large-scale cultivation of immobilized plant cells for biochemical production. For example, plant cell immobilization strategies for *C. roseus* and *Thalictrum rugosum* using glass fibre mats that achieves, 1) relatively rapid, spontaneous cell immobilization, 2) high cell densities, and 3) cultivation of the biomass for extended periods have been developed (33,34). However, biochemical production must be optimized in this system. In using glass fibre mats as the support matrix it was demonstrated that spontaneous cell immobilization of 100% of the inoculum load occurred within 8 hours under physiological conditions within the bioreactor for *T. rugosum*.

Despite advantages of increased product yields during extended cultivation the benefits of immobilization will only be realized if the products are released into the liquid phase. The release of accumulated products does not, however, usually occur; some exceptions include the secreted proteins (e.g. peroxidases, amylases, oxidases) and the alkaloid berberine from *Thalictrum* spp., for example.

The most viable industrial production strategy using immobilized cell systems presently appears to be for biotransformation processes. Biotransformations rely on the ability of cells to alter the chemical structure of some precursor to a more desirable product. Examples of biotransformations include the conversion of codeinone to codeine by *Papaver somniferum* cells immobilized using polyurethane foam (35) and calcium alginate (36), and the transformation of digitoxin to digoxin by immobilized cells *Digitalis lanata* (37,38). Alferman, Spieler and Reinhard (39) described the semi-continuous biotransformation of methyldigitoxin to methyldigoxin by *D. lanata* cells cultivated as suspensions in a 200 l bioreactor; 505 g of methyldigoxin were produced in 3 months.

2.5 Chemical and enzyme-mediated biosynthesis

Certainly, the synthesis of commodity biochemicals via strictly chemical methods and enzyme mediated processes offer industry the most effective and reliable methods for production. For example, gram amounts of strictosidine can be synthesized using immobilized strictosidine synthase isolated from *C. roseus* (40). Strictosidine synthase has been cloned and the protein was expressed in *Escherichia coli*. Indeed, biotechnological production of the plant enzyme in the bacterium, *E. coli* could lead to the availability of large amounts of strictosidine synthase for the industrial production of strictosidine (41). Other examples that will be useful for industrial production of vinblastine and vincristine involve the chemical and enzyme-mediated dimerization of catharanthine and vindoline to yield α-3', 4'-anhydrovinblastine (AVLB), the precursor of vinblastine and vincristine. Vukovic and Goodbody (42) described the $FeCl_3$-catalyzed dimerization of catharanthine and vindoline to yield AVLB and vinblastine (VLB); maximum yields for AVLB and VLB were 68% and 21%, respectively, under specific conditions.

It has also been shown that coupling of catharanthine and vindoline to yield AVLB is mediated by horseradish peroxidase (43), and a crude enzyme fraction from *C. roseus* cell suspensions in the presence of NADH and $MnCl_2$ catalyzes the formation of VLB from AVLB in 12% yield (44). Peroxidase isolated from bioreactor-cultivated *C. roseus* cells also mediates dimerization of catharanthine and vindoline to yield AVLB. The peroxidase mediated processes has an absolute requirement for hydrogen peroxide (H_2O_2), and it was found that excessive concentrations of H_2O_2 reduce the yield of AVLB (45). It is essential to precisely control and supply optimal concentrations of H_2O_2 that would yield significant amounts of AVLB in the coupling assay. The solution to this problem was through the use of a coupled enzyme system designed to produce H_2O_2 *in situ* for peroxidase mediated dimerization of catharanthine and vindoline (Fig. 1). The coupled enzyme system consists of glucose oxidase (GOD) and horseradish peroxidase (POD). The GOD yields gluconate and H_2O_2 in the presence of D-glucose, and O_2. Subsequently, the POD consumes the H_2O_2 to effect dimerization of catharanthine and vindoline in yields of 70%. Optimization of this process is ongoing and will be described in detail elsewhere.

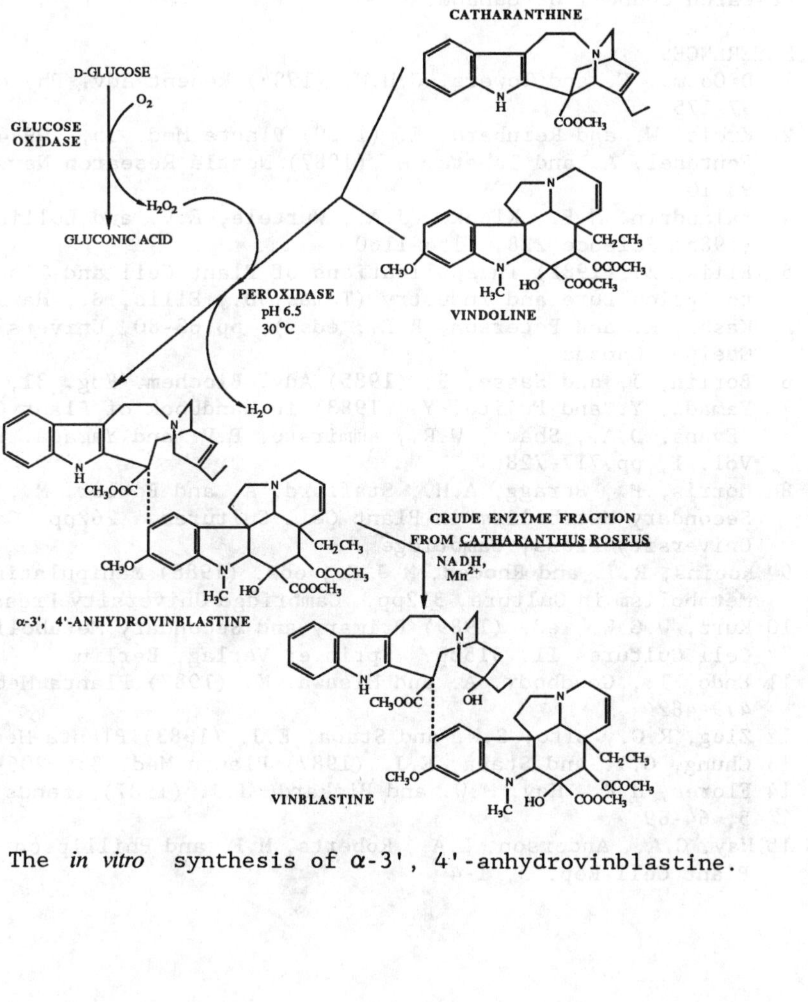

FIGURE 1. The *in vitro* synthesis of α-3', 4'-anhydrovinblastine.

3. Conclusion

The examples presented clearly demonstrate that the industrial production of plant derived biochemicals will eventually yield many successes for astute companies that recognize the immense potential, even though few processes have demonstrated this possibility. Success in this area requires that efforts designed: 1) to discover regulatory controls of product synthesis and how to manipulate these controls, 2) to achieve high density cell cultures expressing high levels of product yield, and 3) to provide efficient down-stream processing.

ACKNOWLEDGEMENTS

F. DiCosmo is grateful to Prof. Dr. H.J.J. Nijkamp and Dr. J. van Aartrijk, and the Scientific Advisory Board of the VII International Congress on Plant Tissue Culture for the opportunity and financial support that made this presentation possible. Ms. J. Bede prepared Fig. 1. Research in F. DiCosmo's laboratory is funded by the Natural Sciences and Engineering Research Council of Canada and the Medical Research Council of Canada.

REFERENCES

1 DiCosmo, F. and Towers, G.H.N. (1984) Recent Adv. Phytochem. 18, 97-175
2 Kreis, W. and Reinhard, E. (1989) Planta Med. 55, 409-416
3 Fontanel, A. and Tabata, M. (1987) Nestlé Research News 1986/87, 93-103
4 Balandrin, M.F., Klocke, J.A., Wurtele, E.V. and Bollinger, W.H. (1985) Science 228, 1154-1160
5 Ellis, B. (1982) in Applications of Plant Cell and Tissue Culture to Agriculture and Industry (Tomes, D., Ellis, B., Harney, P., Kasha, K. and Peterson, R.L., eds.), pp.63-80, University of Guelph, Canada
6 Berlin, J. and Sasse, F. (1985) Adv. Biochem. Eng. 31, 99-132
7 Yamada, Y. and Fujita, Y. (1983) in Handbook of Plant Cell Culture (Evans, D.A., Sharp, W.R., Ammirato, P.V. and Yamada, Y., eds.), Vol. 1, pp.717-728
8 Morris, P., Scragg, A.H., Stafford, A. and Fowler, M., eds. (1986) Secondary Metabolism in Plant Cell Cultures. 262pp. Cambridge University Press, Cambridge
9 Robins, R.J. and Rhodes, M.J.C., eds. (1988) Manipulating Secondary Metabolism in Culture, 312pp. Cambridge University Press, Cambridge
10 Kurz, W.G.W., ed. (1989) Primary and Secondary Metabolism of Plant Cell Cultures II, 315pp. Springer-Verlag, Berlin
11 Endo, T., Goodbody, A. and Misawa, M. (1987) Planta Med. 53, 479-482
12 Zieg, R.G., Zito, S.W. and Staba, E.J. (1983) Planta Med. 48, 88-91
13 Chung, C.T. and Staba, E.J. (1987) Planta Med. 53, 206-210
14 Flores, H.E., Hoy, M.W. and Pickard, J.J. (1987) Trends Biotechnol. 5, 64-69
15 Hay, C.A., Anderson, L.A., Roberts, M.F. and Phillipson, J.D. (1986) Plant Cell Rep. 5, 1-4

16 Parr, A.J., Peerless, A.C.J., Hamill, J.D., Walton, N.J., Robins, R.J., Rhodes, M.J.C. (1988) Plant Cell Rep. 7, 309-312
17 Hashimoto, T. and Yamada, Y. (1983) Planta Med. 47, 195-199
18 Christen, P., Roberts, M.F., Phillipson, J.D. and Evans, W.C. (1989) Planta Cell Rep. 8, 75-77
19 Taya, M., Yohama, A., Nomura, R., Kondo, O., Matsui, C. and Kobayashi, T. (1989) J. Ferm. Bioeng. 67, 31-34
20 Zenk, M.H., El-Shagi, H., Arens, H., Stöckigt, E., Weiler, W. and Deus, B. (1977) in Plant Tissue Culture and its Biotechnological Application (Barz, W., Reinhard, E. and Zenk, M.H., eds.), pp.27-43, Springer-Verlag, Berlin
21 Fujita, Y., Takahashi, S. and Yamada, Y. (1985) Agric. Biol. Chem. 49, 2071-2075
22 Sato, F. and Yamada, Y. (1984) Phytochem. 23, 281-285
23 Morimoto, T., Hara, Y., Kato, Y., Hiratsuka, J., Yoshioka, T., Fujita, Y. and Yamada, Y. (1988) Agric. Biol. Chem. 52, 1835-1836
24 Matsubara, K., Kitani, S., Yoshioka, T., Morimoto, T. and Fujita, Y. (1989) J. Chem. Tech. Biotechnol. 46, 61-69
25 Fujita, Y. and Tabata, M. (1987) in Plant Cell and Tissue Culture (Green, C.E., Somers, D.A., Hackett, W.P. and Biesboer, D.D., eds.), pp.169-185, Alan R. Liss, Inc., New York
26 DiCosmo, F. and Misawa, M. (1985) Trends Biotechnol. 3, 318-322
27 Funk, C., Gugler, K. and Brodelius, P. (1987) Phytochem. 26, 401-405
28 Eilert, U. and Constabel, F. (1986) J. Plant Physiol. 125, 167-172
29 Brodelius, P. and Mosbach, K. (1982) Adv. Appl. Microbiol. 28, 1-26
30 Mosbach, K., ed. (1987) Methods Enz., Vol. 135, 675pp.
31 Mosbach, K., ed. (1987) Methods Enz., Vol. 136, 584pp.
32 Mosbach, K., ed. (1988) Methods Enz., Vol. 137, 767pp.
33 Facchini, P.J. and DiCosmo, F. (1990) Appl. Microbiol. Biotechnol. (in press)
34 Facchini, P.J. and DiCosmo, F. (1990) Biotechnol. Bioeng. (submitted)
35 Corchete, P. and Yeomann, M.M. (1989) Plant Cell Rep. 8, 128-131
36 Furuya, T., Yoshikawa, T. and Taira, M. (1984) Phytochem 5, 999-1001
37 Brodelius, P., Deus, B. and Zenk, M.H. (1979) FEBS Lett. 103, 93-97
38 Spieler, H., Alfermann, A.W. and Reinhard, E. (1985) Appl. Microbiol. Biotechnol. 23, 1-4
39 Alferman, A.W., Spieler, H. and Reinhard, E. (1985) in Primary and Secondary Metabolism of Plant Cell Cultures (Neumann, K.H., Barz, W. and Reinhard, E., eds.), pp.316-322
40 Pfitzner, U. and Zenk, M.H. (1982) Planta Med. 46, 10-14
41 Kutchan, T.M., Hampp, N., Lottspeich, L., Beyreuther, K. and Zenk, M.H. (1988) FEBS Lett. 237, 40-44
42 Vukovic, J. and Goodbody, A.E. (1988) U.S. Patent 4,778,885
43 Goodbody, A.E., Endo, T., Vukovic, J., Kutney, J.P., Choi, L.S.L. and Misawa, M. (1988) Planta Med. 43, 136-140
44 Endo, T., Goodbody, A.E., Vukovic, J. and Misawa, M. (1987) Phytochem. 12, 3233-3234
45 Smith, J.I., Amouzou, E., Yamaguchi, A., McLean, S. and DiCosmo, F. (1988) Biotechnol. Appl. Biochem. 10, 568-575

SELECTION OF HIGH-YIELDING CELL LINES OF *TAGETES* USING FLOW CYTOMETRY.

Paulien Adamse, Centre for Agrobiological Research, P.O. Box 14, 6700 AA Wageningen, The Netherlands.

Selection of high-producing cell lines is an important strategy in developing methods for the industrial production of useful secondary metabolites by cultured plant cells (1,2). Between and within cell lines considerable differences exist in production levels of secondary metabolites. These differences can be attributed either to phenotypical variation or to somaclonal variation. Phenotypical variation - e.g. due to cellular differentiation, extended exposure to various plant growth hormones or variations in environment - is reversible and will not result in a stable high-producing cell line. It is more likely that somaclonal variation will be a source of stable high-producing variants or mutants, especially when additional mutagenic treatments are imposed.

Current methods available for selection of high-yielding cell lines are mainly based on the analysis of clones, either callus tissues or cell suspensions, by various chemical techniques. These procedures are very laborious and time-consuming. Flow cytometry enables a rapid analysis and subsequent sorting of large amounts of cells (Fig. 1). Analysis of the concentration of certain compounds in individual cells by flow cytometry has been reported using the autofluorescence of chlorophyll (3) the fluorescence of metabolites such as serpentine (4,5) and berberine (6) or the fluorescence of special dyes (e.g. 7,8).

FCM selection requires individual cells, for example protoplasts, producing a fluorescent secondary metabolite. Aggregated cell suspensions of *Tagetes patula* produce various thiophenes (natural biocides, 9). When exposed to UV light thiophenes are highly fluorescent. Therefore *T. patula* cell suspensions represent an attractive model-system. From these cultures ca. 4% of the cells yielded protoplasts. HPLC analysis of protoplast-isolation residues (Table 1) indicated that a large fraction of the thiophenes is recovered in the cell wall residue.

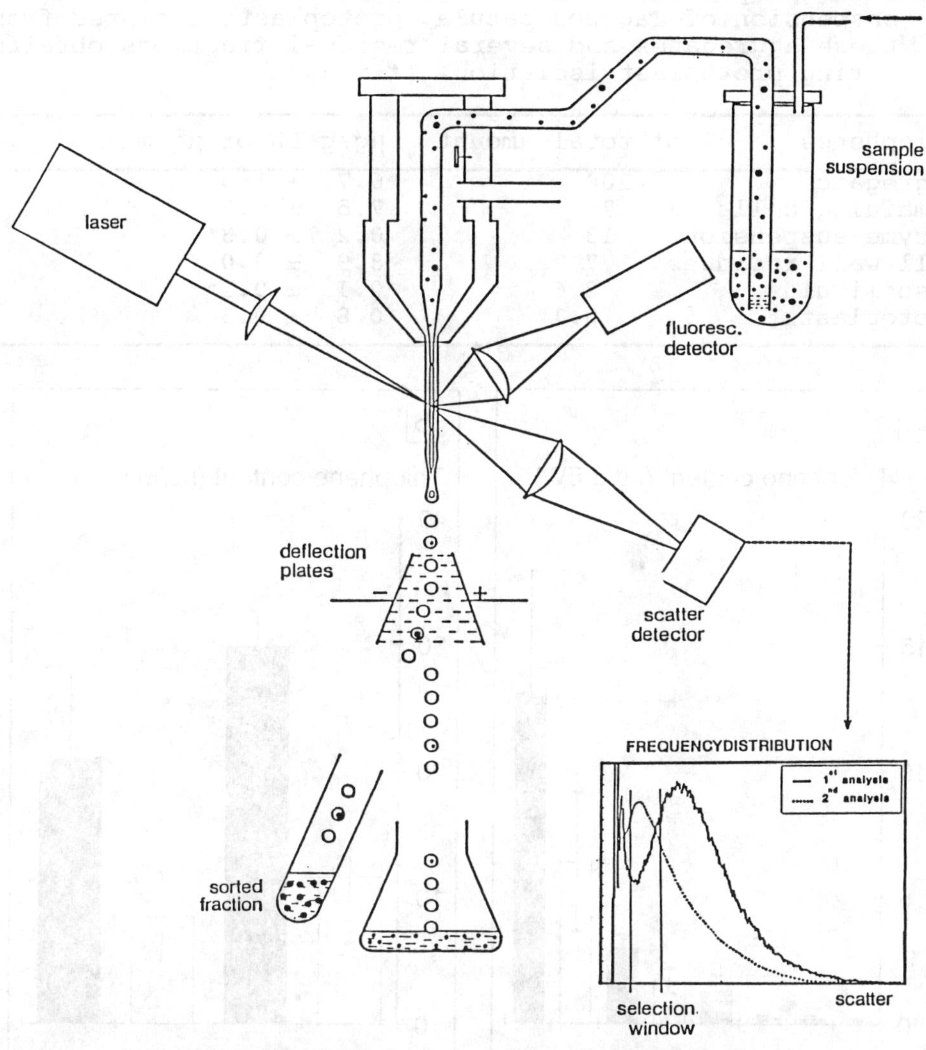

FIGURE 1. Cells or protoplasts are passed single file across a small area irradiated by a laser beam. The fluid jet is broken into droplets by a vibrating piezoelectric crystal. The droplets are either void or contain one cell. The light scatter ant the fluorescence of each cell is detected by light sensors and a frequency distribution is displayed graphically. In this graph criteria can be defined for the desired fraction(s): the sorting window. The measured values are rapidly compared with the preset criteria so that the cell can be charged electrostatically and deflected into the appropriate reservoir.

TABLE 1. Thiophene content of aggregates from a cell suspension of *Tagetes patula*, protoplasts isolated from these aggregates and several residual fractions obtained during protoplast isolation. (* = μg/ml)

Thiophenes	% of total amount	μg/g FW or μg/ml*
aggregates	100	6.7 ± 1.0
remaining cells	75	7.5 ± 0.7
enzyme suspension	13	0.2 ± 0.3*
cell wall residues	7	8.9 ± 0.0
wash liquids	2.6	0.1 ± 0.1*
protoplasts	0.1	0.8 ± 0.3

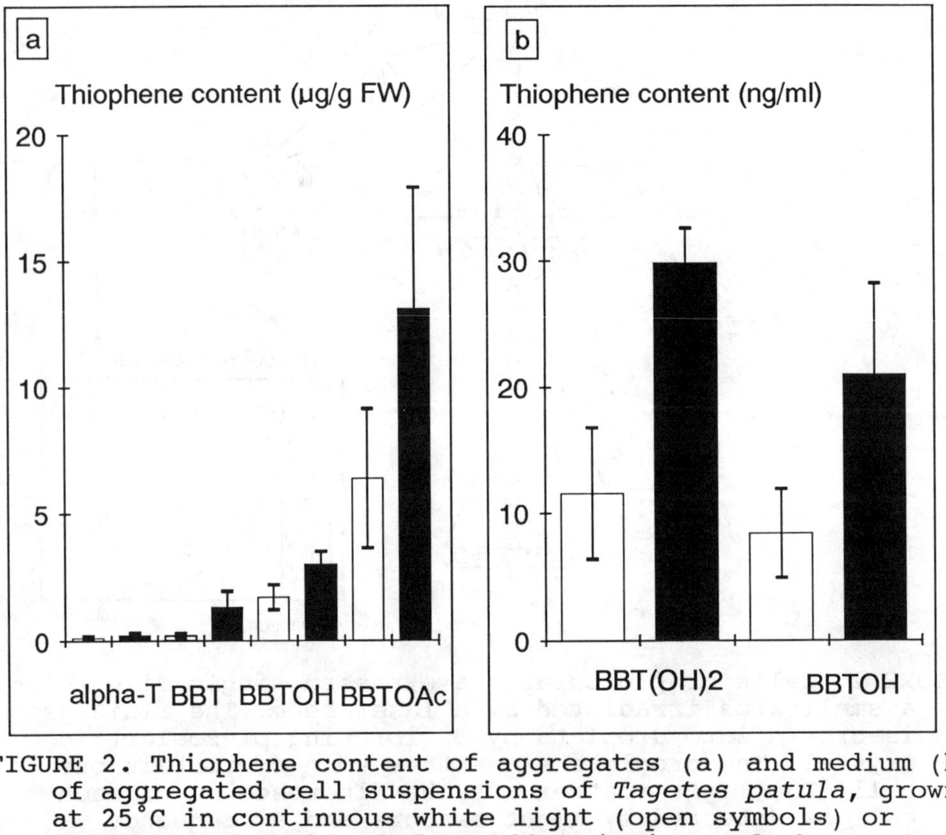

FIGURE 2. Thiophene content of aggregates (a) and medium (b) of aggregated cell suspensions of *Tagetes patula*, grown at 25°C in continuous white light (open symbols) or darkness (closed symbols). Abbreviations of the different thiophenes: alpha-T = α-terthiophene; BBTOH = 5-(4-hydroxy-1-butinyl)-2,2'-bithiophene; BBTOAc = 5-(4-acetoxy-1-butinyl)-2,2'-bithiophene; BBT = 5-(1-butinyl-3-ene)-2,2'-bithiophene.

The thiophene content of protoplasts is very low (ca. 10% of the concentration in whole cells), but can be measured with a flow cytometer. Continuous white light decreased the thiophene content in the aggregates as well as in the medium of *T. patula* cell suspensions (Fig. 2). It is not clear whether this is caused by decreased biosynthesis or increased degradation of the thiophenes.

It has been suggested that enhancement of production is mainly a question of enhancing the number of productive cells within a cell population (6). However, the intensity of blue fluorescence of individual protoplasts as observed with a microscope varied considerably in our system, thus indicating a variability in the thiophene levels of individual cells. Moreover, preliminary flow cytometric analysis of dark- and light-grown cell suspensions (high and low thiophene content, respectively) indicated that the mode of fluorescence distribution shifted towards a higher intensity with the increase of the average thiophene concentration of a cell line (Fig. 3). Also the range of intensities increased.

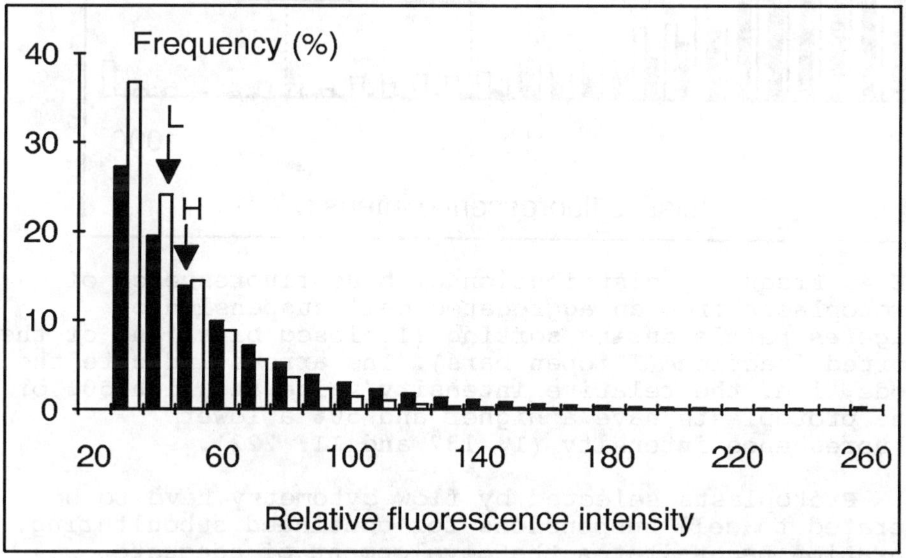

FIGURE 3. Frequency distribution of blue fluorescence of protoplasts from aggregated cell suspensions of *Tagetes patula* with a low (L; open bars) or a high (H; closed bars) thiophene production. Analyzed on a FACS-IV Cell Sorter (Becton Dickinson, Sunnyvale, USA) equipped with a HBO 100 mercury high pressure lamp (360 nm) and a LP-418 filter in the emission beam. The arrows indicate the mode, i.e. the relative intensity value at which 50% of the protoplasts have a higher and 50% a lower fluorescence intensity (L: 34; H: 41).

Preliminary selection of protoplasts with a high thiophene fluorescence showed an increased mode of fluorescence intensity as compared to the original protoplast population. The sorted fraction indeed contained more protoplasts with a high thiophene content (Fig. 4).

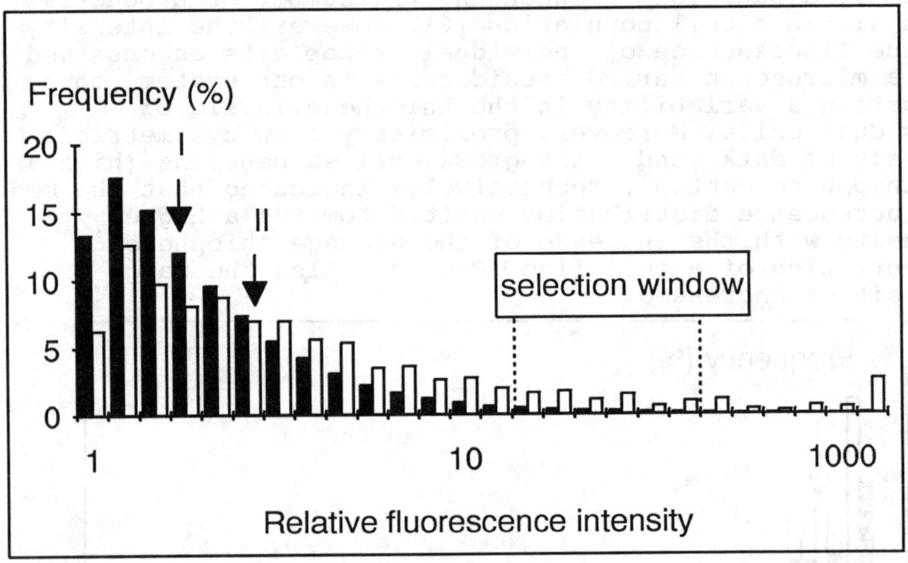

FIGURE 4. Frequency distribution of blue fluorescence of protoplasts from an aggregated cell suspension of *Tagetes patula* during sorting (I:closed bars) and of the sorted fraction (II:open bars). The arrows indicate the mode, i.e. the relative intensity value at which 50% of the protoplasts have a higher and 50% a lower fluorescence intensity (I: 132 and II: 222).

Protoplasts selected by flow cytometry have to be regenerated to cell suspensions for continued subculturing. Regeneration necessitates the development of adequate techniques for the isolation and culture of protoplasts. To our knowledge there is only one reference for the successful regeneration of *Tagetes* protoplasts (10). Unfortunately this reference states no experimental details, therefore we had to develop our own isolation and regeneration method. Successful regeneration of one or a small group of protoplasts to whole cells has resulted in the formation of callus tissue and subsequently of aggregated cell suspensions. However, the regeneration method is not yet reproducible and rather slow.

The cell suspension cultures regenerated from the selected protoplasts will be tested for stability of growth and thiophene production using HPLC and flow cytometry. A slow decrease of the production of serpentine and ajmalicine in a *Catharanthus roseus* cell line, selected for high production, has been reported by Deus-Neumann and Zenk (5). Similar results were obtained for anthocyanin production in cell cultures of carrot (11). Repeated selection is probably necessary to maintain a high yield. Only if the production of a cell line remains stable after several selection cycles a true overproducing mutant can be recognized. However, without thorough knowledge of relevant biosynthesis pathways and their genetic regulation it is difficult to predict the nature of such a mutant. Till so far our results indicate that the time-consuming regeneration of selected protoplasts will remain the bottleneck for flow cytometric selection of high-producing cell lines of *Tagetes*, especially when repeated selecting is envisaged.

REFERENCES

1 Deus, B., and M.H. Zenk (1982). Biotechn. Bioeng. 24:1965-1974.
2 Wersuhn, G. (1988). Biochem. Physiol. Pflanzen 183: 361-369.
3 Harkins, K.R., and D.W. Galbraith (1984). Physiol. Plant. 60: 43-52.
4 Brown, S., J.-P. Renaudin, C. Prévot and J. Guern (1984). Physiol. Vég. 22: 541-554.
5 Deus-Neumann, B, and M.H. Zenk (1984). Planta Med. 50: 427-431.
6 Hara, Y., H. Yamagata, T. Morimoto, J. Hiratsuka, T. Yoshioka, Y. Fujita and Y. Yamada (1989). Planta Med. 55: 151-154.
7 Brown, S. (1984). Physiol. Vég. 22: 341-349.
8 Galbraith, D.W., C.L. Afonso and K.R. Harkins (1984). Plant Cell Rep. 3: 151-155.
9 Ketel, D.H. (1988). Planta Med. 5: 400-405.
10 Binding, H., R. Nehls, R. Kock, J. Finger and G. Mordhorst (1981). Z. Pflanzenphysiol. 101: 119-130.
11 Dougall, D.K., J.M. Johnson and G.H. Whitten (1980). Planta 149: 292-297.

THE EFFECTS OF OVER-EXPRESSING THE YEAST ORNITHINE DECARBOXYLASE GENE UPON THE NICOTINE AND POLYAMINE LEVELS IN TRANSFORMED ROOTS OF <u>NICOTIANA RUSTICA</u>

J. D. HAMILL, R. J. ROBINS, A. J. PARR, D. M. EVANS, J. M. FURZE, E. G. BENT & M. J. C. RHODES, PLANT BIOTECHNOLOGY GROUP, AFRC INSTITUTE OF FOOD RESEARCH, COLNEY LANE, NORWICH, NR4 7UA. UNITED KINGDOM.

1. INTRODUCTION

Putrescine is an important precursor of alkaloids such as nicotine in <u>Nicotiana</u> and scopolamine and hyoscyamine in <u>Datura</u>. It acts at the interface between primary and secondary metabolism and its primary role is as an intermediate in the formation of polyamines which have important roles in the growth of animal and microbial as well as plant cells. In animals and microbes the formation of putrescine and polyamines is regulated by effects on the expression of the enzymes ornithine decarboxylase (ODC) and S-adenosyl methionine decarboxylase (SAM dC)[1] both at the nucleic acid and protein levels. In plants, the situation is complicated by the occurrence of an alternative route of formation of putrescine from arginine and arginine decarboxylase (ADC) as well as from ornithine and ornithine decarboxylase and the presence of large buffering pools of putrescine and polyamine conjugates. Nevertheless, the role of polyamines in growth-related processes in plants is well established[2,3,4]. In tissues where putrescine-derived alkaloids are found, there is clearly competition for putrescine as substrate between the polyamine and alkaloid pathways. Feeding low levels of putrescine (1-5mM) can have a stimulatory effect upon the production of nicotine in <u>N. rustica</u> transformed root cultures[5]. This suggests that increasing the levels of this intermediate, by genetic manipulation techniques, may allow a greater flux down the metabolic pathway leading to nicotine and thus increase the secondary product concentrations of these cultures. We have investigated these interactions by attempting to over-express a yeast ornithine decarboxylase (yODC) gene in transformed roots of <u>Nicotiana rustica</u> to increase internal pools of putrescine and to study its partitioning between polyamines and alkaloids.

2. METHODS

The yeast ODC gene was introduced into transformed roots of <u>N. rustica</u> under the control of the CaMV35S promoter or a derivitive with duplicated enhancer sequences (CaMV35S^{++}) as described by Hamill <u>et al</u>[6]. Control tissues contained a chloramphenicol acetyl transferase gene under the control of the CaMV35S promoter. The levels of nicotine and intermediates and of polyamines were assayed exactly as described in previous papers[7,6]. The extraction and assay of enzymes was as described by Robins et al[8]. Plants were regenerated from transformed roots of N. rustica by culturing on MS medium containing IAA (2 mg l^{-1}) and 6-BAP (1 mg l^{-1}), pH6, 30g l^{-1} sucrose, and were grown to flowering stage in a greenhouse. Pollen was stained with acetocamine as described[9] and the percentage of fully formed grains assessed.

3. RESULTS

The presence of the yeast ODC in the transformed lines and of the CAT gene in the control lines was confirmed by Southern Blotting. The yeast ODC was shown to be integrated in an intact form into the genome of the transformed root lines, but was absent from the control lines. The transformed lines together with several of the control lines were assayed for their content of ODC and the metabolically related enzymes ADC, putrescine methyltransferase (PMT) and methyl putrescine oxidase (MPO). Of the six transformed lines containing the yODC gene under the CaMV35S^{++} promoter with duplicated enhancers which were investigated in detail, three lines, 6, 8 and 15 had ODC levels significantly greater than a range of control roots when measured at day 7 of a culture cycle, when the endogenous ODC levels (Robins, unpublished results) are maximal, and at early stationary phase (day 21) when the endogenous ODC had fallen to low levels (Table 1). A detailed time course of changes in the four enzymic activities in line 8 is shown in Figure 1. It can be seen that the expression of the yODC gene behind the constitutive CaMV35S^{++} promoter completely alters the course of expression of the enzyme. In particular, the expression of the enzyme remains high even late in stationary phase, by which time the endogenous activity has fallen dramatically. In this respect the kinetics of expression are similar to that of CAT under both the CaMV35S^{++} promoter[10] and nopaline synthesis promoter (Robins, unpublished). At day 7 the ODC level in line 8 is about 2-fold higher than in the control, but by day 21 it is more than 20-fold higher.

Northern Blotting of line 8 showed the presence of a 2.1 kb transcript homologous to the yeast ODC and that the transcript could be detected at all stages in the growth cycle (up to 22 days).

The levels of intermediates and end-products of the polyamine and nicotine pathway were assayed in lines containing the yODC gene under the CaMV355^{++} promoter and in control root cultures. The level of putrescine was found to be significantly higher than that of the control roots in two lines (line 6 and 8), particularly at the late stage of the cycle. Other lines showed similar levels of putrescine as controls. In contrast to putrescine, none of the transformed lines showed an enhanced level of either of the polyamine products, spermidine or spermine. In the pathway leading to nicotine biosynthesis, substantual variation in levels of the intermediates, N-methylputrescine and N-methyl pyrrolium were observed among the control lines analysed. The level of N-methylputrescine in line 6 was significantly different ($p<0.01$, <0.05 respectively) for the controls at both the early (day 7) and late (day 21) time point investigations but the levels of these intermediates in the other lines containing the

Line	ODC Activity 7 day	21 day	ADC Activity 7 day	21 day	PMT Activity 7 day	21 day	MPO Activity 7 day	21 day
1	30.2	4.6	3.3	0.5	14.9	0.9	17.0	2.3
3	21.8	8.2	3.2	0.9	10.3	2.5	35.4	0.0
6	36.9	25.8	3.3	0.6	12.8	1.3	30.0	3.6
8	55.3	47.1	3.3	1.1	10.7	4.1	36.0	1.5
14	9.7	2.4	1.7	0.4	6.4	0.0	8.9	0.2
15	54.0	21.7	5.4	1.0	11.3	4.3	39.2	0.0
51	18.1	1.1	4.1	0.4	7.6	0.0	22.1	3.2
58	7.9	1.4	2.2	0.4	5.2	0.3	15.6	6.8
24	20.1	0.4	9.6	1.0	6.1	0.0	39.6	3.9
27	15.6	1.7	3.4	0.5	5.1	2.1	36.0	2.4
29	22.4	4.4	3.1	0.6	12.4	3.0	51.8	2.1
31	9.7	2.2	1.6	0.7	4.4	0.3	21.6	0.2

TABLE 1: The activity (pkat/mg protein) of ODC, ADC, PMT and MPO (means of duplicate samples) in a number of manipulated lines and controls. Lines 1 - 15 contained the yODC gene under control of the CaMV35S promoter with duplicated enhancer sequences, lines 51 + 58 contained the yODC gene under control of the CaMV35S promoter and lines 24 - 31 were controls containing the CAT gene under control of the CaMV35S promoter with duplicated enhancer sequences.

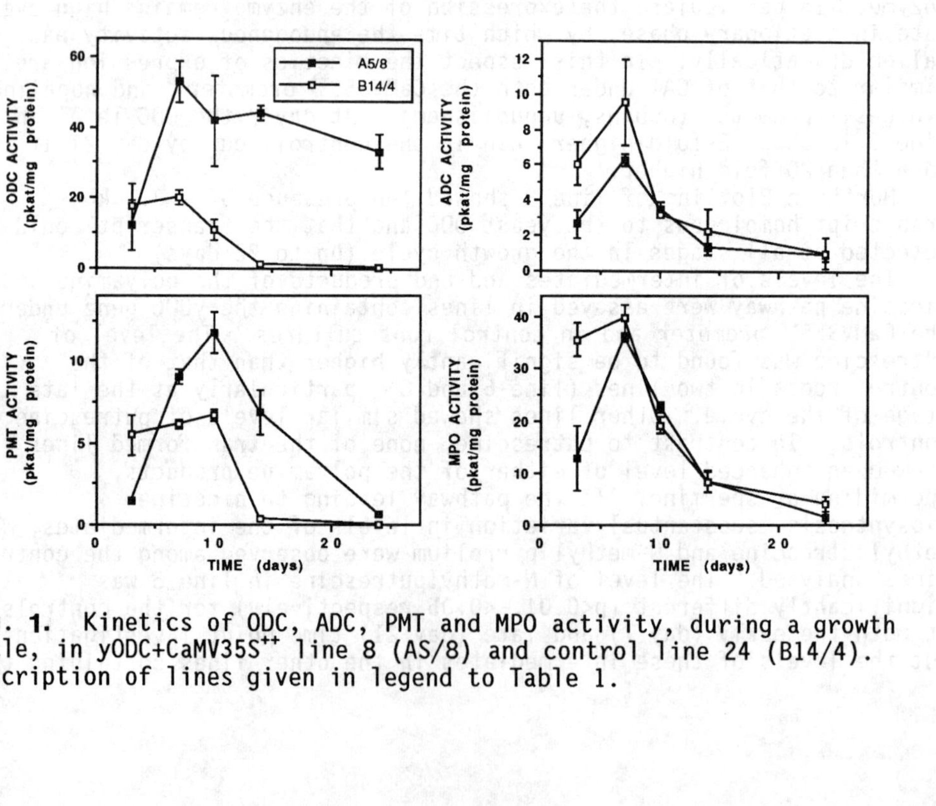

Fig. **1.** Kinetics of ODC, ADC, PMT and MPO activity, during a growth cycle, in yODC+CaMV35S^{++} line 8 (AS/8) and control line 24 (B14/4). Description of lines given in legend to Table 1.

ODC gene fell within the range seen in the control tissues which suggests that minimal stimulation, if any, occurred in the levels of these intermediates.

The effects of introducing the yODC gene, upon the concentration of nicotine, were assessed for lines containing the yODC under either the CaMV35S promoter or the CaMV35S promoter with duplicated enhancer sequences and were compared to control tissues. These results are shown in Figure 2. It would appear that there was no clear effect of introducing the yODC under the control of the CaMV35S promoter upon the nicotine levels as only one line showed a significant difference from control lines (line 53, Figure 2). The mean nicotine content of both populations was not significantly different (2.28±0.22μmol/g fr. wt for controls compared to 2.65±0.4μmol/g fr. wt for CaMV35S yODC lines, $p<0.4$; t-test). However there was an increase in the nicotine content of lines containing the yODC gene under the control of the CaMV35S^{++} promoter with duplicated enhancer sequences (Figure 2), and the mean nicotine content of this population was 4.04±0.48μmol/g fr. wt. which is significantly greater than the levels of control tissues ($p<0.05$, t-test).

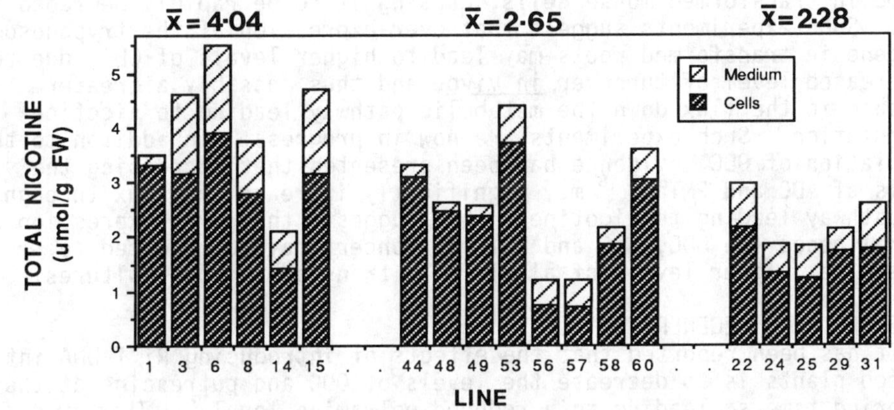

Fig.2

Nicotine levels in lines containing the yODC gene under the CaMV35S^{++} promoter (1-15), the CaMV35S promoter (44-60) and control lines containing the CAT gene under the CaMV35S^{++} promoter (22-31).

Thus it would appear that the nicotine levels in transformed roots of <u>N. rustica</u> can be stimulated (albeit less than twofold on average) by over expression of the yeast ODC gene, but only when this gene is driven by a very strong constitutive promoter. Stimulatory effects upon ODC enzyme levels and upon metabolic intermediates leading to nicotine were seen in some lines but not others. No stimulatory effects upon the level of polyamines spermine and spermidine were seen. Taken together these results suggest that other regulatory factors act to minimise the metabolic perturbation effects of over-expressing yeast

ODC in plant cells. We are not sure at present where these regulatory factors are operating though several possibilities are immediately obvious. For example, post-transcriptional[11] and post-translational mechanisms[12] have been reported which regulate ODC levels and the yODC gene transcript in the transformed roots would presumably not be independent of such regulation. This would tend to minimise the levels of putrescine accumulation and thus partly explain the results reported above. Sequences inherent in the yeast ODC protein may also be important regarding protein turnover and may explain why we were only able to achieve modest increases in the peak of ODC activity whereas other workers have reported much greater stimulation of enzyme levels using the CaMV35S promoter to overexpress genes eg, 40-fold in EPSP synthesis[13] and 35-fold in glutamine synthase[14]. It is of great interest that the ODC enzyme from trypanosomes is much more stable <u>in vivo</u> than that from mammalian cells and that the reasons are due to sequences near the carboxy end of the protein. (PEST regions in the mammalian enzyme which are lacking in the trypanosome enzyme[15]). Recent experiments by Coffino and Wang[16] have in fact shown that replacement of the sequences near the carboxy terminus of the trypanosome ODC gene with those from the mouse gene confers much greater instability to the enzyme in transformed mouse cells, causing it to be rapidly degraded <u>in vivo</u>. Such experiments suggest that over-expression of the trypanosome ODC gene in transformed roots may lead to higher levels of ODC, due to a decreased level of turnover <u>in vivo</u>, and thus possibly a greater increase in the flux down the metabolic pathway leading to nicotine accumulation. Such experiments are now in progress. In additon to the stimulation of ODC, evidence has been presented that increasing the levels of ADC and PMT[8,17,18] may significntly increase the flux through the pathway leading to nicotine. This suggests that over-expression of several genes (ie ODC, ADC and PMT) in concert may be required to achieve yet higher levels of alkaloid in transformed root cultures.

4. EFFECTS ON REGENERATED PLANTS

It has been reported that the effects of introducing Ri T-DNA into tobacco plants is to decrease the levels of ODC and putrescine at the flowering time so leading to a reduced polyamine level[19]. This may partly explain the abnormalities seen in floral development in plants containing Ri T-DNA[20]. In our experience, plants of <u>N. rustica</u> regenerated from transformed roots do show typical Ri T-DNA effects and are self-sterile. We have observed that at least one of the lines containing the yeast ODC gene under the control of the CaMV35S promoter with duplicated enhancer sequences (line 6) has given rise to plants which, though not totally normal, do have high levels of fully formed pollen and are self-fertile. The seeds produced by these plants are fertile (over 95% germination) and we are in the process of studying the segregation of the yODC gene and the Ri T-DNA genes in the R_2 generation. Using the PCR approach outlined in another presentation at this conference (Hamill <u>et al</u>), we are able to screen large numbers of offspring to assess whether there is indeed a correlation between fertility and the presence of the yODC gene in plants containing Ri T-DNA.

REFERENCES

1 Helby, O. and Persson, L. (1990). T. I. B. S. 15, 153-158.
2 Meijer, E. G. M. and Simmonds, J. (1988). J. Expt. Bot. 39, 737 -794.
3 Mengoli, M., Bagni, N., Luccarini, G., Nuti-Ronchi, V., Serafini-Fracassini, D., (1989) J. Plant Physiol. 134, 389-394.
4 Tiburcio, A. F., Kaur-Sewhney, R., Galston, A. W. (1988). Plant Cell Physiol. 29 1241-1249.
5 Walton, N. J., Robins, R. J., and Rhodes, M. J. C. (1988). Plant Sci 54, 125-131.
6 Hamill, J. D., Robins, R. J., Parr, A. J., Evans, D. M., Furze, J. M., Rhodes, M. J. C. (1990). Plant. Mol. Biol. (in press).
7 Parr, A. J. and Hamill, J. D. (1987). Phytochemistry 26: 3241-3245.
8 Robins, R. J, Parr, A. J., Payne, J., Walton, N. J., Rhodes, M. J. C. (1990) Planta (in press).
9 Collins, G. B., (1984) in Nicotiana, procedures for experimental use, USDA, Technical Bulletin, No. 1586.
10 Rhodes, M. J. C., Robins, R. J., Hamill, J. D., Parr, A. J., Hilton, M., Walton, N.J. (1990). In Secondary Products from plant tissue culture, Charlwood, B. and Rhodes, M. J. C. (eds) Oxford University Press, pp.199-223.
11 Kameji, T., and Pegg, A. E., (1987). J. Biol. Chem. 262:2427-2430.
12 Koromilas, A. E. and Kryiakidis, D. A., (1988). Physiol Plantarum 72:718-724.
13 Shah, D. M., Horsch, R. B., Klee, H. J., Kishore, G. M., Winter, J. A., Turner, N. E., Hironaka, C. M., Sanders, P. R., Gasser, C. S., Aykert, S., Siegel, N. R., Rogers, S. G., Fraley, R. T. (1988) Science, 233:478-481.
14 Eckes, P., Schimitt, P., Daub, W., Wegermeyer, F., (1989). Mol Gen Genet. 217:263-268.
15 Coffino, P. (1989) in Ornithine Decarboxylase: Biology, Enzymolgy and Molecular Genetics (Hayashi S, ed) pp 135-144 Pergamon Press.
16 Coffino, P. and Wang, C. C. (1990). Abstract of 634th meeting on Biochemistry of Trypanosomes, Bath University, April 1990.
17 Tiburcio, A. F., Galston, A. W., (1986) Phytochemistry 25, 107-110.
18 Rhodes, M. J. C., Robins, A. J., Aird, E. L. H., Payne, J. Parr, A. J., Walton, N. J. W., (1989). In: Proceedings of the Second International Symposium on Primary and Secondary Metabolism in Plant Cell Culture (Kurz W G W, ed) Springer-Verlag, Berlin, p 58-72.
19 Tepfer, D., (1984) Cell 47:959-967.
20 Martin-Tanguy, J., Tepfer, D., Paynot, M., Burtin, D., Heisler, L., Martin, C., (1990) Plant Physiol.

SEMISYNTHETIC PRODUCTION OF VINBLASTINE INVOLVING CELL CULTURES OF CATHARANTHUS ROSEUS AND CHEMICAL REACTION

YASUHIRO FUJITA[1], YASUHIRO HARA[1], TEIJIRO MORIMOTO[1] & MASANARU MISAWA[2]
1. Bioscience Research Laboratories, Mitsui Petrochemical Ind. Ltd., Waki-cho, Kuga-gun, Yamaguchi 740, Japan; 2. Allelix Inc., 6850 Goreway Drive, Mississauga, Ontario L4V 1P1, Canada (Present address: Bio International Inc., Suite 430, 151 Bloor Street West, Toronto, Ontario, M5S 1S4, Canada)

1. INTRODUCTION

Vinblastine (VLB) is a rare dimeric alkaloid contained in *Catharanthus roseus* and used in treating leukemia and malignant lymphoma. Many researchers have studied the production of VLB by plant cell and tissue cultures. So far, there has been no success in producing VLB by undifferentiated cell cultures, but the monomer precursor, catharanthine, can be produced by this way (1). VLB can be produced in multiple shoot cultures of *C. roseus*, but the VLB yield in the cultures is only about 10% of that of the Catharanthus leaves (2). Furthermore, it is difficult to synthesize VLB chemically due to its intricate structure.

None the less, Misawa, the co-author of this paper, and his colleagues developed semisynthetic production of VLB in which catharanthine, produced by cell suspension cultures of *C. roseus*, and vindoline, extracted from the plants, are coupled by the chemical reaction using Fe^{3+} as a catalyst (3,4). This report explains the results of our studies on industrial production of VLB focusing on the production of catharanthine.

2. MATERIALS AND METHODS

2.1. Cell lines and basic culture conditions

An LBE-1 line that was induced from the anther culture of *C. roseus* var. Little Bright Eye by Dr. Smith, Allelix Inc, was used. Murashige-Skoog medium (5) containing 30 g dm^{-3} sucrose, 1 mg dm^{-3} NAA, and 0.1 mg dm^{-3} kinetin was used as a basic medium. Cell suspensions were cultured at 25°C in the dark, shaken at 100 rpm on a rotary type shaker, and subcultured every week.

2.2. Selection of highly productive cell lines for catharanthine

A small cell-aggregate method was used for selecting highly productive cell lines. Seven day old cells separated from the medium were spread on an agar plate and cultured for 2-3 weeks. The 2-3 mm diameter colonies which formed on the plates were transferred to new plates. After 2-3 more weeks of growth, each resulting callus was inoculated into a liquid medium and subcultured as mentioned above. The alkaloid productivity of each cell line was checked in every subculture, and highly productive cell lines for catharanthine were selected. For the selection of the cell lines resistant to 5-methyltryptophan (5-MT), the small cell-aggregate method was used, but an agar medium containing 0.05 m mol dm^{-3} 5-MT was used for colony and callus formation.

2.3. Scale-up experiments

A jar fermenter equipped with a hollow-paddle stirring wing developed by Matsubara et al. (6). The volume of the medium was 1.7 dm^3 and the medium contained 60 ppm antifoam (TSA-7341, Toshiba Silicone).

2.4. Conditions for extraction and determination of alkaloids produced

The harvested cells were dried at 40°C for 24 h. Alkaloids were extracted by ultrasonic treatment of the dried cells in methanol and determined according to the method of Smith et al. (3).

2.5. VLB production by chemical coupling of catharanthine and vindoline

Reactions were performed according to the method of Vukovic et al. (4). Products of the reaction were separated by silica gel chromatography and identified using ^{1}H-NMR, ^{13}C-NMR and MS by comparison with authentic VLB.

3. RESULTS AND DISCUSSION

3.1. Selection of highly productive cell lines for catharanthine

We tried two selection procedures, the method with small cell-aggregate and the method of resistance to tryptophan analogue.

3.1.1 The method with small cell aggregate: We could collect highly productive cell lines having about 5 times the cellular content of catharanthine after one passage through the aggregate (Table 1.). However, further highly productive cell line could not be obtained even when the process was repeated.

3.1.2 The method utilizing resistance to 5-MT: We investigated the effect of L-tryptophan (Trp) on the catharanthine production and found that the addition of 3 m mol dm^{-3} to the medium gave an increase in the catharanthine yield by 30% of the control. A similar effect of Trp was reported on the production of serpentine and ajmalicine (7). These results suggested that the synthesis of Trp is one of the key steps for indole alkaloid production. Therefore, we tried to acquire the highly productive cell lines for catharanthine from the 5-MT resistant cell

TABLE 1. Enhancement of catharanthine production by selection of highly productive cell lines using a small cell-aggregate method. Each cell line was subcultured every week. The values show the average of 10 subcultures after the lines selected.

Selection	Cell yield (gDW dm^{-3})	Catharanthine content (%)	Catharanthine yield (mg dm^{-3})
Parent	19	0.10	20
1st	20	0.50	100
2nd	19	0.53	100
4th	18	0.67	120

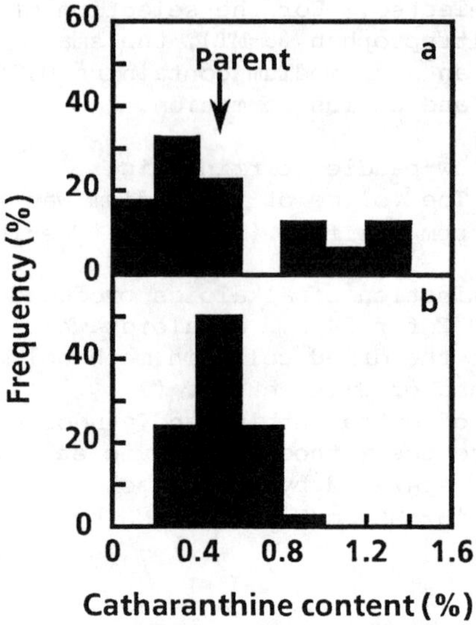

FIGURE 1. Distribution of catharanthine content in offspring obtained from a) the 5-MT method (the number of offspring: 18) and b) the small cell-aggregate method (the number of offspring: 34). All the cells were harvested after 10 days of growth.

lines that should have a capacity to produce excess Trp. Consequently, 18 cell lines resistant to 5-MT were obtained and highly productive cell lines were included at a frequency higher (Fig. 1.a.) than with the normal small cell-aggregate method (Fig. 1.b.). The highest yield of cellular catharanthine was about 1.4%DW.

The parent line contained ajmalicine as well as catharanthine; the cellular content of ajmalicine was about 70% of catharanthine. The selection with the 5-MT method increased the contents of both alkaloids, but in almost all the 5-MT resistant lines, the content of ajmalicine was lowered to about 40% of catharanthine. Conversely, the alkaloid composition in the offspring did not change even when the small cell-aggregate method was carried out. This finding suggests that the 5-MT treatment influenced regulation of the enzymes located on the branches of the Iboga-type (catharanthine) and the Corynanthe-type (ajmalicine) alkaloids.

Scott et al. studied the selection of highly productive cell lines for ajmalicine using the 5-MT resistance method, but the resistant cell lines they obtained contained no target alkaloid although the cells accumulated a large amount of Trp (8). Our 5-MT method seems to differ from their method at the following points: 1) selection of the resistant cell lines with agar medium instead of liquid medium, 2) reduction of the 5-MT concentration to about 20% of that used by Scott et al. and 3) shortening the period for 5-MT treatment. Only 3 resistant lines were obtained from the method of Scott et al., but 18 lines survived in our case. Because we used weaker treatment of 5-MT than they did which enhanced the frequency of the survivals, the probability for obtaining highly productive cell lines may have been increased.

3.2. Effect of medium composition and inoculum size on catharanthine production

The effect of all the components contained in the MS medium were individually investigated, but there were no components that could markedly enhance the catharanthine output. The investigation of the combined effect of the main components showed that reduction of

TABLE 2. Increase in cellular content of catharanthine by modification of medium composition. The 5-MT resistant cell line was used and cells were harvested after 1 week of growth.

Inoculum size ($gFW\ dm^{-3}$)	Concentration of components NAA	Sucrose	N+P*	Catharanthine content (%)	Cell yield ($gDW\ dm^{-3}$)
35	N**	N	N	0.84	19
	L	N	N	0.90	17
	N	L	N	0.54	7
	N	N	L	0.82	14
	L	L	N	0.75	8
	N	L	L	1.47	7
	L	N	L	0.99	13
	L	L	L	1.39	7
105	N	N	N	1.69***	24

* Ammonium, nitrate and phosphate

** N: normal concentration in the basic medium, L: one third of the normal concentration

*** The effect of high inoculation on the production was also reported by Berlin et al. (9).

all the concentrations of sucrose, nitrogen sources and phosphate increased the cellular content of catharanthine (Table 2.). Increasing inoculum size also brought the similar effect on the content without modification of the medium composition. Therefore, increasing inoculum size can enhance the catharanthine yield per unit volume of a fermenter.

3.3. Scale-up of cultures with jar fermenter

Our Catharanthus cells grew forming fine cell aggregates in which the aggregate smaller than 0.8 mm in diameter were 70-80% of the total weight of cells. Cell density in the culture was beyond 300 $gFW\ dm^{-3}$ after the late exponential growth phase; as a result, air bubbles did not disperse well because of an increase in the viscosity of cell suspension. Increases in agitation speed, aeration rate or both, with cell proliferation, were necessary to keep k_La (volumetric oxygen transfer coefficient) constant at about 10 h^{-1} which is the same as shaken flask culture. However, these conditions would increase shear stress to the cells or aggravate blowing cells over the surface of the medium. To avoid these problems, we tried to culture the cells keeping the concentration of dissolved oxygen (DO) in the medium constant by enriching the aeration gas with oxygen (Fig. 2.). We found that catharanthine production was improved with the increase in DO concentration, and the same productivity for catharanthine as the flask culture was obtained above 8 ppm of DO in jar fermenter culture (Table 3.). Furthermore, in this culture system, the cell culture with increased inoculum size produced the expected catharanthine yield.

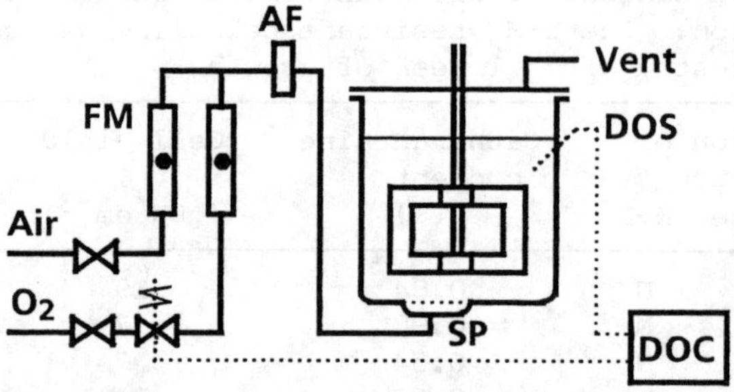

FIGURE 2. Diagram of equipment used for the culture of C. roseus. AF: air filter, DOC: DO controller, DOS: DO sensor, FM: flow meter, SP: sparger

TABLE 3. Effect of DO on cell yield and on catharanthine yield by C. roseus cells. Aeration rates of air and oxygen was 0.05 vvm each and agitation speed was 30 rpm. The 5-MT resistant cells were used and cells were harvested after 1 week of growth.

DO (ppm)	Inoculum size ($gFW\ dm^{-3}$)	Cell yield ($gDW\ dm^{-3}$)	Catharanthine content (%)	Catharanthine yield ($mg\ dm^{-3}$)
not controlled*	35	20	0.23	50
5	35	18	0.45	80
8	35	17	0.83	140
12	35	16	0.88	140
16	35	16	0.97	150
8	70	22	1.07	230

* Aerated without oxygen enriched air.

3.4. Chemical synthesis of VLB

Vukovic et al. found that the dimeric alkaloids including VLB were obtained by the chemical coupling of catharanthine and vindoline in the glycine buffer (pH 2.0) at 4°C using $FeCl_3$ as a catalyst (4). We have studied the mechanisms of this reaction and found that the formation of VLB happens in two steps via anhydroVLB; therefore, we could establish the optimum conditions for each step of the reaction. At the first step which anhydroVLB is formed from the two precursors, the yield was elevated to about 90%, and at the second step which anhydroVLB is converted to VLB, the yield was increased to about 50% (Fig. 3.).

This research was done in collaboration with Allelix and Mitsui Petrochemical; Allelix established the cell line capable of catharanthine production and the method for the chemical coupling and Mitsui Petrochemical improved the technology for industrialization.

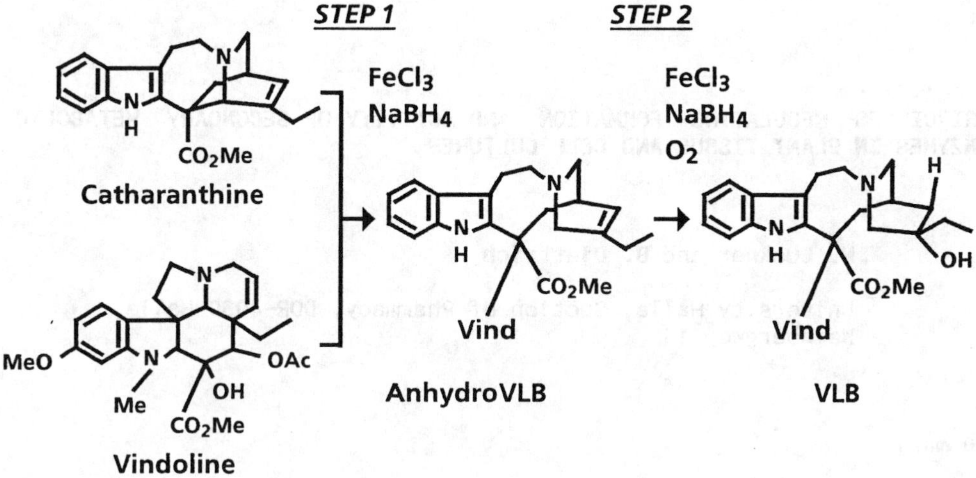

FIGURE 3. VLB synthesis by two step chemical reactions.

ACKNOWLEDGMENTS

We thank Dr. J.I. Smith, Dr. N.J. Smart, Dr. A.E. Goodbody, Dr. J. Vukovic and Dr. T. Endo, Allelix, for their helpful advice. Also we are grateful to Prof. Dr. F. DiCosmo, Toronto Univ., for his useful suggestions for catharanthine production.

REFERENCES

1 Kutney, J.P., Aweryn, B., Chatson, K.B., Choi, L.S.L. and Kurz, W.G.W. (1985) Plant Cell Rep. 4, 259-262
2 Japanese Patent, Tokkaisho 61-254127
3 Smith, J.I., Smart, N.J., Kurz W.G.W. and Misawa, M. (1987) J. Exp. Bot. 194, 1501-1506
4 US Patent 4778885
5 Murashige, T. and Skoog, F. (1962) Physiol. Plant. 15, 473-497
6 Matsubara, K., Kitani, S., Yoshioka, T., Morimoto, T., Fujita, Y. and Yamada, Y. (1989) J. Chem. Tech. Biotechnol. 46, 61-69
7 Zenk, M.H., El-Shagi, H., Arens, H., Stöckigt, J., Weiler, E.W. and Deus, B. (1977) in Plant Tissue Culture and Its Bio-technological Application (Barz, W., Reinhard, E. and Zenk, M.H., eds.) pp.27-43, Springer-Verlag, Berlin Heiderberg New York
8 Scott, A.I., Mizukami, H. and Lee, S.L. (1979) Phytochemistry 18, 795-798
9 Berlin, J., Mollenschott, C. and DiCosmo, F. (1987) Z. Naturforsch. Sect C. 42, 1101-1108

PRINCIPLES REGULATING FORMATION AND ACTIVITY OF SECONDARY METABOLIC ENZYMES IN PLANT TISSUE AND CELL CULTURES

M. Luckner and B. Diettrich

University Halle, Section of Pharmacy, DDR-4050 Halle, Weinbergweg 15

Summary

Regulation of enzyme amount is the most important means in the control of secondary product biosynthesis. In the native plants expression of most secondary enzymes is integrated into the programs of gene expression and development leading to specialized cells and organs. In cell cultures "natural expression" may be achieved as part of these programs. Biosynthesis of secondary products thus is easily triggered in vitro as part of tissue and organ development.

In addition, expression of secondary metabolism may be possible independent of these programs. Formation of several secondary products may be raised, for instance, in unspecialized cells by elicitors. The expression not controlled by natural programs probably is based on multiple sites of regulation at the genetic material encoding secondary metabolic enzymes and the interaction of effectors not formed within the programs with the "additional" sites.

Regulation of enzyme activity is the second important principle in controlling the formation of secondary products. Precursors and intermediates are compartmentalized and channeled to the sites of metabolism. In vivo most enzymes of secondary metabolism work under substrate limitation and do not reach V_{max}. In some cases enzyme activity in vivo may be raised by increase of the supply of endogenous or exogenous precursors.

In most cell cultures, however, even under favorite conditions the rates of secondary product biosynthesis remain low. Due to limited expression of secondary metabolic genes only small amounts of secondary metabolic enzymes are formed. Higher rates may be obtained after fusion of the structural part of the genes with strong promoters. Thus in future the rate of biosynthesis of plant secondary products may be raised by the methods of genetic engineering. For the biotechnological production of secondary compounds the plant cells have the disadvantage of slow growth and of fragility. Therefore the expression of the "improved" genes in microorganisms may be more suitable for secondary product biosynthesis than that in plant cells.

1. The natural expression of secondary metabolic pathways in vitro as part of developmental programs

The expression of secondary metabolism is an aspect of cell differentiation and secondary products are formed in specialized cells [15, 19]. The lack of secondary product biosynthesis in plant cell cultures is in most cases the result of absence or incorrect procedure of cell specialization. Based on this fact the following strategy for the "natural expression" of secondary metabolic pathways in plant cell cultures was proposed [16]:

a) in the first step the programs in which synthesis and accumulation of secondary metabolites are integrated will be elucidated, and

b) in the second step these programs will to be triggered in vitro, i.e., in the cell cultures.

This strategy will be exemplified in the following with the formation of cardenolides in cultures of Digitalis lanata, which in our laboratory was extensively investigated in the last years (summaries s. [16 - 18]).

D. lanata is a medicinal plant which due to the formation of cardenolides is cultivated in different parts of the world. The highest cardenolide contents were found in the leaves. Detailed examination showed that the mesophyll cells were the most important site of cardenolide biosynthesis as well as of cardenolide storage. With the usage of sensitive radioimmunoassays it was found that the leaves contain the highest cardenolide content of all cell types of the D. lanata plants [28]. Protoplasts derived from mesophyll cells were able to incorporate precursors, e.g., progesterone-4-^{14}C, into the cardenolides [16], a result also found with D. purpurea mesophyll protoplasts [10]. Mesophyll cells are therefore a cell type, the formation of which in cell cultures was expected to cause the "natural" triggering of cardenolide biosynthesis and accumulation.

The detailed investigation of the developmental programs causing organogenesis in cell cultures of D. lanata showed that mesophyll cells are formed during the regeneration of shoots as well as during the formation of somatic embryos. In both cases cardenolides were synthesized and accumulated. The formation of somatic embryos was most suitable for the investigation of cardenolide biosynthesis, because the embryo structures forming cardenolides were morphologically very simple and could be obtained in large quantities [25].

The formation of cardenolides began at a relatively early stage of embryo development (transformation of stage-I-globules into stage-II-globules) [25]. In the first step a slight increase in the cardenolide content became measurable, which was independent from external triggers. It was signalizing the competence of the cells to form cardenolides. Cardenolide biosynthesis was influenced to a great extent by light in the competent cells. Red light and blue light were shown to be most suitable [23]. The "natural" expression of cardenolide biosynthesis as part of the developmental program "somatic embryogenesis" resulted in a cardenolide content of about 1 µmol digitoxin equivalents per gram dry weight during a cultivation period of 4 weeks.

2. Triggering of secondary metabolism by "unphysiological" signals

The "natural" expression of secondary product formation as part of the programs of cell differentiation and development includes many morphological and biochemical events which are related only indirectly to secondary metabolism. The pattern of differentiation may be simplified by the "direct" expression of the genes of secondary metabolism apart of the programs in which secondary product formation usually is integrated. In some cell cultures direct triggering of secondary product biosynthesis by unphysiological signals was successful. Flavonoids, naphthoquinones, rosmarinic acid, and several other groups of secondary products may be synthesized in suspensions of "parenchyma-like" cells [2, 4]. The formation of organs or other complex morphological structures was proven to be unnecessary in these cases.

The expression of secondary metabolic pathways apart from the programs of gene expression may be possible due to the existence of different cis-acting regulatory sites (enhancer and/or silencer sequences) connected with the genes of secondary metabolism. At least two cis-regulating control regions were found, for instance, as part of the chalcone synthase gene of Antirrhinum majus (s. [26]). At the different regulatory sequences different regulatory molecules may interact allowing the expression of the genes of secondary metabolism by independent signals.

Many questions still have to be answered with respect to the program-independent expression of secondary metabolic genes, e.g.:

- which genes of secondary metabolism have more than one regulatory site?
- do all genes of a given secondary metabolic pathway possess similar regulatory sequences?
- is there a "master switch", which governs the transcription of all genes of a secondary metabolic pathway and which is regulated by different signals and effectors?

In most experiments expression of secondary metabolism apart from the morphological programs in cultures of parenchyma-like cells was therefore handled empirically. In these experiments different potential signals were tested, e.g., stress factors, like heavy metal ions, media causing starvation, and elicitors [6].

In cultures of parenchyma-like cells of Digitalis lanata stress factors were not very successful in triggering cardenolide biosynthesis. This result was probably caused by the fact that cardenolides emerged during evolution as compounds mediating qualitative defence against herbivores [15]. Cardenolides were less active, however, against microorganisms (unpublished results), which are plant stress factors causing elicitation of many types of secondary products in nature. The only stress originating cardenolide biosynthesis in parenchyma-like cells of D. lanata in vitro was cultivation of submerged grown cells on the surface of solidified nutrient media (stress probably caused by the reduced moisture and the air). Cardenolide biosynthesis was, however, not stable and disappeared after a few subcultivations [16].

3. The manipulation of secondary metabolism of cell cultures on the base of the naturally occurring genetic material

3.1 The simplification of developmental programs

The developmental programs occurring in nature may be simplified in vitro by omitting parts not necessary for the biosynthesis of secondary products. Suitable methods for this type of experiments are:

- the isolation of mutants in which certain areas of the gene expression programs are destroyed or cannot be expressed furthermore, and

- the use of chemicals, which inhibit the proceeding of the programs at certain stages.

Both methods were used with cell cultures of D. lanata. A Digitalis lanata cell strain was isolated by Ohlson et al. [21] from a normal embryogenic strain which formed cardenolides in spite of the fact that the development was simplified. Whereas in the wild-type strains the formation of cardenolides was dependent on illumination which caused the transformation of the plastids present in the stage-I-globules into chloroplasts (s. above), the strain of Ohlson was unable to form chloroplasts and synthesized cardenolides already in the dark. In another type of experiments chloroplast formation in the wild type embryogenic strains was blocked by inhibitors of plastid differentiation, e.g., SAN 9789, under conditions which did not repress cardenolide biosynthesis [24]. These results indicated that chloroplasts are not necessary for cardenolide biosynthesis and in the wild-type strains coupling of cardenolide formation with the transformation of plastids into chloroplasts is accidentally.

3.2 The increase of precursor flux

In a multitude of experimental systems it was shown that the enzymes of secondary metabolism did not work under V_{max} conditions [15] because the concentrations of precursors, cosubstrates and other necessary cell products were too low. An increase of the amounts of precursors and other rate-limiting substances thus may be expected to yield better rates of secondary product biosynthesis. For improved precursor supply either strains may be used which produce these compounds in excess or the substances may be added to the nutrient medium. Also inhibitors of metabolic pathways competing for the rate limiting substances may be applied. In many instances, however, experiments of this type were unsuccessful, probably due to the strict compartmentation and channeling of precursors, cosubstrates and intermediates in the producer cell. Also with D. lanata cell cultures feeding of potential precursors and intermediates yielded no substantial increase in cardenolide biosynthesis (unpublished results). Further experiments will show whether the implantation of additional copies of 3-hydroxy-3-methylglutaryl CoA reductase, probably a key enzyme in the biosynthesis of precursors of all types of isoprenoids, or the selection of overproducing mutants resistant to inhibitors of steroid biosynthesis will increase the formation of cardenolides.

3.3 The significance of storage compartments

With lupine cell cultures it was shown that the absence of suitable storage compartments inhibited the formation of quinolizidine alkaloids [29]. This result is in agreement with experiments which demonstrated that product inhibition may decrease the formation of secondary compounds [15]. Artificial storage compartments built, e.g., by the addition of ion exchange resins for adsorption of alkaloids or of liquid paraffins for solving lipophilic secondary products, yielded higher rates of secondary product biosynthesis in cell cultures releasing secondary products into the nutrient medium [3].

However, most secondary products are stored within the protoplast, for instance, in the vacuoles [15]. In these cases the creation of artificial storage compartments in the cultures is without use. It would make sense if the storage compartments could be permeabilized and the secondary products would be released from the cells. As yet, however, no suitable method has been found for a (partial) permeabilization of plant cells without destroying the viability.

The cardiac glycosides, formed in cell cultures of D. lanata, also are accumulated in the vacuoles [5, 11]. Glucosylated cardenolides were taken up in an energy-dependent reaction. The amount of cardenolides stored depended on the metabolic state (ATP content) of the cells and could not exceed a critical value. If cells were confronted with large amounts of cardenolides an increasing portion was found in the medium surrounding the cells.

Cardenolides which could not be glucosylated could not be stored. This was used in the development of a biotechnological procedure for the 12ß-hydroxylation of cardenolides by cell cultures of D. lanata (transformation of methyldigitoxin to methyldigoxin. In methyldigitoxin the glucosylation site is blocked by methylation). Methyldigoxin was not accumulated in the cells, but released into the nutrient medium [1].

4 "Improvement" of the genetic material of secondary metabolic pathways

In nature secondary products are accumulated in plants during the whole period of vegetation or the entire life span. Even if the rates are small due to the long periods of time available considerable quantities of secondary products may be stored. In vitro the period of cultivation comprises only a few days or weeks. Secondary product formation has therefore to be much faster to reach comparable concentrations. Rate limiting is usually the amount and the in vitro activity of the secondary metabolic enzymes (s. above). Several attempts have been made to overcome this drawback.

4.1 Production and selection of mutants

Changes in the genetic material of cultivated plant cells may be obtained by chemical and physical means. They may be (rarely) accompanied with an increased rate of secondary product biosynthesis. However, mutations often do not result in a changed phenotype because nearly all plant cells (in contrast to most microbial cells) are diploid or polyploid. In addition efficient screening methods for the selection of cells overproducing secondary products are rare. The effort necessary for the induction/selection of mutants overproducing secondary products in plant cell cultures therefore is rather high and in a limited number of cell strains only the selection of mutants resulted in better producing strains.

Relative to D. lanata respective experiments with diploid cells did not yield improved strains. Only in the last time haploid cultures were derived from D. lanata anthers (unpublished results). These cultures are the base for a new mutation program which, however, is still under work.

4.2 The usage of somaclonal variability

Cell strains cultivated for long periods in vitro often contain a multitude of cell lines deviating from each other in their genetic material. These differences arise spontaneously, e.g., by irregular mitoses, by the activation of cryptic transposons, by gene amplifications, etc. [14, 20, 27]. Higher producing cell lines may be selected from these cultures. In most cases, however, the higher productivity was shown to be unstable.

The existence of somaclonal variants was demonstrated also in cultures of the D. lanata cell strain VII, which was cultivated in vitro for about 10 years. The variants showed differences in the rates of cardenolide biosynthesis under stress in parenchyma-like cells (s. above) [16]. Until now, no lines with stable cardenolide biosynthesis could be selected. Also plants regenerated from cells of strain VII showed differences in the cardenolide contents. The deviations exceeded those of D. lanata clone plants (unpublished results) and probably reflected somaclonal variation.

4.3 The combination of complex sets of characteristics by fusion of cells and organells

Hybridization of cells allows the combination of the (positive) characteristics of different cell types. It may be brought about by fusion of protoplasts of different cell strains [8]. The procedure results in the combination of a multitude of genes of both partners, yielding complex changes in the phenotype. It was used successfully, e.g., in the combination of fast growth and a high rate of shikonin production substantiated in different cell lines of Lithospermum erythrorhizon [7]. Also the fusion of protoplasts with organells from other plants has been described. Since organells like nuclei, plastids and mitochondria house own genetic material with this type of experiments new genotypes may be constructed too.

4.4 Change of the regulatory sites of the secondary metabolic genes by the methods of genetic engineering

The rate of synthesis of the enzymes of secondary metabolism depends mainly on the rate of transcription of the respective genes. In this respect the naturally occurring regulatory sequences form a frontier which in nature hinders secondary metabolism to withdraw too much of the resources from primary metabolism. Higher rates of secondary product biosynthesis may be reached if the structural part of the genes becomes isolated and fused by the methods of genetic engineering with regulatory sequences allowing higher rates of expression. This was successfully carried out in some cases. There are promoters available which result in a strong and in several cases also constitutive expression of the genes they are connected with. The effort necessary for isolation and cloning of secondary metabolic genes, however, is still rather high and until now only a few genes were obtained (cf. [12, 13]).

With D. lanata the base was laid for the use of the techniques of genetic engineering in the last years. Methods were worked out allowing the transformation with the Agrobacterium tumefaciens system using wild-type as well as genetically engineered plasmids with bacterial marker and reporter genes (unpublished results). In future experiments genes will be introduced into the cells which, hopefully allow a higher rate of precursor supply and an increased rate of cardenolide biosynthesis.

5. The future of the production of secondary metabolic compounds with plant cell cultures

As yet in a few countries only (Japan, USSR) plant cell cultures are used for the industrial production of secondary compounds. In several other countries (e.g., Germany) respective procedures currently are tested. Also with D. lanata cell cultures most questions were solved connected with the cultivation in fermenters [1, 4, 9]. With most experimental systems it was shown, however, that the rates of production of secondary compounds still are to small for an economical output. Deeper penetration into the problems of plant physiology, enzymatics, and molecular biology is necessary in the future.

As yet it cannot be predicted whether the "natural" rate of secondary product biosynthesis in a considerable number of plant cell strains can be raised sufficiently by manipulation of physiology and genetics. The transfer of plant genes after genetic treatment into organisms easier to handle biotechnologically (e.g., bacteria and yeasts) often may be the better solution. In nearly all fields the biotechnological use of the enzymes of plant secondary metabolism today is still in its infancy. The large multitude of genes and enzymes present in plants, however, forms a source, which is a challenge for the biochemist and may shape biotechnology in the future.

6. References

[1] Alfermann, A. W., Spieler, H., Reinhard, E. (1985) in Primary and Secondary Metabolism of Plant Cell Cultures (Neumann, K. H., Barz, W., Reinhard, E., eds.), pp. 316-322, Springer-Verlag, Berlin.

[2] Bajaj, Y. P. S., ed. (1988) Biotechnology in Agriculture and Forestry, Vol. 4: Medicinal and Aromatic Plants I, Springer-Verlag, Berlin.

[3] Beiderbeck, R., Knoop, B. (1988) in Biotechnology in Agriculture and Forestry, Vol. 4: Medicinal and Aromatic Plants I (Bajaj, Y. P. S., ed.), pp. 123-135, Springer Verlag, Berlin.

[4] Constabel, F., Vasil, I. K., eds. (1988) Cell Culture and Somatic Cell Genetics of Plants, Vol 5: Phytochemicals in Plant Cell Cultures, Academic Press, San Diego.

[5] Diettrich, B., Aster, U., Greidziak, N., Roos, W., Luckner, M. (1987) Biochem. Physiol. Pflanzen 182, 245-255.

[6] Eilert, U., Kurz, W. G. W., Constabel, F. (1987) in Plant Tissue and Cell Culture (Green, C. E., Somers, D. A., Hackett, W. P., Biesboer, D. D., eds.), pp. 213-219, Alan R. Liss Inc., New York.

[7] Fujita, Y. (1986) in 5th Intern. Symp on the Genetics of Industrial Microorganisms, CIM 86, Book of Abstracts p. 79, Minneapolis.

[8] Gleba, Y. Y., Sytnik, K. M. (1984) Protoplast fusion, Springer Verlag, Berlin.

[9] Greidziak, N., Diettrich, B., Luckner, M. (1990) Planta med. 56, 175-178.

[10] Hagimori, M., Matsumoto, T., Mikami, Y. (1984) Plant & Cell Physiol. 25, 947-953.

[11] Kreis, W., Reinhard, E. (1987) J. Plant Physiol. 128, 311-326.

[12] Kutchan, T. M. (1989) FEBS Letters 257, 127-130.

[13] Kutchan, T. M., Hampp, N., Lottspeich, F., Beyreuther, K., Zenk, M. H. (1988) FEBS Letters 237, 40-44.

[14] Lee, M., Phillips, R. L. (1988) Ann. Rev. Plant Physiol. Plant Mol. Biol. 39, 413-437.

[15] Luckner, M. (1990) Secondary Metabolism in Microorganisms, Plants, and Animals, Fischer Verlag and Springer Verlag, Jena and Berlin.

[16] Luckner, M., Diettrich, B. (1987) in Plant Tissue and Cell Culture (Green, C. E., Somers, D. A., Hackett, W. P., Bisboer, D. D., eds.), pp. 187-197, Allan R. Liss Inc., New York.

[17] Luckner, M., Diettrich, B. (1988) in Cell Culture and Somatic Cell Genetics of Plants, Vol. 5: Phytochemicals in Plant Cell Cultures (Vasil, I. K., ed.), pp. 193-212, Academic Press, San Diego.

[18] Luckner, M., Diettrich, B. (1989) in Primary and Secondary Metabolism of Plant Cell Cultures II (Kurz, W. G. W., ed.), pp. 117-124, Springer Verlag, Berlin.

[19] Luckner, M., Nover, L., Böhm, H. (1977) Secondary Metabolism and Cell Differentiation, Springer Verlag, Berlin.

[20] Mantell, S. H. (1986) in Secondary Metabolism in Plant Cell Cultures (Morris, P., Scragg, A. H., Stafford, A., Fowler, M. W, eds.), pp. 208-218, Cambridge University Press, Cambridge.

[21] Ohlsson, A. B., Björk, L., Gatenbeck, S. (1983) Phytochemistry 22, 2447-2450.

[22] Reinbothe, C., Diettrich, B., Luckner, M. (1990) Biochem. Physiol. Pflanzen im Druck, .

[23] Scheibner, H., Björk, L., Schulz, U., Diettrich, B., Luckner, M. (1987) J. Plant Physiol. 130, 211-219.

[24] Scheibner, H., Björk, L., Schulz, U., Neumann, D., Diettrich, B., Luckner, M. (1989) Biochem. Physiol. Pflanzen 184, 63-67.

[25] Scheibner, H., Diettrich, B., Schulz, U., Luckner, M. (1989) Biochem. Physiol. Pflanzen 184, 311-320.

[26] Schell, J. S. (1987) Science 237, 1176-1183.

[27] Stafford, A. (1986) in Secondary Metabolism of Plant Cell Cultures (Scragg, A. H., Stafford, A., Fowler, M. W., eds.), pp. 219-223, Cambridge University Press, Cambridge.

[28] Vogel, E., Luckner, M. (1981) Planta med. 41, 161-165.

[29] Wink, M. (1987) Plant Cell Tissue Organ Culture 8, 103-111.

ENZYMOLOGY OF ALKALOID PRODUCTION IN PLANT CELL CULTURES

E. GALNEDER and M. H. ZENK

LEHRSTUHL FÜR PHARMAZEUTISCHE BIOLOGIE DER UNIVERSITÄT MÜNCHEN, KARLSTRASSE 29, D-8000 MÜNCHEN 2, WEST-GERMANY

INTRODUCTION

The art of plant cell culture has developed in the past 15 years to such an extent that we can now recognize at least four categories of cell or callus cultures with regard to their ability to produce secondary plant products: a) cultures which invariably produce the compounds in question (e.g. anthraquinones in the case of **Morinda, Galium**, etc.), b) cultures which contain variant cells producing compounds in the presence of a majority of non-producing cells (e.g. ajmalicine and solasonine in the case of **Catharanthus** or **Solanum** (3)), c) cultures which invariably do not produce the compounds which are elaborated by the original differentiated plant (e.g. morphine, scopolamine in **Papaver somniferum** (4) and **Datura**, respectively), d) cell cultures which produce compounds which are not found in the differentiated plant (5). The handicap for the commercial use of plant cell cultures is the fact that the majority of medicinal plants yield cell cultures which are unable to produce the desired compounds (e.g. codeine, digoxin, ginkgolides, morphine, pilocarpine, quinidine, scopolamine, sennosides). The plant systems which would yield these highly priced and necessary compounds have been studied extensively in this and other laboratories without success. The most important question is how to achieve in suspension cultures chemical differentiation without morphological differentiation. In non-producing cultures "shortage of precursor will seldom cause the lack of secondary substances" (6,4). However, on the other side, statements have been made that "It becomes more and more evident that the activity of the first enzyme leading into a secondary pathway might play a very decisive role for the capability of cell cultures to synthesize secondary compounds" (7,8). Clearly the regulation of enzymes which are involved in the catalysis of secondary products is not yet understood and deserves attention (3). From a biotechnological view point there is a three fold interest in the enzymes catalysing the formation of secondary metabolites: elucidation of the true precursors for a given end-product, discovery of the rate-limiting steps (if there are any) in a biosynthetic pathway, and use of the isolated enzymes for biomimetic synthesis (9). In the case of alkaloids, the pathway leading to the protoberberine alkaloid, berberine, is the first which has been completely elucidated at the enzyme level (9). Berberine is, therefore, an interesting target molecule of which to study the regulation of biosynthesis at the enzyme level. During our investigation on the biosynthesis of berberine using cell suspension cultures of **Thalictrum glaucum**, we noticed the segregation of this culture into a yellow, berberine-producing strain and a colorless strain which was unable to produce this alkaloid. This system seemed to be an ideal one with which to study how non-producing strains are regulated, a study which may lead to a hypothesis about how to localize and possibly overcome the problem of non-producing cell cultures. Clearly the questions posed above should be answered with this system. Does "that enzyme connecting primary and secondary pathway (really) play a decisive regulatory role in the biosynthesis of secondary metabolites?" (8). Are the precursors rate limiting or are there individual enzymes along the pathway that are repressed?

PROCEDURE

Materials and methods

Biological material: A cell culture strain of **Thalictrum glaucum** has been established about 8 years ago and maintained on Linsmaier and Skoog (LS) medium (10). Suspension cultures were started from callus and maintained in 1 liter Erlenmeyer flasks containing 250 ml LS medium. The cultures were shaken at 100 rpm at 23° $\pm$ 2° in diffuse light (650 lux). The cells were harvested by suction filtration, frozen in liquid N_2 and stored at -20° until they were analysed. After about 2 years in culture, a chance observation was made that one of the subcultures had lost its color. This culture was maintained separately and designated as strain "white" in contrast to the original strain "yellow". The white strain occassionally and spontaneously reverted back to the yellow variety.

Analytical procedure: All enzymes and other parameters were measured as summarized in (11).

Chemical material: All labelled and unlabelled alkaloids were from the collection of this department.

RESULTS AND DISCUSSION

HPLC analysis of the ethanol soluble fraction of the yellow strain of **T. glaucum** revealed that the two major alkaloids present were berberine and columbamine. The yield was 92 mg berberine/liter medium and 23 mg columbamine/liter medium. In the white variety both alkaloids were completely absent, no other protoberberine or tetrahydro derivatives could be detected. The metabolic block was therefore comparable to the situation in economically important cell cultures like **Papaver** or **Datura** where virtually no alkaloids of the pathway under investigation are formed. Using the berberine producing yellow strain of **T. glaucum** the pathway leading to the target molecule was analyzed at the enzyme level. The first crucial enzyme in this pathway is the one condensing stereoselectively dopamine with p-hydroxyphenylacetaldehyde, (S)-norcoclaurine synthase (12); the last aromatizes ring C of the (S)-tetrahydroberberine (canadine) to yield berberine as has been discovered in the family **Ranunculaceae** for the species **Coptis japonica** by Yamada's group (13). The same terminal enzyme also catalyses the transition from (S)-canadine to berberine in **Thalictrum**, however, its molecular weight with 43 + 1 kD is slightly lower than the one from **Coptis** (58 kD) and, further more, it is a monomer while the **Coptis** enzyme is a dimer. A total of 13 enzymic steps are involved transforming two molecules of the primary metabolite L-tyrosine to berberine. Two enzymes of this pathway take part in the formation of dopamine from tyrosine. Another two enzymes first transaminate tyrosine to 4-OH-phenylpyruvate which is subsequently decarboxylated to furnish 4-OH-phenylacetaldehyde (14). The alkaloid pathway involving the 9 enzymatic steps is shown in Fig. 1. While the pathway from the non-alkaloidal precursors to (S)-reticuline is catalyzed by soluble cytosolic enzymes, (S)-reticuline has to enter a specific vesicle which houses the berberine bridge enzyme responsible for the transformation of the universal isoquinoline precursor, (S)-reticuline, to (S)-scoulerine, the parent compound to a multitude of tetrahydroprotoberberine alkaloids and derivatives thereof. It is important to note that we have found now evidence that the subsequent methyltransferase ((S)-scoulerine-9-O-methyltransferase) is not contained in this alkaloid vesicle as we had originally assumed (9) but rather is cytosolic. Careful sucrose density gradient centrifugation and thorough washing of the vesicles clearly indicated that this enzyme occurs in the cytoplasma. Therefore, the metabolite (S)-scoulerine has to leave the vesicle to be transformed to (S)-tetrahydrocolumbamine by action of a SAM dependent O-methyltransferase and this metabolite has to re-enter the vesicle where it is finally processed to columbamine and berberine. The c-ring aromatizing enzyme is exclusively located in the vesicles.

Activities of all the enzymes of this pathway could be measured in **Thalictrum** cell cultures, except for the hydroxylase taking N-methylcoclaurine to 3'-hydroxy-N-methylcoclaurine and the one introducing the methylenedioxy bridge in berberine. These two enzymes could be detected in neither the "yellow" nor in the "white" strain. No doubt technical reasons are responsible for this lack of detectable enzyme activity in **Thalictrum.** Eleven enzymes out of this pathway could, however, be quantitated, especially the complete set of enzymes connecting the primary with the secondary pathway. Fig. 2 shows the comparison of enzyme activities (pkat/mg protein) between the alkaloid producing and non-producing strains. It can be seen that of the 11 enzymes which could be measured in both extracts, the majority shows

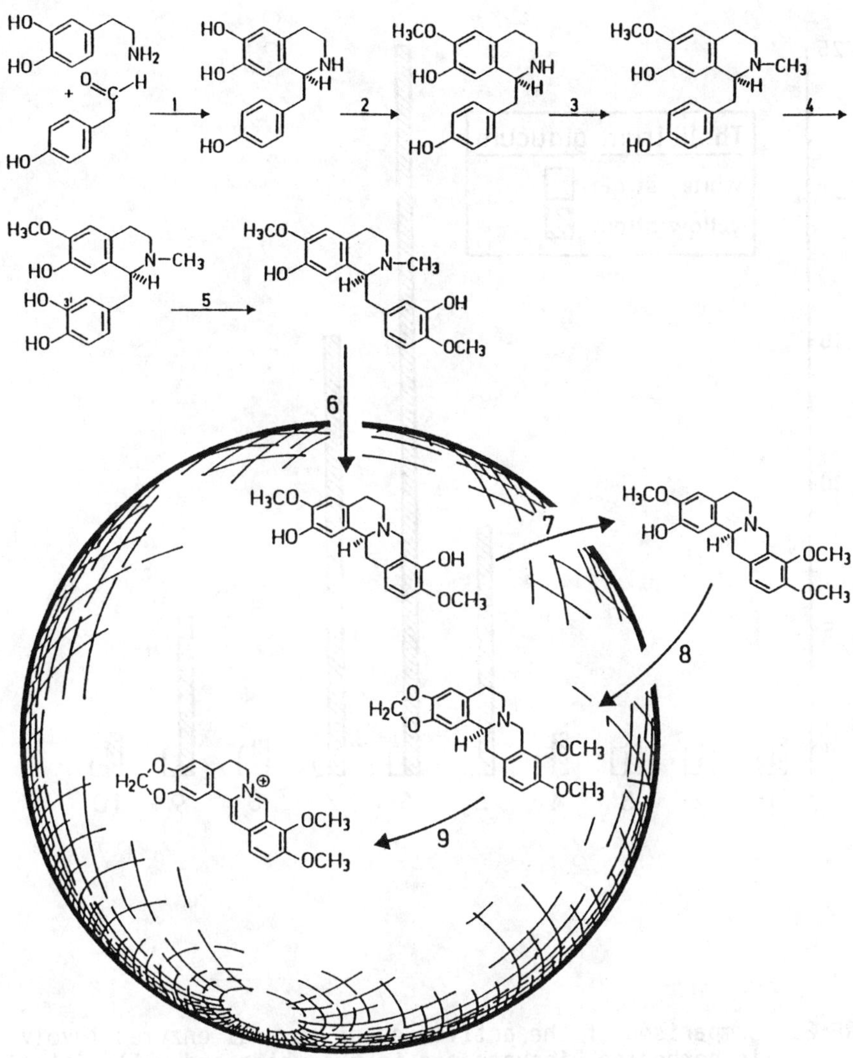

FIGURE 1. Biosynthetic pathway leading from dopamine and 4-OH-phenylacetaldehyde to berberine in **Thalictrum glaucum** cell cultures. The enzymes 1-5 and 7 are located in the cytoplasma while enzymes 6 and 9 are located in vesicles; 1 = (S)-Norcoclaurine synthase; 2 = (R,S)-Norcoclaurine-6-O-Methyltransferase; 3 = (S)-Coclaurine-N-Methyltransferase; 4 = Phenolase; 5 = (S)-3'OH-N-Methylcoclaurine-4'-O-Methyltransferase; 6 = Berberine Bridge Enzyme; 7 = (S)-Scoulerine-9-O-Methyltransferase; 9 = (S)-Canadine Oxidase

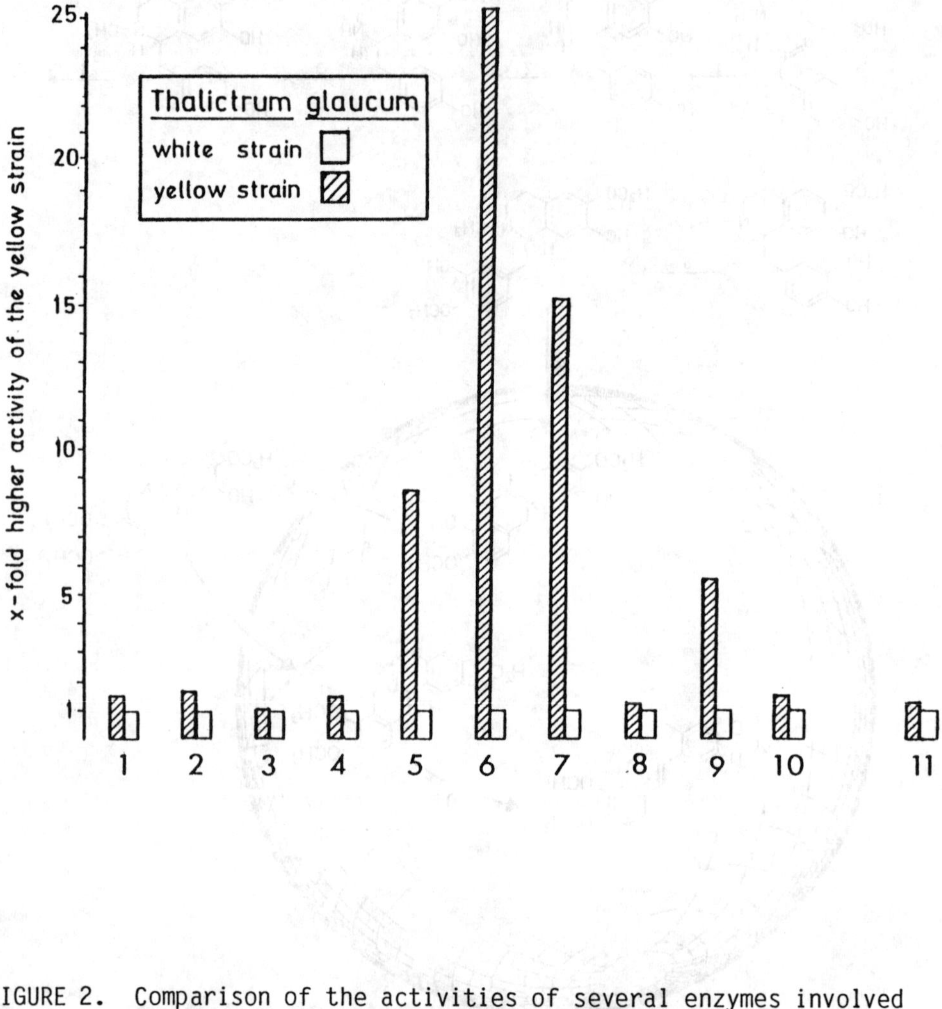

FIGURE 2. Comparison of the activities of several enzymes involved in berberine biosynthesis in the white and yellow strain of **Thalictrum glaucum** (based on pkat/mg protein). 1 = DOPA-Decarboxylase; 2 = Tyrosine-Transaminase; 3 = p-Hydroxyphenylpyruvate-Decarboxylase; 4 = (S)-Norcoclaurine-synthase; 5 = (R,S)-Norcoclaurine-6-O-Methyltransferase; 6 = (S)-Coclaurine-N-Methyltransferase; 7 = (S)-3'-OH-N-Methylcoclaurine-4'-O-Methyltransferase; 8 = Berberine Bridge Enzyme; 9 = (S)-Scoulerine-9-O-Methyltransferase; 10 = (S)-Canadine Oxidase; 11 = Glucose-6-Phosphate-Dehydrogenase

equivalent catalytic activity. There is, however, a clear difference between both strains with regard to all four methyltransferases. Surprisingly, the O- and N-methyltransferase activities in the yellow strain are drastically higher in specific, as well as total (pkat/l medium), activity than in the white strains. We should assume therefore that specifically the three methyltransferases which are found in series in the pathway leading to (S)-reticuline, namely: (R,S)- norcoclaurine-6-O-methyltransferase, (S)-coclaurine-N-methyltransferase and (S)-3'-hydroxy-N-methylcoclaurine-4'-O-methyltransferase are more weakly expressed (9,25,15 fold) in the "white" strain when compared to the "yellow" strain. Computation of these values (under substrate saturation) would indicate that the transformation of norcoclaurine to reticuline proceeds ca. 3300 times faster in the yellow strain compared to the non-producing strain. Assuming that in the white strain the N- and 4'-O-methyltransferase are not under conditions of substrate saturation, this value will be even considerably higher.

On the other hand, the fourth methyltransferase, (S)-scoulerine-9-O-methyltransferase, is only 5 fold higher in the yellow strain, but the total activity is considerably higher than the other 3 methyltransferases and thus should not cause a blockade in the tetrahydroprotoberberine part of the pathway. To prove this assumption, to the "white" strain was added (R,S)-coclaurine in substrate concentrations. No berberine was found as indicated by lack of color formation. However, addition of (S)-reticuline to the medium (0.7 mM final concentration) resulted after a growth period of 8 days in the originally colorless strain turning bright yellow, producing visually detectable amounts of berberine. This experiment clearly demonstrates that the biosynthetic block in the "white" **Thalictrum** strain can be localized by substrate feeding experiments. Precursor added to the culture prior to the block does not abolish the block, however, addition of substrate which is biosynthetically localized after the block fully restores biosynthetic activity. This block could also not be abolished by addition of L-tyrosine, tyramine or dopamine. These experiments were supplemented by feeding experiments using ^{14}C-labelled L-(U-^{14}C)-tyrosine, (S)-(O-$^{14}CH_3$)-coclaurine and (S)-(N-$^{14}CH_3$)-reticuline. While in the "yellow" strain all three precursors were metabolized to yield labelled berberine, in the colorless strain only (S)-reticuline lead to formation of radioactive berberine and columbamine, the other two precursors remaining unmetabolized in the cells as revealed by two-dimensional TLC. This fact clearly indicates that the 5 fold lower (S)-scoulerine-9-O-methyltransferase activity in the colorless strain does not inhibit the pathway from reticuline to berberine.

This excludes also the possibility that simultaneous with the lack of expression of the methyltransferases, maybe a block in SAM synthesis could lead to the block in berberine synthesis, since enough nucleotide cosubstrate is available to methylate substrate amounts of scoulerine. These feeding experiments also show that the enzyme responsible for forming the methylenedioxy bridge in berberine, which could not be measured for technical reasons **in vitro** is present in both strains.

This experiment is the first to demonstrate why a given strain of a plant cell culture is not capable of synthesizing a secondary product. The biochemical block could be pinpointed by both substrate feeding and determination of enzyme activity of the individual steps in the pathway (Fig. 3). Both experimental approaches showed convincingly that the methyltransferases occurring in series in the reticuline part of the pathway determine whether a cell culture produces the secondary metabolite, berberine, or not. Reduction in total activity of a multiplicative factor of at least 3000 is enough to completely stop the flow of precursors through the biosynthetic chain. Furthermore, this experiment clearly demonstrates that the hypothesis that the enzyme connecting primary and secondary pathway "plays a decisive regulatory role in the biosynthesis of secondary metabolites" (8) is incorrect. We will demonstrate later in a second example (benzophenanthridine biosynthesis) that again this hypothesis is untenable. Unlike in microorganisms (15,16), in higher plants obviously the enzyme linking primary to secondary metabolism is <u>not</u> the determining factor in inducing natural products. Attempts to clone those enzymes linking primary to secondary metabolism (e.g. tryptophandecarboxylase (17,18)) and overexpress these genes in non-producing tissues has no hope for success to overcome the metabolic blockade in indole alkaloid metabolism. In order to overcome the repression in dedifferentiated plant cell cultures, the complete pathways leading to the final and often commercially desirable products have to be studied with precision and in detail. Only if the whole set of enzymes leading to a given natural product are known, can we monitor which of the enzymes are expressed poorly or not at all in a non-producing culture. The strategy of the future has to be to localize the rate limiting enzymes of a given pathway, isolate the genes of these enzymes and identify the regulatory elements governing the control of gene expression. Given this information, we can forsee a derepression of the genes of alkaloid biosynthesis in undifferentiated cell cultures. Alternatively we can envisage the expression of portions of or total biosynthetic pathways leading to secondary compounds in microorganisms so that the plant product is synthesized by microbes possessing much better fermentation characteristics than plants. The solution to the problem of non-producing plant cell culture strains will come from a combination of biochemistry with molecular biology.

ACKNOWLEDGEMENT

We thank Dr. T.M. Kutchan for linguistic help in preparing this manuscript. This work was supported by Deutsche Forschungsgemeinschaft, Bonn (SFB 145) and Fonds der Chemischen Industrie.

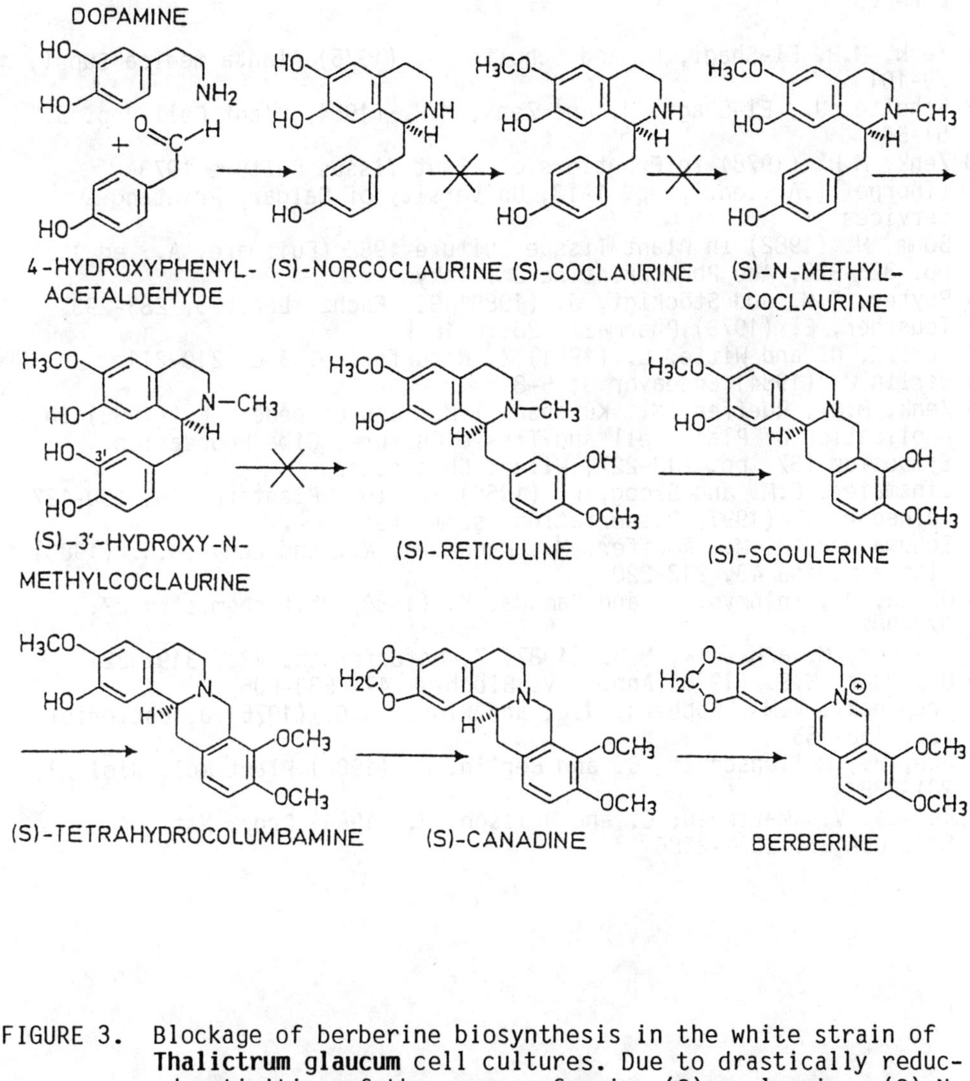

FIGURE 3. Blockage of berberine biosynthesis in the white strain of **Thalictrum glaucum** cell cultures. Due to drastically reduced activities of the enzymes forming (S)-coclaurine, (S)-N-methylcoclaurine and (S)-reticuline, no berberine production can be observed in the cells of the white strain.

REFERENCES

1 Zenk, M.H. El-Shagi, N. and Schulte, U. (1975) Planta medica Suppl. 79-101
2 Schulte, U., El-Shagi, H. and Zenk, M.H. (1984) Plant Cell Rep. 3, 51-54
3 Zenk, M.H. (1978) in Frontiers of Plant Tissue Culture 1978 (Thorpe, T.A., ed.), pp. 1-13, University of Calgary Printing Services
4 Böhm, H. (1982) in Plant Tissue Culture 1982 (Fujiware, A., ed.) pp. 325-328, Abe Photoprinting Co. Tokyo
5 Ruyter, C.M. and Stöckigt, J. (1989) GIT Fachz. Lab. 33, 283-293
6 Teuscher, E. (1973) Pharmazie 28, 6-18
7 Berlin, J. and Witte, L. (1981) Z. Naturforsch. 36c, 210-214
8 Berlin J. (1984) Endeavor 8, 5-8
9 Zenk, M.H., Rueffer, M., Kutchan, T.M., and Galneder, E. (1988) in Application of Plant Cell and Tissue Culture, Ciba Foundation Symposium 137, pp. 213-227, Wiley, Chichester
10 Linsmaier, E.M. and Skoog, F. (1965) Physiol. Plantarum 18, 100-127
11 Galneder, E. (1991) Dissertation, submitted
12 Schumacher, H.-M., Rueffer, M., Nagakura, N., and Zenk, M.H. (1983) Planta medica 48, 212-220
13 Okada, N., Shinmyo, A. and Yamada, Y. (1988) Phytochemistry 27, 979-982
14 Rueffer, M. and Zenk, M.H. (1987) Z. Naturforsch. 42c, 319-332
15 Umbarger, H.E. (1978) Ann. Rev. Biochem. 47, 533-606
16 Krupinski, V.M., Robbers, J.E. and Floss, H.G. (1976) J. Bacteriol. 125, 158-165
17 Noe, W., Mollenschott, C. and Berlin, J. (1984) Plant Mol. Biol. 3, 281-288
18 DeLuca, V., Marineau, C. and Brisson, N. (1989) Proc. Natl. Acad. Sci. USA 86, 2582-2586

AFFECTING SECONDARY PRODUCT FORMATION IN SUSPENSION AND HAIRY ROOT CULTURES - A COMPARISON

J. BERLIN[1,2], C. MOLLENSCHOTT[1,2], N. GREIDZIAK[1], S. ERDOGAN[1] and I. KUZOVKINA[3]

[1]BBA - Institut f.Biochemie, Biol.Bundesanstalt f.Forst- und Landwirtschaft, and [2]GBF - Ges. f.Biotechnol.Forschung, Braunschweig, FRG; [3]Timirjazev Institute Plant Physiology, Acad.Science, Moscow, USSR.

1. INTRODUCTION

Recently, the spectrum of secondary metabolites which might be produced in good levels by tissue culture techniques, has been extended by the use of organ cultures, especially of transformed root cultures. These hairy root cultures accumulate usually the spectrum of secondary metabolites in levels as found in roots of the intact plant. Nevertheless for technological purposes growth and product levels of many root cultures have to be optimized. Thus we wanted to analyze whether secondary pathways of root cultures would be affected by the same factors as known from studies with normal cultures. The ease of manipulating secondary pathways of suspension cultures has provided first insight into their regulatory controls. As activities of enzymes diverting primary metabolites into secondary pathways are often greatly induced under culture conditions favoring secondary metabolism, it has been suggested that these enzymes serve as one of diverse regulatory controls of secondary pathways. Therefore we were interested to see whether such enzymes undergo rapid changes in transformed root or suspension cultures. The availability of cell lines with different productivities let us compare the activities of such enzymes. Here we summarize some of our experiences in altering isoflavonoid levels of *Lupinus* and ß-carbolines and serotonin of *Peganum* by variation of the medium composition.

2. MATERIAL AND METHODS

2.1. Plant material: a) The cell lines LUPO (*L. polyphyllus*) and LUHA (*L. hartwegii*) were established from sterile grown seedlings on MS-medium containing 2 µM 2,4-D. Transformed suspension cultures (LUPO 30150, LUHA 30150) and hairy root cultures (LUPO 15834, LUHA 15834) were obtained by infecting pieces of sterile grown seedlings with *Agrobacterium tumefaciens* strain 30150 (DSM, Braunschweig) or *A. rhizogenes* 15834, respectively. The transformed lines have been maintained for 2-6 years on MS-medium without phytohormone supplement.
b) Establishment and maintenance of the suspension cultures of *Peganum harmala* have been described (1). The transformed root cultures of *Peganum* were obtained either by infection with *A. rhizogenes* A4 (2), 15834 or a strain carrying the reporter gene GUS (pLTC GUS; Commandeur

et al., in preparation). Maintenance of the *Peganum* root cultures has been on phytohormone-free B5-medium.

2.2. Analytics: a) The isoflavonoid glucosides of *Lupinus* cultures were determined on a Lichrosorb RP_{18}-column (250x4 mm, 7μ) using a gradient from H_2O (0.5% CH_3COOH) to MeOH (0.5% CH_3COOH) in 20 min, flow rate 1.5 ml/min. Detection at 280 nm. The isoflavonoid content was expressed as genistein equivalents.

b) Serotonin and ß-carboline alkaloids in cultures of *Peganum* were quantified by fluorescence assays (3) and/or by a modified HPLC method: Lichrosorb RP_{18}-7μ (250x4 mm); solvent A: 20% MeOH (0.05% HCOOH/triethylamine pH 8.5), B: 100% MeOH (pH 8.5); exponential gradient 3: from 0% B to 100% B in 15 min, then 15 min at 100% B; flow rate 1 ml/min; detection for the ß-carboline alkaloids at 330, for serotonin at 302 nm.

2.3. Enzyme assays: Activity of tryptophan decarboxylase (TDC) activity in *Peganum* cultures and phenylalanine ammonia-lyase in *Lupinus* cells was measured as described (1,4).

3. RESULTS AND DISCUSSION

3.1. On the isoflavonoid formation in various cultures of *Lupinus*

The main aim of our work on *Lupinus* cultures is to outfit the cells with a permanently expressed bacterial lysine decarboxylase for studying its effect on quinolizidine alkaloid biosynthesis. As by-product of this project we analyzed a number of *Lupinus* cultures transformed with wild type strains of *Agrobacterium* to see whether transformation itself would have effects on the biosynthetic pathway of quinolizidine alkaloids. This was not the case as neither lysine decarboxylase activity nor any alkaloids were detected in white or green transformed suspension cultures or in hairy root cultures. Main secondary metabolites of normal and transformed lines of *L. polyphyllus* and *L. hartwegii* were various isoflavonoid glucosides (orobol-3',7-O-diglucoside; genistein-4',7-O-diglucoside; orobol-7-O-glucoside; genistein-7-O-glucoside; 6-dimethylallylorobol-3',7-O-diglucoside) (5). As minor components the flavonol kaempferol and its glucosides accounted for 5-10% of the total flavonoid content.

As specific levels (mg/g) of secondary metabolites of suspension cultures depend upon the physiological state of the cells, we wanted to compare the reactions of phytohormone independent transformed suspension and root cultures and normal suspensions in medium variation experiments. It is well accepted that in phytohormone dependent cell cultures removal or change of the phytohormone, reduction of PO_4, increase of the carbon source and altering the ratio NH_4/NO_3 have positive effects on secondary product formation. For such experiments the PO_4 level was varied from 0 - 2 mM, NO_3 and NH_4 from 0 - 40 mM, sucrose from 0 - 10%. From all our experiments with the various *Lupinus* cell lines (in preparation) the most important findings are summarized in Table 1. Phosphate deficiency increased the specific isoflavonoid content in all suspension cultures, while PO_4 had no effect on the specific yields of root cultures. As root cultures of *Lupinus* showed the highest biomass production at 2 mM PO_4, high PO_4 levels helped to increase product yields. Growth of all root cultures of *Lupinus* was greatly impaired in the presence of NH_4, especially when used as sole nitrogen source. The highest biomass increases were achieved when KNO_3 was the only nitrogen source. With increasing concentrations of NH_4 isoflavonoid formation became lower and lower, while an optimum curve was obtained for NO_3, with a maximum at 10 mM. Suspension cultures of *Lupinus* showed a similar behaviour as the root cultures (poor growth and very low production with ammonia as sole nitrogen source).

Table 1: Effects of the culture medium on the isoflavonoid content of a phytohormone dependent (LUPO) and a transformed phytohormone independent (LUPO 30150) suspension culture and a hairy root culture (LUHA 15834) The suspension cultures were harvested after 10d, the root cultures after 14d. DM = Dry mass/flask; ISO = Isoflavonoids; GM = Growth medium

	LUPO		LUPO 30150		LUHA 15834	
	DM mg/flask	ISO mg/g	DM mg/flask	ISO mg/g	DM mg/flask	ISO mg/g
Growth medium	561	2.1	562	16.9	480	14.2
GM - P	140	8.5	137	27.4	110	13.9
GM - N+10mM KCl	186	3.4	192	18.3	170	13.6
GM - N+10mM KCl						
+ 20mM $(NH_4)_2SO_4$	92	0.6	103	5.4	130	2.1
+ 20 mM KNO_3	275	2.7	306	19.8	590	18.3
GM + 8% sucrose	272	2.9	275	22.4	95	2.6

Sucrose levels up to 6% had no effect on the specific isoflavonoid content of root cultures. Higher concentrations caused strong reductions of growth and specific isoflavonoid levels (Table 1). In addition, many media experiments with different phytohormone combinations (2,4-D, NAA, kinetin, BAP) were performed.Neither growth nor the high product levels of the phytohormone independent line LUPO 30150 (Table 1) were affected by typical phytohormone concentrations. In contrast, appearance and growth of root cultures was greatly affected by many different phytohormone combinations. After 4 passages on MS-medium with 2 μM 2,4-D, for example, a slowly growing aggregated suspension culture was obtained. Growth did not improve during the next 5 passages in the presence 2,4-D and product levels were 15-20% of that found in the root state. After 4 passages of 3 weeks on phytohormone-free medium the root character and the original productivity was returned.

PAL-activity was greatly induced in all suspension cultures during the first 24 h after being transferred into fresh medium while it remained roughly at the same level in root cultures (Table 2). As one would have expected from the medium variation experiments, the root cells were also less responsive on the enzyme level to external changes. As long as turnover rates for the isoflavonoid pools have not been determined, a simple correlation between PAL-activity and isoflavonoid levels may not be found (Table 2).

Table 2: Relationship between PAL-activity and accumulation of isoflavonoids in various cell lines of *L. polyphyllus* (LUPO) and *L. hartwegii* (LUHA). The lines with number 30150 and 15834 are transformed suspension and root cultures, respectively.

CELL LINE	PAL-ACTIVITY pkat/g fresh mass					ISOFLAVONOIDS mg/g dry mass	
	0h	24h	48h	96h	168h	96h	216h
LUPO	19	72	31	14	8	4.2	1.8
LUPO 30150	18	244	61	27	13	22.4	31.5
LUPO 15834	12	21	18	23	25	8.3	8.3
LUHA	8	133	60	24	21	12.2	11.5
LUHA 30150	5	192	60	8	2	6.4	4.7
LUHA 15834	22	30	44	31	31	13.5	15.2

3.2. On the formation of serotonin and ß-carbolines in *Peganum harmala*

Cell suspension cultures contain only low levels of ß-carbolines and serotonin. Even this low biosynthetic potential is often lost in long term cultures. By selection, variant lines with high levels of serotonin were recovered while ß-carboline alkaloids remained low (Table 3).

Table 3: Effects of the medium on specific product levels of low producing wild-type cells, a 4-methyltryptophan-tolerant variant line and a hairy root culture. Cells were harvested after 9 days of culture.

Medium	Wild type		Variant		Root Culture	
	ser	ß-carb	ser	ß-carb	ser	ß-carb
			mg/g dry mass			
MS + 2μM 2,4-D	1.2	0.3	14.8	0.8	3.9	9.2
B5 + 5μM 2,4-D	0.9	0.3	9.6	0.7	3.3	7.7
MS + NAA/Kinetin	0.7	0.1	12.2	0.8	5.1	10.3
MSO	4.5	0.4	11.7	0.8	8.6	11.8
B5O	3.7	0.3	12.7	0.8	7.5	15.2
MS - N	3.4	0.3	7.6	0.8	3.3	13.2
MS - N + K	1.9	0.3	8.5	0.9	4.4	17.3
MS - N + K + NH_4	7.8	0.3	10.9	0.9	10.9	18.5
MS - N + KNO_3	2.1	0.4	9.1	0.7	4.4	16.7
MS - P	7.2	0.5	13.1	0.7	8.0	12.7
MS + 2 P	4.2	0.4	11.7	0.8	5.1	11.0
MS + 6% Sucrose	11.5	0.6	9.0	0.7	6.1	9.9

NAA/Kinetin: 10/1 μM; K: 20 mM KCl; NH_4: 20 mM $(NH_4)_2SO_4$; KNO_3: 40 mM; P: 1.2 mM PO_4;

Therefore the finding that hairy root cultures of *Peganum* accumulate high levels of serotonin and harmane alkaloids has opened new avenues for studying this pathway. First, we wanted to know whether product levels of the cultures could be altered by medium variation experiments (Table 3). ß-Carboline levels could not be improved by any variation of the medium while serotonin levels of wild type cells were increased by "stress" factors. When the roots were grown in the presence of phytohormones, first callus formation was seen after 2-3 weeks and dedifferentiated aggregates contained as low ß-carboline/serotonin levels as normal callus cultures. Despite the fact that alkaloid production nearly disappeared, the possibility of changing the root culture into a root/callus culture is of technological interest. On phytohormone-free media the roots were long and interlacing and had to be transferred by a pair of tweezers. In the presence of various phytohormone combinations, the roots became short and could be transferred by a spoon. In addition the root/callus cultures showed better growth than the highly productive roots. Even fine suspension cultures were obtained. However, growth of the suspensions remained rather poor. The root state with its high production was restored after 1-2 cycles on phytohormone-free medium. Thus there exists the possibility of scaling up the transformed root culture as an aggregated callus culture and of reinducing root formation and alkaloid production at the desired time.

The appearance of the roots could be changed by the nitrogen source. On MS-medium (NH_4 high) the roots became thick while on B5-medium (NH_4 low) the roots were very thin. Thin roots produced twice the amounts of ß-carbolines (eg. harmine) as the MS-cultured roots. The high specific content with NH_4 as sole nitrogen source (Table 3) resulted from a complete growth inhibition under that culture condition.

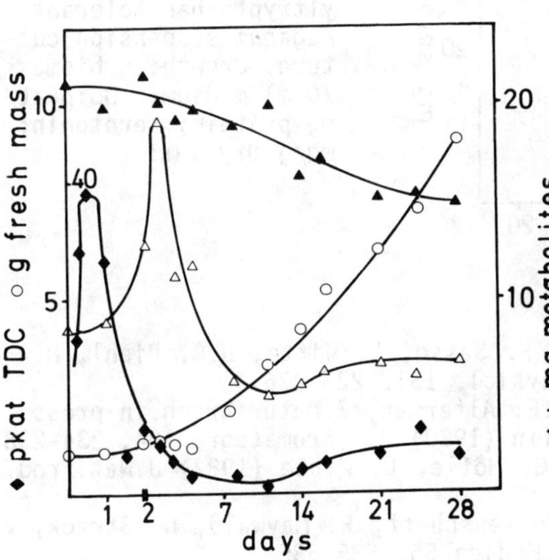

Fig.1: Growth and production kinetic of a transformed root culture of *Peganum* on B50-medium. Growth: g fresh mass/70 ml medium; TDC: pkat/mg protein; metabolites: mg/g dry mass.

Reasonable growth and highest production of the root culture was achieved on B50-medium (Fig. 1). When maintained under standard conditions specific alkaloid levels did not change distinctly, while serotonin levels were often enhanced during the first week (Fig. 1). When transferred to fresh medium, TDC activity was induced within 24 h and fell back to a low level after two days. The extent of induction of TDC depended upon the age and the initial inoculum. So far, however, 40 pkat/mg protein was the highest specific activity measured in transformed root cultures. Despite the fact that the root cultures contained 1.5-3% indole derivatives on dry mass basis, TDC activity was found to be surprisingly low. Normal root cultures (induced from aggregated callus cultures) accumulating substantial amounts of ß-carboline alkaloids, contained also rather low TDC activity (6). In contrast, TDC of 4-methyltryptophan-tolerant suspension cultures containing 1-3% serotonin, was induced to a much greater extent (Fig. 2) when transferred to fresh medium. In conclusion: Despite the fact that their response to external manipulations was rather sluggish, especially root cultures with altered product patterns offer new opportunities for basic and biotechnological studies.

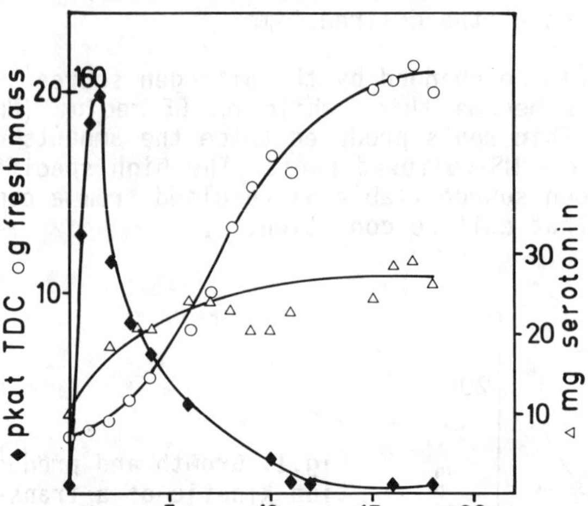

Fig.2: Growth and production kinetic of a 4-methyltryptophan-tolerant *Peganum* suspension culture. Growth: g biomass/ 70 ml medium; TDC:pkat/ mg protein; serotonin: mg/g dry mass.

REFERENCES:

1. Berlin, J., C. Mollenschott, F. Sasse, L. Witte, H.G. Piehl, H. Büntemeyer (1987) J. Plant Physiol. 131, 225-236.
2. Kuzovkina, I.N., A. Gohar, I.E. Alterman, Z.Naturforsch.in press
3. Sasse, F., J. Hammer, J. Berlin (1980) J. Chromatogr. 194, 234-238.
4. Berlin, J., K.-H. Knobloch, G. Höfle, L. Witte (1982) J.Nat.Prod. 45, 83-87.
5. Berlin, J., M. Rippert, C. Mollenschott, F. Maywald, D. Strack, V. Wray, C. Sator (1989) Planta Medica 55, 685-686.
6. Berlin, J. (1988) In: Production of metabolites by plant cell cultures. APRIA, Paris p. 89-98.

PURIFICATION OF CYTOCHROME P-450-DEPENDENT GERANIOL-10-HYDROXYLASE FROM A CELL SUSPENSION CULTURE OF *CATHARANTHUS ROSEUS*

A.H. MEIJER, E.J.M. PENNINGS, A. DE WAAL AND R. VERPOORTE.
BIOTECHNOLOGY DELFT LEIDEN, PROJECT-GROUP PLANT CELL BIOTECHNOLOGY, CENTER FOR BIO-PHARMACEUTICAL SCIENCES, DIVISION OF PHARMACOGNOSY, GORLAEUS LABORATORIA, P.O. BOX 9502, 2300 RA LEIDEN, THE NETHERLANDS

1. INTRODUCTION

The conversion of geraniol to its 10-hydroxy derivative is a possible site for regulatory control in the biosynthesis of secologanin, the precursor of the indole alkaloids formed by the higher plant *Catharanthus roseus* [1,2]. This reaction is catalysed by geraniol-10-hydroxylase (G10H), a membrane-bound enzyme complex, consisting of a cytochrome P-450 monooxygenase, and an NADPH:cytochrome P-450 reductase [3].

Cytochrome P-450 monooxygenases are well investigated in animals, in which they are involved in the formation of endogenous compounds as well as in a detoxification mechanism. A variety of xenobiotics can be metabolised by a microsomal system consisting of several isoforms of cytochrome P-450, that have broad and overlapping substrate specificities [4].

In plants a growing number of specific biosynthetic reactions, mostly in secondary metabolism, are found to be catalysed by cytochrome P-450 monooxygenases. The purification of cytochrome P-450 from plants and the reconstitution of enzyme activity with a combination of cytochrome P-450, NADPH:cytochrome P-450 reductase and lipid has been achieved in only a few cases. A cytochrome P-450 with unknown physiological role was purified from *Tulipa gesneriana* bulbs [5]. A p-chloro-N-methylaniline demethylase, consisting of two nearly identical polypeptides, was purified from *Persea americana* mesocarp [6]. Cinnamate hydroxylase from *Helianthus tuberosis* was purified and reconstituted [7]. Partial purification and reconstitution of digitoxin 12β-hydroxylase from *Digitalis lanata* was described [8]. From *Glycine max* 3,9-dihydroxypterocarpan 6a-hydroxylase was purified, and separated from cinnamate 4-hydroxylase, thus proving for the first time the existence of cytochrome P-450 isoforms with different substrate specificity in plants [9].

Cytochrome P-450-dependent geraniol-10-hydroxylase was isolated from *Catharanthus roseus* seedlings [10]. The enzyme activity was associated with provacuolar membranes, rather than with endoplasmic reticulum like in most cytochrome P-450 monooxygenases [11]. The reductase component was purified to homogeneity, but the P-450 component was only partially purified by an ion-exchange step [12]. Approx. 10% of the solubilised G10H activity could be reconstituted with the enzyme components and a lipid extract [10,12].

Here we describe the isolation of G10H from a *Catharanthus roseus* cell suspension culture, grown on an alkaloid production medium. We describe the purification of both the cytochrome P-450 and the reductase component, and report on the reconstitution of G10H activity.

2. PROCEDURE

2.1. Materials and Methods

2.1.1. Enzyme assays: NADPH:cytochrome P-450 reductase was assayed with cytochrome *c* as a substrate. Geraniol 10-hydroxylase was measured by the formation of [1-^{3}H]10-hydroxygeraniol from [1-^{3}H]geraniol according to [10] with modifications. Following the first chromatographic separation G10H was assayed in a reconstituted system. For this purpose, 0.36 nkat NADPH:cytochrome P-450 reductase (DEAE-Sephacel fraction) and 10 µg *C. roseus* lipids were added to the cytochrome P-450 containing fraction. Before the addition of substrates, the incubation mixture was passed through Extracti-gel D (Pierce) for detergent removal.

2.1.2. Analytical methods: Cytochrome P-450 concentration was determined from the CO-difference spectrum according to [13]. Protein was determined according to [14] with bovine serum albumin as a standard.

2.1.3. Enzyme purification: 5-6 days old *C. roseus* cells (1.35 kg) grown on alkaloid production medium were frozen in liquid N_2, homogenised in a Waring blendor for 1 min, and rapidly thawed in 2 parts of homogenisation buffer. Membranes sedimenting between 20 min 1000 x g and 60 min 20000 x g were solubilised with 1 % sodium cholate. Soluble protein was applied to a DEAE Sephacel (Pharmacia) column and eluted with a KCL gradient. The NADPH:cytochrome P-450 reductase containing fractions were further purified by affinity chromatography on 2',5'-ADP-Sepharose (Pharmacia). The P-450 (G10H) containing fractions were fractionated on an HA-Ultrogel (IBF) column. Elution was with a 10-500 mM KPi gradient. The peak fractions were applied on an ω-aminooctyl agarose (Sigma) column. Elution was with a 0-0.2 % Renex 690 gradient. G10H containing fractions were fractionated on a TSK Phenyl-5PW (LKB) HPLC column. Elution was with a 0.1-0.6 % Renex 690 gradient. All steps were performed in a cold room, except the TSK Phenyl-5PW chromatography, which was performed at room temperature.

3. RESULTS AND DISCUSSION

3.1. Solubilisation of G10H: For solubilisation of G10H CHAPS, CHAPSO, and cholate were tested at various concentrations. In all cases at concentrations above the critical micelle concentration, enzyme activity was only present in the solubilised fraction. The yield for the reductase was about 100 %, whereas G10H activity was never higher than 10 %. The loss of G10H activity could be caused by denaturation of 90 % of the enzyme, or by distribution of the P-450 and the reductase in the solubilised preparation over different micelles. Cholate was selected for the solubilisation of G10H, since this was the least inhibitory to the G10H activity.

3.2. Reconstitution of G10H activity: After separation of the P-450 component and the reductase by ion-exchange chromatography G10H activity can only be measured in a reconstituted system. We determined the dependence of G10H activity on cytochrome P-450 concentration (Fig. 1.a), reductase concentration (Fig. 1.b), and (phospho)lipid concentration and composition (Fig. 2a).

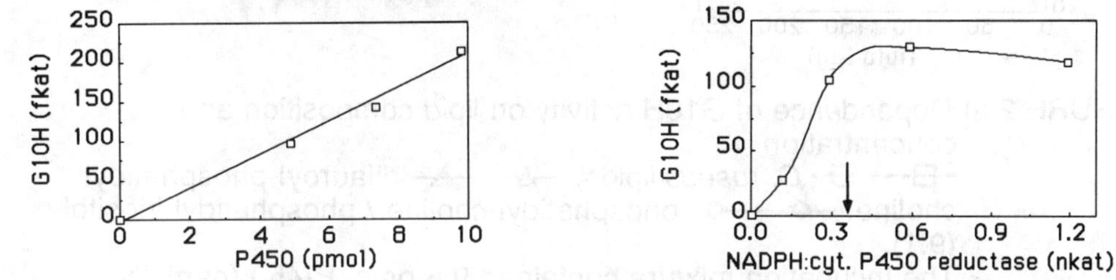

FIGURE 1.a) Dependence of G10H activity on cytochrome P-450 concentration. The incubation mixture contained 0.3 nkat reductase (from DEAE Sephacel) and 100 µg *C. roseus* lipids. P-450 was partially purified by DEAE Sephacel, HA-Ultrogel, and aminooctyl agarose.
b) Dependence of G10H activity on NADPH:cytochrome P-450 reductase concentration.
The incubation mixture contained 100 µg *C. roseus* lipids, and 4.9 pmol P-450 (partially purified by DEAE Sephacel, HA-Ultrogel, and aminooctyl agarose). The arrow indicates the concentration chosen for routine assays.

Reconstitution with *C. roseus* lipids (10-250 µg) stimulated G10H activity. The highest (2-fold) increase was with 10 µg. A stimulation could also be effected with 10 µg of purified phospholipids, but these were inhibitory at higher concentrations. Detergent removal with Extracti-gel D is essential for an efficient reconstitution. Without this step G10H activity can be more than 80 % lower. Addition of 0.1 % Renex 690 to the reconstituted system after detergent removal caused a similar inhibition. The G10H activity is significantly lower in the reconstituted system after the first chromatographic step, as compared to the solubilised enzyme complex. This may reflect a less efficient interaction between the components in the reconstituted system, rather than loss or denaturation of protein, since during all subsequent chromatographic steps the recovery of enzyme activity is remarkably high.

3.3. Purification of the NADPH:cytochrome P-450 reductase component of G10H: The NADPH:cytochrome P-450 reductase purified by anion-exchange and 2',5'-ADP-Sepharose chromatography shows electrophoretic homogeneity on SDS-PAGE and has an M_r of 74000 (Fig. 2.b). The IEP is 4.6. A proteolytic fragment of the enzyme with an M_r of

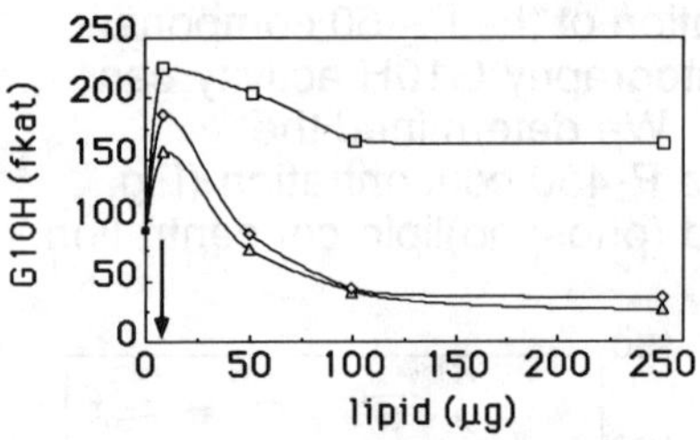

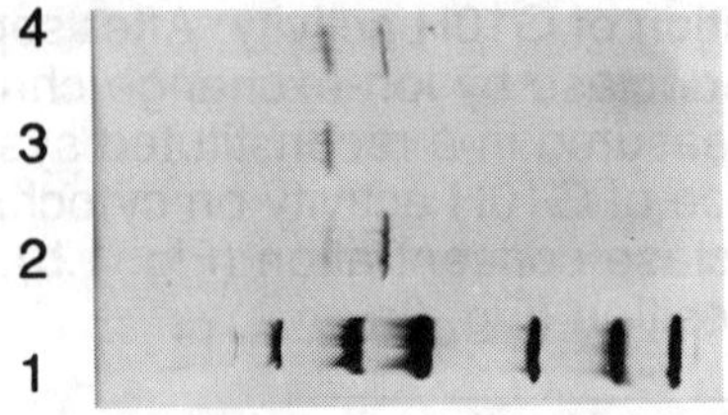

FIGURE 2.a) Dependence of G10H activity on lipid composition and concentration.
□——□ *C. roseus* lipids, △——△ dilauroyl-phosphatidyl choline, ◇——◇ phosphatidyl choline / phosphatidyl inositol (9:1).
The incubation mixture contained 9.8 pmol P-450 (partially purified by DEAE Sephacel, HA-Ultrogel, and aminooctyl agarose) and 0.3 nkat reductase (from DEAE Sephacel). The arrow indicates the concentration of *C. roseus* lipids chosen for routine assays.
b) SDS-PAGE of cytochrome P-450 and NADPH:cytochrome P-450 reductase.
A 10-15% Phastgel (Pharmacia) with SDS bufferstrips was used. Detection was by silverstaining. Lane 1: LMW markers (from left to right 94, 67, 43, 30, 20.1 and 14.4 kDa), lane 2: P-450, lane 3: reductase, lane 4: reductase + P-450.

66000 is separated from the intact protein during the second purification step. The specific activity for cytochrome *c* is 1.2 μkat/mg. The purified reductase is capable of reconstitution of G10H activity, but for routine reconstitution assays we use a partially purified preparation (DEAE Sephacel fraction).

3.4. Purification of the cytochrome P-450 component of G10H: The solubilised cytochrome P-450 component was purified in a 4-step procedure, comprising chromatography on DEAE Sephacel (Fig. 3), Hydroxylapatite Ultrogel, ω-aminooctyl agarose, and TSK Phenyl-5PW (Fig. 4). The purification scheme is given in Table 1. Peak fractions from TSK Phenyl-5PW show a band at 56000 on SDS-PAGE , the intensity of which corresponds to the G10H activity in the fractions. Apart from the first fractions, only one contaminating protein can be detected in the peak fractions (Fig. 2.b shows the pooled fractions). The M_r of this protein and that of the purified reductase are identical (Fig. 2.b). It is possible that some of the reductase has remained associated with the cytochrome P-450. We could, however, not detect any enzyme activity associated with this protein. The cytochrome P-450 has an IEP of 8.3. The cell-free extract also catalyses the hydroxylation of nerol, the cis-isomer of geraniol. The same applies to the purified enzyme.

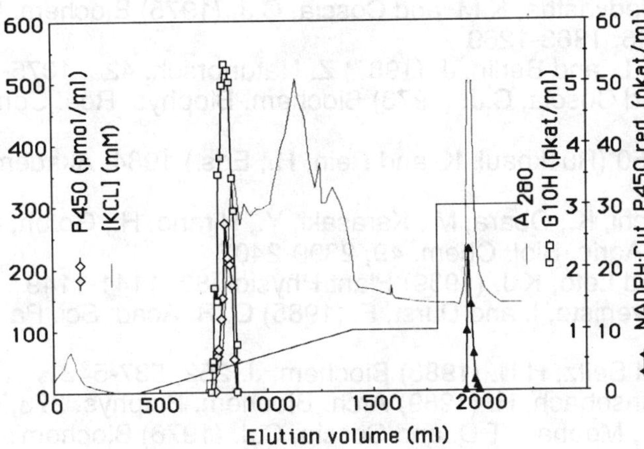

FIGURE 3) Typical elution profile from DEAE Sephacel.

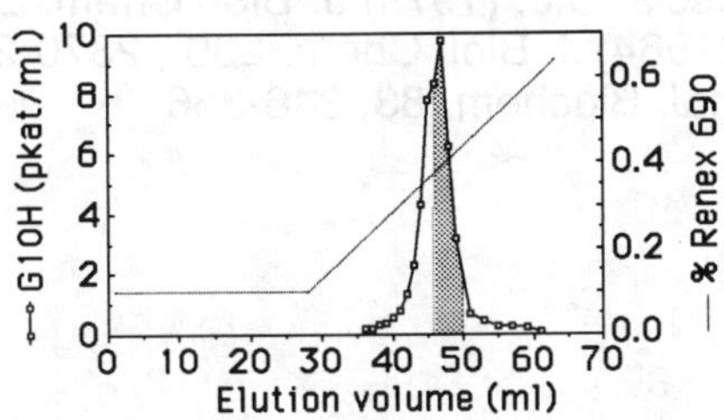

FIGURE 4) Elution profile from TSK-Phenyl 5PW.
Shaded area represents pooled fractions.

TABLE 1 : Purification of geraniol-10-hydroxylase

purification step	protein (mg)	G10H (pkat)	specific activity (pkat/mg)	purifi-. cation (-fold)	recovery (%)
supernatant 1000 x g	34067.0	7887.6	9.3	1.0	100.00
Pellet 20000 x g	2689.0	4865.0	36.1	3.9	65.63
solubilised	269.1	1483.0	5.5	0.6	3.91
DEAE Sephacel	8.9	85.6	9.6	1.0	0.23
HA-Ultrogel	3.7	55.4	14.8	1.6	0.15
ω-aminooctyl agarose	3.4	129.2	37.9	4.1	0.34
TSK Phenyl-5PW	0.09	32.4	344.7	9.6	0.09

REFERENCES

1 McFarlane, J., Madyastha, K.M. and Coscia, C.J. (1975) Biochem. Biophys. Res. Commun. 66, 1363-1269
2 Schiel, O., Witte, L. and Berlin, J. (1987) Z. Naturforsch. 42c, 1075-1081
3 Meehan, T.D. and Coscia, C.J. (1973) Biochem. Biophys. Res. Commun. 53, 1043-1048
4 Cytochrome P-450 (Ruckpaul, K. and Rein, H., Eds.) 1984, Akademie-Verlag, Berlin
5 Higashi, K., Ikeuchi, K., Obara, M., Karasaki, Y., Hirano, H., Gotoh, S. and Koga, Y. (1985) Agric. Biol. Chem. 49, 2399-2405
6 O'Keefe, D.P. and Leto, K.J. (1989) Plant Physiol. 89, 1141-1149
7 Gabriac, B., Benveniste, I. and Durst, F. (1985) C. R. Acad. Sci. Paris 301, 753-758
8 Petersen, M. and Seitz, H.U. (1988) Biochem. J. 252, 537-543
9 Kochs, G. and Grisebach, H. (1989) Arch. Biochem. Biophys. 273, 543-553
10 Madyastha, K.M., Meehan, T.D. and Coscia, C.J. (1976) Biochemistry 15, 1097-1102
11 Madyastha, K.M., Ridgway, J.E., Dwyer, J.G. and Coscia, C.J. (1977) J. Cell. Biol. 72, 302-313
12 Madyastha, K.M. and Coscia, C.J. (1979) J. Biol. Chem. 254, 2419-2427
13 Omura, T. and Sato, R. (1964) J. Biol. Chem. 239 , 2370-2378
14 Peterson, G.L. (1977) Anal. Biochem. 83, 346-356

MOLECULAR CLONING, AND TISSUE- AND CELL-SPECIFIC EXPRESSION OF HYOSCYAMINE 6β-HYDROXYLASE

T. HASHIMOTO, J. MATSUDA, S. OKABE, Y. AMANO, D.J. YUN, A. HAYASHI & Y. YAMADA
Research Center for Cell and Tissue Culture, Faculty of Agriculture, Kyoto University, Kyoto 606, Japan

1. INTRODUCTION

Scopolamine, the 6,7-epoxide of [S]-hyoscyamine, is one of the major alkaloids that accumulate in the family Solanaceae (Fig. 1). The epoxide oxygen of scopolamine is derived from molecular oxygen, this addition of oxygen being catalyzed by a 2-oxoglutarate-dependent dioxygenase, hyoscyamine 6β-hydroxylase (H6H; EC 1.14.11.11)(1). H6H incorporates one atom of dioxygen into the 6β-position of [S]-hyoscyamine as a hydroxyl group while simultaneously using the other dioxygen atom for oxidative decarboxylation of 2-oxoglutarate to succinate and CO_2. This oxidative reaction requires ferrous ion and ascorbate, and probably involves a ferryl enzyme as a highly reactive reaction intermediate (2). Besides hydroxylation of hyoscyamine and its derivatives, H6H catalyzes the epoxidation of 6,7-dehydrohyoscyamine (De-hyos; a synthetic alkaloid) to scopolamine (3) but the unsaturated alkaloid is excluded as a precursor of scopolamine *in vivo* (4). In plants, 6β-hydroxyhyoscyamine (Hyos-OH) is directly converted to scopolamine by 7β-dehydrogenation, thus retaining the 6β-hydroxyl oxygen in scopolamine. Low levels of enzyme activity catalyzing this epoxidation reaction in the presence of co-factors of 2-oxoglutarate-dependent dioxygenases were found in cultured roots of *Hyoscyamus niger*, and co-purified with H6H during partial purification (5). On the basis of this co-purification, it was considered likely that the low epoxidase activity may also be catalyzed by H6H.

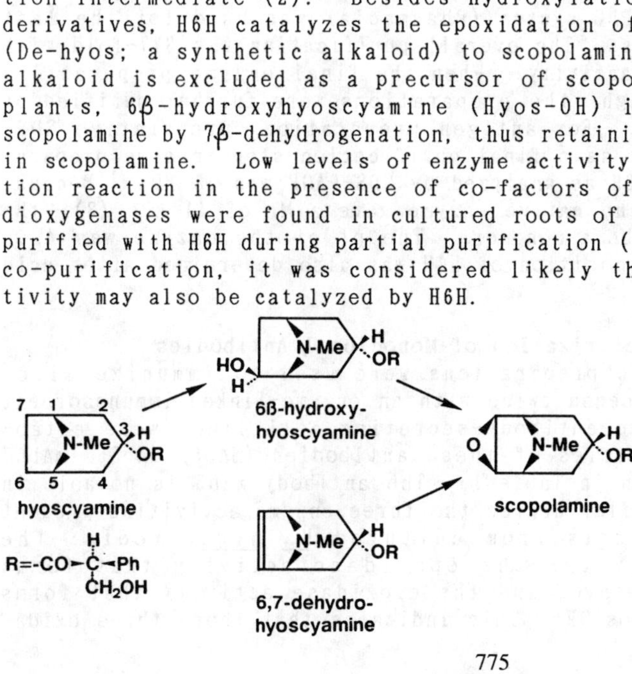

FIGURE 1. Biosynthetic pathway from hyoscyamine to scopolamine. Although 6,7-dehydrohyoscyamine is converted to scopolamine when fed to plants, this alkaloid is not an intermediate in this pathway.

Classical grafting experiments first demonstrated that the major site of tropane alkaloid biosynthesis is the root. Grafting a scion from tropane alkaloid-producing species on a root stock from nonproducing species resulted in no accumulation of alkaloids in the grafted plants; whereas, grafting in the reciprocal combination produced plants that accumulated tropane alkaloids (reviewed in 6). These experiments also indicate that the alkaloids are transported from the root to the aerial parts. More recently, two enzymes of tropane alkaloid biosynthesis, H6H (1) and putrescine N-methyltransferase (7), have been shown to be active in cultured roots but scarcely detectable in cultured cells and cultured shoots.

We here report the purification of H6H to homogeneity from cultured roots of H. niger. The purified H6H protein was used to immunize mice for the production of monoclonal antibodies against H6H. These antibodies were not only useful for studying the various reactions catalyzed by H6H, but also allowed immunohistochemical localization of H6H in the root. Further, information based on the internal amino acid sequences obtained from the homogeneous H6H were used to synthesize oligonucleotide probes for the isolation of H6H cDNA. The predicted amino acid sequence of H6H is homologous to the enzymes involved in the biosynthesis of cephalosporin in microorganisms.

2. RESULTS AND DISCUSSION

2.1. Purification of H6H

H6H was purified to homogeneity from cultured roots of H. niger by precipitation with ammonium sulfate and four consecutive chromatographies on Butyl-, DEAE-, Phenyl- and Hydroxyapatite-columns (8). The FPLC system (Pharmacia) was used for the last two chromatographies. The overall purification was 377-fold with a 0.6% recovery of activity. When the final enzyme preparations were not pure enough, the preparations were further purified by reversed-phase HPLC. For antigen preparation, preparative SDS-PAGE was also used as a final purification step in some preparations. The Mr of H6H as analyzed by SDS-PAGE was 38 kDa. Because gel filtration of the native enzyme gave a Mr of 41 kDa (3), the active enzyme must be a monomer. The pI of the enzyme was 4.6. The amino acid composition of H6H was also determined after acid hydrolysis.

2.2. Production and Characterization of Monoclonal Antibodies

Highly purified H6H preparations were used to immunize mice. Hybridomas were screened twice with an enzyme-linked immunosorbent assay. Four stable antibody-secreting cell lines were established. The properties of these antibodies (mAb5, mAb55, mAb67 and mAb94) are shown in Table 1. The antibody mAb5 is notable in inhibiting to a similar degree the three enzyme activities present in cell-free extracts from cultured H. niger roots: the hydroxylase activity, the epoxidase activity that forms scopolamine from De-hyos and the epoxidase activity that forms scopolamine from Hyos-OH. This indicates that these three oxidation reactions are catalyzed by a single protein, H6H. Immunoaffinity column prepared by covalently coupling mAb5 to cyanogen bromide-activated Sepharose CL-4B was used to purify in a single step enzymatically active H6H from partially purified H6H preparations. Because mAb5 also detects SDS-denatured H6H protein on Western blots, this antibody must bind to both the native and denatured forms of H6H. The other three antibodies bound only to denatured H6H.

TABLE 1. Properties of monoclonal antibodies against H6H

Antibody	Enzyme-linked immunosorbent assay	Immuno-precipitation	Western blot	Inhibition of enzyme activities	Subclass
mAb5	+	+	+	+	IgG1
mAb55	+	-	+	-	IgM
mAb67	+	-	+	-	IgG1
mAb94	+	-	+	-	IgM

+ and - respectively indicate positive and negative reactions.

Crude enzyme extracts from cultured roots of _Atropa belladonna_, _Brassica campestris_, _Duboisia leichhardtii_, _Hyoscyamus albus_, _H. gyorffi_, _H. muticus_, _H. niger_ and _Nicotiana tabacum_ were analyzed by Western blotting after SDS-PAGE. The antibody mAb5 recognized bands of 38-39 kDa in extracts from scopolamine-producing root cultures and the intensities of the bands roughly correlated with the hydroxylase activities in the extracts. The other three antibodies also reacted strongly with the 38 kDa band of _H. niger_, but their reactions with the H6H proteins from _Atropa_ and _Duboisia_ species were relatively weak. These studies show that the H6Hs from various species have Mr of 38-39 kDa and possess a common epitope that is recognized by mAb5. The fact that mAb5 inhibits enzyme activities of H6H indicates that this common epitope is located near the active site of H6H.

2.3. Tissue- and Cell-specific Localization

Western blot analysis of crude cell extracts from various tissues of _H. niger_ showed that the H6H protein is abundant in cultured roots, and present in plant roots but absent in petals, calyxes, leaves, stems, cultured cells and cultured shoots. For the immunohistochemical localization, all cultured roots except _Duboisia_ species were first embedded in paraffin then cut into 10 μm transverse sections, whereas other plant tissues were frozen and cut by a cryostat into 20 μm sections. Immunogold-silver enhancement staining with mAb5 as the primary antibody was used. H6H was detected in a layer of cells near the vascular cylinder in the several scopolamine-producing cultured roots but no signal was observed in the sections from cultured tobacco roots nor in any sections when a preimmune mouse IgG was used as the primary antibody.

Fig. 2 shows immunostaining of cryosections from cultured _D. leichhardtii_ roots using mAb5 or preimmune IgG as the primary antibody. When cryosections from the young tissues of _D. myoporoides_ were immunostained, a strong signal was seen at the periphery of the vascular cylinder of the root but no such strong staining was observed in the stem and leaf sections, indicating that H6H is expressed specifically in the pericycle cells of the root. Proximity of alkaloid-synthesizing cells to the phloem and the xylem may be important for the transport of tropane alkaloids from the root to the aerial parts of plant.

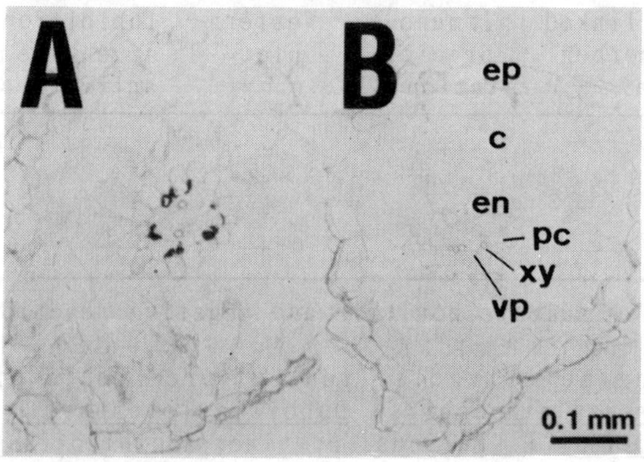

FIGURE 2. Immunohistochemical localization of H6H in transverse cryosections of cultured _Duboisia leichhardtii_ roots.
A) H6H-specific antibody mAb5 was used as the primary antibody in immunogold-silver staining. The pericycle cells are stained.
B) Preimmune mouse IgG was used instead of mAb5. No signal was observed. ep; epidermis, c; cortex, en; endodermis, pc; pericycle, vp; vascular parenchyma, xy; xylem.

2.4. **Molecular Cloning of H6H cDNA**

Direct amino terminal sequencing of the homogeneous H6H protein resulted in negligible yields on an amino acid sequencer, presumably because of amino terminal blockage. Therefore, after digestion with lysylendopeptidase the internal peptides of H6H were separated by HPLC and analyzed by automated Edman degradation on an ABI 470 protein sequencer. Amino acid sequences of 12 peptides were determined, three of which were used to design oligonucleotide probes. A cDNA library was made from mRNA from cultured roots of _H. niger_ by a unidirectional cloning method using a Not I-oligo dT primer. After screening the cDNA library with these three probes, several hybridization positive clones were obtained. All the cDNA clones encoded a polypeptide of Mr 38,999 and only differed in the length of their 5'-untranslated regions and in the polyadenylation sites (three sites were found). The predicted amino acid composition matched well with the composition determined from the purified H6H. The predicted amino acid sequence of H6H showed a relatively low but distinct regions of homology with the oxygenases involved in the biosynthesis of cephalosporin in certain bacteria (e.g. Streptomyces) and filamentous fungi (e.g. Cephalosporium). These enzymes are deacetoxycephalosporin C synthetase (DAOCS; 9, 10) and deacetylcephalosporin C synthetase (DACS; 10), both of which are 2-oxoglutarate-dependent dioxygenases, and isopenicillin N synthase (IPNS; 11) which requires ferrous ion, ascorbate and dioxygen for catalysis. It is intriguing to find several regions in the H6H amino acid sequence that are respectively homologous to both DAOCS/DACS and IPNS, and either to DAOCS/DACS or IPNS. These oxygenases in microorganisms and higher plants may have evolved from a common ancestral protein.

The H6H cDNA was placed under the control of the CaMV 35S promoter with the NOS terminator in pGA482 (Pharmacia) and introduced into tobacco by a Ti plasmid-based binary system. Transgenic tobacco calli expressing the cDNA had H6H enzyme activity and a 38 kDa protein that was recognized by mAb5 on Western blots. This unequivocally demonstrated that the cDNA encodes H6H. We are now introducing the H6H gene construct into several plants that produce various amounts of scopolamine. Nothern hybridization showed that H6H mRNA is abundant in cultured roots, and present in plant roots but absent in leaves, stems and cultured cells of H. niger. Thus, the tissue- and cell-specific expression of H6H is primarily controlled at the transcriptional level.

3. CONCLUSION

Starting from establishing fast-growing, alkaloid-producing root cultures in mid 80's, our research on alkaloid biosynthesis has now come to a molecular stage. The expression of scopolamine biosynthesis turned out to be highly tissue- and cell-specific, and this specificity appears to be controlled at the transcriptional level. Characteization of an H6H genomic clone is now in progress. Because most of the tropane alkaloid-producing plants are easily transformed by Agrobacterium, it should be straightforward to analyze the functions of the H6H promoter in homologous host plants as well as in tobacco. Also, the availability of H6H cDNA provides new opportunities to improve scopolamine production in transgenic medicinal plants.

REFERENCES

1. Hashimoto, T. and Yamada, Y. (1986) Plant Physiol. **81**, 619-625
2. Yamada, Y. and Hashimoto, T. (1989) Proc. Japan Acad. **65**, Ser. B, 156-159
3. Hashimoto, T. and Yamada, Y. (1987) Eur. J. Biochem. **164**, 277-285
4. Hashimoto, T., Kohno, J. and Yamada, Y. (1987) Plant Physiol. **84**, 144-147
5. Hashimoto, T., Kohno, J. and Yamada, Y. (1989) Phytochemistry **28**, 1077-1082
6. Waller, G.R. and Nowacki, E.K. (1978) in Alkaloid Biology and Metabolism in Plants, pp.121-141, Plenum Press, New York
7. Hashimoto, T., Yukimune, Y. and Yamada, Y. (1989) Planta **178**, 123-

130
8. Yamada, Y., Okabe, S. and Hashimoto, T. (1990) Proc. Japan Acad. (in press)
9. Kovacevic, S., Weigel, B.J., Tobin, M.B., Ingolia, T.D. and Miller, J.R. (1989) J. Bacteriol. **171**, 754-760
10. Samson, S.M., Dotzlaf, J.E., Llisz, M.L., Becker, G.W., van Frank, R.M., Veal, L.E., Yeh, W.-K., Miller, J.R., Queener, S.W. and Ingolia, T.D. (1987) Bio/technology **5**, 1207-1214
11. Cohen, G., Shiffman, D., Mevarech, M. and Aharonowitz, Y. (1990) TIBTECH **8**, 105-111

Section D
Biotechnology and Developing Countries

PERSPECTIVES IN PLANT CELL AND TISSUE CULTURE IN DEVELOPING COUNTRIES

Dr A. Ian Robertson, Department of Crop Science
University of Zimbabwe, PO Box MP.167, Mount Pleasant, Harare, Zimbabwe

A Third World Perspective

The Industrial Revolution allowed Europe to colonise and dominate the world. The green Revolution helped Asia at least towards self-sufficiency in food. The oil crises increased fertilizer prices and made it capital-dependent which pushed some of the benefits into the hands of those with ready cash.

The combined Computer/Information/Communication Revolution that we are experiencing also left many countries behind. A quick grasp of the potential and implications allowed the hard-working countries of Asia to employ this micro-chip revolution to promote themselves from poorly developed to increasingly well-developed status. I don't need to point out to a predominantly Nederlander audience who is making most of the costly luxury goods in the shops - and it is not just Japan. However a prominent African scientist, Dr Odiambho(1) of ICIPE has said "there are many Africans who feel that we really have lost a century." As a result, we in Africa and perhaps South America sometimes look on as spectators with frank envy. Can we have a share of all this prosperity?

Some Goals

My hope is that the Gene revolution based on trans-species movement of DNA plus a de novo writing of new DNA programmes will provide the opportunity for us to clamber aboard the fast train to tomorrow.
Can this be possible? Can we participate in the Third World? The organisers may have noted a tiny but significant change in my title: Perspectives in Plant Cell and Tissue Cultures in (not for) Developing Countries. I believe we need to apply with every sophisticated means at our disposal, a kind of "barefoot biotechnology" whose test is "Does it work : do the peasants benefit?" To use Odiambho's words we need to indigenise this new technology. Our first target is dignity and dignity flows from self-sufficiency in food production. Second is security : security comes from exporting enough to buy what you

need. Third is prosperity which incidentally comes from exporting enough to buy what you want - we will leave that to future generations.

The Potential

Plant cell and tissue culture can help traditional breeding to provide the means to those ends. We desperately need crops with a higher yield - there is basically no new land to bring into cultivation so yields must go up.

That is the bad news. The good news is that massive multiplication of clean planting material can double yields of our staple peasant crops which are fortunately clonal - cassava, sweet potato, potato, yams, coco-yams and many other lesser known roots and tubers. Our first task therefore is to make it available to Government, University or commercial laboratories; the second task, much harder, is to ensure its delivery to the remotest village in every developing country. Now that there is no longer any good reason, if there ever was, for the arms race, perhaps East and West can offer their idealists the challenge of helping the Third World off its knees and into the 20th century by inventing new ways to deliver the potential to the peasants. In Mozambique some are already using hovercraft and others are planning on airships! We need intelligent and economic use of all forms of transport and communication to service our disadvantaged. I believe no African country can afford to be without an efficient and effective micropropagation factory for receiving and multiplying lines of those cloned staple crops. As a modified version of the Brandt Commission recommendations might have put it "For the price of one nuclear sub we could provide every African nation with a clonal multiplication factory." Why not do it?

Our second task is to add single traits to already good varieties so that the cost of certain inputs will be reduced or eliminated. As of now few peasants can afford inputs like insecticides, herbicides, fungicides, acaricides, nematicides or bacteriocides so they take losses in yield that vary from a few to one hundred percent.
If we provide a scientific package that confers resistance to caterpillars(2) we can increase African yields by about 20% and reduce costs by another 10%, according to the Chief of Crop Production for Agritex, our Zimbabwean extension service for rural farmers(3). If we can provide a comprehensive range of virus resistance(4) we

would again contain yield losses of probably 50% in clonal crops - that means doubling the actual yields. The benefits are bound to be huge - no one can be seriously scientific about estimating their exact magnitude : the figures are not there. Yet any scientist and every peasant, when made aware of the potential of the gene-transfer revolution cannot help but get excited and hopeful of liberation from poverty and debt.

May I remind you of a well known graph which shows that most Africans of necessity plant crops that are bound to produce less than 50% of their potential. If we plant rubbish, we will harvest rubbish.

```
    100% l   x    ....commercial farmers
         l        x    {return to 100% every 3 years}
Yield    l             x
         l                  x
     50% l                       x    ....peasant
         l                            x    farmers{forever}
         l                                      x         x
         l
      0% l_____________________________________________________

Year...  0   1    2    3    4    5    6    7    8    9    10   11
```

Figure 1. Theoretical reduction in yield of clonal crops when planting material is saved from current crop year by year. Unless the peasant farmer gets fresh, clean material he is condemned forever to low yields.

The difference is the establishment of the aforementioned factories for the production of clean planting material that must be established in every African country.

Current Status

I know very well there are problems along the way: regulatory problems, political problems, many and varied technical problems, ecological problems, but I have confidence in the wit and determination of my colleagues in East and West, North and South such that these problems will be addressed and solved. The truth is that the real problems are those of paralysed democracies, favouritism and rivalries which are common yet specific to all nations and can only be resolved by a change in character and attitude of the people involved. That is beyond my brief at this conference so we will pass on!

<u>Situation Report</u>

Most of us in Africa live with a war on-going in our country or nearby so I will use the war-time jargon and give you a situation report. It is fragmentory as our communications are poor. This latter I hope the African Biotech Network that FAO and IITA have serious plans for (5) will alleviate.

To summarise:
Starting from the North, Morocco has tissue culture facilites in several labs, national, University and commercial, producing bananas and working on dates(6). IITA (Nigeria) is multiplying and distributing germplasm of cassava and sweet potato and plantain and banana plus yams and cocoyam. Good germplasm has gone to many countries(7). Egypt is working on its own Bt gene for use in cotton and potatoes(8). Kenya(9) is multiplying pyrethrum effectively thus challenging synthetic producers with a more natural product. Cameroon is cleaning up local varieties of root and tuber crops(10). Cote d'Ivorie has multiplication and virus-elimination work ongoing.

In Zimbabwe we have recently started adding marker genes to potato and tobacco(11) in addition to providing the National Programmes with "Virus-Eliminated" cassava, potato and strawberries. We have an efficient system for importing new varieties <u>in vitro</u>, growing them out, multiplying them, planting out in variety trials and so selecting what suits our ecological niches(12). Soon we will be in a position to add genes of relevance to agronomic constraints(13). Further south there are laboratories producing clean citrus and sweet potato, bananas and pawpaw. There is also a strong commitment to molecular biology which surely one day will be at the service of the whole continent(14).
My own lab now has eight PhD researchers hard at work on biotechnology for improving potato, sweet potato, strawberry, cassava, yam and cowpeas and coffee. Only one is funded from abroad, the rest are funded by local business or producer groups and the equipment is from our University Research Board and Government funded. These "human resources" will form the indigenous talent for our own gene revolution and we are increasing our links with other African centres particularly Kenya, Nigeria and Botswana.

I should add that I am only familiar with Africa although I am impressed with S America's beginnings(eg15) and more so with Asia's results! We truly are the last in line. That is a sketchy situation report but I hope its very paucity puts you in the picture.

I have said before(16) and I repeat to you : given a bus-load of scientists and a containerload of equipment- and a fax machine! - and we could together, scientists of North and South, solve the food crisis and perhaps the health crisis and therefore in time the population crisis of Africa. Which challenge could be more important and more fun?

Incidentally, I am sure you can see that the risk/benefit or the bad news/good news ratio looks very different from a developing country's perspective. In EEC or USA a neat gene transfer might shift the market share of someone's insecticide or herbicide by a few per cent but the tomato the customer eats is still virtually the same. In Africa the good news could be that some of those who might have quietly died because they did not find enough to eat, might actually live to enjoy life,might survive to tell the tale of how a powerful new technology was put to good use to give them and their family and their village a decent dignified life.

REFERENCES

1. Odiambo, T.R. (1990) ILEIA newsletter 6 (1) p5.
2. Hofke, H. and Whiteley, H.R. (1989) Microbiological Reviews, June 1989, pp242-255
3. P. Johnson, personal communication.
4. Beachy, R. (1989) Proc. of Third Conf. of IPBNet, Nairobi, Kenya.
5. Conclusions and Recommendations of CTA/FAO Symposium "Plant Biotechnologies for Developing Countries" Luxembourg, June 1989 p11 and informal discussions thereat.
6. M. Aaouine, personal communication.
7. IITA Annual Reports. Also, Proceedings of IITA Workshop on the Use of Biotechnology for the Improvement of Cassava, Yams and Plantain in Africa (1987) S.Y.C.Ng, IITA.
8. Kamel A. Ahmed, personal communication.
9. Okioga,D.M., Muriithi,L.M., Gichuru,S.P. and Ottaro,W.G. M. (1989) Abstracts of Third Conf of IPBNet, Nairobi, Kenya.
10. Wutoh, et al (this conference)

11. A. Connolly, personal communication
12. Robertson, A.I., Sakina, K. and Masona, V.M. Symposium on Science & Technology, Harare, January 1988
13. Jefferson, R. (1990) "Preliminary Assessment of the Potential for Agricultural Molecular Biology and Biotechnology in Zimbabwe and Kenya" Consultant Report to FAO.
14. C. Cresswell, personal communication, Conference on Plant Biology, Maputo, Mozambique, February 1990.
15. Tejada,M. Andean Biotech Programme of the Andean Development Corporation (CAF) at CIP Biotech Planning Conf, Lima, Peru, March 1990.
16. Robertson, A.I. and Sakina, E.K. (1989) TIBTECH 7 (1) S14; also Robertson,A.I. Household and National Food Security in Southern Africa. Proceedings of the Fourth Annual Conference on Food Security Research in Southern Africa. Harare, October,1988.

N.B. I apologise for the chauvinism of the references: I am trying to illustrate what is happening in Africa.

COMMERCIALISATION OF AGRICULTURAL BIOTECHNOLOGY IN INDIA: OPPORTUNITIES, PROGRESS AND PROBLEMS

JITENDRA PRAKASH

Biotechnology Division, Indo-American Hybrid Seeds, Post Box No. 7099, Bangalore - 560 070. India.

INTRODUCTION

India is an agribased economy and the role of new technologies in agriculture have more relevance for the country than in other areas. The abundance of genetic base, diverse agroclimatic zones and highly qualified man power offer a rich scope for technological advances in agricultural biotechnology.

India has been a leading exporter of primary produce such as tea, coffee, rubber, cardamom, cotton and various other spices, flavoring, aromatic and medical plants, but has, of late, started losing its premier position to other competing developing countries such as Guatemala, Kenya and Brazil, due to non competitive prices and low productivity. Thus the need for use of cell and tissue culture related techniques in increasing the vertical productivity is immensely felt today than any time before.

On the domestic front agricultural production is governed more by the climatic vagaries and the preharvest scarcity and post harvest glut lead to speculative trading by the producer (1). The shortage of healthy seeds/planting material and lack of disease resistant clones have often been effecting the Indian agricultural economy. Large scale production of elite clones through micropropagation or somatic embryogenesis, production of disease free propagules and meaningful development of plant varieties utilizing cellular, molecular and whole plant breeding techniques hold the promise of alleviating some of these problems in Indian agriculture, horticulture and forestry.

OPPORTUNITIES FOR DEVELOPMENT IN INDIA

Among all the countries in the world, India has a rich resource of land spread over diverse agroclimatic zones that have the potential to produce a wide variety of agricultural and commercially high value crops. It has also a germplasm of rare ornamental and medicinal species that are native to the tropical forests of different states. As a whole, to develop and apply biotechnology to a maximum extent, India offers many advantages over other countries in the world.

The native forests all over the country are well known for their richness and variety in foliage and ornamental plants. These plants can be propagated in large scale by tissue culture methods and can be used either as pot plants or as cut flowers. As commercial growers, breeders and producers are interested in new variety of plants, these plants can be introduced as new products in horticultural markets.

Plants obtained from anywhere in the world can be maintained, propagated and transplanted into soil in India as the country has diversified agroclimatic regions. To economize the procedure, it is advisable to cut down towards purchase and maintenance of the equipment (airconditioning systems) needed for high rates of multiplication and survival, if the appropriate site is located within the country. However, before locating a site for plant tissue culture industry, one must decided the products that are going to be produced.

In India, the scientific talent and the operation skills of the people is on par with the best in the world and also the cost of skilled personnel is low as compared to other countries. As a result a significant portion of production costs would come down whereas the reverse is true in the case of most advanced countries such as Europe, U.S.A. etc. In fact, some of the important discoveries in plant biotechnology are of Indian scientists. For example, anther culture technology, widely practised today throughout the world, was introduced by Indian scientists in 1964 on solanaceous species, from Delhi University. The finding of 'in vitro' pollination to effectively overcome the incompatibility mechanism in many plants and ensuring proper fertilization, good seed set and also the possibility of obtaining rare hybrids, is from the same school.

Embryo culture is yet another novel 'in vitro' technology originated from India that is extensively being used by the plant breeders. This technology can help overcome the hurdles of endosperm-embryo incompatibility and poor endosperm growth encountered in traditional breeding methodologies; rare hybrids can also be obtained by this method. Inter-specific as well as intergeneric hybrids have been successfully regenerated through embryo rescue and culture. Similarly, Indian University and Institutes have pioneered micropropagation procedures for various crops for the first time.

However, a drawback of the Indian Research programme has been that, though the best technology is being evolved here, the implementation/commercialization part is being done in other parts of the world because of lack of proper direction.

Blessed with the rich flora and scientific talents, India is in a better position in compared to other so called cheap labour areas of the world. As we do not have to import technology, we only need to scale them up, protect and patent the technology for the benefit of the nation.

PROGRESS IN COMMERCIALISATION

India ranks among first five countries in the area of space and nuclear research. This was due to foresightness of our late Prime Minister Mrs.Indira Gandhi. It was her again who in 1982 formed National Board of Biotechnology to encourage and promote research and development work in the area of Biotechnology in the country. Since then several well defined programmes of National priorities began in various Government Research Institutes and Universities. In recognition of the need for a focal point in the administrative structure of the Government, the Department of Biotechnology, (DBT) was set up in February 1986, as a separate department in the ministry of science and technology, for planning, promotion and coordination of biotechnological programmes in the country.

Currently DBT is funding 13 research projects in the area of crop biotechnology and expects to fund more programmes in this area. The ongoing projects on rice deal with the use of RFLP and allied techniques for genetics and breeding programme, regeneration of plants from protoplasts of indica rice and wide hybridization with special reference to diversification of cytoplasm and heterosis exploitation. A task force set up by DBT on 'production of biomass (fuel, fodder, timber and commercial/industrial woods) using Tissue Culture Technology' identified a list of fourteen nationally important forest tree species. Based on the recommendations of the task force, two pilot plant units were set up at National Chemical Laboratory, Pune and Tata Energy Research Institute, New Delhi. Seven R & D projects in selected forest tree species have also been commenced for developing and standardising know how/protocols for complete plantlet regeneration using tissue culture techniques from mature elite budwood material/plus trees (2).

It is only in 1987 that commercialization of agricultural biotechnology began in India in real sense when various private sector organisations looked at this technology as a means of increasing the productivity of their primary produce or as a part of their diversification programme. These include A.V. Thomas Group Companies (AVT) Indo American Hybrid Seeds (IAHS), Hindustan Lever, Tata Tea, Unicorn Biotek, Nath seeds, RPG Enterprises, Indian Tobacco Company, Maharastra Hybrid seeds company, Hindustan Agrigenetics Ltd., and many others either preparing or sensing to do so.

The first professionally managed commercial plant tissue culture unit in the country was set up in 1987-88 by AVT, when the author was their Research Director. This unit was set up in Cochin Export processing zone and remained as one of the largest in Asia and only one of its kind in India, until IAHS began production in its Bangalore based unit which not only has forty eight Laminar flow work stations, five US class, hundred growth rooms, each accomodating 400,000 cultures at one time, but also has a green house covering 30,000 Sq. feet (2790 Sq M) with most advanced equipments such as automatic shade system, fogging, blackout and irrigation systems, and movable benches for hardening of tissue culture plants.

Both IAHS and AVT have exported millions of tissue culture plants to Europe and released several hundred thousands bananas and cardamom to farmers at home. In addition IAHS has won several contracts from the Netherlands, UK and Denmark to develop new plant varieties by somatic hybridisation, generation and selection of somaclonal variants, endosperm culture, "in vitro" pollination and embryo rescue, c-DNA, satellite-RNA and antisense RNA technologies in conjunction with the whole plant breeding.

PROBLEMS ASSOCIATED WITH AGRICULTURAL BIOTECHNOLOGY

Biotechnology is highly capital intensive industry and since the delivery of large volumes at targetted time is generally not made possible in most cases, the per plant cost becomes prohibitive.

When the economics of an industry is under consideration, it becomes necessary to scrutinize the factors responsible for its performance over the past years in order to set priority to its objectives. Though an extensive technology has been developed during the last several years, there are still some major difficulties especially those associated with expansion of this technology to a variety of crops. Some of these problems are production cost, seasonal nature and heavy peak demands, personnel/organisation problem and contamination control (3).

Besides, the marketing of biotech products adds another dimension to the problem, restricting the furtherance of agricultural-biotechnology in India. The insensitivity of the farmer, to per plant returns and ultimate potential of his crop, acts as a serious impediment to fasten adoption of the technology in India. He is less receptive to the concept and thus needs a gestation period before he is convinced (1).

On the export scene, lack of variety of packaging material at affordable prices, high air freights, unsuitable air connections, lengthy phytosanitary, custom and other export documentation formalities make this nontraditional highly perishable export commodity much less interesting as compared to potential it has.

CONCLUSION

Not withstanding these problems, the prospects for plant biotechnolgy in India are very many. The sound academic base and scientific manpower strength available in the country could be effectively utilised by the technology, thereby reversing the brain drain. This skilled manpower that is available at a cost effective rate could also be utilised for such off-shore projects of the developed countries, for whom the scientific labour is proving costlier.

The non-seasonability of biotech production could help ease out violent price fluctuations. This can help the farmer to plan his agriculture accordingly and get out of speculative trading. This also has its advantages in microlevel agricultural planning in altering the crop balance.

Further, larger volumes of plant delivery, in time, at affordable prices could be made possible by the technology. The commercial scale-up however has to come from the entrepreneurial R and D. Favourable government policies could develop this in the country. Simultaneously, rapid advancements in the newer and advanced frontiers of plant biotechnology by the academic institutions and their commercial upgradation by entrepreneurial R and D could revolutionise Indian agriculture subsequent to that of the Green Revolution(1).

REFERENCES

1. George E and Prakash J. (1988) Proc. Biotek India, '88, 19 - 20.

2. Annual Report (1989 - 1990) Department of Biotechnology, Ministry of Science and Technology, 6-7.

3. Prakash J (1989) in applications of biotechnology in Forestry and Horticulture (Dhawan, V ed.) pp 265-274, Plenum Publishing Corporation, NEW YORK.

IMPACT OF TISSUE CULTURE TECHNIQUES ON THE DEVELOPMENT OF AGRIBUSINESS IN MEXICO

F. MEDINA-GOMEZ
NOVUM CORPORATIVO SA DE CV, BOSQUE DE CIRUELOS 180, PISO 1, 11700 MEXICO, D.F., MEXICO

ABSTRACT

Some private Mexican agribusinesses have only recently discovered the long term impact of the micropropagation of selected plant species in order to sustain vital overall economic yields, increase the production of secondary plant metabolites of economic importance, increase tolerance to high salinity, lack of water or certain diseases, provide a ready source of germplasm as well as a means of preserving it for the future, and finally offer alternatives for massive improved seed production. The application of these techniques has brought about a revolution in the way Mexican industry perceives biotechnology in general and some specific case examples are presented and discussed in more detail to provide both information and insight to other developing countries.

I. INTRODUCTION

Mexico, unlike other developing countries throughout the world, has a very unique set of problems regarding the development of its agribusiness sector. After all, the Mexican Revolution of 1910, set forth as one of its basic principles an Agrarian Reform that oversaw the redistribution of the large estates to the small communities of peasants, who were organized collectively in cooperatives called **ejidos** where every individual has his own share of land, but is also unable to sell or otherwise legally dispose of it. The Agrarian Reform, therefore, virtually tied down the peasants and their ever increasing number of descendants to a shrinking individual plot of land. Most of these peasants have remained at the subsistance farmer level and even the Green Revolution, with its emphasis on the use of normally expensive agricultural chemicals to increase yields, failed to provide, via technology, a better standard of living for the farmers themselves or even any incentives for the creation of a stronger ejido-based agribusiness sector.

On the other hand, the Agrarian Reform did set a limit to private ownership of land, with a maximum of 150 hectares in the case of prime irrigated land, thus allowing for the relative growth of a small group of independent farmers who quickly adopted both the Green Revolution technologies and a set of specialized crops that commanded higher prices mainly in the export markets. These farmers were to become the initiators of the Mexican agribusiness sector, which later grew to encompass some of the ejidos themselves and some large multinational firms in a strange but effective liaison.

II PRIMARY IMPACTS

Obviously, not all the crops are likely to receive a strong impact from the adoption of tissue culture techniques. Most of the basic cereal crops are of little interest to the Mexican agribusiness sector, because of the lack of economies of scale associated to the small plots of land. Furthermore, the sharp drop in the international prices of certain traditional cash crop commodities, such as sugar, coffee, cocoa, cotton, etc., have also reduced the farmer's interest in both developing and applying sophisticated tissue culture techniques to increase or at least maintain his eroded profit margins.

Thus, the greater impact of these techniques is definitely associated with those crops where the cost of production can be reduced directly, and not with those crops where productivity is increased through yield increases. The development of new crop varieties of high added-value, resistant to pest, disease and environmental stress, saves costs of purchasing agricultural chemicals at little additional expense, as well as decreases operating costs associated to energy, irrigation and labour.

The massive use of pesticides, herbicides, plant-growth regulators, irrigation, fertilizers, and energy, typically increases yields but at an additional expense, the major flaw of the Green Revolution.

The factors which are more likely to induce the adoption of tissue culture techniques are the prices of both agricultural inputs and final products and the absence of economies of scale associated with the size of the plot of land. Thus, the major technological change has taken place in the case of high added-value crops for the export markets, because of the ratio price of final product to price of inputs: 1) cut flowers such as roses, carnations, chrysanthemums, statice, tuberose, gerbera and orchids, and tropical foliage plants such as Dieffenbachia spp, Ficus spp, Philodendron spp, Anthurium spp, Syngonium spp, Guzmania spp, and Spathiphyllum spp; 2) vegetables such as asparragus, broccoli, cauliflower, tomato, cucumber, onion, garlic, chilli pepper, strawberry, etc.; 3) fruits such as avocado, mango, guava, papaya, pineapple, grapes, apples, lemon and oranges; 4) industrial crops such as sugar cane, Agave tequilana and marigold flower; 5) valuable timbers such as red cedar and pine; and 6) fodders such as alfalfa.

All of these crops are excellent candidates for the application of tissue culture techniques, since they already rely heavily on the application of high and conventional technologies.

III CASE EXAMPLES

III.1 CUT FLOWERS AND TROPICAL FOLIAGE PLANTS.

This has been the single most successful case of widespread application of tissue culture techniques in Mexico. At least four private companies have been in operation for the last 5 years, producing via meristem culture techniques, virus and disease-indexed material of both carnation and chrysanthemum, which is then propagated using rapid clonal propagation techniques. Also, various types of useful mutant traits have been selected directly in tissue cultures. Other techniques, such as somatic hybridization or embryo rescue, have been more readily applied in the rather more industrial crops.

The tissue culture techniques have to compete with the more traditional and simple rapid growing procedures such as rooting cuttings, and the slower more sophisticated procedures such as grafting and buddying. In the cases where rooting cuttings are cheaper than cuttings produced from clonally propagated plantlets, there is little economic argument in favour of the latter.

In at least four species of common cut flowers: rose, gerbera, carnation and pompom chrysanthemum, the application of tissue culture techniques has demonstrated a definite economical advantage, even in spite of the absence of economies of scale at the commercial operation level. In one particular case, the flower producers realized that they could apply tissue culture techniques, such as meristem culture, in order to produce in vitro large number of plantlets of economically important clonal lines, which were then successively rooted and transferred in sequence from the first to the third generation greenhouse, after being vegetatively multiplied at every step. It would have been prohibitively expensive to use these plantlets for direct commercial production of cut flowers, but they were excellent starting material for the commercial production of rooting cuttings.

Out of a total of 7 hectares, 3 of which were dedicated to the production of rooting-cuttings and the rest for commercial flower production (one hectare for each of the four species mentioned), the use of tissue culture techniques provided overall savings of about 100,000 USD equivalent to 3.5 % of total sales, which allowed for the recovery of a total investment of 71,000 USD made on a relatively modest micropropagation lab, in less than one year.

The investment in the tissue culture facilities was only 21% of the total fixed assets, including the indexing equipment, and they were only used to about 35 % of their production potential, because of the small plot of land available. The cost of producing each rooted plantlet amounted to about 98 US cents, excluding the cost of the mother material, and labour accounted for 51 % of this cost, compared to 39 % of micropropagation materials. The fact that these techniques are labour-intensive, should make developing countries very competitive. It is important to consider that labour includes the people working at the first, second and third generation greenhouses where the rooting-cuttings are obtained by traditional techniques. The manpower employed by the areas associated with the tissue culture technology is 49 % of the total.

However, clonal propagation techniques also have a greater impact on the final quality of the cut flowers (less than 10% rejected material, as opposed to 30 % without), thus ensuring a better acceptance in the export markets (normally more than 80 % of the total sales, and with prices between 4 - 5 times greater than the domestic prices), plus a greater productivity in terms of numbers of export-quality flowers (in the order of 20% more tonnage for sale). If this is quantified, the benefit derived from tissue culture technology exceeds 15 % of the total anual sales of 2 million USD. This example has now been followed by at least four more companies.

The total sales of rooting cuttings of two of these companies, of carnation and chrysanthemum obtained from clonally propagated material, amounted to roughly 2.1 million US dollars in the past five years. These cuttings eventually transformed themselves into cut flowers, which meant a total export revenue of over 16 million dollars to the flower producers.

III.2 INDUSTRIAL CROPS.

__Agave tequilana__.- The agave plant from which tequila is made was saved from an imminent extinction thanks to the development of tissue culture techniques for the rapid clonal propagation of the agave plants. These agave usually have a life cycle of 7 years, and the whole plant is uprooted and the core utilized to extract the sap which is fermented and distilled into tequila.

Tissue culture techniques did not only guarantee the long term preservation of the agave plantations and, therefore the only source of raw material for the tequila brewing industry, but also made an impact on the agronomical and industrial yields of the plant. Since the raw material cost is close to 65 % of the total cost of production, an increase of 10 % in the amount of sap produced decreases the cost in about 6 %. Considering that Mexico produces more than 63 million litres of tequila, 80 % of which are destined for the export markets, such an impact on costs is equivalent to 15 million USD in savings per year.

The investment made by one of the larger tequila companies was also of the order of 100,000 USD. The internal rate of return attributable to the adoption of tissue culture technology, defined as the increase in profits directly derived from this technology, divided by the overall investment in this technology spread over a period of 10 years, was greater than 270, an extremely attractive figure, compared to the average 50 - 60 rate characteristic of good Mexican investments in technology.

CAROTENOID BIOSYNTHESIS.- One peculiar crop grown in Mexico is the marigold flower, Tagetes erecta, which is transformed into a series of pigmenting products for the poultry industry. T. erecta belongs to the Compositae and thus is characterized by an enormous genetic diversity.

The pigmenting carotenoids are basically present in the flower's petals, so carotenoid productivity is dependent on two main factors: flower heterogeneity and flower loss from disease and/or pests.

Tissue culture techniques have been developed to select a number of clonal lines according to phenotype, where the architecture of the flower can be greatly improved: 1) number of petals in each flower; 2) flowers with more surface on petals, (more solar exposition and thus more carotenoid production); 3) an intense orange colour.

The actual biosynthesis of carotenoids takes place as a result of the plant's metabolism. It is easier to study the metabolism of genetically stable and homogenous lines which already exhibit desirable traits, with the aim of increasing productivity in terms of the number of petals (and therefore weight of each flower), the number of "super" flower (i.e. flowers with more petals) yielding plants, and the carotenoid content in the petals.

The research initiated by a couple of private Mexican companies has involved some agronomical work in the nursery and greenhouse to assess the impact of environmental and physiological conditions that determine flowering behaviour and carotenoid content (i.e. photoperiod) as well as in vitro work on growth regulators to maintain uniformity in plant growth and flowering.

Research in cell culture, in the other hand, has lead to somaclonal variation to produce elite genotypes and phenotypes with carotenoid hyperproducing characteristics and resistant to major agronomical diseases and pests.

Since the carotenoid concentration in the petals accounts for the greatest impact on the companies' return on investment, it is important to develop clonal lines to standardize this concentration and increase profitability.

The economic impact can be easily assessed by estimating an increase to 24 g/kg of carotenoids in dry petals, from the current 20 g/kg, coupled with an increase from 50 to 90 % of "super" flowers. At the 1988 production levels, this would have meant an increase of 77 % more grams, with the same price per kilogram of fresh flowers. There is an obvious benefit for the pigment producer, who can reduce his production costs in about 40 %, but also for the farmer who gets more tons of flowers per hectare (up to 50 % more tonnage with the same land).

III.3 PLANT SECONDARY METABOLITES.

The lack of availability of raw materials to produce certain value-added products has resulted in the necessity of exploring the possibility of producing secondary metabolites in plant cell fermenters.

One such case is that of capsaicin, the compound that gives the chilli pepper, Capsicum annuum, its pungency. The main objective of this agribusiness initiative was to screen different commercially available cultivars of C. annuum for their capsaicin content and once the cultivars of interest were established, manipulate a number of environmental parameters in culture: photoperiodic effects, intensity and light quality effects, as well as the culture media themselves: plant hormones and their ratio, nutrients, etc., to develop a strategy for callus formation, secondary metabolite synthesis in the callus, the transfer to liquid suspension culture and the optimization of the plant cell fermenter at a laboratory scale to develop the bioreactor design suitable for scale-up.

C. annuum yields mainly two products of interest: capsaicin (with a price of 40 USD/g) and a mixture of red and yellow carotenoids at a combined price of 0.52 USD/g. Obviously, the main advantage of this technique is that very large numbers of cells can be grown in a small volume. Cells in suspension can be gradually exposed to different environmental and culture media conditions, as well as to stress, and then the most promising cells are encouraged to regenerate either back into calluses or whole plants which are then screened for the given metabolite content.

A rule of thumb has been developed in Mexican industry that only a plant product of a value higher than 4 USD/g is profitable enough to be produced by a plant cell fermenter with the existing bioreactor technology. However, products of strategic importance, with a price as low as 0.40 USD/g, but with a raw material cost higher than 75 % of the total production cost and a downstream processing cost associated to the product recovery and purification not higher than 10 % of the total cost, can and should be considered as potential candidates on a case by case basis.

This project is still in progress, but the final result might be the decision that it is economically more feasible to rely on standard agricultural methods, but using the lines developed by secondary metabolite technologies, to obtain larger yields.

REFERENCES

Medina-Gomez, F., Biotechnology in Mexican Agriculture. Unpublished.

A MODEL FOR COMPARING AND EVALUATING BIOTECHNOLOGICAL INNOVATIONS FOR SMALL-SCALE FARMERS IN DEVELOPING COUNTRIES

J.E.W. Broerse*, J.F.G. Bunders*, J.D. Byloo** and A. Stolp*
** Department of Biology and Society, Free University Amsterdam, De Boelelaan 1087, 1081 HV Amsterdam, The Netherlands.*
*** National Council for Agricultural Research, Ministry of Agriculture, Nature Management and Fisheries, P.O. Box 20401, 2500 EK The Hague, The Netherlands.*

1. Introduction

Biotechnology* is widely seen as one of the key technologies for generating innovations in agriculture, industry, human health care, etc. (1). This broad field of research is considered specifically as an instrument to promote rural development in the Third World by an improvement of agricultural production and the generation of jobs (1-4). However, recent developments in research are more likely to threaten than to support this improvement. The most important cause is that the present direction of biotechnological research is largely determined by the goals of pharmaceutical and chemical multinationals (5,6). Research in the fields chosen by these firms is strongly supported by Western governmental institutes and universities which closely cooperate with them (6-8). Market orientation hence dictates the focus of research as the substitution of raw agricultural products and the delivery of inputs for capital intensive, well-controlled farming systems (9).

On the basis of experiences with earlier efforts to introduce Western agricultural innovations in developing countries, it can be expected that innovations resulting from such a focus will lead to a decrease of commodity prices and result in a greater instability of the agricultural structure in many developing countries. These developments may lead to a further marginalization of small-scale farmers and to an accelerated migration to the already overcrowded cities, where job opportunities are very limited (6). This analysis suggests that the new biotechnological applications currently being developed will have a mainly negative impact on many developing countries and will be beneficial for a limited group of mainly large-scale farmers in some developing countries only.

---

* The following definition of biotechnology is used by the Netherlands Directorate General for International Cooperation: "Biotechnology is the integrated use of molecular genetics, biochemistry, microbiology and process technology to supply goods and services, employing micro-organisms, parts of micro-organisms, or cells and tissues of higher organisms".

It can be concluded that with current research directions the category of resource-poor, small-scale farmers in developing countries will not benefit from biotechnological developments. This conclusion raises two questions:

1. Does biotechnology really have the potential to contribute to the solution of several important problems of this group of farmers?
2. Are there ways to redirect part of international biotechnological research?

In answer to the first question, it is clear that from a biological point of view the problems of small-scale farmers are related to: drought, poor soils, erosion, poorly adapted varieties of crops, and diseases of crops, animals and human beings. Biotechnology has significant potential for tackling several of these specific problems. It provides new possibilities for controlling biological processes to improve crops, animals and micro-organisms. Furthermore, new processing techniques could provide outputs with higher quality. Hence, from a purely technological viewpoint the claim that biotechnological innovations can make a significant contribution to the solution of the problems of the small-scale farmers seems justified.

Dealing with the second question, it must be noted that public research institutes are not obliged to focus on corporal industrial interests. In fact, these institutes can make an effort to ensure that small-scale farmers benefit from biotechnological developments (supported with some additional funding from donors). However, most public organizations that might be expected and able to deliver a contribution lack information on which they can base such research.

Therefore, the first step in such a redirection is a careful analysis of how and under which conditions biotechnology could contribute to sustainable rural development, and how choices can be made between the various options.

In order to be able to direct biotechnological research to the needs and interests of small-scale farmers the following questions will have to be answered:

1. What are the major problems of small-scale farmers in developing countries?
2. What are the potential contributions of biotechnology to the agricultural sector?
3. Under which conditions are certain biotechnological solutions:
   (a) appropriate, i.e. relevant in terms of expected outputs and absence of negative side-effects;
   (b) feasible, i.e. the expected output has to be achievable in terms of scientific and technical research, and diffusion to and adoption by farmers; and
   (c) have a comparative advantage in terms of implementability (appropriatemess and feasibility), problem-solving capacity and cost-effectiveness?

In answering these questions one has to deal with two major problems: 1) lack of information exchange between different scientific fields, and between different groups involved in rural development such as farmers' organizations, governments, donor organizations and scientists; and 2) lack of an adequate model, based on a coherent set of criteria, to assess and thus to evaluate and compare ideas on the development and introduction of biotechnological innovations.

On the basis of the questions mentioned above, a project named "Biotechnology for small-scale farmers in developing countries: an assessment of opportunities" was executed in 1988-1989 with the following objectives:

1. Collection of information about problems of small-scale farmers in developing countries from literature and resource persons;
2. Collection of information about developments in agricultural research and the opportunities and constraints for developing countries from literature and resource persons;
3. Presentation of a model for information exchange; the "interactive bottom-up approach";
4. Presentation of an assessment model for evaluating and comparing biotechnological innovations; and
5. Application of the models in three developing countries (Zimbabwe, Pakistan and Bolivia) in order to identify and prioritize problem areas of small-scale farmers for which biotechnological solutions might be appropriate and feasible.

In this paper we will present the model for information exchange and the assessment model for comparing and evaluating biotechnological innovations. We hope that these models will provide adequate tools in the structuring and assessing of programmes and projects.

2. Presentation of the models

Different methods can be used to identify research priorities in biotechnological research for developing countries. A well-known procedure has been developed for the Rockefeller Foundation's International Program on Rice Biotechnology. In this approach, the identification of research targets is based on the assessments of molecular biologists only. They identified the research targets for rice crop improvement that promise the highest theoretical yield increase on a world scale at the lowest research costs (measured in research time). The assessment '*excludes ... socio-economic research designed to discover why producers may or may not utilize technology or what the impact of technology is*' (10). The argument used not to involve other experts than scientists in the identification of research priorities is that '*the inclusion of social scientists, extension workers and farmers would likely improve the accuracy with which the responses reflect farm level realities, but they would be less accurate in precisely identifying very specific research targets*' (10).

Instead of focusing exclusively on the possible biotechnological contribution to theoretical yield increase and research costs, we have specifically examined the appropriateness and feasibility of biotechnological innovations. The rationale is that research is costly and that (financial) resources are too scarce to be allotted to projects where the benefits have no chance of reaching the end-user (in this case small-scale farmers). Only after having identified appropriate and feasible solutions does the question of most effective impact become relevant.

Interactive Bottom-up Approach

The "interactive bottom-up approach" is developed on the basis of technology assessment research, and on ethnographic studies of innovations and research priority setting processes (11). This approach is needed to avoid technology-push, and to include the opinions of not only scientists but also that of end-users and the organizations which represent and/or work with them (12,13). Three phases can be distinguished:

1. *Inventory phase.* This phase consists of exploratory activities which are aimed at providing an overview of the literature, and of the ideas and opinions of different groups and organizations involved in rural development on research priorities for small-scale farmers, and of the way in which the various options can be compared and evaluated (assessment model). On the basis of this information, priorities for biotechnological innovations can be identified, and communication and an exchange of information between different groups, like farmers, scientists, policy makers and NGOs, can be established. Through the activities in phase 1 it will become clear which organizations and people feel strongly committed to the idea of using biotechnology for the improvement of the living conditions of the rural poor. This means that an informal team can be formed to link up with relevant power structures and with local institutions through which priorities are set to improve the situation of small-scale farmers. This team can operate as an important element within the social structure by supporting and guiding the implementation of the ideas when they may reach phase 3.
2. *Interaction phase.* In this phase a public debate takes place in which the results of the preceding phase are discussed in order to identify unanticipated negative side-effects and constraints to the development of the innovations. This may result in a change in priorities or adaptation of proposals. Ideas are also discussed on how to realize the conditions necessary for effective implementation.
3. *Phase of institutionalization.* In this phase specific projects are prepared, program studies are initiated, existing institutional frameworks are adjusted, and new institutional frameworks are developed.

Assessment model for evaluating and comparing biotechnological innovations

The assessment model presented in this paper is based on criteria for evaluating the appropriateness, feasibility and comparative advantage of biotechnological innovations. These criteria can be classified in three major categories: the national context (11,14,15), the farming system (14,16-18), and the comparative advantage of the biotechnological innovation (11,14,19). Each category refers to a number of necessary conditions that should be met in order to make the introduction of biotechnological innovations a success.

1. *National Context*

   A. The availability of scientific knowledge about the farming system of small-scale farmers. The agricultural activities of small-scale farmers are highly interrelated. Changes in one kind of agricultural activity may have certain effects on one or more other activities, and in some cases, may jeopardize the improvements being pursued. Before introducing new technological innovations among small-scale farmers it is very important to have reliable knowledge about the structure and functioning of their farming system. Without such information the possible contributions of biotechnological solutions cannot be assessed properly hence making it very difficult to choose rationally between different options.

   B. National/regional organizations for the adaptation and diffusion of (bio-) technological innovations. The introduction of biotechnological innovations is much easier in situations in which this type of organizations already exists and functions adequately. These organizations, which may be extension services or other specialized institutes established by the government of the country, or NGOs, facilitate the adaptation of biotechnological innovations to local conditions and their diffusion to small-scale farmers.

   C. Decision-making structure in and between government bodies and organization involved in research development for small-scale farmers. The introduction of biotechological innovations can only be successful if they are conceived as 'practical' solutions for the problems of the country concerned. Therefore it is necessary that within the country itself relations exist between a variety of organizations and services that make the identification of problems and their solutions possible in a relatively structured way. In addition, it should be possible to identify relevant fields of research. In that way the development of biotechnological innovations is based on a more balanced judgement and endorsed by political support. The lack of a decision-making structure on research development implies that additional efforts are required in order to ensure that the right problems are being addressed, and that proper solutions will be identified and integrated in research and development programmes.

   D. Commercial markets for agricultural products. The existence of a commercial market for agricultural crops, together with other marketing facilities and policy measures to improve the small-scale agricultural sector, is another important aspect of the national context to be taken into account when assessing the potential success of the introduction of biotechnological innovations. In cases where there is no commercial market for certain agricultural products, or where markets are highly unstable, e.g. as a result of no limitations on competing imports nor guaranteed prices, the introduction of biotechnological innovations may be severely restricted. Possibilities are even more limited if transport and storage facilities (road systems, harbour facilities, availability of transport vehicles and field depots) are poor.

E. Infrastructure for biotechnological research. The realization of appropriate biotechnological innovations strongly depends on the existence of a local scientific infrastructure on biotechnology.

2. *Farming System*

The special focus on the needs of the small-scale farming sector in the identification of potential research fields in biotechnology is essential as the different circumstances of the large-scale commercial farming sector often lead to different priorities even if the same crop is grown or the same diseases and pests occur. Five crucial aspects of the farming system are:

A. Severity of the problem being addressed. The impact a biotechnological innovation might have on a given farming system is dependent on the severity of the problem being addressed by the innovation, which can be determined by calculating the amount of farmers affected by the problem and by analysing the relative importance of the problem in terms of output.

B. Input requirements. For a biotechnological innovation to be successfully applied in a given farming system, certain types and amounts of inputs should be available. These input requirements can be divided in two different types:
   1. External: access to capital intensive inputs, credit facilities, education services;
   2. Internal: means of production (land, draught animals, capital, technological means, labour), the adjustability of the land use pattern and distribution of labour, and the adaptability of the socio-cultural pattern.

C. Output characteristics. Four different kinds of output characteristics have to be taken into account:
   1. Income generating effect. In order to realize an income the outputs have to be marketed. For some products the market does not exist or is highly unstable, due to lack of a price mechanism. However, at local markets, the output can also be exchanged for resources other than money (e.g. food and school uniforms).
   2. The quality of the output for marketing. Sometimes processing is an important prerequisite for the successful marketing of a product. This, however, often cannot be realized within the small-scale farming system. An innovation that requires extra processing while the inputs for such a process are lacking is unlikely to be successful without additional support.
   3. Added nutritive value of food/fodder consumption. Biotechnological innovations may change the characteristics of the output in such a way that the amount of food for household consumption and its nutritive value is changed.
   4. Possibilities to use the output for other productive or remunerative purposes within the farming system. The possibility to use the output of a certain agricultural activity for new or already existing productive or remunerative farm activities is an important factor in the adoption of biotechnological innovations. This is especially true for small-scale

farmers who are usually involved in a variety of complementary and interchangeable productive on- and off-farm activities.

D. Robustness

The small-scale farming systems are highly vulnerable. They are based on a very delicate interrelationship between four major elements:

- Biotic elements (plants, animals, micro-organisms);
- Abiotic elements (water, temperature, soil texture and structure, nutrients, etc.);
- Other internal elements (draught power, land, processing facilities, agricultural practices, availability of labour, etc.); and
- External elements (fertilizers, seed, market prices, market relations, institutional linkages).

Changes in any of them can easily affect:

a. Stability of the output (persistence of the output against normal and mainly short term fluctuations to which it is exposed):
   1. Pests, crop and animal diseases, weeds, etc.
   2. Climate (drought, heat or frost periods); and
   3. Internal/external inputs.

b. Sustainability of the farming system (persistence of the basic resources for farming against structural disturbances):
   1. Biotic and abiotic factors like deterioration of gene pools of plant and animal species, or exhaustion of soils and water, salinization and soil erosion;
   2. Economic factors like loss of capital and labour (e.g. overburdening of the women agricultural workers); and
   3. Socio-cultural factors like the desintegration of labour organization and the disappearance of practises of coopertion and mutual help and of traditional knowledge of agricultural practices.

Only a few types of small-scale farming systems in developing countries are stable and sustainable. Most systems are involved in processes of change, caused by e.g. population pressure or technological innovations, rendering them highly unstable and unsustainable. In addition, small-scale farming systems in developing countries are less understood and less predictable being also relatively complex. In consequence, innovations in small-scale farming systems can easily result in negative side-effects.

F. Synergetic and/or antagonistic effects with other changes in the farming system and national context. The introduction of a biotechnological innovation has synergetic effects if it facilitates the preparation or the implementation of another innovation. More reliable and higher yields, resulting in a more constant supply of products, for instance, could make the establishment of an economically feasible transport system much easier. If a biotechnological innovation has strong synergetic effects, its introduction may be much easier compared with innovations that do not have these effects or are accompanied by antagonistic effects which can be caused, for instance, by a reduction of import restrictions, thus influencing the prices of cash crops for local markets. In such a case, the biotechnological improvement of cash crops that are produced for local markets will not be a successful innovation.

3. *Comparative Advantage of Biotechnological Innovation* in terms of:
   A. Implementability (appropriateness and feasibility). This can be determined by comparing:
      1. The assessments based on the national criteria and the farming systems criteria; and
      2. Technical feasibility, which must be measured in terms of state of the art of the compared technologies and the scientific capacity of the country involved. For the realization of a project, a distinction can be made between that part of the research to be carried out abroad and that locally.
   B. Problem-solving capacity. The extent to which a proposed biotechnological innovation can solve a problem in comparison with other relevant innovations should be determined.
   C. Cost-effectiveness. The comparison of financial feasibility requires identification of time-frames for research and development, and the costs involved. Also possible funding by donor organizations and the developing country for which the innovation is intended have to be taken into account.

In order to be able to judge whether a certain biotechnological innovation is appropriate and feasible, and has a comparative advantage the criteria mentioned above should, first, be rated and scored one by one. Secondly, the end score which can be calculated by simple enumeration, gives a general indication of the degree to which the conditions for a successful introduction of a biotechnological innovation already exist and to what extent the necessary conditions have not yet been fulfilled. On the basis of a complete evaluation of appropriateness, feasibility and comparative advantage, priorities for biotechnological research can be established.

Besides determining problem areas for which biotechnological innovations are pre-eminently suitable, such an assessment model can also be used as a manual for the construction and evaluation of project proposals, because it specifies primary conditions relevant to the succes of biotechnology project for the sustainable development of small-scale agriculture. In both cases the assessment model can be used to provide answers to the following questions:

- Will the biotechnological innovation be able to reach the small-scale farmers without many additional requirements? (national context)
- Will the small-scale farmer be able to apply the biotechnological innovation successfully? (farming system)
- Is the proposed biotechnological innovation pre-eminently suitable to solve a specific problem? (comparative advantage of the biotechnological innovation)

3. Conclusions and recommendations

Several important conclusions and recommendations of the project are given below:

1. It is not possible, at a general level, to identify biotechnological innovations which are appropriate, feasible and have comparative advantage for application at the small-scale farm level in developing countries. The appropriateness and feasibility of a biotechnological innovation to promote sustainable rural development strongly depends on the specific context of which the following are important elements: national context, farming system and comparative advantage of the biotechnological innovation.

2. Most biotechnological innovations that are expected to be beneficial for small-scale farmers require additional institutional measures and adaptations of socio-cultural patterns. Only a few biotechnological innovations that are relatively easy to introduce have been identified. Examples include improved nitrogen fixing bacteria in legumes and disease or pest resistant varieties of crops that can be multiplied by farmers themselves. Although the implementation may be difficult, it is important to emphasize that the impact of biotechnological innovations can be considerable even for resource-poor farmers.

3. A detailed examination of the conditions under which the necessary adaptations in the institutional infrastructure and/or socio-cultural patterns are likely to occur is a prerequisite for the successful introduction of any biotechnological innovation.
   A major constraint to the examination of these conditions is the lack of knowledge of the possibilities of changing socio-cultural patterns and/or adapting relevant institutional structures.

4. The context has to be specified carefully to identify the most relevant problem areas and conditions that determine appropriateness, feasibility and comparative advantages. For such an identification, short inventory studies, based on a few interviews with experts or literature reviews, are not sufficient. Data collection has to be more comprehensive; effective exchange of information between policy-makers, scientists and representatives of end-users is crucial. Data should be collected by a team that has access to multidisciplinary expertise; it is not feasible to establish a complete multidisciplinary team.

5. The identification of priorities for biotechnological innovations aiming at sustainable rural development requires specific expertise in the field of Technology Assessment and biotechnology, which is often not present in developing countries. Assistance from experts from industrialized countries may be necessary. Donor organization could consider providing lacking expertise to groups and governments of developing countries which are interested in the identification of priorities.

6. In order to achieve sustainable rural development through biotechnological innovation, the following strategies can be recommended:
   - For countries with very poor scientific infrastructure, advanced research projects could be carried out in Western or international research institutes, or in regional institutes in the form of South-South cooperation. Important is to investigate whether adequate diffusion channels for regional cooperation exist or can be established;
   - For countries with a reasonably developed scientific infrastructure, this infrastructure could be further developed by the implemetation of e.g. MSc. programmes and/or by supporting existing biotechnological research programmes directed towards rural development.

7. Biotechnology as a strategy should not be considered in isolation but as an instrument integrated into general policies of developing countries and donor organization in order to facilitate synergetic effects.

References

1. Persley, G. (1989) AgBiotech News and Information 1989, Vol.1, no.1, pp.23-26.
2. Ahmed, I.F.T. (1988) ILO Studies, International Labour Review, Vol.127, no.1.
3. Swaminathan, M.S. (1982) Science, Vol.218, pp. 967-972.
4. Okereke, G.U. (1988) World Employment Programme Research, Working Paper. ILO, Geneva.
5. Dembo, D. and Morehouse, W. (1987) Technology Trends Series No.6, UNIDO.
6. Bunders, J. (1988) Trends in Biotechnolology, Vol.6, pp. 173-180.
7. Kennedy, M. (1986) Yale University Press.
8. Gotsch, N. and Rieder, P. (1989) TIBTECH, Vol.7, pp. 29-34.
9. Ruivenkamp, G. (1989) PhD Thesis.
10. Toenniessen, G.H. and Herdt, R.W. (1988) Presented at the USAID sponsored conference on Strengthening Collaboration in Biotechnology: Research and the Private Sector.
11. Bunders, J., Broerse, J. and Stolp, A. (1989) TIBTECH, Vol. 7, pp. S16-S24.
12. NRLO (1988) NRLO, 1988, no.13.
13. Chambers, R., Pacey, A. and Thrupp, L.A. (1989) Intermediate Technology Publications, London.
14. Bunders, J. (1988) TIBTECH, Vol.6, pp. 173-180.
15. Radnor, M. and Kaufman, S. (1988) In : A. Ward (ed), 'Science, Technology and Development', Westview press, Colorado, USA.
16. Fresco, L.O. (1988) Tropical crops Communication no. 13, Dept. of Tropical Crop Science, Agricultural Univesity Wageningen.
17. Norman, D.W. (1980) Michigan State University.
18. Ellis, F. (1988) Wye studies in agricultural and rural development, Cambridge University Press.
19. Kennedy, J.O.S. (1986) Elsevier Applied Science Publishers Ltd.

Current Plant Science and Biotechnology in Agriculture

1. H.J. Evans, P.J. Bottomley and W.E. Newton (eds.): *Nitrogen Fixation Research Progress.* Proceedings of the 6th International Symposium on Nitrogen Fixation (Corvallis, Oregon, 1985). 1985 ISBN 90-247-3255-7
2. R.H. Zimmerman, R.J. Griesbach, F.A. Hammerschlag and R.H. Lawson (eds.): *Tissue Culture as a Plant Production System for Horticultural Crops.* Proceedings of a Conference (Beltsville, Maryland, 1985). 1986 ISBN 90-247-3378-2
3. D.P.S. Verma and N. Brisson (eds.): *Molecular Genetics of Plant-microbe Interactions.* Proceedings of the 3rd International Symposium on this subject (Montréal, Québec, 1986). 1987 ISBN 90-247-3426-6
4. E.L. Civerolo, A. Collmer, R.E. Davis and A.G. Gillaspie (eds.): *Plant Pathogenic Bacteria.* Proceedings of the 6th International Conference on this subject (College Park, Maryland, 1985). 1987 ISBN 90-247-3476-2
5. R.J. Summerfield (ed.): *World Crops: Cool Season Food Legumes.* A Global Perspective of the Problems and Prospects for Crop Improvement in Pea, Lentil, Faba Bean and Chickpea. Proceedings of the International Food Legume Research Conference (Spokane, Washington, 1986). 1988 ISBN 90-247-3641-2
6. P. Gepts (ed.): *Genetic Resources of* Phaseolus *Beans.* Their Maintenance, Domestication, Evolution, and Utilization. 1988 ISBN 90-247-3685-4
7. K.J. Puite, J.J.M. Dons, H.J. Huizing, A.J. Kool, M. Koorneef and F.A. Krens (eds.): *Progress in Plant Protoplast Research.* Proceedings of the 7th International Protoplast Symposium (Wageningen, The Netherlands, 1987). 1988 ISBN 90-247-3688-9
8. R.S. Sangwan and B.S. Sangwan-Norreel (eds.): *The Impact of Biotechnology in Agriculture.* Proceedings of the International Conference 'The Meeting Point between Fundamental and Applied *in vitro* Culture Research' (Amiens, France, 1989). 1990. ISBN 0-7923-0741-0

KLUWER ACADEMIC PUBLISHERS – DORDRECHT / BOSTON / LONDON

